Introduction to the Fast Multipole Method

T0076718

Introduction to the Fast Multipole Method

Topics in Computational Biophysics, Theory, and Implementation

Victor Anisimov
James J.P. Stewart

CRC Press is an imprint of the
Taylor & Francis Group, an **informa** business

CRC Press
Taylor & Francis Group
6000 Broken Sound Parkway NW, Suite 300
Boca Raton, FL 33487-2742

First issued in paperback 2022

© 2020 by Taylor & Francis Group, LLC
CRC Press is an imprint of Taylor & Francis Group, an Informa business

No claim to original U.S. Government works

ISBN-13: 978-1-439-83905-8 (hbk)
ISBN-13: 978-1-03-233740-1 (pbk)
DOI: 10.1201/9780429063862

This book contains information obtained from authentic and highly regarded sources. Reasonable efforts have been made to publish reliable data and information, but the author and publisher cannot assume responsibility for the validity of all materials or the consequences of their use. The authors and publishers have attempted to trace the copyright holders of all material reproduced in this publication and apologize to copyright holders if permission to publish in this form has not been obtained. If any copyright material has not been acknowledged, please write and let us know so we may rectify in any future reprint.

Except as permitted under U.S. Copyright Law, no part of this book may be reprinted, reproduced, transmitted, or utilized in any form by any electronic, mechanical, or other means, now known or hereafter invented, including photocopying, microfilming, and recording, or in any information storage or retrieval system, without written permission from the publishers.

For permission to photocopy or use material electronically from this work, please access www.copyright.com (http://www.copyright.com/) or contact the Copyright Clearance Center, Inc. (CCC), 222 Rosewood Drive, Danvers, MA 01923, 978-750-8400. CCC is a not-for-profit organization that provides licenses and registration for a variety of users. For organizations that have been granted a photocopy license by the CCC, a separate system of payment has been arranged.

Trademark Notice: Product or corporate names may be trademarks or registered trademarks, and are used only for identification and explanation without intent to infringe.

Publisher's Note

The publisher has gone to great lengths to ensure the quality of this reprint but points out that some imperfections in the original copies may be apparent.

Visit the Taylor & Francis Web site at
http://www.taylorandfrancis.com

and the CRC Press Web site at
http://www.crcpress.com

Contents

Preface

The Fast Multipole Method (FMM) is an efficient procedure developed by Greengard and Rokhlin for computing the potential energy in systems that obey the inverse-square law. Since the amount of work required in computing interaction energies scales quadratically with the number of interacting particles, the computational cost of direct summation quickly becomes prohibitive. The FMM addresses that challenge by making the amount of work scale linearly, $O(N)$, with system size N.

Originally developed for astrophysics, the FMM algorithm quickly gained attention as a promising method for the treatment of electrostatic interactions in molecular modeling applications. This field has traditionally been dominated by simpler and computationally less-demanding Particle Mesh Ewald methods (PME), which scale as $O(N \log N)$. However, the advent of exascale computing (10^{18} floating point operations per second), combined with the trend in molecular modeling to simulate larger and more realistic systems, pushes the number of particles in the system close to a billion, and that steadily increases the value of the true $O(N)$ method.

Starting from the discovery of the FMM algorithm, several seminal works published by the two teams lead by Martin Head-Gordon and Gustavo Scuseria marked significant milestones in the development of the FMM method for biophysics applications. Although FMM theory is well understood, its complexity and the difficulty in efficient computer implementation are two major factors that limit the adoption of the method. Attempting to address these issues, our intent is to provide a complete, self-contained explanation of the math of the FMM method, so that anyone having an undergraduate grasp of calculus should be able to follow the presented material.

FMM theory can be regarded as an umbrella that covers numerous topics in mathematical physics, and, as a result, a comprehensive introduction to FMM theory is virtually non-existent in the literature. Despite an abundance of literature on those subjects, many publications are often either too concise or too advanced to be used in self-study. This book responds to the need by providing a simple and easy-to-understand introduction to every important topic, while conducting a self-contained complete reconstruction of the FMM theory from the ground up. The reader learns the method by rederiving the equations involved and turning them into computer code.

This project aims at a wide readership. Some, particularly those with an extensive physics background, may find the presented material useful in the teaching process. Many, who have no time or resources to gather the pieces of information from the distributed science publications, will be able to study the method from a single source. Others may find complementary insights about such topics as angular momentum, addition theorems, or Wigner matrix, which go beyond their utility for the FMM method.

Finding the best way to present all the required information in order to explain FMM requires undertaking a separate research work. This textbook presents an attempt to logically and concisely introduce the FMM method as if deriving it from first principles. Of necessity, many advanced topics about FMM have had to be excluded due to space constraints. It is our hope that the material presented will give the reader the necessary background to continue their study of the FMM theory on an independent journey.

In this book, Gauss units are employed because of their simplicity and their visual clarity in the equations. Readers who want to should have no difficulty converting them to SI units.

Victor Anisimov
James J.P. Stewart

1

Legendre Polynomials

Electrostatics is a branch of physics that deals with determining the electrostatic potential created by a system of stationary or slowly moving electric charges. Elementary electrostatic theory for a system of point charges is rather straightforward. It consists of the Coulomb law and the superposition principle, summing over the independent pairwise interactions of individual particles. The Coulomb law describes the force, $\mathbf{F}_i$, acting on particle i having charge q_i arising from other particles having charge q_j:

$$\mathbf{F}_i = q_i \sum_j \frac{q_j}{|\mathbf{r}_i - \mathbf{r}_j|^2} \frac{\mathbf{r}_i - \mathbf{r}_j}{|\mathbf{r}_i - \mathbf{r}_j|}, \tag{1.1}$$

where $\mathbf{r}_i = (x_i, y_i, z_i)$ represents a coordinate vector of particle i. Summation over the particles reflects the superposition principle according to which each pairwise interaction is essentially independent of the other ones.

Integration of Equation 1.1 leads to the definition of electrostatic potential, Φ_i:

$$\Phi_i = \sum_j \frac{q_j}{|\mathbf{r}_i - \mathbf{r}_j|}, \tag{1.2}$$

which, if multiplied by particle charge, q_i, gives the electrostatic energy, $U_i = q_i \, \Phi_i$ that particle i has in the presence of other charge sources j. Although, historically, the theory of electrostatics begins from the Coulomb law, in practice, however, it is easier to obtain Equation 1.1 by differentiation of Equation 1.2, which is a subject of Chapter 9.

A fundamental problem in electrostatics is the evaluation of the electrostatic potential. Equations of the type that in Equation 1.2 are difficult to numerically compute when the number of particles becomes very large, because that leads to the computational cost growing quadratically with the number of particles in the system. The problematic part is the inverse distance function, $|\mathbf{r}_1 - \mathbf{r}_2|^{-1}$ that mixes up the coordinates of two particles. Solution to that problem called factorization redefines inverse distance function in the form of product of coordinates of individual particles.

A number of powerful mathematical factorization techniques are available, depending on how the source and field points are placed relative to each other. Among those techniques, Legendre polynomials being a solution of the Laplace's equation take a prominent position. Legendre polynomials are especially useful when studying systems that have azimuthal symmetry, for example, in finding the electrostatic potential of a point charge placed on the z-axis. In such cases, the presence of azimuthal symmetry greatly simplifies the expansion of the inverse distance function.

1.1 Potential of a Point Charge Located on the *z*-Axis

Consider a positive unit charge placed on the z-axis at the position of coordinate vector $\mathbf{a} = (x_a, y_a, z_a)$ with x_a and y_a components being zero, as illustrated in Figure 1.1. This particle creates an electrostatic potential $|\mathbf{r} - \mathbf{a}|^{-1}$ at a point P, defined by coordinate vector $\mathbf{r} = (x_r, y_r, z_r)$, and having angle θ subtended between the vectors $\mathbf{a}$ and $\mathbf{r}$.

$$\frac{1}{|\mathbf{r} - \mathbf{a}|} = \frac{1}{\sqrt{(x_r - x_a)^2 + (y_r - y_a)^2 + (z_r - z_a)^2}}. \tag{1.3}$$

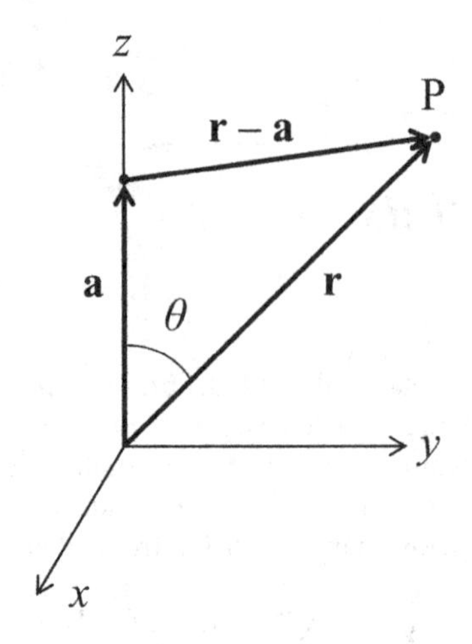

FIGURE 1.1 Electrostatic potential created at the field point, P by a charge source placed on the z-axis.

The distance vector $\mathbf{r} - \mathbf{a}$ points from the source charge toward the point P of measurement of electrostatic potential, and has the length $|\mathbf{r} - \mathbf{a}|$. The point P is called a field point because it senses the electric field produced by the charge sources.

Since the inverse distance function, and the electrostatic potential in general, is a scalar function, it is possible to rewrite $|\mathbf{r} - \mathbf{a}|^{-1}$ in the form of a product of two scalars, $|\mathbf{r}| = r = \sqrt{x_r^2 + y_r^2 + z_r^2}$ and $|\mathbf{a}| = a = \sqrt{x_a^2 + y_a^2 + z_a^2}$. In practice, those scalars will be present in the form of a ratio rather than as a product. This is a valid operation because division is equivalent to a product, as in $a/r = ar^{-1}$, so the elementary definition of factorization as a product of two terms still holds.

In the situation when the point charge is placed on the z-axis as in Figure 1.1, the cosine theorem can be used to express the distance function *via* the cosine of the angle subtended between the two vectors:

$$|\mathbf{r} - \mathbf{a}|^2 = r^2 + a^2 - 2\,ar\cos\theta = r^2\left[1 + \left(\frac{a}{r}\right)^2 - 2\frac{a}{r}\cos\theta\right] = r^2\left[1 + \left(\frac{a}{r}\right)\left(\frac{a}{r} - 2\cos\theta\right)\right]. \tag{1.4}$$

Likewise, the inverse distance function becomes:

$$\frac{1}{|\mathbf{r} - \mathbf{a}|} = \frac{1}{r}\left[1 + \left(\frac{a}{r}\right)\left(\frac{a}{r} - 2\cos\theta\right)\right]^{-\frac{1}{2}}. \tag{1.5}$$

After making the substitution

$$t = \left(\frac{a}{r}\right)\left(\frac{a}{r} - 2\cos\theta\right), \tag{1.6}$$

the expression in square brackets in Equation 1.5 can be expanded in powers of t about zero by using the Taylor theorem:

$$f(t) = (1 + t)^{-1/2} = \sum_{k=0}^{\infty} \frac{1}{k!} t^k f^{(k)}(0), \tag{1.7}$$

where $f^{(k)}(0)$ is k-th derivative of function $f(t)$ at the point $t = 0$. All $f^{(k)}(0) = 1$, therefore:

$$\frac{1}{|\mathbf{r}-\mathbf{a}|} = \frac{1}{r}(1+t)^{-1/2} = \frac{1}{r}\left(1-\frac{1}{2}t+\frac{3}{8}t^2-\frac{5}{16}t^3+\cdots\right). \tag{1.8}$$

Substituting Equation 1.6 back into Equation 1.8 and gathering terms (a/r) of the same power gives the following power series:

$$\frac{1}{|\mathbf{r}-\mathbf{a}|} = \frac{1}{r}\left[1+\left(\frac{a}{r}\right)\cos\theta+\left(\frac{a}{r}\right)^2(3\cos^2\theta-1)/2+\left(\frac{a}{r}\right)^3(5\cos^3\theta-3\cos\theta)/2+\cdots\right]. \tag{1.9}$$

In this expansion the trigonometric coefficients are the Legendre polynomials, $P_l(\mu)$, where $\mu = \cos\theta$ and l is the degree of the polynomial. This allows the series to be written in a compact form:

$$\frac{1}{|\mathbf{r}-\mathbf{a}|} = \sum_{l=0}^{\infty}\frac{a^l}{r^{l+1}}P_l(\mu). \tag{1.10}$$

This quite remarkable result gives rise to a number of improvements over the original inverse distance function. First, the expansion in Equation 1.10 separates particle coordinates into a ratio of two scalar numbers a and r, which are the individual particle distances from the center of the coordinate system. Second, the angle subtended between the radius-vectors of the source and field particles are expressed as Legendre polynomials, a very versatile mathematical tool. This equation now provides a first glimpse into the possibility that the coordinates in the electrostatic potential function can be separated; an operation of central significance in electrostatics.

Legendre polynomials have values in the range of $[-1: +1]$, and the angle θ is restricted to values in the range $[0: \pi]$. Based on that, for the series in Equation 1.8 to be convergent, $|t|$ should be less than 1, that is., that $|r| > |a|$ in Equation 1.10. This constraint requires that the source particle be located closer to the origin of the coordinate system than the field point (see Figure 1.1). However, due to the symmetry of the problem, the larger of a and r can arbitrarily be put into the denominator and the smaller into the numerator. Indeed the electrostatic potential in point $\mathbf{r}$ due to charge q in point $\mathbf{a}$ is the same as the potential in point $\mathbf{a}$ when charge q is placed in point $\mathbf{r}$.

1.2 Laplace's Equation

Performing a Taylor expansion of the inverse distance when the particle is located on the z-axis is one of the ways to obtain Legendre polynomials. The use of these polynomials shows a promising way toward factorization of the inverse distance function, and it would be desirable to maintain and, if possible, widen that progress. It turns out that the more general approach to perform the factorization, in which the Legendre polynomials appear as a particular case, comes from the analysis of mathematical properties of potential function *via* its derivatives.

In a mathematical analysis of a function of a single variable, $f(x)$, its second derivative $\partial^2 f/\partial x^2$ describes the curvature of the function. At points where the second derivative is zero, $\partial^2 f/\partial x^2 = 0$, the function f is linear and no maxima or minima are present, except on the boundaries. A general solution of this second-order differential equation is $f = ax + b$, where a and b are constants. In a linear function, as shown in Figure 1.2, the extreme points are always located on the boundaries of the function.

Extending this concept to 3D, where $\Phi(x, y, z)$ is a scalar function, leads to the second derivative in the form:

$$\nabla^2\Phi = \frac{\partial^2\Phi}{\partial x^2}+\frac{\partial^2\Phi}{\partial y^2}+\frac{\partial^2\Phi}{\partial z^2}. \tag{1.11}$$

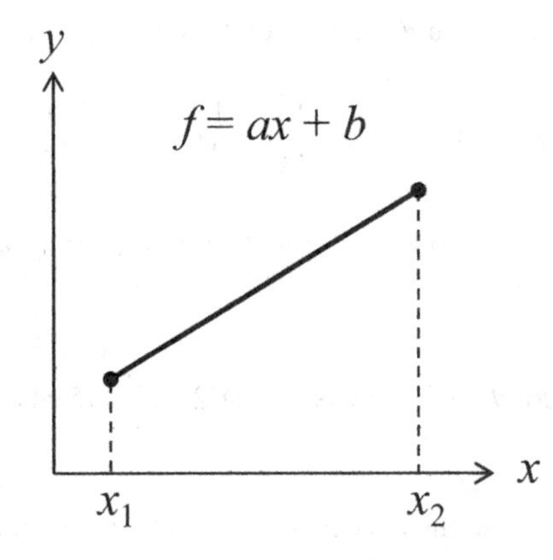

FIGURE 1.2 A linear function f that is defined on the interval $[x_1: x_2]$ has a zero second derivative over the entire interval, and its minimum and maximum values are located at the boundaries.

Setting this derivative to zero introduces Laplace's second-order partial differential equation:

$$\nabla^2\Phi = 0. \tag{1.12}$$

For a function Φ to have a vanishing second derivative, certain conditions must be fulfilled. One of these is that, similar to the 1D case, function Φ in Equation 1.12 has to be a smooth analytic function that cannot have any minima or maxima except on the boundary points. Working with a 3D space makes it possible to drop the uninteresting cases of a linear function where each individual second derivative is trivially equal to zero. Now the partial second derivatives, when combined, can cancel out so that their sum becomes zero. This cancellation should hold over the entire range of function Φ, where it is defined, and within the specified boundaries. This makes the problem much more interesting, in that function Φ can have a curvy profile, but those curves still have to be smooth in the sense that neither minima nor maxima would occur within the defined boundaries.

Laplace's equation typically appears in problems that involve deriving a function Φ for predefined boundary conditions. Of course, on its own, Equation 1.12 has an infinite number of solutions. Matching those to the boundary condition excludes all but a single unique solution. This is the essence of the uniqueness theorem for Laplace's equation. The boundary conditions can be presented by the known value of function, Φ, or of its first derivative, defined at each point of the boundary.

Satisfying Laplace's equation is a general property of any potential function, and that includes the electrostatic potential function, $\Phi(x, y, z) \equiv \Phi(\mathbf{r})$, where each coordinate point in space is defined by a radius-vector $\mathbf{r} = (x, y, z)$. This function has the extreme values of $-\infty$ and $+\infty$ at the positions of negative and positive source charges, respectively; however, these values are obviously of no practical interest. Therefore, in order to define the potential function Φ, Laplace's equation requires the origins of all source charges to be located away from the region where the electrostatic potential is measured. The physical meaning of that requirement will now be described.

Laplace's equation states that, in any closed 3D volume including no source charges, the incoming flow of electric field is equal to the amount of electric field that leaves the boundary of the region, so the net flux is zero (Figure 1.3). Due to this property, the electrostatic potential gradually changes between the points on the boundary surface where the electric field enters and leaves the object; therefore, the resulting electrostatic potential function has neither minima nor maxima. Correspondingly, the electrostatic potential assumes its extreme values only at points on the boundary surface.

Indeed, if there were source charges in the region of definition of electrostatic potential function this would introduce a non-zero flux of electric field over the boundaries of the system, which would invalidate the Laplace's equation.

The simplest form of potential function is the one generated by a point charge. One can check by direct substitution that the inverse distance function, $1/r$, being the electrostatic potential function of a unit point

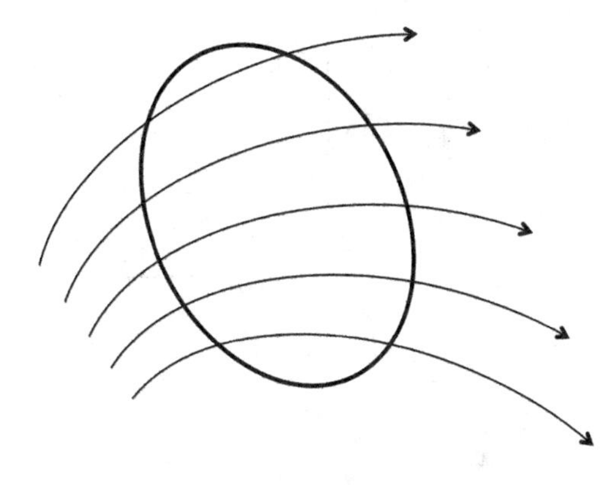

FIGURE 1.3 Zero net flux of electric field for 2D object away from charge sources.

charge placed at the origin of coordinate system, satisfies this equation. Second-order differentiation of $1/r$ along coordinate x gives:

$$\frac{\partial^2}{\partial x^2}\left[(x^2+y^2+z^2)^{-\frac{1}{2}}\right]=\frac{\partial}{\partial x}\left[-x\,(x^2+y^2+z^2)^{-\frac{3}{2}}\right]$$
$$=-(x^2+y^2+z^2)^{-\frac{3}{2}}+3x^2(x^2+y^2+z^2)^{-\frac{5}{2}}. \tag{1.13}$$

The other two partial derivatives along coordinates y and z are obtained in the same way.

$$\begin{aligned}\frac{\partial^2}{\partial y^2}\left[(x^2+y^2+z^2)^{-\frac{1}{2}}\right]&=-(x^2+y^2+z^2)^{-\frac{3}{2}}+3y^2(x^2+y^2+z^2)^{-\frac{5}{2}}\\ \frac{\partial^2}{\partial z^2}\left[(x^2+y^2+z^2)^{-\frac{1}{2}}\right]&=-(x^2+y^2+z^2)^{-\frac{3}{2}}+3z^2(x^2+y^2+z^2)^{-\frac{5}{2}}\end{aligned} \tag{1.14}$$

Summing up the resulting partial derivatives leads to:

$$\left(\frac{\partial^2}{\partial x^2}+\frac{\partial^2}{\partial y^2}+\frac{\partial^2}{\partial z^2}\right)\frac{1}{r}=-3(x^2+y^2+z^2)^{-\frac{3}{2}}+3(x^2+y^2+z^2)(x^2+y^2+z^2)^{-\frac{5}{2}}=0. \tag{1.15}$$

This example illustrates the fact that if the electrostatic potential function were known it would indeed satisfy Laplace's equation. However the utility of Laplace's equation extends far beyond such simple validation. The differential equation Equation 1.12, when solved, gives the general analytic solution for function Φ. When combined with the known boundary conditions, such a solution is unique, as it can be seen in Figure 1.2 for the 1D case. The ability of Laplace's equation to provide a detailed characterization of potential function makes it the most important equation in electrostatic theory.

1.3 Solution of Laplace's Equation in Cartesian Coordinates

Laplace's equation has a very simple form when it is given in Cartesian coordinates:

$$\frac{\partial^2\Phi}{\partial x^2}+\frac{\partial^2\Phi}{\partial y^2}+\frac{\partial^2\Phi}{\partial z^2}=0. \tag{1.16}$$

In practical terms, Equation 1.16 can only be solved if there is a way to separate the variables in Φ. Hence the strategy for solving it depends on the nature of the boundary conditions. The simplest boundary

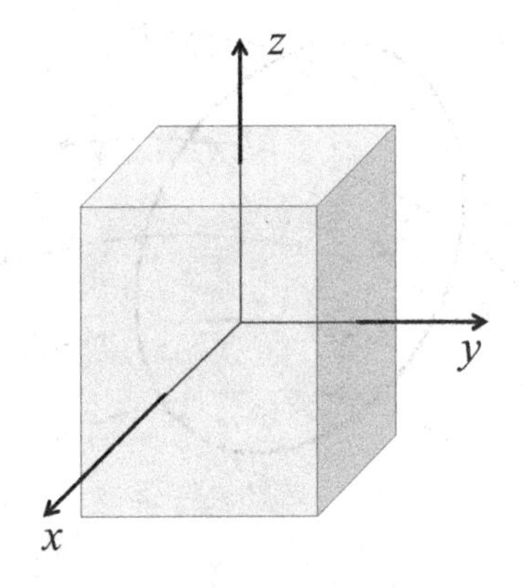

FIGURE 1.4 Rectangular boundary conditions.

condition is represented by a rectangular cuboid as shown in Figure 1.4 in which the boundary planes are positioned perpendicular to the coordinate axes.

Due to the uniqueness theorem, function Φ can be expressed in the form that is most convenient for the solution process. Thus, for the problem illustrated in Figure 1.4, separation of coordinates can be most easily achieved if function Φ is factorized into the product of independent functions of x, y, and z coordinates, in the form:

$$\Phi(x,y,z) = X(x)\,Y(y)\,Z(z). \tag{1.17}$$

Substituting Equation 1.17 into Equation 1.16 gives:

$$\frac{\partial^2 X(x)}{\partial x^2}Y(y)\,Z(z) + X(x)\frac{\partial^2 Y(y)}{\partial y^2}Z(z) + X(x)\,Y(y)\frac{\partial^2 Z(z)}{\partial z^2} = 0. \tag{1.18}$$

Dividing the each term by the product $X(x)$ $Y(y)$ $Z(z)$ leads to:

$$\frac{1}{X(x)}\frac{\partial^2 X(x)}{\partial x^2} + \frac{1}{Y(y)}\frac{\partial^2 Y(y)}{\partial y^2} + \frac{1}{Z(z)}\frac{\partial^2 Z(z)}{\partial z^2} = 0. \tag{1.19}$$

Since $X(x)$ is function of x only, $Y(y)$ is function of y only, and $Z(z)$ is function of z only, each of the three terms in Equation 1.19 must be constant in order for their sum to become zero. This gives three equations:

$$\frac{1}{X(x)}\frac{\partial^2 X(x)}{\partial x^2} = \alpha^2 \qquad \frac{1}{Y(y)}\frac{\partial^2 Y(y)}{\partial y^2} = \beta^2 \qquad \frac{1}{Z(z)}\frac{\partial^2 Z(z)}{\partial z^2} = \gamma^2, \tag{1.20}$$

which can be solved independently under the condition $\alpha^2 + \beta^2 + \gamma^2 = 0$, where α, β, and γ are complex constants. Taking the first equation from Equation 1.20 one can re-write it in the form:

$$\frac{\partial^2 X(x)}{\partial x^2} - \alpha^2 X(x) = 0. \tag{1.21}$$

This is the equation of a harmonic oscillator and has a general solution of the form:

$$X(x) = A_1 e^{\alpha x} + A_2 e^{-\alpha x}. \tag{1.22}$$

Correspondingly, the solution of Equation 1.19 becomes:

$$\Phi(x,y,z) = X(x)\,Y(y)\,Z(z) = (A_1e^{\alpha x} + A_2e^{-\alpha x})(B_1e^{\beta y} + B_2e^{-\beta y})(C_1e^{\gamma z} + C_2e^{-\gamma z}), \qquad (1.23)$$

where A_1, A_2, B_1, B_2, C_1, and C_2 are arbitrary constants which can be determined from the supplied boundary conditions. Since Laplace's equation is linear, any linear combination of functions in the form of Equation 1.23, including their product with a constant, is also a solution.

This analysis illustrates that Laplace's equation can be solved by factorization of the target function Φ into the product of individual functions, each of which depends only on one coordinate.

1.4 Laplace's Equation in Spherical Polar Coordinates

When solving Laplace's equation, the specific choice of coordinate system is dictated by the geometry of the boundary conditions, thus Cartesian coordinates would be used for the boundary having the shape of a rectangular cuboid. Conversely, because the potential generated by point charges has spherical symmetry, the coordinate system of choice would be spherical polar rather than Cartesian. Indeed, for point charges, the use of spherical polar coordinates (r, θ, ϕ) leads to a powerful solution of Laplace's equation.

Like the Cartesian coordinate system, the spherical polar coordinate system is orthogonal, as illustrated in Figure 1.5. Because of this, the function Φ can be factorized into a product of three independent functions:

$$\Phi(r,\theta,\phi) = R(r)\,P(\theta)\,Z(\phi), \qquad (1.24)$$

where functions R, P, and Z denote radial, angular, and azimuthal dependencies respectively, and where each function depends on only one coordinate.

The first step, then, is to convert Laplace's equation into spherical polar coordinates. This requires establishing the relationship between the partial derivatives of the corresponding coordinate systems. To illustrate the procedure, one can differentiate function $\Phi(r, \theta, \phi)$ over coordinate x by using the chain rule. This gives:

$$\frac{\partial \Phi}{\partial x} = \frac{\partial \Phi}{\partial r}\frac{\partial r}{\partial x} + \frac{\partial \Phi}{\partial \theta}\frac{\partial \theta}{\partial x} + \frac{\partial \Phi}{\partial \phi}\frac{\partial \phi}{\partial x}. \qquad (1.25)$$

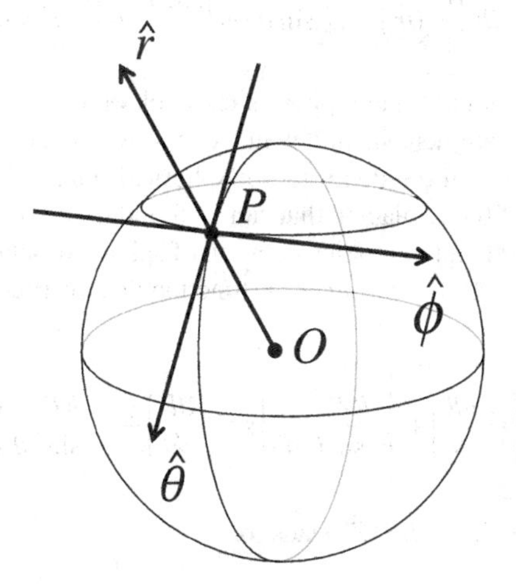

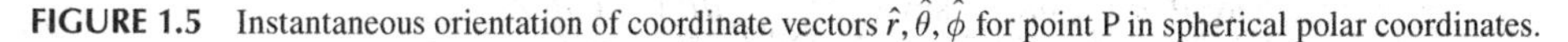
FIGURE 1.5 Instantaneous orientation of coordinate vectors $\hat{r}, \hat{\theta}, \hat{\phi}$ for point P in spherical polar coordinates.

After that, Equation 1.25 should be differentiated once more to produce a second-order partial derivative:

$$\frac{\partial^2\Phi}{\partial x^2}=\frac{\partial}{\partial x}\left(\frac{\partial\Phi}{\partial r}\frac{\partial r}{\partial x}\right)+\frac{\partial}{\partial x}\left(\frac{\partial\Phi}{\partial\theta}\frac{\partial\theta}{\partial x}\right)+\frac{\partial}{\partial x}\left(\frac{\partial\Phi}{\partial\phi}\frac{\partial\phi}{\partial x}\right). \tag{1.26}$$

Applying the differentiation to the expressions in brackets in Equation 1.26 in the form of a product of two functions leads to:

$$\frac{\partial^2\Phi}{\partial x^2}=\frac{\partial}{\partial x}\left(\frac{\partial\Phi}{\partial r}\right)\frac{\partial r}{\partial x}+\frac{\partial\Phi}{\partial r}\frac{\partial^2 r}{\partial x^2}+\frac{\partial}{\partial x}\left(\frac{\partial\Phi}{\partial\theta}\right)\frac{\partial\theta}{\partial x}+\frac{\partial\Phi}{\partial\theta}\frac{\partial^2\theta}{\partial x^2}+\frac{\partial}{\partial x}\left(\frac{\partial\Phi}{\partial\phi}\right)\frac{\partial\phi}{\partial x}+\frac{\partial\Phi}{\partial\phi}\frac{\partial^2\phi}{\partial x^2}. \tag{1.27}$$

Once again, the derivative of the function placed in brackets can be found by using the chain rule. Applying this recipe to the first term in right-hand side of Equation 1.27 gives:

$$\frac{\partial}{\partial x}\left(\frac{\partial\Phi}{\partial r}\right)\frac{\partial r}{\partial x}=\left(\frac{\partial^2\Phi}{\partial r^2}\frac{\partial r}{\partial x}\right)\frac{\partial r}{\partial x}+\left(\frac{\partial^2\Phi}{\partial r\partial\theta}\frac{\partial\theta}{\partial x}\right)\frac{\partial r}{\partial x}+\left(\frac{\partial^2\Phi}{\partial r\partial\phi}\frac{\partial\phi}{\partial x}\right)\frac{\partial r}{\partial x}. \tag{1.28}$$

From this point on the procedure is rather straightforward. The remaining partial derivatives of spherical polar coordinates over Cartesian coordinates can be found from differentiation of the following relations existing between these two coordinate systems:

$$\begin{aligned} r&=\sqrt{x^2+y^2+z^2},\\ x^2+y^2&=r^2\sin\theta,\\ \cos\theta&=\frac{z}{r},\\ \frac{y}{x}&=\tan\phi. \end{aligned} \tag{1.29}$$

The first-order partial derivatives of spherical polar coordinates over Cartesian coordinates are obtained in Appendix A.1. Continuing this bulky and somewhat tedious procedure, rearranging, and cancelling out cross-terms eventually produces the Laplacian, ∇^2, in spherical polar coordinates:

$$\nabla^2=\frac{1}{r^2}\frac{\partial}{\partial r}\left(r^2\frac{\partial}{\partial r}\right)+\frac{1}{r^2\sin\theta}\frac{\partial}{\partial\theta}\left(\sin\theta\frac{\partial}{\partial\theta}\right)+\frac{1}{r^2\sin^2\theta}\frac{\partial^2}{\partial\phi^2}. \tag{1.30}$$

Equation 1.30 can be easily found in any physics textbook on electrostatics; so its derivation is only sketched here. One important comment should be made, however. Due to physicists and mathematicians employing different conventions in the definition of spherical polar coordinates, the reader should be aware of two different forms of the Laplacian that can be found in physics and mathematics literature. In our case, Equation 1.30 is based on the physics notation of spherical polar coordinates.

The next step is to derive the factorized form of function Φ. Substituting Equation 1.24 instead of Φ into Equation 1.12 gives:

$$\frac{PZ}{r^2}\frac{d}{dr}\left(r^2\frac{dR}{dr}\right)+\frac{RZ}{r^2\sin\theta}\frac{d}{d\theta}\left(\sin\theta\frac{dP}{d\theta}\right)+\frac{RP}{r^2\sin^2\theta}\frac{d^2Z}{d\phi^2}=0. \tag{1.31}$$

Multiplying each term by $r^2\sin^2\theta/(R\,P\,Z)$ leads to:

$$\frac{\sin^2\theta}{R}\frac{d}{dr}\left(r^2\frac{dR}{dr}\right)+\frac{\sin\theta}{P}\frac{d}{d\theta}\left(\sin\theta\frac{dP}{d\theta}\right)+\frac{1}{Z}\frac{d^2Z}{d\phi^2}=0. \tag{1.32}$$

Since, in Equation 1.32, only the last term depends on azimuthal angle ϕ, this term must be a constant in order for the overall sum to be zero. For reasons that will be clarified in Chapter 2, this term can be equated to $-m^2$, with parameter m assuming integer values in the range from $-\infty$ to $+\infty$. Although, in general, parameter m can be a decimal or even a complex number, the common physics applications of Laplace's equation require m to have integer values. With that choice Equation 1.32 turns into two equations, each of which can be solved separately:

$$\frac{\sin^2\theta}{R}\frac{d}{dr}\left(r^2\frac{dR}{dr}\right)+\frac{\sin\theta}{P}\frac{d}{d\theta}\left(\sin\theta\frac{dP}{d\theta}\right)=m^2, \tag{1.33}$$

$$\frac{1}{Z}\frac{d^2Z}{d\phi^2}=-m^2. \tag{1.34}$$

Equation 1.34 is the equation of a harmonic oscillator, and has a general solution of the form:

$$Z=C\,e^{\pm im\phi}, \tag{1.35}$$

where C is a constant. The validity of Equation 1.35 can be checked by direct substitution. Next is to solve Equation 1.33. A full solution of Equation 1.33 for $m \neq 0$ is presented in Chapter 2. In systems that have azimuthal symmetry, $m = 0$. This condition is also known as cylindrical symmetry when the main axis of the cylinder coincides with the z-axis. Under that condition the value of function Z along the z-axis does not depend on coordinate ϕ, and the derivative in Equation 1.34 is equal to zero. In that case Equation 1.33 simplifies to:

$$\frac{\sin^2\theta}{R}\frac{d}{dr}\left(r^2\frac{dR}{dr}\right)+\frac{\sin\theta}{P}\frac{d}{d\theta}\left(\sin\theta\frac{dP}{d\theta}\right)=0. \tag{1.36}$$

Dividing both terms in Equation 1.36 by $\sin^2\theta$ one obtains:

$$\frac{1}{R}\frac{d}{dr}\left(r^2\frac{dR}{dr}\right)+\frac{1}{P\sin\theta}\frac{d}{d\theta}\left(\sin\theta\frac{dP}{d\theta}\right)=0. \tag{1.37}$$

Once again, the variables in Equation 1.37 are fully separated so that both its summands must be equal to a constant, and in order for the solution to be physically meaningful that constant must be of integer type. For that purpose the constant can be set to $l(l+1)$, where l is a non-negative integer number that leads to two equations that can be solved separately:

$$\frac{1}{R}\frac{d}{dr}\left(r^2\frac{dR}{dr}\right)=l(l+1), \tag{1.38}$$

and

$$\frac{1}{P\sin\theta}\frac{d}{d\theta}\left(\sin\theta\frac{dP}{d\theta}\right)=-l(l+1). \tag{1.39}$$

The reason for the peculiar choice of the constant in the form $l(l+1)$ will be explained shortly. To find the solution of Equation 1.38 it should be rearranged. This gives:

$$\frac{d}{dr}\left(r^2\frac{dR}{dr}\right)-l(l+1)\,R=0. \tag{1.40}$$

Applying differentiation in the left-most term in Equation 1.40 leads to the Euler-Cauchy linear homogeneous ordinary differential equation:

$$r^2 \frac{d^2R}{dr^2} + 2r\frac{dR}{dr} - l\,(l+1)\,R = 0. \tag{1.41}$$

A solution of this equation has the form $R = r^k$, so this will be used as a trial function. The first and second derivatives of this function are:

$$\frac{dR}{dr} = k\,r^{k-1}, \quad \frac{d^2R}{dr^2} = k\,(k-1)r^{k-2}. \tag{1.42}$$

Substituting these values into Equation 1.41 gives:

$$r^2\left[k(k-1)r^{k-2}\right] + 2r\left[k\,r^{k-1}\right] - l(l+1)r^k = 0. \tag{1.43}$$

After simplification of Equation 1.43 one obtains:

$$k\,(k-1)r^k + 2k\,r^k - l(l+1)r^k = 0. \tag{1.44}$$

To resolve Equation 1.44 one needs to find roots of the quadratic equation:

$$k(k-1) + 2k - l(l+1) = 0. \tag{1.45}$$

After simplification this gives:

$$k^2 + k - l(l+1) = 0. \tag{1.46}$$

Since Equation 1.46 is a quadratic equation over k it has two roots. The square root of the discriminant, D, of this equation is:

$$\sqrt{D} = \sqrt{1 + 4l(l+1)} = \sqrt{4l^2 + 4l + 1} = 2l + 1. \tag{1.47}$$

With that, the roots of Equation 1.46 are:

$$k_1 = \frac{-1 + 2l + 1}{2} = l, \qquad k_2 = \frac{-1 - 2l - 1}{2} = -l - 1. \tag{1.48}$$

Now the reason for choosing a constant in the form of $l\,(l+1)$ in Equation 1.38 is revealed: a constant in this form allows the square root of the discriminant in Equation 1.47 to be neatly resolved so that the parameter k in Equation 1.46 can remain of integer type.

Since any linear combination of individual solutions r^l and r^{-l-1} is also a solution of Equation 1.38, the general form of the radial function is:

$$R = A_l r^l + B_l r^{-l-1}, \tag{1.49}$$

where A_l and B_l are constants of integration. These can be derived from the value of the electrostatic potential specified at the boundary points.

Equation 1.49 provides a family of solutions to Equation 1.38 based on the ability of index l to assume any integer value in the range from 0 to $+\infty$. From a mathematical standpoint, negative values of index l are also legitimate solutions but, because they do not have any physical meaning, these can safely be ignored.

All that is left is to determine the dependence of the angular coordinate θ in Equation 1.39. This part of Laplace's equation can be rearranged to give:

$$\frac{d}{d\theta}\left(\sin\theta\frac{dP}{d\theta}\right)+l(l+1)P\sin\theta=0. \tag{1.50}$$

Applying the substitution $\mu = \cos\theta$ and $d\mu = \sin\theta\, d\theta$ one obtains:

$$\sin\theta\frac{d}{d\mu}\left(\sin^2\theta\frac{dP}{d\mu}\right)+l(l+1)P\sin\theta=0. \tag{1.51}$$

Dividing Equation 1.51 by $\sin\theta$ then gives the ordinary Legendre differential equation:

$$\frac{d}{d\mu}\left[(1-\mu^2)\frac{dP_l(\mu)}{d\mu}\right]+l(l+1)P_l(\mu)=0. \tag{1.52}$$

The solutions to Equation 1.52 are the Legendre polynomials $P_l(\mu)$, which were previously derived in Equation 1.10. As shown in Equation 1.9, analytic expressions for Legendre polynomials can be obtained using a Taylor expansion, but such a procedure is somewhat clumsy. Instead, a direct and elegant representation of Legendre polynomials is provided by Rodrigues' formula:

$$P_l(\mu)=\frac{1}{2^l l!}\frac{d^l}{d\mu^l}(\mu^2-1)^l, \tag{1.53}$$

where the coefficient $1/(2^l l!)$ is added to satisfy the condition that $P_l(1) = 1$ for each polynomial term.

Following the choice previously made that when azimuthal symmetry is present, that is, the solution does not depend on azimuthal angle, then $m = 0$, and combining the formulae in Equations 1.49 and 1.53, the solution of Laplace's equation now becomes:

$$\Phi(r,\mu)=\sum_{l=0}^{\infty}\left[A_l r^l + B_l r^{-l-1}\right]P_l(\mu). \tag{1.54}$$

This is a general form of the electrostatic potential function for the case of azimuthal symmetry. Before proceeding further to take advantage of the factorization of the electrostatic potential provided by Equation 1.54 it is necessary to briefly review the major mathematical properties of Legendre polynomials.

1.5 Orthogonality and Normalization of Legendre Polynomials

Two important properties of Legendre polynomials are their orthogonality and normalization conditions; these allow the expansion of any arbitrary function as a series of Legendre polynomials. The Legendre polynomials are defined on the interval $-1 \le \mu \le 1$ (recall that $\mu = \cos\theta$), which determines the integration limits. The orthogonality condition states that

$$\int_{-1}^{+1} P_n(\mu)\,P_l(\mu)\,d\mu=0, \qquad \text{for any } n \neq l. \tag{1.55}$$

To prove this statement one can take the Legendre differential equation for $P = P_l(\mu)$, given by Equation 1.52,

$$\frac{d}{d\mu}\left[(1-\mu^2)\frac{dP_l(\mu)}{d\mu}\right]+l\,(l+1)\,P_l(\mu)=0, \tag{1.52}$$

and left-multiply it by $P_n(\mu)$, and integrate. This leads to:

$$\int_{-1}^{+1} P_n(\mu)\left\{\frac{d}{d\mu}\left[(1-\mu^2)\frac{dP_l(\mu)}{d\mu}\right]+l(l+1)\,P_l(\mu)\right\}d\mu=0. \tag{1.56}$$

Equation 1.56 can be rewritten as a sum of two integrals:

$$\int_{-1}^{+1} P_n(\mu)\left\{\frac{d}{d\mu}\left[(1-\mu^2)\frac{dP_l(\mu)}{d\mu}\right]\right\}d\mu+l(l+1)\int_{-1}^{+1} P_n(\mu)\,P_l(\mu)\,d\mu=0. \tag{1.57}$$

The first integral in Equation 1.57 can be solved using integration by parts:

$$\int u\,dv=uv-\int v\,du. \tag{1.58}$$

For that purpose, the following assignments can be made:

$$u=P_n(\mu),\quad \text{and}\quad dv=\left\{\frac{d}{d\mu}\left[(1-\mu^2)\frac{dP_l(\mu)}{d\mu}\right]\right\}d\mu. \tag{1.59}$$

From Equation 1.59 it follows that:

$$du=\frac{dP_n(\mu)}{d\mu}d\mu,\quad \text{and}\quad v=(1-\mu^2)\frac{dP_l(\mu)}{d\mu}. \tag{1.60}$$

Applying integration by parts (Equations 1.58 through 1.60) to the first integral in Equation 1.57 leads to:

$$\int_{-1}^{+1} P_n(\mu)\frac{d}{d\mu}\left[(1-\mu^2)\frac{dP_l(\mu)}{d\mu}\right]d\mu=P_n(\mu)(1-\mu^2)\frac{dP_l(\mu)}{d\mu}\bigg|_{-1}^{+1}-\int_{-1}^{+1}(1-\mu^2)\frac{dP_l(\mu)}{d\mu}\frac{dP_n(\mu)}{d\mu}d\mu. \tag{1.61}$$

Since $(1-\mu^2)$ is zero for the present integration limits, Equation 1.61 reduces to:

$$\int_{-1}^{+1} P_n(\mu)\frac{d}{d\mu}\left[(1-\mu^2)\frac{dP_l(\mu)}{d\mu}\right]d\mu=\int_{-1}^{+1}(\mu^2-1)\frac{dP_l(\mu)}{d\mu}\frac{dP_n(\mu)}{d\mu}d\mu. \tag{1.62}$$

Based on this reduction Equation 1.57 assumes the form:

$$\int_{-1}^{+1}\left[(\mu^2-1)\frac{dP_l(\mu)}{d\mu}\frac{dP_n(\mu)}{d\mu}+l(l+1)\,P_l(\mu)\,P_n(\mu)\right]d\mu=0. \tag{1.63}$$

In Equation 1.63 both $P_l(\mu)$ and $P_n(\mu)$ are now present on the equal footing. Due to this fact one can interchange indices l and n in Equation 1.63 and the equality to zero will still hold. This leads to another equation:

$$\int_{-1}^{+1}\left[(\mu^2-1)\frac{dP_n(\mu)}{d\mu}\frac{dP_l(\mu)}{d\mu}+n\,(n+1)\,P_n(\mu)\,P_l(\mu)\right]d\mu=0. \tag{1.64}$$

Subtracting Equation 1.64 from Equation 1.63 one obtains:

$$\left[l(l+1)-n\,(n+1)\right]\int_{-1}^{+1}P_n(\mu)\,P_l(\mu)\,d\mu=0. \tag{1.65}$$

Since l and n are arbitrary indices this equality can only be satisfied if the Legendre polynomials are in fact orthogonal. This proves Equation 1.55.

When $l=n$ the integral in Equation 1.55 assumes a finite value, N_l, representing the normalization condition:

$$\int_{-1}^{+1}P_l(\mu)\,P_l(\mu)\,d\mu=N_l. \tag{1.66}$$

This integral can be found with help of Rodrigues' formula, Equation 1.53. Substituting the explicit expression of Legendre polynomials given by Rodrigues' formula into Equation 1.66 gives:

$$N_l=\int_{-1}^{+1}P_l(\mu)\,P_l(\mu)\,dx=\frac{1}{2^{2l}(l!)^2}\int_{-1}^{+1}\left[\frac{d^l}{d\mu^l}(\mu^2-1)^l\right]\left[\frac{d^l}{d\mu^l}(\mu^2-1)^l\right]d\mu. \tag{1.67}$$

The first step in solving this integral is to eliminate one of the differentiation operations inside the integral; this can be achieved by using integration by parts. To assist in this, the following variable substitutions are introduced:

$$u=\frac{d^l}{d\mu^l}(\mu^2-1)^l,\quad\text{and}\quad dv=\frac{d^l}{d\mu^l}(\mu^2-1)^l\,d\mu. \tag{1.68}$$

From this it follows that:

$$du=\frac{d^{l+1}}{d\mu^{l+1}}(\mu^2-1)^l,\quad\text{and}\quad v=\frac{d^{l-1}}{d\mu^{l-1}}(\mu^2-1)^l. \tag{1.69}$$

Additionally, for $(\mu^2-1)^l$, every successive derivative of up to l-th degree will include the term (μ^2-1) in the result. This leads to:

$$uv\Big|_{-1}^{+1}=0. \tag{1.70}$$

Applying once the integration by parts to Equation 1.67 gives:

$$N_l=\frac{-1}{2^{2l}(l!)^2}\int_{-1}^{+1}\left[\frac{d^{l-1}}{d\mu^{l-1}}(\mu^2-1)^l\right]\left[\frac{d^{l+1}}{d\mu^{l+1}}(\mu^2-1)^l\right]d\mu. \tag{1.71}$$

Repeating this procedure to the total of l times leads to:

$$N_l = \frac{(-1)^l}{2^{2l}(l!)^2} \int_{-1}^{+1} (\mu^2 - 1)^l \left[\frac{d^{2l}}{d\mu^{2l}} (\mu^2 - 1)^l \right] d\mu. \tag{1.72}$$

The remaining differentiation operation in Equation 4.71 can be eliminated in a similar way *via* integration by parts. The following substitutions help simplify this process:

$$u = (\mu^2 - 1)^l, \qquad dv = \frac{d^{2l}}{d\mu^{2l}} (\mu^2 - 1)^l \, d\mu. \tag{1.73}$$

Applying integration by parts $2l$ times to Equation 1.72 based on Equation 1.58 brings in the coefficient $(-1)^{2l} = 1$, and reduces the term v in Equation 1.58 to $(\mu^2 - 1)^l$. All that remains is to differentiate the term u, an operation that will need to be performed $2l$ times. To assist in this task it is convenient to rewrite u using the binomial theorem, and differentiate. This gives:

$$\frac{d^{2l}}{d\mu^{2l}} (\mu^2 - 1)^l = \frac{d^{2l}}{d\mu^{2l}} \sum_{k=0}^{l} \frac{l!}{k!(l-k)!} \mu^{2k} (-1)^{l-k} = \frac{d^{2l}}{d\mu^{2l}} \mu^{2l} = (2l)!. \tag{1.74}$$

A careful examination of the binomial expansion reveals that only the term $k = l$ in the sum survives the differentiation. Based on the steps just performed, the normalization integral takes the form:

$$N_l = \frac{(-1)^l (2l)!}{2^{2l}(l!)^2} \int_{-1}^{+1} (\mu^2 - 1)^l \, d\mu = \frac{(2l)!}{2^{2l}(l!)^2} \int_{-1}^{+1} (1 - \mu^2)^l \, d\mu. \tag{1.75}$$

This integral can be solved in trigonometric form with $\mu = \cos\theta$ and $d\mu = -\sin\theta \, d\theta$. When that substitution is made, the integration limit $\mu = 1$ becomes $\theta = 0$, and $\mu = -1$ becomes $\theta = \pi$. This leads to:

$$\int_{-1}^{+1} (1 - \mu^2)^l \, d\mu = -\int_{\pi}^{0} (1 - \cos^2\theta)^l \sin\theta \, d\theta = \int_{0}^{\pi} \sin^{2l+1}\theta \, d\theta. \tag{1.76}$$

This is a standard table integral and has the solution:

$$\int_{0}^{\pi} \sin^n\theta \, d\theta = -\frac{\sin^{n-1}\theta \cdot \cos\theta}{n} \Bigg|_{0}^{\pi} + \frac{n-1}{n} \int_{0}^{\pi} \sin^{n-2}\theta \, d\theta, \tag{1.77}$$

where $n \geq 1$. Since the first term in the right-hand side in Equation 1.77 is zero for the present integration limits, Equation 1.77 reduces to:

$$\int_{0}^{\pi} \sin^n\theta \, d\theta = \frac{n-1}{n} \int_{0}^{\pi} \sin^{n-2}\theta \, d\theta. \tag{1.78}$$

A second recursive application of Equation 1.78 reduces the power of the sine function by another 2 and gives:

$$\int_{0}^{\pi} \sin^n\theta \, d\theta = \frac{n-1}{n} \int_{0}^{\pi} \sin^{n-2}\theta \, d\theta = \frac{(n-1)}{n} \frac{(n-3)}{(n-2)} \int_{0}^{\pi} \sin^{n-4}\theta \, d\theta. \tag{1.79}$$

For $n = 2l + 1$, repeating the application of Equation 1.78 to the total of l-times leads to the following result:

$$\int_0^{\pi} \sin^{2l+1}\theta \, d\theta = \frac{2l}{(2l+1)} \frac{(2l-2)}{(2l-1)} \frac{(2l-4)}{(2l-3)} \cdots \frac{2}{3} 2. \tag{1.80}$$

The final 2 in Equation 1.80 appears because:

$$\int_0^{\pi} \sin\theta \, d\theta = -\cos\theta \Big|_0^{\pi} = 2. \tag{1.81}$$

The numerator in Equation 1.80 can be simplified by separating the multiplier 2 from each multiplication term. There are l such terms. This leads to:

$$2l(2l-2)(2l-4)\ldots 2 = 2l\, 2(l-1)2(l-2)\ldots 2 = 2^l l!. \tag{1.82}$$

The denominator in Equation 1.80 can be rearranged in a similar way. Note that in it, each subsequent term is different from the preceding one by −2. This product can be turned into a factorial by multiplying and simultaneously dividing it by the missing terms. The extra terms, which are clearly identifiable in the divisor, happen to be equal to the expression in Equation 1.82. These operations lead to:

$$(2l+1)(2l-1)(2l-3)\ldots 3 = (2l+1)\frac{2l(2l-1)(2l-2)(2l-3)(2l-4)\ldots 1}{2l(2l-2)(2l-4)\ldots 1} = (2l+1)\frac{(2l)!}{2^l l!}. \tag{1.83}$$

Incorporating Equations 1.82 and 1.83 into Equation 1.80 gives:

$$\int_0^{\pi} \sin^{2l+1}\theta \, d\theta = \frac{2}{(2l+1)} \frac{2^{2l}(l!)^2}{(2l)!}. \tag{1.84}$$

Finally the normalization integral given by Equation 1.75 becomes:

$$N_l = \int_{-1}^{+1} P_l(\mu) P_l(\mu) dx = \frac{(2l)!}{2^{2l}(l!)^2} \frac{2}{2l+1} \frac{2^{2l}(l!)^2}{(2l)!} = \frac{2}{2l+1}. \tag{1.85}$$

1.6 Expansion of an Arbitrary Function in Legendre Series

Given that the Legendre polynomials $P_l(\mu)$ form a complete normalized and orthogonal set of basis functions in a Hilbert space for the interval $-1 \leq \mu \leq 1$ and $0 \leq \theta \leq \pi$, it follows that they can be used for expanding any arbitrary function. That is, any smooth and continuous function $f(\mu)$ defined for the domain [−1: 1] can be expanded in a Legendre series. This expansion takes the form:

$$f(\mu) = \sum_{l=0}^{\infty} c_l P_l(\mu). \tag{1.86}$$

The derivative of a product of two functions μ and $(\mu^2 - 1)^{l-1}$ in square brackets in Equation 1.101 can be found by applying the Leibnitz formula (see Appendix B.1). This leads to:

$$\frac{d}{d\mu}P_l(\mu)=\frac{2l}{2^l l!}\sum_{k=0}^{1}\frac{l!}{k!(l-k)!)}\left[\frac{d^k}{d\mu^k}\mu\right]\left[\frac{d^{l-k}}{d\mu^{l-k}}(\mu^2-1)^{l-1}\right]. \tag{1.102}$$

Note that index k in the summation in Equation 1.102 has the upper value of 1. Beyond that the sum terminates, because the derivative enclosed in the first square brackets becomes zero. This makes it possible to write the two terms in the sum explicitly:

$$\frac{d}{d\mu}P_l(\mu)=\frac{2l}{2^l l!}\left[\mu\frac{d^l}{d\mu^l}(\mu^2-1)^{l-1}+l\frac{d^{l-1}}{d\mu^{l-1}}(\mu^2-1)^{l-1}\right]. \tag{1.103}$$

In Equation 1.103 the constant value of $2l$ can be canceled out in the numerator and in the denominator. In addition to that the first term in the square brackets can be rearranged to the form of Rodrigues' formula. This gives:

$$\frac{d}{d\mu}P_l(\mu)=\frac{1}{2^{l-1}(l-1)!}\left[\mu\frac{d}{d\mu}\frac{d^{l-1}}{d\mu^{l-1}}(\mu^2-1)^{l-1}+l\frac{d^{l-1}}{d\mu^{l-1}}(\mu^2-1)^{l-1}\right]. \tag{1.104}$$

In Equation 1.104, the coefficient in front of the square brackets and the derivatives inside then compose a Legendre polynomial of degree $l - 1$, as defined by Rodrigues' formula. This means that:

$$\frac{d}{d\mu}P_l(\mu)=\mu\frac{d}{d\mu}P_{l-1}(\mu)+l\,P_{l-1}(\mu). \tag{1.105}$$

Considering the fact that parameter l in the above equation can take any value in the permitted range of $[0\colon \infty]$, it can be substituted by $l + 1$. This substitution turns Equation 1.97 into Equation 1.86, and completes the proof for Equation 1.94.

The next recurrence relation, Equation 1.95, can be obtained from the Legendre differential equation given by Equation 1.52:

$$\frac{d}{d\mu}\left[\left(1-\mu^2\right)\frac{dP_l(\mu)}{d\mu}\right]+l(l+1)\,P_l(\mu)=0. \tag{1.52}$$

Integrating Equation 1.52 leads to:

$$\int\frac{d}{d\mu}\left[(1-\mu^2)\frac{dP_l(\mu)}{d\mu}\right]d\mu+\int l(l+1)\,P_l(\mu)\,d\mu=0. \tag{1.106}$$

This expression simplifies to:

$$(1-\mu^2)\frac{d}{d\mu}P_l(\mu)+l(l+1)\int P_l(\mu)\,d\mu=0. \tag{1.107}$$

Further progress on solving Equation 1.107 depends on a knowledge of the integral over Legendre polynomial. Using Rodrigues' form of Legendre polynomials, the integral takes the form:

$$\int P_l(\mu)d\mu=\int\frac{d}{d\mu}\left[\frac{1}{2^l l!}\frac{d^{l-1}}{d\mu^{l-1}}(\mu^2-1)^l\right]d\mu=\frac{1}{2^l l!}\frac{d^{l-1}}{d\mu^{l-1}}(\mu^2-1)^l. \tag{1.108}$$

To find out where the integral in Equation 1.108 leads, we can start with a polynomial of degree $l+1$ *via* Rodrigues' formula, and apply the differentiation inside it once:

$$P_{l+1}(\mu)=\frac{1}{2^{l+1}(l+1)!}\frac{d^{l+1}}{d\mu^{l+1}}(\mu^2-1)^{l+1}=\frac{2(l+1)}{2^{l+1}(l+1)!}\frac{d^l}{d\mu^l}\left[\mu(\mu^2-1)^l\right]. \tag{1.109}$$

Using the Leibnitz formula, the expression in the square brackets can be differentiated to give:

$$P_{l+1}(\mu)=\frac{1}{2^l l!}\sum_{k=0}^{1}\frac{l!}{k!(l-k)!}\left[\frac{d^k}{d\mu^k}\mu\right]\left[\frac{d^{l-k}}{d\mu^{l-k}}(\mu^2-1)^l\right] \tag{1.110}$$

Since the sum in Equation 1.110 has the upper limit of 1 the summands can again be written explicitly. This leads to:

$$P_{l+1}(\mu)=\frac{1}{2^l l!}\left[\mu\frac{d^l}{d\mu^l}(\mu^2-1)^l+l\frac{d^{l-1}}{d\mu^{l-1}}(\mu^2-1)^l\right] \tag{1.111}$$

A brief inspection of Equation 1.111 reveals that it contains a Legendre polynomial of degree l in its first term. The second term can be written in an integral form. With that Equation 1.111 becomes:

$$P_{l+1}(\mu)=\mu\,P_l(\mu)+\frac{l}{2^l l!}\int\frac{d^l}{d\mu^l}(\mu^2-1)^l d\mu. \tag{1.112}$$

Since the integral in Equation 1.112 is a Legendre polynomial of degree l, Equation 1.112 can be rearranged to produce:

$$l\int P_l(\mu)\,d\mu=P_{l+1}(\mu)-\mu\,P_l(\mu). \tag{1.113}$$

Substituting Equation 1.113 into Equation 1.107 produces Equation 1.95, and that proves the corresponding recurrence relation.

The next recurrence relation to derive is Equation 1.96. We start the derivation by multiplying both sides of Equation 1.94 by $(1-\mu^2)$. This gives:

$$(1-\mu^2)\frac{d}{d\mu}P_{l+1}(\mu)=\mu\,(1-\mu^2)\frac{d}{d\mu}P_l(\mu)+(l+1)(1-\mu^2)\,P_l(\mu). \tag{1.114}$$

Next, substitute the first term in the right-hand side of Equation 1.114 by its value from Equation 1.95. This leads to:

$$(1-\mu^2)\frac{d}{d\mu}P_{l+1}(\mu)=(l+1)\,\mu^2\,P_l(\mu)-(l+1)\mu\,P_{l+1}(\mu)+(l+1)(1-\mu^2)\,P_l(\mu). \tag{1.115}$$

Equation 1.115 simplifies to give:

$$(1-\mu^2)\frac{d}{d\mu}P_{l+1}(\mu)=(l+1)P_l(\mu)-(l+1)\mu\,P_{l+1}(\mu). \tag{1.116}$$

In Equation 1.116, substituting parameter l by $l-1$ generates Equation 1.96.

To prove the next recurrence relation, Equation 1.97, one can equate the right-hand sides of Equations 1.95 and 1.96. This gives:

$$(l+1)\mu\, P_l(\mu) - (l+1)P_{l+1}(\mu) = l\, P_{l-1}(\mu) - l\mu\, P_l(\mu). \tag{1.117}$$

In this equation, one can move the Legendre polynomial of degree l from the left to the right-hand side, and multiply both sides by -1. This leads to:

$$(l+1)P_{l+1}(\mu) = (l+1)\mu\, P_l\,(\mu) + l\mu\, P_l\,(\mu) - l\, P_{l-1}(\mu). \tag{1.118}$$

After combining like terms one obtains Equation 1.97:

$$(l+1)P_{l+1}(\mu) = (2l+1)\mu\, P_l\,(\mu) - lP_{l-1}(\mu). \tag{1.97}$$

Equation 1.97 is Bonnet's recurrence relation, which can be used for computing successive terms of Legendre polynomials given a knowledge of the two preceding terms. If needed, the entire set of Legendre polynomials can be recursively computed starting from the first two terms, $P_0(\mu) = 1$ and $P_1(\mu) = \mu$.

Replacing l by $l - 1$ in Equation 1.97 gives a more convenient form of Bonnet's recursion formula for digital computing where the desired element l depends on the value of two previous terms, $l - 1$ and $l - 2$:

$$P_l(\mu) = \frac{1}{l}\left[(2l-1)\mu P_{l-1}(\mu) - (l-1)P_{l-2}(\mu)\right]. \tag{1.119}$$

The following recurrence relation, Equation 1.98, can also be derived using Bonnet's equation. Differentiating Equation 1.97 leads to:

$$(l+1)\frac{d}{d\mu}P_{l+1}(\mu) = (2l+1)P_l(\mu) + (2l+1)\mu\frac{d}{d\mu}P_l(\mu) - l\frac{d}{d\mu}P_{l-1}(\mu). \tag{1.120}$$

Subtracting Equation 1.94 from Equation 1.120 gives:

$$l\frac{d}{d\mu}P_{l+1}(\mu) = l\, P_l(\mu) + 2l\mu\frac{d}{d\mu}P_l(\mu) - l\frac{d}{d\mu}P_{l-1}(\mu). \tag{1.121}$$

Canceling the common coefficient l and rearranging terms leads to:

$$\frac{d}{d\mu}P_{l+1}(\mu) = 2\mu\frac{d}{d\mu}P_l(\mu) - \frac{d}{d\mu}P_{l-1}(\mu) + P_l(\mu). \tag{1.122}$$

To simplify Equation 1.122 and eliminate the $(d/d\mu)P_l(\mu)$ term, take Equation 1.94 and multiply it by 2. This gives:

$$2\frac{d}{d\mu}P_{l+1}(\mu) = 2\mu\frac{d}{d\mu}P_l(\mu) + 2(l+1)P_l(\mu). \tag{1.123}$$

After that, subtracting Equation 1.122 from Equation 1.123 gives the sought Equation 1.98:

$$\frac{d}{d\mu}P_{l+1}(\mu) = \frac{d}{d\mu}P_{l-1}(\mu) + (2l+1)P_l(\mu). \tag{1.98}$$

To prove the last recurrence relation (Equation 1.99), one can equate the right-hand side of Equations 1.94 and 1.98. This gives:

$$\mu \frac{d}{d\mu} P_l(\mu) + (l+1)P_l = \frac{d}{d\mu} P_{l-1}(\mu) + (2l+1)P_l(\mu). \tag{1.124}$$

By rearranging and combining like terms we get:

$$\frac{d}{d\mu} P_{l-1}(\mu) = \mu \frac{d}{d\mu} P_l(\mu) - l\, P_l(\mu). \tag{1.99}$$

The recurrence relations for Legendre polynomials, which are derived in this section, will be used in Chapter 2 to derive the recurrence relations for associated Legendre functions.

1.8 Analytic Expressions for First Few Legendre Polynomials

In mathematical manipulations, it is sometimes necessary to have access to analytic expressions of Legendre polynomials of degree l. Such expressions can theoretically be obtained by continuously applying differentiation in the Rodrigues' formula:

$$P_l(\mu) = \frac{1}{2^l l!} \frac{d^l}{d\mu^l} (\mu^2 - 1)^l, \tag{1.53}$$

where $\mu = \cos\,\theta$. It is indeed acceptable to use Equation 1.53 to derive the beginning terms in the Legendre series, for example, $P_0(\mu)$, $P_1(\mu)$, and the like. Beyond that, the direct differentiation approach quickly becomes too cumbersome. The more practical technique is to derive the successive terms in degree l by using Bonnet's recurrence relation, Equation 1.119. Table 1.1 lists the first few Legendre polynomials derived in this way.

If the numerical values of Legendre polynomials are needed in a calculation, the analytic expressions from Table 1.1 should not be used because numerical precision errors increase rapidly as the degree l increases. These errors arise from the cancellation of large numbers in the analytic expressions; note the alternating signs in the individual terms of the Legendre polynomials. Instead, whenever numerical values of Legendre polynomials are needed in a computer program, they can be recursively computed by using Bonnet's recurrence relation, Equation 1.119.

The analytic expressions in Table 1.1 are, however, useful in testing the correctness of a computer code that implements the recursive computation of Legendre polynomials.

1.9 Symmetry Properties of Legendre Polynomials

The Rodrigues' formula undoubtedly provides the most frequently used representation of Legendre polynomials in analytical manipulations. However, due to the presence of the differentiation operator in the formula, it is difficult to plot this function, inspect the function value in specific points, or determine its symmetry properties. Therefore it would be useful to derive an alternative representation of Legendre polynomials that is better suited for function analysis.

The next target to find is the value of Legendre polynomials at the point $\mu = 1$. Neither Rodrigues' equation nor Equation 1.130 are suitable for that purpose. As in the previous case, the presence of the differentiation operator in Rodrigues' equation obscures the interpretation, and for $\mu = 1$, Equation 1.130 leads to a sum of coefficients that is difficult to simplify. A different representation of Legendre polynomials than that in Equation 1.130 is needed in order to reduce the number of summands in the sum to a single summand when $\mu = 1$. The first step in deriving such an equation starts with considering the equality:

$$(\mu^2-1)^l = (\mu-1)^l(\mu+1)^l = 2^l(\mu-1)^l\left(1+\frac{\mu-1}{2}\right)^l. \tag{1.134}$$

By using the binomial theorem, the second multiplier in Equation 1.134 can be expanded in a power series. This gives:

$$(\mu^2-1)^l = 2^l(\mu-1)^l\sum_{k=0}^{l}\frac{l!}{k!(l-k)!}\left(\frac{\mu-1}{2}\right)^k. \tag{1.135}$$

The term $(\mu - 1)^l$ that is located in front of the sum can then be merged with the corresponding term inside the sum. Similarly, the coefficient $2l$ can also be migrated into the sum. These transformations produce:

$$(\mu^2-1)^l = \sum_{k=0}^{l}\frac{2^{l-k}l!}{k!(l-k)!}(\mu-1)^{l+k}. \tag{1.136}$$

Substituting Equation 1.136 into Rodrigues' equation, and cancelling out common factors leads to:

$$P_l(\mu) = \frac{1}{2^l l!}\frac{d^l}{d\mu^l}\left[\sum_{k=0}^{l}\frac{2^{l-k}l!}{k!(l-k)!}(\mu-1)^{l+k}\right] = \frac{d^l}{d\mu^l}\left[\sum_{k=0}^{l}\frac{1}{2^k k!(l-k)!}(\mu-1)^{l+k}\right]. \tag{1.137}$$

Now the expression in square brackets can easily be differentiated. Differentiation of the term $(\mu - 1)^{l+k}$ l times reduces its power from $l + k$ to k, and generates the coefficient:

$$(l+k)(l+k-1)\ldots(l+k-l+1) = \frac{(l+k)!}{k!}. \tag{1.138}$$

With that, the result of differentiation is:

$$P_l(\mu) = \sum_{k=0}^{l}\frac{(l+k)!}{(k!)^2(l-k)!}\left(\frac{\mu-1}{2}\right)^k. \tag{1.139}$$

This provides another expression, in addition to Equation 1.130 for Legendre polynomials, as an alternative to Rodrigues' formula.

Equation 1.139 makes it easy to find the value of Legendre polynomials in the point $\mu = 1$. This condition results in all summands, except the first one, in the sum in Equation 1.139 being zero. For the first one, that is, when $k = 0$ the coefficient is equal to:

$$\frac{l!}{(0!)^2 l!} = 1. \tag{1.140}$$

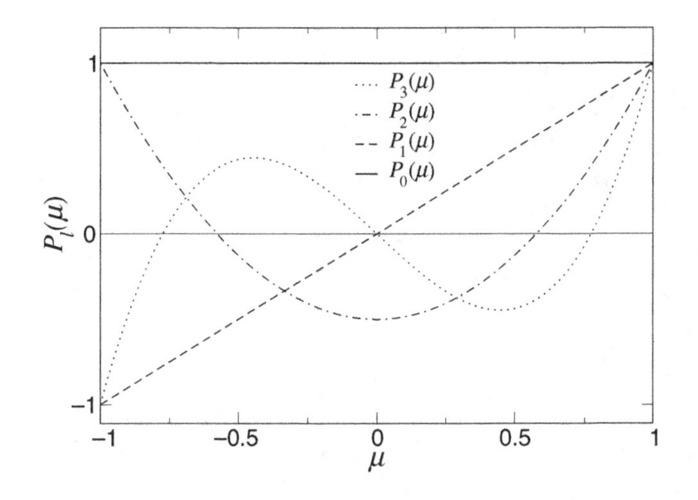

FIGURE 1.6 Legendre polynomials.

This finding leads to an interesting conclusion that the value of Legendre polynomial in the point $\mu = 1$ is always unity regardless of the value of index l, that is:

$$P_l(1) = 1. \tag{1.141}$$

In concluding the above analysis, it is illuminating to visualize a few of the Legendre polynomials in order to see how they behave inside the function domain. Select Legendre polynomials are graphed in Figure 1.6.

The graphs plotted in Figure 1.6 visually illustrate the conclusions derived in this section about the symmetry in the value of Legendre polynomials between the negative and positive ends of the argument. The plots demonstrate the value of Legendre polynomials in the points $\mu = -1, 0, 1$, and show how many times the function crosses the horizontal line of zero. Remarkably, that number coincides with the value of index l for the corresponding Legendre polynomial. This is consistent with Legendre function $P_l(\mu)$ being a polynomial of degree l that has the highest power of l in the argument μ. This is a re-statement of the fact that a polynomial of degree l has exactly l roots.

Despite their interesting properties and overall significance, Legendre polynomials have limited utility in the Fast Multipole Method. Therefore, the next step in building up the mathematical tool of Fast Multipole Method is to look at more general functions, which incorporate Legendre polynomials as a special case.

2

Associated Legendre Functions

A fundamental problem in electrostatics is finding the electrostatic potential function, Φ, for a system of electric charges. Although the exact form of this function depends on specific details of the electrostatic system, much information about the general properties of potential function may be obtained from the solution of Laplace's equation, $\nabla^2\Phi = 0$, which the potential function must obey. Having led to the introduction of Legendre polynomials in Chapter 1, Laplace's equation guides the concept of factorization of the electrostatic potential function, Φ, further toward finding a general solution, with the associated Legendre functions being the next milestone on this path. Developing that solution is the purpose of the present chapter.

The most elegant solution of Laplace's equation emerges in spherical polar coordinates (r, θ, ϕ) that leads to factorization of the electrostatic potential function Φ into a product of radial, $R(r)$, angular, $P(\theta)$, and azimuthal, $Z(\phi)$ functions,

$$\Phi(r,\ \theta,\ \phi) = R(r)\,P(\theta)\,Z(\phi), \tag{2.1}$$

so that each function in the product can be found independently. When written in spherical polar coordinates, Laplace's equation takes the form:

$$\left[\frac{1}{r^2}\frac{\partial}{\partial r}\left(r^2\frac{\partial}{\partial r}\right) + \frac{1}{r^2\sin\theta}\frac{\partial}{\partial\theta}\left(\sin\theta\frac{\partial}{\partial\theta}\right) + \frac{1}{r^2\sin^2\theta}\frac{\partial^2}{\partial\phi^2}\right]\Phi = 0. \tag{2.2}$$

Substituting the factorized potential function from Equation 2.1 into Equation 2.2 and multiplying the resulting equation by $r^2\sin^2\theta/(R\,P\,Z)$ leads to two equations, each of which can then be solved separately:

$$\frac{1}{Z}\frac{d^2Z}{d\phi^2} = -m^2, \tag{2.3}$$

and

$$\frac{\sin^2\theta}{R}\frac{d}{dr}\left(r^2\frac{dR}{dr}\right) + \frac{\sin\theta}{P}\frac{d}{d\theta}\left(\sin\theta\frac{dP}{d\theta}\right) = m^2, \tag{2.4}$$

where m is a parameter that can assume integer values in the positive and negative range, including zero.

Equation 2.3 describes the motion of harmonic oscillator. Its solution is:

$$Z = C\,e^{\pm im\phi}, \tag{2.5}$$

where C is an arbitrary constant, and i is a complex number. This leaves Equation 2.4 to work on.

The method of solution of Equation 2.4 depends on the value of parameter m. The simplest solution appears in the case of azimuthal symmetry when $m = 0$. In that case, the radial, r, and angular, θ, coordinates are easily separated. This turns Equation 2.4 into two independent equations, which are coupled *via* parameter $l\ (l + 1)$:

$$\frac{1}{R}\frac{d}{dr}\left(r^2\frac{dR}{dr}\right) = l(l+1), \tag{2.6}$$

and

$$\frac{1}{P\sin\theta}\frac{d}{d\theta}\left(\sin\theta\frac{dP}{d\theta}\right)=-l(l+1). \tag{2.7}$$

The solution for radial function, R, in Equation 2.6 is:

$$R=A_l r^l+B_l r^{-l-1}, \tag{2.8}$$

where A_l and B_l are arbitrary constants. The solutions for angular function, P, in Equation 2.7 are the Legendre polynomials. Defined by means of Rodrigues' formula, Legendre polynomials have the form:

$$P_l(\mu)=\frac{1}{2^l l!}\frac{d^l}{d\mu^l}(\mu^2-1)^l, \tag{2.9}$$

where $\mu=\cos\theta$, and l is a non-negative integer number that assumes values in the range $[0: +\infty]$. Combining Equations 2.5, 2.8, and 2.9 according to Equation 2.1 gives the electrostatic potential function solution of Laplace's equation for the case of azimuthal symmetry.

All the above steps and the conclusions derived from them are of significance in electrostatics. However, solving Laplace's equation by implying azimuthal symmetry limits the range of its application. In practice, few real cases satisfy that requirement, as the majority of real systems do not have symmetry. Therefore it is important to find a general solution of Laplace's equation when no symmetry is present in the position of source charges. The present chapter addresses that task.

2.1 Generalized Legendre Equation

The general solution of Equation 2.4 requires considering the case of $m\neq 0$. To start with, this equation can be rewritten in a more convenient form. Dividing it by $\sin^2\theta$ and rearranging the terms leads to:

$$\frac{1}{P\sin\theta}\frac{d}{d\theta}\left(\sin\theta\frac{dP}{d\theta}\right)-\frac{m^2}{\sin^2\theta}+\frac{1}{R}\frac{d}{dr}\left(r^2\frac{dR}{dr}\right)=0. \tag{2.10}$$

Since the radial, R and angular, P functions in Equation 2.10 are independent on each other this equation can only be satisfied if the radial and angular terms are equal to a constant and mutually cancel out. This leads to a similar variable separation that has been previously applied to Equation 2.4. Equating the radial and angular terms in Equation 2.10 to a constant, $l(l+1)$ leads to Equation 2.6 for the radial term, and to Equation 2.11 for the angular part:

$$\frac{1}{P\sin\theta}\frac{d}{d\theta}\left(\sin\theta\frac{dP}{d\theta}\right)-\frac{m^2}{\sin^2\theta}=-l(l+1). \tag{2.11}$$

The solution of Equation 2.6 is already known and is given by Equation 2.8. All that remains is to find the solution for Equation 2.11. Multiplying Equation 2.11 by P and rearranging the terms leads to:

$$\frac{1}{\sin\theta}\frac{d}{d\theta}\left(\sin\theta\frac{dP}{d\theta}\right)+\left[l(l+1)-\frac{m^2}{\sin^2\theta}\right]P=0. \tag{2.12}$$

Next, introducing variable substitution $\mu=\cos\theta$ and $d\mu=-\sin\theta\, d\theta$ produces a generalized Legendre equation:

$$\frac{d}{d\mu}\left[(1-\mu^2)\frac{dP}{d\mu}\right]+\left[l(l+1)-\frac{m^2}{1-\mu^2}\right]P=0. \tag{2.13}$$

This is a very important differential equation in mathematical physics. The name *generalized* emphasizes the fact that this equation applies to the general case when no azimuthal symmetry is present. In this equation, setting $m = 0$ leads back to the ordinary Legendre differential equation:

$$\frac{d}{d\mu}\left[(1-\mu^2)\frac{dP_l}{d\mu}\right]+l(l+1)P_l=0, \tag{2.14}$$

which corresponds to the case of azimuthal symmetry and has the Legendre polynomials $P_l(\mu)$ as its solution. This connection between the general and ordinary Legendre equations will help in finding a solution of Equation 2.13 based on the known solution of Equation 2.14. To reveal the transformation path that establishes the connection between these equations it is necessary to first rearrange Equation 2.13.

Implementing this strategy, the solution of Equation 2.13 can be sought in the form:

$$P=(1-\mu^2)^{\frac{m}{2}}u, \tag{2.15}$$

where u is a function to be determined later on. Substituting the expression for P from Equation 2.15 into Equation 2.13 gives:

$$\frac{d}{d\mu}\left[(1-\mu^2)\frac{d}{d\mu}\left[(1-\mu^2)^{\frac{m}{2}}u\right]\right]+\left[l(l+1)-\frac{m^2}{1-\mu^2}\right](1-\mu^2)^{\frac{m}{2}}u=0. \tag{2.16}$$

Applying differentiation inside the double enclosed square brackets leads to:

$$\frac{d}{d\mu}\left[(1-\mu^2)^{\frac{m+2}{2}}\frac{du}{d\mu}-m\mu(1-\mu^2)^{\frac{m}{2}}u\right]+\left[l(l+1)-\frac{m^2}{1-\mu^2}\right](1-\mu^2)^{\frac{m}{2}}u=0. \tag{2.17}$$

The expression inside the left-most square brackets after differentiation becomes:

$$(1-\mu^2)^{\frac{m+2}{2}}\frac{d^2u}{d\mu^2}-(m+2)\mu(1-\mu^2)^{\frac{m}{2}}\frac{du}{d\mu}-m(1-\mu^2)^{\frac{m}{2}}u+m^2\mu^2(1-\mu^2)^{\frac{m-2}{2}}u-m\mu(1-\mu^2)^{\frac{m}{2}}\frac{du}{d\mu}. \tag{2.18}$$

It can be simplified by combining the second and fifth terms. This operation gives:

$$(1-\mu^2)^{\frac{m+2}{2}}\frac{d^2u}{d\mu^2}-2(m+1)\mu(1-\mu^2)^{\frac{m}{2}}\frac{du}{d\mu}-m(1-\mu^2)^{\frac{m}{2}}u+m^2\mu^2(1-\mu^2)^{\frac{m-2}{2}}u. \tag{2.19}$$

The second summand in Equation 2.17 can be re-written as:

$$l(l+1)(1-\mu^2)^{\frac{m}{2}}u-m^2(1-\mu^2)^{\frac{m-2}{2}}u. \tag{2.20}$$

Note that the sum of Equations 2.19 and 2.20 is equal to the left-hand side of Equation 2.17. Before adding those parts together, a further simplification can be made by combining the fourth term in Equation 2.19 with second term in Equation 2.20. This gives:

$$m^2\mu^2(1-\mu^2)^{\frac{m-2}{2}}u-m^2(1-\mu^2)^{\frac{m-2}{2}}u=-m^2(1-\mu^2)^{\frac{m}{2}}u. \tag{2.21}$$

Now everything is ready to re-assemble Equation 2.17 from the prepared parts. Adding Equations 2.19 and 2.20, and taking into account Equation 2.21 leads to:

$$(1-\mu^2)^{\frac{m+2}{2}}\frac{d^2u}{d\mu^2}-2(m+1)\mu(1-\mu^2)^{\frac{m}{2}}\frac{du}{d\mu}-m(1-\mu^2)^{\frac{m}{2}}u+l(l+1)(1-\mu^2)^{\frac{m}{2}}u-m^2(1-\mu^2)^{\frac{m}{2}}u=0. \tag{2.22}$$

The term $(1 - \mu^2)^{m/2}$ cannot equal zero—if it did the search for function u would be meaningless. With that, all terms in Equation 2.22 can be divided by $(1 - \mu^2)^{m/2}$ This leads to:

$$(1-\mu^2)\frac{d^2u}{d\mu^2}-2(m+1)\mu\frac{du}{d\mu}-mu+l(l+1)u-m^2u=0. \tag{2.23}$$

This equation can be written in a more compact form:

$$(1-\mu^2)\frac{d^2u}{d\mu^2}-2(m+1)\mu\frac{du}{d\mu}+(l-m)(l+m+1)u=0. \tag{2.24}$$

Equation 2.24 is a transformed form of the generalized Legendre equation, which was previously defined by Equation 2.13. Unlike Equation 2.13 that is written for function P, Equation 2.24 is provided in terms of an unknown function u still to be determined. This accomplishes the first part of the transformation that aims to connect the generalized and ordinary Legendre equations. The second part of the transformation requires making a modification to the ordinary Legendre equation.

The idea behind this portion of transformation is as follows. Since the solution of the ordinary Legendre equation is known, if we could transform the ordinary Legendre equation into the generalized Legendre equation *via* a series of steps, it would mean that the application of the same steps to Legendre polynomials would likewise turn them into solutions of the generalized Legendre equation. Recall that Legendre polynomials are the solutions of the ordinary Laplace's equation. Implementing this strategy, the ordinary Legendre equation given by Equation 2.14 can be differentiated m-times. This gives:

$$\frac{d^{m+1}}{d\mu^{m+1}}\left[(1-\mu^2)\frac{d}{d\mu}P_l\right]+l(l+1)\frac{d^m}{d\mu^m}P_l=0. \tag{2.25}$$

The expression in square brackets is a product of two functions. Based on that, the Leibnitz formula (see Appendix B.1) can be used to perform the differentiation. This leads to:

$$\sum_{k=0}^{m+1}\frac{(m+1)!}{k!(m+1-k)!}\left[\frac{d^{m+1-k}}{d\mu^{m+1-k}}(1-\mu^2)\right]\frac{d^{k+1}}{d\mu^{k+1}}P_l+l(l+1)\frac{d^m}{d\mu^m}P_l=0. \tag{2.26}$$

The expression in square brackets can be differentiated exactly twice because all other terms are automatically zero. The terms surviving differentiation have the index $k = m + 1$, m, and $m - 1$. These are:

$$\frac{d^0}{d\mu^0}(1-\mu^2)=1-\mu^2, \qquad \frac{d}{d\mu}(1-\mu^2)=-2\mu, \qquad \frac{d^2}{d\mu^2}(1-\mu^2)=-2, \tag{2.27}$$

respectively. Writing those terms in the sum in Equation 2.26 explicitly turns the equation into:

$$(1-\mu^2)\frac{d^{m+2}}{d\mu^{m+2}}P_l-2\frac{(m+1)!}{m!}\mu\frac{d^{m+1}}{d\mu^{m+1}}P_l-\frac{(m+1)!}{(m-1)!}\frac{d^m}{d\mu^m}P_l+l(l+1)\frac{d^m}{d\mu^m}P_l=0. \tag{2.28}$$

After canceling out factorials, combining like terms and taking into account that

$$l(l+1)-m(m+1)=(l-m)(l+m+1), \tag{2.29}$$

we obtain the final transformed form of the ordinary Legendre equation:

$$(1-\mu^2)\frac{d^{m+2}}{d\mu^{m+2}}P_l-2(m+1)\mu\frac{d^{m+1}}{d\mu^{m+1}}P_l+(l-m)(l+m+1)\frac{d^m}{d\mu^m}P_l=0. \tag{2.30}$$

At this point it is easy to see that for $m=0$ this equation reverts back to the ordinary Legendre equation:

$$(1-\mu^2)\frac{d^2}{d\mu^2}P_l-2\mu\frac{d}{d\mu}P_l+l(l+1)P_l=0, \tag{2.31}$$

which is the expanded form of Equation 2.14 that is obtained after applying differentiation to the term in the square brackets in Equation 2.14. The smooth backward transition from Equations 2.30 to 2.31 verifies that the differentiation was performed correctly.

Now that both generalized and ordinary Legendre equations have undergone individual transformations that resulted in obtaining Equations 2.24 and 2.30, respectively, it is time to show that the latter two equations are equivalent, and that the transformations merged into the common point. To show the equivalence of Equations 2.24 and 2.30, the latter equation can be rewritten as:

$$(1-\mu^2)\frac{d^2}{d\mu^2}\left[\frac{d^m}{d\mu^m}P_l\right]-2(m+1)\mu\frac{d}{d\mu}\left[\frac{d^m}{d\mu^m}P_l\right]+(l-m)(l+m+1)\left[\frac{d^m}{d\mu^m}P_l\right]=0. \tag{2.32}$$

From a comparison of Equations 2.32 and 2.24, it becomes apparent that the unknown function u in Equation 2.24 has the form:

$$u=\frac{d^m}{d\mu^m}P_l. \tag{2.33}$$

Substituting the value of function u from Equation 2.33 into Equation 2.15 gives the solution of the generalized Legendre equation, Equation 2.13:

$$P=C(1-\mu^2)^{\frac{m}{2}}\frac{d^m}{d\mu^m}P_l, \tag{2.34}$$

where C is an arbitrary constant, normally set to $(-1)^m$. With that substitution, the solution of the generalized Legendre equation, Equation 2.13, becomes:

$$P_l^m=(-1)^m(1-\mu^2)^{\frac{m}{2}}\frac{d^m}{d\mu^m}P_l. \tag{2.35}$$

where $P_l^m(\mu)$ or simply P_l^m are referred to as associated Legendre functions.

Having resolved the angular dependence in Laplace's equation for the general case, that is, when no symmetry is present in the position of charge sources, it is now possible to write a completely general form of the electrostatic potential function that was sketched by Equation 2.1. Combining the individual solutions for radial, angular, and azimuthal functions presented by Equations 2.8, 2.35, and 2.5, respectively, in the form of their product leads to:

$$\Phi(r,\theta,\phi)=\sum_{l=0}^{\infty}\sum_{m=-l}^{l}[A_l^m r^l+B_l^m r^{-l-1}]P_l^m e^{im\phi}, \tag{2.36}$$

where A_l^m and B_l^m are linear coefficients. An exact form of this expansion will be determined in subsequent chapters.

According to Equation 2.36, the expansion of a potential function in a series of products of radial, angular, and azimuthal functions has an infinite number of terms. This would obviously make the use of Equation 2.36 unsuitable for digital computation. However, in practical applications, the series quickly converges, and the expansion can be truncated to a finite number of terms with only negligible rounding error.

2.2 Associated Legendre Functions

Associated Legendre functions are important mathematical functions that find numerous applications in mathematics, physics, geodesy, engineering, and computer graphics. They are an essential part of the mathematical apparatus of the Fast Multipole Method. Therefore an understanding of the key mathematical properties of these remarkable functions is essential.

Associated Legendre functions are defined by Equation 2.35 for the argument $\mu = \cos\theta$, where θ is the polar angle in a spherical polar coordinate system. Substituting the expression for Legendre polynomials, $P_l(\mu)$, in the form defined by Rodrigues' formula into Equation 2.35 gives the differential form of associated Legendre functions:

$$P_l^m(\mu) = \frac{(-1)^m}{2^l l!}(1-\mu^2)^{\frac{m}{2}} \frac{d^{l+m}}{d\mu^{l+m}}(\mu^2-1)^l. \tag{2.37}$$

Each associated Legendre function is characterized by integer indices l and m. Conventionally, index l is referred to as the degree and index m as the order of the function. Index l is defined in the interval $[0: +\infty]$, and index m has the values in the range $-l \leq m \leq l$. For $m > l$, the expression in Equation 2.37 becomes zero due to differentiation. When $m = 0$, the associated Legendre function becomes a Legendre polynomial of degree l, $P_l(\mu)$. For $m < -l$, the associated Legendre function is non-zero but, because there are no applications in physics that use that range, functions of these orders are of no interest.

The constant of $(-1)^m$ that stands in front of associated Legendre functions is a phase factor that comes from the Condon-Shortley phase convention. This phase factor is necessary for a consistent treatment of atomic spectra and to reproduce early works in quantum mechanics. Specifically, the choice of phase factor affects the tabulated values of the Gaunt integral of the triple product of associated Legendre functions that appears in atomic spectroscopy.

At this point, it might be useful to clarify the naming difference between associated Legendre functions, $P_l^m(\mu)$ and Legendre polynomials, $P_l(\mu)$. Unlike Legendre polynomials the solutions of generalized Legendre equation are not referred to as polynomials; instead they are called associated Legendre functions. This is because in order for a function to be a polynomial it should have the argument raised only to integer powers. In $P_l^m(\mu)$ this condition is satisfied only for even values of m. For odd values of m, $P_l^m(\mu)$ is not a polynomial due to the presence of term $(1 - \mu^2)^{m/2}$ in the function.

Continuing the overview of associated Legendre functions, it is worth mentioning that there are many ways to represent them analytically; see Appendices B.2 and B.4 for a few examples. Among the various forms of associated Legendre functions the Rodrigues' representation given by Equation 2.37 is the most popular. It conveniently generates the correct associated Legendre functions for both positive and negative range of m. From now onward, for consistency and clarity in notation we will adopt a convention of treating parameter m as a positive number unless otherwise noted, while explicitly showing the negative sign outside the letter m.

The first useful property to review concerns the symmetry behavior of index m. For example, the associated Legendre functions for m and $-m$ are identical except for a proportionality coefficient. This statement can be proved and the proportionality coefficient found by converting the differentiation part inside the associated Legendre function into its polynomial form. The derivation starts from considering the positive values of m. The differentiation term of associated Legendre function can be represented in the following form:

$$\frac{d^{l+m}}{d\mu^{l+m}}(\mu^2-1)^l = \frac{d^{l+m}}{d\mu^{l+m}}\left[(\mu-1)^l(\mu+1)^l\right]. \tag{2.38}$$

Using the Leibnitz formula, a differentiation can be performed on the product of the two functions inside the square brackets, this gives:

$$\frac{d^{l+m}}{d\mu^{l+m}}(\mu^2-1)^l = \sum_{k=0}^{l+m} \frac{(l+m)!}{k!(l+m-k)!}\left[\frac{d^k}{d\mu^k}(\mu-1)^l\right]\left[\frac{d^{l+m-k}}{d\mu^{l+m-k}}(\mu+1)^l\right]. \tag{2.39}$$

For the result of the differentiation to be non-zero, the index k must obey the condition $m \leq k \leq l$; this then defines the lower index in the sum. Applying differentiation to the expressions in square brackets leads to:

$$\frac{d^{l+m}}{d\mu^{l+m}}(\mu^2-1)^l = \sum_{k=m}^{l} \frac{(l+m)!}{k!(l+m-k)!}\frac{l!}{(l-k)!}(\mu-1)^{l-k}\frac{l!}{(k-m)!}(\mu+1)^{k-m}. \tag{2.40}$$

Next consider the term with negative m, that is, $-m$. Differentiation now leads to:

$$\frac{d^{l-m}}{d\mu^{l-m}}(\mu^2-1)^l = \sum_{j=0}^{l-m} \frac{(l-m)!}{j!(l-m-j)!}\frac{l!}{(l-j)!}(\mu-1)^{l-j}\frac{l!}{(j+m)!}(\mu+1)^{j+m}. \tag{2.41}$$

Equation 2.41 can be rearranged in two steps to give it a more convenient look for the further analysis. In the first step, we take a common multiplier out of the sum. This gives:

$$\frac{d^{l-m}}{d\mu^{l-m}}(\mu^2-1)^l = (l-m)!(\mu^2-1)^m \sum_{j=0}^{l-m} \frac{l!}{j!(l-m-j)!(l-j)!}(\mu-1)^{l-j-m}\frac{l!}{(j+m)!}(\mu+1)^{j}. \tag{2.42}$$

After that, rearranging the terms inside the sum leads to:

$$\frac{d^{l-m}}{d\mu^{l-m}}(\mu^2-1)^l = (l-m)!(\mu^2-1)^m \sum_{j=0}^{l-m} \frac{l!}{(j+m)!(l-j)!(l-j-m)!}(\mu-1)^{l-j-m}\frac{l!}{j!}(\mu+1)^{j}. \tag{2.43}$$

Now we can go back to Equation 2.40 and apply the substitution $k = j + m$. With that the lower limit of index $k = m$ in the sum turns into $j = 0$. The upper limit in the sum gets reduced from l to $l - m$ so that the total number of elements in the sum remains unchanged. This gives:

$$\frac{d^{l+m}}{d\mu^{l+m}}(\mu^2-1)^l = \sum_{j=0}^{l-m} \frac{(l+m)!}{(j+m)!(l-j)!}\frac{l!}{(l-j-m)!}(\mu-1)^{l-j-m}\frac{l!}{j!}(\mu+1)^{j} \tag{2.44}$$

From a visual comparison of Equations 2.43 and 2.44 it becomes apparent that these two equations are proportional so that Equation 2.43 takes the form:

$$\frac{d^{l-m}}{d\mu^{l-m}}(\mu^2-1)^l = \frac{(l-m)!}{(l+m)!}(\mu^2-1)^m\frac{d^{l+m}}{d\mu^{l+m}}(\mu^2-1)^l. \tag{2.45}$$

This finishes the preparation, and we are finally ready to derive the relation between associated Legendre functions with positive and negative values of index m. Using the explicit form of the associated Legendre functions given in Equation 2.37, the expression for the negative range of m is:

$$P_l^{-m}(\mu) = \frac{(-1)^{-m}}{2^l l!}(1-\mu^2)^{-m/2}\frac{d^{l-m}}{d\mu^{l-m}}(\mu^2-1)^l. \tag{2.46}$$

Substituting the expression from Equation 2.45 into Equation 2.46 and using the fact that $(-1)^{-m}$ is equal to $(-1)^m$ gives:

$$P_l^{-m}(\mu) = \frac{(-1)^m}{2^l l!}(1-\mu^2)^{-m/2}\frac{(l-m)!}{(l+m)!}(\mu^2-1)^m\frac{d^{l+m}}{d\mu^{l+m}}(\mu^2-1)^l. \quad (2.47)$$

Taking into account that $(\mu^2-1)^m = (-1)^m(1-\mu^2)^m$, Equation 2.47 simplifies to:

$$P_l^{-m}(\mu) = (-1)^m\frac{(l-m)!}{(l+m)!}\frac{(-1)^m}{2^l l!}(1-\mu^2)^{m/2}\frac{d^{l+m}}{d\mu^{l+m}}(\mu^2-1)^l. \quad (2.48)$$

From this it follows that:

$$P_l^{-m}(\mu) = (-1)^m\frac{(l-m)!}{(l+m)!}P_l^m(\mu). \quad (2.49)$$

Equation 2.49 establishes the symmetry relation between associated Legendre functions of opposite sign in index m in the range $-l \le m \le l$. This equation is often used to compute $P_l^{-m}(\mu)$ from the known value of $P_l^m(\mu)$.

From here on, in order to reduce the clutter in manipulations, it is convenient to drop the argument μ from associated Legendre functions and refer to it simply as P_l^m whenever its argument is trivially obvious.

2.3 Orthogonality and Normalization of Associated Legendre Functions

Associated Legendre functions, introduced by Equation 2.37, form a complete basis set in the lower index l for any fixed value of the upper index m, and are orthogonal when index m is the same in both functions, that is:

$$\int_{-1}^{1} P_l^m P_n^m d\mu = 0, \qquad \text{when } l \neq n. \quad (2.50)$$

Their normalization integral is:

$$\int_{-1}^{1} P_l^m P_l^m d\mu = \frac{2}{2l+1}\frac{(l+m)!}{(l-m)!}. \quad (2.51)$$

To prove the first integral in Equation 2.50, we shall consider the non-negative range of index m, that is, $m \ge 0$. The negative range is handled *via* symmetry relation defined by Equation 2.49. Substituting the expression for associated Legendre functions from Equation 2.35, written *via* Legendre polynomials, into Equation 2.50 gives:

$$\int_{-1}^{1} P_l^m P_n^m \, d\mu = \int_{-1}^{1} (1-\mu^2)^m \left[\frac{d^m}{d\mu^m}P_l\right]\left[\frac{d^m}{d\mu^m}P_n\right] d\mu. \quad (2.52)$$

As the result of this substitution, the term $(-1)^{2m}$ vanishes and power of the term $(1 - \mu^2)$ changes from $m/2$ to m. Having obtained the integer power of the multiplication factor $(1 - \mu^2)$, the next step is to change the sign in it in order to unify the latter with the similar term $(\mu^2 - 1)$ located inside the Legendre polynomial, Equation 2.9. Applying the sign change and substituting the Legendre polynomials in their Rodrigues' form leads to:

$$\int_{-1}^{1} P_l^m P_n^m \, d\mu = \frac{1}{2^{2l}(l!)^2}(-1)^m \int_{-1}^{1} (\mu^2-1)^m \left[\frac{d^{l+m}}{d\mu^{l+m}}(\mu^2-1)^l\right]\left[\frac{d^m}{d\mu^m}(\mu^2-1)^n\right] d\mu. \tag{2.53}$$

Equation 2.53 can be solved using integration by parts, which uses the formula:

$$\int_{-1}^{1} u\,dv = uv\Big|_{-1}^{1} - \int_{-1}^{1} v\,du. \tag{2.54}$$

This involves introducing the following substitutions:

$$u = (\mu^2-1)^m \left[\frac{d^{l+m}}{d\mu^{l+m}}(\mu^2-1)^l\right], \tag{2.55}$$

and

$$dv = \left[\frac{d^{n+m}}{d\mu^{n+m}}(\mu^2-1)^n\right] d\mu. \tag{2.56}$$

These substitutions are chosen in order to make the term u easily differentiable and the term dv easily integrable. They rely on the assumption that $l < n$. Since $l \neq n$, and because l and n are arbitrary parameters, it is not particularly important which one out of two indices is smaller than the other one as long as we keep track of their relation. As it will soon be apparent, associating the variable u with index l makes Equation 2.55 differentiable fewer times than the number of times Equation 2.56 can be integrated. The reason for this and the purpose of that condition will be revealed during the course of discussion. Now everything is ready to perform integration by parts. This operation will be performed $l+m$ times to reduce Equation 2.53 to the point where the remaining integral can be easily resolved. Differentiating Equation 2.55 $l+m$ times gives:

$$\frac{d^{l+m}}{d\mu^{l+m}} u = \frac{d^{l+m}}{d\mu^{l+m}}\left[(\mu^2-1)^m \frac{d^{l+m}}{d\mu^{l+m}}(\mu^2-1)^l\right]. \tag{2.57}$$

Differentiation of the right-hand side in Equation 2.57 requires the application of the Leibniz formula. This gives:

$$\frac{d^{l+m}}{d\mu^{l+m}} u = \sum_{k=0}^{l+m} \frac{(l+m)!}{k!(l+m-k)!}\left[\frac{d^k}{d\mu^k}(\mu^2-1)^m\right]\left[\frac{d^{2l+2m-k}}{d\mu^{2l+2m-k}}(\mu^2-1)^l\right], \tag{2.58}$$

Analysis of Equation 2.58 reveals that the terms in square brackets become zero for a certain range of index k, that is:

$$\frac{d^k}{d\mu^k}(\mu^2-1)^m = 0, \quad \text{when } k > 2m; \tag{2.59}$$

and

$$\frac{d^{2l+2m-k}}{d\mu^{2l+2m-k}}(\mu^2-1)^l = 0, \quad \text{when } k < 2m. \tag{2.60}$$

This tells that only the term with index $k = 2m$ in the sum in Equation 2.58 is non-zero. Based on that Equation 2.58 simplifies to:

$$\frac{d^{l+m}}{d\mu^{l+m}}u = \frac{(l+m)!}{(2m)!(l-m)!}\left[\frac{d^{2m}}{d\mu^{2m}}(\mu^2-1)^m\right]\left[\frac{d^{2l}}{d\mu^{2l}}(\mu^2-1)^l\right]. \tag{2.61}$$

To find the remaining derivatives consider the fact that only the polynomial term of the highest power in $(\mu^2 - 1)^m$ will survive the differentiation $d^{2m}/d\mu^{2m}$, that is, the derivative $(d^{2m}/d\mu^{2m})(\mu^2 - 1)^m$ reduces to $(d^{2m}/d\mu^{2m})\mu^{2m}$ and $(d^{2m}/d\mu^{2m})\mu^{2m} = (2m)!$. With that, Equation 2.61 becomes:

$$\frac{d^{l+m}}{d\mu^{l+m}}u = \frac{(l+m)!}{(2m)!(l-m)!}(2m)!(2l)! = \frac{(l+m)!}{(l-m)!}(2l)! \tag{2.62}$$

Having resolved the differentiation portion of integration by parts, the next step is to process the integration component given by Equation 2.56. This term will be integrated $l + m$ times during the course of solution of Equation 2.53. Each time the integration is applied it simply reduces the order of differentiation by a factor of 1 inside the square brackets. In the end the result of integration becomes:

$$v = \frac{d^{n-l}}{d\mu^{n-l}}(\mu^2-1)^n. \tag{2.63}$$

Equation 2.63 explains the purpose of the previously stated condition $l < n$, to keep the differentiation order in Equation 2.63 positive. Since the basic shape of integration component presented by Equation 2.56 remains the same after each integration step the result of that operation does not add anything into the differentiation portion of the next integration by parts that is to follow up, and that keeps the overall process of integration by parts fairly simple. This property turns the repetitive $l + m$ times application of integration by parts into separate $l + m$ times differentiation of Equation 2.55 and separate $l + m$ times integration of Equation 2.56.

Performing integration by parts i times, where $i > 0$, produces the integrated component $u^{(i-1)}v^{[i]}\Big|_{-1}^{1}$. In this notation $u^{(i-1)}$ indicates the $i-1$-th differentiation of term u, and $v^{[i]}$ indicates the i-th integration of term dv, respectively. Differentiating Equation 2.55 $i - 1$ times gives:

$$u^{(i-1)} = \frac{d^{i-1}}{d\mu^{i-1}}\left[(\mu^2-1)^m\frac{d^{l+m}}{d\mu^{l+m}}(\mu^2-1)^l\right]. \tag{2.64}$$

Integrating Equation 2.56 i times produces:

$$v^{[i]} = \frac{d^{n+m-i}}{d\mu^{n+m-i}}(\mu^2-1)^n. \tag{2.65}$$

A brief inspection of Equation 2.64 reveals that $u^{(i-1)}$ contains at least one common multiplier $\mu^2 - 1$ in the result of differentiation for the range $1 \leq i \leq m$. Similarly, it follows from Equation 2.65 that $v^{[i]}$ contains at least one common multiplier $\mu^2 - 1$ in the result of integration for the range $m < i \leq n + m$. This tells that each integrated term vanishes for the integration limits of ± 1 due to the presence of at least of one common multiplier $\mu^2 - 1$ in it, that is:

$$u^{(i-1)}v^{[i]}\Big|_{-1}^{1} = 0, \tag{2.66}$$

for any i in the range $1 \leq i \leq n + m$.

With that everything is ready to assemble the result of integration by parts, that has been applied to Equation 2.53. Combining Equations 2.62, 2.63, and 2.66 leads to:

$$\int_{-1}^{1} P_l^m P_n^m d\mu = \frac{(-1)^m}{2^{2l}(l!)^2} \frac{(l+m)!}{(l-m)!} (2l)!\,(-1)^{l+m} \int_{-1}^{1} \left[\frac{d^{n-l}}{d\mu^{n-l}} (\mu^2-1)^n \right] d\mu. \tag{2.67}$$

Note that application of integration by parts $l+m$ times introduces the term $(-1)^{l+m}$ into Equation 2.67. The remaining integral in Equation 2.67 can be easily found. The integral is:

$$\int_{-1}^{1} \left[\frac{d^{n-l}}{d\mu^{n-l}} (\mu^2-1)^n \right] d\mu = \left. \frac{d^{n-l-1}}{d\mu^{n-l-1}} (\mu^2-1)^n \right|_{-1}^{1} = 0. \tag{2.68}$$

This vanishes because at least one term $\mu^2 - 1$ survives the differentiation in the integrated part in Equation 2.68 for the integration limits of ± 1, and that means that Equation 2.67 has the value zero. This outcome proves the orthogonality condition defined by Equation 2.50.

The derivations performed above also help in finding the normalization integral defined by Equation 2.51. Setting the condition $l = n$ in Equation 2.67 simplifies the expression to:

$$\int_{-1}^{1} P_l^m P_l^m d\mu = \frac{(-1)^m}{2^{2l}(l!)^2} \frac{(l+m)!}{(l-m)!} (2l)!\,(-1)^{l+m} \int_{-1}^{1} (\mu^2-1)^l d\mu. \tag{2.69}$$

The integral that appears in the right-hand side of Equation 2.69 has already been found in Chapter 1. Combining Equations 1.75 and 1.83 from that chapter gives:

$$\int_{-1}^{+1} (\mu^2-1)^l d\mu = (-1)^l \frac{2}{(2l+1)} \frac{2^{2l}(l!)^2}{(2l)!}. \tag{2.70}$$

Substituting Equation 2.70 into Equation 2.69 and cancelling out common terms leads to:

$$\int_{-1}^{1} P_l^m P_l^m d\mu = \frac{2}{2l+1} \frac{(l+m)!}{(l-m)!}. \tag{2.51}$$

which is the normalization integral.

2.4 Recurrence Relations for Associated Legendre Functions

Associated Legendre functions frequently appear in theoretical manipulations in electrostatics, including the Fast Multipole Method. This requires having an efficient procedure for their computation. There are many different formulas for numeric computation of the associated Legendre functions; however, most of them are of little practical use as algorithms because they are prone to rounding errors that accumulate rapidly. Instead of using the formulae directly, a useful strategy for computing associated Legendre functions in computer code is to obtain them *via* recurrence relations. This is accomplished using the following numerically stable recurrence relations:

$$(l-m) P_l^m = (2l-1)\,\mu P_{l-1}^m - (l+m-1)\,P_{l-2}^m, \tag{2.71}$$

$$P_l^{l-1} = (2l-1)\,\mu P_{l-1}^{l-1}. \tag{2.72}$$

$$P_l^l = -(2l-1)(1-\mu^2)^{1/2} P_{l-1}^{l-1}, \tag{2.73}$$

$$(1-\mu^2)^{\frac{1}{2}} P_l^{m+1} = (l-m+1)P_{l+1}^m - (l+m+1)\mu P_l^m. \tag{2.74}$$

The derivation of the above equations relies on the well-established relations for Legendre polynomials that are derived in Chapter 1. One of those to be used now is Bonnet's formula:

$$l P_l = (2l-1)\mu P_{l-1} - (l-1)P_{l-2}. \tag{1.118}$$

The fact that differentiation of an ordinary Legendre equation turns it into a generalized Legendre equation suggests that, by analogy, differentiation of Bonnet's formula should convert it into a corresponding relation for associated Legendre functions. Differentiating Equation 1.118 m-times gives:

$$l \frac{d^m}{d\mu^m} P_l = (2l-1)\frac{d^m}{d\mu^m}\left[\mu P_{l-1}\right] - (l-1)\frac{d^m}{d\mu^m} P_{l-2}. \tag{2.75}$$

To continue further, one needs to find the derivative of the expression in square brackets. Using the Leibnitz formula (see Appendix B.1) for differentiation of a product of two functions leads to:

$$\frac{d^m}{d\mu^m}\left[\mu P_{l-1}\right] = \sum_{k=0}^{m} \frac{m!}{k!(m-k)!}\left[\frac{d^k}{d\mu^k}\mu\right]\left[\frac{d^{m-k}}{d\mu^{m-k}} P_{l-1}\right]. \tag{2.76}$$

Since the derivative in the first square brackets becomes zero for $k > 1$, the sum in Equation 2.76 has only two terms, which can be written out explicitly. This gives:

$$\frac{d^m}{d\mu^m}\left[\mu P_{l-1}\right] = \mu \frac{d^m}{d\mu^m} P_{l-1} + m \frac{d^{m-1}}{d\mu^{m-1}} P_{l-1}, \tag{2.77}$$

where $m \geq 1$.

Substituting Equation 2.77 into Equation 2.75 gives:

$$l \frac{d^m}{d\mu^m} P_l = (2l-1)\mu \frac{d^m}{d\mu^m} P_{l-1} + (2l-1)m \frac{d^{m-1}}{d\mu^{m-1}} P_{l-1} - (l-1)\frac{d^m}{d\mu^m} P_{l-2}. \tag{2.78}$$

In this equation, the presence of derivatives of order $m-1$ is inconvenient since all other terms are of order m. Ideally the order of differentiation in this term should be raised from $m-1$ to m. The necessary instrument for achieving this will now be derived. The derivation starts from a recurrence relation for Legendre polynomial that has previously been derived in Chapter 1:

$$\frac{d}{d\mu} P_l = \mu \frac{d}{d\mu} P_{l-1} + l P_{l-1}. \tag{1.93}$$

The slight difference between Equation 1.93 in Chapter 1 and its present form is due to replacement of index $l+1$ by l. This equation should be rearranged to collect the differentiation terms in the right-hand side. This leads to:

$$l P_{l-1} = \frac{d}{d\mu} P_l - \mu \frac{d}{d\mu} P_{l-1}. \tag{2.79}$$

Following the previously established recipe, this equation should be differentiated $m - 1$ times. This gives:

$$l\frac{d^{m-1}}{d\mu^{m-1}}P_{l-1} = \frac{d^m}{d\mu^m}P_l - \frac{d^{m-1}}{d\mu^{m-1}}\left[\mu\frac{d}{d\mu}P_{l-1}\right]. \tag{2.80}$$

As before, the Leibnitz formula helps in differentiating the expression in square brackets. With that one obtains:

$$\frac{d^{m-1}}{d\mu^{m-1}}\left[\mu\frac{d}{d\mu}P_{l-1}\right] = \sum_{k=0}^{m-1}\frac{(m-1)!}{k!(m-1-k)!}\left[\frac{d^k}{d\mu^k}\mu\right]\left[\frac{d^{m-k}}{d\mu^{m-k}}P_{l-1}\right]. \tag{2.81}$$

Since the derivative $(d^k/d\mu^k)\mu$ becomes zero for $k > 1$, the sum in Equation 2.81 has only two terms, which can be written out explicitly:

$$\frac{d^{m-1}}{d\mu^{m-1}}\left[\mu\frac{d}{d\mu}P_{l-1}\right] = \mu\frac{d^m}{d\mu^m}P_{l-1} + (m-1)\frac{d^{m-1}}{d\mu^{m-1}}P_{l-1}. \tag{2.82}$$

With that, Equation 2.80 takes the form:

$$l\frac{d^{m-1}}{d\mu^{m-1}}P_{l-1} = \frac{d^m}{d\mu^m}P_l - \mu\frac{d^m}{d\mu^m}P_{l-1} - (m-1)\frac{d^{m-1}}{d\mu^{m-1}}P_{l-1}. \tag{2.83}$$

In it, combining like terms gives:

$$(l+m-1)\frac{d^{m-1}}{d\mu^{m-1}}P_{l-1} = \frac{d^m}{d\mu^m}P_l - \mu\frac{d^m}{d\mu^m}P_{l-1}. \tag{2.84}$$

Equation 2.84 is the recurrence relation for Legendre polynomials and shows how to express a derivative of order $m - 1$ *via* the derivatives of order m. This result can be substituted into Equation 2.78. This leads to:

$$l\frac{d^m}{d\mu^m}P_l = (2l-1)\mu\frac{d^m}{d\mu^m}P_{l-1} + \frac{2l-1}{l+m-1}m\frac{d^m}{d\mu^m}P_l - \frac{2l-1}{l+m-1}m\mu\frac{d^m}{d\mu^m}P_{l-1} - (l-1)\frac{d^m}{d\mu^m}P_{l-2}. \tag{2.85}$$

Combining like terms produces:

$$\frac{l(l+m-1)-(2l-1)m}{l+m-1}\frac{d^m}{d\mu^m}P_l = \frac{(2l-1)(l+m-1)-(2l-1)m}{l+m-1}\mu\frac{d^m}{d\mu^m}P_{l-1} - (l-1)\frac{d^m}{d\mu^m}P_{l-2}. \tag{2.86}$$

Since all terms in Equation 2.86 now have the differentiation order m, the Legendre polynomials in it can be converted into corresponding associated Legendre functions by multiplying both sides of the equation by $(-1)^m(1-\mu^2)^{m/2}$. This gives:

$$\frac{l(l+m-1)-(2l-1)m}{l+m-1}P_l^m = \frac{(2l-1)(l+m-1)-(2l-1)m}{l+m-1}\mu P_{l-1}^m - (l-1)P_{l-2}^m. \tag{2.87}$$

In this equation, one may see that $l(l+m-1)-(2l-1)m = l^2 - lm - l + m = (l-1)(l-m)$. Replacing two instances of this term leads to:

$$\frac{(l-1)(l-m)}{l+m-1}P_l^m = \frac{(l-1)(l+m-1)+(l-1)(l-m)}{l+m-1}\mu P_{l-1}^m - (l-1)P_{l-2}^m. \tag{2.88}$$

Multiplying both parts of Equation 2.88 by $(l+m-1)/(l-1)$ gives the desired recurrence relation:

$$(l-m)P_l^m = (2l-1)\,\mu P_{l-1}^m - (l+m-1)P_{l-2}^m. \tag{2.71}$$

This recurrence relation allows the computation of any associated Legendre function of degree l given the value of the two previously computed functions of degree $l-1$ and $l-2$. For Equation 2.71 to be valid, it is necessary that $|m| \leq l-2$.

The next recurrence relation to derive is Equation 2.72. This follows directly from Equation 2.71. Indeed, for $m = l-1$, $P_{l-2}^{l-1} = 0$. This reduces Equation 2.71 into Equation 2.72:

$$P_l^{l-1} = (2l-1)\,\mu P_{l-1}^{l-1}. \tag{2.72}$$

The collective span of index m in Equations 2.71 and 2.72 is $|m| \leq l-1$. All that remains is to find a relation for the last case of $|m| = l$ to completely cover the allowable range of index m. Computation of the associated Legendre function P_l^l requires finding a relation that connects the function of order $m+1$ with previously computed functions of order m. The derivation starts from the recurrence relation for Legendre polynomials that was previously derived in Chapter 1:

$$\frac{d}{d\mu}P_{l+1} = \frac{d}{d\mu}P_{l-1} + (2l+1)P_l. \tag{1.97}$$

This equation should be differentiated m times in order to convert it to the corresponding relation for associated Legendre functions. The differentiation produces:

$$\frac{d^{m+1}}{d\mu^{m+1}}P_{l+1} = \frac{d^{m+1}}{d\mu^{m+1}}P_{l-1} + (2l+1)\frac{d^m}{d\mu^m}P_l. \tag{2.89}$$

Multiplying Equation 2.89 by $(-1)^{m+1}(1-\mu^2)^{(m+1)/2}$ leads to:

$$P_{l+1}^{m+1} = P_{l-1}^{m+1} - (2l+1)(1-\mu^2)^{\frac{1}{2}}P_l^m. \tag{2.90}$$

This is one of the many recurrence relations that exist for associated Legendre functions. To adapt it to the present goal of computing P_l^l it is necessary to replace index m by l. This makes the term $P_{l-1}^{l+1} = 0$ due to differentiation. The next step is to replace $l+1$ by l. This gives the desired recurrence relation:

$$P_l^l = -(2l-1)(1-\mu^2)^{\frac{1}{2}}P_{l-1}^{l-1}. \tag{2.73}$$

The derived Equations 2.71 through 2.73 represent a complete set of numerically stable recurrence relations that are needed for the calculation of the associated Legendre functions.

The final recurrence relation, Equation 2.74, which still remains to be derived, is a prerequisite for differentiation of associated Legendre functions. Its derivation starts from the corresponding recurrence relation for Legendre polynomials that was previously derived in Chapter 1:

$$\frac{d}{d\mu}P_{l-1} = \mu\frac{d}{d\mu}P_l - lP_l. \tag{1.98}$$

To turn it into a recurrence relation for associated Legendre functions it needs to be differentiated. Differentiating this equation m times gives:

$$\frac{d^{m+1}}{d\mu^{m+1}}P_{l-1} = \frac{d^m}{d\mu^m}\left[\mu\frac{d}{d\mu}P_l\right] - l\frac{d^m}{d\mu^m}P_l. \tag{2.91}$$

Applying the Leibnitz formula to the derivative of $\mu(d/d\mu)P_l$ leads to:

$$\frac{d^m}{d\mu^m}\left[\mu\frac{d}{d\mu}P_l\right]=\sum_{k=0}^{m}\frac{m!}{k!(m-k)!}\left[\frac{d^k}{d\mu^k}\mu\right]\left[\frac{d^{m-k+1}}{d\mu^{m-k+1}}P_l\right]. \tag{2.92}$$

Since the derivative $(d^k/d\mu^k)\mu$ becomes zero for $k > 1$, the sum in Equation 2.92 has only two terms, which can be written out explicitly:

$$\frac{d^m}{d\mu^m}\left[\mu\frac{d}{d\mu}P_l\right]=\mu\frac{d^{m+1}}{d\mu^{m+1}}P_l+m\frac{d^m}{d\mu^m}P_l. \tag{2.93}$$

Substituting Equation 2.93 into Equation 2.91 leads to:

$$\frac{d^{m+1}}{d\mu^{m+1}}P_{l-1}=\mu\frac{d^{m+1}}{d\mu^{m+1}}P_l+m\frac{d^m}{d\mu^m}P_l-l\frac{d^m}{d\mu^m}P_l. \tag{2.94}$$

Combining like terms and resolving about $(d^m/d\mu^m)P_l$ gives:

$$(l-m)\frac{d^m}{d\mu^m}P_l=\mu\frac{d^{m+1}}{d\mu^{m+1}}P_l-\frac{d^{m+1}}{d\mu^{m+1}}P_{l-1}. \tag{2.95}$$

To proceed further it is necessary to lower the degree in $(d^{m+1}/d\mu^{m+1})P_l$ from l to $l-1$ in this equation. This conversion can be accomplished using the previously derived Equation 2.84. Replacing index m by $m+1$ in Equation 2.84 produces:

$$(l+m)\frac{d^m}{d\mu^m}P_{l-1}=\frac{d^{m+1}}{d\mu^{m+1}}P_l-\mu\frac{d^{m+1}}{d\mu^{m+1}}P_{l-1}. \tag{2.96}$$

Rearranging terms to resolve $(d^{m+1}/d\mu^{m+1})P_l$ leads to:

$$\frac{d^{m+1}}{d\mu^{m+1}}P_l=\mu\frac{d^{m+1}}{d\mu^{m+1}}P_{l-1}+(l+m)\frac{d^m}{d\mu^m}P_{l-1}. \tag{2.97}$$

Equation 2.97 reduces the degree of derivative of Legendre polynomial from l to $l-1$. Substituting the obtained equation into Equation 2.95 gives:

$$(l-m)\frac{d^m}{d\mu^m}P_l=\mu^2\frac{d^{m+1}}{d\mu^{m+1}}P_{l-1}+(l+m)\,\mu\frac{d^m}{d\mu^m}P_{l-1}-\frac{d^{m+1}}{d\mu^{m+1}}P_{l-1}. \tag{2.98}$$

Combining like terms in Equation 2.98 leads to:

$$(l-m)\frac{d^m}{d\mu^m}P_l=(l+m)\,\mu\frac{d^m}{d\mu^m}P_{l-1}-(1-\mu^2)\frac{d^{m+1}}{d\mu^{m+1}}P_{l-1}. \tag{2.99}$$

With that everything is ready now to convert Equation 2.99 from a recurrence relation for derivatives of the Legendre polynomials into a recurrence relation for associated Legendre functions. Multiplying both sides of Equation 2.99 by $(-1)^m(1-\mu^2)^{m/2}$ and applying Equation 2.35 gives:

$$(l-m)P_l^m=(l+m)\,\mu P_{l-1}^m-(-1)^m(1-\mu^2)^{\frac{m}{2}}(1-\mu^2)\frac{d^{m+1}}{d\mu^{m+1}}P_{l-1}. \tag{2.100}$$

In the right-most term in Equation 2.100 the derivative is of order $m+1$. This suggests converting the other individual components in the product to the order of $m+1$ as well. This leads to:

$$(l-m)P_l^m = (l+m)\mu P_{l-1}^m + (-1)^{m+1}(1-\mu^2)^{\frac{m+1}{2}}(1-\mu^2)^{\frac{1}{2}}\frac{d^{m+1}}{d\mu^{m+1}}P_{l-1}, \tag{2.101}$$

which, after application of Equation 2.35 to the right-most term, becomes:

$$(l-m)P_l^m = (l+m)\mu P_{l-1}^m + (1-\mu^2)^{\frac{1}{2}}P_{l-1}^{m+1}. \tag{2.102}$$

The last step in the derivation is to replace the degree l by $l+1$ and resolve the equation about P_l^{m+1}. This gives the result needed:

$$(1-\mu^2)^{\frac{1}{2}}P_l^{m+1} = (l-m+1)P_{l+1}^m - (l+m+1)\mu P_l^m. \tag{2.74}$$

This completes the derivation of recurrence relations for associated Legendre functions that will be used in the Fast Multipole Method.

2.5 Derivatives of Associated Legendre Functions

The solution of Laplace's equation given by Equation 2.36 provides a general form of the electrostatic potential function Φ. However, a knowledge of the electrostatic potential at a particular point of space is not sufficient to completely characterize the state of a particle located in that point. The complementary information is supplied by the derivative of the potential, that is, the electric field, E, and is defined by the equation:

$$E = -\nabla\Phi, \tag{2.103}$$

where ∇ is a gradient.

Differentiation of potential function, Equation 2.36, requires a knowledge of the derivative of the associated Legendre function. This differentiation can be performed either over the variable $\mu = \cos\theta$ or directly over the angle θ by using the chain rule:

$$\frac{dP_l^m}{d\theta} = \frac{dP_l^m}{d\mu}\frac{d\mu}{d\theta} = -\sin\theta\frac{dP_l^m}{d\mu}. \tag{2.104}$$

In Equation 2.104, the derivative over angle θ is simply a product of the derivative over variable μ times the negative sine of θ. The derivative over variable μ is obtained by differentiating the associated Legendre function, Equation 2.35:

$$P_l^m = (-1)^m(1-\mu^2)^{\frac{m}{2}}\frac{d^m}{d\mu^m}P_l. \tag{2.35}$$

Inspection of Equation 2.35 suggests that its differentiation can be performed as the product of two functions:

$$\frac{d}{d\mu}P_l^m = (-1)^m\frac{d}{d\mu}\left[(1-\mu^2)^{\frac{m}{2}}\right]\frac{d^m}{d\mu^m}P_l + (-1)^m(1-\mu^2)^{\frac{m}{2}}\frac{d^{m+1}}{d\mu^{m+1}}P_l. \tag{2.105}$$

In this equation, differentiation of Legendre polynomial simply increases the differential order of P_l from $d^m/d\mu^m$ to $d^{m+1}/d\mu^{m+1}$. The derivative of the term in square brackets is:

$$\frac{d}{dx}(1-\mu^2)^{\frac{m}{2}} = \frac{m}{2}(-2\mu)(1-\mu^2)^{\frac{m}{2}-1} = -\frac{m\,\mu}{1-\mu^2}(1-\mu^2)^{\frac{m}{2}}. \tag{2.106}$$

Substituting Equation 2.106 into Equation 2.105 leads to:

$$\frac{d}{d\mu}P_l^m = -(-1)^m \frac{m\,\mu}{1-\mu^2}(1-\mu^2)^{\frac{m}{2}} \frac{d^m}{d\mu^m}P_l + (-1)^m(1-\mu^2)^{\frac{m}{2}} \frac{d^{m+1}}{d\mu^{m+1}}P_l. \tag{2.107}$$

The first summand in Equation 2.107 contains an associated Legendre function of order m. The second summand can be transformed to an associated Legendre function of order $m + 1$. This gives:

$$\frac{d}{d\mu}P_l^m = -\frac{m\,\mu}{1-\mu^2}P_l^m - (-1)^{m+1}(1-\mu^2)^{\frac{m+1}{2}}(1-\mu^2)^{-\frac{1}{2}} \frac{d^{m+1}}{d\mu^{m+1}}P_l. \tag{2.108}$$

Gathering the terms in the right-hand side into an associated Legendre function of order $m + 1$ and multiplying both sides by $1 - \mu^2$ leads to:

$$(1-\mu^2)\frac{d}{d\mu}P_l^m = -m\,\mu\,P_l^m - (1-\mu^2)^{\frac{1}{2}}P_l^{m+1}. \tag{2.109}$$

Equation 2.109 describes the derivative of the associated Legendre function over variable $\mu = \cos\theta$. However, it is not optimal for digital computation. This equation can be transformed in two steps into a more efficient expression that requires fewer floating-point operations. The first step is to eliminate the square root in Equation 2.109. This can be accomplished using the recurrence relation 2.74. Substituting Equation 2.74 in Equation 2.109 leads to:

$$(1-\mu^2)\frac{d}{d\mu}P_l^m = -m\,\mu\,P_l^m - (l-m+1)P_{l+1}^m + (l+m+1)\mu\,P_l^m. \tag{2.110}$$

Combining like terms in $\mu\,P_l^m$ gives:

$$(1-\mu^2)\frac{d}{d\mu}P_l^m = (l+1)\,\mu\,P_l^m - (l+1-m)P_{l+1}^m. \tag{2.111}$$

Note that in Equation 2.111 all terms have the same order of m. The remaining problem is the presence of term P_{l+1}^m; this has the higher degree of $l + 1$ than that in the sought derivative $(d/d\mu)P_l^m$. It means that if the expansion series for electrostatic potential in Equation 2.36 were to be truncated at the degree l, Equation 2.111 would require accessing the unavailable term of degree $l + 1$. The last step in the transformation is to eliminate the dependence of derivative of the associated Legendre function of degree l on the function of degree $l + 1$ by using the recurrence relation 2.71. Replacing index l by $l + 1$ in Equation 2.71 leads to:

$$(l+1-m)P_{l+1}^m = (2l+1)\,\mu\,P_l^m - (l+m)P_{l-1}^m. \tag{2.112}$$

Substituting Equation 2.112 into Equation 2.111 gives:

$$(1-\mu^2)\frac{d}{d\mu}P_l^m = (l+1)\,\mu\,P_l^m - (2l+1)\,\mu\,P_l^m + (l+m)P_{l-1}^m, \tag{2.113}$$

and combining like terms in $\mu\, P_l^m$ leads to:

$$(1-\mu^2)\frac{d}{d\mu}P_l^m = (l+m)P_{l-1}^m - l\,\mu\, P_l^m. \tag{2.114}$$

Equation 2.114 is the recurrence relation for efficient numerical computation of derivatives of associated Legendre functions.

2.6 Analytic Expression for First Few Associated Legendre Functions

There are several ways to produce analytic expression for associated Legendre functions. Among those, Rodrigues' form given by Equation 2.37 is the most frequently used. Here it is reproduced for convenience:

$$P_l^m(\mu) = \frac{(-1)^m}{2^l l!}(1-\mu^2)^{\frac{m}{2}}\frac{d^{l+m}}{d\mu^{l+m}}(\mu^2-1)^l. \tag{2.37}$$

Using Equation 2.37 to obtain associated Legendre functions quickly becomes tedious as index l grows, because of that it is practical to carry out the differentiation in Equation 2.37 only for relatively small values of l. Associated Legendre functions of higher degree of l are usually obtained *via* recurrence relations. Section 2.4 of this chapter presented a complete set of recurrence relations needed for computation of associated Legendre function. The analytic expressions for first few associated Legendre functions, which are obtained according to the above procedure, are presented in Table 2.1. Functions in the range $0 \le l \le 1$ are computed from Equation 2.37, whereas all other higher-degree functions follow from Equations 2.71 through 2.73.

As Table 2.1 illustrates, analytic expressions for many associated Legendre functions contain sums involving large numbers, which essentially cancel out due to alternating sign of the terms. An implication of this is that the computation of numerical values of associated Legendre functions from their analytic expression is subject to accumulation of rounding errors. The solution to that problem is similar to the way the analytic expressions in Table 2.1 are obtained. When computing the numerical values of associated Legendre functions for a given value of argument μ, the numerical values for functions of degree $l = 0$ and 1 are obtained from their analytic expressions given in Table 2.1, whereas the numerical values of associated Legendre functions for higher degree are computed recursively based on Equations 2.71 through 2.73 using numerical values of the previously computed functions.

The analytic expressions from Table 2.1 can be useful to verify the validity of the computer code performing numerical computation of associated Legendre functions based on recurrence relations.

2.7 Symmetry Properties of Associated Legendre Functions

Associated Legendre functions $P_l^m(\mu)$ exhibit a number of useful symmetry properties which are indispensible in the computation of these functions. Equation 2.49 is one such relationship that applies to the sign of index m. Another symmetry relationship that can be obtained pertains to the effect of a change in the sign of the argument μ; values of which are limited to the range $\mu = [-1{:}\ 1]$ since $\mu = \cos\theta$.

The Rodrigues' formula, Equation 2.37, which is frequently used to express associated Legendre functions in analytical manipulations, is of little use for predicting the outcome of a change in sign of μ due to the presence of an unresolved differentiation operation in the Rodrigues' formula. Instead, what is needed is a different type of expression for associated Legendre functions that would avoid the use of a differentiation operator.

TABLE 2.1
Analytic Expressions for the First Few Associated Legendre Functions

$P_0^0 = 1$

$P_1^0 = \cos\theta$

$P_1^1 = -\sin\theta$

$P_2^0 = \frac{1}{2}\left[3\cos^2\theta - 1\right]$

$P_2^1 = -3\sin\theta\cos\theta$

$P_2^2 = 3\sin^2\theta$

$P_3^0 = \frac{1}{2}\left[5\cos^3\theta - 3\cos\theta\right]$

$P_3^1 = -\frac{3}{2}\sin\theta[5\cos^2\theta - 1]$

$P_3^2 = 15\sin^2\theta\cos\theta$

$P_3^3 = -15\sin^3\theta$

$P_4^0 = \frac{1}{8}[35\cos^4\theta - 30\cos^2\theta + 3]$

$P_4^1 = -\frac{5}{2}\sin\theta[7\cos^3\theta - 3\cos\theta]$

$P_4^2 = \frac{15}{2}\sin^2\theta[7\cos^2\theta - 1]$

$P_4^3 = -105\sin^3\theta\cos\theta$

$P_4^4 = 105\sin^4\theta$

$P_5^0 = \frac{1}{8}[63\cos^5\theta - 70\cos^3\theta + 15\cos\theta]$

$P_5^1 = -\frac{15}{8}\sin\theta[21\cos^4\theta - 14\cos^2\theta + 1]$

$P_5^2 = \frac{105}{2}\sin^2\theta\cos\theta[3\cos^2\theta - 1]$

$P_5^3 = -\frac{105}{2}\sin^3\theta[9\cos^2\theta - 1]$

$P_5^4 = 945\sin^4\theta\cos\theta$

$P_5^5 = -945\sin^5\theta$

$P_6^0 = \frac{1}{16}[231\cos^6\theta - 315\cos^4\theta + 105\cos^2\theta - 5]$

$P_6^1 = -\frac{21}{8}\sin\theta\cos\theta[33\cos^4\theta - 30\cos^2\theta + 5]$

$P_6^2 = \frac{105}{8}\sin^2\theta[33\cos^4\theta - 18\cos^2\theta + 1]$

$P_6^3 = -\frac{315}{2}\sin^3\theta\cos\theta[11\cos^2\theta - 3]$

$P_6^4 = \frac{945}{2}\sin^4\theta[11\cos^2\theta - 1]$

$P_6^5 = -10{,}395\sin^5\theta\cos\theta$

$P_6^6 = 10{,}395\sin^6\theta$

$P_7^0 = \frac{1}{16}[429\cos^7\theta - 693\cos^5\theta + 315\cos^3\theta - 35\cos\theta]$

$P_7^1 = -\frac{7}{16}\sin\theta[429\cos^6\theta - 495\cos^4\theta + 135\cos^2\theta - 5]$

(*Continued*)

TABLE 2.1 (*Continued*)

Analytic Expressions for the First Few Associated Legendre Functions

$$P_7^2 = \frac{63}{8}\sin^2\theta\cos\theta[143\cos^4\theta - 110\cos^2\theta + 115]$$

$$P_7^3 = -\frac{315}{8}\sin^3\theta[143\cos^4\theta - 66\cos^2\theta + 3]$$

$$P_7^4 = \frac{3465}{2}\sin^4\theta\cos\theta[13\cos^2\theta - 3]$$

$$P_7^5 = -\frac{10{,}395}{2}\sin^5\theta[13\cos^2\theta - 1]$$

$$P_7^6 = 135{,}135\sin^6\theta\cos\theta$$

$$P_7^7 = -135{,}135\sin^7\theta$$

$$P_8^0 = \frac{1}{128}[6435\cos^8\theta - 12{,}012\cos^6\theta + 6930\cos^4\theta - 1260\cos^2\theta + 35]$$

$$P_8^1 = -\frac{9}{16}\sin\theta\cos\theta[715\cos^6\theta - 1001\cos^4\theta + 385\cos^2\theta - 35]$$

$$P_8^2 = \frac{315}{16}\sin^2\theta[143\cos^6\theta - 143\cos^4\theta + 33\cos^2\theta - 1]$$

$$P_8^3 = -\frac{3465}{8}\sin^3\theta\cos\theta[39\cos^4\theta - 26\cos^2\theta + 3]$$

$$P_8^4 = \frac{10{,}395}{8}\sin^4\theta[65\cos^4\theta - 26\cos^2\theta + 1]$$

$$P_8^5 = -\frac{135{,}135}{2}\sin^5\theta\cos\theta[5\cos^2\theta - 1]$$

$$P_8^6 = \frac{135{,}135}{2}\sin^6\theta[15\cos^2\theta - 1]$$

$$P_8^7 = -2{,}027{,}025\sin^7\theta\cos\theta$$

$$P_8^8 = 2{,}027{,}025\sin^8\theta$$

$$P_9^0 = \frac{1}{128}[12{,}155\cos^9\theta - 25{,}740\cos^7\theta + 18{,}018\cos^5\theta - 4620\cos^3\theta + 315\cos\theta]$$

$$P_9^1 = -\frac{45}{128}\sin\theta[2431\cos^8\theta - 4004\cos^6\theta + 2002\cos^4\theta - 308\cos^2\theta + 7]$$

$$P_9^2 = \frac{3465}{112}\sin^2\theta\cos\theta[221\cos^6\theta - 273\cos^4\theta + 91\cos^2\theta - 7]$$

$$P_9^3 = -\frac{3465}{16}\sin^3\theta[221\cos^6\theta - 195\cos^4\theta + 39\cos^2\theta - 1]$$

$$P_9^4 = \frac{135{,}135}{8}\sin^4\theta\cos\theta[17\cos^4\theta - 10\cos^2\theta + 1]$$

$$P_9^5 = -\frac{135{,}135}{8}\sin^5\theta[85\cos^4\theta - 30\cos^2\theta + 1]$$

$$P_9^6 = \frac{675{,}675}{2}\sin^6\theta\cos\theta[17\cos^2\theta - 3]$$

$$P_9^7 = -\frac{2{,}027{,}025}{2}\sin^7\theta[17\cos^2\theta - 1]$$

$$P_9^8 = 34{,}459{,}425\sin^8\theta\cos\theta$$

$$P_9^9 = -34{,}459{,}425\sin^9\theta$$

$$P_{10}^0 = \frac{1}{256}[46189\cos^{10}\theta - 109{,}395\cos^8\theta + 90{,}090\cos^6\theta - 30{,}030\cos^4\theta + 3465\cos^2\theta - 63]$$

$$P_{10}^1 = -\frac{55}{128}\sin\theta\cos\theta[4199\cos^8\theta - 7956\cos^6\theta + 4914\cos^4\theta - 1092\cos^2\theta + 63]$$

$$P_{10}^2 = \frac{495}{128}\sin^2\theta[4199\cos^8\theta - 6188\cos^6\theta + 2730\cos^4\theta - 364\cos^2\theta + 7$$

(*Continued*)

TABLE 2.1 (*Continued*)

Analytic Expressions for the First Few Associated Legendre Functions

$$P_{10}^{3} = -\frac{6435}{16}\sin^3\theta\cos\theta[323\cos^6\theta - 357\cos^4\theta + 105\cos^2\theta - 7]$$

$$P_{10}^{4} = \frac{45{,}045}{16}\sin^4\theta[323\cos^6\theta - 255\cos^4\theta + 45\cos^2\theta - 1]$$

$$P_{10}^{5} = -\frac{135{,}135}{8}\sin^5\theta\cos\theta[323\cos^4\theta - 170\cos^2\theta + 15]$$

$$P_{10}^{6} = \frac{675{,}675}{8}\sin^6\theta[323\cos^4\theta - 102\cos^2\theta + 3]$$

$$P_{10}^{7} = -\frac{11{,}486{,}475}{2}\sin^7\theta\cos\theta[19\cos^2\theta - 3]$$

$$P_{10}^{8} = \frac{34{,}459{,}425}{2}\sin^8\theta[19\cos^2\theta - 1]$$

$$P_{10}^{9} = -654{,}729{,}075\sin^9\theta\cos\theta$$

$$P_{10}^{10} = 654{,}729{,}075\sin^{10}\theta$$

In order to transform the Rodrigues' formula, it is necessary to obtain a different representation of the term $(\mu^2 - 1)^l$ that would be easier to repeatedly differentiate. A suitable solution can be obtained from a binomial expansion of $(\mu^2 - 1)^l$. This leads to the following expression:

$$(\mu^2 - 1)^l = \sum_{k=0}^{l} \frac{l!}{k!(l-k)!}(-1)^k \mu^{2l-2k}. \tag{2.115}$$

Equation 2.115 is now easily differentiable. Differentiating the right-hand side of this equation $l - m$ times reduces the power of μ from $2l - 2k$ to $l - m - 2k$. Simultaneously, the product of coefficients, which emerges due to $l - m$ successive differentiations of the argument μ^{2l-2k}, transforms into a ratio of factorials:

$$(2l-2k)(2l-2k-1)\ldots(2l-2k-l-m+1) = \frac{(2l-2k)!}{(l-m-2k)!}. \tag{2.116}$$

Taking into account the above steps the result of differentiation becomes:

$$\frac{d^{l+m}}{\mu^{l+m}}(\mu^2 - 1)^l = \sum_{k=0}^{N}(-1)^k \frac{l!}{k!(l-k)!}\frac{(2l-2k)!}{(l-m-2k)!}\mu^{l-m-2k}, \tag{2.117}$$

where the upper summation limit, N is defined from the condition that the power of argument μ should be a non-negative integer number, that is, $l - m - 2k \geq 0$. Specifically, if the value of $l - m$ is even, the upper value of index k is $(l - m)/2$. If the value of $l - m$ is odd, then the upper value of index k should be $(l - m - 1)/2$. A convenient way to manage both of these conditions is to use the Fortran function mod($l - m$, 2). This function, in the way it is invoked, returns the remainder after division of $l - m$ by 2; using it, the upper limit of summation can then be concisely defined *via* the expression: $N = [l - m - \text{mod}(l - m, 2)]/2$.

Inserting Equation 2.117 into Rodrigues' formula, Equation 2.37, and cancelling out the common factor $l!$ gives the desired expression for associated Legendre functions:

$$P_l^m(\mu) = (1-\mu^2)^{m/2} \sum_{k=0}^{[l-m-\text{mod}(l-m,2)]/2} (-1)^{m+k} \frac{(2l-2k)!}{2^l k!(l-k)!(l-m-2k)!}\mu^{l-m-2k}, \tag{2.118}$$

which is valid for both positive and negative values of index m.

Equation 2.118 is a good starting point for the derivation of a couple of useful symmetry relations for associated Legendre functions. It immediate follows from the above expansion that coordinate inversion, that is, the change in the sign of argument from μ to $-\mu$, has the following effect:

$$P_l^m(-\mu)=(-1)^{l-m}P_l^m(\mu). \tag{2.119}$$

Next, Equation 2.118 makes it possible to determine the value of associated Legendre function when the argument $\mu=0$. If the index $l-m$ is odd, each summand in the sum in Equation 2.118 includes the argument μ; therefore, the entire associated Legendre function vanishes. Resolving the alternative case requires a bit more effort. If the index $l-m$ is even, all summands except the last one will vanish as well. The last summand will have a non-zero value due to $\mu^0=1$. With that, the value of associated Legendre function in point $\mu=0$ becomes:

$$(l-m)_{even}:\ \ P_l^m(0)=(-1)^{m+k}\frac{(2l-2k)!}{2^l k!(l-k)!(l-m-2k)!}. \tag{2.120}$$

This equation can be further simplified. Since $l-m$ is even, the upper value of the summation index is $k=(l-m)/2$. Substituting this value into Equation 2.120 gives:

$$(l-m)_{even}:\ \ P_l^m(0)=(-1)^{\frac{l+m}{2}}\frac{(2l-l+m)!}{2^l\left(\frac{1-m}{2}\right)!\left(\frac{(1+m)}{2}\right)!(l-m-l+m)!}. \tag{2.121}$$

After simplification of Equation 2.121 by taking into account that $2^l((l-m)/2)!((l+m)/2)!=(l-m)!!(l+m)!!$ and $(l+m)!/(l+m)!!=(l+m-1)!!$, the final result describing both odd and even cases assumes the following form:

$$\begin{aligned}&(l-m)_{odd}:\ \ P_l^m(0)=0,\\&(l-m)_{even}:\ \ P_l^m(0)=(-1)^{\frac{l+m}{2}}\frac{(l+m-1)!!}{(l-m)!!}.\end{aligned} \tag{2.122}$$

Another special point of interest in the coordinate space is $\mu=1$. All associated Legendre functions that have $m>0$ are equal to zero at that point, that is, $P_l^m(1)=0$, due to multiplier $(\mu^2-1)^{m/2}$ in Equation 2.118 becomes zero. For $m=0$, associated Legendre functions turn into Legendre polynomials, and these all have the same value when $\mu=1$:

$$P_l(1)=1. \tag{1.140}$$

Thus, depending on index m, associated Legendre functions become zero, when $m>0$, or unity, when $m=0$ in the coordinate point $\mu=1$. This summarizes to:

$$\begin{aligned}m>0:\ \ P_l^m(1)=0,\\m=0:\ \ P_l^m(1)=1.\end{aligned} \tag{2.123}$$

The behavior of associated Legendre functions in the function domain $\mu=[-1:1]$ is graphically depicted in Figure 2.1. The plots demonstrate the non-trivial character of the function in the points $\mu=-1, 0, 1$ due to azimuthal index m.

Unlike Legendre polynomials, which have the function range limited to unity, associated Legendre functions defined by Equation 2.38 have no fixed boundaries. Scaling the functions by the inverse square root of the normalization integral, Equation 2.51, can limit the function range to a constant value, if

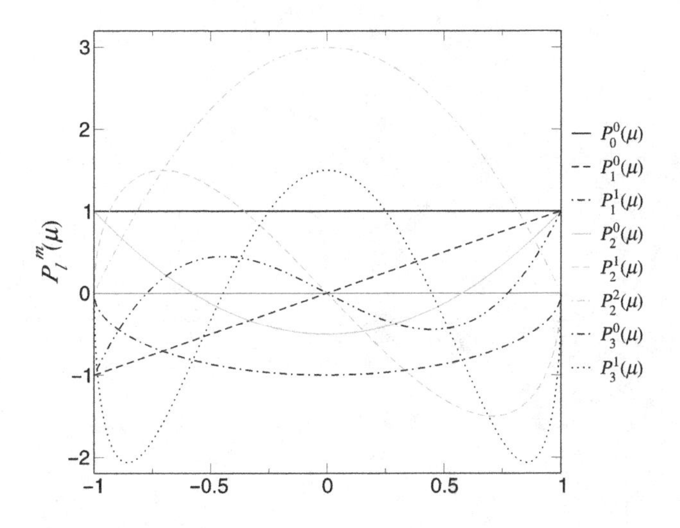

FIGURE 2.1 Associated Legendre functions.

desired. Since associated Legendre functions appear in the Fast Multipole Method as a component of the spherical harmonics, a separate normalization of the associated Legendre functions is not necessary.

Associated Legendre functions include Legendre polynomials as a special case of index $m = 0$. With the introduction of the azimuthal index m, associated Legendre functions represent a significant step toward breaking free from the requirement of cylindrical symmetry in the applications. However, on their own, they are insufficient for managing a rotational symmetry of finite order. Additional generalizations are necessary in order to completely remove this limitation. The necessary tools will be developed in the next chapter.

3

Spherical Harmonics

Solutions of Laplace's equation $\nabla^2\Phi = 0$ comprise a very important class of functions called harmonic functions. This name traces its roots to the family of higher-frequency overtones, also called harmonics, which accompany the fundamental tone in the sound spectrum of a musical instrument.

Harmonic functions are solutions of Laplace's equation. Their common mathematical property is that they can be expanded in a series of appropriate harmonic functions, which form a complete orthonormal set. Each function in the latter set is of course also a solution of Laplace's equation.

Since the electrostatic potential function Φ satisfies Laplace's equation, it is a harmonic function, and it too can be expanded in a series of suitable harmonic functions. This provides an alternative way of determining the potential function of a physical system. Finding the coefficients of such an expansion is often a less formidable task than seeking a conventional solution for the potential function. To implement that approach in a practical way requires a closer understanding of harmonic functions.

While there exists an infinite variety of various harmonic functions, the most important family of harmonic functions forming a complete orthonormal set that makes them suitable to serve as basis functions in the expansion series comes from the general solution of Laplace's equation in spherical polar coordinates (r, θ, ϕ) sought in the form of a product of three independent functions $R(r)$, $P(\theta)$, and $Z(\phi)$:

$$\Phi(r,\theta,\phi) = R(r)\,P(\theta)\,Z(\phi), \tag{3.1}$$

where each individual function depends only on a single coordinate. Equation 3.1 corresponds to separation of variables. Each function in Equation 3.1 is defined separately by solving a corresponding ordinary differential equation. Giving a brief overview of the discussion started in Chapters 1 and 2, the solution for radial function, $R(r)$ is:

$$R(r) = A_l\, r^l + B_l\, r^{-l-1}, \tag{3.2}$$

where A_l and B_l are arbitrary constants, and l is an integer number satisfying the condition $l \geq 0$.

The function $P(\theta)$, which describes the polar angle dependence, is an associated Legendre function:

$$P_l^m(\mu) = \frac{(-1)^m}{2^l l!}(1-\mu^2)^{m/2}\frac{d^{l+m}}{d\mu^{l+m}}(\mu^2-1)^l, \tag{3.3}$$

where $\mu = \cos\theta$, and l and m are integers with $l \geq 0$, and $-l \leq m \leq l$.

The third function, $Z(\phi)$, in the product in Equation 3.1 describes the azimuthal dependence. Its solution is a complex exponential function:

$$Z(\phi) = e^{\pm im\phi}. \tag{3.4}$$

The three individual functions, $R(r)$, $P(\theta)$, and $Z(\phi)$, link together *via* common parameters to form a single function Φ; $R(r)$ and $P(\theta)$ share a parameter l, and the functions $P(\theta)$ and $Z(\phi)$ couple through a parameter m.

3.1 Spherical Harmonics Functions

Due to the innate spherical symmetry of the spherical polar coordinate system, it is beneficial to combine the two angular terms in Equation 3.1, which depend on polar θ and azimuthal ϕ angles, into a single function $Y(\theta, \phi)$:

$$Y(\theta,\phi) = P(\theta)\,Z(\phi). \tag{3.5}$$

Functions of this type are called spherical harmonics, and are defined thus:

$$Y_{lm}(\theta,\phi) = \sqrt{\frac{(2l+1)}{4\pi}\frac{(l-m)!}{(l+m)!}}P_l^m(\cos\theta)e^{im\phi}, \tag{3.6}$$

where the square root term is a normalization coefficient. The index l in Equation 3.6 is referred to as the degree and index m is the order of the spherical harmonic function. These indices span the intervals $l \geq 0$ and $-l \leq m \leq l$. Note that, unlike the associated Legendre functions P_l^m that have their indices l and m placed at the bottom and at the top of the function symbol, respectively, spherical harmonic functions Y_{lm} have both indices placed at the bottom. This is a consequence of the fact that most spherical harmonics are complex functions; the place of the upper index is reserved for the star symbol to indicate complex conjugate, that is, Y_{lm}^*.

Having both polar and azimuthal angular dependencies combined into a single function, Y reduces Equation 3.1 to:

$$\Phi = R(r)Y(\theta,\phi). \tag{3.7}$$

Additionally, fixing the distance parameter r at the constant value $r=1$ leads to $R(r)=1$, and that results in the polar θ and azimuthal ϕ coordinates spanning the surface of a unit sphere. This condition equates Φ to Y in Equation 3.7. Using this convention, it is customary to refer to the spherical harmonics as describing the electrostatic potential on the surface of a unit sphere.

This introduction of spherical harmonics during the solution of Laplace's equation leads to the important topic of 3D rotations to be introduced in Chapter 4. Therefore, a brief description of the background is in order.

Given that Laplace's equation $\nabla^2\Phi = 0$ in spherical polar coordinates is:

$$\frac{1}{r^2}\frac{\partial}{\partial r}\left(r^2\frac{\partial \Phi}{\partial r}\right) + \frac{1}{r^2\sin\theta}\frac{\partial}{\partial \theta}\left(\sin\theta\frac{\partial \Phi}{\partial \theta}\right) + \frac{1}{r^2\sin^2\theta}\frac{\partial^2 \Phi}{\partial \phi^2} = 0, \tag{3.8}$$

the first step to learn its implication is to differentiate the potential function, Φ. Since Φ is a product of R and Y, its partial derivatives are:

$$\frac{\partial}{\partial r}(RY) = Y\frac{\partial}{\partial r}R \quad \frac{\partial}{\partial \theta}(RY) = R\frac{\partial}{\partial \theta}Y \quad \frac{\partial}{\partial \phi}(RY) = R\frac{\partial}{\partial \phi}Y, \tag{3.9}$$

note that function arguments are omitted to avoid clutter. In Equation 3.9, the function in the product that does not depend on the differentiation variable is moved outside the differentiation operator. Second-order derivatives of Φ are similar, the only change being the operator of first-order differentiation being replaced by the corresponding operator of second-order differentiation.

Substituting the electrostatic potential Φ given by Equation 3.7 into Equation 3.8, and using the results of differentiation given by Equation 3.9 leads to:

$$\frac{Y}{r^2}\frac{\partial}{\partial r}\left(r^2\frac{\partial R}{\partial r}\right) + R\left[\frac{1}{r^2\sin\theta}\frac{\partial}{\partial \theta}\left(\sin\theta\frac{\partial Y}{\partial \theta}\right) + \frac{1}{r^2\sin^2\theta}\frac{\partial^2 Y}{\partial \phi^2}\right] = 0. \tag{3.10}$$

Since R and Y are independent functions, Equation 3.10 can be multiplied by r^2 and simultaneously divided by RY. The result of that operation is:

$$\frac{1}{R}\frac{\partial}{\partial r}\left(r^2\frac{\partial R}{\partial r}\right)+\frac{1}{Y}\left[\frac{1}{\sin\theta}\frac{\partial}{\partial\theta}\left(\sin\theta\frac{\partial Y}{\partial\theta}\right)+\frac{1}{\sin^2\theta}\frac{\partial^2 Y}{\partial\phi^2}\right]=0, \tag{3.11}$$

in this expression the terms depending on R and Y are now fully separated.

Due to the fact that the two summands in Equation 3.11 are completely independent, their sum is equal to zero only if they cancel each other out. It happens that the only solution to Equation 3.11 has the first and second summands being equal to $l(l+1)$ and $-l(l+1)$, respectively, that is:

$$\frac{1}{R}\frac{d}{dr}\left(r^2\frac{dR}{dr}\right)=l(l+1), \tag{3.12}$$

$$\frac{1}{Y}\left[\frac{1}{\sin\theta}\frac{\partial}{\partial\theta}\left(\sin\theta\frac{\partial Y}{\partial\theta}\right)+\frac{1}{\sin^2\theta}\frac{\partial^2 Y}{\partial\phi^2}\right]=-l(l+1). \tag{3.13}$$

Solution of Equation 3.12 leads to Equation 3.2. Solutions of Equation 3.13 are the spherical harmonics, as provided by Equation 3.6. This is a restatement of facts that have already been discussed earlier in Chapters 1 and 2. Before proceeding further in examining Equation 3.13, it is helpful to perform a rearrangement. Multiplying both sides of the equation by $-Y$, and then moving the function Y to the right side of the square brackets leads to:

$$-\left[\frac{1}{\sin\theta}\frac{\partial}{\partial\theta}\left(\sin\theta\frac{\partial}{\partial\theta}\right)+\frac{1}{\sin^2\theta}\frac{\partial^2}{\partial\phi^2}\right]Y_{lm}=l(l+1)Y_{lm}. \tag{3.14}$$

The lower indices lm are reattached to function Y in order to explicitly show the dependence of the function on those indices.

Close inspection of Equation 3.14 reveals that it is similar to the fundamental equation for angular momentum:

$$L^2Y_{lm}=l(l+1)\hbar^2Y_{lm}, \tag{3.15}$$

where L^2 is an operator for the square of the magnitude of angular momentum:

$$L^2=-\hbar^2\left[\frac{1}{\sin\theta}\frac{\partial}{\partial\theta}\left(\sin\theta\frac{\partial}{\partial\theta}\right)+\frac{1}{\sin^2\theta}\frac{\partial^2}{\partial\phi^2}\right]. \tag{3.16}$$

Equation 3.15 implies that spherical harmonics Y_{lm} are eigenfunctions of the angular momentum operator and have eigenvalues $l(l+1)$. The eigenvalues $l(l+1)$ are integer numbers that have the units of $\hbar^2$. The transition from Laplace's equation to the eigenvalue problem for the square of the magnitude of angular momentum operator, L^2, establishes a connection between the theory of harmonic functions and the theory of angular momentum. The latter thus providing a tool for performing 3D rotations of spherical harmonics. More on this topic will be given in Chapter 4; at present, it is necessary to establish a number of useful relations for spherical harmonics.

3.2 Orthogonality and Normalization of Spherical Harmonics

Spherical harmonics $Y_{lm}(\theta,\phi)$ form a complete set of orthonormal, that is, orthogonal and normalized, functions. A set of functions is regarded as complete when there cannot exist another function in the coordinate space (θ,ϕ) that would be simultaneously orthogonal to all currently existing functions in the set. A complete set of spherical harmonic functions consists of an infinite number of functions, all of which are defined in the domain (θ,ϕ), and have unique pairs of indices l and m. Based on completeness of the set, any function $f(\theta,\phi)$ defined on the surface of a unit sphere can be expanded in a series of spherical harmonics:

$$f(\theta,\phi)=\sum_{l=0}^{\infty}\sum_{m=-l}^{l}A_{lm}Y_{lm}(\theta,\phi), \tag{3.17}$$

where A_{lm} are the coefficients of expansion to be determined. In general, the index l in Equation 3.17 goes from zero to infinity. No problem arises as a result of the infinite number of members in the sum because this series quickly converges, and only a finite number of summation elements is required to obtain a convergent solution. Equation 3.17 reads as a function $f(\theta,\phi)$ being expanded in the basis of spherical harmonic functions; therefore, it is customary to call the expansion functions *basis functions.*

The completeness of the function set relies on the fact that spherical harmonics are orthogonal functions in the indices l and m. This means that the integral over a product of two spherical harmonic functions vanishes unless the indices of both functions are equal to each other:

$$\int Y^*_{l'm'}(\theta,\phi)Y_{lm}(\theta,\phi)d\Omega=\delta_{l'l}\,\delta_{m'm}, \tag{3.18}$$

where $d\Omega=d(\cos\theta)\,d\phi$ indicates integration over the surface of the unit sphere. Since spherical harmonics are complex functions, the integral in Equation 3.18 requires one of the functions to be in its complex-conjugate form; that is, indicated by the star. Kronecker's symbol $\delta_{l'l}$ is equal to one if the two indices are the same; otherwise it is zero.

Proof of Equation 3.18 follows from the fact that spherical harmonics are a product of polar, $P(\theta)$, and azimuthal, $Z(\phi)$, functions, which are independent of each other. Therefore the orthogonality condition can be analyzed for each coordinate θ and ϕ separately. First consider the azimuthal function:

$$Z(\phi)=e^{\pm im\phi}. \tag{3.19}$$

The argument ϕ in this function varies from 0 to 2π. The azimuthal function is orthogonal in index m. Using a second azimuthal function, with k in place of m, this can be easily verified for the case of $m\neq k$. Since $Z(\phi)$ is a complex function, the first function in the integral is taken in a complex conjugate form. The corresponding orthogonality integral is:

$$\int_0^{2\pi} Z^*_m Z_k d\phi=\int_0^{2\pi} e^{\mp im\phi}e^{\pm ik\phi}d\phi=\int_0^{2\pi} e^{\pm i(k-m)\phi}d\phi=\frac{1}{\pm i(k-m)}e^{\pm i(k-m)\phi}\Big|_0^{2\pi}=0. \tag{3.20}$$

The integral in Equation 3.20 resolves *via* the standard differentiation rule of exponential functions, that is, $(d/d\phi)e^{\pm i(k-m)\phi}=\pm i(k-m)e^{\pm i(k-m)\phi}$. Since the azimuthal function is periodic with the period of 2π, this leads to the equality $e^{\pm i(k-m)2\pi}=e^{\pm i(k-m)0}$. Because of that, the upper and lower integration limits cancel out, and the integral vanishes. This proves that azimuthal functions having different azimuthal index, $m\neq k$, are orthogonal to each other, so that the integral of the product of such functions is equal to zero.

In addition to being orthogonal, the basis functions used in the expansion series should be normalized to unity. Normalization of azimuthal functions is determined by their normalization integral:

$$\int_0^{2\pi} Z^*_m Z_m d\phi=\int_0^{2\pi} e^{-im\phi}e^{im\phi}d\phi=\int_0^{2\pi} d\phi=2\pi, \tag{3.21}$$

which differs from the orthogonality integral by the condition $m=k$. For normalized functions this integral must equal unity. In order to achieve such a result, it is necessary to scale the azimuthal function by the inverse square root of the outcome of integration in Equation 3.21. Implementing the above normalization condition leads to:

$$Z(\phi)_{normalized} = \frac{1}{\sqrt{2\pi}} e^{\pm im\phi}. \tag{3.22}$$

The next function in Equation 3.5 to examine for orthogonality is $P(\theta)$, which describes the dependence on polar angle θ. This function is represented by the associated Legendre functions, P_l^m, as defined in Equation 3.3. Unlike the above case of azimuthal dependence functions, $Z(\phi)$, associated Legendre functions exhibit only conditional orthogonality. Specifically, associated Legendre functions are orthogonal in index l only when they have the same value of upper index m. This condition is described by the orthogonality integral:

$$\int_{-1}^{1} P_l^m P_n^m d\mu = 0, \quad \text{when } n \neq l. \tag{2.50}$$

The companion case of $n=l$ leads to a normalization integral:

$$\int_{-1}^{1} P_l^m P_l^m d\mu = \frac{2}{(2l+1)} \frac{(l+m)!}{(l-m)!}. \tag{2.51}$$

Proof for both of these equations is given in Chapter 2. Note that because the coordinate of integration is $\mu = \cos\theta$, the limits of integration are –1 to 1.

Since the $P(\theta)$ are components of spherical harmonics functions, and since the latter need to be normalized in order for them to be applicable as basis functions in expansion series, it is necessary for the functions $P(\theta)$ to be normalized. The integral in Equation 2.51 provides the normalization coefficient. Scaling the non-normalized functions taken from Equation 3.3 by the inverse square root of the normalization integral gives the normalized form of the associated Legendre functions:

$$P(\theta)_{normalized} = \sqrt{\frac{(2l+1)}{2} \frac{(l-m)!}{(l+m)!}} P_l^m(\cos\theta). \tag{3.23}$$

The product of azimuthal (Equation 3.22) and polar functions (Equation 3.23) in their normalized forms result in the desired spherical harmonic functions (Equation 3.6). The normalization coefficient of the spherical harmonics is simply a product of normalization coefficients of the individual functions. Since the azimuthal and polar components of the spherical harmonics functions are individually orthonormal in the domain of coordinates θ and ϕ, respectively, and because they are independent of each other, their product, Equation 3.6 is also orthonormal in the same coordinate space. This completes the proof that spherical harmonics, as stated in Equation 3.18, are orthonormal functions.

As an apparent benefit, the switch from the individual azimuthal and polar dependence functions to their product improves the orthogonality property of the resulting spherical harmonics over that of the separate associated Legendre functions. Indeed, associated Legendre functions P_l^m that have different index m are not orthogonal in index l. Moreover, they are not even orthogonal in index m regardless of the value that index l assumes. These factors limit the usefulness of stand-alone associated Legendre functions as basis functions in any series expansion. Merging azimuthal and polar functions into a single function removes this limitation. As a result, all spherical harmonics functions, Y_{lm}, which likewise depend on indices l and m, are orthogonal in indices l and m independently. Indeed, Equation 3.20 guarantees the orthogonality of spherical harmonics functions of index m regardless of the value of index l. If index m is the same in two spherical harmonic functions, they remain orthogonal in index l due to Equation 2.50. The resulting independent orthogonality property in indices l and m makes spherical harmonic functions particularly useful for generating series expansions.

3.3 Symmetry Properties of Spherical Harmonics

Spherical harmonics functions exhibit a number of symmetry relations in their indices l and m that make mathematical manipulation easier. The most important and frequently used symmetry property applies to the azimuthal index, m. This index has the range $-l \leq m \leq l$. Spherical harmonic functions having negative value of index m are connected to the corresponding terms having positive value of index m *via* a simple relation:

$$Y_{l,-m} = (-1)^m Y_{lm}^*, \tag{3.24}$$

where star indicates complex conjugate.

The symmetry relation introduced by Equation 3.24 can be obtained by replacing index m by $-m$ in Equation 3.6. This gives:

$$Y_{l,-m}(\theta,\phi) = \sqrt{\frac{(2l+1)}{4\pi}\frac{(l+m)!}{(l-m)!}} P_l^{-m}(\cos\theta)\, e^{-im\phi}. \tag{3.25}$$

The next step relies on a corresponding property of the associated Legendre functions that was obtained in Chapter 2:

$$P_l^{-m}(\cos\theta) = (-1)^m \frac{(l-m)!}{(l+m)!} P_l^m(\cos\theta). \tag{2.49}$$

Substituting the value of P_l^{-m} from Equation 2.49 into Equation 3.25 leads to:

$$Y_{l,-m}(\theta,\phi) = \sqrt{\frac{(2l+1)}{4\pi}\frac{(l+m)!}{(l-m)!}}(-1)^m \frac{(l-m)!!}{(l+m)!} P_l^m(\cos\theta)\, e^{-im\phi}. \tag{3.26}$$

Cancelling out common factors, and noting that the negative sign in front of complex number i corresponds to a complex conjugate form of the spherical harmonic function, leads to the desired result:

$$Y_{l,-m}(\theta,\phi) = (-1)^m \sqrt{\frac{(2l+1)}{4\pi}\frac{(l-m)!}{(l+m)!}} P_l^m(\cos\theta)\, e^{-im\phi} = (-1)^m Y_{lm}^*(\theta,\phi). \tag{3.27}$$

This proves Equation 3.24.

The next symmetry relation involves index l and describes the change in spherical harmonic functions resulting from coordinate inversion, that is, a change in sign of Cartesian coordinates from (x, y, z) to $(-x, -y, -z)$. Figure 3.1 illustrates coordinate inversion acting on particle P resulting in it being translated to point P′, which is located in the opposite octant relative to that of point P.

Although the change in Cartesian coordinates caused by inversion is trivially simple due to the similarity between all three coordinate axes, the equivalent change in a spherical polar coordinate system (r, θ, ϕ) upon coordinate inversion is a bit more involved. Since inversion does not change the distance from the particle to the center of coordinate system the radial term, r remains unchanged. The polar angle, θ, is not so simple. Figure 3.2 illustrates the change in polar angle that results from coordinate inversion.

In spherical polar coordinates, the magnitude of polar angle θ spans the range 0 to π. Inversion of the coordinates of point P translates it into point P′. The resulting point P′ makes an angle $\pi+\theta$ with the z-axis. Due to the trigonometric relationships $\cos\theta = -\cos(\pi+\theta) = -\cos(\pi-\theta)$, there is flexibility in

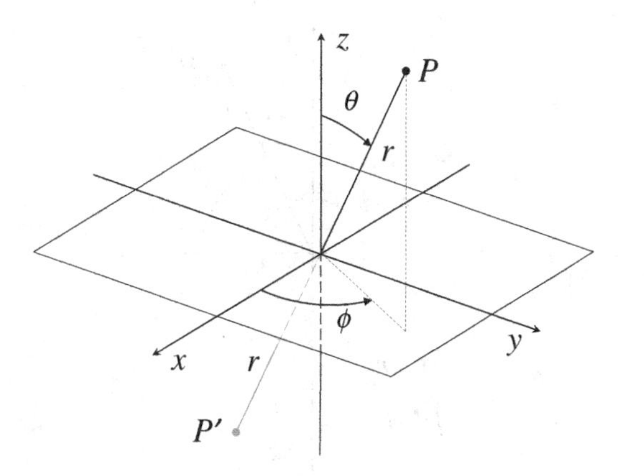

FIGURE 3.1 Translation of particle P into P′ upon coordinate inversion.

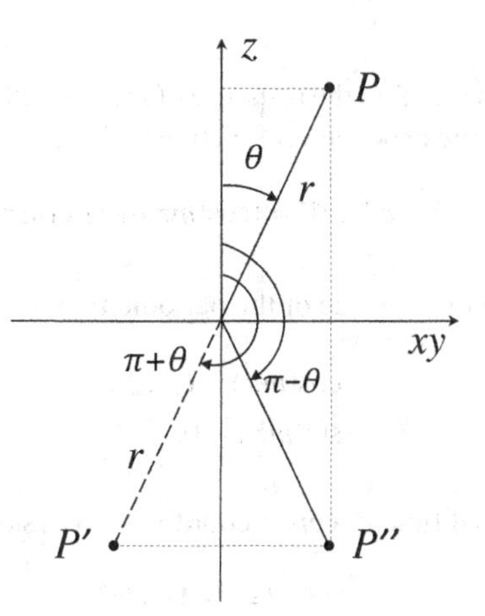

FIGURE 3.2 Change in the polar angle upon coordinate inversion.

the choice of angles to be used. For convenience, the choice is made in favor of the smaller angle, $\pi - \theta$ rather than the larger one, $\pi + \theta$. This means that upon coordinate inversion the angle θ changes to $\pi - \theta$.

The remaining coordinate to analyze on the nature of its change upon coordinate inversion is azimuthal angle, ϕ. In spherical polar coordinates, its value spans the range 0 to 2π. This full 360° scan is necessary in order to cover all four quadrants in the xy-plane. Based on the way azimuthal angle is measured, coordinate inversion adds π to the value of angle, ϕ. Figure 3.3 illustrates how azimuthal angle of a particle changes from ϕ to $\pi + \phi$ upon coordinate inversion.

Putting all the pieces about each individual coordinate together results in the rule for inversion of spherical polar coordinates. This rule states that upon inversion spherical polar coordinates change from (r, θ, ϕ) to $(r, \pi - \theta, \pi + \phi)$. In the associated Legendre functions and spherical harmonics the polar angle θ occurs in the form of the cosine of the angle, so the coordinates transform is from $(r, \cos\theta, \phi)$ to $(r, -\cos\theta, \pi + \phi)$.

Having the details of transformation of spherical polar coordinates resolved makes it possible to work out how the coordinate inversion affects spherical harmonics. As before, the solution comes from an analysis of individual polar and azimuthal functions that constitute spherical harmonics. Upon coordinate inversion, the associated Legendre functions undergo the following transformation:

$$P_l^m(-\mu) = (-1)^{l-m} P_l^m(\mu). \tag{2.119}$$

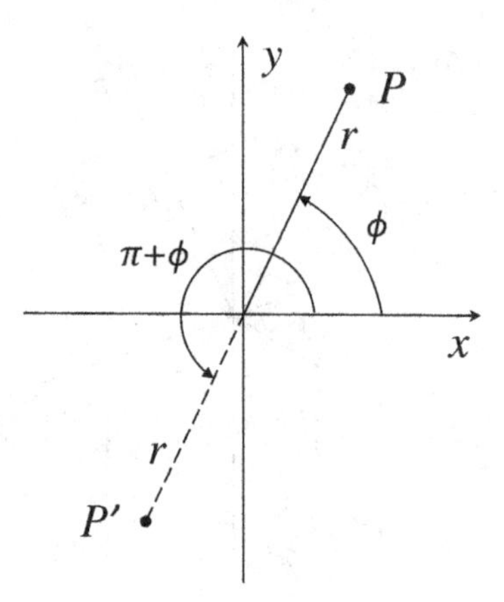

FIGURE 3.3 Change in the azimuthal angle upon coordinate inversion.

where $\mu = \cos\,\theta$. All that remains is to find the response of the azimuthal function to coordinate inversion. Substituting angle ϕ by $\pi + \phi$ in the exponential function and expanding $e^{im\pi}$ *via* Euler's formula gives:

$$e^{im(\pi+\phi)} = e^{im\pi}e^{im\phi} = [\cos(m\pi) + i\sin(m\pi)]e^{im\phi}. \tag{3.28}$$

Since m is an integer number, a consequence of the periodicity of trigonometric functions is that:

$$\begin{aligned}\cos(m\pi) &= (-1)^m,\\ \sin(m\pi) &= 0.\end{aligned} \tag{3.29}$$

With that, the change in azimuthal function upon coordinate inversion becomes:

$$e^{im(\pi+\phi)} = (-1)^m e^{im\phi}. \tag{3.30}$$

The results presented in Equation 2.119 and Equation 3.30 can now be combined. Replacing the spherical polar coordinates (θ, ϕ) in spherical harmonics by their transformed values, ($\pi - \theta$, $\pi + \phi$), and using the fact that $\cos(\pi - \theta) = -\cos\,\theta$ gives:

$$Y_{lm}(\pi - \theta, \pi + \phi) = \sqrt{\frac{(2l+1)}{4\pi}\frac{(l-m)!}{(l+m)!}}P_l^m(-\cos\theta)e^{im(\pi+\phi)}. \tag{3.31}$$

Substituting the conclusions from Equation 2.119 and Equation 3.30 into Equation 3.31 leads to:

$$Y_{lm}(\pi - \theta, \pi + \phi) = \sqrt{\frac{(2l+1)}{4\pi}\frac{(l-m)!}{(l+m)!}}(-1)^{l-m}P_l^m(\cos\theta)(-1)^m e^{im\phi}. \tag{3.32}$$

Upon multiplication of $(-1)^{l-m}$ and $(-1)^m$ terms, Equation 3.32 becomes:

$$Y_{lm}(\pi - \theta, \pi + \phi) = (-1)^l Y_{lm}(\theta, \phi). \tag{3.33}$$

From this analysis, the conclusion can be made that coordinate inversion changes spherical harmonics by a phase factor of $(-1)^l$.

Both symmetry relations Equations 3.24 and 3.33, which are derived in this chapter, help in finding the value of spherical harmonic functions based on the previously computed ones; that helps to reduce the amount of computation that would otherwise be necessary. However, the bulk of computation of spherical harmonics requires the use of more powerful techniques. A suitable tool for this will now be developed.

3.4 Recurrence Relations for Spherical Harmonics

Spherical harmonics have many recurrence relations. Some apply to index l or m individually, others increment both indices at the same time. Recurrence relations provide the most efficient approach to the evaluation of spherical harmonics. However, it is important to use the right recurrence relation since some are numerically unstable. The following three recurrence relations are numerically stable, and together cover the full range of indices l and m:

$$Y_{l,m} = \sqrt{\frac{(2l-1)(2l+1)}{(l-m)(l+m)}}\,\mu Y_{l-1,m} - \sqrt{\frac{(2l+1)(l-m-1)(l+m-1)}{(2l-3)(l-m)(l+m)}}\,Y_{l-2,m}, \tag{3.34}$$

$$Y_{l,l-1} = \sqrt{2l-3}\,\mu Y_{l-1,l-1}, \tag{3.35}$$

$$Y_{l,l} = -\sqrt{\frac{2l+1}{2l}}(1-\mu^2)^{\frac{1}{2}} e^{i\phi}\, Y_{l-1,l-1}. \tag{3.36}$$

In order to acquire a full understanding of those relations it is helpful to see how they are derived. Since the associated Legendre functions represent the most complicated part of spherical harmonics, it is reasonable to begin the derivation of recurrence relations for spherical harmonics by starting with the numerically stable recurrence relations for associated Legendre functions.

The first recurrence relation for spherical harmonics to be derived begins with the previously discussed relation for associated Legendre functions:

$$(l-m)P_l^m = (2l-1)\mu P_{l-1}^m - (l+m-1)P_{l-2}^m. \tag{2.71}$$

Each term in Equation 2.71 needs a normalization coefficient and a corresponding azimuthal function in order to become a spherical harmonic function. Multiplying both sides of this equation by

$$\frac{1}{(l-m)}\sqrt{\frac{2l+1}{4\pi}\frac{(l-m)!}{(l+m)!}}\,e^{im\phi}$$

produces the desirable effect, and changes the left side of Equation 2.71 into a spherical harmonic of degree l and order m:

$$Y_{l,m} = \mu\frac{(2l-1)}{(l-m)}\sqrt{\frac{2l+1}{4\pi}\frac{(l-m)!}{(l+m)!}}P_{l-1}^m e^{im\phi} - \frac{(l+m-1)}{(l-m)}\sqrt{\frac{2l+1}{4\pi}\frac{(l-m)!}{(l+m)!}}P_{l-2}^m e^{im\phi}. \tag{3.37}$$

A similar transformation also needs to be performed in the right-hand side of Equation 3.37. However, some additional modifications are necessary before the remaining two terms can be converted into spherical harmonics. The exponential functions, which are required for the transformation, are already in place. This is not true for the normalization coefficients. These coefficients do not quite match the ones required for spherical harmonics of degree $l-1$ and $l-2$:

$$Y_{l-1,m} = \sqrt{\frac{2l-1}{4\pi}\frac{(l-m-1)!}{(l+m-1)!}}P_{l-1}^m e^{im\phi}, \tag{3.38}$$

$$Y_{l-2,m} = \sqrt{\frac{2l-3}{4\pi}\frac{(l-m-2)!}{(l+m-2)!}}P_{l-2}^{m}e^{im\phi}. \tag{3.39}$$

The first coefficient, which is in front of P_{l-1}^{m} in Equation 3.37, should be modified to the form shown in Equation 3.38. To do this, it is necessary to separate off the leading term in the first factorial in Equation 3.37. This gives:

$$\frac{(2l-1)}{(l-m)}\sqrt{\frac{2l+1}{4\pi}\frac{(l-m)!}{(l+m)!}} = \frac{(2l-1)}{(l-m)}\sqrt{\frac{2l+1}{4\pi}\frac{(l-m)(l-m-1)!}{(l+m)(l+m-1)!}}. \tag{3.40}$$

Cancelling out a square root of $(l-m)$ and rearranging the terms in the right-hand side of Equation 3.40 leads to:

$$\frac{(2l-1)}{(l-m)}\sqrt{\frac{2l+1}{4\pi}\frac{(l-m)!}{(l+m)!}} = \sqrt{\frac{(2l-1)(2l+1)}{(l-m)(l+m)}}\sqrt{\frac{2l-1}{4\pi}\frac{(l-m-1)!}{(l+m-1)!}}. \tag{3.41}$$

The second normalization coefficient in Equation 3.41 now matches the one in Equation 3.38.

The second coefficient from Equation 3.37 can be transformed in a similar way to produce a normalization coefficient that is required for spherical harmonics of degree $l-2$. The target form of the coefficient is the one shown in Equation 3.39. This time it is necessary to separate off the two leading terms in the factorials. This gives:

$$\frac{(l+m-1)}{(l-m)}\sqrt{\frac{2l+1}{4\pi}\frac{(l-m)!}{(l+m)!}} = \frac{(l+m-1)}{(l-m)}\sqrt{\frac{2l+1}{4\pi}\frac{(l-m)(l-m-1)(l-m-2)!}{(l+m)(l+m-1)(l+m-2)!}}. \tag{3.42}$$

Cancelling out square roots of $(l+m-1)$ and $(l-m)$ and rearranging the terms in the right-hand side of Equation 3.42 leads to:

$$\frac{(l+m-1)}{(l-m)}\sqrt{\frac{2l+1}{4\pi}\frac{(l-m)!}{(l+m)!}} = \sqrt{\frac{(2l+1)(l-m-1)(l+m-1)}{(2l-3)(l-m)(l+m)}}\sqrt{\frac{2l-3}{4\pi}\frac{(l-m-2)!}{(l+m-2)!}}. \tag{3.43}$$

The resulting expression now contains the normalization coefficient for spherical harmonics of degree $l-2$.

Substituting Equations 3.41 and 3.43 into Equation 3.37 gives:

$$\begin{aligned} Y_{l,m} = {} & \mu\sqrt{\frac{(2l-1)(2l+1)}{(l-m)(l+m)}}\sqrt{\frac{2l-1}{4\pi}\frac{(l-m-1)!}{(l+m-1)!}}P_{l-1}^{m}e^{im\phi} \\ & -\sqrt{\frac{(2l+1)(l-m-1)(l+m-1)}{(2l-3)(l-m)(l+m)}}\sqrt{\frac{2l-3}{4\pi}\frac{(l-m-2)!}{(l+m-2)!}}P_{l-2}^{m}e^{im\phi}. \end{aligned} \tag{3.44}$$

Finally, replacing the product of normalization coefficient, associated Legendre function, and azimuthal function by the corresponding spherical harmonics produces the desired recurrence relation:

$$Y_{l,m} = \sqrt{\frac{(2l-1)(2l+1)}{(l-m)(l+m)}}\mu Y_{l-1,m} - \sqrt{\frac{(2l+1)(l-m-1)(l+m-1)}{(2l-3)(l-m)(l+m)}}Y_{l-2,m}. \tag{3.34}$$

This recurrence relation allows the calculation of the value of spherical harmonics Y_{lm} of degree l given the value of two spherical harmonics $Y_{l-1,m}$ and $Y_{l-2,m}$ of degree $l-1$ and $l-2$, respectively, while keeping index m constant. Although this equation works for both positive and negative values of the index m, the absolute value of the index m is limited by the condition $|m| \leq l-2$ because of the presence of the spherical harmonic of degree $l-2$ in Equation 3.34. Therefore the upper limit of m is $l-2$, so only the functions $Y_{l,0} \ldots Y_{l,l-2}$ can appear in the left-hand side of Equation 3.34.

Literature on spherical harmonics typically presents this recurrence relation in a slightly different form:

$$\mu Y_{l,m} = \sqrt{\frac{(l-m+1)(l+m+1)}{(2l+1)(2l+3)}} Y_{l+1,m} + \sqrt{\frac{(l-m)(l+m)}{(2l-1)(2l+1)}} Y_{l-1,m}, \tag{3.45}$$

which differs from Equation 3.34 only in the arrangement of index l. Since the ability to go back and forth between Equations 3.34 and 3.45 helps to better understand this recurrence relation, it is useful at this point to derive Equation 3.45 as well. To perform the derivation it is necessary to substitute all occurrences of l by $l+1$ in Equation 3.34. This gives:

$$Y_{l+1,m} = \sqrt{\frac{(2l+1)(2l+3)}{(l-m+1)(l+m+1)}} \mu Y_{l,m} - \sqrt{\frac{(2l+3)(l-m)(l+m)}{(2l-1)(l-m+1)(l+m+1)}} Y_{l-1,m}. \tag{3.46}$$

Next, Equation 3.46 is rearranged so that the term $Y_{l,m}$ appears in the left-hand side and the term $Y_{l+1,m}$ goes into the right-hand side.

$$\sqrt{\frac{(2l+1)(2l+3)}{(l-m+1)(l+m+1)}} \mu Y_{l,m} = Y_{l+1,m} + \sqrt{\frac{(2l+3)(l-m)(l+m)}{(2l-1)(l-m+1)(l+m+1)}} Y_{l-1,m}. \tag{3.47}$$

Multiplying both sides of Equation 3.47 by

$$\sqrt{\frac{(l-m+1)(l+m+1)}{(2l+1)(2l+3)}}$$

leads to Equation 3.45.

Now that two representations of the same recurrence relation have been derived, it is possible to critically evaluate their utility. Despite being better known, Equation 3.45 is not suitable for digital computation because it requires the knowledge of term $l+1$ before the term l is computed. Equation 3.34 is free from that limitation, and it can be directly coded into a computer program.

The recurrence relation given by Equation 3.34 limits the range of the azimuthal index to $|m| = [0: l-2]$, as a result, computation of spherical harmonics $Y_{l,l-1}$ and $Y_{l,l}$ requires the application of different recurrence relations. The recurrence relation for computation of $Y_{l,l-1}$ directly follows from Equation 3.34. Setting $m = l-1$ makes the term $Y_{l-1,l-2}$ vanish since the corresponding associated Legendre function P_{l-2}^{l-1} vanishes due to the differentiation present in the Rodrigues' formula. With that Equation 3.34 reduces to:

$$Y_{l,l-1} = \sqrt{\frac{(2l-3)(2l-1)}{(2l-1)}} \mu Y_{l-1,l-1}, \tag{3.48}$$

which after simplification becomes:

$$Y_{l,l-1} = \sqrt{2l-3}\, \mu Y_{l-1,l-1}. \tag{3.35.1}$$

This gives the second recurrence relationship required for the calculation of spherical harmonics. The third and final recurrence relation needed is to calculate $Y_{l,l}$, that is, the spherical harmonic with the highest value of azimuthal index $|m| = l$. This time the recurrence has to take place in the index m as well as in index l. The starting point for the derivation is to use the corresponding relation for associated Legendre functions:

$$P_l^l = -(2l-1)(1-\mu^2)^{\frac{1}{2}} P_{l-1}^{l-1}. \tag{2.73}$$

The next step in the derivation requires the left and right sides of Equation 2.73 to be converted into the form:

$$Y_{l,l} = \sqrt{\frac{2l+1}{4\pi}\frac{1}{(2l)!}}P_l^l e^{il\phi}, \quad (3.49)$$

and

$$Y_{l-1,l-1} = \sqrt{\frac{2l-1}{4\pi}\frac{1}{(2l-2)!}}P_{l-1}^{l-1} e^{i(l-1)\phi}, \quad (3.50)$$

respectively. For that purpose, both sides of Equation 2.73 need to be multiplied by

$$\sqrt{\frac{2l+1}{4\pi}\frac{1}{(2l)!}}e^{il\phi}.$$

This leads to:

$$Y_{l,l} = -(2l-1)\sqrt{\frac{2l+1}{4\pi}\frac{1}{(2l)!}}(1-\mu^2)^{\frac{1}{2}}P_{l-1}^{l-1}e^{il\phi} \quad (3.51)$$

where the left-hand side is converted to the spherical harmonic function, $Y_{l,l}$ according to Equation 3.49. The coefficient in Equation 3.51 needs to be modified in order to put it into the same form as the normalization coefficient from Equation 3.50. As with the second recurrence procedure, this step starts by taking two leading terms off the factorial in the denominator. The next step is to move $2l-1$ into the square root. This gives:

$$-(2l-1)\sqrt{\frac{2l+1}{4\pi}\frac{1}{(2l)!}} = -\sqrt{\frac{2l+1}{4\pi}\frac{(2l-1)(2l-1)}{(2l)(2l-1)(2l-2)!}}. \quad (3.52)$$

After cancelling out the common terms $2l-1$, the remaining terms on the right-hand side of Equation 3.52 can be rearranged to separate out the coefficient necessary for spherical harmonics functions of the type $Y_{l-1,l-1}$. This leads to:

$$-(2l-1)\sqrt{\frac{2l+1}{4\pi}\frac{1}{(2l)!}} = -\sqrt{\frac{2l+1}{2l}}\sqrt{\frac{2l-1}{4\pi}\frac{1}{(2l-2)!}}. \quad (3.53)$$

Substituting Equation 3.53 into Equation 3.51, and splitting up the exponential function gives:

$$Y_{l,l} = -\sqrt{\frac{2l+1}{2l}}\sqrt{\frac{2l-1}{4\pi}\frac{1}{(2l-2)!}}(1-\mu^2)^{\frac{1}{2}}P_{l-1}^{l-1}e^{i(l-1)\phi}e^{i\phi}. \quad (3.54)$$

With that, Equation 3.54 is now fully prepared for the replacement of the parts in its right-hand side by the corresponding spherical harmonics function $Y_{l-1,l-1}$. This leads to the third and final recurrence relation for spherical harmonics:

$$Y_{l,l} = -\sqrt{\frac{2l+1}{2l}}(1-\mu^2)^{\frac{1}{2}}e^{i\phi}Y_{l-1,l-1}. \quad (3.36)$$

This equation is suitable for use in calculating spherical harmonic functions that have the highest value of azimuthal index, $|m| = l$.

With the completion of the derivation of recurrence relations for spherical harmonics, all the tools necessary for generating analytic expressions for the spherical harmonics have now been constructed.

3.5 Analytic Expression for the First Few Spherical Harmonics

Earlier, we saw that the general expression for spherical harmonic functions, Equation 3.6, is unsuitable as an analytic expression for individual spherical harmonic functions because of the presence of the differentiation operator in Rodrigues' formula in the expression for associated Legendre functions:

$$Y_{lm}(\theta,\phi) = \sqrt{\frac{(2l+1)}{4\pi}\frac{(l-m)!}{(l+m)!}}\,P_l^m(\cos\theta)\,e^{im\phi}, \tag{3.6}$$

$$P_l^m(\mu) = \frac{(-1)^m}{2^l l!}(1-\mu^2)^{m/2}\frac{d^{l+m}}{d\mu^{l+m}}(\mu^2-1)^l, \tag{2.37}$$

$$Y_{lm}(\theta,\phi) = \sqrt{\frac{(2l+1)}{4\pi}\frac{(l-m)!}{(l+m)!}}\,\frac{(-1)^m}{2^l l!}(1-\mu^2)^{\frac{m}{2}}\frac{d^{l+m}}{d\mu^{l+m}}(\mu^2-1)^l e^{im\phi}, \tag{3.37}$$

where $\mu = \cos\theta$.

Theoretically, spherical harmonics can be obtained with little effort from analytic expressions for associated Legendre functions, if those are available, by substituting them into Equation 3.6. Since associated Legendre functions may not always be available, we will proceed on the premise that they are not. In that case, computation of spherical harmonics from their recurrence relations is the only feasible approach.

Due to the availability of recurrence relations, only the first two terms (l,m) referred to as (0,0) and (1,0) need to be directly computed from Equation 3.37; that is a straightforward task. All the remaining spherical harmonic functions can then be computed recursively from Equations 3.34 through 3.36 for the non-negative range of index m. Spherical harmonics that have azimuthal index $m < 0$ can then be computed from the corresponding functions having $m > 0$ by using Equation 3.24. Analytic expressions for first few spherical harmonics are presented in Table 3.1.

These analytic expressions may be useful when an individual function is needed for some reason. However, the above expressions are less than ideal for generating the numerical values of a series of spherical harmonic functions due to the rapid accumulation of rounding errors resulting from manipulation of large coefficients. Once again, the numerical values of spherical harmonics functions should be computed by using the recurrence relations just developed in Section 3.4.

3.6 Nodal Properties of Spherical Harmonics

Access to analytic expressions for the individual spherical harmonics provides an opportunity to closely study their spatial properties. There are three major types of spherical harmonics, depending on their nodal properties. Being a product of polar and azimuthal functions, spherical harmonics include the Legendre polynomials for the special case when index $m = 0$. This condition reduces the equation for spherical harmonics Equation 3.6 to:

$$Y_{l,0}(\theta) = \sqrt{\frac{(2l+1)}{4\pi}}\,P_l(\cos\theta). \tag{3.38}$$

TABLE 3.1
Analytic Expressions for the First Few Spherical Harmonics

$$Y_{0,0} = \frac{1}{2}\sqrt{\frac{1}{\pi}}$$

$$Y_{1,0} = \frac{1}{2}\sqrt{\frac{3}{\pi}}\cos\theta$$

$$Y_{1,1} = -\frac{1}{2}\sqrt{\frac{3}{2\pi}}e^{i\phi}\sin\theta$$

$$Y_{1,-1} = \frac{1}{2}\sqrt{\frac{3}{2\pi}}e^{-i\phi}\sin\theta$$

$$Y_{2,0} = \frac{1}{4}\sqrt{\frac{5}{\pi}}(3\cos^2\theta - 1)$$

$$Y_{2,1} = -\frac{1}{2}\sqrt{\frac{15}{2\pi}}e^{i\phi}\sin\theta\cos\theta$$

$$Y_{2,2} = \frac{1}{4}\sqrt{\frac{15}{2\pi}}e^{2i\phi}\sin^2\theta$$

$$Y_{3,0} = \frac{1}{4}\sqrt{\frac{7}{\pi}}\cos\theta(5\cos^2\theta - 3)$$

$$Y_{3,1} = -\frac{1}{8}\sqrt{\frac{21}{\pi}}e^{i\phi}\sin\theta(5\cos^2\theta - 1)$$

$$Y_{3,2} = \frac{1}{4}\sqrt{\frac{105}{2\pi}}e^{2i\phi}\sin^2\theta\cos\theta$$

$$Y_{3,3} = -\frac{1}{8}\sqrt{\frac{35}{\pi}}e^{3i\phi}\sin^3\theta$$

$$Y_{4,0} = \frac{3}{16}\sqrt{\frac{1}{\pi}}(35\cos^4\theta - 30\cos^2\theta + 3)$$

$$Y_{4,1} = -\frac{3}{8}\sqrt{\frac{5}{\pi}}e^{i\phi}\sin\theta\cos\theta(7\cos^2\theta - 3)$$

$$Y_{4,2} = \frac{3}{8}\sqrt{\frac{5}{2\pi}}e^{2i\phi}\sin^2\theta(7\cos^2\theta - 1)$$

$$Y_{4,3} = -\frac{3}{8}\sqrt{\frac{35}{\pi}}e^{3i\phi}\sin^3\theta\cos\theta$$

$$Y_{4,4} = \frac{3}{16}\sqrt{\frac{35}{2\pi}}e^{4i\phi}\sin^4\theta$$

$$Y_{5,0} = \frac{1}{16}\sqrt{\frac{11}{\pi}}(63\cos^5\theta - 70\cos^3\theta + 15\cos\theta)$$

$$Y_{5,1} = -\frac{1}{16}\sqrt{\frac{165}{2\pi}}e^{i\phi}\sin\theta(21\cos^4\theta - 14\cos^2\theta + 1)$$

$$Y_{5,2} = \frac{1}{8}\sqrt{\frac{1155}{2\pi}}e^{2i\phi}\sin^2\theta\cos\theta(3\cos^2\theta - 1)$$

$$Y_{5,3} = -\frac{1}{32}\sqrt{\frac{385}{\pi}}e^{3i\phi}\sin^3\theta(9\cos^2\theta - 1)$$

(Continued)

TABLE 3.1 (*Continued*)

Analytic Expressions for the First Few Spherical Harmonics

$$Y_{5,4} = \frac{3}{16}\sqrt{\frac{385}{2\pi}}\, e^{4i\phi} \sin^4\theta\cos\theta$$

$$Y_{5,5} = -\frac{3}{32}\sqrt{\frac{77}{\pi}}\, e^{5i\phi} \sin^5\theta$$

$$Y_{6,0} = \frac{1}{32}\sqrt{\frac{13}{\pi}}(231\cos^6\theta - 315\cos^4\theta + 105\cos^2\theta - 5)$$

$$Y_{6,1} = -\frac{1}{16}\sqrt{\frac{273}{2\pi}}\, e^{i\phi} \sin\theta\cos\theta(33\cos^4\theta - 30\cos^2\theta + 5)$$

$$Y_{6,2} = \frac{1}{64}\sqrt{\frac{1365}{\pi}}\, e^{2i\phi} \sin^2\theta(33\cos^4\theta - 18\cos^2\theta + 1)$$

$$Y_{6,3} = -\frac{1}{32}\sqrt{\frac{1365}{\pi}}\, e^{3i\phi} \sin^3\theta\cos\theta(11\cos^2\theta - 3)$$

$$Y_{6,4} = \frac{3}{32}\sqrt{\frac{91}{2\pi}}\, e^{4i\phi} \sin^4\theta(11\cos^2\theta - 1)$$

$$Y_{6,5} = -\frac{3}{32}\sqrt{\frac{1001}{\pi}}\, e^{5i\phi} \sin^5\theta\cos\theta$$

$$Y_{6,6} = \frac{1}{64}\sqrt{\frac{3003}{\pi}}\, e^{6i\phi} \sin^6\theta$$

$$Y_{7,0} = \frac{1}{32}\sqrt{\frac{15}{\pi}}(429\cos^7\theta - 693\cos^5\theta + 315\cos^3\theta - 35\cos\theta)$$

$$Y_{7,1} = -\frac{1}{64}\sqrt{\frac{105}{2\pi}}\, e^{i\phi} \sin\theta(429\cos^6\theta - 495\cos^4\theta + 135\cos^2\theta - 5)$$

$$Y_{7,2} = \frac{3}{64}\sqrt{\frac{35}{\pi}}\, e^{2i\phi} \sin^2\theta\cos\theta(143\cos^4\theta - 110\cos^2\theta + 15)$$

$$Y_{7,3} = -\frac{3}{64}\sqrt{\frac{35}{2\pi}}\, e^{3i\phi} \sin^3\theta(143\cos^4\theta - 66\cos^2\theta + 3)$$

$$Y_{7,4} = \frac{3}{32}\sqrt{\frac{385}{2\pi}}\, e^{4i\phi} \sin^4\theta\cos\theta(13\cos^2\theta - 3)$$

$$Y_{7,5} = -\frac{3}{64}\sqrt{\frac{385}{2\pi}}\, e^{5i\phi} \sin^5\theta(13\cos^2\theta - 1)$$

$$Y_{7,6} = \frac{3}{64}\sqrt{\frac{5005}{\pi}}\, e^{6i\phi} \sin^6\theta\cos\theta$$

$$Y_{7,7} = -\frac{3}{64}\sqrt{\frac{715}{2\pi}}\, e^{7i\phi} \sin^7\theta$$

$$Y_{8,0} = \frac{1}{256}\sqrt{\frac{17}{\pi}}(6435\cos^8\theta - 12{,}012\cos^6\theta + 6930\cos^4\theta - 1260\cos^2\theta + 35)$$

$$Y_{8,1} = -\frac{3}{64}\sqrt{\frac{17}{2\pi}}\, e^{i\phi} \sin\theta\cos\theta(715\cos^6\theta - 1001\cos^4\theta + 385\cos^2\theta - 35)$$

$$Y_{8,2} = \frac{3}{128}\sqrt{\frac{595}{\pi}}\, e^{2i\phi} \sin^2\theta(143\cos^6\theta - 143\cos^4\theta + 33\cos^2\theta - 1)$$

(*Continued*)

TABLE 3.1 (*Continued*)
Analytic Expressions for the First Few Spherical Harmonics

$$Y_{8,3} = -\frac{1}{64}\sqrt{\frac{19,635}{2\pi}}\, e^{3i\phi}\sin^3\theta\cos\theta(39\cos^4\theta - 26\cos^2\theta + 3)$$

$$Y_{8,4} = \frac{3}{128}\sqrt{\frac{1309}{2\pi}}\, e^{4i\phi}\sin^4\theta(65\cos^4\theta - 26\cos^2\theta + 1)$$

$$Y_{8,5} = -\frac{3}{64}\sqrt{\frac{17,017}{2\pi}}\, e^{5i\phi}\sin^5\theta\cos\theta(5\cos^2\theta - 1)$$

$$Y_{8,6} = \frac{1}{128}\sqrt{\frac{7293}{\pi}}\, e^{6i\phi}\sin^6\theta(15\cos^2\theta - 1)$$

$$Y_{8,7} = -\frac{3}{64}\sqrt{\frac{12,155}{2\pi}}\, e^{7i\phi}\sin^7\theta\cos\theta$$

$$Y_{8,8} = \frac{3}{256}\sqrt{\frac{12,155}{2\pi}}\, e^{8i\phi}\sin^8\theta$$

$$Y_{9,0} = \frac{1}{256}\sqrt{\frac{19}{\pi}}(12,155\cos^9\theta - 25,740\cos^7\theta + 18,018\cos^5\theta - 4620\cos^3\theta + 315\cos\theta)$$

$$Y_{9,1} = -\frac{3}{256}\sqrt{\frac{95}{2\pi}}\, e^{i\phi}\sin\theta(2431\cos^8\theta - 4004\cos^6\theta + 2002\cos^4\theta - 308\cos^2\theta + 7)$$

$$Y_{9,2} = \frac{3}{128}\sqrt{\frac{1045}{\pi}}\, e^{2i\phi}\sin^2\theta\cos\theta(221\cos^6\theta - 273\cos^4\theta + 91\cos^2\theta - 7)$$

$$Y_{9,3} = -\frac{1}{256}\sqrt{\frac{21,945}{\pi}}\, e^{3i\phi}\sin^3\theta(221\cos^6\theta - 195\cos^4\theta + 39\cos^2\theta - 1)$$

$$Y_{9,4} = \frac{3}{128}\sqrt{\frac{95,095}{2\pi}}\, e^{4i\phi}\sin^4\theta\cos\theta(17\cos^4\theta - 10\cos^2\theta + 1)$$

$$Y_{9,5} = -\frac{3}{256}\sqrt{\frac{2717}{\pi}}\, e^{5i\phi}\sin^5\theta(85\cos^4\theta - 30\cos^2\theta + 1)$$

$$Y_{9,6} = \frac{1}{128}\sqrt{\frac{40,755}{\pi}}\, e^{6i\phi}\sin^6\theta\cos\theta(17\cos^2\theta - 3)$$

$$Y_{9,7} = -\frac{3}{512}\sqrt{\frac{13,585}{\pi}}\, e^{7i\phi}\sin^7\theta(17\cos^2\theta - 1)$$

$$Y_{9,8} = \frac{3}{256}\sqrt{\frac{230,945}{2\pi}}\, e^{8i\phi}\sin^8\theta\cos\theta$$

$$Y_{9,9} = -\frac{1}{512}\sqrt{\frac{230,945}{\pi}}\, e^{9i\phi}\sin^9\theta$$

$$Y_{10,0} = \frac{1}{512}\sqrt{\frac{21}{\pi}}(46,189\cos^{10}\theta - 109,395\cos^8\theta + 90,090\cos^6\theta - 30,030\cos^4\theta + 3465\cos^2\theta - 63)$$

$$Y_{10,1} = -\frac{1}{256}\sqrt{\frac{1155}{2\pi}}\, e^{i\phi}\sin\theta\cos\theta(4199\cos^8\theta - 7956\cos^6\theta + 4914\cos^4\theta - 1092\cos^2\theta + 63)$$

$$Y_{10,2} = \frac{3}{512}\sqrt{\frac{385}{2\pi}}\, e^{2i\phi}\sin^2\theta(4199\cos^8\theta - 6188\cos^6\theta + 2730\cos^4\theta - 364\cos^2\theta + 7)$$

$$Y_{10,3} = -\frac{3}{256}\sqrt{\frac{5005}{\pi}}\, e^{3i\phi}\sin^3\theta\cos\theta(323\cos^6\theta - 357\cos^4\theta + 105\cos^2\theta - 7)$$

(Continued)

TABLE 3.1 (*Continued*)

Analytic Expressions for the First Few Spherical Harmonics

$$Y_{10,4} = \frac{3}{256}\sqrt{\frac{5005}{2\pi}}\, e^{4i\phi}\sin^4\theta(323\cos^6\theta - 255\cos^4\theta + 45\cos^2\theta - 1)$$

$$Y_{10,5} = -\frac{3}{256}\sqrt{\frac{1001}{\pi}}\, e^{5i\phi}\sin^5\theta\cos\theta(323\cos^4\theta - 170\cos^2\theta + 15)$$

$$Y_{10,6} = \frac{3}{1024}\sqrt{\frac{5005}{\pi}}\, e^{6i\phi}\sin^6\theta(323\cos^4\theta - 102\cos^2\theta + 3)$$

$$Y_{10,7} = -\frac{3}{512}\sqrt{\frac{85{,}085}{\pi}}\, e^{7i\phi}\sin^7\theta\cos\theta(19\cos^2\theta - 3)$$

$$Y_{10,8} = \frac{1}{512}\sqrt{\frac{255{,}255}{2\pi}}\, e^{8i\phi}\sin^8\theta(19\cos^2\theta - 1)$$

$$Y_{10,9} = -\frac{1}{512}\sqrt{\frac{4{,}849{,}845}{\pi}}\, e^{9i\phi}\sin^9\theta\cos\theta$$

$$Y_{10,10} = \frac{1}{1024}\sqrt{\frac{969{,}969}{\pi}}\, e^{10i\phi}\sin^{10}\theta$$

Figure 3.4A shows a typical example of such a function. In it, the z-axis goes from the south to the north pole of the sphere, and the center of the sphere is placed in the center of the coordinate system. Also known as Legendre polynomials, the functions defined by Equation 3.38 are called *zonal* harmonics. The name zonal indicates that the horizontally drawn parallels (latitudes) divide the sphere into zones. The lines of latitude represent points on the sphere where the value of the function is zero, that is, where it changes sign. The number of latitudes in the zonal harmonics is equal to the value of index l. In the drawings in Figure 3.4, gray indicates positive areas; negative areas are shown in white. The function changes sign when the azimuthal angle crosses a latitude line. This creates a striping pattern of alternating signs.

Pure Legendre polynomials (a.k.a. zonal harmonics) have rotational symmetry of infinite order. This very high symmetry rarely occurs in reality; therefore, the range of practical application of Legendre polynomials is very limited. The introduction of spherical harmonics as a product of polar and azimuthal functions in Equation 3.5 accomplishes the goal of creating functions that are much more useful in solving practical applications.

Figures 3.4B and 3.4C illustrate other two types of spherical harmonics. In a special case of $m=l$ (or $m=-l$), the vertical lines of meridians, or longitude, divide the surface of the sphere into sectors. Because of that, these functions, illustrated in Figure 3.4B, are called *sectorial* harmonics. Each line of meridians represents points on the surface where the function is zero, that is, where it changes sign.

Derivation of analytic expression for sectorial harmonics begins from the polynomial-style expression for associated Legendre functions:

$$P_l^m(\mu) = (1-\mu^2)^{m/2} \sum_{k=0}^{[l-m-\mathrm{mod}(l-m,2)]/2} (-1)^{m+k} \frac{(2l-2k)!}{2^l k!(l-k)!(l-m-2k)!}\mu^{l-m-2k}. \tag{2.118}$$

Setting $m=l$ in Equation 2.118 reduces the summation in index k to a single term, $k=0$. Taking into account that $(1-\mu^2)^{l/2} = \sin^l\theta$ further simplifies Equation 2.118 to:

$$P_l^l = (-1)^l \frac{(2l)!}{2^l l!}\sin^l\theta. \tag{3.39}$$

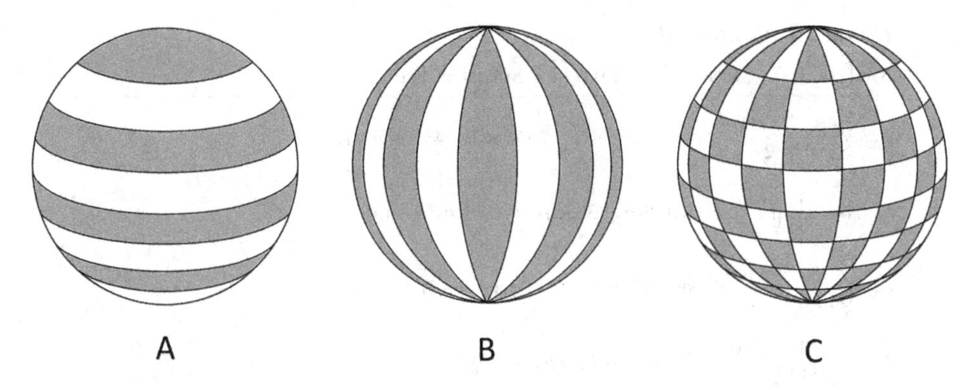

FIGURE 3.4 Types of spherical harmonics. A: zonal, B: sectorial, C: tesseral.

Substituting Equation 3.39 into Equation 3.6 leads to an analytic expression for sectorial harmonics:

$$Y_{l,l}(\theta,\phi) = (-1)^l \frac{1}{2^l l!}\sqrt{\frac{(2l+1)(2l)!}{4\pi}}\, e^{il\phi} \sin^l \theta. \tag{3.40}$$

Equation 3.40 helps us to understand the specific pattern of sectorial harmonics, which are illustrated in Figure 3.4B. Indeed, having $m = l$ reduces the dependence of the polar angle θ to powers of sin θ. Since sin θ and all of its powers are positive in the entire domain of θ, which varies from 0 to π, no horizontal nodal lines appear in sectorial spherical harmonics. Correspondingly, all vertical nodal lines in sectorial harmonics are due to the periodicity of the azimuthal function. The number of vertically positioned nodal circles, that is, where the function changes sign, is equal to the value $m = l$. The zonal function in Equation 3.40 thus describes a system that has m-fold rotational symmetry.

The third and last type, Figure 3.4C, represents *tesseral* spherical harmonics. They obtain their name from the surface of the sphere being cut into tesserae by intersecting horizontal and vertical nodal lines. Tesseral spherical harmonics have l nodal lines in total. Out of those l lines, m nodal lines are vertical (meridians), and $l - m$ lines are horizontal (latitudes).

Figure 3.5 helps in understanding the origin of the nodal properties of the spherical harmonics $Y_{0,0}$, $Y_{1,0}$, and $Y_{1,1}$. Since $Y_{0,0}$ is a constant, it has no nodal lines. The positive value of $Y_{0,0}$ on the surface of a unit sphere is shown in gray. $Y_{1,0}$ has no azimuthal dependence because $m = 0$, therefore it is a zonal spherical harmonic function. The polar angle dependence in $Y_{1,0}$ is described by cos θ, which is positive in the northern hemisphere (shown in gray; θ varies from 0 to $\pi/2$) and is negative in the southern hemisphere (shown in white; θ varies from $\pi/2$ to π). A horizontal nodal line appears at the equator due to $\cos(\pi/2) = 0$.

The third spherical harmonic function that is displayed in Figure 3.5 is $Y_{1,1}$. It is a sectorial function, and is slightly more elaborate than the two previous ones due to its dependency on both polar and azimuthal angles. In the diagrams, the negative sign of the spherical harmonic function is attached to the polar function. With that, the polar dependence in $Y_{1,1}$ is described by function $-\sin\theta$. This function is negative in the entire domain from 0 to π, excluding the boundary points (poles), at which point it goes to zero. The azimuthal function, $e^{i\phi}$, consists of real (*Re*) and imaginary (*Im*) components; these are shown separately in the diagram. The real component is cos ϕ, and the imaginary component is sin ϕ. Note that the coordinate system used here is the standard right-handed one. The real part of the azimuthal function, cos ϕ assumes positive values on that portion of the sphere that is located in the positive section of the x-axis. Similarly, sin ϕ assumes positive values in that region where y-axis is positive. The product of the polar and azimuthal functions gives $Y_{1,1}$, so the sign of the spherical harmonic function at any particular point of the unit sphere is determined by multiplying the sign of the individual polar and azimuthal functions. In this color scheme, white multiplied by white gives gray; white multiplied by gray gives white; and gray multiplied by gray gives gray. Because $m = 1$, $Y_{1,1}$ has one vertical nodal line,

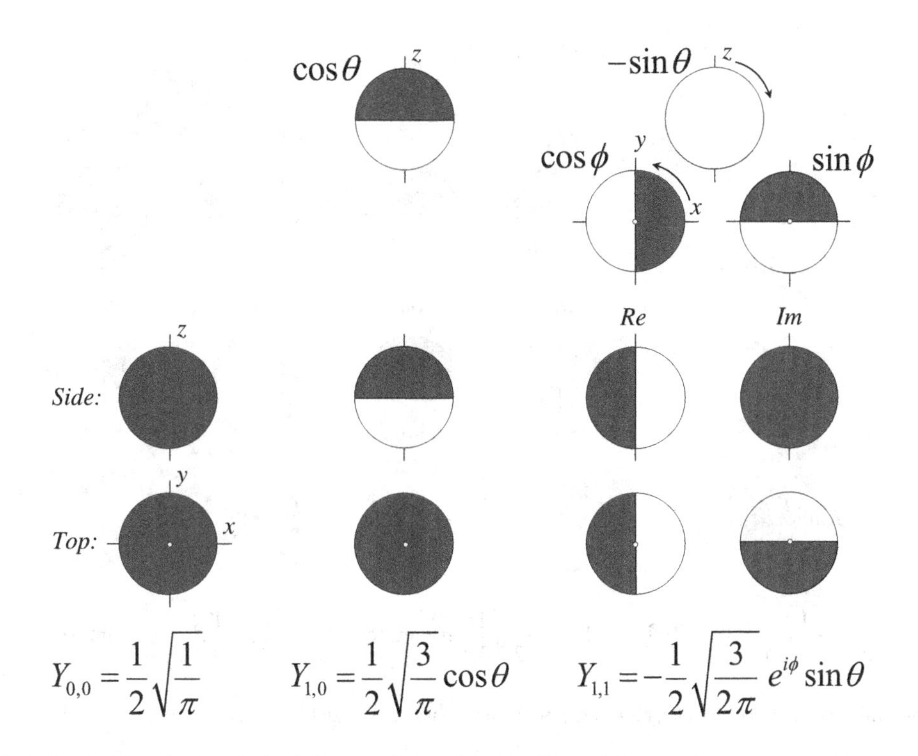

FIGURE 3.5 Nodal properties of spherical harmonic functions $Y_{0,0}$, $Y_{1,0}$, and $Y_{1,1}$.

which separates positive and negative values of the real part of the function. The imaginary part of $Y_{1,1}$ is obtained by rotating the corresponding real part by 90° in the standard counter clockwise direction about the z-axis when looking at the xy-plane from the tip of the z-axis.

The complexity of construction of nodal diagrams for spherical harmonics quickly increases as the value of indices l and m becomes larger. Figure 3.6 shows the nodal properties of three of the spherical harmonic functions of degree $l=2$. Function $Y_{2,0}$ is a zonal spherical harmonic; it has no azimuthal dependence. Its polar function is $3\cos^2\theta-1$; this has two roots, one at $\theta=54.7°$ and one at $\theta=125.3°$. These are responsible for the two horizontal nodal lines on the surface of the unit sphere. In the range $0°\leq\theta<54.7°$, function $3\cos^2\theta-1$ is positive (shown in gray); it is negative (shown in white) in the range $54.7°<\theta<125.3°$; and is positive once again in the range $125.3°<\theta\leq 180°$.

Spherical harmonic function $Y_{2,1}$ has both polar and azimuthal dependencies. The polar dependence function $-\sin\theta\cos\theta$ divides the entire sphere by an equatorial nodal line into north and south hemispheres. This function is negative in the northern hemisphere, is zero at the equatorial nodal line, and is positive in the southern hemisphere. The azimuthal dependence function, $e^{i\phi}$, is the same as that in $Y_{1,1}$; it creates one vertical nodal line with its exact position depending on whether it is located in a real or complex component of the azimuthal function. The product of polar and azimuthal functions has one horizontal and one vertical nodal line that makes $Y_{2,1}$ a tesseral spherical harmonic. Once again, the imaginary component of spherical harmonic $Y_{2,1}$ is obtained by 90° counter clockwise rotation of the real component around z-axis.

The third and the last function in this series, $Y_{2,2}$, is a sectorial spherical harmonic. Since $l-m=0$ it has no horizontal nodal lines because its polar dependence function, $\sin^2\theta$, is positive in the entire domain $0°<\theta<180°$, except at the poles where it goes to zero. The angular dependence function $e^{2i\phi}$ is represented by a real $\cos^2\theta$ and imaginary $\sin^2\theta$ components, each exhibiting a two-fold rotational symmetry. The product of polar and azimuthal functions splits the unit sphere into four vertically cut slices. Nodal properties of spherical harmonics of higher degree than $l=2$ can be worked out by using the same techniques as were used in Figures 3.5 and 3.6.

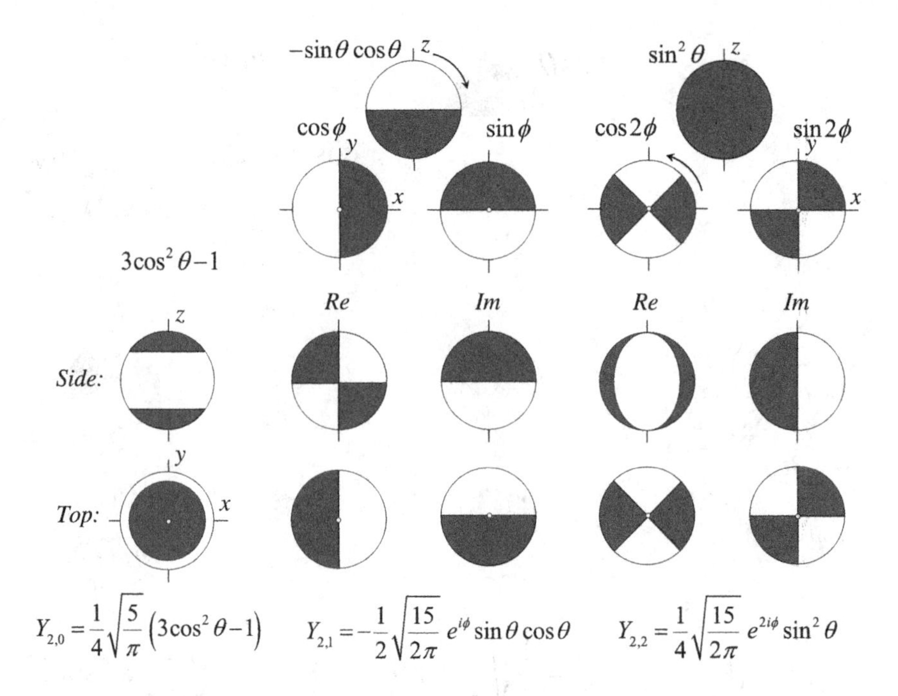

FIGURE 3.6 Nodal properties of spherical harmonic functions $Y_{2,0}$, $Y_{2,1}$, and $Y_{2,2}$.

Spherical harmonics are robust mathematical functions that can be used as basis functions in expansion series when the target function is defined on the surface of a unit sphere. The target function can be entirely asymmetric, which, as mentioned earlier, is a powerful advantage over the series expansion into Legendre polynomials. There is an important caveat in the application of spherical harmonics, though. Electrostatic energy, as well as many other physical properties of real objects, does not depend on the choice of coordinate system. However, spherical harmonics that are used to describe those properties depend on the choice of coordinate system. Because of this, their numerical values change as the coordinate system is rotated. Explaining the transformation of spherical harmonics under spatial rotation is a subject of the following chapter.

4

Angular Momentum

The concept of rotation of a body in 3D space is one of the pillars of classical mechanics. The general model of orbital rotation describes a particle, P, attached to the center of a coordinate system by a weightless and non-stretchable string, rotating around the center, O, as illustrated in Figure 4.1. Transitioning from that model to the rotation of a solid body requires summing up through all particles constituting the body; and that corresponds to taking the integral over the volume.

Classical mechanics defines rotation as an orbital motion of an object around a center of rotation. Quantum mechanics introduces an additional degree of freedom, called spin, which is an intrinsic property of any quantum object. Since spin has no classical equivalent, it cannot be visualized. However, because spin and orbital rotation share a great deal of the common theoretical ground, these two topics naturally appear together.

Any mathematical description of rotation requires paying close attention to the choice of the reference frame of rotation. This assumes aligning the center of coordinate system with the center of rotation. If this is not done, then the equations describing rotation would become extremely cumbersome. By having a physical body defined using a local coordinate frame, any rotation within this system can be described in two different ways, depending on the point of view. In the most intuitive approach, the local coordinate system is held fixed and the body is rotated. Since the body is doing the actual rotation, this is called *active rotation*. This framework would be the natural choice when describing the spinning parts of machinery. In an alternative view, the body is held fixed and the local coordinate system is rotated. This is called *passive rotation*, which reflects the independence of physical properties on the choice of coordinate system.

While the meaning of active rotation is intuitively clear, the purpose of passive rotation needs additional clarification. With a few exceptions, including coordinate-independent properties such as energy, the values of the functions that describe the state of a physical system often depend on the choice of coordinate system. If the coordinate system is rotated while the object is held fixed in space, the point of evaluation of a physical property, such as electrostatic potential, would rotate together with the coordinate system, so that the values of the function recorded in the point before and after rotation will be different. Since the initial position and value of a function before rotation are typically known, the theory of rotation deals with determining the value of function in specific points after the coordinate system is rotated.

Finding the outcome of rotation for a general function sounds like an almost impossible task, but an analytic solution can be achieved if the function in question could be expanded in a series of other functions that have known transformation properties. Recall that electrostatic potential function Φ may be expanded in a series of spherical harmonic functions, Y_{lm}, on the surface of a unit sphere:

$$\Phi(\theta,\phi) = \sum_{l=0}^{\infty} \sum_{m=-l}^{l} C_{lm} Y_{lm}(\theta,\phi), \tag{4.1}$$

where C_{lm} are coefficients of the expansion, and θ and ϕ are polar and azimuthal angles, respectively, in spherical polar coordinates. Spherical harmonics functions exhibit elegant properties when the coordinate frame is rotated, and exploring these will be the subject of the present chapter. Briefly, rotation transformations of a spherical harmonics function in 3D space correspond to a virtual rotation of the function in the space of its azimuthal degrees of freedom, so the viewpoint of passive rotation provides the most suitable formalism for operations of this type. Since spherical harmonics functions happen to be eigenfunctions of the angular momentum operator, this makes the theory of rotation, which is otherwise known as the theory of angular momentum, an integral part of electrostatics.

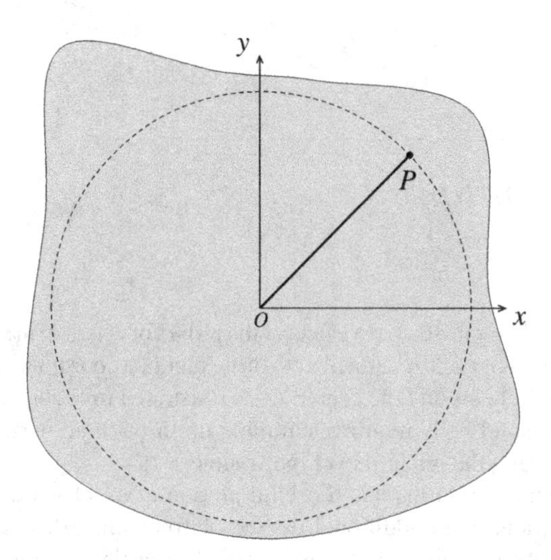

FIGURE 4.1 Rotation of a physical body and a particle P belonging to it around the center of coordinate system, O.

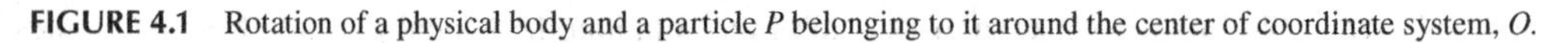

4.1 Rotation Matrices

The Cartesian coordinate system describes the position of a particle, P, in space by expressing it in the basis of three mutually orthogonal unit coordinate vectors, $\hat{\mathbf{x}}, \hat{\mathbf{y}}, \hat{\mathbf{z}}$, as illustrated in Figure 4.2.

Projecting the particle position onto the coordinate axes produces three coordinate vectors: $\mathbf{x} = x\hat{\mathbf{x}}$, $\mathbf{y} = y\hat{\mathbf{y}}$, and $\mathbf{z} = z\hat{\mathbf{z}}$, where x, y, z are scalar coefficients for the corresponding unit vectors. The components x, y, z are conventionally arranged in a column vector, $\mathbf{r}$:

$$\mathbf{r} = \begin{pmatrix} x \\ y \\ z \end{pmatrix} \quad \begin{matrix} \hat{\mathbf{x}} \\ \hat{\mathbf{y}}, \\ \hat{\mathbf{z}} \end{matrix} \tag{4.2}$$

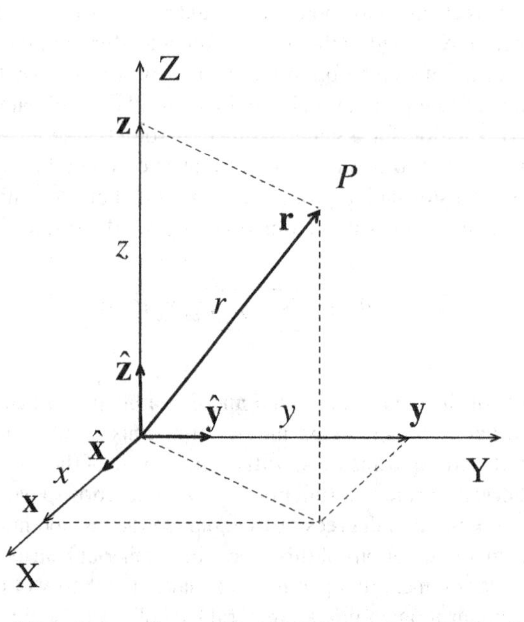

FIGURE 4.2 Representation of Cartesian coordinates of particle P in terms of unit vectors, $\hat{\mathbf{x}}, \hat{\mathbf{y}}, \hat{\mathbf{z}}$.

so that the radius vector of the particle becomes a sum of vector projections along the coordinate axes, $\mathbf{r} = \mathbf{x} + \mathbf{y} + \mathbf{z} = x\hat{\mathbf{x}} + y\hat{\mathbf{y}} + z\hat{\mathbf{z}}$. The matrix notation of the vector in Equation 4.2 omits the basis components $\hat{\mathbf{x}}, \hat{\mathbf{y}}, \hat{\mathbf{z}}$ for simplicity.

If the coordinate system is rotated around the z-axis by angle ϕ, as shown in Figure 4.3, the radius vector $\mathbf{r} = (x, y, z)$ of particle P changes to $\mathbf{r}' = (x', y', z')$ in the new coordinate system according to the equation:

$$\mathbf{r}' = R(\phi)\mathbf{r}, \tag{4.3}$$

where $R(\phi)$ is the rotation operator. A positive value of the rotation angle ϕ corresponds to a counter-clockwise direction of rotation when looking from the tip of the z-axis toward the xy-plane. As a reminder, whenever the topic of rotation appears in the discussion, any rotation of the coordinate system means a rotation of the local coordinate system only, leaving the particle's position in the global coordinate system unchanged.

A mathematical expression for the rotation transformation of the point P may be deduced from the analysis of geometric relations that connect the rotated frame with the initial one. The initial coordinate system comes with three basis vectors $\hat{\mathbf{x}}, \hat{\mathbf{y}}, \hat{\mathbf{z}}$ that determine the orientation of the coordinate axes. Rotation of the coordinate system changes the initial basis set, $\hat{\mathbf{x}}, \hat{\mathbf{y}}, \hat{\mathbf{z}}$, into a new one, $\hat{\mathbf{x}}', \hat{\mathbf{y}}', \hat{\mathbf{z}}'$, thus, $\hat{\mathbf{z}} = \hat{\mathbf{z}}'$ represents the outcome of rotation around the z-axis, and the actual rotation occurs in the xy-plane. Using the diagram in Figure 4.3, the coordinate transformation $x, y \rightarrow x', y'$ for particle P from the initial to the rotated coordinate frame occurs according to the trigonometric relations:

$$\begin{aligned} x' &= x \cos\phi + y \sin\phi, \\ y' &= -x \sin\phi + y \cos\phi, \\ z' &= z. \end{aligned} \tag{4.4}$$

The trigonometric coefficients that appear in the rotation transformation in Equation 4.4 can be arranged in a table form that comprises a rotation matrix, $\mathbf{R}_z(\phi)$:

$$\begin{array}{cc} & \begin{array}{ccc} \hat{\mathbf{x}} & \hat{\mathbf{y}} & \hat{\mathbf{z}} \end{array} \\ \mathbf{R}_z(\phi) = & \begin{pmatrix} \cos\phi & \sin\phi & 0 \\ -\sin\phi & \cos\phi & 0 \\ 0 & 0 & 1 \end{pmatrix} \end{array} \begin{array}{c} \\ \hat{\mathbf{x}}' \\ \hat{\mathbf{y}}' \\ \hat{\mathbf{z}}' \end{array} \tag{4.5}$$

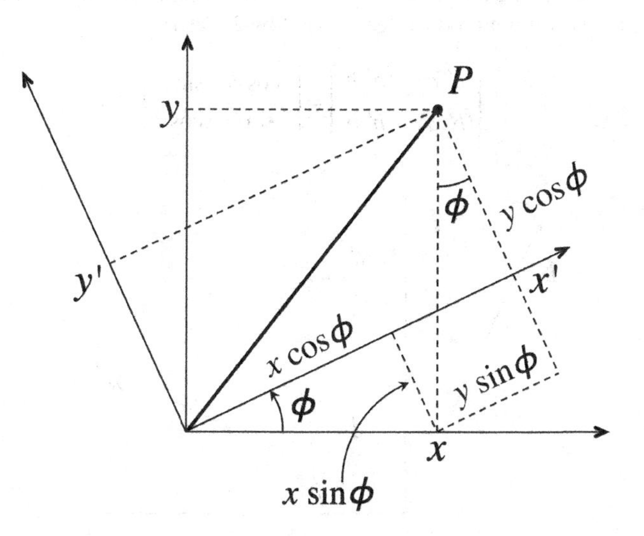

FIGURE 4.3 Counter-clockwise rotation of the right-handed coordinate system about the z-axis, in which the z-axis points toward the viewer.

Table $\mathbf{R}_z(\phi)$ is a matrix representation of the operator $R(\phi)$ in Cartesian space. The columns of this matrix represent a projection of the basis vectors $\hat{\mathbf{x}}, \hat{\mathbf{y}}, \hat{\mathbf{z}}$ of the initial coordinate system onto the basis vectors $\hat{\mathbf{x}}', \hat{\mathbf{y}}', \hat{\mathbf{z}}'$ of the transformed frame. Likewise, the rows of the matrix are projections of the new basis vectors $\hat{\mathbf{x}}', \hat{\mathbf{y}}', \hat{\mathbf{z}}'$ onto the initial ones, $\hat{\mathbf{x}}, \hat{\mathbf{y}}, \hat{\mathbf{z}}$. Using the introduced rotation matrix, a particle's coordinates in the new coordinate system resulting from a rotation of the coordinate system about the z-axis through the angle ϕ can be expressed as a matrix equation:

$$\begin{pmatrix} x' \\ y' \\ z' \end{pmatrix} = \begin{pmatrix} \cos\phi & \sin\phi & 0 \\ -\sin\phi & \cos\phi & 0 \\ 0 & 0 & 1 \end{pmatrix} \begin{pmatrix} x \\ y \\ z \end{pmatrix} \tag{4.6}$$

In symbolic matrix form, Equation 4.6 reads as $\mathbf{r}' = \mathbf{R}\,\mathbf{r}$, where $\mathbf{R}$ is a rotation matrix. This appears to be similar to Equation 4.3. The subtle difference that exists between the matrix $\mathbf{R}$ and the operator $R(\phi)$ comes from the matrix $\mathbf{R}$ being a representation of the operator $R(\phi)$ in a specific basis set, whereas $R(\phi)$ is an abstract operator. Due to the equivalence of the matrices and operator equations, both forms will be used interchangeably.

In addition to trigonometric relations, which define the rotation transformation in Equation 4.4, a rotation of the coordinate system can also be expressed in terms of angles subtended between the axes of the initial and rotated coordinate systems. To aid in this analysis, Figure 4.4 provides an example of mutual disposition of two coordinate systems in 2D. The axes a and b serve for any two coordinate axes from the right-handed Cartesian coordinate system. The rotation axis points toward the viewer, and is perpendicular to the plane ab.

Rotation $ab \to a'b'$ through angle ϕ creates three angles, ϕ, $(\pi/2) - \phi$, and $(\pi/2) + \phi$, which are subtended between the axes aa', $a'b$, and ab', respectively. Cosines of these angles are:

$$\begin{aligned} \phi &: \ \cos\phi, \\ \frac{\pi}{2}+\phi &: \ \cos\left(\frac{\pi}{2}+\phi\right) = -\sin\phi, \\ \frac{\pi}{2}-\phi &: \ \cos\left(\frac{\pi}{2}-\phi\right) = \sin\phi. \end{aligned} \tag{4.7}$$

Arranging these cosines into a table form by placing the axis vectors of the transformed coordinate system $a'b'$ in the rows while positioning the axes of the original coordinate system ab in the columns, and using bra-ket notation for the vector dot product, leads to a 2-by-2 matrix:

$$\begin{pmatrix} \langle a'|a\rangle & \langle a'|b\rangle \\ \langle b'|a\rangle & \langle b'|b\rangle \end{pmatrix} = \begin{pmatrix} \cos\phi & \sin\phi \\ -\sin\phi & \cos\phi \end{pmatrix}. \tag{4.8}$$

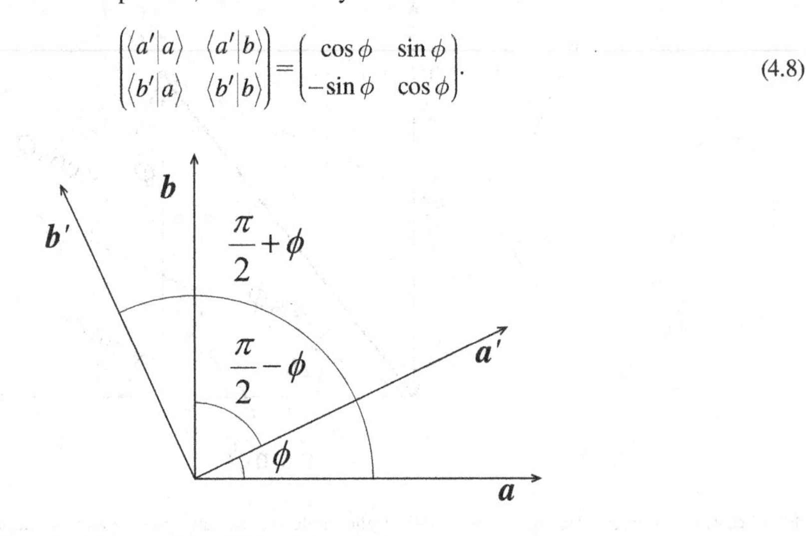

FIGURE 4.4 Rotation of coordinate system $ab \to a'b'$ through angle ϕ.

The matrix thus obtained (Equation 4.8) is identical to the matrix in Equation 4.5. Therefore, elements of the rotation matrix are indeed the cosines of the angle subtended between the axes of the initial and rotated coordinate systems. This result applies to rotation matrices of any dimension. With that, the general matrix for 3D rotation is:

$$\mathbf{R} = \begin{pmatrix} \langle \hat{\mathbf{x}}' | \hat{\mathbf{x}} \rangle & \langle \hat{\mathbf{x}}' | \hat{\mathbf{y}} \rangle & \langle \hat{\mathbf{x}}' | \hat{\mathbf{z}} \rangle \\ \langle \hat{\mathbf{y}}' | \hat{\mathbf{x}} \rangle & \langle \hat{\mathbf{y}}' | \hat{\mathbf{y}} \rangle & \langle \hat{\mathbf{y}}' | \hat{\mathbf{z}} \rangle \\ \langle \hat{\mathbf{z}}' | \hat{\mathbf{x}} \rangle & \langle \hat{\mathbf{z}}' | \hat{\mathbf{y}} \rangle & \langle \hat{\mathbf{z}}' | \hat{\mathbf{z}} \rangle \end{pmatrix}. \tag{4.9}$$

The theory developed for 2D rotation about the z-axis naturally extends to the x- and y-axes being the rotation axes. Figure 4.5 illustrates all three such possible cases. In the picture, the rotation axis points out from the rotation plane toward the viewer. All similar 2D rotations follow the same transformation rules described in Figure 4.3 with the only difference being the name of the rotation axis and the names of the axes composing the plane. Because of these similarities, rotation in 2D about different coordinate axes can be generalized into a single principle. Arranging the matrix elements from Equation 4.8 into a column for each rotation axis while placing the label of rotation axis on the top, and replacing the letters a and b by the pair of the axes from the rotation plane, produces a table:

$$\begin{array}{ccccccccc} & & (z) & & (x) & & (y) & & \\ \langle a' | a \rangle & = & \langle x' | x \rangle & = & \langle y' | y \rangle & = & \langle z' | z \rangle & = & \cos\phi \\ \langle b' | a \rangle & = & \langle y' | x \rangle & = & \langle z' | y \rangle & = & \langle x' | z \rangle & = & -\sin\phi \\ \langle a' | b \rangle & = & \langle x' | y \rangle & = & \langle y' | z \rangle & = & \langle z' | x \rangle & = & \sin\phi \\ \langle b' | b \rangle & = & \langle y' | y \rangle & = & \langle z' | z \rangle & = & \langle x' | x \rangle & = & \cos\phi \end{array} \tag{4.10}$$

The case of rotation about the z-axis corresponds to the situation when the axes x and y replace the labels a and b, respectively. Rotation about the x-axis emerges from the replacement of the labels ab with the pair yz. Similarly, replacement of the labels ab by zx gives the rotation around the y-axis. Equation 4.10, thus provides four of the nine matrix elements needed for the construction of a 2D rotation matrix in Equation 4.9 about a particular rotation axis. Evaluation of the remaining matrix elements is straightforward: the numeric value of the diagonal matrix element corresponding to the rotation axis is 1.0, since the coordinate value along this axis does not change on rotation. All off-diagonal matrix elements along the rows and columns that include the index of the rotation axis are zero because the coordinate components along the rotation axis are orthogonal with the coordinate components from the rotation plane. Of course, these simple rules only work when the rotation axis coincides with one of the coordinate axes.

The order of replacement of the axes in these 2D rotations corresponds to a cyclic permutation: $x \to y \to z \to x$, as can be seen in Figure 4.5. Conveniently, this cyclic permutation also preserves the

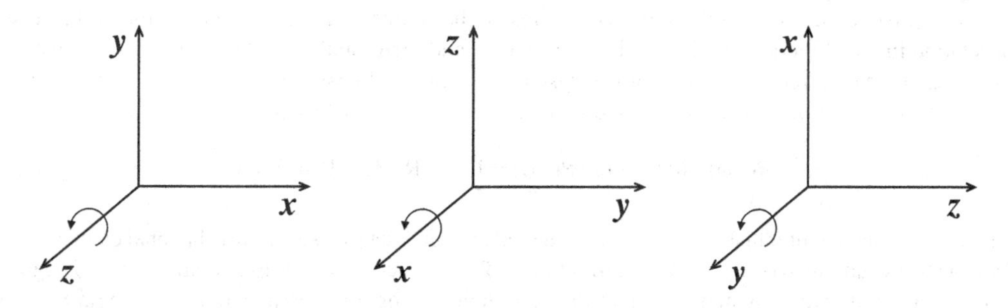

FIGURE 4.5 2D rotation of the coordinate system around different coordinate axes.

right-handedness of the coordinate system. Based on Equation 4.10, the matrices for rotation around the x- and y-axes become:

$$\mathbf{R}_x(\gamma)=\begin{pmatrix}1 & 0 & 0\\ 0 & \cos\gamma & \sin\gamma\\ 0 & -\sin\gamma & \cos\gamma\end{pmatrix},\qquad \mathbf{R}_y(\beta)=\begin{pmatrix}\cos\beta & 0 & -\sin\beta\\ 0 & 1 & 0\\ \sin\beta & 0 & \cos\beta\end{pmatrix}, \tag{4.11}$$

respectively.

Thus far, the rotations considered have dealt only with positive angles of rotation, that is, counter-clockwise rotations. However, rotations through a negative angle $-\phi$, clockwise rotations, are also valid. A clockwise rotation is described by the operator $R(-\phi)$, which is the inverse $R^{-1}(\phi)$ of the regular rotation operator, $R(\phi)$, given by Equation 4.3. The matrix representation for clockwise rotation from Equation 4.5 takes advantage of the parity of trigonometric functions: $\cos(-\phi) = \cos(\phi)$ and $\sin(-\phi) = -\sin(\phi)$. The matrix representation for operator $R(-\phi)$ for rotation around the z-axis is:

$$\mathbf{R}_z^{-1}(\phi)=\begin{pmatrix}\cos(-\phi) & \sin(-\phi) & 0\\ -\sin(-\phi) & \cos(-\phi) & 0\\ 0 & 0 & 1\end{pmatrix}=\begin{pmatrix}\cos\phi & -\sin\phi & 0\\ \sin\phi & \cos\phi & 0\\ 0 & 0 & 1\end{pmatrix}. \tag{4.12}$$

Matrix representations for rotation about other coordinate axes can be derived from Equation 4.11 in a similar manner:

$$\mathbf{R}_x(-\gamma)=\begin{pmatrix}1 & 0 & 0\\ 0 & \cos\gamma & -\sin\gamma\\ 0 & \sin\gamma & \cos\gamma\end{pmatrix},\qquad \mathbf{R}_y(-\beta)=\begin{pmatrix}\cos\beta & 0 & \sin\beta\\ 0 & 1 & 0\\ -\sin\beta & 0 & \cos\beta\end{pmatrix}. \tag{4.13}$$

The next step in the analysis of the properties of the rotation operator is to combine two or more rotations. Consecutive rotations may be applied either around the same axis or around different axes. In both cases, the second rotation is performed on the state that results from the first rotation. Given that, a sequence of rotations can be viewed as a successive step-by-step application of the rotation operators to the result of the previous rotation. This process can be expressed in operator form as:

$$\mathbf{r}' = R_2(R_1\mathbf{r}) = R_2R_1\mathbf{r}. \tag{4.14}$$

Equation 4.14 describes two consecutive rotations of the radius vector $\mathbf{r}$, first by the operator R_1, and then by the operator R_2. Since the vector $\mathbf{r}$ that is acted upon stays on the right side of the rotation operator, the order of application of the rotation operators reads from right to left. This is important when dealing with rotations about different axes. Conversely, the order of rotations is not important when multiple rotations are to be performed about the same axis.

Consecutive rotations about the same axis represent the simplest case of multiple rotations. The result of rotation in one direction, either positive or negative, and then rotation backward about the same axis by the same amount returns the coordinate system to its original position. The corresponding equation that describes this process when expressed in the operator or matrix form is:

$$R(-\phi)\,R(\phi) = R(\phi)\,R(-\phi) = 1,\qquad \mathbf{R}^{-1}\mathbf{R} = \mathbf{R}\,\mathbf{R}^{-1} = \mathbf{I}, \tag{4.15}$$

where $\mathbf{I}$ is a diagonal unit matrix that has all diagonal elements set to 1 and all off-diagonal elements to 0. Obviously, the unit matrix represents a null rotation. That is, it corresponds to a rotation by zero degrees.

Rotation about z-axis by angle ϕ_1, and then by angle ϕ_2 corresponds to rotation by angle $\phi_1 + \phi_2$. In matrix form, the total rotation is $\mathbf{r}' = \mathbf{R}(\phi_2)\,\mathbf{R}(\phi_1)\,\mathbf{r}$. When both rotations are made about the same axis, the order of matrix multiplications is not important. This is because the total rotation angle is the sum of individual rotation angles, $\phi = \phi_1 + \phi_2$, and because scalar addition is a commutative operation. Therefore,

$$\mathbf{R}_z(\phi) = \mathbf{R}_z(\phi_1)\,\mathbf{R}_z(\phi_2) = \mathbf{R}_z(\phi_2)\,\mathbf{R}_z(\phi_1) = \mathbf{R}_z(\phi_1 + \phi_2). \tag{4.16}$$

Proof that this relation holds can be obtained by multiplying two rotation matrices:

$$\begin{aligned}
\mathbf{R}_z(\phi_2)\mathbf{R}_z(\phi_1) &= \begin{pmatrix} \cos\phi_2 & \sin\phi_2 \\ -\sin\phi_2 & \cos\phi_2 \end{pmatrix}\begin{pmatrix} \cos\phi_1 & \sin\phi_1 \\ -\sin\phi_1 & \cos\phi_1 \end{pmatrix} \\
&= \begin{pmatrix} \cos\phi_1\cos\phi_2 - \sin\phi_1\sin\phi_2 & \sin\phi_1\cos\phi_2 + \cos\phi_1\sin\phi_2 \\ -\cos\phi_1\sin\phi_2 - \sin\phi_1\cos\phi_2 & -\sin\phi_1\sin\phi_2 + \cos\phi_1\cos\phi_2 \end{pmatrix} \\
&= \begin{pmatrix} \cos(\phi_1+\phi_2) & \sin(\phi_1+\phi_2) \\ -\sin(\phi_1+\phi_2) & \cos(\phi_1+\phi_2) \end{pmatrix} = \mathbf{R}_z(\phi_1+\phi_2).
\end{aligned} \tag{4.17}$$

The derivation in Equation 4.17 uses the formulas of angle addition from trigonometry:

$$\begin{aligned}
\cos\phi_1\cos\phi_2 - \sin\phi_1\sin\phi_2 &= \cos(\phi_1+\phi_2), \\
\sin\phi_1\cos\phi_2 + \cos\phi_1\sin\phi_2 &= \sin(\phi_1+\phi_2).
\end{aligned} \tag{4.18}$$

It is important to remember that after a sequence of operations is performed, not all rotations add up as shown in Equation 4.16. For example, if the rotations are performed about different rotation axes, the final result of the rotation will depend on the order in which the operations were performed. This can be illustrated by first rotating the coordinate system 90° about the z-axis, and then rotating it 90° about the x'-axis. The resulting transformation is $x,y,z \rightarrow y,-x,z \rightarrow y,z,x$. Alternatively, first rotating the coordinate system 90° about the x-axis, and then rotating 90° about the z'-axis gives the transformation $x,y,z \rightarrow x,z,-y \rightarrow z,-x,-y$. Clearly, these two transformations end up producing different results.

Matrices that describe rotation, whether it be a 2D or 3D or higher dimensional rotation, all have common properties. The nature of that unifying principle will now be explored.

4.2 Unitary Matrices

The theory of rotation in vector space naturally extends to the theory of rotation in function space, where a complete set of orthonormal basis functions serves the same role as the base vectors in Cartesian space. This transition utilizes the similarity that exists between vectors and functions. The basis functions, φ_k, chosen for that purpose are typically the eigenvectors of a Hermitian operator; this guarantees the orthonormality and completeness of the basis set. Orthonormality implies that any pair of functions satisfies the equation

$$\langle \varphi_i | \varphi_j \rangle = \delta_{ij}. \tag{4.19}$$

Completeness of the set means that there exists no other function, ψ, which is simultaneously orthogonal to all the basis functions φ_k. The reason for using orthonormal and complete functions is that any target function can then be expanded into a linear combination of these basis functions. An aid in understanding the abstract concepts of rotation in function space is to imagine that the functions have been substituted by vectors, and to think about the multi-dimensional operations as if they were happening in 3D space.

Important properties of the rotation operator R and its matrix representation $\mathbf{R}$ follow from the standard properties of the rotation operation. It is intuitively obvious that rotation, as a physical process, preserves angles and lengths of the rotated object. Rotation in the function space also conforms to the same principle; the metric of the space, that is, distances and angles, are preserved upon rotation. This requires the dot product of two functions, which are handled in the same way as vectors, to be unaffected by the rotation, R:

$$\langle R\psi | R\varphi \rangle = \langle \psi | \varphi \rangle. \tag{4.20}$$

The left-hand side in this relation reads:

$$\langle R\psi | R\varphi \rangle = \int (R\psi)^* R\varphi \, dr = \int \psi^* R^+ R\varphi \, dr. \tag{4.21}$$

Recall that taking a complex conjugate of an operator (matrix) gives its adjoint; this is indicated by the presence of cross in the upper index of the operator R^+. In matrix terms, adjoint implies interchanging rows by columns and taking the complex conjugate of each matrix element. Comparison of Equation 4.21 with Equation 4.20, reveals that, in order for Equation 4.20 to hold, it is necessary to have:

$$R^+R = 1 \quad \text{and} \quad R\,R^+ = 1, \tag{4.22}$$

where 1 represents the unit operator, that is, the diagonal unit matrix. The second relation on the right in Equation 4.22 is generated by taking the complex conjugate of the first relation. It is obvious that the diagonal unit matrix would be unchanged under such an operation.

In terms of matrix elements, Equation 4.22 requires that:

$$\sum_{i=1} |R_{ij}|^2 = 1 \quad \text{and} \quad \sum_{j=1} |R_{ij}|^2 = 1, \tag{4.23}$$

where the relation on the left is a sum of squares over column elements, and the relation on the right is that over elements of the row. The complementary relations for off-diagonal elements are:

$$\sum_{i=1} R_{ij}^* R_{ik} = 0 \text{ where } j \neq k \quad \text{and} \quad \sum_{j=1} R_{ij}^* R_{kj} = 0 \text{ where } i \neq k, \tag{4.24}$$

when summed up over column and row matrix elements, respectively. Combining together Equations 4.23 and 4.24 produces the orthonormality conditions when applied to rows and columns:

$$\sum_{k=1} R_{ki}^* R_{kj} = \delta_{ij} \quad \text{and} \quad \sum_{k=1} R_{ik}^* R_{jk} = \delta_{ij}, \tag{4.25}$$

where δ is a Kronecker delta that has the value:

$$\delta_{ij} = \begin{cases} 0 & \text{for } i \neq j \\ 1 & \text{for } i = j \end{cases}. \tag{4.26}$$

Matrices that satisfy Equations 4.22 and 4.25 are called unitary matrices and are often labeled with the letter **U** in order to distinguish them from other matrices. This analysis leads to the important conclusion that the rotation operator R and its matrix representation **R** are unitary. The labels **U** and **R** will therefore be used interchangeably when designating unitary matrices.

As has already been stated, a unitary matrix has the property that any given pair of rows or columns is orthonormal. From that, other properties can be derived easily. Given that the absolute value of the determinant of a unitary matrix is 1.0, there must exist a matrix inverse, $\mathbf{R}^{-1}$ so that:

$$\mathbf{R}^{-1}\mathbf{R} = 1 \quad \text{and} \quad \mathbf{R}\,\mathbf{R}^{-1} = 1. \tag{4.27}$$

Combining Equation 4.27 with Equation 4.22 leads to another important property of unitary matrices:

$$\mathbf{R}^{+} = \mathbf{R}^{-1}. \tag{4.28}$$

This relationship implies that the Hermitian adjoint (or transpose, in the case of real matrices) of a unitary matrix is its inverse. The same conclusion also applies to unitary operators.

Based on these properties of the unitary matrix, it is possible to find the numerical value of the determinant of the matrix. Application of the theorem of determinants to the left- and right-hand sides of Equation 4.27, by evaluating the determinant of each term in the expression, leads to:

$$\det(\mathbf{R}^{+})\det(\mathbf{R}) = \det(1) = 1, \qquad \det(\mathbf{R}^{-1})\det(\mathbf{R}) = \det(1) = 1, \tag{4.29}$$

where the determinant of a unit number is obviously the number itself. Since matrix inversion does not change the value of the determinant, and since $\det(\mathbf{R}^{+}) = \det(\mathbf{R})^*$, it follows that:

$$\det(\mathbf{R})^2 = 1; \qquad \text{therefore,} \qquad \det(\mathbf{R}) = 1. \tag{4.30}$$

From this, it follows that the determinant of a unitary matrix is unity. This result corresponds to a regular rotation operation, or "pure rotation." The other possible value of -1 for the determinant corresponds to rotation followed by inversion about the center of coordinate system. Further discussion in this chapter will be solely focused on pure rotations.

A few more useful properties of unitary matrices follow easily from the theoretical framework just developed. First, it follows from Equation 4.27 that the inverse of a unitary matrix is a unitary matrix. Second, a product of any number of unitary matrices is also a unitary matrix. Indeed, if $\mathbf{U}_1$ and $\mathbf{U}_2$ are unitary, the Hermitian adjoint of their product $\mathbf{U} = \mathbf{U}_1\mathbf{U}_2$ is $\mathbf{U}^{+} = \mathbf{U}_2^{+}\mathbf{U}_1^{+}$. The product of the matrix with its own adjoint is $\mathbf{U}\mathbf{U}^{+} = \mathbf{U}_1\mathbf{U}_2\mathbf{U}_2^{+}\mathbf{U}_1^{+} = \mathbf{U}_1\mathbf{U}_1^{+} = 1$, from which it follows that $\mathbf{U} = \mathbf{U}_1\mathbf{U}_2$ is also unitary.

The theory developed so far will now be briefly summarized before proceeding further. Rotation operations can be expressed as unitary transformations and represented by unitary matrices. Rotation is performed in a space defined by the basis vectors φ_k, the size of which could either be finite or infinite. If manipulating entities in spaces of infinite dimensionality poses conceptual difficulties, it might help to mentally map the infinite space to a regular 3D space, while temporarily ignoring the fact that the basis index goes to infinity. Continuing on, the completeness of a function set implies that there does not exist any other function, ψ, which is orthogonal to all the basis functions φ_k. From the completeness of a basis set it follows that for a different complete set of orthogonal basis functions ψ_k in the space that is initially defined by functions φ_k, any given function from the set ψ_k can be represented by a linear combination of functions φ_k. The connection between the two basis sets φ_k and ψ_k constitutes a unitary transformation that can be expressed in matrix form as:

$$\psi_i = \sum_{j=1} U_{ij}\varphi_j. \tag{4.31}$$

Comparison with a regular 3D space might help in understanding the meaning of Equation 4.31, which shows that rotation of a coordinate system, defined by three base vectors $\hat{\mathbf{x}}, \hat{\mathbf{y}}, \hat{\mathbf{z}}$, produces a rotated coordinate system, now defined by a new set of orthonormal base vectors $\hat{\mathbf{x}}', \hat{\mathbf{y}}', \hat{\mathbf{z}}'$. The rotation transformation from the old vectors $\hat{\mathbf{x}}, \hat{\mathbf{y}}, \hat{\mathbf{z}}$ to the new ones $\hat{\mathbf{x}}', \hat{\mathbf{y}}', \hat{\mathbf{z}}'$ is performed according to the equation $\mathbf{r}' = R\,\mathbf{r}$, where $\mathbf{r}$ is a vector in the old coordinate system, and $\mathbf{r}'$ is a vector in the new system. Since rotation is a unitary operation, the rotation equation can also be written in a matrix form, $\mathbf{r}' = \mathbf{U}\,\mathbf{r}$.

Matrix elements U_{ij} are determined in a standard way by multiplying both sides of Equation 4.31 by φ_j^* from the left, and integrating. This step along with orthonormality of the initial and transformed basis sets leads to the relation:

$$U_{ij} = \langle \varphi_j | \psi_i \rangle, \tag{4.32}$$

which determines the matrix elements, and represents a dot product of two functions φ_j and ψ_i.

Within the topic of rotation in function space, unitary matrices are universally used as a tool for rotating a representation from one basis set to another. This technique requires understanding of how basis functions transform under rotation. Once again, this topic may be best introduced by analyzing the rotation of Cartesian base vectors. Concerning transformation of basis vectors, Figure 4.3 illustrates a rotation from the basis set $\hat{\mathbf{x}}, \hat{\mathbf{y}}$ to $\hat{\mathbf{x}}', \hat{\mathbf{y}}'$. This transformation corresponds to a unitary operation:

$$\hat{\mathbf{x}}' = \mathbf{U}\hat{\mathbf{x}} \quad \text{and} \quad \hat{\mathbf{y}}' = \mathbf{U}\hat{\mathbf{y}}, \tag{4.33}$$

where $\hat{\mathbf{x}} = (1,0,0)$ and $\hat{\mathbf{y}} = (0,1,0)$ are column vectors, and $\mathbf{U}$ is a unitary transformation. The analytic form of the transformation of base vectors follows from the trigonometric relations that exist between the vectors, and are visualized in Figure 4.3:

$$\hat{\mathbf{x}}' = \hat{\mathbf{x}}\cos\phi + \hat{\mathbf{y}}\sin\phi \quad \text{and} \quad \hat{\mathbf{y}}' = -\hat{\mathbf{x}}\sin\phi + \hat{\mathbf{y}}\cos\phi. \tag{4.34}$$

These relations determine the matrix $\mathbf{U}$:

$$\mathbf{U} = \begin{pmatrix} \cos\phi & -\sin\phi & 0 \\ \sin\phi & \cos\phi & 0 \\ 0 & 0 & 1 \end{pmatrix}. \tag{4.35}$$

With that, the basis vector transformation $\hat{\mathbf{x}} \to \hat{\mathbf{x}}'$ becomes, in matrix form:

$$\hat{\mathbf{x}}' = \begin{pmatrix} \cos\phi \\ \sin\phi \\ 0 \end{pmatrix} = \begin{pmatrix} \cos\phi & -\sin\phi & 0 \\ \sin\phi & \cos\phi & 0 \\ 0 & 0 & 1 \end{pmatrix} \begin{pmatrix} 1 \\ 0 \\ 0 \end{pmatrix} = \mathbf{U}\,\hat{\mathbf{x}}. \tag{4.36}$$

The matrix $\mathbf{U}$ in Equation 4.35 represents a rotation of the basis vectors of a Cartesian coordinate system around the z-axis. As a rotation matrix around the z-axis has already been defined in Equation 4.5 for matrix $\mathbf{R}_z$, matrix $\mathbf{U}$ is therefore the transpose of matrix $\mathbf{R}_z$, that is, rows and columns are swapped around. Since both matrices are unitary, they both correspond to a rotation. The subtle difference that exists between these matrices reflects the different application of those matrices. Matrix $\mathbf{R}_z$ describes a transformation of a radius-vector $\mathbf{r}$ upon rotation of the coordinate system, so that the coordinates of the rotated vector $\mathbf{r}'$ are given in the basis of the rotated basis vectors $\hat{\mathbf{x}}', \hat{\mathbf{y}}', \hat{\mathbf{z}}'$. Matrix $\mathbf{U}$, which is a transpose of matrix $\mathbf{R}_z$, describes the rotated basis vectors $\hat{\mathbf{x}}', \hat{\mathbf{y}}', \hat{\mathbf{z}}'$ in terms of the original basis vectors $\hat{\mathbf{x}}, \hat{\mathbf{y}}, \hat{\mathbf{z}}$. Since these two matrices describe the rotated vector about different coordinate systems before and after rotation, $\mathbf{U}$ is a Hermitian adjoint of $\mathbf{R}_z$, that is, $\mathbf{U} = \mathbf{R}_z^+$, and $\mathbf{U}^+ = \mathbf{R}_z$. The rotation matrix can be labeled using either $\mathbf{U}$ or $\mathbf{R}$, so both these letters can be used interchangeably. An important conclusion is that the choice of a particular form, $\mathbf{U}$ or $\mathbf{U}^+$, of the rotation matrix determines either the initial or the rotated basis set as the basis set for the outcome of the rotation transformation.

The basic principle of the basis set transformation by means of a unitary matrix that is derived for Cartesian space directly applies to rotation in function space. Switching from Cartesian to function space provides some important benefits in that it is often simpler to obtain a matrix representation of an operator in function space than in Cartesian space. The matrix representation of an operator A in basis set φ is:

$$\langle \varphi_i \,|\, A\varphi_j \rangle \equiv \langle \varphi_i \,|\, A \,|\, \varphi_j \rangle \equiv \langle i \,|\, A \,|\, j \rangle = A_{ij}, \tag{4.37}$$

where A_{ij} is a matrix element. Equation 4.37 presents three popular bra-ket notations for the matrix element. The left-most expression underlines the notion that the operator acts on the function placed on its right side. Having a vertical bar added after the operator symbol in the second form helps to distinguish the expression for the operator from the dot product of functions. The third expression in Equation 4.37

offers a simplified notation of the matrix element by referring to basis functions by their indices while omitting the symbol of the function.

A few important properties of the matrix representation readily follow from its definition. If the basis functions φ_k are eigenvectors of operator A, and a_k are their eigenvalues, so that the eigenvalue equation $A\varphi_k = a_k\varphi_k$ is satisfied, the matrix representation **A** of the operator A will be a diagonal matrix with the eigenvalues a_k placed along the diagonal according to:

$$A_{kk} = \langle \varphi_k | A\varphi_k \rangle = \langle \varphi_k | a_k\varphi_k \rangle = a_k \langle \varphi_k | \varphi_k \rangle = a_k. \quad (4.38)$$

Under that condition, all off-diagonal elements in the matrix representation will be zero, due to the orthonormality condition $\langle \varphi_i | \varphi_j \rangle = \delta_{ij}$ that exists between the eigenvectors. Conversely, if the functions φ_k are not eigenvectors of the operator A, the matrix representation of the operator A in this basis will not be diagonal.

Rotation of the basis set changes the matrix representation of the operator as follows: assume that a unitary transformation U of a basis set φ, which has a k-th member φ_k, generates a rotated basis set ψ. Since the initial basis set is complete, the rotated basis set has the same number of elements as the initial basis set. The members of the initial and rotated basis sets are connected *via* a unitary transformation $\psi_k = U\varphi_k$. Multiplying both sides of this relation by U^{-1} from the left, and taking advantage of the unitary property of the operator U, provides a reverse transformation $\varphi_k = U^{-1}\psi_k$. Replacing the basis functions φ_k in the matrix representation $\langle \varphi_i | A | \varphi_j \rangle$ of operator A by their reverse transformation $\varphi_k = U^{-1}\psi_k$ produces the equality:

$$\langle \varphi_i | A | \varphi_j \rangle = \langle U^{-1}\psi_i | A | U^{-1}\psi_j \rangle. \quad (4.39)$$

Application of the property $(U^{-1}\psi_i)^* \equiv (U^+\psi_i)^* = \psi_i U$ transforms the above expression to:

$$\langle \varphi_i | A | \varphi_j \rangle = \langle \psi_i | UAU^{-1} | \psi_j \rangle = \langle \psi_i | A' | \psi_j \rangle, \quad (4.40)$$

where

$$A' = UAU^{-1} \equiv UAU^+. \quad (4.41)$$

Therefore, if the matrix **A** is a representation of the operator A in basis φ, the matrix $\mathbf{UAU^+}$ is the representation of that operator in basis ψ, where U is the unitary transformation from the basis set φ to basis ψ. In a particular case of functions φ being eigenfunctions of operator A, matrix **A** is diagonal in the basis φ, whereas matrix $\mathbf{UAU^+}$ is diagonal in the basis ψ.

The property that makes unitary operators a particularly powerful tool is their ability to be represented in exponential form. The general form of the exponential operator e^A, expanded as a Taylor series, is:

$$e^A = \sum_{k=0} \frac{1}{k!} A^k, \quad (4.42)$$

where A is an operator. The upper index k in A^k indicates the k'th application of the operator A, that is, $A^k = A \times A \times \cdots \times A$. In general, the expansion series in Equation 4.42 contains an infinite number of terms.

The exponential operator has the remarkable property that its complex form e^{iA} is unitary if the operator A is Hermitian, that is, when $A^+ = A$. To prove this unitary property it is necessary to find the adjoint, $(e^{iA})^+$:

$$(e^{iA})^+ = \sum_{k=0} \frac{1}{k!} [(iA)^k]^+ = \sum_{k=0} \frac{1}{k!} (-iA^+)^k = \sum_{k=0} \frac{1}{k!} (-iA)^k = e^{-iA}. \quad (4.43)$$

From this it follows that $(e^{iA})^{+}(e^{iA}) = e^{-iA+iA} = 1$. Therefore the operator e^{iA} is unitary, and $U = e^{iA}$ and $U^{-1} = e^{-iA}$. Note that the exponents add up in the usual way,

$$e^{iA}e^{iB} = e^{iB}e^{iA} = e^{i(A+B)}, \tag{4.44}$$

only when operators A and B commute, that is, $AB = BA$. This requirement becomes apparent when replacing the exponents in the product by their corresponding Taylor expansions. The product of two exponential operators $e^{iA}e^{iB}$ is always unitary when the operators A and B are Hermitian.

4.3 Rotation Operator

Having developed the abstract notion of a rotation operator R, and demonstrating how it can be expressed as a matrix, an analytical mathematical expression for it will now be derived.

Rotation of the coordinate system through angle ϕ about the axis along a unit vector $\mathbf{n}$ turns the function ψ into function ψ'. This rotation corresponds to a unitary transformation:

$$\psi' = R(\mathbf{n},\phi)\psi. \tag{4.45}$$

In order to maintain the unitary property of $R(\mathbf{n},\phi)$ while conducting the search for the explicit form of this operator, it is convenient to write the operator in the exponential form:

$$R(\mathbf{n},\phi) = e^{-iA(\mathbf{n},\phi)}. \tag{4.46}$$

Expressed in this way, the search for the operator $R(\mathbf{n},\phi)$ is transformed into a search for the proper Hermitian operator A. The choice of the negative sign in the power of the exponent can be justified as follows: both positive and negative signs satisfy the requirement for a unitary operator since the phase factor of a unitary operator is undefined. A positive sign in the power of the exponent corresponds to a unitary operator that returns the coordinates of the rotated vector relative to the rotated coordinate system. The matrix representation of that operator is provided in Equation 4.5. The negative case corresponds to a rotation operation that returns the coordinates of the rotated vector relative to the initial coordinate system. An example would be the rotation of a basis set. Equation 4.35 provides the necessary matrix representation for the corresponding rotation operator. The adjoint of Equation 4.35 is Equation 4.5; this corresponds to the change of sign in the power of the exponential operator, resulting in the negative sign in Equation 4.46.

A useful condition in the search for an analytic expression of operator $A(\mathbf{n},\phi)$ is that the rotation operator $R(\mathbf{n},\phi)$ must become unity in the limit $\phi \to 0$. This means that $A(\mathbf{n},\phi) \to 0$ for $\phi \to 0$, and that operator A is proportional to the value of angle ϕ. Because of the condition $\phi \to 0$, second- and higher-order terms in the Taylor expansion of the exponential operator become vanishingly small and can therefore be safely discarded. This reduces the expansion to first-order terms:

$$R(\mathbf{n},\phi) = 1 - iA(\mathbf{n},\phi). \tag{4.47}$$

This equation satisfies the requirement that the rotation operator R becomes unity if $A(\mathbf{n},\phi) \to 0$.

The mathematical form of the rotation operator is determined by the transformation properties of the system under study. An intuitive perception of real 3D space suggests that for infinitesimally small values of angle ϕ, the transformation caused by the rotation operator is linearly proportional to the value of ϕ. Equation 4.34 describes how the coordinates transform when the coordinate system is rotated about the z-axis through angle ϕ. In it, the condition $\phi \to 0$ leads to $\lim_{\phi\to 0}(\cos\phi) = 1$ and $\lim_{\phi\to 0}(\sin\phi) = 0$. Therefore, infinitesimal rotation $d\phi$ about the z-axis transforms the Cartesian components according to:

$$\begin{aligned} x' &= x + y\,d\phi, \\ y' &= y - x\,d\phi, \\ z' &= z. \end{aligned} \tag{4.48}$$

These relations may be rewritten in a more general form. For that, consider the effect of the operator $(y(\partial/\partial x) - x(\partial/\partial y))$ on the Cartesian coordinates:

$$\left(y\frac{\partial}{\partial x} - x\frac{\partial}{\partial y}\right)x = y, \quad \left(y\frac{\partial}{\partial x} - x\frac{\partial}{\partial y}\right)y = -x, \quad \left(y\frac{\partial}{\partial x} - x\frac{\partial}{\partial y}\right)z = 0. \tag{4.49}$$

Combining Equations 4.48 and 4.49 leads to:

$$\begin{aligned} x' &= x + d\phi\left(y\frac{\partial}{\partial x} - x\frac{\partial}{\partial y}\right)x, \\ y' &= y + d\phi\left(y\frac{\partial}{\partial x} - x\frac{\partial}{\partial y}\right)y, \\ z' &= z + d\phi\left(y\frac{\partial}{\partial x} - x\frac{\partial}{\partial y}\right)z. \end{aligned} \tag{4.50}$$

This gives rise to the remarkable result that the same operator $d\phi(y(\partial/\partial x) - x(\partial/\partial y))$ acts upon each Cartesian coordinate component x, y, z in an entirely symmetric way. With that, the equation for rotation of basis vectors can now be extended to the rotation of functions. By analogy with Equation 4.50, a transformation of a scalar function $\psi(x, y, z)$, where each point in space has a single value, gives the rotated function $\psi(x', y', z')$:

$$\psi(x', y', z') = \psi(x, y, z) - d\phi\left(x\frac{\partial}{\partial y} - y\frac{\partial}{\partial x}\right)\psi(x, y, z). \tag{4.51}$$

The partial derivative in Cartesian coordinates, shown in Equations 4.50 and 4.51, is known as the orbital angular momentum operator:

$$L_z = -i\hbar\left(x\frac{\partial}{\partial y} - y\frac{\partial}{\partial x}\right), \tag{4.52}$$

or, to be more exact, as the projection of the angular momentum operator onto the z-axis.

Angular momentum $\mathbf{L}$ is present in any particle that has circular motion, as illustrated in Figure 4.6. Angular momentum is a vector quantity and has three Cartesian components:

$$\mathbf{L} = \mathbf{L}_x + \mathbf{L}_y + \mathbf{L}_z. \tag{4.53}$$

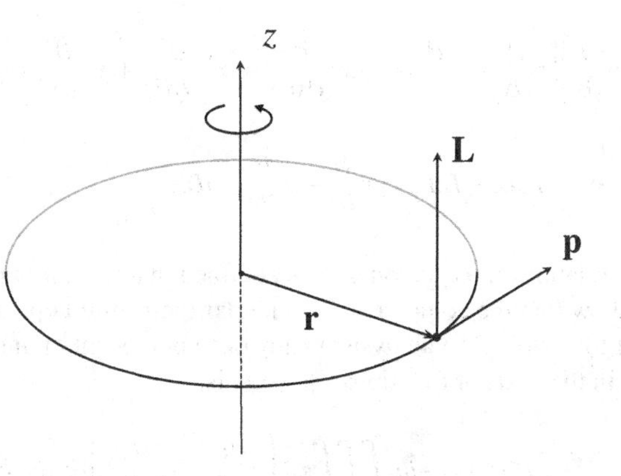

FIGURE 4.6 Angular momentum **L** of a particle undergoing circular motion.

In classical physics, angular momentum is a vector product of the particle radius vector $\mathbf{r}$ and the momentum $\mathbf{p}$. It is defined as a determinant of the matrix:

$$\mathbf{L} = \mathbf{r} \times \mathbf{p} = \det \begin{pmatrix} \hat{\mathbf{x}} & \hat{\mathbf{y}} & \hat{\mathbf{z}} \\ x & y & z \\ p_x & p_y & p_z \end{pmatrix} = (yp_z - zp_y)\hat{\mathbf{x}} + (zp_x - xp_z)\hat{\mathbf{y}} + (xp_y - yp_x)\hat{\mathbf{z}}. \tag{4.54}$$

Quantum mechanics defines the components of momentum $\mathbf{p}$ in operator form:

$$p_x = -i\hbar \frac{\partial}{\partial x}, \qquad p_y = -i\hbar \frac{\partial}{\partial y}, \qquad p_z = -i\hbar \frac{\partial}{\partial z}. \tag{4.55}$$

Substituting Equation 4.55 into Equation 4.54 leads to the operator form of the angular momentum:

$$\mathbf{L} = -i\hbar\left(y\frac{\partial}{\partial z} - z\frac{\partial}{\partial y}\right)\hat{\mathbf{x}} - i\hbar\left(z\frac{\partial}{\partial x} - x\frac{\partial}{\partial z}\right)\hat{\mathbf{y}} - i\hbar\left(x\frac{\partial}{\partial y} - y\frac{\partial}{\partial x}\right)\hat{\mathbf{z}}. \tag{4.56}$$

The angular momentum operator has the units of energy, which explains the occurrence of the reduced Planck's constant, $\hbar$. The presence of complex number $-i$ in Equations 4.52 and 4.56 is required in order to make the angular momentum operator and its components Hermitian; this will be elaborated on shortly. Cartesian components of the angular momentum operator follow from Equation 4.56:

$$L_x = -i\left(y\frac{\partial}{\partial z} - z\frac{\partial}{\partial y}\right) \qquad L_y = -i\left(z\frac{\partial}{\partial x} - x\frac{\partial}{\partial z}\right) \qquad L_z = -i\left(x\frac{\partial}{\partial y} - y\frac{\partial}{\partial x}\right). \tag{4.57}$$

These may also be obtained from Equation 4.52 by applying the cyclic permutation $z \to x \to y \to z$. An alternative approach is to repeat the steps for the derivation of Equation 4.52, but starting the analysis by examining the effect of a rotation about the x- and y-axes. The components of the angular momentum operator are linked to each other through the relations:

$$L_yL_z - L_zL_y = iL_x, \qquad L_zL_x - L_xL_z = iL_y, \qquad L_xL_y - L_yL_x = iL_z, \tag{4.58}$$

which are related by a cyclic permutation. The validity of the relations in Equation 4.58 can be verified by hand. For instance,

$$L_yL_z = -\left(z\frac{\partial}{\partial x} - x\frac{\partial}{\partial z}\right)\left(x\frac{\partial}{\partial y} - y\frac{\partial}{\partial x}\right) = -z\frac{\partial}{\partial y} - xz\frac{\partial^2}{\partial x\partial y} + yz\frac{\partial^2}{\partial x^2} + x^2\frac{\partial^2}{\partial y\partial z} - xy\frac{\partial^2}{\partial x\partial z}, \tag{4.59}$$

$$L_zL_y = -\left(x\frac{\partial}{\partial y} - y\frac{\partial}{\partial x}\right)\left(z\frac{\partial}{\partial x} - x\frac{\partial}{\partial z}\right) = -xz\frac{\partial^2}{\partial x\partial y} + x^2\frac{\partial^2}{\partial y\partial z} + yz\frac{\partial^2}{\partial x^2} - y\frac{\partial}{\partial z} - xy\frac{\partial^2}{\partial x\partial z}, \tag{4.60}$$

$$L_yL_z - L_zL_y = y\frac{\partial}{\partial z} - z\frac{\partial}{\partial y} = iL_x. \tag{4.61}$$

Proof for the other two relations in Equation 4.58 is obtained in a similar manner.

The next task is to show that the components of angular momentum operator are Hermitian. For this purpose, it is sufficient to perform the analysis on any one component of angular momentum. A matrix element of operator L_x in the basis of functions ψ_i and ψ_j is:

$$\langle \psi_i | L_x \psi_j \rangle = -i\hbar \iiint \psi_i^* \left(y\frac{\partial \psi_j}{\partial z} - z\frac{\partial \psi_j}{\partial y}\right) dx\, dy\, dz. \tag{4.62}$$

The integral in Equation 4.62 consists of two terms enclosed in brackets, which may be treated independently. Applying integration by parts,

$$d(uv) = v\,du + u\,dv, \qquad \int d(uv) = \int v\,du + \int u\,dv, \qquad \int v\,du = uv\Big|_{-\infty}^{\infty} - \int u\,dv, \tag{4.63}$$

to the first term $\iiint \psi_i^* y \frac{\partial \psi_j}{\partial z} dx\,dy\,dz$ over coordinate z, and denoting

$$du = \frac{\partial \psi_j}{\partial z} dz, \qquad u = \psi_j, \qquad v = \psi_i^* y\,dx\,dy, \qquad dv = y\frac{\partial \psi_i^*}{\partial z} dx\,dy\,dz, \tag{4.64}$$

where the derivative of $\psi_i^* y$ over coordinate z is $y(\partial\psi_i^*/\partial z)$ since y and z are independent coordinates, leads to:

$$\iiint \psi_i^* y \frac{\partial \psi_j}{\partial z} dx\,dy\,dz = \iint \left(\psi_j\Big|_{-\infty}^{\infty}\right)\psi_i^* y\,dx\,dy - \iiint \psi_j y \frac{\partial \psi_i^*}{\partial z} dx\,dy\,dz. \tag{4.65}$$

Since any physically meaningful function is normalized and must go to zero at infinity, the first integral in the right-hand side vanishes:

$$\iint \left(\psi_j\Big|_{-\infty}^{\infty}\right)\psi_i^* y\,dx\,dy = 0, \tag{4.66}$$

therefore:

$$\iiint \psi_i^* y \frac{\partial \psi_j}{\partial z} dx\,dy\,dz = -\iiint \psi_j y \frac{\partial \psi_i^*}{\partial z} dx\,dy\,dz. \tag{4.67}$$

The second term in Equation 4.62 behaves in the same way. Therefore,

$$\langle \psi_i | L_x \psi_j \rangle = -i\hbar \iiint \psi_i^* \left(y\frac{\partial \psi_j}{\partial z} - z\frac{\partial \psi_j}{\partial y}\right) dx\,dy\,dz = i\hbar \iiint \psi_j \left(y\frac{\partial \psi_i^*}{\partial z} - z\frac{\partial \psi_i^*}{\partial y}\right) dx\,dy\,dz = \langle L_x\psi_i | \psi_j \rangle. \tag{4.68}$$

Note that the complex conjugation $(L_x\psi_i)^*$ implicit in $\langle L_x\psi_i | \psi_j\rangle$ consumes the negative sign in Equation 4.68. Since all terms in the integrals for other components of angular momentum work in the same way, this gives:

$$\langle \psi_i | L_x\psi_j \rangle = \langle L_x\psi_i | \psi_j \rangle, \qquad \langle \psi_i | L_y\psi_j \rangle = \langle L_y\psi_i | \psi_j \rangle, \qquad \langle \psi_i | L_z\psi_j \rangle = \langle L_z\psi_i | \psi_j \rangle. \tag{4.69}$$

These relationships satisfy the definition of a Hermitian operator, and therefore prove that the components of angular momentum are Hermitian. Since any sum of Hermitian operators is also Hermitian, it follows that the angular momentum operator $L = L_x + L_y + L_z$ is also Hermitian.

Equation 4.51 can now be rewritten to include the projection of the angular momentum operator on the z-axis into the formula. This leads to:

$$\psi(x', y', z') = (1 - i\,d\phi\,L_z)\psi(x, y, z). \tag{4.70}$$

In its general form, the rotation equation is $\psi(x', y', z') = R\psi(x, y, z)$, where R is a rotation operator. Equation 4.70, together with the symmetry relationships revealed in Equation 4.50, allows rotation around the x, y, and z-axes to be expressed in general form as:

$$R_x\psi = (1 - i\,d\phi\,L_x)\psi, \qquad R_y\psi = (1 - i\,d\phi\,L_y)\psi, \qquad R_z\psi = (1 - i\,d\phi\,L_z)\psi, \tag{4.71}$$

where R_x, R_y, and R_z are Cartesian components of the rotation operator R. A few details in Equation 4.71 still need to be resolved before the derivation for a mathematical formula of operator R can be completed. The analysis of those issues continues.

Equation 4.71 defines each component of the rotation operator R *via* Cartesian components L_x, L_y, and L_z of angular momentum operator. These latter components are Hermitian operators that comply with the requirements imposed on the operator $A(\mathbf{n},\phi)$ in Equation 4.46 in order to keep operator R unitary. Operators L_x, L_y, and L_z correspond to the canonical case of the rotation axis being aligned with a primary coordinate axis. These operators form a complete set in the operator space, so that the operator that corresponds to rotation about an arbitrary rotation axis becomes a linear combination of these three canonical operators. For any arbitrary axis of rotation, aligned with a unit vector **n**, the general form of the rotation transformation is:

$$R\psi = (1 - i\,d\phi(\mathbf{n}\cdot\mathbf{L}))\psi, \tag{4.72}$$

where $\mathbf{n}\cdot\mathbf{L} = n_xL_x + n_yL_y + n_zL_z$. In this notation, n_k are the Cartesian components of **n**, and L_k are the Cartesian components of the angular momentum operator **L**. For instance, if the rotation axis is aligned with z-axis, the product becomes $\mathbf{n}\cdot\mathbf{L} \equiv L_z$.

A Taylor expansion of the rotation operator R can be used in constructing the differential form of the operator A. The Taylor expansion form is only needed during this construction, and, once it is completed, the series expansion of the rotation operator R can be abandoned. Converting the rotation operator back to the exponential form gives $R = \exp(-i d\phi(\mathbf{n}\cdot\mathbf{L}))$. In this expression, the rotation angle is still in infinitesimal form, $d\phi$, not in the finite angle form ϕ. This problem can be resolved based on the fact that operator L is independent of the angle or rotation. Transition from an infinitesimal $d\phi$ to a finite rotation relies on the common property of rotations that motion though an angle $\alpha + \beta$ is equivalent to a rotation through angle α followed by a rotation through angle β, about the same rotation axis. In operator form, this corresponds to a product of individual operators, $R(\alpha + \beta) = R(\beta)R(\alpha)$. Likewise, rotation on a finite angle ϕ is equivalent to a succession of infinitesimal rotations on angle $d\phi$. In operator form, this reads:

$$R(\phi) = R(d\phi)R(d\phi)\ldots R(d\phi) = \exp(-i[d\phi + d\phi + \cdots + d\phi](\mathbf{n}\cdot\mathbf{L})) = \exp(-i\phi(\mathbf{n}\cdot\mathbf{L})). \tag{4.73}$$

Since the powers of exponents add up in the product, the rotation operator has the same form whether it employs infinitesimal $d\phi$ or finite ϕ angle of rotation. This analysis leads to the general form of the rotation operator R:

$$R(\phi) = e^{-i\phi(\mathbf{n}\cdot\mathbf{L})}. \tag{4.74}$$

This analytical expression for the rotation operator R shows that a rotation of coordinate bases can be expressed mathematically in terms of the angular momentum operator **L**. Because of its importance, a good understanding of some of the properties of **L** is essential before continuing the work with rotation operators.

4.4 Commutative Properties of the Angular Momentum

The preceding section described the derivation of the angular momentum operator using the rotation properties of the physical system. In addition to orbital angular momentum, L, for which there exists a corresponding classical counterpart, there also exists spin angular momentum, S, which is an intrinsic property of quantum systems and has no classical counterpart. Angular momenta of the system can be added, so the total angular momentum of a system is $J = L + S$; if the system has no spin, that is, $S = 0$, then $J \equiv L$.

The analytic expression for orbital angular momentum operator L was developed in the previous section, but because it has no classical analog this approach would not work for spin angular momentum S. Fortunately, however, no analytic expression for the spin operator is required. Everything of value that pertains to spin can be derived using only the properties of the abstract spin operator. To keep the following analysis as general as possible, further reference to angular momentum will use index J in order to emphasize that the formalism applies to both the orbital and the spin parts of the angular momentum, whereas the index L will be used when only the orbital angular momentum operator is involved. Although spin is not an intrinsic part of the Fast Multipole Method, some mathematical techniques that refer to spin also happen to be useful in FMM theory. Therefore, before continuing with the broader overview of angular momentum, a brief introduction to spin will be given.

The first important subject in this review is the derivation of commutation rules. Since rotation does not alter any physical properties of the system described by the Hamiltonian H, a rotation operation can be used to regenerate a Hamiltonian that has undergone a basis set transformation. This transformation is:

$$RHR^{+} = e^{-i\phi(\mathbf{n}\cdot\mathbf{J})} H e^{i\phi(\mathbf{n}\cdot\mathbf{J})} = H, \tag{4.75}$$

where $R = e^{-i\phi(\mathbf{n}\cdot\mathbf{J})}$ is a unitary operator that rotates the basis set. Multiplying this expression by R from the right gives:

$$e^{-i\phi(\mathbf{n}\cdot\mathbf{J})} H = H\, e^{-i\phi(\mathbf{n}\cdot\mathbf{J})}. \tag{4.76}$$

The relationship described in Equations 4.75 and 4.76 rests on the assumption that operators R and H commute, that is, they can be applied to the wavefunction in any order without changing the end result. Expressed differently: in an experiment, pairs of physical quantities can be simultaneously measured if and only if the operators that represent them commute.

A Taylor expansion of the rotation operator helps to reveal the underlying mechanism of commutation of the rotation operator with the Hamiltonian. Considering a projection of the angular momentum on the z-axis, and keeping only the first-order terms in the expansion being substituted into Equation 4.76 gives:

$$(1 - i\phi J_z)H = H(1 - i\phi J_z). \tag{4.77}$$

Canceling out the terms in common leads to:

$$[J_k, H] \equiv J_k H - H J_k = 0, \tag{4.78}$$

where square brackets denote the commutator, and index k represents one of the components x, y, or z. This relation reveals that the commutation of the rotation operator with the Hamiltonian is based on commutation of the angular momentum operator J with the Hamiltonian. Proof for this commutator follows; and that in turn will prove Equation 4.75.

Before proceeding to the proof, it is convenient to obtain one more useful relation that follows from Equation 4.78. If the commutator of the angular momentum operator with the Hamiltonian holds, that is, $[J, H] = 0$, any power of J would also commute with the Hamiltonian. In order to show it for $[J^2,H]$, one may add and subtract $J\,H\,J$ into the commutator, and rearrange the terms. This gives:

$$[J^2,H] \equiv J^2 H - H J^2 = J(J\,H) - J(H\,J) + (J\,H)J - (H\,J)J. \tag{4.79}$$

Therefore,

$$[J^2,H] \equiv J[J,H] + [J,H]J = 0. \tag{4.80}$$

For higher powers of J, the derivation of the commutator progresses recursively by employing the results from the previously obtained commutators. For example, for J^3 the commutator is:

$$[J^3,H] = J^2(J\,H) - J^2(H\,J) + (J^2 H)J - (H\,J^2)J = J^2[J,H] + [J^2,H]J = 0. \tag{4.81}$$

This argument proves that if J commutes with H, then any higher power of J that appears in the Taylor expansion of rotation operator $R = e^{-i\phi(\mathbf{n}\cdot\mathbf{J})} = \sum_{k=0} 1/k![-i\phi(\mathbf{n}\cdot\mathbf{J})]^k$ will also commute with the Hamiltonian, which in turn proves the commutation relation in Equation 4.76 for the whole rotation operator R.

The above derivation rests on the assumption that $[J, H] = 0$. This is the last major step in the chain leading back to the commutation of the rotation and the Hamiltonian operators that need to be proven. In order to verify that the relation $[J, H] = 0$ holds (that the total angular momentum operator commutes with the Hamiltonian), it is sufficient to show that each component of the angular momentum operator satisfies that condition independently. The following derivation employs the z-component, J_z of the orbital angular momentum, and a time-independent Hamiltonian for an electron in a hydrogen atom; this is sufficient for the purpose of the analysis. The Hamiltonian is:

$$H = -\frac{\hbar^2}{2m}\nabla^2 - \frac{1}{r}, \tag{4.82}$$

where m is the mass of the particle, the Laplacian ∇^2 is the kinetic energy operator, and the inverse distance $1/r$ is the potential function.

The analysis of commutation relation of the angular momentum operator with the Hamiltonian consists of two steps. In the first step, it is necessary to show that the angular momentum operator and the Laplacian commute, that is, $[J_z,\nabla^2] = 0$. Later on, the second step will be to prove that $[J_z, r^{-1}] = 0$.

In the first step, it is convenient to express the Laplacian and angular momentum operators by a momentum operator, p:

$$\nabla^2 = -\frac{1}{\hbar^2}(p_x^2 + p_y^2 + p_z^2) \qquad J_z = xp_y - yp_x \qquad p_x = -i\hbar\frac{\partial}{\partial x}. \tag{4.83}$$

Also, for the purpose of this analysis, it is useful to derive the commutation relationship between Cartesian coordinates and the momentum operator. Making the commutator act on function ψ, in order to facilitate taking the derivative, leads to:

$$[x, p_x]\psi = -i\hbar\, x\frac{\partial}{\partial x}\psi + i\hbar\frac{\partial}{\partial x}x\psi = -i\hbar\, x\frac{\partial}{\partial x}\psi + i\hbar\psi + i\hbar\, x\frac{\partial}{\partial x}\psi, \tag{4.84}$$

Therefore,

$$[x, p_x] = i\hbar \quad \text{and} \quad [p_x, x] = -i\hbar. \tag{4.85}$$

From these, a few more commutators immediately follow:

$$[p_x, yp_x] = p_x y p_x - yp_x p_x = yp_x p_x - yp_x p_x = 0, \tag{4.86}$$

$$[p_y, xp_y] = p_y x p_y - xp_y p_y = xp_y p_y - xp_y p_y = 0, \tag{4.87}$$

$$[p_x, xp_y] = (p_x x)p_y - (xp_x)p_y = [p_x, x]p_y = -i\hbar\, p_y, \tag{4.88}$$

$$[p_y, yp_x] = (p_y y)p_x - (yp_y)p_x = [p_y, y]p_x = -i\hbar\, p_x, \tag{4.89}$$

$$[p_x, L_z] = [p_x, x\, p_y - y\, p_x] = [p_x, x\, p_y] - [p_x, y\, p_x] = -i\hbar\, p_y, \tag{4.90}$$

$$[p_y, L_z] = [p_y, x\, p_y - y\, p_x] = [p_y, x\, p_y] - [p_y, y\, p_x] = i\hbar\, p_x. \tag{4.91}$$

These intermediate relations are needed in the derivation that follows.

When the Laplacian is expressed in terms of momentum operators, the commutator becomes:

$$[\nabla^2, J_z] = -\frac{1}{\hbar^2}[p_x^2 + p_y^2 + p_z^2, J_z]. \tag{4.92}$$

This splits the original commutator $[\nabla^2, J_z]$ into three simpler commutators $[p_x^2, J_z]$, $[p_y^2, J_z]$, and $[p_z^2, J_z]$, each of which can be found independently. Since operators acting on different spaces commute, this immediately leads to:

$$[p_z^2, J_z] = [p_z^2, x\,p_y - y\,p_x] = 0. \tag{4.93}$$

The other two commutators for the p_y and p_x components can be found after a minor additional manipulation. The first step is to reduce the power of the momentum operator in the commutator by using the corresponding property of the commutators. This results in:

$$[p_y^2, J_z] = p_y(p_y J_z) - p_y(J_z p_y) + (p_y J_z)p_y - (J_z p_y)p_y = p_y[p_y, J_z] + [p_y, J_z]p_y. \tag{4.94}$$

After that, substituting the value of the previously derived intermediate commutator $[p_y, J_z]$ into the above equation leads to:

$$[p_y^2, J_z] = p_y i\hbar\, p_x + i\hbar\, p_x p_y = 2i\hbar\, p_x p_y. \tag{4.95}$$

Similar manipulations produce the third commutator:

$$[p_x^2, J_z] = p_x[p_x, J_z] - [p_x, J_z]p_x = -2i\hbar\, p_x p_y. \tag{4.96}$$

Combining the results from these steps gives the commutator of the Laplacian ∇^2 with the angular momentum operator, J_z:

$$[\nabla^2, J_z] = -\frac{1}{\hbar^2}([p_x^2, J_z] + [p_y^2, J_z] + [p_z^2, J_z]) = 0, \quad \text{and} \quad [J_z, \nabla^2] = 0. \tag{4.97}$$

This completes the proof that the z-component of the angular momentum operator commutes with the kinetic energy operator.

The second step is to show that the operator J_z commutes with the potential energy operator, r^{-1}, where the distance r is a square root of the squares of Cartesian components x, y, and z of the particle:

$$r = \sqrt{x^2 + y^2 + z^2}. \tag{4.98}$$

Since $J_z = xp_y - yp_x$, the commutators $[p_x, r^{-1}]$ and $[p_y, r^{-1}]$ need to be evaluated. The first of these is:

$$[p_x, r^{-1}]\psi = p_x r^{-1}\psi - r^{-1}p_x\psi = -i\hbar\frac{\partial}{\partial x}(x^2 + y^2 + z^2)^{-\frac{1}{2}}\psi - r^{-1}p_x\psi. \tag{4.99}$$

Differentiation leads to:

$$[p_x, r^{-1}]\psi = -i\hbar x(x^2 + y^2 + z^2)^{-\frac{3}{2}}\psi + r^{-1}p_x\psi - r^{-1}p_x\psi. \tag{4.100}$$

From this it follows that:

$$[p_x, r^{-1}] = i\hbar\frac{x}{r^3} \quad \text{and} \quad [p_y, r^{-1}] = i\hbar\frac{y}{r^3}. \tag{4.101}$$

Combining the intermediate results leads to:

$$[J_z,r^{-1}]=x[p_y,r^{-1}]-y[p_x,r^{-1}]=i\hbar\frac{xy}{r^3}-i\hbar\frac{xy}{r^3}=0. \quad (4.102)$$

This confirms that the z-component of the angular momentum operator J_z commutes with the potential energy operator, r^{-1}. Since the operator J_z individually commutes with the kinetic ∇^2 and potential energy r^{-1} operators from the Hamiltonian, this leads to $[J_z,H]=0$. Repeating the above step would show that the Hamiltonian commutes with the other two components of the angular momentum operator, J_x and J_y. This completes the proof that the entire angular momentum operator J commutes with the Hamiltonian, that $[J, H]=0$.

Although the component operators J_x, J_y, and J_z individually commute with the Hamiltonian, the physical quantities that correspond to them cannot be measured simultaneously because those operators do not commute with each other. This is consistent with the previous conclusion that the result of rotation about any two given axes depends on the order of rotations. For a set of physical properties to be simultaneously observable their corresponding operators must simultaneously commute in all possible pair combinations. Because of the commutation issue, it is conventionally assumed that only the expectation value of the operator J_z can always be measured with the unrestricted precision in the experiment, whereas the expectation values for operators J_x and J_y are generally not available.

This restriction does not exhaust the list of observable properties of the angular momentum, however. There is one more form of the angular momentum operator that simultaneously commutes with J_z and H operators: the square of the angular momentum operator, J^2.

$$J^2=\mathbf{J}\cdot\mathbf{J}=J_x^2+J_y^2+J_z^2. \quad (4.103)$$

This is the vector dot product of the angular momentum operator with itself. Since operator J commutes with the Hamiltonian, the same is true about any power of J according to the previously provided proof. Specifically:

$$[J^2,H]\equiv J^2H-H\,J^2=0. \quad (4.104)$$

The commutator of the square of the angular momentum with the projection along the z-axis can be found in the following steps. Expanding operator J^2 *via* the sum of its components, and noticing that the components representing a projection on the same axis automatically commute, $[J_z^2,J_z]=0$, leads to:

$$[J^2,J_z]=[J_x^2+J_y^2+J_z^2,J_z]=[J_x^2+J_y^2,J_z]=[J_x^2,J_z]+[J_y^2,J_z]. \quad (4.105)$$

The next step is to apply the rule for reducing the power of the components in the last two commutators. This substitution leads to the sum of four elementary commutators:

$$[J^2,J_z]=J_x[J_x,J_z]+[J_x,J_z]J_x+J_y[J_y,J_z]+[J_y,J_z]J_y. \quad (4.106)$$

The solution for these commutators has already been worked out:

$$[J_y,J_z]=i\,J_x, \qquad [J_z,J_x]=i\,J_y, \qquad [J_x,J_y]=i\,J_z. \quad (4.107)$$

Substituting these commutators into the above equation gives:

$$[J^2,J_z]=-iJ_xJ_y-iJ_yJ_x+iJ_yJ_x+iJ_xJ_y=0. \quad (4.108)$$

Repeating the above steps to find the commutators of J^2 with other two Cartesian coordinate projections, J_x and J_y, leads to the conclusion that operator J^2 simultaneously commutes with all three components, J_x,

J_y, and J_z, of the angular momentum operator and also with the Hamiltonian. To summarize, the following commutators characterize the angular momentum operator:

$$[J_z,H]=0, \qquad [J^2,H]=0, \qquad [J^2,J_z]=0. \tag{4.109}$$

That the operators J^2 and J_z commute is not unexpected, since the two operators describe different aspects of the angular momentum. Specifically, J^2 is a scalar quantity, a vector length, whereas J_z is a projection of the vector J on to the z-axis.

4.5 Eigenvalues of the Angular Momentum

Angular momentum theory provides a mathematical tool for performing a rotation of the potential energy function in the Fast Multipole Method. Developing this tool requires an excursion into the eigenvalue problem of the angular momentum operator.

An operator is an abstract mathematical device characterized by its eigenfunctions and eigenvalues that performs a certain mathematical operation on the function it acts upon. In its simplest form, the eigenvalue problem for the angular momentum operators J^2 and J_z can be written as:

$$J^2\psi = n\psi, \qquad J_z\psi = m\psi, \tag{4.110}$$

where n and m are eigenvalues, which represent experimentally observable physical quantities, and ψ is the eigenfunction of the operator.

When an operator is represented in the basis of its eigenfunctions, it takes the form of a simple diagonal matrix in which the diagonal elements are the eigenvalues. In that form, the action of the operator on its eigenfunction corresponds to multiplying the eigenfunction by the corresponding eigenvalue.

In operator language, the value of each unique, experimentally observable, physical quantity is the eigenvalue of its associated operator. A necessary condition for a set of different physical quantities to be simultaneously observable is that their corresponding operators must share a common set of eigenfunctions. Since, by definition, operators are diagonal in the basis of their eigenfunctions, an alternative way of saying that operators commute is to say that they share the same set of eigenfunctions. For instance, for the angular momentum values n and m to be simultaneously measured in an experiment together with the energy e of the system, which is a solution of the Schrödinger equation, $H\psi = e\psi$, the eigenfunction ψ of the Hamiltonian H must also be the eigenfunction of the angular momentum operators J^2 and J_z.

An experimental measurement of a system in a state described by the function ψ_{jm} would give results that correspond to the eigenvalues n and m. That is, the experimental results correspond to solving the eigenvalue problems:

$$J^2\psi_{jm} = n\psi_{jm}, \qquad J_z\psi_{jm} = m\psi_{jm}. \tag{4.111}$$

Mathematically, the next task is to find the eigenvalues n and m of the angular momentum operators J^2 and J_z.

Because ψ_{jm} is an eigenfunction of the angular momentum operators, these operators can be represented by a diagonal matrix, and the diagonal elements are the eigenvalues n and m, respectively. Subtraction of two diagonal matrices J^2 and J_z^2, which have ψ_{jm} as their eigenfunction, produces another diagonal matrix:

$$J_x^2 + J_y^2 = J^2 - J_z^2, \tag{4.112}$$

that has ψ_{jm} as its eigenfunction, since operator $J_x^2 + J_y^2$ satisfies the eigenvalue equation:

$$(J_x^2 + J_y^2)\psi_{jm} = (J^2 - J_z^2)\psi_{jm} = (n - m^2)\psi_{jm}. \tag{4.113}$$

The right-hand side follows from $J_z(J_z\psi_{jm}) = J_z(m\psi_{jm}) = m(J_z\psi_{jm}) = m^2\psi_{jm}$ and from the fact that m is a constant, so that it may be moved to the left of the operator.

Since J_x and J_y are each Hermitian, the matrix elements of J_x^2 and J_y^2, which reside on the diagonal, should be greater than or equal to zero, according to the general properties of Hermitian matrices:

$$(A^2)_{kk} = \sum_i A_{ki}A_{ik} = \sum_i A_{ik}^*A_{ik} = \sum_i |A_{ik}|^2 \geq 0. \tag{4.114}$$

Of course, this property is satisfied for all regular Hermitian matrices, not just diagonal ones. Since eigenvalues are the diagonal elements of the matrix representation of an operator in the basis of its eigenfunctions, the eigenvalues of squared operators $J_x^2 + J_y^2$ must be greater than or equal to zero:

$$n - m^2 \geq 0. \tag{4.115}$$

This relation establishes a range limit on the eigenvalues n and m.

In order to determine the numerical values of eigenvalue m, it is helpful to introduce non-Hermitian operators J_- and J_+ *via* the definitions:

$$J_- = J_x - iJ_y \qquad J_+ = J_x + iJ_y. \tag{4.116}$$

The operators J_- and J_+ are called lowering and raising (also termed shift) operators, respectively, due to the effect that these operators produce when they act on the eigenfunctions $\psi_{j,m}$ of the angular momentum operator. In this notation of the eigenfunction, the letters j and m in the lower index are additionally separated by a comma in order to make them more distinct. The lowering operator changes the eigenfunction $\psi_{j,m}$ to $\psi_{j,m-1}$, and the raising operator changes the eigenfunction $\psi_{j,m}$ to $\psi_{j,m+1}$. This operation is accomplished by the equation:

$$J_\pm\psi_{j,m} = N_\pm\psi_{j,m\pm 1}, \tag{4.117}$$

where $N_\pm$ is a constant still to be determined. Note that this equation is not an eigenvalue problem, since the eigenfunctions in the left and right are different. This can be verified by probing the product $(J_\pm\psi_{j,m})$ with the operators J^2 and J_z.

In order to use the lowering and raising operators J_- and J_+ together with the operators J^2 and J_z, the commutation rules for these operators need to be established. Since operator J^2 commutes with every Cartesian component of the angular momentum operator, it immediately leads to:

$$[J^2, J_\pm] = 0. \tag{4.118}$$

Finding the commutator of $J_\pm$ with operator J_z involves expressing the former operator in terms of Cartesian components. Substituting the values of the commutators already known into the resulting equation leads to:

$$[J_z, J_\pm] = [J_z, J_x \pm iJ_y] = [J_z, J_x] \pm i[J_z, J_y] = iJ_y \pm J_x = \pm J_\pm. \tag{4.119}$$

These relationships greatly simplify the analysis of the shift operators, thus, since J^2 and $J_\pm$ commute, the application of operator J^2 to the product $(J_\pm\psi_{j,m})$ produces:

$$J^2(J_\pm\psi_{j,m}) = J_\pm(J^2\psi_{j,m}) = n(J_\pm\psi_{j,m}). \tag{4.120}$$

The application of operator J_z to the product $(J_\pm\psi_{j,m})$ proceeds in a like manner, employing the commutation rule for operators J_z and $J_\pm$, and yields:

$$J_z(J_\pm\psi_{j,m}) = (J_\pm J_z + [J_z, J_\pm])\psi_{j,m} = (J_\pm J_z \pm J_\pm)\psi_{j,m} = J_\pm(J_z \pm 1)\psi_{j,m}. \tag{4.121}$$

Since $J_z\psi_{j,m} = m\psi_{j,m}$,

$$J_z(J_\pm\psi_{j,m}) = (m \pm 1)(J_\pm\psi_{j,m}). \tag{4.122}$$

This proves that the application of lowering or raising operator $J_\pm$ to eigenfunction $\psi_{j,m}$ leads to another eigenfunction, denoted as $(J_\pm\psi_{j,m})$, which has the eigenvalue of $m \pm 1$.

Since the values of m are bounded by the condition $n - m^2 \geq 0$, the repeated application of the operator $J_\pm$ must eventually reach the end in both directions. The shift would stop on the boundary values m_{min} and m_{max}, beyond which the lowering and raising operations cannot proceed:

$$J_-\psi_{j,m_{min}} = 0 \qquad J_+\psi_{j,m_{max}} = 0. \tag{4.123}$$

At these limits, the effect of the shift operators on the eigenfunction is to annihilate the eigenfunction. No further operations are meaningful, and the relationships expressed in Equation 4.123 will still hold. For example, applying operators J_+ and J_- from the left side to the above equations leads to:

$$J_+J_-\psi_{j,m_{min}} = 0 \qquad J_-J_+\psi_{j,m_{max}} = 0. \tag{4.124}$$

The product of the lowering and raising operators turns into a familiar form when expressed in Cartesian components:

$$J_+J_- = (J_x + iJ_y)(J_x - iJ_y) = J_x^2 + iJ_yJ_x - iJ_xJ_y + J_y^2, \tag{4.125}$$

$$J_-J_+ = (J_x - iJ_y)(J_x + iJ_y) = J_x^2 - iJ_yJ_x + iJ_xJ_y + J_y^2. \tag{4.126}$$

After identifying and replacing the commutator $[J_x,J_y] = iJ_z$ by its value, these equations become:

$$J_+J_- = J_x^2 + J_y^2 - i[J_x,J_y] = J_x^2 + J_y^2 + J_z, \tag{4.127}$$

$$J_-J_+ = J_x^2 + J_y^2 + i[J_x,J_y] = J_x^2 + J_y^2 - J_z. \tag{4.128}$$

Substituting these results back into the operator equation leads to:

$$J_+J_-\psi_{j,m_{min}} = (J_x^2 + J_y^2 + J_z)\psi_{j,m_{min}} = (n - m_{min}^2 + m_{min})\psi_{j,m_{min}} = 0, \tag{4.129}$$

$$J_-J_+\psi_{j,m_{max}} = (J_x^2 + J_y^2 - J_z)\psi_{j,m_{max}} = (n - m_{min}^2 - m_{min})\psi_{j,m_{max}} = 0. \tag{4.130}$$

Since the eigenfunctions $\psi_{j,m_{min}}$ and $\psi_{j,m_{max}}$ cannot be zero, it means that the sum of linear coefficients, which are arranged in brackets, is equal to zero in each equation. With that, the expressions inside the brackets can be equated, which gives:

$$n - m_{min}^2 + m_{min} = n - m_{max}^2 - m_{max}. \tag{4.131}$$

Canceling out n on both sides and factoring out the common multiplier leads to:

$$-m_{min}^2 + m_{min} = -m_{max}^2 - m_{max}, \tag{4.132}$$

$$m_{min}(m_{min} - 1) = m_{max}(m_{max} + 1). \tag{4.133}$$

Based on the fact that $m_{max} \geq m_{min}$, and since lowering and raising operators shift the m index of the eigenfunction in opposite directions, the only possible solution for the above equation is:

$$m_{min} = -m_{max}. \tag{4.134}$$

From this it follows that the negative and positive values of m are symmetric in their range and magnitude in both the positive and negative directions. Assigning the boundary values of m to a constant j gives:

$$m_{\min} = -j \qquad m_{\max} = j. \tag{4.135}$$

Analysis of the possible values of m using the shift operators reveals that adjacent values of the eigenvalue m differ by unity and span a total $2j + 1$ values. For a given j, the permitted values of m are in the range $-j \le m \le j$, so that:

$$m = -j, -j+1, -j+2, \ldots, j-2, j-1, j. \tag{4.136}$$

These requirements are satisfied by index j being either a whole or a half integer, that is,:

$$j = 0, \tfrac{1}{2}, 1, \tfrac{3}{2}, 2, \ldots. \tag{4.137}$$

Spin angular momentum values of j can be half integer or whole numbers, while orbital angular momentum values are limited to whole numbers only.
Inserting the maximum value of m into $n - m_{\max}^2 - m_{\max} = 0$ gives $n - j^2 - j = 0$, which leads to:

$$n = j(j+1). \tag{4.138}$$

This neatly provides a solution to the eigenvalue problem for the square of the angular momentum operator:

$$J^2 \psi_{jm} = j(j+1)\psi_{jm}, \tag{4.139}$$

where $j(j + 1)$ is the eigenvalue of the operator J^2, and the number j is the angular momentum of the eigenfunction ψ_{jm}, when expressed in units of $\hbar$.

All that remains is to find the value of the constant $N_\pm$ that appears in Equation 4.117, $J_\pm \psi_{j,m} = N_\pm \psi_{j,m\pm1}$. Multiplying the function $(J_\pm \psi_{j,m})$ by its complex conjugate $(J_\pm \psi_{j,m})^*$ and integrating the product leads to:

$$\left\langle J_\pm \psi_{j,m} \middle| J_\pm \psi_{j,m} \right\rangle = |N_\pm|^2 \left\langle \psi_{j,m+1} \middle| \psi_{j,m+1} \right\rangle = |N_\pm|^2. \tag{4.140}$$

Since the $\psi_{j,m+1}$ form an orthonormal set, the integral $\langle \psi_{j,m+1} | \psi_{j,m+1} \rangle$ is equal to unity. The product in the left-hand side of the integral can also be transformed though the identity $(J_\pm \psi_{j,m})^* = \psi_{j,m}^* J_\pm^+$. This leads to:

$$\left\langle J_\pm \psi_{j,m} \middle| J_\pm \psi_{j,m} \right\rangle = \left\langle \psi_{j,m} \middle| J_\pm^+ J_\pm \psi_{j,m} \right\rangle. \tag{4.141}$$

Using the complex conjugate transpose of operator $J_\pm$,

$$J_\pm^+ = (J_x \pm i J_y)^+ = J_x \mp i J_y = J_\mp, \tag{4.142}$$

which incidentally proves that it is not Hermitian. Inserting Equation 4.142 back into the integral in Equation 4.141 and substituting the product of the shift operators by the previously derived relationship $J_\mp J_\pm = J_x^2 + J_y^2 \mp J_z$ gives:

$$\left\langle J_\pm \psi_{j,m} \middle| J_\pm \psi_{j,m} \right\rangle = \left\langle \psi_{j,m} \middle| J_\mp J_\pm \psi_{j,m} \right\rangle = \left\langle \psi_{j,m} \middle| \left(J_x^2 + J_y^2 \mp J_z\right) \psi_{j,m} \right\rangle. \tag{4.143}$$

The solution to the eigenvalue problem in the right-hand side has already been obtained. Replacing the operators by their corresponding eigenvalues gives:

$$\left\langle J_\pm \psi_{j,m} \middle| J_\pm \psi_{j,m} \right\rangle = \left\langle \psi_{j,m} \middle| [j(j+1) - m(m\pm1)]\, \psi_{j,m} \right\rangle = \left\langle \psi_{j,m} \middle| \psi_{j,m} \right\rangle [j(j+1) - m(m\pm1)]. \tag{4.144}$$

Again, because the eigenfunctions $\psi_{j,m}$ are orthonormal, their normalization integral $\langle \psi_{j,m} | \psi_{j,m} \rangle$ is equal to unity. Combining this and the previously obtained integral leads to:

$$\left\langle J_{\pm}\psi_{j,m} \,|\, J_{\pm}\psi_{j,m} \right\rangle = j(j+1) - m(m\pm 1) = |N_{\pm}|^2 . \tag{4.145}$$

With that, the linear coefficient $N_{\pm}$ for the shift operators is:

$$N_{\pm} = [j(j+1) - m(m\pm 1)]^{\frac{1}{2}} = [j^2 - m^2 + j \mp m]^{\frac{1}{2}} = [(j\mp m)(j\pm m+1)]^{\frac{1}{2}}. \tag{4.146}$$

Therefore, the application of the shift operator to an eigenfunction gives:

$$J_{+}\psi_{j,m} = [(j-m)(j+m+1)]^{\frac{1}{2}}\psi_{j,m+1}, \tag{4.147}$$

$$J_{-}\psi_{j,m} = [(j+m)(j-m+1)]^{\frac{1}{2}}\psi_{j,m-1}. \tag{4.148}$$

This result makes it straightforward to verify that the application of the shift operators to the corresponding terminal eigenfunctions does indeed lead to zero due to vanishing linear coefficients. Therefore,

$$J_{+}\psi_{j,j} = 0, \tag{4.149}$$

$$J_{-}\psi_{j,-j} = 0. \tag{4.150}$$

Up to this point, knowledge of the general properties of the angular momentum operator has been sufficient for the derivation of all details of the eigenvalues without the need to express the eigenfunctions in their explicit form. Although the explicit form of eigenfunctions is not essential for most of the theory of angular momentum, these functions will be instrumental during the development of the theory of the Fast Multipole Method.

4.6 Angular Momentum Operator in Spherical Polar Coordinates

The orbital angular momentum operator $\mathbf{L} = L_x\,\hat{\mathbf{x}} + L_y\,\hat{\mathbf{y}} + L_z\,\hat{\mathbf{z}}$ being a vector quantity has the following components defined in Cartesian coordinates:

$$L_x = -i\hbar\left(y\frac{\partial}{\partial z} - z\frac{\partial}{\partial y}\right), \qquad L_y = -i\hbar\left(z\frac{\partial}{\partial x} - x\frac{\partial}{\partial z}\right), \qquad L_z = -i\hbar\left(x\frac{\partial}{\partial y} - y\frac{\partial}{\partial x}\right). \tag{4.151}$$

The simple appearance of these formulae is the direct consequence of using the Cartesian coordinate system, but that advantage quickly disappears when an attempt is made to obtain the formula for the square of the orbital angular momentum operator, $L^2 = \mathbf{L}\cdot\mathbf{L} = L_x^2 + L_y^2 + L_z^2$. When expressed in Cartesian coordinates, this operator becomes extremely bulky and is of little use. An alternative, the spherical polar coordinate system with its innate symmetry, provides a much better match for the physics problem of modeling angular momentum. Switching from the Cartesian to the spherical polar coordinate system helps illustrate important additional physical insights about angular momentum.

Conversion from Cartesian to spherical polar coordinates involves replacing coordinates x, y, z and the partial derivatives $\partial/\partial x$, $\partial/\partial y$, $\partial/\partial z$ in the angular momentum components by their spherical polar counterparts. The replacement rule for coordinates x, y, z is defined by the relation between Cartesian and spherical polar coordinates:

$$x = r\sin\theta\cos\phi, \qquad y = r\sin\theta\sin\phi, \qquad z = r\cos\theta. \tag{4.152}$$

Conversion of partial derivatives from one coordinate system to another requires the application of the chain rule. The partial derivatives in Cartesian coordinates are:

$$\begin{aligned}\frac{\partial}{\partial x}&=\frac{\partial r}{\partial x}\frac{\partial}{\partial r}+\frac{\partial \theta}{\partial x}\frac{\partial}{\partial \theta}+\frac{\partial \phi}{\partial x}\frac{\partial}{\partial \phi},\\ \frac{\partial}{\partial y}&=\frac{\partial r}{\partial y}\frac{\partial}{\partial r}+\frac{\partial \theta}{\partial y}\frac{\partial}{\partial \theta}+\frac{\partial \phi}{\partial y}\frac{\partial}{\partial \phi},\\ \frac{\partial}{\partial z}&=\frac{\partial r}{\partial z}\frac{\partial}{\partial r}+\frac{\partial \theta}{\partial z}\frac{\partial}{\partial \theta}+\frac{\partial \phi}{\partial z}\frac{\partial}{\partial \phi}.\end{aligned} \tag{4.153}$$

These formulae include cross-derivatives of spherical polar coordinates over Cartesian coordinates. Those are:

$$\begin{aligned}&\frac{\partial r}{\partial x}=\sin\theta\cos\phi, &&\frac{\partial r}{\partial y}=\sin\theta\sin\phi, &&\frac{\partial r}{\partial z}=\cos\theta,\\ &\frac{\partial \theta}{\partial x}=\frac{\cos\theta\cos\phi}{r}, &&\frac{\partial \theta}{\partial y}=\frac{\cos\theta\sin\phi}{r}, &&\frac{\partial \theta}{\partial z}=-\frac{\sin\theta}{r},\\ &\frac{\partial \phi}{\partial x}=-\frac{\sin\phi}{r\sin\theta}, &&\frac{\partial \phi}{\partial y}=\frac{\cos\phi}{r\sin\theta}, &&\frac{\partial \phi}{\partial z}=0.\end{aligned} \tag{4.154}$$

Chapter 2 provided a detailed account of the derivation of Equation 4.154. Substituting the cross derivatives into the expression for partial derivatives of Cartesian coordinates gives:

$$\begin{aligned}\frac{\partial}{\partial x}&=\sin\theta\cos\phi\frac{\partial}{\partial r}+\frac{\cos\theta\cos\phi}{r}\frac{\partial}{\partial \theta}-\frac{\sin\phi}{r\sin\theta}\frac{\partial}{\partial \phi},\\ \frac{\partial}{\partial y}&=\sin\theta\sin\phi\frac{\partial}{\partial r}+\frac{\cos\theta\sin\phi}{r}\frac{\partial}{\partial \theta}+\frac{\cos\phi}{r\sin\theta}\frac{\partial}{\partial \phi},\\ \frac{\partial}{\partial z}&=\cos\theta\frac{\partial}{\partial r}-\frac{\sin\theta}{r}\frac{\partial}{\partial \theta}.\end{aligned} \tag{4.155}$$

As the result of this transformation, Cartesian derivatives are now completely expressed in terms of spherical polar coordinates.

Returning to the angular momentum operator, the x-component can be converted as follows: first, the Cartesian coordinates are replaced by the corresponding spherical polar coordinates,

$$L_x=-i\hbar\left(y\frac{\partial}{\partial z}-z\frac{\partial}{\partial y}\right)=-i\hbar\left(r\sin\theta\sin\phi\frac{\partial}{\partial z}-r\cos\theta\frac{\partial}{\partial y}\right), \tag{4.156}$$

then, replacement of the partial derivatives using Equation 4.154 gives:

$$L_x=-i\hbar\left(r\sin\theta\sin\phi\left[\cos\theta\frac{\partial}{\partial r}-\frac{\sin\theta}{r}\frac{\partial}{\partial \theta}\right]-r\cos\theta\left[\sin\theta\sin\phi\frac{\partial}{\partial r}+\frac{\cos\theta\sin\phi}{r}\frac{\partial}{\partial \theta}+\frac{\cos\phi}{r\sin\theta}\frac{\partial}{\partial \phi}\right]\right). \tag{4.157}$$

After canceling out terms that involve derivatives of the type $\partial/\partial r$, the remaining terms are:

$$L_x=-i\hbar\left(-\sin^2\theta\sin\phi\frac{\partial}{\partial \theta}-\cos^2\theta\sin\phi\frac{\partial}{\partial \theta}-\frac{\cos\theta}{\sin\theta}\cos\phi\frac{\partial}{\partial \phi}\right). \tag{4.158}$$

This expression simplifies to:

$$L_x = -i\hbar\left(-\sin\phi\frac{\partial}{\partial\theta} - \frac{\cos\theta}{\sin\theta}\cos\phi\frac{\partial}{\partial\phi}\right). \tag{4.159}$$

Similar operations can now be applied to the y-component of the angular momentum operator. The replacement of coordinates leads to:

$$L_y = -i\hbar\left(z\frac{\partial}{\partial x} - x\frac{\partial}{\partial z}\right) = -i\hbar\left(r\cos\theta\frac{\partial}{\partial x} - r\sin\theta\cos\phi\frac{\partial}{\partial z}\right). \tag{4.160}$$

This is followed by replacement of the partial derivatives:

$$L_y = -i\hbar\left(r\cos\theta\left[\sin\theta\cos\phi\frac{\partial}{\partial r} + \frac{\cos\theta\cos\phi}{r}\frac{\partial}{\partial\theta} - \frac{\sin\phi}{r\sin\theta}\frac{\partial}{\partial\phi}\right] - r\sin\theta\cos\phi\left[\cos\theta\frac{\partial}{\partial r} - \frac{\sin\theta}{r}\frac{\partial}{\partial\theta}\right]\right). \tag{4.161}$$

Once again, terms involving $\partial/\partial r$ cancel out. This leads to:

$$L_y = -i\hbar\left(\cos^2\theta\cos\phi\frac{\partial}{\partial\theta} - \frac{\cos\theta}{\sin\theta}\sin\phi\frac{\partial}{\partial\phi} + \sin^2\theta\cos\phi\frac{\partial}{\partial\theta}\right). \tag{4.162}$$

After simplification, the y-component of the angular momentum operator becomes:

$$L_y = -i\hbar\left(\cos\phi\frac{\partial}{\partial\theta} - \frac{\cos\theta}{\sin\theta}\sin\phi\frac{\partial}{\partial\phi}\right). \tag{4.163}$$

Conversion of the z-component of the angular momentum operator from Cartesian to spherical polar coordinates is performed in a similar manner. As before, the replacement of coordinates goes first, and leads to:

$$L_z = -i\hbar\left(x\frac{\partial}{\partial y} - y\frac{\partial}{\partial x}\right) = -i\hbar\left(r\sin\theta\cos\phi\frac{\partial}{\partial y} - r\sin\theta\sin\phi\frac{\partial}{\partial x}\right). \tag{4.164}$$

Next follows the replacement of Cartesian partial derivatives by their spherical polar counterparts:

$$\begin{aligned} L_z = &-i\hbar\left(r\sin\theta\cos\phi\left[\sin\theta\sin\phi\frac{\partial}{\partial r} + \frac{\cos\theta\sin\phi}{r}\frac{\partial}{\partial\theta} + \frac{\cos\phi}{r\sin\theta}\frac{\partial}{\partial\phi}\right]\right) \\ &-i\hbar\left(-r\sin\theta\sin\phi\left[\sin\theta\cos\phi\frac{\partial}{\partial r} + \frac{\cos\theta\cos\phi}{r}\frac{\partial}{\partial\theta} - \frac{\sin\phi}{r\sin\theta}\frac{\partial}{\partial\phi}\right]\right). \end{aligned} \tag{4.165}$$

In Equation 4.165, terms that involve derivatives $\partial/\partial r$ and $\partial/\partial\theta$ cancel out. Summing up the remaining terms and simplifying leads to:

$$L_z = -i\hbar\left(\cos^2\phi\frac{\partial}{\partial\phi} + \sin^2\phi\frac{\partial}{\partial\phi}\right) = -i\hbar\frac{\partial}{\partial\phi}. \tag{4.166}$$

The considerably simpler look of the operator L_z over that of L_x and L_y in spherical polar coordinates provides additional justification for the decision to prefer the component L_z over L_x and L_y in the choice of which observable operator to use along with the operator L^2.

To summarize the derivation, the components of orbital angular momentum in spherical polar coordinates are:

$$L_x = -i\hbar\left(-\sin\phi\frac{\partial}{\partial\theta} - \frac{\cos\theta}{\sin\theta}\cos\phi\frac{\partial}{\partial\phi}\right), \tag{4.167}$$

$$L_y = -i\hbar\left(\cos\phi\frac{\partial}{\partial\theta} - \frac{\cos\theta}{\sin\theta}\sin\phi\frac{\partial}{\partial\phi}\right), \tag{4.168}$$

$$L_z = -i\hbar\frac{\partial}{\partial\phi}. \tag{4.169}$$

Having the individual components of orbital angular momentum expressed in spherical polar coordinates provides a good starting point for the derivation of the corresponding expression for the orbital angular momentum operator, $\mathbf{L} = L_x\,\hat{\mathbf{x}} + L_y\,\hat{\mathbf{y}} + L_z\,\hat{\mathbf{z}}$. Due to vector components L_x, L_y, and L_z being operators, it is important to remember that these operators act on the components x, y, and z of the unit vectors $\hat{\mathbf{x}}$, $\hat{\mathbf{y}}$, and $\hat{\mathbf{z}}$, and not on the unit vectors themselves. Because of that, this expression is frequently presented in the form $\mathbf{L} = \hat{\mathbf{x}}\,L_x + \hat{\mathbf{y}}\,L_y + \hat{\mathbf{z}}\,L_z$.

Since $\mathbf{L}$ is a vector quantity, in addition to formulating its components in the new coordinate system, it is also necessary to express the Cartesian unit vectors as their spherical polar counterparts. The required equations are:

$$\begin{aligned}\hat{\mathbf{x}} &= \sin\theta\cos\phi\,\hat{\mathbf{r}} + \cos\theta\cos\phi\,\hat{\theta} - \sin\phi\,\hat{\phi},\\ \hat{\mathbf{y}} &= \sin\theta\sin\phi\,\hat{\mathbf{r}} + \cos\theta\sin\phi\,\hat{\theta} + \cos\phi\,\hat{\phi},\\ \hat{\mathbf{z}} &= \cos\theta\,\hat{\mathbf{r}} - \sin\theta\,\hat{\theta}.\end{aligned} \tag{4.170}$$

The derivation of these equations was described in Chapter 2. Substituting the various components and unit vectors into the equation for the orbital angular momentum leads to:

$$\begin{aligned}\mathbf{L} = &-i\hbar(\sin\theta\cos\phi\,\hat{\mathbf{r}} + \cos\theta\cos\phi\,\hat{\theta} - \sin\phi\,\hat{\phi})\left(-\sin\phi\frac{\partial}{\partial\theta} - \frac{\cos\theta}{\sin\theta}\cos\phi\frac{\partial}{\partial\phi}\right)\\ &-i\hbar(\sin\theta\sin\phi\,\hat{\mathbf{r}} + \cos\theta\sin\phi\,\hat{\theta} + \cos\phi\,\hat{\phi})\left(\cos\phi\frac{\partial}{\partial\theta} - \frac{\cos\theta}{\sin\theta}\sin\phi\frac{\partial}{\partial\phi}\right)\\ &-i\hbar(\cos\theta\,\hat{\mathbf{r}} - \sin\theta\,\hat{\theta})\frac{\partial}{\partial\phi}.\end{aligned} \tag{4.171}$$

Each of the Cartesian unit vectors $\hat{\mathbf{x}}$, $\hat{\mathbf{y}}$, and $\hat{\mathbf{z}}$ depends on the various spherical polar unit vectors. Because of this, when the angular momentum is expressed in spherical polar coordinates, the various components contain contributions from the Cartesian L_x, L_y, and L_z. To determine the new vector components it is convenient to sum up the contributions separately. For the radial component L_r the individual contributions are:

$$\begin{aligned}L_r = &-i\hbar\left(-\sin\theta\sin\phi\cos\phi\,\hat{\mathbf{r}}\frac{\partial}{\partial\theta} - \cos\theta\cos^2\phi\,\hat{\mathbf{r}}\frac{\partial}{\partial\phi}\right)\\ &-i\hbar\left(\sin\theta\sin\phi\cos\phi\,\hat{\mathbf{r}}\frac{\partial}{\partial\theta} - \cos\theta\sin^2\phi\,\hat{\mathbf{r}}\frac{\partial}{\partial\phi}\right)\\ &-i\hbar\frac{\partial}{\partial\phi}\cos\theta\,\hat{\mathbf{r}} = 0.\end{aligned} \tag{4.172}$$

That all terms cancel out is a very important result. It means that angular momentum in spherical polar coordinates has a zero component in the direction of radial unit vector $\hat{\mathbf{r}}$. This lack of dependence of angular momentum on radial coordinate implies that momentum is conserved, which is the quantum theoretical equivalent of the corresponding law of classical mechanics. The effect of this law may be observed on an object undergoing orbital rotation around the center of a coordinate system. If its radius of rotation is changed suddenly, then its speed of rotation must also change in order to keep the angular momentum constant. Reducing the radius of rotation of the body results in an increase in rotation speed, and increasing the radius of rotation correspondingly reduces the speed of rotation of the body.

Having established that the radial component of angular momentum is zero, that leaves only the angular coordinates, $\hat{\theta}$ and $\hat{\phi}$, in the equation:

$$\begin{aligned}\mathbf{L} = & -i\hbar(\cos\theta\cos\phi\,\hat{\theta} - \sin\phi\,\hat{\phi})\left(-\sin\phi\frac{\partial}{\partial\theta} - \frac{\cos\theta}{\sin\theta}\cos\phi\frac{\partial}{\partial\phi}\right)\\ & -i\hbar(\cos\theta\sin\phi\,\hat{\theta} + \cos\phi\,\hat{\phi})\left(\cos\phi\frac{\partial}{\partial\theta} - \frac{\cos\theta}{\sin\theta}\sin\phi\frac{\partial}{\partial\phi}\right)\\ & -i\hbar(-\sin\theta\,\hat{\theta})\frac{\partial}{\partial\phi}\end{aligned} \tag{4.173}$$

to be determined.

Opening the brackets, and moving the unit vector symbols to the left gives:

$$\begin{aligned}\mathbf{L} = & -i\hbar\left(-\hat{\theta}\cos\theta\sin\phi\cos\phi\frac{\partial}{\partial\theta} - \hat{\theta}\frac{\cos^2\theta}{\sin\theta}\cos^2\phi\frac{\partial}{\partial\phi} + \hat{\phi}\sin^2\phi\frac{\partial}{\partial\theta} + \hat{\phi}\frac{\cos\theta}{\sin\theta}\sin\phi\cos\phi\frac{\partial}{\partial\phi}\right)\\ & -i\hbar\left(\hat{\theta}\cos\theta\sin\phi\cos\phi\frac{\partial}{\partial\theta} - \hat{\theta}\frac{\cos^2\theta}{\sin\theta}\sin^2\phi\frac{\partial}{\partial\phi} + \hat{\phi}\cos^2\phi\frac{\partial}{\partial\theta} - \hat{\phi}\frac{\cos\theta}{\sin\theta}\sin\phi\cos\phi\frac{\partial}{\partial\phi}\right)\\ & -i\hbar\left(-\hat{\theta}\sin\theta\frac{\partial}{\partial\phi}\right).\end{aligned} \tag{4.174}$$

After canceling out the common terms the remaining terms are:

$$\mathbf{L} = -i\hbar\left(-\hat{\theta}\frac{\cos^2\theta}{\sin\theta}\cos^2\phi\frac{\partial}{\partial\phi} - \hat{\theta}\frac{\cos^2\theta}{\sin\theta}\sin^2\phi\frac{\partial}{\partial\phi} - \hat{\theta}\sin\theta\frac{\partial}{\partial\phi} + \hat{\phi}\sin^2\phi\frac{\partial}{\partial\theta} + \hat{\phi}\cos^2\phi\frac{\partial}{\partial\theta}\right). \tag{4.175}$$

Using trigonometric identities reduces this to:

$$\mathbf{L} = -i\hbar\left(-\hat{\theta}\frac{\cos^2\theta}{\sin\theta}\frac{\partial}{\partial\phi} - \hat{\theta}\sin\theta\frac{\partial}{\partial\phi} + \hat{\phi}\frac{\partial}{\partial\theta}\right). \tag{4.176}$$

Which finally simplifies to:

$$\mathbf{L} = -i\hbar\left(-\hat{\theta}\frac{1}{\sin\theta}\frac{\partial}{\partial\phi} + \hat{\phi}\frac{\partial}{\partial\theta}\right). \tag{4.177}$$

This is the equation for orbital angular momentum in spherical polar coordinates. Its simplicity illustrates the advantage of using the spherical polar coordinate system for physics problems that involve angular momentum.

Equation 4.177 provides a good starting point for the derivation of the square of the orbital angular momentum in spherical polar coordinates. The operator for this is the vector dot product of the angular momentum operator with itself:

$$L^2 = \mathbf{L}\cdot\mathbf{L} = -\hbar^2\left(-\hat{\theta}\frac{1}{\sin\theta}\frac{\partial}{\partial\phi}+\hat{\phi}\frac{\partial}{\partial\theta}\right)\left(-\hat{\theta}\frac{1}{\sin\theta}\frac{\partial}{\partial\phi}+\hat{\phi}\frac{\partial}{\partial\theta}\right). \tag{4.178}$$

In this equation, the operators in the left brackets act on the entire content enclosed in the right brackets, including the unit vectors. Using the rules for differentiation of unit vectors,

$$\frac{\partial\hat{\theta}}{\partial\theta} = -\hat{\mathbf{r}},\quad \frac{\partial\hat{\phi}}{\partial\theta} = 0,\quad \frac{\partial\hat{\theta}}{\partial\phi} = \cos\theta\,\hat{\phi},\quad \frac{\partial\hat{\phi}}{\partial\phi} = -\sin\theta\,\hat{\mathbf{r}} - \cos\theta\,\hat{\theta}, \tag{4.179}$$

and opening the brackets gives:

$$L^2 = -\hbar^2\left[\hat{\theta}\frac{1}{\sin\theta}\frac{\partial}{\partial\phi}\left(\hat{\theta}\frac{1}{\sin\theta}\frac{\partial}{\partial\phi}\right)-\hat{\theta}\frac{1}{\sin\theta}\frac{\partial}{\partial\phi}\left(\hat{\phi}\frac{\partial}{\partial\theta}\right)-\hat{\phi}\frac{\partial}{\partial\theta}\left(\hat{\theta}\frac{1}{\sin\theta}\frac{\partial}{\partial\phi}\right)+\hat{\phi}\frac{\partial}{\partial\theta}\left(\hat{\phi}\frac{\partial}{\partial\theta}\right)\right]. \tag{4.180}$$

For convenience, each term in square brackets may be evaluated independently. The first term is:

$$\begin{aligned}\hat{\theta}\frac{1}{\sin\theta}\frac{\partial}{\partial\phi}\left(\hat{\theta}\frac{1}{\sin\theta}\frac{\partial}{\partial\phi}\right) &= \hat{\theta}\frac{1}{\sin^2\theta}\left(\left[\frac{\partial\hat{\theta}}{\partial\phi}\right]\frac{\partial}{\partial\phi}+\hat{\theta}\left[\frac{\partial^2}{\partial\phi^2}\right]\right)\\ &= (\hat{\theta}\cdot\hat{\phi})\frac{\cos\theta}{\sin^2\theta}\frac{\partial}{\partial\phi}+(\hat{\theta}\cdot\hat{\theta})\frac{1}{\sin^2\theta}\frac{\partial^2}{\partial\phi^2} = \frac{1}{\sin^2\theta}\frac{\partial^2}{\partial\phi^2}.\end{aligned} \tag{4.181}$$

Resolving the second term involves similar steps:

$$\begin{aligned}\hat{\theta}\frac{1}{\sin\theta}\frac{\partial}{\partial\phi}\left(\hat{\phi}\frac{\partial}{\partial\theta}\right) &= \hat{\theta}\frac{1}{\sin\theta}\left(\frac{\partial\hat{\phi}}{\partial\phi}\frac{\partial}{\partial\theta}+\hat{\phi}\frac{\partial^2}{\partial\phi\,\partial\theta}\right)\\ &= \hat{\theta}\frac{1}{\sin\theta}(-\sin\theta\,\hat{\mathbf{r}}-\cos\theta\,\hat{\theta})\frac{\partial}{\partial\theta}+(\hat{\theta}\cdot\hat{\phi})\frac{1}{\sin\theta}\frac{\partial^2}{\partial\phi\,\partial\theta}\\ &= \hat{\theta}\frac{1}{\sin\theta}(-\sin\theta\,\hat{\mathbf{r}}-\cos\theta\,\hat{\theta})\frac{\partial}{\partial\theta} = -(\hat{\theta}\cdot\hat{\mathbf{r}})\frac{\partial}{\partial\theta}-(\hat{\theta}\cdot\hat{\theta})\frac{\cos\theta}{\sin\theta}\frac{\partial}{\partial\theta} = -\frac{\cos\theta}{\sin\theta}\frac{\partial}{\partial\theta}.\end{aligned} \tag{4.182}$$

The third term is:

$$\hat{\phi}\frac{\partial}{\partial\theta}\left(\hat{\theta}\frac{1}{\sin\theta}\frac{\partial}{\partial\phi}\right) = \hat{\phi}\frac{1}{\sin\theta}\left(\frac{\partial\hat{\theta}}{\partial\theta}\frac{\partial}{\partial\phi}+\hat{\theta}\frac{\partial^2}{\partial\theta\,\partial\phi}\right) = -(\hat{\phi}\cdot\hat{\mathbf{r}})\frac{1}{\sin\theta}\frac{\partial}{\partial\phi}+(\hat{\phi}\cdot\hat{\theta})\frac{1}{\sin\theta}\frac{\partial^2}{\partial\theta\,\partial\phi} = 0. \tag{4.183}$$

The fourth and final term is:

$$\hat{\phi}\frac{\partial}{\partial\theta}\left(\hat{\phi}\frac{\partial}{\partial\theta}\right) = \hat{\phi}\frac{\partial\hat{\phi}}{\partial\theta}\frac{\partial}{\partial\theta}+(\hat{\phi}\cdot\hat{\phi})\frac{\partial^2}{\partial\theta^2} = \frac{\partial^2}{\partial\theta^2}. \tag{4.184}$$

These terms can now be combined, which leads to:

$$L^2 = -\hbar^2\left[\frac{1}{\sin^2\theta}\frac{\partial^2}{\partial\phi^2}+\frac{\cos\theta}{\sin\theta}\frac{\partial}{\partial\theta}+\frac{\partial^2}{\partial\theta^2}\right]. \tag{4.185}$$

Rearrangement gives the conventional form of the orbital angular momentum operator:

$$L^2 = -\hbar^2 \left[\frac{1}{\sin\theta}\frac{\partial}{\partial\theta}\left(\sin\theta\frac{\partial}{\partial\theta}\right) + \frac{1}{\sin^2\theta}\frac{\partial^2}{\partial\phi^2} \right]. \tag{4.186}$$

Having obtained the explicit equations for angular momentum operators L_z and L^2 in spherical polar coordinates, the next step is to find their eigenfunctions.

4.7 Eigenvectors of the Angular Momentum Operator

Eigenfunctions $\psi_{l,m}$, also denoted as $Y_{l,m}$, are solutions of the eigenvalue equation:

$$L^2 Y_{l,m} = l(l+1)\hbar^2 Y_{l,m}, \tag{4.187}$$

where $l(l+1)$ is an eigenvalue of the operator L^2 in units of $\hbar^2$. Section 4.4 explained how this eigenvalue was derived. The indices l and m in the eigenfunction indicate the dependence of the function on those numbers.

Before attempting to solve the eigenvalue equation, a little simplification may prove useful. Substituting the expression for the operator L^2, and canceling out $\hbar^2$ on both sides, gives:

$$-\left[\frac{1}{\sin\theta}\frac{\partial}{\partial\theta}\left(\sin\theta\frac{\partial}{\partial\theta}\right) + \frac{1}{\sin^2\theta}\frac{\partial^2}{\partial\phi^2} \right] Y_{l,m} = l(l+1)Y_{l,m}. \tag{4.188}$$

This equation appears in the process of solution of Laplace's equation, as described in Chapter 3. The solution of Laplace's equation leads to the spherical harmonics:

$$Y_{l,m}(\theta,\phi) = \sqrt{\frac{(2l+1)}{4\pi}\frac{(l-m)!}{(l+m)!}} P_l^m(\cos\theta)e^{im\phi}, \tag{4.189}$$

where P_l^m are the associated Legendre functions.

Tracing the steps from the solution of Laplace's equation back to the eigenvalue equation shows that spherical harmonic functions $Y_{l,m}$ are eigenfunctions of the square of the angular momentum operator, L^2, and have the eigenvalue $l(l+1)$. The index l in spherical harmonics $Y_{l,m}$ indicates the relationship of the function to its eigenvalue $l(l+1)$.

Spherical harmonics are also eigenfunctions of the operator L_z. This can be verified by direct application of the operator to the function $Y_{l,m}$:

$$L_z Y_{lm} = -i\hbar\frac{\partial}{\partial\phi}\left[\sqrt{\frac{(2l+1)}{4\pi}\frac{(l-m)!}{(l+m)!}} P_l^m(\cos\theta)e^{im\phi} \right] = m\hbar Y_{lm}. \tag{4.190}$$

The operator acts only on the exponent part of the function, which depends only on coordinate ϕ. The multiplier resulting from differentiation of $e^{im\phi}$ is im. This cancels with the leading $-i$ to give m, showing that the application of the operator L_z on spherical harmonics $Y_{l,m}$ is equivalent to scaling that function by the coefficient m. Therefore, $Y_{l,m}$ is an eigenfunction of operator L_z and has the eigenvalue m, in units of $\hbar$. The presence of index m in $Y_{l,m}$ reflects the relation between the eigenfunction and its eigenvalue m. In literature, it is common to omit Planck's constant in the eigenvalue equation for the sake of brevity. In such cases, the presence of Planck's constant is simply assumed.

When the eigenvalues and eigenfunctions of the angular momentum operators are known, the numbers l and m, which appear in the eigenvalue equations and in the indices of eigenfunction $Y_{l,m}$, have their proper physical interpretation. In the conventional language, eigenvalue is the magnitude of the physical quantity that is represented by the operator. Based on that, the interpretation of the number m is straightforward. Since it is an eigenvalue of operator L_z, the number m is the magnitude of the projection of the angular momentum $\mathbf{L}$ on the z-axis, which is called a quantization axis for that reason.

The meaning of number l is more complicated. In the classical mechanical limit, it is the magnitude of the angular momentum. It cannot be, however, an eigenvalue of the operator $\mathbf{L}$ because the latter is a vector operator and all three components cannot be determined simultaneously. The number l is determined indirectly, in the form of the expression $l(l + 1)$, which is the eigenvalue of operator L^2. In the classical limit of large l, the value of $l(l + 1)$ becomes indistinguishable from l^2, so that the classical limit of l for the magnitude of the angular momentum $\mathbf{L}$ is met. The extra "+1" in the magnitude of the square of the angular momentum is a purely quantum mechanical effect.

The usual names for numbers l and m are orbital and azimuthal (or magnetic) quantum numbers, respectively. The equivalent names in spherical harmonic functions, $Y_{l,m}$, are the orbital and azimuthal indices. Correspondingly, if a quantum particle happens to be in the state described by eigenfunction $Y_{l,m}$, that particle has the magnitude $l(l + 1)$ for the square of its angular momentum, and the magnitude m for the projection of the angular momentum onto the z-axis.

The conclusion that spherical harmonic functions are eigenfunctions of the angular momentum operator has important implications. Eigenfunctions are the ideal candidates to serve as basis functions in series expansions, because they form a complete orthonormal set. Since any electrostatic potential function may be expanded in a series of spherical harmonics, the problem of rotation of a potential function in 3D translates into the simpler task of rotation of spherical harmonic functions, which is a well-studied mathematical problem. The following section further formalizes the subject of rotation in 3D space, thus laying the groundwork for the theory of rotation of spherical harmonics.

4.8 Characteristic Vectors of the Rotation Operator

Rotation is a process that changes the relative position of a point around a rotation axis and that preserves the angles and distances between the rotated points themselves and those between the points and the rotation axis. In the general form, a rotation operator $R(\hat{\mathbf{n}},\phi)$ is defined by specifying a rotation axis with help of a unit vector $\hat{\mathbf{n}}$, and the angle or rotation ϕ. The direction of the vector $\hat{\mathbf{n}}$ defines the direction of rotation, which is conventionally set to be positive for a counter-clockwise course when looking from the tip of the vector toward its base.

The general form of the rotation operator does not depend on the choice of the coordinate system, and can be easily deduced from the basic geometric principles. Figure 4.7 depicts the rotation of a point

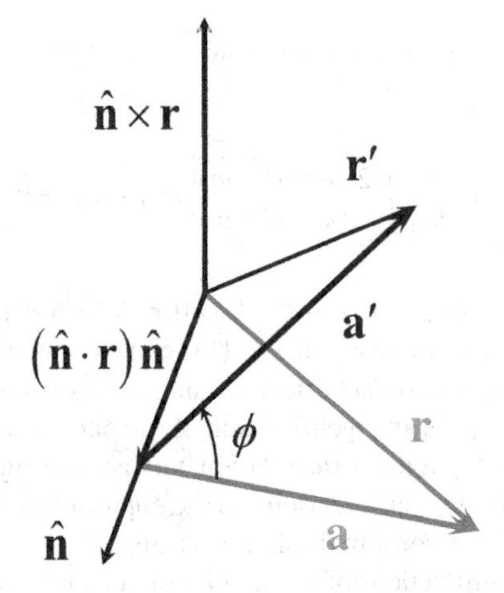

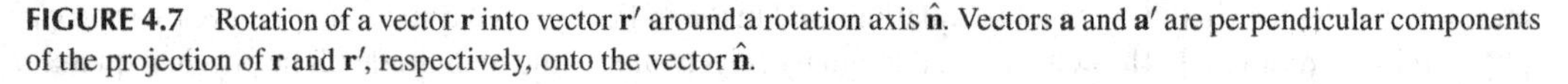
FIGURE 4.7 Rotation of a vector $\mathbf{r}$ into vector $\mathbf{r}'$ around a rotation axis $\hat{\mathbf{n}}$. Vectors $\mathbf{a}$ and $\mathbf{a}'$ are perpendicular components of the projection of $\mathbf{r}$ and $\mathbf{r}'$, respectively, onto the vector $\hat{\mathbf{n}}$.

defined by a radius vector $\mathbf{r}$ into the radius vector $\mathbf{r}'$ around a rotation axis $\hat{\mathbf{n}}$ through angle ϕ. Rotation preserves the length of the vector, $|\mathbf{r}| = |\mathbf{r}'|$.

The component of $\mathbf{r}$ along $\hat{\mathbf{n}}$, which is $(\hat{\mathbf{n}}\cdot\mathbf{r})\hat{\mathbf{n}}$, does not undergo rotation. In that vector, the dot product $\hat{\mathbf{n}}\cdot\mathbf{r}$ gives the length of the component, and the unit vector $\hat{\mathbf{n}}$ gives the direction. The perpendicular component that undergoes rotation is $\mathbf{a} = \mathbf{r} - (\hat{\mathbf{n}}\cdot\mathbf{r})\hat{\mathbf{n}}$. The third vector that establishes the rotation frame is $\hat{\mathbf{n}}\times\mathbf{r}$, which is perpendicular to vectors $\hat{\mathbf{n}}$ and $\mathbf{r}$. Together, vectors $\hat{\mathbf{n}}$, $\mathbf{a}$, and $\hat{\mathbf{n}}\times\mathbf{r}$ form a right-handed reference frame.

Rotation about the axis $\hat{\mathbf{n}}$ through angle ϕ turns vector $\mathbf{a}$ into $\mathbf{a}'$, so that $|\mathbf{a}'| = |\mathbf{a}| = |\hat{\mathbf{n}}\times\mathbf{r}|$. The latter portion of this equality comes from $|\hat{\mathbf{n}}\times\mathbf{r}| = |\hat{\mathbf{n}}||\mathbf{r}|\sin\theta = |\mathbf{r}|\sin\theta = |\mathbf{a}|$, where θ is the angle subtended between vectors $\hat{\mathbf{n}}$ and $\mathbf{r}$. The resulting vector $\mathbf{a}'$ is a linear combination of the orthogonal base vectors $\mathbf{a}$ and $\hat{\mathbf{n}}\times\mathbf{r}$, which mix through standard trigonometric relations, as illustrated in Figure 4.8.
Therefore,

$$\mathbf{a}' = \mathbf{a}\cos\phi + (\hat{\mathbf{n}}\times\mathbf{r})\sin\phi. \tag{4.191}$$

Since $\mathbf{r}' = (\hat{\mathbf{n}}\cdot\mathbf{r})\hat{\mathbf{n}} + \mathbf{a}'$, this leads to:

$$R(\hat{\mathbf{n}},\phi)\mathbf{r} = (\hat{\mathbf{n}}\cdot\mathbf{r})\,\hat{\mathbf{n}} + \mathbf{a}\cos\phi + (\hat{\mathbf{n}}\times\mathbf{r})\sin\phi. \tag{4.192}$$

Therefore, taking into account that

$$\mathbf{a} = \mathbf{r} - (\hat{\mathbf{n}}\cdot\mathbf{r})\hat{\mathbf{n}}:$$

$$\mathbf{r}' = R(\hat{\mathbf{n}},\phi)\,\mathbf{r} = \mathbf{r}\cos\phi + (\hat{\mathbf{n}}\cdot\mathbf{r})\,\hat{\mathbf{n}}(1-\cos\phi) + (\hat{\mathbf{n}}\times\mathbf{r})\sin\phi. \tag{4.193}$$

This equation describes the rotation of an arbitrary vector $\mathbf{r}$ by the operator $R(\hat{\mathbf{n}},\phi)$. It also leads to the definition of the characteristic vectors of a rotation $\hat{\mathbf{p}} + i\hat{\mathbf{q}}$, $\hat{\mathbf{n}}$, and $\hat{\mathbf{p}} - i\hat{\mathbf{q}}$, in which $\hat{\mathbf{p}}$, $\hat{\mathbf{q}}$, and $\hat{\mathbf{n}}$ are unit vectors that are mutually perpendicular, so that, taken together, they form a right-handed triad $(\hat{\mathbf{n}}, \hat{\mathbf{p}}, \hat{\mathbf{q}})$ of unit vectors. These vectors are unusual in that they do not change their orientation after application of the rotation operator, but only get scaled by a phase factor, so that:

$$\begin{aligned} &R(\hat{\mathbf{n}},\phi)(\hat{\mathbf{p}} + i\,\hat{\mathbf{q}}) = e^{-i\phi}(\hat{\mathbf{p}} + i\,\hat{\mathbf{q}}), \\ &R(\hat{\mathbf{n}},\phi)\hat{\mathbf{n}} = \hat{\mathbf{n}}, \\ &R(\hat{\mathbf{n}},\phi)(\hat{\mathbf{p}} - i\,\hat{\mathbf{q}}) = e^{i\phi}(\hat{\mathbf{p}} - i\,\hat{\mathbf{q}}). \end{aligned} \tag{4.194}$$

Eigenvalue equations are recognizable in these formulae, in that they resemble the change in index m in $e^{im\phi}$ from negative, through zero, to positive integer numbers, and in the construction of complex spherical harmonics from Cartesian coordinates, which will be described shortly.

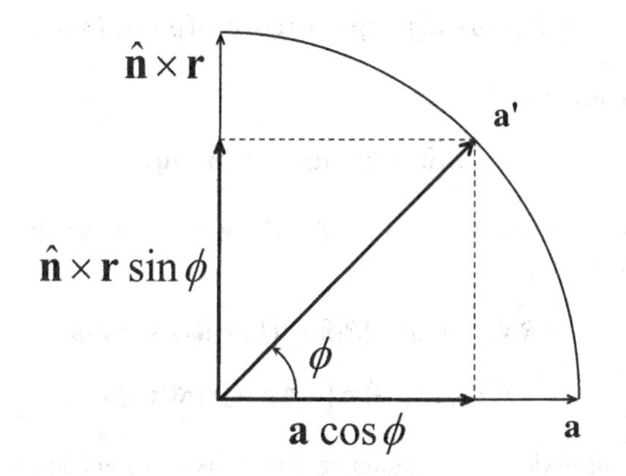

FIGURE 4.8 Rotation of vector $\mathbf{a}$ into $\mathbf{a}'$ produces a linear combination of basis vectors.

A proof for these equations follows from the application of direct substitution. Substitution of $\hat{\mathbf{p}}+i\hat{\mathbf{q}}$ for $\mathbf{r}$ in the rotation equation leads to:

$$R(\hat{\mathbf{n}},\phi)(\hat{\mathbf{p}}+i\hat{\mathbf{q}}) = (\hat{\mathbf{p}}+i\hat{\mathbf{q}})\cos\phi + [\hat{\mathbf{n}}\cdot(\hat{\mathbf{p}}+i\hat{\mathbf{q}})]\hat{\mathbf{n}}(1-\cos\phi) + [\hat{\mathbf{n}}\times(\hat{\mathbf{p}}+i\hat{\mathbf{q}})]\sin\phi. \tag{4.195}$$

Since vectors $\hat{\mathbf{n}}$, $\hat{\mathbf{p}}$, $\hat{\mathbf{q}}$ are mutually orthogonal, their dot product is zero, and the cross products are $\hat{\mathbf{n}}\times\hat{\mathbf{p}} = \hat{\mathbf{q}}$ and $\hat{\mathbf{n}}\times\hat{\mathbf{q}} = -\hat{\mathbf{p}}$, thus:

$$R(\hat{\mathbf{n}},\phi)(\hat{\mathbf{p}}+i\hat{\mathbf{q}}) = (\hat{\mathbf{p}}+i\hat{\mathbf{q}})\cos\phi + (\hat{\mathbf{q}}-i\hat{\mathbf{p}})\sin\phi. \tag{4.196}$$

The right-most term can be rearranged so that:

$$R(\hat{\mathbf{n}},\phi)(\hat{\mathbf{p}}+i\hat{\mathbf{q}}) = (\hat{\mathbf{p}}+i\hat{\mathbf{q}})\cos\phi - (\hat{\mathbf{p}}+i\hat{\mathbf{q}})i\sin\phi. \tag{4.197}$$

Applying the substitution $\cos\phi - i\sin\phi = e^{-i\phi}$ leads to the expected result:

$$R(\hat{\mathbf{n}},\phi)(\hat{\mathbf{p}}+i\hat{\mathbf{q}}) = \mathrm{e}^{-i\phi}(\hat{\mathbf{p}}+i\hat{\mathbf{q}}). \tag{4.198}$$

The proof for the characteristic vector $\hat{\mathbf{n}}$ follows from substituting $\hat{\mathbf{n}}$ for $\mathbf{r}$ in the rotation equation. This gives:

$$R(\hat{\mathbf{n}},\phi)\hat{\mathbf{n}} = \hat{\mathbf{n}}\cos\phi + (\hat{\mathbf{n}}\cdot\hat{\mathbf{n}})\,\hat{\mathbf{n}}(1-\cos\phi) + (\hat{\mathbf{n}}\times\hat{\mathbf{n}})\sin\phi. \tag{4.199}$$

The dot product of $\hat{\mathbf{n}}$ with itself is unity and the cross product is zero. This leads to:

$$R(\hat{\mathbf{n}},\phi)\hat{\mathbf{n}} = \hat{\mathbf{n}}\cos\phi + \hat{\mathbf{n}}(1-\cos\phi) = \hat{\mathbf{n}}, \tag{4.200}$$

which agrees with the common-sense notion that points located on the rotation axis do not undergo rotation.

A similar proof exists for the third characteristic vector, $\hat{\mathbf{p}}-i\hat{\mathbf{q}}$. Substitution of $\hat{\mathbf{p}}-i\hat{\mathbf{q}}$ for $\mathbf{r}$ in the rotation equation leads to:

$$R(\hat{\mathbf{n}},\phi)(\hat{\mathbf{p}}-i\hat{\mathbf{q}}) = (\hat{\mathbf{p}}-i\hat{\mathbf{q}})\cos\phi + [\hat{\mathbf{n}}\cdot(\hat{\mathbf{p}}-i\hat{\mathbf{q}})]\,\hat{\mathbf{n}}(1-\cos\phi) + [\hat{\mathbf{n}}\times(\hat{\mathbf{p}}-i\hat{\mathbf{q}})]\sin\phi. \tag{4.201}$$

This simplifies through:

$$R(\hat{\mathbf{n}},\phi)(\hat{\mathbf{p}}-i\hat{\mathbf{q}}) = (\hat{\mathbf{p}}-i\hat{\mathbf{q}})\cos\phi + (\hat{\mathbf{q}}+i\hat{\mathbf{p}})\sin\phi, \tag{4.202}$$

to

$$R(\hat{\mathbf{n}},\phi)(\hat{\mathbf{p}}-i\hat{\mathbf{q}}) = (\hat{\mathbf{p}}-i\hat{\mathbf{q}})\cos\phi + (\hat{\mathbf{p}}-i\hat{\mathbf{q}})i\sin\phi, \tag{4.203}$$

which results in the final equation:

$$R(\hat{\mathbf{n}},\phi)(\hat{\mathbf{p}}-i\hat{\mathbf{q}}) = e^{i\phi}(\hat{\mathbf{p}}-i\hat{\mathbf{q}}). \tag{4.204}$$

Using the explicit form of the rotation operator $R(\hat{\mathbf{n}},\phi)$, one can verify that the following relations hold for any two vectors $\mathbf{r}$ and $\mathbf{s}$:

$$R(\hat{\mathbf{n}},\phi)(\mathbf{r}\cdot\mathbf{s}) = [R(\hat{\mathbf{n}},\phi)\mathbf{r}]\cdot[R(\hat{\mathbf{n}},\phi)\mathbf{s}] = \mathbf{r}\cdot\mathbf{s}, \tag{4.205}$$

$$R(\hat{\mathbf{n}},\phi)(\mathbf{r}\times\mathbf{s}) = [R(\hat{\mathbf{n}},\phi)\mathbf{r}]\times[R(\hat{\mathbf{n}},\phi)\mathbf{s}]. \tag{4.206}$$

These relations reiterate the axiom that rotation preserves distances and angles.

4.9 Rotation of Eigenfunctions of Angular Momentum

FMM computation requires rotation of multipole expansion to be performed in 3D space in order to properly align the multipoles, which, like vectors, depend on the choice of the coordinate system. The fundamental basics needed for multipole rotation will be worked out in this section. Due to multipole expansions being made in the basis of spherical harmonics, it is necessary to deduce the basic principles of rotation of the functions involved. Since spherical harmonics, denoted $\psi_{j,m}$ or $Y_{l,m}$, are eigenfunctions of the angular momentum operator, the goal changes to the study of the rotation properties of the eigenfunctions of angular momentum. In the notation used here, index j applies to all eigenfunctions of angular momentum, including spin, whereas index l denotes only integer values of index j since only integer numbers occur in spherical harmonic functions. Although FMM theory is intended for use with spherical harmonics only, it is worthwhile developing the more general theory of rotation for eigenfunctions $\psi_{j,m}$ of angular momentum rather than for those limited to the case of $j = l$, because both sets of functions rotate according to exactly the same theory.

The fact that spherical harmonics are eigenfunctions of the angular momentum operator leads to an important theorem that greatly helps in construction of the theory of rotation of these functions. According to this theorem, the eigenvalue of the square of the angular momentum is unchanged by rotation. The theorem states that if $\psi_{j,m}$ is an eigenvalue of operator J^2, such that

$$J^2\psi_{j,m} = j(j+1)\psi_{j,m}, \tag{4.207}$$

where Planck's constant is omitted for brevity, then the rotated function $R\psi_{j,m}$ is also an eigenfunction of J^2 with the same eigenvalue $j(j + 1)$:

$$J^2[R\psi_{j,m}] = j(j+1)[R\psi_{j,m}]. \tag{4.208}$$

In order to prove this theorem, it is necessary to find a commutator of J^2 with $R = e^{-i\phi(\mathbf{n}\cdot\mathbf{J})}$. Expanding the rotation operator inside the commutator in a Taylor series shows that these operators commute:

$$[J^2, R] = \left[J^2, \sum_{k=0}^{\infty}\frac{1}{k!}(-i\phi\,\mathbf{n}\cdot\mathbf{J})^k\right] = \sum_{k=0}^{\infty}\frac{1}{k!}(-i\phi)^k[J^2, (\mathbf{n}\cdot\mathbf{J})^k] = 0. \tag{4.209}$$

Indeed, since for an arbitrary axis of rotation, the product $\mathbf{n}\cdot\mathbf{J}$ is a linear combination of components J_x, J_y, and J_z, and, because operator J^2 commutes with each of them, and with any power of them, the commutator of J^2 with R is equal to zero. Since operators J^2 and R commute, it implies that:

$$J^2[R\psi_{j,m}] = R[J^2\psi_{j,m}] = R[j(j+1)\psi_{j,m}] = j(j+1)[R\psi_{j,m}], \tag{4.210}$$

and therefore the theorem is proved.

The independence of the eigenvalue of the angular momentum on any particular orientation of the eigenfunction corresponds to the well-known fact that the total angular momentum does not depend on the choice of coordinate system.

A consequence of the above theorem is that rotation of eigenfunctions of angular momentum does not mix functions that have different indices j. This significantly reduces the complexity of the problem. The only remaining degree of freedom that is left in spherical harmonics for transformations due to rotation is associated with index m. The rotation operator mixes functions in index m because the rotated function is not an eigenfunction of operator J_z. The latter follows from the fact that operators J_z and R do not commute. This can be proved by the presence of components J_x and J_y in the product $\mathbf{n}\cdot\mathbf{J}$ in the Taylor expansion of the rotation operator, R. Since those components do not commute with J_z, it follows that neither does the rotation operator.

Recalling that rotation always leads to a linear combination of the orthogonal basis function (compare it with 2D case), and taking into account that eigenfunctions having the same index j are mutually orthogonal in index m, leads to the general form of the rotation operation:

$$R\psi_{j,m} = \sum_{k=-j}^{j} C_{k,m}\psi_{j,k}, \tag{4.211}$$

where index k runs through the complete azimuthal space $-j \le k \le j$ of the j-multiplet, and $C_{k,m}$ is a linear coefficient still to be determined. This equation indicates that the rotated function $R\psi_{j,m}$ is a linear combination of the functions with the same orbital index j and different azimuthal index k. Coefficient $C_{k,m}$ is determined by the standard rule of multiplying both sides of the equation by $\psi^*_{j,k}$ from the left, and integrating. Due to the orthogonality of the eigenfunctions, the linear coefficient is:

$$C_{k,m} = \left\langle \psi_{j,k} \,|\, e^{-i\phi(\mathbf{n}\cdot\mathbf{J})} \,|\, \psi_{j,m} \right\rangle = \left\langle jk \,|\, e^{-i\phi(\mathbf{n}\cdot\mathbf{J})} \,|\, jm \right\rangle. \tag{4.212}$$

As shown in Equation 4.212, symbols of the function are frequently omitted, and are replaced by their indices for the sake of brevity. The iterating index k applies to the left-most eigenfunction that is in the complex conjugate state in the integral. With that, the equation that describes rotation of a spherical harmonic function is:

$$R\psi_{j,m} = \sum_{k=-j}^{j} \left\langle jk \left| e^{-i\phi(\mathbf{n}\cdot\mathbf{J})} \right| jm \right\rangle \psi_{j,k}. \tag{4.213}$$

Rotation about the quantization axis has a particularly simple form, in which the product $\mathbf{n}\cdot\mathbf{J}$ is equal to J_z, and the rotation equation becomes:

$$R_z\psi_{j,m} = \sum_{k=-j}^{j} \left\langle jk \,|\, e^{-i\phi J_z} \,|\, jm \right\rangle \psi_{j,k}. \tag{4.214}$$

Expanding the exponential operator in Taylor series, and taking into account that $\psi_{j,m}$ is an eigenfunction of operator J_z with the eigenvalue m, leads to:

$$R_z\psi_{j,m} = \sum_{k=-j}^{j} \left\langle jk \left| \sum_{t=0}^{\infty} \frac{1}{t!}(-i\phi J_z)^t \right| jm \right\rangle \psi_{j,k} = \sum_{k=-j}^{j} \left\langle jk \left| \sum_{t=0}^{\infty} \frac{1}{t!}(-i\phi m)^t \right| jm \right\rangle \psi_{j,k}. \tag{4.215}$$

This expression can be rearranged by taking the constant multiplier outside the integral, so that it becomes:

$$R_z\psi_{j,m} = \left[\sum_{t=0}^{\infty} \frac{1}{t!}(-i\phi m)^t\right] \left[\sum_{k=-j}^{j} \left\langle jk \,|\, jm \right\rangle \psi_{j,k}\right]. \tag{4.216}$$

Due to the orthonormality of eigenfunctions, the integral $\langle jk|jm\rangle$ is equal to $\delta_{k,m}$, so it is finite and equal to unity only when $k = m$. Thus,

$$R_z\psi_{j,m} = \left[\sum_{t=0}^{\infty} \frac{1}{t!}(-i\phi m)^t\right] \psi_{j,m}. \tag{4.217}$$

The remaining sum can be converted back to the exponential form, which leads to:

$$R_z\psi_{j,m} = e^{-i\phi m}\psi_{j,m}. \tag{4.218}$$

This equation demonstrates that rotation of an eigenfunction $\psi_{l,m}$ of angular momentum about the z-axis corresponds to multiplication of the function by a phase factor, $e^{-i\phi m}$, which, by definition, has a modulus of one:

$$|e^{-i\phi m}| = e^{i\phi m}e^{-i\phi m} = e^{0} = 1. \tag{4.219}$$

Rotation about the quantization axis is interesting because of its simplicity. It is also present in the general case of rotation around an arbitrary axis, since a general rotation can be cast into a series of rotations around individual coordinate axes. The most elegant and convenient approach for defining a general 3D rotation is to make use of Euler angles, which is introduced in the next chapter.

5

Wigner Matrix

This chapter continues the discussion of theory of rotation started in Chapter 4. It details the parametric form of the rotation operator, and derives the general form of rotation matrix for spherical harmonics.

5.1 The Euler Angles

A rotation in 3D space requires specifying three parameters. This is reflected in the dependence of the unitary transformation operator $R(\mathbf{n}, \phi)$ on three angles. There is considerable freedom in the way in which these angles may be defined. For instance, from the geometric perspective, two angles may be used to define the position and direction of the rotation axis $\mathbf{n}$ about the coordinate system, and the third angle ϕ would define the magnitude of rotation. Among possible alternatives, Euler's angles provide the most elegant approach to the definition of rotation.

The Euler's parametric form defines 3D rotation *via* a sequence of 2D rotations around the coordinate axes:

$$R = R(\gamma)R(\beta)R(\alpha), \tag{5.1}$$

where α, β, and γ are Euler's angles. Figure 5.1 illustrates the definition of those angles, and Figure 5.2a shows the rotation in a sequence of steps. The domain of definition of Euler's angles is:

$$0 \leq \alpha < 2\pi, \quad 0 \leq \beta \leq \pi, \quad 0 \leq \gamma < 2\pi. \tag{5.2}$$

The individual 2D rotations in the Euler's parametric form are defined as follows.

First, a rotation is made around the z-axis through angle α. It corresponds to the right-most operator $R(\alpha)$ in the operator product in Equation 5.1. In this rotation, the z-axis stays in place, so that the original z and the resulting z' axes coincide, and the axes x and y transform into the new coordinate axes x' and y', respectively.

The second rotation is made around the y'-axis through angle β, and corresponds to the operator $R(\beta)$ in Equation 5.1. As the result of this rotation, the y'-axis remains in its original position so that the axes y' and y'' coincide, and the axes x' and z' transform into new coordinate axes x'' and z'', respectively.

The third and the final 2D rotation is around the z''-axis, through the angle γ; this rotation corresponds to operator $R(\gamma)$ in Equation 5.1. After completion of the rotation, the z''-axis coincides with the z'''-axis, and the axes x'' and y'' transform into axes x''' and y''', respectively.

Euler's angles represent a completely general definition of rotation in 3D space. This approach replaces the previously introduced rotation operator $R(\mathbf{n}, \phi)$ by a product of three operators:

$$R(\mathbf{n}, \phi) = e^{-i\phi(\mathbf{n}\cdot\mathbf{J})} \equiv R_\gamma R_\beta R_\alpha, \tag{5.3}$$

where lower indices α, β, and γ indicate a 2D rotation around the axis associated with the corresponding Euler's angle. These two definitions of 3D rotation are equivalent, and therefore there must exist a relationship that connects the angle ϕ and rotation axis $\mathbf{n}$ on one side, and the angles α, β, and γ on the other. The explicit form of that dependence, however, will not be needed in the development of FMM theory, and thus it will not be analyzed. From this point onward, all 3D rotations will be specified by the three Euler's angles so there will no longer be any need for the former, less practical, definition.

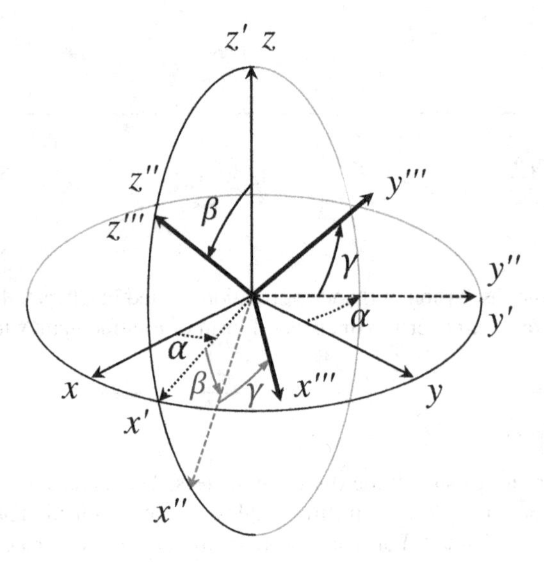

FIGURE 5.1 Rotations of coordinate system xyz through three Euler's angles, α, β and γ in three successive steps $xyz \to x'y'z' \to x''y''z'' \to x'''y'''z'''$.

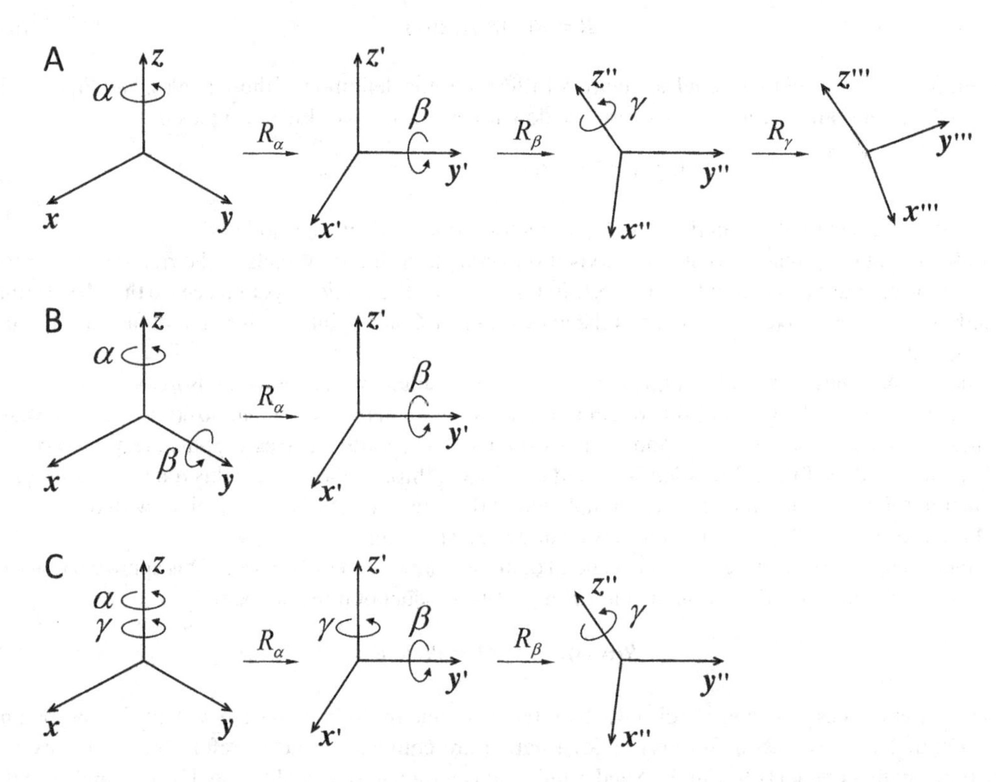

FIGURE 5.2 Rotations using Euler's angles. Panel A: Schematic representation of 3D rotation in a sequence of 2D rotation steps: $xyz \to x'y'z' \to x''y''z'' \to x'''y'''z'''$. Panel B: Conversion of the operator $e^{-i\beta J_y}$, which rotates around the y-axis, into the operator R_β, which rotates about the y'-axis, with help of a unitary transformation R_α. Panel C: Two-step conversion of the operator $e^{-i\gamma J_z}$ about the z-axis into the operator R_γ about the z''-axis *via* unitary transformations R_β and R_α.

Operators R_α, R_β, and R_γ represent successive rotations around the z, y', and z'' axes, respectively. These steps are illustrated in Figure 5.2A. Each of the individual rotation operators can be written in the usual exponential form. Thus,

$$R_\gamma R_\beta R_\alpha = e^{-i\gamma J_{z''}} e^{-i\beta J_{y'}} e^{-i\alpha J_z} \tag{5.4}$$

where J_z, $J_{y'}$, and $J_{z''}$ are the components of operator $\mathbf{J}$ along the axes z, y', and z'', respectively. Rotation around the z-axis represented by operator R_α is a known procedure, and the explicit form of component J_z has already been established. The other two rotations R_β and R_γ will now be determined.

Operator R_β performs a rotation around the y'-axis through the angle β. The axis y' originates from the y-axis due to rotation of the coordinate system xyz around the z-axis by using the operator $R_\alpha = e^{-i\alpha J_z}$. This step is illustrated in Figure 5.2B. In a specific case of $\alpha = 0$, operator R_β becomes a standard rotation operator $e^{-i\beta J_y}$ for rotation around the y-axis. This observation provides a hint to the idea that operator R_β is a unitary transform of the standard rotation operator, $e^{-i\beta J_y}$. The following analysis explains that conjecture. A unitary transformation, U, transforms an operator A into UAU^{-1}. Since rotation is a unitary operation, this implies that the unitary operator R_α can transform the operator $e^{-i\beta J_y}$, which performs rotation around the y-axis on angle β, into the operator R_β:

$$R_\beta = e^{-i\beta J_{y'}} = R_\alpha e^{-i\beta J_y} R_\alpha^{-1}. \tag{5.5}$$

A similar argument applies to the operator R_γ, but now the operator is a result of two consecutive unitary transformations. Tracing the route from right to left in Figure 5.2C reveals that R_γ is a unitary transform of the rotation operator $e^{-i\gamma J_{z'}}$ by means of a unitary operator R_β, so that:

$$R_\gamma = e^{-i\gamma J_{z''}} = R_\beta e^{-i\gamma J_{z'}} R_\beta^{-1}. \tag{5.6}$$

Operator $e^{-i\gamma J_{z'}}$ describes rotation around the z'-axis through angle γ in coordinate system $x'y'z'$. In turn, this operator is a unitary transform of the operator $e^{-i\gamma J_z}$, which describes rotation around the z-axis through angle γ, by means of a unitary operator R_α:

$$e^{-i\gamma J_{z'}} = R_\alpha e^{-i\gamma J_z} R_\alpha^{-1}. \tag{5.7}$$

Combining both unitary transformations into a single equation gives:

$$R_\gamma = R_\beta R_\alpha e^{-i\gamma J_z} R_\alpha^{-1} R_\beta^{-1}. \tag{5.8}$$

The expression for R_γ may now be substituted into the main equation, Equation 5.3. Taking into account that $U^{-1}U = 1$ simplifies the equation to:

$$R_\gamma R_\beta R_\alpha = R_\beta R_\alpha e^{-i\gamma J_z} R_\alpha^{-1} R_\beta^{-1} R_\beta R_\alpha = R_\beta R_\alpha e^{-i\gamma J_z}. \tag{5.9}$$

Substituting the earlier expression for operator R_β from Equation 5.5 into this equation leads to:

$$R_\gamma R_\beta R_\alpha = R_\alpha e^{-i\beta J_y} R_\alpha^{-1} R_\alpha e^{-i\gamma J_z}. \tag{5.10}$$

To give the final result:

$$R_\gamma R_\beta R_\alpha = e^{-i\alpha J_z} e^{-i\beta J_y} e^{-i\gamma J_z}, \tag{5.11}$$

which defines a 3D rotation around the original axes x, y, and z in terms of Euler's angles.

Unlike the initial Equation 5.4 for rotation R, which applies the consecutive 2D rotation to the transformed coordinate axes y' and z'', Equation 5.11 performs all three 2D rotations around the coordinate axes of the original unmodified coordinate system xyz. All that is necessary for the 2D rotations to be performed in the original coordinate system is to apply the rotations in an inverted order: γ, β, and α. With that, the first 2D rotation is performed around the z-axis though angle γ while keeping the original coordinate system xyz fixed in place. Then, the resulting coordinate system $x'y'z'$ is rotated around the original y-axis through angle β. Finally, the new coordinate system $x''y''z''$ is once again rotated around the original z-axis through the angle α.

This definition of the rotation operator R in terms of Euler's angles leads to a new form of the linear coefficients of the rotation transformation:

$$R\psi_{j,m} = \sum_{k=-j}^{j} \left\langle jk \,|e^{-i\alpha J_z} e^{-i\beta J_y} e^{-i\gamma J_z}|\, jm \right\rangle \psi_{j,k}. \tag{5.12}$$

These coefficients are known as the Wigner matrix elements, D^j_{km}:

$$D^j_{km} = \left\langle jk \,|e^{-i\alpha J_z} e^{-i\beta J_y} e^{-i\gamma J_z}|\, jm \right\rangle. \tag{5.13}$$

Using these matrix elements the rotation equation assumes the form:

$$R\psi_{j,m} = \sum_{k=-j}^{j} D^j_{km}\psi_{j,k}. \tag{5.14}$$

The Wigner matrix is a 3-dimensional table that depends on indices j, k, and m. It also can be viewed as a set of regular 2-dimensional matrices; one such matrix for each value of j. Each 2-dimensional matrix is of the size $(2j + 1)$-by-$(2j + 1)$ since there are $2j + 1$ values of m for the given value of j. By convention, the Wigner matrix elements are arranged in a table such that the index m goes through the list of values starting from its maximum positive value first: $m = j, j - 1, \ldots, -j$.

Representation of rotation by three Euler angles α, β, and γ provides a tremendous aid in finding the matrix elements D^j_{km}. The part of rotation around the z-axis is particularly simple due to $\psi_{j,m}$ being the eigenfunctions of operator J_z. The expression for Wigner matrix elements has two such rotations, which leads to a useful simplification. Letting $e^{-i\gamma J_z}$ operate on the right-side eigenfunction $|jm\rangle$, gives

$$e^{-i\gamma J_z}\left|jm\right\rangle = e^{-i\gamma m}\left|jm\right\rangle, \tag{5.15}$$

and allowing $e^{-i\alpha J_z}$ to operate on the left-side eigenfunction $\langle jk|$, so that

$$\left\langle jk\right| e^{-i\alpha J_z} = \psi^*_{j,k} e^{-i\alpha J_z} = \left(e^{i\alpha J_z}\psi_{j,k}\right)^* = \left(e^{i\alpha k}\psi_{j,k}\right)^* = \psi^*_{j,k} e^{-i\alpha k} = e^{-i\alpha k}\left\langle jk\right|, \tag{5.16}$$

leading to:

$$D^j_{km} = e^{-i\alpha k}\left\langle jk\right| e^{-i\beta J_y}\left|jm\right\rangle e^{-i\gamma m}. \tag{5.17}$$

Because $e^{-i\alpha k}$ is a constant it can be moved to the left.

Since functions $\psi_{j,m}$ are not eigenfunctions of the operator J_y, resolving the remaining part for $e^{-i\beta J_y}$ requires a considerable effort. For ease of reference those terms are denoted as matrix elements:

$$d^j_{km}(\beta) = \left\langle jk\right| e^{-i\beta J_y}\left|jm\right\rangle. \tag{5.18}$$

Therefore,

$$D^j_{km}(\alpha\beta\gamma) = e^{-i\alpha k} d^j_{k,m}(\beta)\, e^{-i\gamma m}. \tag{5.19}$$

Evidently,

$$D^j_{km}(0,\beta,0) = d^j_{k,m}(\beta). \tag{5.20}$$

As a useful observation, if a rotation $R(\phi)$ is to be performed on angle ϕ, then the inverse R^{-1} is equivalent to a rotation in opposite direction, $R(-\phi)$, since rotation in one direction and then in the reverse direction returns the system back into the initial position:

$$R(\phi)R^{-1}(\phi) = R^{-1}(\phi)R(\phi) = R(\phi)R(-\phi) = R(-\phi)R(\phi). \tag{5.21}$$

Similarly, if a rotation is defined using Euler's angles, $R(\alpha,\beta,\gamma) = e^{-i\alpha J_z}e^{-i\beta J_y}e^{-i\gamma J_z}$ then its inverse is:

$$R^{-1} = R(-\gamma,-\beta,-\alpha) = e^{i\gamma J_z}e^{i\beta J_y}e^{i\alpha J_z}, \tag{5.22}$$

which is a product of inverse operators in inverse order. Finding the inverse R^{-1} relies on the unitary property of the rotation operator. For unitary operators, the inverse of the operator is identical to its Hermitian adjoint, $R^{-1} = R^{+}$. In terms of matrix elements, this means that:

$$R^{-1}_{k,m} = \langle jk \,|\, R^{-1} \,|\, jm \rangle = (R_{k,m})^{+} = R^{*}_{m,k} = \langle jm \,|\, R \,|\, jk \rangle^{*}, \tag{5.23}$$

which involves transposing the matrix $\mathbf{R}$ and taking the complex conjugate of all its elements. In Wigner matrix element notation, the inverse rotation corresponds to:

$$D^j_{k,m}(-\gamma,-\beta,-\alpha) = e^{ik\gamma}\, d^j_{k,m}(-\beta)\, e^{im\alpha}. \tag{5.24}$$

Due to the symmetry property of the Wigner matrix explained in Section 7.1, and because the exponents can easily be moved from one place to another, this relation becomes:

$$D^j_{k,m}(-\gamma,-\beta,-\alpha) = e^{im\alpha}\, d^j_{m,k}(\beta)\, e^{ik\gamma} = D^{*j}_{m,k}(\alpha,\beta,\gamma). \tag{5.25}$$

A detailed analysis of the Wigner matrix continues in the sections that follow.

5.2 Wigner Matrix for $j = 1$

This section presents the derivation of the Wigner matrix elements following the steps outlined by Rose. The Wigner matrix $\mathbf{D}^j$ performs a rotational transformation of the eigenfunctions of angular momentum, which corresponds to a rotation of a coordinate system in which eigenfunctions play the role of basis functions. There are three indices present in the Wigner matrix element D^j_{km}: index j is the total angular momentum, that is, the sum of angular and spin quantum numbers $j = l + s$, while the indices k and m are azimuthal quantum numbers having values in the range $-j \le k \le j$ and $-j \le m \le j$. For a given value of index j, $\mathbf{D}^j$ is a 2-dimensional matrix of the form:

$$\begin{matrix} & \rightarrow \; m \; \rightarrow \\ \begin{matrix}\downarrow\\ k\\ \downarrow\end{matrix} & \begin{pmatrix} D^j_{j,j} & D^j_{j,j-1} & \cdots & D^j_{j,-j} \\ D^j_{j-1,j} & D^j_{j-1,j-1} & \cdots & D^j_{j-1,j-1} \\ \cdots & \cdots & \cdots & \cdots \\ D^j_{-j,j} & D^j_{-j,j-1} & \cdots & D^j_{-j,-j} \end{pmatrix} \end{matrix} \tag{5.26}$$

The azimuthal indices k and m begin from the value of j and continue to the value of $-j$; by convention, the index k runs in a column, and the index m varies in a row of the matrix.

Obtaining the Wigner matrix for a general value of index j is non-trivial, so it is useful to consider a few specific cases first. The easiest to derive are the Wigner matrix elements for the value of angular momentum $j = 1$. In this case, the basis set used to construct a matrix representation of the rotation operator consists of three eigenfunctions. These are three spherical harmonic functions, Y_{lm}:

$$\begin{aligned} Y_{1,1} &= -\frac{1}{2}\sqrt{\frac{3}{2\pi}}\, e^{i\phi} \sin\theta, \\ Y_{1,0} &= \frac{1}{2}\sqrt{\frac{3}{\pi}} \cos\theta, \\ Y_{1,-1} &= \frac{1}{2}\sqrt{\frac{3}{2\pi}}\, e^{-i\phi} \sin\theta. \end{aligned} \tag{5.27}$$

where $l = 1$ and $m = 1, 0, -1$ (Chapter 3 explains the derivation of these functions). The functions in Equation 5.27 may easily be mapped to the three Cartesian coordinates, x, y, z. The elements of the Wigner matrix for angular momentum $j = 1$ may be derived by directly rotating these eigenfunctions in Cartesian space.

The conversion of spherical harmonics functions in Equation 5.27 to Cartesian form can be accomplished with help of the relations $z = r\cos\theta$ and $x \pm iy = e^{\pm i\phi} \sin\theta$. This gives:

$$\begin{matrix} Y_{1,1} \\ Y_{1,0} \\ Y_{1,-1} \end{matrix} \;=\; \sqrt{\frac{3}{4\pi}}\frac{1}{r} \begin{cases} -\dfrac{1}{\sqrt{2}}(x+iy), \\ z, \\ \dfrac{1}{\sqrt{2}}(x-iy). \end{cases} \tag{5.28}$$

The coefficient $\sqrt{3/4\pi}\,(1/r)$ is separated from the other parts of the spherical harmonic functions because it is invariant under rotation. Only the parts that transform need to be included in the following analysis. Those parts are shown on the right of the curly bracket in Equation 5.28.

The strategy for finding the matrix elements D^1_{km} consists of formulating a rotational transformation of an arbitrary vector in Cartesian coordinates under a rotation of the coordinate system, then finding the proper rotation matrix, and finally converting that matrix into the basis of spherical harmonics. An arbitrary point has the radius vector $\mathbf{r}$ and $\mathbf{r}'$ in the Cartesian coordinate system before and after rotation of the coordinate system, respectively:

$$\mathbf{r} = \begin{pmatrix} x \\ y \\ z \end{pmatrix}, \quad \mathbf{r}' = \begin{pmatrix} x' \\ y' \\ z' \end{pmatrix}. \tag{5.29}$$

A rotation operation that translates the radius vector $\mathbf{r}$ into $\mathbf{r}'$ corresponds to the matrix equation:

$$\mathbf{r}' = \mathbf{R}\,\mathbf{r}, \tag{5.30}$$

where $\mathbf{R}$ is a 3×3 rotation matrix. Written in component form, this rotation transformation is:

$$r_i' = \sum_{k=1}^{3} R_{ik} r_k. \tag{5.31}$$

Note that matrix $\mathbf{R}$ is not yet a Wigner matrix since radius vectors are not eigenfunctions of angular momentum.

Previously, the analytic form of the rotation matrix was derived, and matrix **R** was shown to be a product of three rotation matrices:

$$\mathbf{R}(\alpha\beta\gamma) = \mathbf{R}(\gamma)\,\mathbf{R}(\beta)\,\mathbf{R}(\alpha), \tag{5.32}$$

where α, β, and γ are Euler's angles. In Equation 5.32, first the coordinate frame is rotated around the z-axis through angle α, the resulting coordinate frame is rotated through angle β around the original y-axis, and, finally, the frame is rotated around the original z-axis through angle γ. These are familiar 2D rotations. The individual matrices performing those transformations are:

$$\mathbf{R}(\alpha) = \begin{pmatrix} \cos\alpha & \sin\alpha & 0 \\ -\sin\alpha & \cos\alpha & 0 \\ 0 & 0 & 1 \end{pmatrix}, \quad \mathbf{R}(\beta) = \begin{pmatrix} \cos\beta & 0 & -\sin\beta \\ 0 & 1 & 0 \\ \sin\beta & 0 & \cos\beta \end{pmatrix},$$
$$\mathbf{R}(\gamma) = \begin{pmatrix} \cos\gamma & \sin\gamma & 0 \\ -\sin\gamma & \cos\gamma & 0 \\ 0 & 0 & 1 \end{pmatrix}. \tag{5.33}$$

Substitution of the individual rotation matrices from Equation 5.33 into Equation 5.32 produces the total rotation matrix. Note that the order of multiplication is important. The multiplication can be accomplished in two steps. In the first step, multiplication of $\mathbf{R}(\beta)$ and $\mathbf{R}(\alpha)$ gives the product:

$$\mathbf{R}(\beta)\,\mathbf{R}(\alpha) = \begin{pmatrix} \cos\alpha\cos\beta & \sin\alpha\cos\beta & -\sin\beta \\ -\sin\alpha & \cos\alpha & 0 \\ \cos\alpha\sin\beta & \sin\alpha\sin\beta & \cos\beta \end{pmatrix} \tag{5.34}$$

then, multiplying $\mathbf{R}(\gamma)$ to $\mathbf{R}(\beta)\mathbf{R}(\alpha)$ produces the total rotation matrix in Cartesian coordinates:

$$\mathbf{R}(\alpha\beta\gamma) = \begin{pmatrix} \cos\alpha\cos\beta\cos\gamma - \sin\alpha\sin\gamma & \sin\alpha\cos\beta\cos\gamma + \cos\alpha\sin\gamma & -\sin\beta\cos\gamma \\ -\cos\alpha\cos\beta\sin\gamma - \sin\alpha\cos\gamma & -\sin\alpha\cos\beta\sin\gamma + \cos\alpha\cos\gamma & \sin\beta\sin\gamma \\ \cos\alpha\sin\beta & \sin\alpha\sin\beta & \cos\beta \end{pmatrix}. \tag{5.35}$$

The rotation matrix **R** presented in Equation 5.35 is obtained in the basis of Cartesian base vectors $\hat{\mathbf{x}},\hat{\mathbf{y}},\hat{\mathbf{z}}$. Deriving the Wigner matrix requires converting the matrix **R** into the basis of spherical harmonic functions defined in Equation 5.28. For that purpose, it is necessary to establish the relationship between the coordinate of a point in Cartesian coordinate system and its counterpart in the basis of spherical harmonics. The position of an arbitrary point in the basis of spherical harmonic functions Y_{1m}, where index $m = 1, 0, -1$, is described by a column matrix:

$$\mathbf{r}_s = \begin{vmatrix} -\frac{1}{\sqrt{2}}(x+iy) \\ z \\ \frac{1}{\sqrt{2}}(x-iy) \end{vmatrix}, \quad \mathbf{r}_s' = \begin{vmatrix} -\frac{1}{\sqrt{2}}(x'+iy') \\ z' \\ \frac{1}{\sqrt{2}}(x'-iy') \end{vmatrix}, \tag{5.36}$$

where $\mathbf{r}_s$ and $\mathbf{r}_s'$ are the coordinate vectors of a point before and after rotation, respectively. The subscript s points to the underlying basis of spherical harmonics to distinguish these radius vectors from vector $\mathbf{r}$ and $\mathbf{r}'$, which are in the basis of Cartesian base vectors. Compare Equations 5.36 through 5.29 and Equation 5.28.

The transition from $\mathbf{r}$ to $\mathbf{r}_s$, and from $\mathbf{r}'$ to $\mathbf{r}_s'$ that corresponds to the replacement in the basis set should preserve the norm of the radius vectors, therefore the norm of the radius vectors before and after transformation must be the same. The choice of radius vectors in Equation 5.36 satisfies this requirement:

$$|\mathbf{r}_s| = \sqrt{\frac{1}{2}|x+iy|^2 + z^2 + \frac{1}{2}|x-iy|^2} = \sqrt{x^2+y^2+z^2} = |\mathbf{r}|. \tag{5.37}$$

As is customary with basis set transformations, the transformation from a Cartesian to a spherical harmonics basis is accomplished using a unitary matrix $\mathbf{U}$. For any radius vector in the old coordinate system, the matrix $\mathbf{U}$ establishes a one-to-one relationship with the radius vector in the new coordinate system. Therefore, matrix $\mathbf{U}$ transforms $\mathbf{r}$ into $\mathbf{r}_s$, and $\mathbf{r}'$ into $\mathbf{r}_s'$:

$$\mathbf{r}_s = \mathbf{U}\,\mathbf{r}, \quad \mathbf{r}_s' = \mathbf{U}\,\mathbf{r}'. \tag{5.38}$$

The backward translation from $\mathbf{r}_s$ to $\mathbf{r}$ can be obtained by multiplying both sides of the above equation with $\mathbf{U}^{-1}$:

$$\mathbf{U}^{-1}\,\mathbf{r}_s = \mathbf{U}^{-1}\,\mathbf{U}\,\mathbf{r}. \tag{5.39}$$

Therefore,

$$\mathbf{r} = \mathbf{U}^{-1}\,\mathbf{r}_s. \tag{5.40}$$

With that, the conversion from the coordinate $\mathbf{r}_s$ to the rotated $\mathbf{r}_s'$ is:

$$\mathbf{r}_s' = \mathbf{U}\,\mathbf{r}' = \mathbf{U}\,\mathbf{R}\,\mathbf{r} = \mathbf{U}\,\mathbf{R}\,\mathbf{U}^{-1}\mathbf{r}_s. \tag{5.41}$$

This leads to the formulation of rotation matrix $\mathbf{R}_s$ in the basis of spherical harmonics:

$$\mathbf{R}_s = \mathbf{U}\,\mathbf{R}\,\mathbf{U}^{-1}. \tag{5.42}$$

This equation is the standard unitary transformation of a matrix from one basis set to another.

The next step is to find the matrix $\mathbf{U}$ that transforms the radius vector $\mathbf{r}$ into $\mathbf{r}_s$. This can be accomplished in three steps, starting with a reasonable assumption, with each step being represented by a matrix, so that the total transformation is:

$$\mathbf{U} = \mathbf{c}\,\mathbf{b}\,\mathbf{a}. \tag{5.43}$$

The purpose of the first matrix, $\mathbf{a}$, is to change the y coordinate into iy in the radius vector $\mathbf{r}$ by incorporating a complex number i. No coordinate mixing is expected, therefore matrix $\mathbf{a}$ must be diagonal:

$$\mathbf{a} = \begin{pmatrix} 1 & 0 & 0 \\ 0 & i & 0 \\ 0 & 0 & 1 \end{pmatrix}. \tag{5.44}$$

Multiplication of matrix $\mathbf{a}$ to vector $\mathbf{r}$ produces the anticipated result:

$$\mathbf{a}\,\mathbf{r} = \begin{pmatrix} 1 & 0 & 0 \\ 0 & i & 0 \\ 0 & 0 & 1 \end{pmatrix}\begin{pmatrix} x \\ y \\ z \end{pmatrix} = \begin{pmatrix} x \\ iy \\ z \end{pmatrix}. \tag{5.45}$$

The purpose of the second matrix, $\mathbf{b}$, is to create the linear combinations $-(1/\sqrt{2})x-(1/\sqrt{2})iy$ and $(1/\sqrt{2})x-(1/\sqrt{2})iy$, present in Equation 5.36, out of x and iy components from Equation 5.45, and to preserve intact the component z. This involves placing the coefficient $(1/\sqrt{2})$ with the appropriate sign into the 2×2 block of matrix **b** in order to properly mix the components x and iy. Placing a unit value in the diagonal element in position z preserves that component unmodified. Placing zeros in the off-diagonal elements of the coordinate z prevents it from admixing into the components x and y. Implementing these steps gives matrix **b**:

$$\mathbf{b} = \begin{pmatrix} -\frac{1}{\sqrt{2}} & -\frac{1}{\sqrt{2}} & 0 \\ \frac{1}{\sqrt{2}} & -\frac{1}{\sqrt{2}} & 0 \\ 0 & 0 & 1 \end{pmatrix}. \tag{5.46}$$

Application of the transformation **b** to the result of the previous transformation **a r** leads to:

$$\mathbf{b\,a\,r} = \begin{pmatrix} -\frac{1}{\sqrt{2}} & -\frac{1}{\sqrt{2}} & 0 \\ \frac{1}{\sqrt{2}} & -\frac{1}{\sqrt{2}} & 0 \\ 0 & 0 & 1 \end{pmatrix} \begin{pmatrix} x \\ iy \\ z \end{pmatrix} = \begin{pmatrix} -\frac{1}{\sqrt{2}}(x+iy) \\ \frac{1}{\sqrt{2}}(x-iy) \\ z \end{pmatrix}. \tag{5.47}$$

The transform to interchange the second and third components in the column vector in Equation 5.47 requires another matrix, **c**:

$$\mathbf{c} = \begin{pmatrix} 1 & 0 & 0 \\ 0 & 0 & 1 \\ 0 & 1 & 0 \end{pmatrix}. \tag{5.48}$$

Application of matrix **c** to the matrix product **b a r** leads to:

$$\mathbf{U\,r} = \mathbf{c\,b\,a\,r} = \begin{pmatrix} 1 & 0 & 0 \\ 0 & 0 & 1 \\ 0 & 1 & 0 \end{pmatrix} \begin{pmatrix} -\frac{1}{\sqrt{2}}(x+iy) \\ \frac{1}{\sqrt{2}}(x+iy) \\ z \end{pmatrix} = \begin{pmatrix} -\frac{1}{\sqrt{2}}(x+iy) \\ z \\ \frac{1}{\sqrt{2}}(x+iy) \end{pmatrix} = \mathbf{r}_s, \tag{5.49}$$

which produces the desired column vector $\mathbf{r}_s$. All that remains is to assemble the matrix **U** from the various components. Multiplying matrix **b** with matrix **a** gives:

$$\mathbf{b\,a} = \begin{pmatrix} -\frac{1}{\sqrt{2}} & -\frac{1}{\sqrt{2}} & 0 \\ \frac{1}{\sqrt{2}} & -\frac{1}{\sqrt{2}} & 0 \\ 0 & 0 & 1 \end{pmatrix} \begin{pmatrix} 1 & 0 & 0 \\ 0 & i & 0 \\ 0 & 0 & 1 \end{pmatrix} = \begin{pmatrix} -\frac{1}{\sqrt{2}} & -\frac{i}{\sqrt{2}} & 0 \\ \frac{1}{\sqrt{2}} & -\frac{i}{\sqrt{2}} & 0 \\ 0 & 0 & 1 \end{pmatrix}. \tag{5.50}$$

Then left-multiplying the matrix in Equation 5.50 by matrix **c** leads to:

$$\mathbf{c\,b\,a}=\begin{pmatrix}1&0&0\\0&0&1\\0&1&0\end{pmatrix}\begin{pmatrix}-\frac{1}{\sqrt{2}}&-\frac{i}{\sqrt{2}}&0\\\frac{1}{\sqrt{2}}&-\frac{i}{\sqrt{2}}&0\\0&0&1\end{pmatrix}=\begin{pmatrix}-\frac{1}{\sqrt{2}}&-\frac{i}{\sqrt{2}}&0\\0&0&1\\\frac{1}{\sqrt{2}}&-\frac{i}{\sqrt{2}}&0\end{pmatrix}. \tag{5.51}$$

Finally, taking the common factor outside the matrix gives the matrix **U**:

$$\mathbf{U}=\frac{1}{\sqrt{2}}\begin{pmatrix}-1&-i&0\\0&0&\sqrt{2}\\1&-i&0\end{pmatrix}. \tag{5.52}$$

It is necessary to prove that this matrix is unitary. Taking the complex transpose of matrix **U** gives:

$$\mathbf{U}^{+}=\frac{1}{\sqrt{2}}\begin{pmatrix}1&0&1\\i&0&i\\0&\sqrt{2}&0\end{pmatrix}. \tag{5.53}$$

Multiplying matrix **U** with its complex transpose $\mathbf{U}^{+}$ leads to:

$$\mathbf{U}\,\mathbf{U}^{+}=\frac{1}{2}\begin{pmatrix}-1&-i&0\\0&0&\sqrt{2}\\1&-i&0\end{pmatrix}\begin{pmatrix}-1&0&1\\i&0&i\\0&\sqrt{2}&0\end{pmatrix}=\begin{pmatrix}1&0&0\\0&1&0\\0&0&1\end{pmatrix}, \tag{5.54}$$

which is a diagonal matrix, therefore matrix **U** is unitary. From this, it follows that $\mathbf{U}^{+}=\mathbf{U}^{-1}$.

The matrix $\mathbf{R}_s$ can now be obtained from computing the product $\mathbf{U}\mathbf{R}\mathbf{U}^{-1}$. The partial product $\mathbf{R}\mathbf{U}^{-1}$ is:

$$\frac{1}{\sqrt{2}}\begin{pmatrix}\cos\alpha\cos\beta\cos\gamma-\sin\alpha\sin\gamma & \sin\alpha\cos\beta\cos\gamma+\cos\alpha\sin\gamma & -\sin\beta\cos\gamma\\ -\cos\alpha\cos\beta\sin\gamma-\sin\alpha\cos\gamma & -\sin\alpha\cos\beta\sin\gamma+\cos\alpha\cos\gamma & \sin\beta\sin\gamma\\ \cos\alpha\sin\beta & \sin\alpha\sin\beta & \cos\beta\end{pmatrix}\begin{pmatrix}-1&0&1\\i&0&i\\0&\sqrt{2}&0\end{pmatrix}$$

$$=\frac{1}{\sqrt{2}}\begin{pmatrix}-\cos\alpha\cos\beta\cos\gamma+\sin\alpha\sin\gamma+i\sin\alpha\cos\beta\cos\gamma+i\cos\alpha\sin\gamma\\ -\sqrt{2}\sin\beta\cos\gamma\\ \cos\alpha\cos\beta\cos\gamma-\sin\alpha\sin\gamma+i\sin\alpha\cos\beta\cos\gamma+i\cos\alpha\sin\gamma\\ \cos\alpha\cos\beta\cos\gamma+\sin\alpha\sin\gamma-i\sin\alpha\cos\beta\cos\gamma+i\cos\alpha\sin\gamma\\ \sqrt{2}\sin\beta\sin\gamma\\ -\cos\alpha\cos\beta\sin\gamma-\sin\alpha\cos\gamma-i\sin\alpha\cos\beta\sin\gamma+i\cos\alpha\cos\gamma\\ -\cos\alpha\sin\beta+i\sin\alpha\sin\beta\\ \sqrt{2}\cos\beta\\ \cos\alpha\sin\beta+i\sin\alpha\sin\beta\end{pmatrix}. \tag{5.55}$$

The matrix in Equation 5.55 may be simplified using the Euler formula:

$$\cos\alpha \pm i\sin\alpha = e^{\pm i\alpha}. \tag{5.56}$$

A number of useful relations follow from this equation:

$$\begin{aligned} &-\cos\alpha + i\sin\alpha = -e^{-i\alpha},\\ &i\cos\alpha + \sin\alpha = i(\cos\alpha - i\sin\alpha) = ie^{-i\alpha},\\ &i\cos\alpha - \sin\alpha = i(\cos\alpha + i\sin\alpha) = ie^{i\alpha}. \end{aligned} \tag{5.57}$$

Application of Equation 5.57 to Equation 5.55 simplifies the product of matrices $\mathbf{R}\,\mathbf{U}^{-1}$ to:

$$\mathbf{R}\,\mathbf{U}^{-1} = \frac{1}{\sqrt{2}}\begin{pmatrix} -e^{-i\alpha}\cos\beta\cos\gamma + ie^{-i\alpha}\sin\gamma & -\sqrt{2}\sin\beta\cos\gamma & e^{i\alpha}\cos\beta\cos\gamma + ie^{i\alpha}\sin\gamma \\ e^{-i\alpha}\cos\beta\sin\gamma + ie^{-i\alpha}\cos\gamma & \sqrt{2}\sin\beta\sin\gamma & -e^{i\alpha}\cos\beta\sin\gamma + ie^{i\alpha}\cos\gamma \\ -e^{-i\alpha}\sin\beta & \sqrt{2}\cos\beta & e^{i\alpha}\sin\beta \end{pmatrix}. \tag{5.58}$$

Multiplication of Equation 5.58 by matrix $\mathbf{U}$ from the left leads to $\mathbf{U}\mathbf{R}\mathbf{U}^{-1}$:

$$\begin{aligned} \mathbf{U}\mathbf{M}\mathbf{U}^{-1} &= \frac{1}{2}\begin{pmatrix} -1 & -i & 0 \\ 0 & 0 & \sqrt{2} \\ 1 & -i & 0 \end{pmatrix}\begin{pmatrix} -e^{-i\alpha}\cos\beta\cos\gamma + ie^{-i\alpha}\sin\gamma & -\sqrt{2}\sin\beta\cos\gamma & e^{i\alpha}\cos\beta\cos\gamma + ie^{i\alpha}\sin\gamma \\ e^{-i\alpha}\cos\beta\sin\gamma + ie^{-i\alpha}\cos\gamma & \sqrt{2}\sin\beta\sin\gamma & -e^{i\alpha}\cos\beta\sin\gamma + ie^{i\alpha}\cos\gamma \\ -e^{-i\alpha}\sin\beta & \sqrt{2}\cos\beta & e^{i\alpha}\sin\beta \end{pmatrix} \\ &= \frac{1}{2}\begin{pmatrix} \begin{matrix} e^{-i\alpha}\cos\beta\cos\gamma - ie^{-i\alpha}\sin\gamma - ie^{-i\alpha}\cos\beta\sin\gamma + e^{-i\alpha}\cos\gamma \\ \sqrt{2}\sin\beta\cos\gamma - i\sqrt{2}\sin\beta\sin\gamma \\ -e^{i\alpha}\cos\beta\cos\gamma - ie^{i\alpha}\sin\gamma + ie^{i\alpha}\cos\beta\sin\gamma + e^{i\alpha}\cos\gamma \end{matrix} \\ \\ -\sqrt{2}e^{-i\alpha}\sin\beta \qquad 2\cos\beta \qquad \sqrt{2}e^{i\alpha}\sin\beta \\ \\ \begin{matrix} -e^{-i\alpha}\cos\beta\cos\gamma + ie^{-i\alpha}\sin\gamma - ie^{-i\alpha}\cos\beta\sin\gamma + e^{-i\alpha}\cos\gamma \\ -\sqrt{2}\sin\beta\cos\gamma - i\sqrt{2}\sin\beta\sin\gamma \\ e^{i\alpha}\cos\beta\cos\gamma + ie^{i\alpha}\sin\gamma + ie^{i\alpha}\cos\beta\sin\gamma + e^{i\alpha}\cos\gamma \end{matrix} \end{pmatrix}. \end{aligned} \tag{5.59}$$

Application of the Euler formula from Equation 5.57 simplifies this matrix to:

$$\frac{1}{2}\begin{pmatrix} e^{-i\alpha}\cos\beta e^{-i\gamma} + e^{-i\alpha}e^{-i\gamma} & \sqrt{2}\sin\beta e^{-i\gamma} & -e^{i\alpha}\cos\beta e^{-i\gamma} + e^{i\alpha}e^{-i\gamma} \\ -\sqrt{2}e^{-i\alpha}\sin\beta & 2\cos\beta & \sqrt{2}e^{i\alpha}\sin\beta \\ -e^{-i\alpha}\cos\beta e^{i\gamma} + e^{-i\alpha}e^{i\gamma} & -\sqrt{2}\sin\beta e^{i\gamma} & e^{i\alpha}\cos\beta e^{i\gamma} + e^{i\alpha}e^{i\gamma} \end{pmatrix}. \tag{5.60}$$

Rearranging and simplifying this matrix leads to the final form of the rotation matrix $\mathbf{R}_s$ in the basis of spherical harmonics:

$$\mathbf{R}_s = \mathbf{U}\mathbf{M}\mathbf{U}^{-1} = \begin{pmatrix} e^{-i\alpha}\dfrac{1+\cos\beta}{2}e^{-i\gamma} & \dfrac{\sin\beta}{\sqrt{2}}e^{-i\gamma} & e^{i\alpha}\dfrac{1-\cos\beta}{2}e^{-i\gamma} \\ -e^{-i\alpha}\dfrac{\sin\beta}{\sqrt{2}} & \cos\beta & e^{i\alpha}\dfrac{\sin\beta}{\sqrt{2}} \\ e^{-i\alpha}\dfrac{1-\cos\beta}{2}e^{i\gamma} & -\dfrac{\sin\beta}{\sqrt{2}}e^{i\gamma} & e^{i\alpha}\dfrac{1+\cos\beta}{2}e^{i\gamma} \end{pmatrix}. \tag{5.61}$$

The remaining difference between the matrix $\mathbf{R}_s$ and the Wigner matrix $\mathbf{D}$ involves the viewpoints of rotation that these matrices employ. Where matrix $\mathbf{R}_s$ returns the coordinates of the vector relative to the rotated coordinate system, matrix $\mathbf{D}$ describes a rotation of the basis functions themselves. $\mathbf{D}$, therefore, returns the coordinates of the rotated vector relative to the initial coordinate system. A consequence of that difference is that $\mathbf{D}$ is the transpose of $\mathbf{R}_s$. Application of the transpose to Equation 5.61 leads to:

$$D^1(\alpha\beta\gamma)=\begin{pmatrix} e^{-i\alpha}\frac{1+\cos\beta}{2}e^{-i\gamma} & -e^{-i\alpha}\frac{\sin\beta}{\sqrt{2}} & e^{-i\alpha}\frac{1-\cos\beta}{2}e^{i\gamma} \\ \frac{\sin\beta}{\sqrt{2}}e^{-i\gamma} & \cos\beta & -\frac{\sin\beta}{\sqrt{2}}e^{i\gamma} \\ e^{i\alpha}\frac{1-\cos\beta}{2}e^{-i\gamma} & e^{i\alpha}\frac{\sin\beta}{\sqrt{2}} & e^{i\alpha}\frac{1+\cos\beta}{2}e^{i\gamma} \end{pmatrix}. \tag{5.62}$$

Equation 5.62 presents Wigner matrix for the specific case of angular momentum $j = 1$. This equation rotates spherical harmonic functions of the degree $l = 1$ according to:

$$Y_{1m} = \sum_{k=-1}^{1} D^1_{km}(\alpha\beta\gamma)Y_{1k}. \tag{5.63}$$

The rotation transformation in Equation 5.63 employs the Euler parametric form and rotates the coordinate frame through the angles α, β, and γ. Two out of the three elementary rotations take place around the z-axis through the angles α and γ, and therefore have the simple exponential form $e^{-i\alpha k}$ and $e^{-i\gamma m}$ as a part of the Wigner matrix element $D^j_{km}(\alpha\beta\gamma) = e^{-i\alpha k}d^j_{km}(\beta)e^{-i\gamma m}$. With that, the remaining step in determining the rotation matrix $D^1(\alpha\beta\gamma)$ involves finding the matrix elements $d^1_{km}(\beta)$, representing the rotation through angle β around the y-axis. Setting $\alpha = 0$ and $\gamma = 0$ in Equation 5.62 gives the matrix $d^j(\beta)$ for the angular momentum $j = 1$:

$$\mathbf{d}^1(\beta)=\begin{pmatrix} \frac{1+\cos\beta}{2} & -\frac{\sin\beta}{\sqrt{2}} & \frac{1-\cos\beta}{2} \\ \frac{\sin\beta}{\sqrt{2}} & \cos\beta & -\frac{\sin\beta}{\sqrt{2}} \\ \frac{1-\cos\beta}{2} & \frac{\sin\beta}{\sqrt{2}} & \frac{1+\cos\beta}{2} \end{pmatrix}. \tag{5.64}$$

Although, in principle, this procedure could be applied to the derivation of the Wigner matrix elements for higher values of j, the complexity of representing and manipulating spherical harmonic functions in Cartesian coordinates increases rapidly and makes such an approach impractical. Therefore, an alternative procedure needs to be developed for working with Wigner matrix elements. The basic principles for this new procedure will now be introduced, and that requires obtaining the Wigner matrix having half-value in the orbital index.

5.3 Wigner Matrix for $j = 1/2$

Another Wigner matrix of particular value describes the special case of angular momentum $j = 1/2$ which is the equivalent of spin quantum number $s = 1/2$. Half-integer numbers of angular momentum belong to the class of functions called spinors, which are introduced by Cartan. They appear in math with the introduction of an *isotropic* vector $\mathbf{n} = (x, y, z)$, which has zero length, so that:

$$x^2 + y^2 + z^2 = 0, \tag{5.65}$$

where x, y, and z are complex numbers. This condition may be satisfied by rewriting the vector components as two complex numbers λ_1 and λ_2, such that:

$$x = \lambda_1^2 - \lambda_2^2, \quad y = i(\lambda_1^2 + \lambda_2^2), \quad z = -2\lambda_1\lambda_2. \tag{5.66}$$

With that, the sum of the squares of Cartesian components indeed becomes zero:

$$x^2 + y^2 + z^2 = (\lambda_1^2 - \lambda_2^2)^2 - (\lambda_1^2 + \lambda_2^2)^2 + 4\lambda_1^2\lambda_2^2 = 0. \tag{5.67}$$

Conversely, the complex numbers λ_1 and λ_2 can be defined in terms of the real components x, y, z through the relations:

$$x + iy = \lambda_1^2 - \lambda_2^2 - \lambda_1^2 - \lambda_2^2 = -2\lambda_2^2, \tag{5.68}$$

and

$$x - iy = \lambda_1^2 - \lambda_2^2 + \lambda_1^2 + \lambda_2^2 = 2\lambda_1^2, \tag{5.69}$$

so that

$$\lambda_1 = \pm\sqrt{\tfrac{1}{2}(x - iy)}, \quad \lambda_2 = \pm\sqrt{-\tfrac{1}{2}(x + iy)}. \tag{5.70}$$

The complex components make up a 2-dimensional vector, called a spinor:

$$\lambda = \begin{pmatrix} \lambda_1 \\ \lambda_2 \end{pmatrix}, \tag{5.71}$$

which is defined in a complex 2-dimensional space as:

$$\lambda = \lambda_1 \begin{pmatrix} 1 \\ 0 \end{pmatrix} + \lambda_2 \begin{pmatrix} 0 \\ 1 \end{pmatrix}. \tag{5.72}$$

At this point, the sign of the individual components λ_1 and λ_2 is still undefined. All that can be known on this topic is the relative sign, which follows from the relation $z = -2\lambda_1\lambda_2$ introduced in Equation 5.66. This relation also helps to establish the ratio λ_1/λ_2. Substituting $\lambda_1 = (-(1/2)z/\lambda_2)$ into the ratio λ_1/λ_2, and expressing λ_2 *via* the components of the isotropic vector defined in Equation 5.70 leads to:

$$\frac{\lambda_1}{\lambda_2} = \frac{-\frac{1}{2}z}{\lambda_2^2} = \frac{-\frac{1}{2}z}{-\frac{1}{2}(x + iy)} = \frac{z}{x + iy}. \tag{5.73}$$

Similarly, substituting $\lambda_2 = (-(1/2)z/\lambda_1)$ into λ_1/λ_2, and expressing λ_1 *via* the components of the isotropic vector gives:

$$\frac{\lambda_1}{\lambda_2} = \frac{\lambda_1^2}{-\frac{1}{2}z} = \frac{\frac{1}{2}(x - iy)}{-\frac{1}{2}z} = -\frac{x - iy}{z}. \tag{5.74}$$

Therefore,

$$\frac{\lambda_1}{\lambda_2} = -\frac{x - iy}{z} = \frac{z}{x + iy}. \tag{5.75}$$

Equation 5.75 establishes the ratio λ_1/λ_2 that will be used in the development of some very powerful mathematical tools for manipulating spinor matrices. In order to progress in that direction, Equation 5.75 can be re-written in the form of a system of two equations by applying cross-multiplication of the terms. This gives:

$$\begin{cases} z\lambda_1 + (x - iy)\lambda_2 = 0, \\ (x + iy)\lambda_1 - z\lambda_2 = 0. \end{cases} \tag{5.76}$$

Equation 5.76 may be re-cast in matrix form as:

$$\begin{pmatrix} z & x - iy \\ x + iy & -z \end{pmatrix} \begin{pmatrix} \lambda_1 \\ \lambda_2 \end{pmatrix} = 0. \tag{5.77}$$

Replacing all the Cartesian components with their corresponding spinor equivalents defined by Equation 5.66, gives an equivalent matrix equation:

$$\begin{pmatrix} -2\lambda_1\lambda_2 & 2\lambda_1^2 \\ -2\lambda_2^2 & 2\lambda_1\lambda_2 \end{pmatrix} \begin{pmatrix} \lambda_1 \\ \lambda_2 \end{pmatrix} = 0, \tag{5.78}$$

but now composed only of spinor components.

The 2×2 matrix in Equation 5.77 introduces the first significant milestone about spinors. Breaking down the matrix into a sum of three matrices leads to:

$$\begin{pmatrix} z & x - iy \\ x + iy & -z \end{pmatrix} = x\begin{pmatrix} 0 & 1 \\ 1 & 0 \end{pmatrix} + y\begin{pmatrix} 0 & -i \\ i & 0 \end{pmatrix} + z\begin{pmatrix} 1 & 0 \\ 0 & -1 \end{pmatrix}, \tag{5.79}$$

where

$$\sigma_x = \begin{pmatrix} 0 & 1 \\ 1 & 0 \end{pmatrix}, \quad \sigma_y = \begin{pmatrix} 0 & -i \\ i & 0 \end{pmatrix}, \quad \sigma_z = \begin{pmatrix} 1 & 0 \\ 0 & -1 \end{pmatrix} \tag{5.80}$$

are the Pauli matrices, which were originally introduced in quantum mechanics in the description of electron spin.

Using the Pauli matrices, the ratio λ_1/λ_2, which is expressed in Equation 5.77, may now be written in matrix form as

$$(\mathbf{n} \cdot \sigma)\,\lambda = 0, \tag{5.81}$$

where $\mathbf{n} \cdot \boldsymbol{\sigma}$ denotes $n_x\sigma_x + n_y\sigma_y + n_z\sigma_z$, and $\mathbf{n}$ is the vector associated with a spinor λ. The matrix

$$\mathbf{n} \cdot \sigma = \begin{pmatrix} z & x - iy \\ x + iy & -z \end{pmatrix} \tag{5.82}$$

provides a general technique for associating any arbitrary vector $\mathbf{n}$ with a complex 2×2 matrix.

The theory of spinors just developed can be applied to the mathematical description of atomic particles that have spin $\mathbf{S} = 1/2$. Equation 5.82 leads to the definition of the spin operator $\mathbf{S}$ for angular momentum $j = 1/2$:

$$\mathbf{S} = \frac{1}{2}\begin{pmatrix} \hat{\mathbf{z}} & \hat{\mathbf{x}} - i\hat{\mathbf{y}} \\ \hat{\mathbf{x}} + i\hat{\mathbf{y}} & -\hat{\mathbf{z}} \end{pmatrix}, \tag{5.83}$$

$$\lambda_{-\frac{1}{2}} = d^{\frac{1}{2}}_{\frac{1}{2},-\frac{1}{2}}\psi_{\frac{1}{2}} + d^{\frac{1}{2}}_{-\frac{1}{2},-\frac{1}{2}}\psi_{-\frac{1}{2}} = \begin{pmatrix} -\sin\theta \\ 0 \end{pmatrix} + \begin{pmatrix} 0 \\ \cos\theta \end{pmatrix} = \begin{pmatrix} -\sin\theta \\ \cos\theta \end{pmatrix}. \tag{5.112}$$

The rotated functions λ_m are eigenfunctions of the spin operator $\mathbf{S}_t$. Therefore, the relation $\mathbf{S}_t \lambda_m = m\,\lambda_m$ must hold:

$$\lambda_{\frac{1}{2}}: \quad \frac{1}{2}\begin{pmatrix} \cos\beta & \sin\beta \\ \sin\beta & -\cos\beta \end{pmatrix}\begin{pmatrix} \cos\theta \\ \sin\theta \end{pmatrix} = \frac{1}{2}\begin{pmatrix} \cos\theta \\ \sin\theta \end{pmatrix}, \tag{5.113}$$

$$\lambda_{-\frac{1}{2}}: \quad \frac{1}{2}\begin{pmatrix} \cos\beta & \sin\beta \\ \sin\beta & -\cos\beta \end{pmatrix}\begin{pmatrix} -\sin\theta \\ \cos\theta \end{pmatrix} = -\frac{1}{2}\begin{pmatrix} -\sin\theta \\ \cos\theta \end{pmatrix}. \tag{5.114}$$

Solving either of these equations leads to the same result, and resolves the parameter θ in terms of the rotation angle β. The matrix equation for $\lambda_{1/2}$ corresponds to the following system of equations:

$$\begin{cases} \cos\beta\cos\theta + \sin\beta\sin\theta = \cos\theta, \\ \sin\beta\cos\theta - \cos\beta\sin\theta = \sin\theta. \end{cases} \tag{5.115}$$

Moving the terms with $\cos\beta$ from the left to the right side leads to:

$$\begin{cases} \sin\beta\sin\theta = (1-\cos\beta)\cos\theta, \\ \sin\beta\cos\theta = (1+\cos\beta)\sin\theta. \end{cases} \tag{5.116}$$

Separating the terms that include angle θ in the left side and gathering the terms that include angle β in the right side of these equations leads to:

$$\tan\theta = \frac{1-\cos\beta}{\sin\beta} = \tan\left(\frac{\beta}{2}\right), \qquad \tan\theta = \frac{\sin\beta}{1+\cos\beta} = \tan\left(\frac{\beta}{2}\right). \tag{5.117}$$

Appendix A.3 provides the derivation for the tangent of a half angle. Equating the angles on the opposite sides of the above equations resolves the parameter θ:

$$\theta = \frac{\beta}{2}, \tag{5.118}$$

which, in turn, resolves the Wigner matrix for the angular momentum $j = 1/2$:

$$d^{\frac{1}{2}}(\beta) = \begin{pmatrix} \cos\frac{\beta}{2} & -\sin\frac{\beta}{2} \\ \sin\frac{\beta}{2} & \cos\frac{\beta}{2} \end{pmatrix}. \tag{5.119}$$

Having the parameter θ resolved also determines the functions λ_m. These functions,

$$\lambda_{\frac{1}{2}} = \begin{pmatrix} \cos\frac{\beta}{2} \\ \sin\frac{\beta}{2} \end{pmatrix} \quad \text{and} \quad \lambda_{-\frac{1}{2}} = \begin{pmatrix} -\sin\frac{\beta}{2} \\ \cos\frac{\beta}{2} \end{pmatrix}, \tag{5.120}$$

as previously noted, are the eigenfunctions of the spin operator along an arbitrary vector **t**:

$$\mathbf{S}_t = \frac{1}{2}\begin{pmatrix} \cos\beta & \sin\beta \\ \sin\beta & -\cos\beta \end{pmatrix}. \tag{5.121}$$

The eigenfunctions $\lambda_{1/2}$ and $\lambda_{-1/2}$ are spinors. The dependence on half angle $\beta/2$ is the characteristic feature of both spinors and their rotation matrices $d^{1/2}(\beta)$. Setting the angle β to 4π in Equation 5.119 produces the identity matrix:

$$d^{\frac{1}{2}}(4\pi) = \begin{pmatrix} 1 & 0 \\ 0 & 1 \end{pmatrix}. \tag{5.122}$$

Therefore, unlike the usual vectors, spinors require the rotation angle of 4π in order to return back into themselves. Spinor functions appear with the half-values of the angular momentum j. Spherical harmonic functions that correspond to the integral values of the angular momentum j rotate into themselves through the usual angle of 2π.

Having the matrix $d^{1/2}(\beta)$ defined in Equation 5.119 resolves the complete Wigner matrix for the angular momentum $j = 1/2$ through manually adding the dependence on other two angles α and γ:

$$D^{\frac{1}{2}}_{km}(\alpha\beta\gamma) = e^{-ik\alpha} d^{\frac{1}{2}}_{km}(\beta) e^{-im\gamma}. \tag{5.123}$$

Due to the periodicity of $2\pi k$ in the angles α and γ, where k is an integer number, the exponential terms $e^{-ik\alpha}$ and $e^{-im\gamma}$ automatically accommodate the periodicity of 4π that comes with the angle β, so that the periodicity of $D^{1/2}_{km}(\alpha\beta\gamma)$ is also 4π.

The Wigner matrix for half-values of the angular momentum j cannot be directly used in the Fast Multipole Method, because only the integer values of the angular momentum are associated with spherical harmonics. However, the matrix $d^{1/2}(\beta)$ will be needed in the derivation of other important intermediate relations. Also, derivation of the Wigner matrix for $j = 1/2$ illuminates many important qualities of the rotation matrix, and that aspect of the material makes this section essential for building the larger picture of the Fast Multipole Method.

5.4 General Form of the Wigner Matrix Elements

The Wigner matrix provides a general mechanism for rotation of spherical harmonics. Therefore, in addition to a few previously derived specific cases it is useful to derive a general form of this matrix that would apply to all spherical harmonics. The derivation process also sheds additional light on the meaning of the Wigner matrix. In its general form, the Wigner matrix D^j is a 2-dimensional $(2j + 1) \times (2j + 1)$ table that describes a rotation of the eigenfunction of the angular momentum j. A remarkable aspect of the rotation transformation is that, although it is described as occurring in real 3D space, the result of the rotation is mathematically a linear combination of the virtual degrees of freedom associated with index m, where $m = j, j - 1, \ldots, -j$. The internal degrees of freedom in the index m may be loosely called m-dimensional spinors since mathematical manipulation of these degrees of freedom is similar to those for the 2-dimensional spinors for angular momentum $j = 1/2$.

An outline of the derivation strategy proposed by Biedenharn and Louck follows. Spinors provide a convenient mathematical technique to map a point $\mathbf{r} = (x, y, z)$ in 3D space to a complex 2-dimensional matrix $\mathbf{T}$ defined in spinor space. Spinors may be viewed as entities in a $2j + 1$ component space, so matrix $\mathbf{T}$ has the size $(2j + 1) \times (2j + 1)$. Section 5.3 showed an example of the transformation of a 2-component spinor to a 2×2 matrix for $j = 1/2$. The matrix $\mathbf{T}$ can be rotated using a unitary operator $\mathbf{U}$, so that the result of transformation becomes $\mathbf{T}' = \mathbf{U}\mathbf{T}\mathbf{U}^{-1}$, where $\mathbf{U}^{-1}$ is a matrix inverse of $\mathbf{U}$. The matrix $\mathbf{T}'$ may then be transformed back into a vector $\mathbf{r}' = (x', y', z')$ by applying the spinor mapping technique in the backward order. This gives two vectors $\mathbf{r}$ and its rotated counterpart $\mathbf{r}'$ that can be analyzed further. By definition, any vector $\mathbf{r}$ in 3D space can be transformed into another vector $\mathbf{r}'$ using a 3×3 matrix $\mathbf{R}$, so that $\mathbf{r}' = \mathbf{R}\mathbf{r}$. The spinor mapping technique postulates that rotation of a spinor matrix $\mathbf{T}$ in $2j + 1$ component space by the unitary transformation $\mathbf{U}$ corresponds to a vector rotation in 3D space by the matrix $\mathbf{R}$. However, before accepting this claim at its face value, a proof that matrix $\mathbf{R}$ is indeed a rotation matrix, and not some other sort of a linear transformation, is needed. This proof will now be given.

According to the technique illustrated in Section 5.3, any vector $\mathbf{r} = (x, y, z)$ defined in 3D space can be mapped to a complex 2×2 matrix $\mathbf{T}$, so that

$$\mathbf{T} = T_0\mathbf{I} + x\sigma_x + y\sigma_y + z\sigma_z = \begin{pmatrix} T_0 + z & x - iy \\ x + iy & T_0 - z \end{pmatrix} = \begin{pmatrix} T_{11} & T_{12} \\ T_{21} & T_{22} \end{pmatrix}, \tag{5.124}$$

where σ are Pauli spin matrices, and $\mathbf{I}$ is a diagonal unit matrix. The addition of constant T_0 makes the matrix more general. The mapping rule in Equation 5.124 establishes a one-to-one correspondence between the elements of matrix $\mathbf{T}$ and the vector coordinates x, y, z:

$$\begin{aligned} x &= \tfrac{1}{2}(T_{21} + T_{12}), \\ y &= \tfrac{1}{2i}(T_{21} - T_{12}), \\ z &= \tfrac{1}{2}(T_{11} - T_{22}), \\ T_0 &= \tfrac{1}{2}(T_{11} + T_{22}). \end{aligned} \tag{5.125}$$

Next, a unitary matrix $\mathbf{U}$ that rotates matrix $\mathbf{T}$ must be defined.

The unitary matrix $\mathbf{U}$ transforms a 2-component spinor λ into λ' according to matrix equation $\lambda' = \mathbf{U}\lambda$:

$$\begin{pmatrix} \lambda_1' \\ \lambda_2' \end{pmatrix} = \begin{pmatrix} u_{11} & u_{12} \\ u_{21} & u_{22} \end{pmatrix} \begin{pmatrix} \lambda_1 \\ \lambda_2 \end{pmatrix}. \tag{5.126}$$

In its most general form, this unitary matrix, which has previously been derived, is

$$\mathbf{U} = \begin{pmatrix} a & b \\ -b^* & a^* \end{pmatrix}, \tag{5.127}$$

where

$$a = e^{i\xi}\cos\theta, \quad b = e^{i\eta}\sin\theta. \tag{5.128}$$

This definition guarantees the positive unit value of the determinant, so that the unitary matrix $\mathbf{U}$ corresponds to rotation without inversion. The inverse of this matrix is

$$\mathbf{U}^{-1} = \mathbf{U}^{+} = \begin{pmatrix} a^* & -b \\ b^* & a \end{pmatrix}. \tag{5.129}$$

Matrix $\mathbf{T}$ undergoes a unitary transformation following the standard matrix multiplication rule $\mathbf{T}' = \mathbf{U}\mathbf{T}\mathbf{U}^{-1}$. Once again, the rotated 2×2 matrix $\mathbf{T}'$ can be analyzed from the point of view of the spinor mapping rule and decomposed into its vector components. With that,

$$\mathbf{T}' = \begin{pmatrix} T_{11}' & T_{12}' \\ T_{21}' & T_{22}' \end{pmatrix} = \begin{pmatrix} T_0' + z' & x' - iy' \\ x' + iy' & T_0' - z' \end{pmatrix} = T_0'\mathbf{I} + x'\sigma_x + y'\sigma_y + z'\sigma_z. \tag{5.130}$$

Since a unitary transformation does not change the trace of the matrix, i.e., the sum of the diagonal elements $Tr(\mathbf{T}) = 2T_0$, it requires that $T_0' = T_0$. In addition to that, a unitary transformation does not change the value of the determinant, that is, $\det(\mathbf{T}) = \det(\mathbf{T}')$:

$$\det(\mathbf{T}) = T_0^2 - x^2 - y^2 - z^2, \quad \det(\mathbf{T}') = T_0'^2 - x'^2 - y'^2 - z'^2. \tag{5.131}$$

Therefore,

$$x^2 + y^2 + z^2 = x'^2 + y'^2 + z'^2. \tag{5.132}$$

Equation 5.132 implies that the vector components x, y, z and x', y', z' are linearly related *via* the matrix equation $\mathbf{r}' = \mathbf{R}\mathbf{r}$, and that $\mathbf{R}$ is a transformation matrix. Whether $\mathbf{R}$ is a rotation transformation or something else still has to be worked out.

In order to determine the elements of matrix $\mathbf{R}$ it is necessary to explicitly obtain the matrix $\mathbf{T}'$ from the matrix equation $\mathbf{T}' = \mathbf{U}\mathbf{T}\mathbf{U}^{-1}$:

$$\mathbf{T}' = \begin{pmatrix} a & b \\ -b^* & a^* \end{pmatrix} \begin{pmatrix} T_0 + z & x - iy \\ x + iy & T_0 - z \end{pmatrix} \begin{pmatrix} a^* & -b \\ b^* & a \end{pmatrix}. \tag{5.133}$$

Multiplication of the two matrices in the left leads to:

$$\mathbf{T}' = \begin{pmatrix} aT_0 + az + bx + iby & ax - iay + bT_0 - bz \\ -b^*T_0 - b^*z + a^*x + ia^*y & -b^*x + ib^*y + a^*T_0 - a^*z \end{pmatrix} \begin{pmatrix} a^* & -b \\ b^* & a \end{pmatrix}. \tag{5.134}$$

The second multiplication gives:

$$\mathbf{T}' = \begin{pmatrix} aa^*T_0 + aa^*z + a^*bx + ia^*by + ab^*x - iab^*y + bb^*T_0 - bb^*z \\ \qquad - abT_0 - abz - bbx - ibby + aax - iaay + abT_0 - abz \\ -a^*b^*T_0 - a^*b^*z + a^*a^*x + ia^*a^*y - b^*b^*x + ib^*b^*y + a^*b^*T_0 - a^*b^*z \\ \qquad bb^*T_0 + bb^*z - a^*bx - ia^*by - ab^*x + iab^*y + aa^*T_0 - aa^*z \end{pmatrix}. \tag{5.135}$$

All the terms T_0 in the off-diagonal elements cancel out, and that leads to the final form of the matrix:

$$\mathbf{T}' = \begin{pmatrix} aa^*T_0 + aa^*z + a^*bx + ia^*by + ab^*x - iab^*y + bb^*T_0 - bb^*z \\ \qquad - abz - bbx - ibby + aax - iaay - abz \\ -a^*b^*z + a^*a^*x + ia^*a^*y - b^*b^*x + ib^*b^*y - a^*b^*z \\ \qquad bb^*T_0 + bb^*z - a^*bx - ia^*by - ab^*x + iab^*y + aa^*T_0 - aa^*z \end{pmatrix}. \tag{5.136}$$

This matrix can now be converted to vector components. Applying the matrix-to-vector transformation leads to the components of transformed vector:

$$\begin{aligned} x' &= \tfrac{1}{2}(T_{21}' + T_{12}) = \tfrac{1}{2}\left[(aa + a^*a^* - bb - b^*b^*)x + i(a^*a^* - aa - bb + b^*b^*)y - 2(ab + a^*b^*)z\right], \\ y' &= \tfrac{1}{2i}(T_{21}' - T_{12}) = \tfrac{1}{2i}\left[(a^*a^* - aa + bb - b^*b^*)x + i(a^*a^* + aa + bb + b^*b^*)y + 2(ab - a^*b^*)z\right], \\ z' &= \tfrac{1}{2i}(T_{11}' - T_{22}) = \tfrac{1}{2}\left[2(a^*b + ab^*)x + 2i(a^*b - ab^*)y + 2(aa^* - b^*b^*)z\right]. \end{aligned} \tag{5.137}$$

The transformed vector components x', y', z' are related to the initial ones x, y, z by the matrix equation:

$$\begin{pmatrix} x' \\ y' \\ z' \end{pmatrix} = \begin{pmatrix} R_{xx} & R_{xy} & R_{xz} \\ R_{yx} & R_{yy} & R_{yz} \\ R_{zx} & R_{zy} & R_{zz} \end{pmatrix} \begin{pmatrix} x \\ y \\ z \end{pmatrix}. \tag{5.138}$$

Taking the elements a and b of the unitary matrix from Equation 5.128, and taking into account that

$$\begin{aligned} x' &= R_{xx}x + R_{xy}y + R_{xz}z, \\ y' &= R_{yx}x + R_{yy}y + R_{yz}z, \\ z' &= R_{zx}x + R_{zy}y + R_{zz}z. \end{aligned} \tag{5.139}$$

provides everything that is needed in order to determine the elements of matrix $\mathbf{R}$. In order, these are

$\underline{R_{xx}}$:
Employing relations

$$e^{i2\xi} + e^{-i2\xi} = 2\cos(2\xi), \quad e^{i2\eta} + e^{-i2\eta} = 2\cos(2\eta) \tag{5.140}$$

leads to:

$$\begin{aligned} R_{xx} &= \tfrac{1}{2}[aa + a^*a^* - bb - b^*b^*] = \tfrac{1}{2}[e^{i2\xi}\cos^2\theta + e^{-i2\xi}\cos^2\theta - e^{i2\eta}\sin^2\theta - e^{-i2\eta}\sin^2\theta] \\ &= \cos(2\xi)\cos^2\theta - \cos(2\eta)\sin^2\theta. \end{aligned} \tag{5.141}$$

$\underline{R_{xy}}$:
Using

$$-e^{i2\xi} + e^{-i2\xi} = -2i\sin(2\xi), \quad \text{and} \quad -e^{i2\eta} + e^{-i2\eta} = -2i\sin(2\eta) \tag{5.142}$$

$$\begin{aligned} R_{xy} &= \tfrac{i}{2}[-aa + a^*a^* - bb + b^*b^*] = \tfrac{i}{2}[-e^{i2\xi}\cos^2\theta + e^{-i2\xi}\cos^2\theta - e^{i2\eta}\sin^2\theta + e^{-i2\eta}\sin^2\theta] \\ &= \sin(2\xi)\cos^2\theta + \sin(2\eta)\sin^2\theta. \end{aligned} \tag{5.143}$$

$\underline{R_{xz}}$:
Using

$$e^{i(\xi+\eta)} + e^{-i(\xi+\eta)} = 2\cos(\xi+\eta), \quad \text{and} \quad 2\cos\theta\sin\theta = \sin(2\theta) \tag{5.144}$$

$$R_{xz} = -[ab + a^*b^*] = -\left[e^{i(\xi+\eta)}\cos\theta\sin\theta + e^{-i(\xi+\eta)}\cos\theta\sin\theta\right] = -\cos(\xi+\eta)\sin(2\theta). \tag{5.145}$$

$\underline{R_{yx}}$:
Using

$$-e^{i2\xi} + e^{-i2\xi} = -2i\sin(2\xi), \quad \text{and} \quad e^{i2\eta} - e^{-i2\eta} = 2i\sin(2\eta) \tag{5.146}$$

$$\begin{aligned} R_{yx} &= \tfrac{1}{2i}[-aa + a^*a^* + bb - b^*b^*] = \tfrac{1}{2i}[-e^{i2\xi}\cos^2\theta + e^{-i2\xi}\cos^2\theta + e^{i2\eta}\sin^2\theta - e^{-i2\eta}\sin^2\theta] \\ &= -\sin(2\xi)\cos^2\theta + \sin(2\eta)\sin^2\theta. \end{aligned} \tag{5.147}$$

R_{yy}:

In R_{yy}, the exponents sum to:

$$e^{i2\xi} + e^{-i2\xi} = 2\cos(2\xi), \quad \text{and} \quad e^{i2\eta} + e^{-i2\eta} = 2\cos(2\eta). \tag{5.148}$$

Therefore,

$$\begin{aligned} R_{yy} &= \tfrac{1}{2}[aa + a^*a^* + bb + b^*b^*] = \tfrac{1}{2}[e^{i2\xi}\cos^2\theta + e^{-i2\xi}\cos^2\theta + e^{i2\eta}\sin^2\theta + e^{-i2\eta}\sin^2\theta] \\ &= \cos(2\xi)\cos^2\theta + \cos(2\eta)\sin^2\theta. \end{aligned} \tag{5.149}$$

R_{yz}:

The relations

$$e^{i(\xi+\eta)} - e^{-i(\xi+\eta)} = 2i\sin(\xi+\eta), \quad \text{and} \quad 2\cos\theta\sin\theta = \sin(2\theta) \tag{5.150}$$

give:

$$R_{yz} = \tfrac{1}{i}[ab - a^*b^*] = \tfrac{1}{i}[e^{i(\xi+\eta)}\cos\theta\sin\theta - e^{-i(\xi+\eta)}\cos\theta\sin\theta] = \sin(\xi+\eta)\sin(2\theta) \tag{5.151}$$

R_{zx}:

The relations

$$e^{i(\xi-\eta)} + e^{-i(\xi-\eta)} = 2\cos(\xi-\eta), \quad \text{and} \quad 2\cos\theta\sin\theta = \sin(2\theta) \tag{5.152}$$

lead to:

$$R_{zx} = ab^* + a^*b = e^{i(\xi-\eta)}\cos\theta\sin\theta + e^{-i(\xi-\eta)}\cos\theta\sin\theta = \cos(\xi-\eta)\sin(2\theta). \tag{5.153}$$

R_{zy}:

The relations

$$e^{-i(\xi-\eta)} - e^{i(\xi-\eta)} = -2i\sin(\xi-\eta), \quad \text{and} \quad 2\cos\theta\sin\theta = \sin(2\theta), \tag{5.154}$$

leads to:

$$R_{zy} = i[a^*b - ab^*] = i\left[e^{-i(\xi-\eta)}\cos\theta\sin\theta - e^{i(\xi-\eta)}\cos\theta\sin\theta\right] = \sin(\xi-\eta)\sin(2\theta). \tag{5.155}$$

R_{zz}:

There are no exponents, therefore

$$R_{zz} = aa^* - bb^* = \cos^2\theta - \sin^2\theta = \cos(2\theta). \tag{5.156}$$

Assembling all the elements in table form gives the matrix **R**:

$$\mathbf{R} = \begin{pmatrix} \cos(2\xi)\cos^2\theta - \cos(2\eta)\sin^2\theta & \sin(2\xi)\cos^2\theta + \sin(2\eta)\sin^2\theta & -\cos(\xi+\eta)\sin(2\theta) \\ -\sin(2\xi)\cos^2\theta + \sin(2\eta)\sin^2\theta & \cos(2\xi)\cos^2\theta + \cos(2\eta)\sin^2\theta & \sin(\xi+\eta)\sin(2\theta) \\ \cos(\xi-\eta)\sin(2\theta) & \sin(\xi-\eta)\sin(2\theta) & \cos(2\theta) \end{pmatrix}. \tag{5.157}$$

This procedure for calculation of matrix elements **R** provides an opportunity to investigate the type of transformation this matrix actually represents. A previous analysis of matrices **T** and **T′** revealed that matrix **R** has a unit determinant. However, the sign of the determinant remained undetermined. This issue can now be definitely resolved. Application of an identity transformation to matrix **T** leads to the equality $\mathbf{T} = \mathbf{T}'$. The identity transformation

$$\mathbf{U} = \begin{pmatrix} 1 & 0 \\ 0 & 1 \end{pmatrix}$$

corresponds to the choice of parameters $a = 1$ and $b = 0$. Using these parameters in the calculation of **R** produces a diagonal unit matrix:

$$\mathbf{R} = \begin{pmatrix} 1 & 0 & 0 \\ 0 & 1 & 0 \\ 0 & 0 & 1 \end{pmatrix}. \tag{5.158}$$

The determinant of this matrix is positive unity, supporting the supposition that matrix **R** represents a rotation. The determinant of a rotation matrix is a smooth function of parameters a and b, therefore, it remains the same across the entire parameter space. This rules out the possibility that the determinant could suddenly acquire a negative sign.

The final proof that **R** is a rotation matrix follows from converting it to the form that can be recognized as a genuine rotation matrix in the parametric form of Euler angles. This can be illustrated in two different ways. Choosing the parameter values $\theta = 0$ and $b = 0$, and repeating the procedure for calculation of matrix elements **R** produces the following result:

$$R_{xx} = \tfrac{1}{2}[aa + a^*a^* - bb - b^*b^*] = \tfrac{1}{2}[e^{i2\xi} + e^{-i2\xi}] = \cos(2\xi), \tag{5.159}$$

$$R_{xy} = \tfrac{i}{2}[-aa + a^*a^* - bb + b^*b^*] = \tfrac{i}{2}[-e^{i2\xi} + e^{-i2\xi}] = \sin(2\xi), \tag{5.160}$$

$$R_{xz} = -[ab + a^*b^*] = 0, \tag{5.161}$$

$$R_{yx} = \tfrac{1}{2i}[-aa + a^*a^* + bb - b^*b^*] = \tfrac{1}{2i}[-e^{i2\xi} + e^{-i2\xi}] = -\sin(2\xi), \tag{5.162}$$

$$R_{yy} = \tfrac{1}{2}[aa + a^*a^* + bb + b^*b^*] = \tfrac{1}{2}[e^{i2\xi} + e^{-i2\xi}] = \cos(2\xi), \tag{5.163}$$

$$R_{yz} = \tfrac{1}{i}[ab - a^*b^*] = 0, \tag{5.164}$$

$$R_{zx} = ab^* + a^*b = 0, \tag{5.165}$$

$$R_{zy} = i[a^*b - ab^*] = 0, \tag{5.166}$$

$$R_{zz} = aa^* - bb^* = \cos^2\theta = 1. \tag{5.167}$$

Gathering the matrix elements together leads to a familiar matrix:

$$\begin{pmatrix} \cos(2\xi) & \sin(2\xi) & 0 \\ -\sin(2\xi) & \cos(2\xi) & 0 \\ 0 & 0 & 1 \end{pmatrix}, \tag{5.168}$$

which represents rotation around the z-axis through angle 2ξ.

In a second example, setting $\xi = \eta = 0$ produces:

$$R_{xx} = \tfrac{1}{2}[aa + a^*a^* - bb - b^*b^*] = \tfrac{1}{2}[2\cos^2\theta - 2\sin^2\theta] = \cos(2\theta), \tag{5.169}$$

$$R_{xy} = \tfrac{i}{2}[-aa + a^*a^* - bb + b^*b^*] = 0, \tag{5.170}$$

$$R_{xz} = -[ab + a^*b^*] = -2\cos\theta\sin\theta = -\sin(2\theta), \tag{5.171}$$

$$R_{yx} = \tfrac{1}{2i}[-aa + a^*a^* + bb - b^*b^*] = 0, \tag{5.172}$$

$$R_{yy} = \tfrac{1}{2}[aa + a^*a^* + bb + b^*b^*] = \tfrac{1}{2}[2\cos^2\theta + 2\sin^2\theta] = 1, \tag{5.173}$$

$$R_{yz} = \tfrac{1}{i}[ab - a^*b^*] = 0, \tag{5.174}$$

$$R_{zx} = ab^* + a^*b = 2\cos\theta\sin\theta = \sin(2\theta), \tag{5.175}$$

$$R_{zy} = i[a^*b - ab^*] = 0, \tag{5.176}$$

$$R_{zz} = aa^* - bb^* = \cos^2\theta - \sin^2\theta = \cos(2\theta). \tag{5.177}$$

These matrix elements give rise to:

$$\begin{pmatrix} \cos(2\theta) & 0 & -\sin(2\theta) \\ 0 & 1 & 0 \\ \sin(2\theta) & 0 & \cos(2\theta) \end{pmatrix}, \tag{5.178}$$

which corresponds to rotation around the y-axis through angle 2θ. Together, Equations 5.168 and 5.178 prove that $\mathbf{R}$ is a rotation matrix.

The conclusion that $\mathbf{R}$ is a rotation matrix establishes an important connection between the rotation of spinors in virtual space and 3D rotations in real space. The spinor transformation maps a vector $\mathbf{r}$ onto a 2×2 complex matrix $\mathbf{T}$ *via* the equation $\mathbf{T} = T_0\mathbf{I} + \mathbf{r}\cdot\sigma$. Because T_0 is irrelevant here it can be removed from the equation, so that the expression simplifies to $\mathbf{T} = \mathbf{r}\cdot\sigma$. For every $\mathbf{T}$ there is a unique 2×2 unitary matrix $\mathbf{U}$, which transforms matrix $\mathbf{T}$ into $\mathbf{T}'$, and uniquely relates to the matrix $\mathbf{R}$, which corresponds to a 3D rotation of the vector $\mathbf{r}$, according to:

$$\mathbf{T}' = \mathbf{U}\,\mathbf{T}\,\mathbf{U}^{-1} = \mathbf{r}'\cdot\sigma = (\mathbf{R}\,\mathbf{r})\cdot\sigma, \tag{5.179}$$

where $\mathbf{r}' = \mathbf{R}\,\mathbf{r}$. This leads to the conclusion that rotation of the vector in a 3D space corresponds to its rotation in a spinor space, so that one can be replaced for the other, at will. For spherical harmonics Y_{lm}, if they need to be rotated in 3D, this task is equivalent to performing the rotation in its spinor space spanned by index m.

In order to apply the isomorphism, which was established between rotation in real space and that in spinor space, to the derivation of Wigner matrix elements, it is necessary to extend the theory of rotation of spinors to $(2j + 1)$-dimensional function space. For each value of angular momentum j, the number of unique values of eigenvalue m determines the size of the spinor space. For the angular momentum $j = 1/2$, the spinor space consists of two basis functions, which may be called the state "up" and the state "down", having the eigenvalue $m = 1/2$ and $m = -(1/2)$, respectively. Spinor components λ_1 and λ_2, which represent such 2-dimensional eigenfunction, are described in Equation 5.70. The presence of Cartesian components in the definition of spinors makes it possible to express the rotation of spinors in terms of rotation of their Cartesian components. Cartesian components x, y, and z composing a vector $\mathbf{r}$ rotate according to the standard rule $\mathbf{r}' = \mathbf{R}\,\mathbf{r}$, so that

$$\begin{aligned} x' &= R_{11}x + R_{12}y + R_{13}z, \\ y' &= R_{21}x + R_{22}y + R_{23}z, \\ z' &= R_{31}x + R_{32}y + R_{33}z, \end{aligned} \tag{5.180}$$

where R_{ij} are the elements of rotation matrix. With that, the square, λ_1^2, of the first spinor component defined in Equation 5.70 transforms to $\lambda_1'^2$ according to:

$$\lambda_1'^2 = \tfrac{1}{2}[x' - iy'] = \tfrac{1}{2}\big[(R_{11} - iR_{21})x + (R_{12} - iR_{22})y + (R_{13} - iR_{23})z\big], \tag{5.181}$$

where rotated Cartesian components x' and y' are replaced using the elements of the rotation matrix according to Equation 5.180. Furthermore, replacing the Cartesian components x, y, z in Equation 5.181 by their spinor counterparts λ_1 and λ_2 defined in Equation 5.66 leads to:

$$\lambda_1'^2 = \tfrac{1}{2}\big[(R_{11} - iR_{21})(\lambda_1^2 - \lambda_2^2) + (R_{12} - iR_{22})i(\lambda_1^2 + \lambda_2^2) + (R_{13} - iR_{23})(-2\lambda_1\lambda_2)\big]. \tag{5.182}$$

Gathering coefficients with the same spinor component gives:

$$\lambda_1'^2 = \frac{1}{2}\big[(R_{11} - iR_{21} + iR_{12} + R_{22})\lambda_1^2 - 2(R_{13} - iR_{23})\lambda_1\lambda_2 + (-R_{11} + iR_{21} + iR_{12} + R_{22})\lambda_2^2\big]. \tag{5.183}$$

In order to simplify this relation it is helpful to consider a product $(R_{11} - iR_{21} + iR_{12} + R_{22})$ of coefficients standing in front of λ_1^2 and λ_2^2. This product is:

$$\begin{aligned}
&(R_{11} - iR_{21} + iR_{12} + R_{22})(-R_{11} + iR_{21} + iR_{12} + R_{22}) \\
&= \big([iR_{12} + R_{22}] + [R_{11} - iR_{21}]\big)\big([iR_{12} + R_{22}] - [R_{11} - iR_{21}]\big) \\
&= (iR_{12} + R_{22})^2 - (R_{11} - iR_{21})^2 \\
&= -R_{12}^2 + 2iR_{12}R_{22} + R_{22}^2 - R_{11}^2 + 2iR_{11}R_{21} + R_{21}^2.
\end{aligned} \tag{5.184}$$

The terms in this expression may be rearranged, so that:

$$\begin{aligned}
&(R_{11} - iR_{21} + iR_{12} + R_{22})(-R_{11} + iR_{21} + iR_{12} + R_{22}) \\
&= -R_{11}^2 - R_{12}^2 + 2i(R_{11}R_{21} + R_{12}R_{22}) + R_{21}^2 + R_{22}^2.
\end{aligned} \tag{5.185}$$

Due to the orthonormality of the rotation matrix **R**, the following relations exist between the various matrix elements:

$$\begin{aligned}
R_{11}^2 + R_{12}^2 &= 1 - R_{13}^2, \\
R_{21}^2 + R_{22}^2 &= 1 - R_{23}^2, \\
R_{11}R_{21} + R_{12}R_{22} &= -R_{13}R_{23}.
\end{aligned} \tag{5.186}$$

Application of the relations from Equation 5.186 to Equation 5.185 leads to:

$$(R_{11} - iR_{21} + iR_{12} + R_{22})(-R_{11} + iR_{21} + iR_{12} + R_{22}) = R_{13}^2 - 2iR_{13}R_{23} - R_{23}^2 = (R_{13} - iR_{23})^2. \tag{5.187}$$

Based on this result, Equation 5.183 may be rewritten in quadratic form:

$$\lambda_1'^2 = \tfrac{1}{2}\left[(R_{11} - iR_{21} + iR_{12} + R_{22})^{\frac{1}{2}}\lambda_1 - (-R_{11} + iR_{21} + iR_{12} + R_{22})^{\frac{1}{2}}\lambda_2\right]^2, \tag{5.188}$$

where the minus sign between the two terms is necessary in order to generate $-2(R_{13} - iR_{23})\lambda_1\lambda_2$ when the expression is squared.

Taking the square root from both sides of Equation 5.188 leads to:

$$\lambda_1' = \tfrac{1}{\sqrt{2}}\left[(R_{11} - iR_{21} + iR_{12} + R_{22})^{\frac{1}{2}}\lambda_1 - (-R_{11} + iR_{21} + iR_{12} + R_{22})^{\frac{1}{2}}\lambda_2\right], \tag{5.189}$$

which determines the transformed component of the spinor within the uncertainty of the phase factor.

The expression for the second spinor component λ_2' is obtained in a similar way. In terms of the transformed Cartesian coordinates, it is:

$$\lambda_2' = -\tfrac{1}{2}(x' + iy') = -\tfrac{1}{2}\left[(R_{11} + iR_{21})x + (R_{12} + iR_{22})y + (R_{13} + iR_{23})z\right]. \tag{5.190}$$

Substituting the Cartesian components by their corresponding spinor components leads to:

$$\lambda_2' = -\tfrac{1}{2}\left[(R_{11} + iR_{21})(\lambda_1^2 - \lambda_2^2) + (R_{12} + iR_{22})i(\lambda_1^2 + \lambda_2^2) + (R_{13} + iR_{23})(-2\lambda_1\lambda_2)\right]. \tag{5.191}$$

Gathering the terms with the same spinor component gives:

$$\lambda_2' = -\tfrac{1}{2}\left[(R_{11} + iR_{21} + iR_{12} - R_{22})\lambda_1^2 - 2(R_{13} + iR_{23})\lambda_1\lambda_2 + (-R_{11} - iR_{21} + iR_{12} - R_{22})\lambda_2^2\right]. \tag{5.192}$$

Distributing the negative sign that stands in front of the square brackets between the elements inside the brackets changes the equation to:

$$\lambda_2' = \tfrac{1}{2}\left[(-R_{11} - iR_{21} - iR_{12} + R_{22})\lambda_1^2 + 2(R_{13} + iR_{23})\lambda_1\lambda_2 + (R_{11} + iR_{21} - iR_{12} + R_{22})\lambda_2^2\right]. \tag{5.193}$$

As with the previous case, it is useful to consider a product $(-R_{11} - iR_{21} - iR_{12} + R_{22})(R_{11} + iR_{21} - iR_{12} + R_{22})$ of the terms standing in front of the components λ_1^2 and λ_2^2. The product is:

$$\begin{aligned}(-R_{11} - iR_{21} - iR_{12} + R_{22})(R_{11} + iR_{21} - iR_{12} + R_{22}) &= (-iR_{12} + R_{22})^2 - (R_{11} + iR_{21})^2 \\ &= -R_{12}^2 - 2iR_{12}R_{22} + R_{22}^2 - R_{11}^2 - 2iR_{11}R_{21} + R_{21}^2.\end{aligned} \tag{5.194}$$

Because the rotation matrix is orthonormal, the following relations hold:

$$\begin{aligned}-R_{11}^2 - R_{12}^2 &= -1 + R_{13}^2, \\ R_{21}^2 + R_{22}^2 &= 1 - R_{23}^2, \\ -R_{11}R_{21} - R_{12}R_{22} &= R_{13}R_{23}.\end{aligned} \tag{5.195}$$

Therefore,

$$(-R_{11} - iR_{21} - iR_{12} + R_{22})(R_{11} + iR_{21} - iR_{12} + R_{22}) = R_{13}^2 + 2iR_{13}R_{23} - R_{23}^2 = (R_{13} + 2iR_{13})^2. \tag{5.196}$$

Based on Equation 5.196, the expression for the second component of the rotated spinor takes on the final form:

$$\lambda_2' = \frac{1}{\sqrt{2}}\left[(-R_{11} - iR_{21} - iR_{12} + R_{22})^{\frac{1}{2}}\lambda_1 + (R_{11} + iR_{21} - iR_{12} + R_{22})^{\frac{1}{2}}\lambda_2\right]. \tag{5.197}$$

Combining the expressions for transformation of spinor components leads to a unitary matrix **U**:

$$\mathbf{U} = \frac{1}{\sqrt{2}}\begin{pmatrix}(R_{11} - iR_{21} + iR_{12} + R_{22})^{\frac{1}{2}} & -(-R_{11} + iR_{21} + iR_{12} + R_{22})^{\frac{1}{2}} \\ (-R_{11} - iR_{21} - iR_{12} + R_{22})^{\frac{1}{2}} & (R_{11} + iR_{21} - iR_{12} + R_{22})^{\frac{1}{2}}\end{pmatrix}, \tag{5.198}$$

which resembles the standard general 2 × 2 unitary matrix:

$$\mathbf{U} = \begin{pmatrix} a & b \\ -b^* & a^* \end{pmatrix}. \tag{5.199}$$

The result obtained in Equation 5.198 confirms that spinors transform linearly according to the matrix equation $\lambda' = \mathbf{U}\,\lambda$ with the elements of matrix **U** being composed of the elements of rotation matrix **R** for the isotropic 3D vector.

A 3D rotation of a function, which is defined in $2j + 1$ spinor space, will likewise be accomplished by a unitary matrix **U** of dimensionality $2j + 1$. The next task is to find the matrix **U** for the general case. The general set of functions of variables λ_1 and λ_2, which undergo unitary transformation when λ_1 and λ_2 are transformed as $\lambda' = \mathbf{U}\lambda$, is provided by the monomial:

$$\lambda_1^{j+m}\lambda_2^{j-m}, \quad \text{where} - j \leq m \leq j. \tag{5.200}$$

In the simplest example of $j = 1/2$ when $m = 1/2$ and $m = -(1/2)$, the two functions composing the complete set are the variables λ_1 and λ_2 themselves. This case has already been addressed.

Before a multi-dimensional spinor function Λ_{jm} can transform unitarily, it first has to be normalized, which requires adding a normalization coefficient N_{jm}:

$$\Lambda_{jm} = N_{jm}\lambda_1^{j+m}\lambda_2^{j-m}. \tag{5.201}$$

The normalization coefficient can be determined from an analysis of the rotated function:

$$\Lambda'_{jm} = N_{jm}\lambda_1'^{j+m}\lambda_2'^{j-m}, \tag{5.202}$$

in which the rotated function Λ'_{jm} is composed of the rotated components λ'_1 and λ'_2.

Normalization implies the condition:

$$\sum_{m=-j}^{j}\Lambda'_{jm}\Lambda'^{*}_{jm} = \sum_{m=-j}^{j}\Lambda_{jm}\Lambda^{*}_{jm}. \tag{5.203}$$

To understand the meaning of Equation 5.203, compare it to the simplest case of $j = 1/2$ when the normalization condition assumes the form:

$$\left|\Lambda'_{\frac{1}{2},\frac{1}{2}}\right|^2 + \left|\Lambda'_{\frac{1}{2},-\frac{1}{2}}\right|^2 = \left|\Lambda_{\frac{1}{2},\frac{1}{2}}\right|^2 + \left|\Lambda_{\frac{1}{2},-\frac{1}{2}}\right|^2, \tag{5.204}$$

or

$$\lambda'_1\lambda'^{*}_1 + \lambda'_2\lambda'^{*}_2 = \lambda_1\lambda^{*}_1 + \lambda_2\lambda^{*}_2, \tag{5.205}$$

where index 1 corresponds to $j = 1/2$ and $m = 1/2$, and index 2 represents $j = 1/2$ and $m = -(1/2)$. This relation satisfies the requirement that the length of a two-component vector λ would remain unchanged during a rotation. Similarly, the normalization condition for a $(2j + 1)$-dimensional vector Λ requires that the sum of complex-conjugate squares of all its components must not change upon rotation.

The 2-dimensional case presented in Equation 5.205 requires no normalization coefficient. This is equivalent to the normalization coefficient having unit value. Finding the normalization coefficient N_{jm} for a general case relies on the following argument. Since

$$|\lambda_1|^2 + |\lambda_2|^2 = |\lambda'_1|^2 + |\lambda'_2|^2, \tag{5.206}$$

it must also be true that:

$$\left(|\lambda_1|^2 + |\lambda_2|^2\right)^{2j} = \left(|\lambda'_1|^2 + |\lambda'_2|^2\right)^{2j}. \tag{5.207}$$

Expanding Equation 5.207 using the Binomial theorem gives:

$$\left(|\lambda_1|^2+|\lambda_2|^2\right)^{2j}=\sum_{k=0}^{2j}\frac{(2j)!}{k!(2j-k)!}\left(|\lambda_1|^2\right)^k\left(|\lambda_2|^2\right)^{2j-k}. \tag{5.208}$$

Introducing the substitution $k=j+m$ and adjusting the limits $m=-j$ for $k=0$ and $m=j$ for $k=2j$ turns Equation 5.208 to:

$$\left(|\lambda_1|^2+|\lambda_2|^2\right)^{2j}=\sum_{m=-j}^{j}\frac{(2j)!}{(j+m)!(j-m)!}\left(|\lambda_1|^2\right)^{j+m}\left(|\lambda_2|^2\right)^{j-m}. \tag{5.209}$$

Since the term (2*j*)! does not depend on index *m*, it can be moved outside the sum. With that, the sums for the function Λ_{jm} before and after rotation can be equated, so that:

$$(2j)!\sum_{m=-j}^{j}\frac{1}{(j+m)!(j-m)!}\left(|\lambda_1|^2\right)^{j+m}\left(|\lambda_2|^2\right)^{j-m}=(2j)!\sum_{m=-j}^{j}|\Lambda_{jm}|^2=(2j)!\sum_{m=-j}^{j}|\Lambda'_{jm}|^2. \tag{5.210}$$

This relation leads to the normalization coefficient:

$$N_{jm}=\left[(j+m)!(j-m)!\right]^{-\frac{1}{2}}. \tag{5.211}$$

With that, everything is ready for the resolution of the unitary matrix for rotation of a multi-dimensional function Λ.

Since Λ depends on two variables λ_1 and λ_2, according to Equation 5.201, it is necessary to determine how those variables change under rotation. In matrix form, the unitary transformation of the 2-component spinor λ, using the unitary matrix from Equation 5.199, is:

$$\begin{pmatrix}\lambda'_1\\ \lambda'_2\end{pmatrix}=\begin{pmatrix}a & b\\ -b^* & a^*\end{pmatrix}\begin{pmatrix}\lambda_1\\ \lambda_2\end{pmatrix}. \tag{5.212}$$

This relation corresponds to the system of equations:

$$\begin{aligned}\lambda'_1&=a\lambda_1+b\lambda_2,\\ \lambda'_2&=-b^*\lambda_1+a^*\lambda_2,\end{aligned} \tag{5.213}$$

and that helps to fund the transformed function Λ'. Substituting Equation 5.213 into Equation 5.202 gives:

$$\Lambda'_{jm}=N_{jm}\lambda_1'^{\,j+m}\lambda_2'^{\,j-m}=N_{jm}(a\lambda_1+b\lambda_2)^{j+m}(-b^*\lambda_1+a^*\lambda_2)^{j-m}. \tag{5.214}$$

Application of the Binomial theorem to Equation 5.214 produces an expansion:

$$\Lambda'_{jm}=\left[(j+m)!(j-m)!\right]^{-\frac{1}{2}}\sum_{s=0}^{j+m}\sum_{t=0}^{j-m}\frac{(j+m)!(j-m)!}{s!(j+m-s)!t!(j-m-t)!}(a\lambda_1)^{j+m-s}(b\lambda_2)^s(-b^*\lambda_1)^{j-m-t}(a^*\lambda_2)^t. \tag{5.215}$$

Cancelling out $[(j+m)!(j-m)!]^{-1/2}$, and rearranging to combine like terms gives:

$$\Lambda'_{jm} = \sum_{s=0}^{j+m} \sum_{t=0}^{j-m} (-1)^{j-m-t} \frac{[(j+m)!(j-m)!]^{\frac{1}{2}}}{s!(j+m-s)!t!(j-m-t)!} a^{j+m-s}(a^*)^t b^s (b^*)^{j-m-t} \lambda_1^{2j-s-t} \lambda_2^{s+t}. \tag{5.216}$$

In this, it is useful to introduce a variable substitution $k = j - s - t$ in order to reveal the relationship between the indices. This substitution leads to the following changes in the indices:

$$\begin{aligned} &j-m-t = j-m-j+s+k = k+s-m, \\ &t = j-s-k, \\ &2j-s-t = 2j-s-j+s+k = j+k, \\ &s+t = s+j-s-k = j-k. \end{aligned} \tag{5.217}$$

Due to the substitution, the limits in the sum over index t become $k = j - s$ for $t = 0$, and $k = m - s$ for $t = j - m$. Since the order of summation is unimportant, it is convenient to reverse the lower and upper limits in the sum over index k. After all these changes are made, Equation 5.216 becomes:

$$\Lambda'_{jm} = \sum_{s=0}^{j+m} \sum_{k=m-s}^{j-s} \frac{(-1)^{k+s-m}[(j+m)!(j-m)!]^{1/2}}{s!(j+m-s)!(j-k-s)!(k+s-m)!} a^{j+m-s}(a^*)^{j-s-k} b^s (b^*)^{k+s-m} \lambda_1^{j+k} \lambda_2^{j-k}, \tag{5.218}$$

where the lower bound of index k is $k = -j$, when $s = j + m$, and the higher bound is $k = j$ when $s = 0$.

Equation 5.218 contains the terms $\lambda_1^{j+k}\lambda_2^{j-k}$ in the sum, which correspond to the function Λ_{jk} in the state before the transformation:

$$\Lambda_{jk} = \frac{\lambda_1^{j+k}\lambda_2^{j-k}}{\sqrt{(j+m)!(j-m)!}},$$

or

$$\lambda_1^{j+k}\lambda_2^{j-k} = \sqrt{(j+m)!(j-m)!}\,\Lambda_{jk}. \tag{5.219}$$

Using Equation 5.219, $\lambda_1^{j+k}\lambda_2^{j-k}$ in Equation 5.218 can be replaced by $\sqrt{(j+m)!(j-m)!}\Lambda_{jk}$ to give:

$$\Lambda'_{jm} = \sum_{s=0}^{j+m} \sum_{k=m-s}^{j-s} \frac{(-1)^{k+s-m}\sqrt{(j+m)!(j-m)!}\sqrt{(j+k)!(j-k)!}}{s!(j+m-s)!(j-k-s)!(k+s-m)!} a^{j+m-s}(a^*)^{j-s-k} b^s (b^*)^{k+s-m} \Lambda_{jk}. \tag{5.220}$$

Since the unitary transformation from Λ_{jm} to Λ'_{jm} can be performed using the standard equation:

$$\Lambda'_{jm} = \sum_{k=-j}^{j} U^j_{km} \Lambda_{jk}, \tag{5.221}$$

the combination of Equation 5.220 with Equation 5.221 helps to determine the matrix elements U^j_{km}, where the upper index j indicates that it is a $(2j + 1) \times (2j + 1)$ matrix. Recall from the analysis of Equation 5.218 that index k actually varies from $-j$ to j, so the summation over index k can be moved outside the matrix element U^j_{km} in Equation 5.221, while the remaining sum over index s in Equation 5.220 has to remain inside. Therefore,

$$U^j_{km} = \sum_{s=0}^{j+m} \frac{(-1)^{k+s-m}\sqrt{(j+m)!(j-m)!}\sqrt{(j+k)!(j-k)!}}{s!(j+m-s)!(j-k-s)!(k+s-m)!} a^{j+m-s}(a^*)^{j-s-k} b^s (b^*)^{k+s-m}. \quad (5.222)$$

The coefficients a and b that are needed for this equation have previously been determined in Section 5.3 by Equations 5.119 and 5.123, in the course of devising the rule for rotation of spinors. These are

$$a = e^{-i\alpha/2} \cos\frac{\beta}{2}\, e^{-i\gamma/2} \quad \text{and} \quad b = e^{i\alpha/2} \sin\frac{\beta}{2}\, e^{-i\gamma/2}, \quad (5.223)$$

where α, β and γ are Euler angles. After substitution of Equation 5.223 into Equation 5.222, the portion of the equation that includes coefficients a becomes:

$$a^{j+m-s}(a^*)^{j-s-k} = e^{-i(j+m-s)\frac{\alpha}{2}} \left(\cos\frac{\beta}{2}\right)^{j+m-s} e^{-i(j+m-s)\frac{\gamma}{2}} e^{i(j-s-k)\frac{\alpha}{2}} \left(\cos\frac{\beta}{2}\right)^{j-s-k} e^{i(j-s-k)\frac{\gamma}{2}}. \quad (5.224)$$

After combining similar terms, this expression simplifies to:

$$a^{j+m-s}(a^*)^{j-s-k} = e^{-i(m+k)\frac{\alpha}{2}} \left(\cos\frac{\beta}{2}\right)^{2j+m-k-2s} e^{-i(m+k)\frac{\gamma}{2}}. \quad (5.225)$$

The coefficients for b are obtained in the same way: substituting Equation 5.223 into Equation 5.222 gives:

$$b^s(b^*)^{k+s-m} = e^{is\frac{\alpha}{2}} \left(\sin\frac{\beta}{2}\right)^{s} e^{-is\frac{\gamma}{2}} e^{-i(k+s-m)\frac{\alpha}{2}} \left(\sin\frac{\beta}{2}\right)^{k+s-m} e^{i(k+s-m)\frac{\gamma}{2}}. \quad (5.226)$$

Simplification of Equation 5.226 reduces this expression to:

$$b^s(b^*)^{k+s-m} = e^{-i(k-m)\frac{\alpha}{2}} \left(\sin\frac{\beta}{2}\right)^{k-m+2s} e^{-i(m-k)\frac{\gamma}{2}}. \quad (5.227)$$

Combining together the expressions for coefficients a and b gives the intermediate term:

$$\begin{aligned} a^{j+m-s}(a^*)^{j-s-k} b^s (b^*)^{k+s-m} = {} & e^{-i(m+k)\frac{\alpha}{2}} \left(\cos\frac{\beta}{2}\right)^{2j+m-k-2s} \\ & e^{-i(m+k)\frac{\gamma}{2}} e^{-i(k-m)\frac{\alpha}{2}} \left(\sin\frac{\beta}{2}\right)^{k-m+2s} e^{-i(m-k)\frac{\gamma}{2}}, \end{aligned} \quad (5.228)$$

which simplifies to:

$$a^{j+m-s}(a^*)^{j-s-k} b^s (b^*)^{k+s-m} = e^{-ik\alpha} e^{-im\gamma} \left(\cos\frac{\beta}{2}\right)^{2j+m-k-2s} \left(\sin\frac{\beta}{2}\right)^{k-m+2s}. \quad (5.229)$$

Finally, inserting Equation 5.229 into Equation 5.222, and taking into account that $U^j_{km} \equiv D^j_{km}$, leads to the general expression for the Wigner matrix element:

$$D^j_{km} = e^{-ik\alpha} e^{-im\gamma} \sum_{s=0}^{j+m} \frac{(-1)^{k+s-m} \sqrt{(j+m)!(j-m)!(j+k)!(j-k)!}}{s!(j+m-s)!(j-k-s)!(k+s-m)!} \left(\cos \frac{\beta}{2} \right)^{2j+m-k-2s} \left(\sin \frac{\beta}{2} \right)^{k-m+2s} . \quad (5.230)$$

The process of arriving to Equation 5.230 illustrates the physics behind the Wigner matrix. This equation is too cumbersome to be of practical use in the computation of matrix elements, though. Further studies continued in Chapter 6 and 7 will eventually address that limitation.

5.5 Addition Theorem for Spherical Harmonics

One way to come up with a better method for computation of the Wigner matrix elements is to study the relationship between the Wigner matrix and the spherical harmonics. This relation leads to the expansion of Legendre polynomials into a product of two spherical harmonic functions, known as the addition theorem for spherical harmonics, which is to be derived now.

Spherical harmonics $Y_{lm}(\theta,\phi)$ are eigenfunctions of the angular momentum operator for integral values of angular momentum, j. The use of index l instead of index j in function $Y_{lm}(\theta,\phi)$ emphasizes that condition. Spherical harmonics satisfy the eigenvalue equation

$$\mathbf{J}^2 Y_{lm}(\theta,\phi) = l(l+1)\,\hbar Y_{lm}(\theta,\phi), \quad (5.231)$$

and rotate in 3D according to the prescription of the Wigner matrix. Rotation of the coordinate system through the Euler angles α, β and γ transforms a spherical harmonic function $Y_{lm}(\theta,\phi)$ to $Y_{lm}(\theta',\phi')$, where prime denotes spherical polar coordinates in the rotated frame. The rotation transformation decomposes the resulting spherical harmonic $Y_{lm}(\theta',\phi')$ into a linear combination of the functions $Y_{lk}(\theta,\phi)$ from their common spinor space $(2l + 1)$.

$$Y_{lm}(\theta',\phi') = \sum_{k=-l}^{l} D^j_{km}(\alpha\beta\gamma) Y_{lk}(\theta,\phi). \quad (5.232)$$

Azimuthal index k spans the entire spinor space, $-l \leq k \leq l$, determined by the index l, and iterates through the elements of the m-th column of matrix D^j.

Knowing how spherical harmonics transform as a result of rotation of the coordinate system is helpful in the derivation of an addition theorem for spherical harmonics. This theorem states that the sum over the azimuthal index of a product of two spherical harmonics,

$$g = \sum_{m=-l}^{l} Y^*_{lm}(\theta_1,\phi_1)\, Y_{lm}(\theta_2,\phi_2), \quad (5.233)$$

where g is a function to be determined, is rotationally invariant, that is, does not change under rotation. Spherical polar coordinates θ_1, ϕ_1 and θ_2, ϕ_2 are those describing the radius vectors $\mathbf{r}_1$ and $\mathbf{r}_2$, respectively, as illustrated in Figure 5.4.

This theorem can be proved as follows:

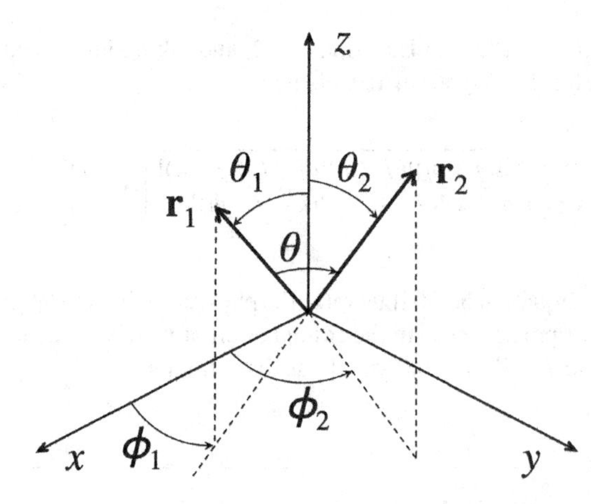

FIGURE 5.4 Two vectors $\mathbf{r}_1(\theta_1, \phi_1)$ and $\mathbf{r}_2(\theta_2, \phi_2)$ defined in spherical polar coordinates, and having angle θ subtended between them.

Rotation of the coordinate system through the Euler angles α, β and γ transforms spherical harmonics $Y_{lm}(\theta_1,\phi_1) \rightarrow Y_{lm}(\theta_1',\phi_1')$ and $Y_{lm}(\theta_2,\phi_2) \rightarrow Y_{lm}(\theta_2',\phi_2')$ according to the following matrix equations:

$$Y_{lm}(\theta_1',\phi_1') = \mathbf{R}\, Y_{lm}(\theta_1,\phi_1) \equiv Y_{lm}(\theta_1',\phi_1') = \sum_{k=-l}^{l} D_{km}^{j}(\alpha\beta\gamma) Y_{lk}(\theta_1,\phi_1), \tag{5.234}$$

and

$$Y_{lm}(\theta_2',\phi_2') = \mathbf{R}\, Y_{lm}(\theta_2,\phi_2) \equiv Y_{lm}(\theta_2',\phi_2') = \sum_{t=-l}^{l} D_{tm}^{j}(\alpha\beta\gamma) Y_{lt}(\theta_2,\phi_2), \tag{5.235}$$

where prime denotes spherical polar coordinates of the vectors $\mathbf{r}_1$ and $\mathbf{r}_2$ in the rotated coordinate system, and $\mathbf{R}$ represents an abstract rotation matrix. The right portions of Equations 5.234 and 5.235 show the actual transformation involving the Wigner matrix.

The inverse transformation $Y_{lm}(\theta',\phi') \rightarrow Y_{lm}(\theta,\phi)$ follows from Equations 5.234 and 5.235 by multiplying both sides of those equations by $\mathbf{R}^{-1}$, and resolving against the original functions $Y_{lm}(\theta,\phi)$. The actual inverse transformation that involves the Wigner matrix directly follows from that. In obtaining matrix $\mathbf{D}^{-1}$ from $\mathbf{D}$, the unitary property of the Wigner matrix requires replacing columns by rows in $\mathbf{D}$ and taking a complex conjugate of the elements of the transposed matrix. This leads to:

$$Y_{lm}(\theta_1,\phi_1) = \mathbf{R}^{-1} Y_{lm}(\theta_1',\phi_1') \equiv Y_{lm}(\theta_1,\phi_1) = \sum_{k=-l}^{l} D_{mk}^{j*}(\alpha\beta\gamma) Y_{lk}(\theta_1',\phi_1'), \tag{5.236}$$

and

$$Y_{lm}(\theta_2,\phi_2) = \mathbf{R}^{-1} Y_{lm}(\theta_2',\phi_2') \equiv Y_{lm}(\theta_2,\phi_2) = \sum_{t=-l}^{l} D_{mt}^{j*}(\alpha\beta\gamma) Y_{lt}(\theta_2',\phi_2'). \tag{5.237}$$

These expressions for $Y_{lm}(\theta_1,\phi_1)$ and $Y_{lm}(\theta_2,\phi_2)$ can now be substituted back into the equation for g, giving:

$$g = \sum_{m=-l}^{l} \left[\sum_{k=-l}^{l} D_{mk}^{j*}(\alpha\beta\gamma) Y_{lk}(\theta_1',\phi_1') \right]^{*} \left[\sum_{t=-l}^{l} D_{mt}^{j*}(\alpha\beta\gamma) Y_{lt}\left(\theta_2',\phi_2'\right) \right], \tag{5.238}$$

which becomes:

$$g=\sum_{m=-l}^{l}\left[\sum_{k=-l}^{l}D^{j}_{mk}(\alpha\beta\gamma)Y^{*}_{lk}(\theta'_1,\phi'_1)\right]\left[\sum_{t=-l}^{l}D^{j*}_{mt}(\alpha\beta\gamma)Y_{lt}(\theta'_2,\phi'_2)\right]. \tag{5.239}$$

A quick inspection of the summation indices in the above equation reveals that index m runs only through the Wigner matrix elements. Therefore, it is helpful to rearrange the order of summation so that the sum involving index m is done first. This gives:

$$g=\sum_{k=-l}^{l}\sum_{t=-l}^{l}\left[\sum_{m=-l}^{l}D^{j}_{mk}D^{j*}_{mt}\right]Y^{*}_{lk}(\theta'_1,\phi'_1)Y_{lt}(\theta'_2,\phi'_2). \tag{5.240}$$

Because the Wigner matrix is orthonormal, for each pair of indices k and t:

$$\sum_{m=-l}^{l}D^{j}_{mk}D^{j*}_{mt}=\delta_{kt}. \tag{5.241}$$

Therefore, only the terms that involve $k=t$ remain non-zero in Equation 5.240. This reduces the triple sum in Equation 5.240 to a single sum:

$$g=\sum_{k=-l}^{l}Y^{*}_{lk}(\theta'_1,\phi'_1)\,Y_{lk}(\theta'_2,\phi'_2). \tag{5.242}$$

Replacing index k with symbol m gives this equation a more conventional appearance. Comparison of Equation 5.242 with Equation 5.233 shows that the sum g of the product of spherical harmonic functions is rotationally invariant, thus proving the theorem:

$$g=\sum_{m=-l}^{l}Y^{*}_{lm}(\theta_1,\phi_1)\,Y_{lm}(\theta_2,\phi_2)=\sum_{m=-l}^{l}Y^{*}_{lm}(\theta'_1,\phi'_1)\,Y_{lm}(\theta'_2,\phi'_2). \tag{5.243}$$

That g is rotationally invariant implies that any coordinate system can be used, which, in turn, simplifies the task of finding the functional form of g. Rotating the coordinate system so that the z-axis becomes aligned with the vector $\mathbf{r}_1$, and orientating the xz-plane so that the vector $\mathbf{r}_2$ is in the plane, produces the spherical polar coordinates $\theta'_1=0$, $\phi'_1=0$ and $\theta'_2=\theta$, $\phi'_2=0$ of the two vectors in the rotated coordinate frame, respectively. This arrangement is illustrated in Figure 5.5.

For vector $\mathbf{r}_1$, the corresponding spherical harmonics function in the rotated frame is:

$$Y^{*}_{lm}(\theta'_1,\phi'_1)=Y^{*}_{lm}(0,0)=\begin{cases}0, & \text{for } m\neq 0,\\ Y_{l0}(0,0), & \text{for } m=0.\end{cases} \tag{5.244}$$

This result follows from the analytic expressions for spherical harmonics discussed in Chapter 3. The value of zero for $m\neq 0$ in Equation 5.244 is a consequence of the spherical harmonic function $Y^{*}_{lm}(\theta'_1,\phi'_1)$ being proportional to $\sin\theta'_1$, and when $\theta'_1=0$, $\sin\theta'_1=0$. For $m=0$, $Y^{*}_{lm}(\theta'_1,\phi'_1)$ is non-zero because only then does the function not include a $\sin\theta'_1$ term. Therefore,

$$Y^{*}_{lm}(\theta'_1,\phi'_1)=Y^{*}_{lm}(0,0)=\delta_{m0}\sqrt{\frac{2l+1}{4\pi}}. \tag{5.245}$$

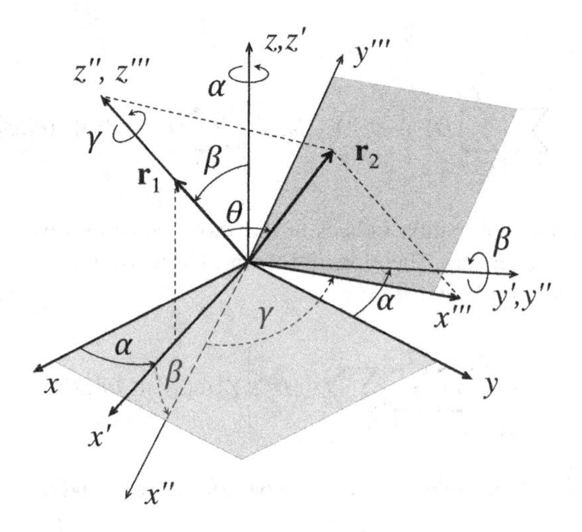

FIGURE 5.5 Rotation of the coordinate frame xyz through Euler angles $\alpha\,\beta\gamma$. The first rotation about the z-axis through angle α places the vector $\mathbf{r}_1$ in the new $x'z'$-plane. In a second rotation, the new coordinate frame $x'y'z'$ rotates about the y'-axis through angle β and aligns the z''-axis with vector $\mathbf{r}_1$. Finally, in a third rotation, the new coordinate frame $x''y''z''$ is rotated about the z'''-axis through angle γ, placing the vector $\mathbf{r}_2$ in the $x'''z'''$-plane.

With that, Equation 5.243 reduces to:

$$g=\sum_{m=-l}^{l} Y_{lm}^{*}(\theta_1',\phi_1')\,Y_{lm}(\theta_2',\phi_2')=\sum_{m=-l}^{l}\delta_{m0}\sqrt{\frac{2l+1}{4\pi}}\,Y_{lm}(\theta_2',\phi_2'). \tag{5.246}$$

The presence of a Kronecker delta in the product of spherical harmonics annihilates all terms in the sum in Equation 5.246, leaving only a single term when $m=0$, at which point the spherical harmonics function $Y_{lm}(\theta_2',\phi_2')$ for vector $\mathbf{r}_2$ reduces to $Y_{l0}(\theta,0)$:

$$Y_{lm}(\theta_2',\phi_2')=Y_{l0}(\theta,0). \tag{5.247}$$

Substituting Equation 5.247 into Equation 5.233 simplifies the function g:

$$g=\sum_{m=-l}^{l} Y_{lm}^{*}(\theta_1,\phi_1)\,Y_{lm}(\theta_2,\phi_2)=\sqrt{\frac{2l+1}{4\pi}}\,Y_{l0}(\theta,0). \tag{5.248}$$

With that, the addition theorem for spherical harmonics takes the form:

$$Y_{l0}(\theta,0)=\sqrt{\frac{4\pi}{2l+1}}\sum_{m=-l}^{l} Y_{lm}^{*}(\theta_1,\phi_1)\,Y_{lm}(\theta_2,\phi_2). \tag{5.249}$$

This can be simplified further. Since

$$Y_{l0}(\theta,0)=\sqrt{\frac{2l+1}{4\pi}}\,P_l(\cos\theta), \tag{5.250}$$

where $P_l(\cos\theta)$ is a Legendre polynomial, this immediately leads to:

$$P_l(\cos\theta)=\frac{4\pi}{2l+1}\sum_{m=-l}^{l} Y_{lm}^{*}(\theta_1,\phi_1)\,Y_{lm}(\theta_2,\phi_2). \tag{5.251}$$

In this form, Equation 5.251 represents the most popular expression for the addition theorem for spherical harmonics.

The addition theorem has many useful applications. For instance, it leads to the derivation of an important equation for the expansion of electrostatic potential as a series of spherical harmonics. This derivation builds on a specific case of finding the electrostatic potential of a source charge placed on the z-axis, as described in Chapter 1. Figure. 1.1 illustrates that particular arrangement of the charge source and the point of measurement of electrostatic potential. Section 1.1 delivers the following expression for electrostatic potential Φ:

$$\Phi = \frac{1}{|\mathbf{r}-\mathbf{a}|} = \sum_{l=0}^{\infty} \frac{a^l}{r^{l+1}} P_l(\cos\theta), \tag{5.252}$$

where a is the distance from the origin to the charge source, r is the distance from the origin to the point of measurement of the electrostatic potential, and θ is the angle between the radius vectors for the charge source and the measurement point. In order for the series to be convergent, this equation requires that $a < r$.

The inconvenience of Equation 5.252 is that it requires the knowledge of angle θ subtended between two radius vectors. Computing that angle for every pair of charges in the system is computationally expensive. It is much faster to have each point and their radius vectors independently defined in spherical polar coordinates, as illustrated in Figure 5.4. If the spherical polar coordinates for any two radius vectors representing a charge source and a measurement point are θ_1, ϕ_1 and θ_2, ϕ_2, respectively, then substituting the Legendre polynomial $P_l(\cos\theta)$ in Equation 5.252 with the product of spherical harmonics from Equation 5.251 leads to:

$$\Phi = \frac{1}{|\mathbf{a}-\mathbf{r}|} = \sum_{l=0}^{\infty}\sum_{m=-l}^{l} \frac{a^l}{r^{l+1}} \frac{4\pi}{2l+1} Y_{lm}^{*}(\theta_1,\phi_1)\, Y_{lm}(\theta_2,\phi_2), \tag{5.253}$$

where, as before, the convergence of the series requires that $a < r$. Since electrostatic potential is fully symmetric upon interchanging the positions of the charge source and the measurement point, it is sufficient to put the smaller of the two distances in the numerator and the larger in the denominator in order to make the expansion series convergent.

The analysis of addition theorem for spherical harmonics also leads to a useful relation between the Wigner matrix and spherical harmonics that could be used for evaluation of the matrix elements. Consider an example of two spherical harmonic functions $Y_{lm}(\theta_1\phi_1)$ and $Y_{lm}(\theta_2\phi_2)$ being defined with spherical polar coordinates θ_1, ϕ_1 and θ_2, ϕ_2, which correspond to two radius vectors $\mathbf{r}_1$ and $\mathbf{r}_2$, respectively. The rotation transformation of $Y_{lm}(\theta_2\phi_2)$ under a rotation of the coordinate system is defined by the standard equation:

$$Y_{lm}(\theta_2',\phi_2') = \sum_{k=-l}^{l} D_{km}^{l}(\alpha\beta\gamma)\, Y_{lk}(\theta_2,\phi_2), \tag{5.254}$$

where prime indicates spherical polar coordinates in the rotated frame. If the vectors $\mathbf{r}_1$, $\mathbf{r}_2$, and the z-axis are initially arranged to reside in one plane, then no rotation through the angle γ will be necessary in order to have the vectors $\mathbf{r}_1$ and $\mathbf{r}_2$ placed into the $x''z''$-plane of the final rotated frame shown in Figure 5.6, that is, $\gamma = 0$. This is most easily seen if rotation is performed about the axes of the original coordinate system according to Equation 5.11. This mode of operation starts the rotation of the coordinate frame about the original z-axis through angle γ, then the transformed frame is rotated about the original y-axis through angle β, and finally, the outcome of that rotation is rotated about the original z-axis through angle α. Since $\gamma = 0$, the other two Euler angles, which are defined in the rotation matrix element $D_{k,m}^{j}(\alpha,\beta,0)$ in Equation 5.254, become spherical polar coordinates of the rotated z''-axis, as shown in Figure 5.3.

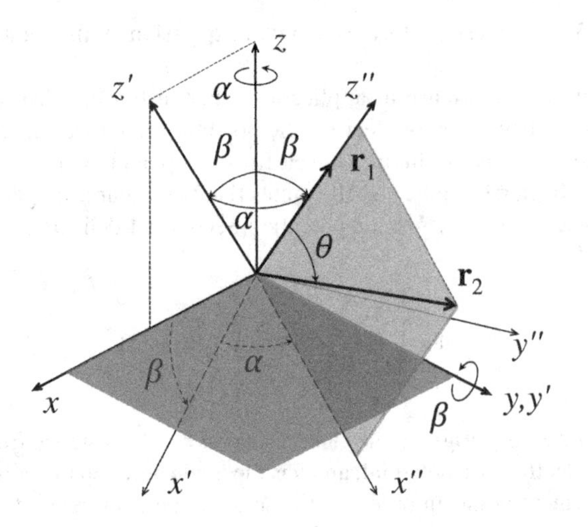

FIGURE 5.6 Rotation of the coordinate frame about the axes of the original coordinate system. Since $\gamma = 0$, the first rotation goes about the y-axis through angle β. The second rotation takes place about the z-axis through angle α. Since the vectors $\mathbf{r}_1$, $\mathbf{r}_2$, and the z-axis reside in one plane, this two-step rotation places the vector $\mathbf{r}_2$ in the $x''z''$-plane of the rotated frame.

The angle β becomes the angle θ_1, and the angle α becomes the angle ϕ_1. Moreover, the angle θ_2' becomes the angle θ that is subtended between the radius vectors $\mathbf{r}_1$ and $\mathbf{r}_2$. These details convert Equation 5.254 to:

$$Y_{lm}(\theta,\phi_2') = \sum_{k=-l}^{l} D_{km}^{l}(\phi_1,\theta_1,0)\, Y_{lk}(\theta_2,\phi_2). \tag{5.255}$$

Another important result follows from the fact that point $\mathbf{r}_2$ resides in the rotated $x''z''$-plane. When $\mathbf{r}_2$ is in the $x''z''$-plane, $\phi_2' = 0$, and that allows m to be zero on both sides of Equation 5.255. Therefore,

$$Y_{l0}(\theta,0) = \sum_{k=-l}^{l} D_{k0}^{l}(\phi_1,\theta_1,0)\, Y_{lk}(\theta_2,\phi_2). \tag{5.256}$$

Comparing this relation to Equation 5.249 for the addition theorem of spherical harmonics establishes the following equality:

$$Y_{l0}(\theta,0) = \sqrt{\frac{4\pi}{2l+1}} \sum_{m=-l}^{l} Y_{lm}^{*}(\theta_1,\phi_1)\, Y_{lm}(\theta_2,\phi_2) = \sum_{k=-l}^{l} D_{k0}^{l}(\phi_1,\theta_1,0)\, Y_{lk}(\theta_2,\phi_2). \tag{5.257}$$

Setting $k = m$ leads to:

$$D_{m0}^{l}(\phi_1,\theta_1,0) = \sqrt{\frac{4\pi}{2l+1}}\, Y_{lm}^{*}(\theta_1,\phi_1). \tag{5.258}$$

Reverting back the assignment $\beta = \theta_1$ and $\alpha = \phi_1$ produces the more conventionally looking equation:

$$D_{m0}^{l}(\alpha,\beta,0) = \sqrt{\frac{4\pi}{2l+1}}\, Y_{lm}^{*}(\beta,\alpha), \tag{5.259}$$

which may be used for evaluation of the respective Wigner matrix element.

6

Clebsch–Gordan Coefficients

Derivation of the general form of the Wigner matrix conducted in Chapter 5 primarily serves the purpose of illuminating the physical meaning of this tool. Understanding the physics of the Wigner matrix provides the basis for development of a form, which would be more suitable for practical computation. To accomplish that transition, it is necessary to make an additional detour to the theory of angular momentum that leads to the introduction of Clebsch–Gordan coefficients. The purpose of this chapter is to provide a mathematical foundation for the efficient computation of Wigner matrix elements.

6.1 Addition of Angular Momenta

Chapter 4 presented the theory of angular momentum of a single particle. However, in reality, the majority of problems of practical interest refers to systems of particles, so it is important to extend the theory to multiple components. The physics of two non-interacting particles, labeled 1 and 2, can be described in terms of two independent wavefunctions $\Psi(1)$ and $\Psi(2)$. Each of these wavefunctions encompasses a unique set of degrees of freedom represented by basis functions $\varphi_{j_1m_1}$ and $\varphi_{j_2m_2}$, respectively. The reason for labeling those basis functions φ_{jm} rather than ψ_{jm}, as in the previous chapter, will become clear later on in this section. The indices j and m are quantum numbers, and their subindices 1 and 2 identify them as being associated with a particular particle. Since these particles are independent of each other, they exist in their own independent coordinate space. Under these conditions, the measurement of angular momentum on each particle mathematically corresponds to an independent set of eigenvalue equations:

$$\begin{matrix} \mathbf{J}_1^2\varphi_{j_1m_1} = j_1(j_1+1)\varphi_{j_1m_1} \\ \mathbf{J}_{1z}\varphi_{j_1m_1} = m_1\varphi_{j_1m_1} \end{matrix}, \quad \text{and} \quad \begin{matrix} \mathbf{J}_2^2\varphi_{j_2m_2} = j_2(j_2+1)\varphi_{j_2m_2} \\ \mathbf{J}_{2z}\varphi_{j_2m_2} = m_2\varphi_{j_2m_2} \end{matrix}, \tag{6.1}$$

where $\mathbf{J}_1^2$, $\mathbf{J}_{1z}$ and $\mathbf{J}_2^2$, $\mathbf{J}_{2z}$ are angular momentum operators of particles 1 and 2.

The formalism of independent coordinate space would be incomplete without providing a specification of how those individual spaces coexist. A natural way of treating a system of several independent particles is to regard them as elements of a single composite space, which is, in turn, the union of the individual spaces of the particles. Thus, for a two-particle system, the eigenfunction would be the product $\varphi_{j_1m_1}\varphi_{j_2m_2}$ of the individual eigenfunctions. Since the particles in such a system are fully independent, the eigenfunction $\varphi_{j_1m_1}\varphi_{j_2m_2}$ represents an uncoupled state of the system in which the particles do not interact. Measurement of the angular momenta of a particle in such a system by application of the uncoupled angular momentum operators $\mathbf{J}_1^2$, $\mathbf{J}_{1z}$ and $\mathbf{J}_2^2$, $\mathbf{J}_{2z}$ to the eigenfunction $\varphi_{j_1m_1}\varphi_{j_2m_2}$ produces, as before, a set of quantum numbers j_1, m_1, and j_2, m_2, respectively, because each operator acts only on its own space and ignores the degrees of freedom of the other particle. For instance, the operator representing the projection of the angular momentum onto the quantization axis of particle 1 acting on the eigenfunction of the uncoupled system would give

$$\mathbf{J}_{1z}\varphi_{j_1m_1}\varphi_{j_2m_2} = m_1\varphi_{j_1m_1}\varphi_{j_2m_2}, \tag{6.2}$$

which is a re-statement of the fact that the product $\varphi_{j_1m_1}\varphi_{j_2m_2}$ is an eigenfunction of the operator $\mathbf{J}_{1z}$ and its eigenvalue is m_1.

The mathematical formalism of the uncoupled state serves as a starting point for dealing with more complex states of the system of particles. It is imperative, therefore, to review the formalism of uncoupled state in more details.

If the states $\varphi_{j_1m_1}$ and $\varphi_{j_2m_2}$ were the only possible states for two independent particles 1 and 2, the eigenfunction of the composite system would be just a product of those individual states, $\varphi_{j_1m_1}\varphi_{j_2m_2}$. However, in reality, the number of possible states for a particle is infinite, and a particle may exist in any distinct state described by a pair of quantum numbers j and m having values in the range $j = 0$, (1/2), 1, (3/2), 2, (5/2), ..., ∞ and $-j \leq m \leq j$. The eigenfunctions φ_{jm} that are labeled with those indices constitute the Hilbert space of the particle. From here on, when a particle space is mentioned in the text, it is the Hilbert space of the particle that is being referred to. The eigenfunctions $\varphi_{j_1m_1}$ and $\varphi_{j_2m_2}$ are basis functions, also known as basis vectors in the independent Hilbert spaces of the particles 1 and 2.

The union of two independent particle spaces is obtained by constructing a direct product, $\varphi_{j_1m_1} \otimes \varphi_{j_2m_2}$, of their corresponding eigenfunctions $\varphi_{j_1m_1}$ and $\varphi_{j_2m_2}$. For each pair of fixed values j_1 and j_2, the direct product generates a matrix:

$$\begin{matrix} \varphi_{j_1j_1}\ \varphi_{j_2j_2} & \varphi_{j_1j_1}\ \varphi_{j_2j_2-1} & \cdots & \varphi_{j_1j_1}\ \varphi_{j_2,-j_2} \\ \varphi_{j_1j_1-1}\varphi_{j_2j_2} & \varphi_{j_1j_1-1}\varphi_{j_2j_2-1} & \cdots & \varphi_{j_1j_1-1}\varphi_{j_2,-j_2} \\ \vdots & \vdots & \vdots & \vdots \\ \varphi_{j_1,-j_1}\varphi_{j_2j_2} & \varphi_{j_1,-j_1}\varphi_{j_2j_2-1} & \cdots & \varphi_{j_1,-j_1}\ \varphi_{j_2,-j_2} \end{matrix} \tag{6.3}$$

Each column in this table features the azimuthal quantum number m_1 of particle 1 taking on the values from j_1 to $-j_1$, and each row has the azimuthal quantum number m_2 of particle 2 taking on the values from j_2 to $-j_2$. For the brevity in notation, each element in this matrix may also be written in bra-ket notation as $\varphi_{j_1m_1}\varphi_{j_2m_2} \equiv |\, j_1m_1; j_2m_2\rangle$.

Since the angular momenta j_1 and j_2 are variable parameters, each pair of numbers j_1 and j_2 would be represented by a unique table in the form of Equation 6.3. The subscripts 1 and 2 represents particle space, so, in order to distinguish between the different values of angular momentum for the same particle, the quantum numbers j will be augmented with a superscript index as in j^a. In that notation, the function space of particle 1 is represented by the set of quantum numbers $j_1^a, j_1^b, j_1^c, \ldots$, and so on. Each pair of indices j_1^a and j_2^b leads to a new table like the one in Equation 6.3 which spans all possible combinations of the azimuthal quantum numbers m_1^a and m_2^b.

Up to this point, the analysis has been concerned with dealing with a system of non-interacting particles. If, however, the particles can interact with each other, their state becomes coupled. In the coupled state, a measurement of the angular momentum of the system no longer produces the values j_1, m_1 and j_2, m_2 corresponding to the independent particles. Instead, the measurement will produce a new set of quantum numbers j and m, which will eventually need to be determined. The coupled state of the system of particles can be denoted by the eigenfunction ψ_{jm} that, for the operators $\mathbf{J}^2$ and $\mathbf{J}_z$, has the eigenvalues $j(j+1)$ and m. The existence of a separate angular momentum of a system of particles has the same origin as the existence of angular momentum of a single particle. As soon as the physical description of the system becomes rotationally invariant, it complies with the conservation of angular momentum. This is the quantum theoretical analogue of the classical mechanics equivalent.

Before proceeding further with the analysis of the topic it is necessary to make a brief note about the notation that will be used throughout this section. The functions φ represent a uncoupled state of the system, whereas functions ψ will be used to denote a coupled state of the system. This designation adopts the convention originating from electronic structure theory, according to which the functions ψ are perceived as a linear combination of functions φ. The present notation for eigenfunctions of angular momentum of the coupled system retains the consistency with that used in Chapter 4.

The first goal is to determine the coupled eigenfunction ψ_{jm} and its dependence on the eigenfunctions $\varphi_{j_1m_1}$ and $\varphi_{j_2m_2}$ of independent particles. Since functions $\varphi_{j_1m_1}$ and $\varphi_{j_2m_2}$, as well as their direct product $\varphi_{j_1m_1}\varphi_{j_2m_2}$, form a complete basis set, and are orthonormal, that is,

$$\langle j^a m^a \,|\, j^b m^b \rangle = \delta_{j^a j^b}\delta_{m^a m^b}, \tag{6.4}$$

the function ψ_{jm} may be expanded in the basis of functions $\varphi_{j_1m_1}\varphi_{j_2m_2}$. In the expansion, the coupled state ψ_{jm} occurs as a linear combination of the uncoupled states $\varphi_{j_1m_1}\varphi_{j_2m_2}$:

$$\psi_{jm} = \sum_{m_1=-j_1}^{j_1}\sum_{m_2=-j_2}^{j_2} C(j_1j_2j;m_1m_2m)\varphi_{j_1m_1}\varphi_{j_2m_2}, \tag{6.5}$$

that weights each term in the uncoupled state with a linear coefficient $C(j_1j_2j;m_1m_2m)$, which depends on the indices j_1, j_2, j, m_1, m_2, m of the functions $\varphi_{j_1m_1}\varphi_{j_2m_2}$ and ψ_{jm}. The nature of the coefficient C is revealed by multiplying both side of the above equation by $\varphi^*_{j_1m_1}\varphi^*_{j_2m_2}$ from the left and integrating, producing:

$$C(j_1j_2j;m_1m_2m) = \langle j_1m_1; j_2m_2 | jm\rangle. \tag{6.6}$$

This relation introduces $C(j_1j_2j;m_1m_2m)$ as an overlap integral between the uncoupled state $\varphi_{j_1m_1}\varphi_{j_2m_2}$ and the coupled state ψ_{jm}, and is commonly referred to as a Clebsch–Gordan coefficient, after the mathematicians who pioneered its use. By convention, Clebsch–Gordan coefficients are treated as real numbers since their generalization to complex numbers does not bring any additional value.

The double sum over indices m_1 and m_2 in Equation 6.5 can be symbolically rewritten as a single sum over the combined index m_1m_2.

$$\psi_{jm} = \sum_{m_1m_2} C(j_1j_2j;m_1m_2m)\varphi_{j_1m_1}\varphi_{j_2m_2}. \tag{6.7}$$

Solving this equation begins by using one of the principles of angular momentum that states that, for the eigenfunction ψ_{jm}, the angular and azimuthal quantum numbers are solutions of the eigenvalue problem for the angular momentum operator $\mathbf{J}$.

The angular momentum operator of the coupled state $\mathbf{J}$ is the vector sum of the angular momentum operators of the uncoupled state, i.e. $\mathbf{J} = \mathbf{J}_1 + \mathbf{J}_2$. Based on that, the total operator of projection of the angular momentum onto the quantization axis is $\mathbf{J}_z = \mathbf{J}_{1z} + \mathbf{J}_{2z}$. The additive nature of the angular momentum operator makes finding the result of application of the total projection operator $\mathbf{J}_z$ onto the coupled state ψ_{jm} straightforward. Since $\mathbf{J}_z\psi_{jm} = m\psi_{jm}$ and $(\mathbf{J}_{1z} + \mathbf{J}_{2z})\varphi_{j_1m_1}\varphi_{j_2m_2} = (m_1 + m_2)\varphi_{j_1m_1}\varphi_{j_2m_2}$:

$$m\psi_{jm} = \mathbf{J}_z\psi_{jm} = \mathbf{J}_z\sum_{m_1m_2} C(j_1j_2j;m_1m_2m)\varphi_{j_1m_1}\varphi_{j_2m_2} = \sum_{m_1m_2} C(j_1j_2j;m_1m_2m)(m_1+m_2)\varphi_{j_1m_1}\varphi_{j_2m_2}. \tag{6.8}$$

From this equation, a dependence of m on m_1 and m_2 can easily be determined. Multiplying both sides of Equation 6.7 by m gives:

$$m\psi_{jm} = \sum_{m_1m_2} C(j_1j_2j;m_1m_2m)\, m\,\varphi_{j_1m_1}\varphi_{j_2m_2}. \tag{6.9}$$

Subtracting Equation 6.8 from Equation 6.9 leads to:

$$\sum_{m_1m_2} C(j_1j_2j;m_1m_2m)(m - m_1 - m_2)\varphi_{j_1m_1}\varphi_{j_2m_2} = 0, \tag{6.10}$$

which, for general $\varphi_{j_1m_1}$ and $\varphi_{j_2m_2}$, can only be satisfied if $m - m_1 - m_2 = 0$. Therefore, the azimuthal number m of the coupled state is an algebraic sum $m = m_1 + m_2$ of azimuthal numbers m_1 and m_2 of the uncoupled states, and therefore all Clebsch–Gordan coefficients are zero unless $m = m_1 + m_2$. For the uncoupled states $m_1 \le j_1$ and $m_2 \le j_2$, consequently the maximum value of m for all j is $j_1 + j_2$. Thus,

$$j_{\max} = j_1 + j_2. \tag{6.11}$$

The minimum value of the angular quantum number $j_{\min}$ in ψ_{jm} may be determined from the requirement that the size of the coupled space must be equal to that of the uncoupled space. The size of the uncoupled space is simply the product of the number of rows and the number of columns in the table $\varphi_{j_1m_1} \otimes \varphi_{j_2m_2}$, where azimuthal quantum numbers m_1 and m_2 assume values in the ranges $-j_1 \le m_1 \le j_1$ and $-j_2 \le m_2 \le j_2$, respectively, so the size of the uncoupled space is $(2j_1+1)(2j_2+1)$. The size of the coupled space originates from the allowed range of quantum numbers j and m in ψ_{jm}. Finding that size requires summing up the sizes of each individual azimuthal multiplet in the coupled state ψ_{jm}. Assuming that the value of j changes from $j_{\min}$ to $j_{\max}$, and taking into account that the size of each multiplet is $(2j+1)$, leads to the equality:

$$\sum_{j=j_{\min}}^{j_{\max}} (2j+1) = (2j_1+1)(2j_2+1). \tag{6.12}$$

This sum represents an arithmetic series, and resolves to:

$$\sum_{j=j_{\min}}^{j_{\max}} (2j+1) = (j_{\max} - j_{\min} + 1)\frac{(2j_{\min}+1)+(2j_{\max}+1)}{2} = (j_{\max} - j_{\min} + 1)(j_{\max} + j_{\min} + 1). \tag{6.13}$$

The value of $j_{\max}$ is already known, $j_1 + j_2$. Therefore,

$$\sum_{j=j_{\min}}^{j_{\max}} (2j+1) = (j_1 + j_2 + 1 - j_{\min})(j_1 + j_2 + 1 + j_{\min}) = (j_1 + j_2 + 1)^2 - j_{\min}^2. \tag{6.14}$$

Equating the size of the coupled space to that of the uncoupled space leads to:

$$(j_1 + j_2 + 1)^2 - j_{\min}^2 = (2j_1 + 1)(2j_2 + 1). \tag{6.15}$$

Starting from Equation 6.15 a sequence of simplifications resolves the unknown value of $j_{\min}$:

$$(j_1 + j_2)^2 + 2(j_1 + j_2) + 1 - j_{\min}^2 = 4j_1j_2 + 2(j_1 + j_2) + 1, \tag{6.16}$$

$$(j_1 + j_2)^2 - j_{\min}^2 = 4j_1j_2, \tag{6.17}$$

$$j_{\min}^2 = (j_1 - j_2)^2, \tag{6.18}$$

Since, by definition, $j_{\min} \ge 0$,

$$j_{\min} = |j_1 - j_2|. \tag{6.19}$$

By convention, the larger of the two angular quantum numbers is labeled j_1, and the smaller one j_2, so the vertical bars denoting the absolute value in Equation 6.19 can be dropped, leading to:

$$j_{\min} = j_1 - j_2. \tag{6.20}$$

Having determined the limits of indices j and m in the coupled eigenfunction ψ_{jm}, it is now possible to represent the Clebsch–Gordan coefficients $C(j_1j_2j;m_1m_2m)$ as the elements of a 3-dimensional matrix. Since j_1 and j_2 are constants, one of the dimensions in matrix C is index j, which spans the range from $j_1 - j_2$ to $j_1 + j_2$. The other two dimensions in the matrix are the azimuthal indices m_1 and m_2; these span the ranges $-j_1 \le m_1 \le j_1$ and $-j_2 \le m_2 \le j_2$. However, because $m_1 + m_2 = m$, it is customary to use the indices m_1 and m, and to compute m_2 from $m_2 = m - m_1$. This leads to a popular notation $C(j_1j_2j;m_1m_2m) \equiv C(j_1j_2j;m_1, m - m_1)$. The variable indices j, m_1, and m constitute the 3-dimensional matrix of Clebsch–Gordan coefficients.

It is convenient to look at the matrix of Clebsch–Gordan coefficients as a set of slices of regular 2-dimensional matrices $m_1 \times m$ or $m_1 \times m_2$ that are cut off from the 3-dimensional matrix $j \times m_1 \times m$ by index j, although other forms of representation of that intricate matrix are equally viable. As the index j varies, the range of the indices m_1 and m changes as well, and therefore the size of the slices also vary in size, so, except for the trivial case of $j_1 = j_2 = 0$, the 3-dimensional matrix is not a cuboid. In an alternative view, the individual matrices may be padded with zeroes in order to give the entire structure a cuboid shape.

Now it is time to explain the nature of the double sum in the Clebsch–Gordan equation, Equation 6.5. In this sum, the value of index m is fixed by the choice of function ψ_{jm}, and the indices m_1 and m_2 are iterated. Due to the condition $m = m_1 + m_2$, it is only necessary to vary one of the two parameters. Typically, the sum runs over index m_1, in which case the other index m_2 is automatically determined by the relation $m_2 = m - m_1$. This decision provides a more conventional form of the Clebsch–Gordan equation:

$$\psi_{jm} = \sum_{m_1=-j_1}^{j_1} C(j_1j_2j;m_1m_2m)\varphi_{j_1m_1}\varphi_{j_2m_2}. \tag{6.21}$$

The general orthonormality property of eigenfunctions $\varphi_{j_1m_1}$, $\varphi_{j_2m_2}$, and ψ_{jm} can be helpful in deriving a number of useful orthogonality relations for Clebsch–Gordan coefficients. For the purpose of these derivations, the original expansion involving Clebsch–Gordan coefficients may be rewritten in order to highlight their dependence on a particular set of indices j^a and m^a. This gives:

$$\psi_{j^am^a} = \sum_{m_1^am_2^a} C(j_1^aj_2^aj^a;m_1^am_2^am^a)\varphi_{j_1^am_1^a}\varphi_{j_2^am_2^a}. \tag{6.22}$$

The use of the old schematic notation in Equation 6.22 showing that the sum depends on both indices m_1^a and m_2^a provides extra clarity at this point. Multiplying both sides of Equation 6.22 by $\psi^*_{j^bm^b} \equiv \langle j^bm^b|$ from the left and integrating leads to:

$$\langle j^bm^b|j^am^a\rangle = \sum_{m_1^am_2^a} C(j_1^aj_2^aj^a;m_1^am_2^am^a)\langle j^bm^b|j_1^am_1^a;j_2^am_2^a\rangle. \tag{6.23}$$

The left side of this equation simplifies to $\delta_{j^aj^b}\delta_{m^am^b}$. The integral $\langle j^bm^b|j_1^am_1^a;j_2^am_2^a\rangle$ on the right-hand side still requires some additional transformation. Expanding $\langle j^bm^b|$ into an additional series of Clebsch–Gordan coefficients gives:

$$\sum_{m_1^am_2^a}\sum_{m_1^bm_2^b} C(j_1^aj_2^aj^a;m_1^am_2^am^a)C(j_1^bj_2^bj^b;m_1^bm_2^bm^b)\langle j_1^bm_1^b;j_2^bm_2^b\,|\,j_1^am_1^a;j_2^am_2^a\rangle = \delta_{j^aj^b}\delta_{m^am^b}. \tag{6.24}$$

The remaining integrals in the sum consists of two independent integrals over the independent function spaces for particles 1 and 2, and these, too, have their own orthonormality conditions:

$$\langle j_1^bm_1^b;j_2^bm_2^b\,|\,j_1^am_1^a;j_2^am_2^a\rangle = \langle j_1^bm_1^b|j_1^am_1^a\rangle\langle j_2^bm_2^b|j_2^am_2^a\rangle = \delta_{j_1^aj_1^b}\delta_{m_1^am_1^b}\delta_{j_2^aj_2^b}\delta_{m_2^am_2^b}. \tag{6.25}$$

Since indices j_1 and j_2 are not iterated under the sum, and have fixed values, it implies that $j_1^a = j_1^b = j_1$ and $j_2^a = j_2^b = j_2$. Therefore,

$$\langle j_1^bm_1^b;j_2^bm_2^b\,|\,j_1^am_1^a;j_2^am_2^a\rangle = \delta_{m_1^am_1^b}\delta_{m_2^am_2^b}. \tag{6.26}$$

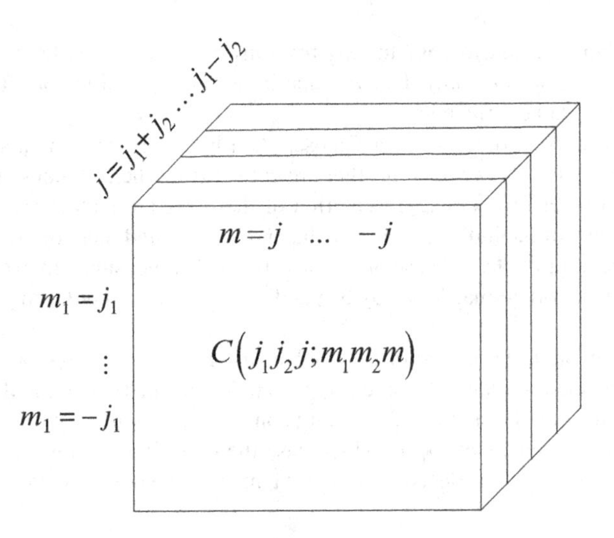

FIGURE 6.1 Three-dimensional matrix of Clebsch–Gordan coefficients.

The presence of this integral in Equation 6.24 zeroes the elements of the sum, unless $m_1^a = m_1^b = m_1$ and $m_2^a - m_2^b = m_2$. These conditions automatically enforce the equality $m^a = m^b = m$ because of the requirement $m = m_1 + m_2$. With those new conditions, Equation 6.24 simplifies to:

$$\sum_{m_1=-j_1}^{j_1} C(j_1 j_2 j^a; m_1 m_2 m) C(j_1 j_2 j^b; m_1 m_2 m) = \delta_{j^a j^b}. \tag{6.27}$$

This important relation shows that the Clebsch–Gordan coefficients are orthogonal for different values of the angular quantum number j of the coupled state ψ_{jm}.

This orthogonality condition for Clebsch–Gordan coefficients provides the means to prove the inverse expansion:

$$\varphi_{j_1 m_1} \varphi_{j_2 m_2} = \sum_{j=j_1-j_2}^{j_1+j_2} C(j_1 j_2 j; m_1 m_2 m) \psi_{jm}. \tag{6.28}$$

This expansion shows that the uncoupled (basis) state $\varphi_{j_1 m_1} \varphi_{j_2 m_2}$ can be obtained from the linear combination of the coupled states ψ_{jm} by summing up all possible states in the quantum number j of the coupled system. In 3-dimensional representation, this corresponds to fixing indices m_1 and m (or m_2) in the matrix slice in Figure 6.1, and weighting the functions ψ_{jm} with the corresponding Clebsch–Gordan coefficients located in row j. Proof of this statement will now be given.

Multiplying both sides of Equation 6.28 by $C(j_1 j_2 j^a; m_1 m_2 m)$ and summing up over index m_1 gives:

$$\sum_{m_1=-j_1}^{j_2} C(j_1 j_2 j^a; m_1 m_2 m) \varphi_{j_1 m_1} \varphi_{j_2 m_2} = \sum_{m_1=-j_1}^{j_2} C(j_1 j_2 j^a; m_1 m_2 m) \sum_{j=j_1-j_2}^{j_1+j_2} C(j_1 j_2 j; m_1 m_2 m) \psi_{jm}. \tag{6.29}$$

Rearranging the order of summation in the right-hand side leads to:

$$\sum_{m_1=-j_1}^{j_1} C(j_1 j_2 j^a; m_1 m_2 m) \varphi_{j_1 m_1} \varphi_{j_2 m_2} = \sum_{j=j_1-j_2}^{j_1+j_2} \left[\sum_{m_1=-j_1}^{j_1} C(j_1 j_2 j^a; m_1 m_2 m) C(j_1 j_2 j; m_1 m_2 m) \right] \psi_{jm}. \tag{6.30}$$

The expression in square brackets is simply δ_{jj^a}, according to Equation 6.27. The orthogonality condition reduces the sum over j in Equation 6.30 to a single term $j = j^a$. With that, Equation 6.30 becomes:

$$\sum_{m_1=-j_1}^{j_1} C(j_1 j_2 j^a; m_1 m_2 m)\,\varphi_{j_1 m_1}\varphi_{j_2 m_2} = \psi_{jm}, \tag{6.31}$$

which is the standard Clebsch–Gordan expansion. Since Equation 6.31 is correct by definition, it proves that the starting Equation 6.28 of this derivation must be true as well.

The inverse expansion reveals another orthogonality property of Clebsch–Gordan coefficients. Multiplying both sides of Equation 6.28 by $\varphi^*_{j_1 m_1^a}\varphi^*_{j_2 m_2^a}$, and integrating, leads to:

$$\langle j_1 m_1^a; j_2 m_2^a | j_1 m_1; j_2 m_2 \rangle = \sum_{j=j_1-j_2}^{j_1+j_2} C(j_1 j_2 j; m_1 m_2 m)\langle j_1 m_1^a; j_2 m_2^a | jm \rangle. \tag{6.32}$$

Due to the eigenfunctions being orthogonal, the left-hand side reduces to:

$$\langle j_1 m_1^a; j_2 m_2^a | j_1 m_1; j_2 m_2 \rangle = \langle j_1 m_1^a | j_1 m_1 \rangle \langle j_2 m_2^a | j_2 m_2 \rangle = \delta_{m_1 m_1^a}\delta_{m_2 m_2^a}. \tag{6.33}$$

The eigenfunction $\psi_{jm} \equiv |\,jm\rangle$ of the coupled state located in the right-hand side of Equation 6.32 may be expanded in the basis of uncoupled states $\varphi_{j_1 m_1}\varphi_{j_2 m_2}$ *via* the Clebsch–Gordan equation. Doing that, and reversing the sides of the equation, leads to:

$$\sum_{j=j_1-j_2}^{j_1+j_2}\sum_{m_1^b=-j_1}^{j_2} C(j_1 j_2 j; m_1 m_2 m)\,C(j_1 j_2 j; m_1^b m_2^b m)\langle j_1 m_1^a; j_2 m_2^a | j_1 m_1^b; j_2 m_2^b \rangle = \delta_{m_1 m_1^a}\delta_{m_2 m_2^a}. \tag{6.34}$$

For the integral under the sum in Equation 6.34 to be non-zero, it is required that $m_1^b = m_1^a$ and $m_2^b = m_2^a$; and that eliminates the summation over index m_1^b. What remains is:

$$\sum_{j=j_1-j_2}^{j_1+j_2} C(j_1 j_2 j; m_1 m_2 m)\,C(j_1 j_2 j; m_1^a m_2^a m) = \delta_{m_1 m_1^a}\delta_{m_2 m_2^a}. \tag{6.35}$$

In this equation, the azimuthal quantum numbers m_1 and m_1^a are both members of the same multiplet $-j_1 \le m_1 \le j_1$. Similarly, the quantum numbers m_2 and m_2^a are members of the second multiplet $-j_2 \le m_2 \le j_2$. Because of that, and due to conditions $m = m_1 + m_2$ and $m^a = m_1^a + m_2^a$ which allow the use of indices m and m^a instead of m_2 and m_2^a, when the values of m_1 and m_1^a are fixed, the pair $\delta_{m_1 m_1^a}\delta_{m_2 m_2^a}$ may be equivalently presented as $\delta_{m_1 m_1^a}\delta_{m m^a}$. This leads to:

$$\sum_{j=j_1-j_2}^{j_1+j_2} C(j_1 j_2 j; m_1 m_2 m)\,C(j_1 j_2 j; m_1^a m_2^a m) = \delta_{m_1 m_1^a}\delta_{m m^a}, \tag{6.36}$$

which is another orthonormality condition of the Clebsch–Gordan coefficients.

6.2 Evaluation of Clebsch–Gordan Coefficients

Section 6.1 introduced the basic principles about the Clebsch–Gordan coefficients. The next step is to find the numerical values of those coefficients. The Clebsch–Gordan equation comprises two important principles.

According to the first rule, the construction of the coupled state from the uncoupled ones follows the principle that the addition of angular momenta j_1 and j_2 yields the total angular momenta j with j having the values $j = j_1 + j_2, j_1 + j_2 - 1, \ldots, j_0, \ldots, j_1 - j_2$, with each value occurring only once. The second rule states that, for a given value of the quantum number $j = j_0$, when a particular eigenvalue $m = m_0$ of angular momentum $\mathbf{J}_z$ occurs in the coupled state $\psi_{j_0 m_0}$, all members of the multiplet $m = j_0, j_0 - 1, \ldots, m_0, \ldots, -j_0$ must occur together with m_0, so that if the state $\psi_{j_0 m_0}$ exists, it is always accompanied by all members of the j_0-multiplet. These two rules provide the basis for computation of Clebsch–Gordan coefficients. A convenient point to start the analysis is at the maximum value of $j = j_1 + j_2$.

The coupled eigenfunction ψ_{jm} is a linear combination of uncoupled states $\varphi_{j_1 m_1}$ and $\varphi_{j_2 m_2}$ that have the angular quantum numbers j_1 and j_2, respectively, and are members of the multiplets $j_1 \le m_1 \le j_1$ and $-j_2 \le m_2 \le j_2$. A direct product $\varphi_{j_1 m_1} \otimes \varphi_{j_2 m_2}$ of these individual basis functions creates a complete basis set of the uncoupled state of the system. For the maximum values of indices $j = j_1 + j_2$ and $m = j_1 + j_2$ the equation for the coupled state ψ_{jm} assumes an especially simple form since the sum over index m_1 in Equation 6.31 reduces to a single term:

$$\psi_{j_1+j_2, j_1+j_2} = C(j_1, j_2, j_1 + j_2; j_1, j_2, j_1 + j_2)\varphi_{j_1 j_1}\varphi_{j_2 j_2}, \tag{6.37}$$

where $m_1 = j_1$ and $m_2 = j_2$. The normalization condition for the coupled state ψ_{jm} requires

$$\langle \psi_{j_1+j_2, j_1+j_2} | \psi_{j_1+j_2, j_1+j_2} \rangle = C^2 \langle \varphi_{j_1 j_1}\varphi_{j_2 j_2} | \varphi_{j_1 j_1}\varphi_{j_2 j_2} \rangle = 1. \tag{6.38}$$

Since the eigenfunctions in the uncoupled state $\varphi_{j_1 m_1}\varphi_{j_2 m_2}$ are normalized, this leads to:

$$C(j_1, j_2, j_1 + j_2; j_1, j_2, j_1 + j_2) = 1. \tag{6.39}$$

The positive unit value in Equation 6.39 follows the convention that Clebsch–Gordan coefficients are real numbers.

The unit value of the Clebsch–Gordan coefficient $C(j_1, j_2, j_1 + j_2; j_1, j_2, j_1 + j_2)$ implies that $\psi_{j_1+j_2, j_1+j_2} = \varphi_{j_1 j_1}\varphi_{j_2 j_2}$. Since ψ_{jm} is an eigenfunction of the square of the angular momentum operator $\mathbf{J}^2$ of the composite system, the same must be true about function $\varphi_{j_1 j_1}\varphi_{j_2 j_2}$. This may be verified by applying the operator $\mathbf{J}^2$ to function $\varphi_{j_1 j_1}\varphi_{j_2 j_2}$. Before doing that, it is necessary to express $\mathbf{J}^2$ in terms of the operators acting onto the individual particle states. The square of the angular momentum operator of the composite system is:

$$\mathbf{J}^2 = (\mathbf{J}_1 + \mathbf{J}_2)^2 = \mathbf{J}_1^2 + 2\mathbf{J}_1 \cdot \mathbf{J}_2 + \mathbf{J}_2^2. \tag{6.40}$$

The operators $\mathbf{J}_1^2$ and $\mathbf{J}_2^2$ and their effects have already been worked out, except the action of the vector dot product $\mathbf{J}_1 \cdot \mathbf{J}_2$ on $\varphi_{j_1 j_1}\varphi_{j_2 j_2}$. However, the operator product is a sum of products of Cartesian operators:

$$\mathbf{J}_1 \cdot \mathbf{J}_2 = \mathbf{J}_{1x}\mathbf{J}_{2x} + \mathbf{J}_{1y}\mathbf{J}_{2y} + \mathbf{J}_{1z}\mathbf{J}_{2z}. \tag{6.41}$$

The next step is to convert the right-hand side of Equation 6.41 into a pair of lowering and raising operators $\mathbf{J}_{-1}\mathbf{J}_{+2}$ and $\mathbf{J}_{-2}\mathbf{J}_{+1}$. Since each operator in the product acts within its own single particle space, the sequence in which the lowering and raising operators are arranged is unimportant. Assigning the index 1 to the lowering operator and index 2 to the raising operator gives:

$$\mathbf{J}_{-1}\mathbf{J}_{+2} = (\mathbf{J}_{1x} - i\mathbf{J}_{1y})(\mathbf{J}_{2x} + i\mathbf{J}_{2y}) = \mathbf{J}_{1x}\mathbf{J}_{2x} - i\mathbf{J}_{1y}\mathbf{J}_{2x} + i\mathbf{J}_{1x}\mathbf{J}_{2y} + \mathbf{J}_{1y}\mathbf{J}_{2y}. \tag{6.42}$$

Interchanging the particle indices gives the complementary operator product:

$$\mathbf{J}_{-2}\mathbf{J}_{+1} = (\mathbf{J}_{2x} - i\mathbf{J}_{2y})(\mathbf{J}_{1x} + i\mathbf{J}_{1y}) = \mathbf{J}_{2x}\mathbf{J}_{1x} - i\mathbf{J}_{2y}\mathbf{J}_{1x} + i\mathbf{J}_{2x}\mathbf{J}_{1y} + \mathbf{J}_{2y}\mathbf{J}_{1y}. \tag{6.43}$$

Note that the operators $\mathbf{J}_1$ and $\mathbf{J}_2$ in the product of type $\mathbf{J}_1\mathbf{J}_2$ in Equations 6.42 and 6.43 commute, since they act on different particle spaces. Therefore, the position of operators in the product may be freely changed. Using that freedom, the addition of Equations 6.42 and 6.43 followed by cancellation of the complex terms produces:

$$\mathbf{J}_{-1}\mathbf{J}_{+2} + \mathbf{J}_{-2}\mathbf{J}_{+1} = 2\mathbf{J}_{1x}\mathbf{J}_{2x} + 2\mathbf{J}_{1y}\mathbf{J}_{2y}. \quad (6.44)$$

The sum of products of the commuting lowering and raising operators in the left-hand side also eliminates the original uncertainty in the assignment of particle index to a particular shift operator.

Substituting Equation 6.44 into Equation 6.41 helps to get rid of the inconvenient operators $\mathbf{J}_1$ and $\mathbf{J}_2$, and turns Equation 6.41 to:

$$2\mathbf{J}_1 \cdot \mathbf{J}_2 = \mathbf{J}_{-1}\mathbf{J}_{+2} + \mathbf{J}_{-2}\mathbf{J}_{+1} + 2\mathbf{J}_{1z}\mathbf{J}_{2z}. \quad (6.45)$$

Substituting this result back to Equation 6.40 for the square of angular momentum operator of the composite system leads to:

$$\mathbf{J}^2 = \mathbf{J}_1^2 + \mathbf{J}_{-1}\mathbf{J}_{+2} + \mathbf{J}_{-2}\mathbf{J}_{+1} + 2\mathbf{J}_{1z}\mathbf{J}_{2z} + \mathbf{J}_2^2. \quad (6.46)$$

At this point, the effect of all the terms when they act upon the eigenfunction of a non-coupled state is known, so application of operator $\mathbf{J}^2$ to function $\varphi_{j_1j_1}\varphi_{j_2j_2}$ gives:

$$\begin{aligned}\mathbf{J}^2\varphi_{j_1j_1}\varphi_{j_2j_2} &= \left(\mathbf{J}_1^2 + \mathbf{J}_{-1}\mathbf{J}_{+2} + \mathbf{J}_{-2}\mathbf{J}_{+1} + 2\mathbf{J}_{1z}\mathbf{J}_{2z} + \mathbf{J}_2^2\right)\varphi_{j_1j_1}\varphi_{j_2j_2} \\ &= j_1(j_1+1)\,\varphi_{j_1j_1}\varphi_{j_2j_2} + 0 + 0 + 2j_1j_2\,\varphi_{j_1j_1}\varphi_{j_2j_2} + j_2(j_2+1)\,\varphi_{j_1j_1}\varphi_{j_2j_2}.\end{aligned} \quad (6.47)$$

The coefficients in front of $\varphi_{j_1j_1}\varphi_{j_2j_2}$ may be rearranged to give:

$$j_1(j_1+1) + 2j_1j_2 + j_2(j_2+1) = j_1^2 + 2j_1j_2 + j_2^2 + j_1 + j_2 = (j_1+j_2)(j_1+j_2+1), \quad (6.48)$$

and the resulting equation becomes:

$$\mathbf{J}^2\varphi_{j_1j_1}\varphi_{j_2j_2} = (j_1+j_2)(j_1+j_2+1)\,\varphi_{j_1j_1}\varphi_{j_2j_2}. \quad (6.49)$$

This confirms that $\varphi_{j_1j_1}\varphi_{j_2j_2}$ is an eigenfunction of the operator $\mathbf{J}^2$ and has the eigenvalue (j_1+j_2) (j_1+j_2+1).

The next illustrative example to consider after $m_{\max} = j_1 + j_2$ in the process of evaluation of Clebsch–Gordan coefficients is $m = j_1 + j_2 - 1$. Due to the general requirement that $m \le j$, the azimuthal quantum number $m = j_1 + j_2 - 1$ may be found only in two coupled states, $\psi_{j_1+j_2,j_1+j_2-1}$ and $\psi_{j_1+j_2-1,j_1+j_2-1}$. The contributing uncoupled states may be identified from an analysis of the possible range of azimuthal quantum numbers m_1 and m_2. The addition rule $m = m_1 + m_2$ restricts the values of m_1 and m_2 to two possible combinations: one $m_1 = j_1 - 1$ and $m_2 = j_2$, and the other $m_1 = j_1$ and $m_2 = j_2 - 1$. Therefore, only the uncoupled states $\varphi_{j_1,j_1-1}\varphi_{j_2,j_2}$ and $\varphi_{j_1,j_1}\varphi_{j_2,j_2-1}$ may contribute to the coupled state that has $m = j_1 + j_2 - 1$, and consequently each of the coupled states $\psi_{j_1+j_2,j_1+j_2-1}$ and $\psi_{j_1+j_2-1,j_1+j_2-1}$ is a separate linear combination of two uncoupled states, $\varphi_{j_1,j_1-1}\varphi_{j_2,j_2}$ and $\varphi_{j_1,j_1}\varphi_{j_2,j_2-1}$.

This analysis may be continued until the lower bound of index m is reached. The frequency of encountering a particular value of azimuthal number m occurring in all coupled states ψ_{jm}, where j varies in the range from $j_1 + j_2$ to $j_1 - j_2$, may be determined by counting the number of stars in the row:

$$\begin{array}{cccccc} & j_1+j_1 & j_1+j_2-1 & j_1+j_2-2 & \cdots & j_1-j_2 \\ & * & & & & \\ & * & * & & & \\ m & * & * & * & & \\ & \cdots & \cdots & \cdots & \cdots & \cdots \\ & * & * & * & \cdots & * \end{array} \quad (6.50)$$

In Equation 6.50, the value $m = j_1 + j_2$ appears precisely once, the value $m = j_1 + j_2 - 1$ is present in two coupled states $j_1 + j_2$ and $j_1 + j_2 - 1$, and the value $m = j_1 - j_2$ is present in every coupled state.

Evaluation of the Clebsch–Gordan coefficients, which scale the weight of each uncoupled state $\varphi_{j_1m_1}\varphi_{j_2m_2}$ in the coupled state ψ_{jm}, is performed in two steps. In the first step, the upper bound, i.e., $m = j$ eigenfunction, ψ_{jj} is obtained for each j-multiplet, where $j_1 + j_2 \leq j \leq |j_1 - j_2|$. The process of finding ψ_{jj} utilizes a property of multiplets, that the raising operator $\mathbf{J}_+ = \mathbf{J}_{1+} + \mathbf{J}_{2+}$ annihilates the upper-bound eigenfunction:

$$\mathbf{J}_+\psi_{jj} = 0. \tag{6.51}$$

In the second step, successive application of the lowering operator $\mathbf{J}_- = \mathbf{J}_{1-} + \mathbf{J}_{2-}$ starting from ψ_{jj} generates all members of the j-multiplet until the last member $\psi_{j,-j}$ of the multiplet is reached, at which point the eigenfunction is annihilated:

$$\mathbf{J}_-\psi_{j,-j} = 0. \tag{6.52}$$

Evaluation of the Clebsch–Gordan coefficients relies on the rule that these coefficients are overlap integrals between the corresponding coupled and uncoupled eigenfunctions of the system. This integral is obtained by multiplying both sides of the Clebsch–Gordan equation, Equation 6.31 by $\varphi^*_{j_1m_1^a}\varphi^*_{j_2m_2^a}$, from the left, and integrating. The orthonormality of the basis eigenvectors $\langle j_1m_1^a | j_1m_1\rangle\langle j_2m_2^a | j_2m_2\rangle = \delta_{m_1^am_1}\delta_{m_2^am_2}$ reduces the sum to a single term:

$$C(j_1j_2j;m_1m_2m) = \langle j_1m_1;j_2m_2 | jm\rangle. \tag{6.53}$$

where $m_1^a = m_1$ and $m_2^a = m_2$.

The first step in evaluating the Clebsch–Gordan coefficients involves determining those coefficients, which constitute the upper-bound coupled eigenfunction ψ_{jj} for each j-multiplet. This uses a recursion relation derived from the application of raising operator $\mathbf{J}_+$ to the coupled eigenfunction ψ_{jj} present in the integral $C(j_1j_2j;m_1m_2j) \equiv \langle j_1m_1;j_2m_2|jj\rangle$. The action of the raising operator results in:

$$\langle j_1m_1;j_2m_2|\mathbf{J}_+|jj\rangle = 0. \tag{6.54}$$

This integral vanishes due to Equation 6.51.

Continuing the derivation, the integral in Equation 6.54 may be transformed into an equivalent representation by having the shift operator act on the uncoupled eigenfunction located in the left side of the integral. Consider the Hermitian adjoint of the integral in Equation 6.54, and recall that the Hermitian adjoint of the shift operator is $(\mathbf{J}_\pm)^+ = \mathbf{J}_\mp$. Thus,

$$\langle j_1m_1;j_2m_2|\mathbf{J}_+|jj\rangle = \langle jj|\mathbf{J}_-|j_1m_1;j_2m_2\rangle^+ = \langle jj|\mathbf{J}_{1-} + \mathbf{J}_{2-}|j_1m_1;j_2m_2\rangle^+. \tag{6.55}$$

Application of the lowering operator to the uncoupled eigenstate under the integral leads to:

$$\begin{aligned}&\langle jj|\mathbf{J}_{1-} + \mathbf{J}_{2-}|j_1m_1;j_2m_2\rangle^+ \\ &\quad = \left[(j_1+m_1)(j_1-m_1+1)\right]^{\frac{1}{2}}\langle jj|j_1m_1-1;j_2m_2\rangle^+ + \left[(j_2+m_2)(j_2-m_2+1)\right]^{\frac{1}{2}}\langle jj|j_1m_1;j_2m_2-1\rangle.\end{aligned} \tag{6.56}$$

Finally, returning the integral back from its Hermitian adjoint gives:

$$\begin{aligned}\langle jj|\mathbf{J}_{1-} + \mathbf{J}_{2-}|j_1m_1;j_2m_2\rangle^+ &= \langle j_1m_1;j_2m_2|\mathbf{J}_+|jj\rangle \\ &= \left[(j_1+m_1)(j_1-m_1+1)\right]^{\frac{1}{2}}\langle jj|j_1m_1-1;j_2m_2\rangle \\ &+ \left[(j_2+m_2)(j_2-m_2+1)\right]^{\frac{1}{2}}\langle j_1m_1;j_2m_2-1|jj\rangle^+.\end{aligned} \tag{6.57}$$

Since the integral in the left-hand side of Equation 6.57 is equal to zero:

$$\left[(j_1+m_1)(j_1-m_1+1)\right]^{\frac{1}{2}}\langle j_1m_1-1;j_2m_2\mid jj\rangle+\left[(j_2+m_2)(j_2-m_2+1)\right]^{\frac{1}{2}}\langle j_1m_1;j_2m_2-1\mid jj\rangle=0. \tag{6.58}$$

Continuing the derivation using the present notation risks that the equations becoming too crowded to be readable. In order to improve their readability, it is necessary to introduce a simplified notation for the integrals. Observing that angular quantum numbers j_1, j_2, and j remain constant for the duration of the derivation, it is convenient to temporarily drop them from the integral notation and keep only the indices of azimuthal quantum number, such as,

$$\langle j_1m_1-1;j_2m_2\,|\,jj\rangle\equiv\langle m_1-1;m_2\,|\,j\rangle. \tag{6.59}$$

Using this notation, Equation 6.58 reduces to:

$$\left[(j_1+m_1)(j_1-m_1+1)\right]^{\frac{1}{2}}\langle m_1-1;m_2\,|\,j\rangle+\left[(j_2+m_2)(j_2-m_2+1)\right]^{\frac{1}{2}}\langle m_1;m_2-1\,|\,j\rangle=0. \tag{6.60}$$

This equation may now be simplified. Keeping the integral $\langle m_1-1;m_2\,|\,j\rangle$ on the left side of the equation, and moving everything else to the right side leads to:

$$\langle m_1-1;m_2\,|\,j\rangle=-\frac{\left[(j_2+m_2)(j_2-m_2+1)\right]^{\frac{1}{2}}}{\left[(j_1+m_1)(j_1-m_1+1)\right]^{\frac{1}{2}}}\langle m_1;m_2-1\,|\,j\rangle. \tag{6.61}$$

This establishes the relation between two Clebsch–Gordan coefficients

$$C(j_1j_2j;m_1-1,m_2,j)\equiv\langle m_1-1;m_2\,|\,j\rangle, \tag{6.62}$$

and

$$C(j_1j_2j;m_1,m_2-1,j)\equiv\langle m_1;m_2-1\,|\,j\rangle. \tag{6.63}$$

Both coefficients must obey the condition $m_1-1+m_2\equiv m_1+m_2-1=m=j$ in order for them to be non-zero. This translates to $m_2=j-m_1+1$. Given that azimuthal indices have the maximum values $m_1=j_1$ and $m_2=j_2$, the maximum value for index j in the recursion becomes j_1+j_2-1. Substituting $j-m_1+1$ for m_2 in Equation 6.61 gives:

$$\langle m_1-1;j-m_1+1\,|\,j\rangle=-\frac{\left[(j_2+j-m_1+1)(j_2-j+m_1)\right]^{\frac{1}{2}}}{\left[(j_1+m_1)(j_1-m_1+1)\right]^{\frac{1}{2}}}\langle m_1;j-m_1\,|\,j\rangle. \tag{6.64}$$

This is the desired recurrence relationship between Clebsch–Gordan coefficients for the upper-bound coupled state ψ_{jj}, where j spans the range from j_1+j_2-1 to j_1-j_2.

In order to use Eq. 6.64 in the derivation of a general Clebsch–Gordan coefficient, it is necessary to sum up multiple recursion steps into a single equation. The recursion computation begins from the upper bound $m_1=j_1$. The first recursion step down gives:

$$\langle j_1-1;j-j_1+1\,|\,j\rangle=-\frac{\left[(j_2+j-j_1+1)(j_2-j+j_1)\right]^{\frac{1}{2}}}{\left[(2j_1)(j_1-j_1+1)\right]^{\frac{1}{2}}}\langle j_1;j-j_1\,|\,j\rangle. \tag{6.65}$$

A second recursion step down starts from $m_1 = j_1 - 1$:

$$\langle j_1-2; j-j_1+2|j\rangle = -\frac{\left[(j_2+j-j_1+2)(j_2-j+j_1-1)\right]^{\frac{1}{2}}}{\left[(2j_1-1)(j_1-j_1+2)\right]^{\frac{1}{2}}}\langle j_1-1; j-j_1+1|j\rangle. \tag{6.66}$$

At the kth step, when $m_1 = j_1 - k + 1$, the equation becomes:

$$\langle j_1-k; j-j_1+k|j\rangle = -\frac{\left[(j_2+j-j_1+k)(j_2-j+j_1-k+1)\right]^{\frac{1}{2}}}{\left[(2j_1-k+1)(j_1-j_1+k)\right]^{\frac{1}{2}}}\langle j_1-k+1; j-j_1+k-1|j\rangle. \tag{6.67}$$

Observe that each step employs the integral computed in the previous step. This makes it possible to jump from the first step straight to the k-th step without the need to compute the intermediate integrals. All that is required is to multiply the coefficients:

$$\left[\frac{(j_2+j-j_1+1)(j_2-j+j_1)}{(2j_1)1}\frac{(j_2+j-j_1+2)(j_2-j+j_1-1)}{(2j_1-1)2}\cdots\frac{(j_2+j-j_1+k)(j_2-j+j_1-k+1)}{(2j_1-k+1)k}\right]^{\frac{1}{2}}. \tag{6.68}$$

In order to simplify this long product, it is necessary to gather related terms in separate groups so they can be treated together. The first group is $(j_2+j-j_1+1)(j_2+j-j_1+2)\cdots(j_2+j-j_1+k)$. Arranging the terms in the product in backward order, $(j_2+j-j_1+k)(j_2+j-j_1+k-1)\cdots(j_2+j-j_1+1)$, reveals that this product is a factorial $(j_2+j-j_1+k)!$, which is truncated at the element (j_2+j-j_1). Therefore,

$$(j_2+j-j_1+1)(j_2+j-j_1+2)\cdots(j_2+j-j_1+k) = \frac{(j_2+j-j_1+k)!}{(j_2+j-j_1)!}. \tag{6.69}$$

Similarly, the second group, $(j_2-j+j_1)(j_2-j+j_1-1)\cdots(j_2-j+j_1-k+1)$, corresponds to a factorial $(j_2-j+j_1)!$, which is truncated at (j_2-j+j_1-k). With that,

$$(j_2-j+j_1)(j_2-j+j_1-1)\cdots(j_2-j+j_1-k+1) = \frac{(j_2-j+j_1)!}{(j_2-j+j_1-k)!}. \tag{6.70}$$

The third group of related terms simplifies to:

$$(2j_1)(2j_1-1)\cdots(2j_1-k+1) = \frac{(2j_1)!}{(2j_1-k)!}. \tag{6.71}$$

The remaining terms compose the fourth and the final factorial, $k!$. Assembling the various parts together completes the reduction of Equation 6.68 to factorial form:

$$\left[\frac{(j_2+j-j_1+k)!}{(j_2+j-j_1)!}\frac{(j_2-j+j_1)!}{(j_2-j+j_1-k)!}\frac{(2j_1-k)!}{(2j_1)!}\frac{1}{k!}\right]^{\frac{1}{2}}. \tag{6.72}$$

With that, the k-th consecutive recursion yields:

$$\langle j_1-k; j-j_1+k|j\rangle = (-1)^k\left[\frac{(j_2+j-j_1+k)!}{(j_2+j-j_1)!}\frac{(j_2-j+j_1)!}{(j_2-j+j_1-k)!}\frac{(2j_1-k)!}{(2j_1)!}\frac{1}{k!}\right]^{\frac{1}{2}}\langle j_1; j-j_1|j\rangle. \tag{6.73}$$

This equation can now be converted into a form suitable for calculating the Clebsch–Gordan coefficients. Given that, in the upper-bound eigenfunction, the azimuthal quantum number of the coupled state has its maximum value of $m = j$, then replacing k by $j_1 - m_1$, and replacing $j - j_1 + k = j - m_1$ by m_2 leads to:

$$\langle m_1;m_2|j\rangle = (-1)^{j_1-m_1}\left[\frac{(j_2+j-m_1)!}{(j_2+j-j_1)!}\frac{(j_2-j+j_1)!}{(j_2-j+m_1)!}\frac{(j_1+m_1)!}{(2j_1)!(j_1-m_1)!}\right]^{\frac{1}{2}}\langle j_1;j-j_1|j\rangle, \tag{6.74}$$

or, after switching back to the full notation for the integrals, to:

$$\langle j_1m_1;j_2m_2|jj\rangle = (-1)^{j_1-m_1}\left[\frac{(j_2+j-m_1)!}{(j_2+j-j_1)!}\frac{(j_2-j+j_1)!}{(j_2-j+m_1)!}\frac{(j_1+m_1)!}{(2j_1)!(j_1-m_1)!}\right]^{\frac{1}{2}}\langle j_1j_1;j_2,j-j_1|jj\rangle. \tag{6.75}$$

Note that, in this recurrence relation, j can only assume values in the range $j_1 + j_2 - 1$ to $j_1 - j_2$, therefore $j_2 \neq j - j_1$.

There are still two undefined integrals in Equation 6.75, one of which needs to be eliminated. Integral $\langle j_1j_1;j_2,j-j_1|jj\rangle$, on the right-hand side of Equation 6.75, can be evaluated by employing the orthonormality condition of Clebsch–Gordan coefficients described in Equation 6.27. For $j^a = j$, the orthonormality condition reduces to the normalization condition:

$$\sum_{m_1=-j_1}^{j_1} C(j_1j_2j;m_1m_2m)^2 = 1. \tag{6.76}$$

Since $C(j_1j_2j;m_1m_2j) \equiv \langle j_1m_1;j_2m_2|jj\rangle$, substituting Equation 6.75 into Equation 6.76, and moving the terms which do not depend on the summation index outside the sum, leads to:

$$\langle j_1j_1;j_2,j-j_1|jj\rangle^2\frac{(j_2-j+j_1)!}{(j_2+j-j_1)!(2j_1)!}\sum_{m_1=-j_1}^{j_1}\frac{(j_2+j-m_1)!(j_1+m_1)!}{(j_2-j+m_1)!(j_1-m_1)!} = 1. \tag{6.77}$$

Employing the equality explained in Appendix A.7:

$$\sum_{m_1=-j_1}^{j_1}\frac{(j_2+j-m_1)!(j_1+m_1)!}{(j_2-j+m_1)!(j_1-m_1)!} = \frac{(j_1+j_2+j+1)!(j_2-j_1+j)!(j_1-j_2+j)!}{(2j+1)!(j_1+j_2-j)!}, \tag{6.78}$$

and canceling out common terms simplifies Equation 6.77 to:

$$\langle j_1j_1;j_2,j-j_1|jj\rangle^2\frac{(j_1+j_2+j+1)!(j_1-j_2+j)!}{(2j_1)!(2j+1)!} = 1, \tag{6.79}$$

or, rearranging:

$$\langle j_1j_1;j_2,j-j_1|jj\rangle^2 = \frac{(2j_1)!(2j+1)!}{(j_1+j_2+j+1)!(j_1-j_2+j)!}. \tag{6.80}$$

The phase convention for Clebsch–Gordan coefficients defines this integral as being positive, therefore,

$$C(j_1,j_2,j;j_1,j-j_1,j) \equiv \langle j_1j_1;j_2,j-j_1|jj\rangle = \left[\frac{(2j_1)!(2j+1)!}{(j_1+j_2+j+1)!(j_1-j_2+j)!}\right]^{\frac{1}{2}}. \tag{6.81}$$

Substituting Equation 6.81 into Equation 6.75, and canceling out $(2j_1)!$ leads to:

$$\langle j_1m_1;j_2m_2|jj\rangle=(-1)^{j_1-m_1}\left[\frac{(j_2+j-m_1)!\;(j_2-j+j_1)!\;(j_1+m_1)!}{(j_2+j-j_1)!\;(j_2-j+m_1)!\;(j_1-m_1)!}\frac{(2j+1)!}{(j_1+j_2+j+1)!(j_1-j_2+j)!}\right]^{\frac{1}{2}}. \quad (6.82)$$

Since $C(j_1j_2j;m_1m_2j)\equiv\langle j_1m_1;j_2m_2|jj\rangle$ for $m=j$, the integral in the left-hand side of Equation 6.82 may be replaced by the corresponding Clebsch–Gordan coefficient. In addition to that, restoring the index $m_2=j-m_1$ gives the first equation for evaluation of the Clebsch–Gordan coefficients:

$$C(j_1j_2j;m_1m_2j)=\delta_{m_1+m_2,j}(-1)^{j_1-m_1}\left[\frac{(j_1+m_1)!(j_2+m_2)!\;(j_2-j+j_1)!}{(j_1-m_1)!(j_2-m_2)!\;(j_2+j-j_1)!}\frac{(2j+1)!}{(j_1+j_2+j+1)!(j_1-j_2+j)!}\right]^{\frac{1}{2}}. \quad (6.83)$$

which determines the upper-bound coupled eigenfunction ψ_{jj} in the expansion:

$$\psi_{jj}=\sum_{m_1=-j_1}^{j_1}C(j_1j_2j;m_1m_2j)\,\varphi_{j_1m_1}\varphi_{j_2m_2}, \quad (6.84)$$

where $m_1+m_2=m=j$. The term $\delta_{m_1+m_2,j}$ in Equation 6.83, explicitly enforces the latter condition. As usual, m must equal m_1+m_2, and the value of index j spans the range from j_1+j_2-1 to j_1-j_2.

Before moving on to step two in the evaluation of Clebsch–Gordan coefficients, it is helpful to briefly review the working of the lowering and raising operators. Application of the lowering operator $\mathbf{J}_-$ to the eigenfunction $|jm\rangle$ has the following effect:

$$\mathbf{J}_-|jm\rangle=\left[(j+m)(j-m+1)\right]^{\frac{1}{2}}|jm-1\rangle. \quad (6.85)$$

Starting from the upper-bound eigenfunction $|jj\rangle$, and performing k consecutive applications of the lowering operator produces:

$$\begin{aligned}
&1)\quad \mathbf{J}_-|jj\rangle=\left[(2j)\cdot 1\right]^{\frac{1}{2}}|jj-1\rangle,\\
&2)\quad \mathbf{J}_-|jj-1\rangle=\left[(2j-1)\cdot 2\right]^{\frac{1}{2}}|jj-2\rangle,\\
&3)\quad \mathbf{J}_-|jj-2\rangle=\left[(2j-2)\cdot 3\right]^{\frac{1}{2}}|jj-3\rangle,\\
&\;\vdots\qquad\quad \vdots\qquad\qquad\qquad\quad \vdots\\
&k)\quad \mathbf{J}_-|jj-k+1\rangle=\left[(2j-k+1)\cdot k\right]^{\frac{1}{2}}|jj-k\rangle.
\end{aligned} \quad (6.86)$$

The k consecutive steps combine into a single equation:

$$\mathbf{J}_-^k|jj\rangle=\left[\frac{(2j)!}{(2j-k)!}k!\right]^{\frac{1}{2}}|jj-k\rangle, \quad (6.87)$$

where $(2j)\cdot1\cdot(2j-1)\cdot2\ldots(2j-k+1)\cdot k=\dfrac{(2j)!}{(2j-k)!}k!$.

Applying the substitution $k = j - m$, which correspond to $j - m$ consecutive applications of the lowering operator, gives:

$$\mathbf{J}_{-}^{j-m} | jj\rangle = \left[\frac{(2j)!(j-m)!}{(j+m)!}\right]^{\frac{1}{2}} | jm\rangle. \tag{6.88}$$

This equation, when rearranged as

$$| jm\rangle = \left[\frac{(j+m)!}{(2j)!(j-m)!}\right]^{\frac{1}{2}} \mathbf{J}_{-}^{j-m} | jj\rangle, \tag{6.89}$$

provides a convenient mechanism for obtaining the mth member of the j-multiplet from the upper-bound eigenfunction $|jj\rangle$.

The effect of the raising operator $\mathbf{J}_+$ on the eigenfunction $|jm\rangle$ is complementary to the effect produced by the lowering operator, so the effect of k consecutive applications of the raising operator is:

$$\begin{aligned}
&1)\quad \mathbf{J}_+ | jm\rangle = \left[(j-m)(j+m+1)\right]^{\frac{1}{2}} | jm+1\rangle, \\
&2)\quad \mathbf{J}_+ | jm+1\rangle = \left[(j-m-1)(j+m+2)\right]^{\frac{1}{2}} | jm+2\rangle, \\
&3)\quad \mathbf{J}_+ | jm+2\rangle = \left[(j-m-2)(j+m+3)\right]^{\frac{1}{2}} | jm+3\rangle, \\
&\vdots \qquad \vdots \qquad\qquad \vdots \\
&k)\quad \mathbf{J}_+ | jm+k-1\rangle = \left[(j-m-k+1)(j+m+k)\right]^{\frac{1}{2}} | jm+k\rangle.
\end{aligned} \tag{6.90}$$

Combining all k steps into a single equation leads to:

$$\mathbf{J}_+^k | jm\rangle = \left[\frac{(j-m)!}{(j-m-k)!}\frac{(j+m+k)!}{(j+m)!}\right]^{\frac{1}{2}} | jm+k\rangle, \tag{6.91}$$

where

$$\begin{aligned}
&(j-m-k+1)(j+m+k)\cdot(j-m-k)(j+m+k-1)\cdots(j-m)(j+m+1) \\
&= \frac{(j-m)!}{(j-m-k)!}\frac{(j+m+k)!}{(j+m)!}.
\end{aligned} \tag{6.92}$$

The upper-bound member $|jj\rangle$ of the j-multiplet appears when $k = j - m$. Therefore, applying substitution $k = j - m$ leads to:

$$\mathbf{J}_+^{j-m} | jm\rangle = \left[\frac{(2j)!(j-m)!}{(j+m)!}\right]^{\frac{1}{2}} | jj\rangle. \tag{6.93}$$

Rearranging the obtained equation gives:

$$| jj\rangle = \left[\frac{(j+m)!}{(2j)!(j-m)!}\right]^{\frac{1}{2}} \mathbf{J}_+^{j-m} | jm\rangle. \tag{6.94}$$

This equation provides the mechanism for generating the members of the j-multiplet all the way to $|jj\rangle$ by application of the raising operator $\mathbf{J}_+$ to the eigenfunction $|jm\rangle$. These relations for the lowering and raising operators provide the necessary tool to complete the evaluation of the Clebsch–Gordan coefficients.

The final step two in the computation of the Clebsch–Gordan coefficients deals with determining the members of the j-multiplet ψ_{jm}, which includes the functions $\psi_{j,j}$, $\psi_{j,j-1}$, …, $\psi_{j,-j}$, and where index j as before varies in the range from $j_1 + j_2$ to $j_1 - j_2$. Again, a comma separator will occasionally be used between the angular and azimuthal indices, as in $\psi_{j,j-1}$, in order to avoid confusion when the indices become complicated. All upper-bound coupled eigenfunctions ψ_{jj} and their expansion coefficients C have already been derived in step one. The remaining members ψ_{jm} of the j-multiplet can be derived from the standard Clebsch–Gordan expansion:

$$\psi_{jm} = \sum_{m_1=-j_1}^{j_1} C(j_1 j_2 j; m_1 m_2 m)\, \varphi_{j_1 m_1} \varphi_{j_2 m_2}. \tag{6.95}$$

Multiplying both sides of this equation by $\varphi^*_{j_1 m_1} \varphi^*_{j_2 m_2}$ from the left, and integrating, resolves the coefficient C as the overlap integral between the uncoupled and coupled states:

$$C(j_1 j_2 j; m_1 m_2 m) = \langle j_1 m_1; j_2 m_2 \,|\, jm\rangle. \tag{6.96}$$

Further analysis focuses on finding this integral.

A similar integral was resolved in the first step of the evaluation of the Clebsch–Gordan coefficients:

$$C(j_1 j_2 j; m_1 m_2 j) = \langle j_1 m_1; j_2 m_2 \,|\, jj\rangle, \tag{6.97}$$

for the upper-bound eigenfunction ψ_{jj}, which obeys the condition $m_1 + m_2 = m = j$. The Clebsch–Gordan coefficient for the next member, $\psi_{j,j-1}$, of the j-multiplet emerges as the result of application of the lowering operator $\mathbf{J}_-$ to the coupled function ψ_{jj}:

$$C(j_1 j_2 j; m_1 m_2 j-1) = \langle j_1 m_1; j_2 m_2 \,|\, jj-1\rangle = \frac{1}{\sqrt{2j}} \langle j_1 m_1; j_2 m_2 |\mathbf{J}_-|\, jj\rangle. \tag{6.98}$$

The coefficient $(1/\sqrt{2j})$ in Equation 6.98 is the inverse of the normalization factor of the lowering operator. Its purpose is to cancel out the regular normalization coefficient $\sqrt{2j}$, which will appear following application of $\mathbf{J}_-$ to ψ_{jj} according to Equation 6.86. The solution for the integral on the right-hand side of Equation 6.98 is obtained by taking its Hermitian adjoint. The Hermitian adjoint changes the lowering operator into a raising one, and that leads to:

$$\begin{aligned} &\langle j_1 m_1; j_2 m_2 \,|\, \mathbf{J}_- |\, jj\rangle = \langle jj \,|\, \mathbf{J}_+ |\, j_1 m_1; j_2 m_2\rangle^+ = \langle jj|\mathbf{J}_{1+} + \mathbf{J}_{2+}|\, j_1 m_1; j_2 m_2\rangle^+ \\ &= \left[(j_1 - m_1)(j_1 + m_1 + 1)\right]^{\frac{1}{2}} \langle jj|\, j_1, m_1 + 1; j_2 m_2\rangle^+ + \left[(j_2 - m_2)(j_2 + m_2 + 1)\right]^{\frac{1}{2}} \langle jj|\, j_1 m_1; j_2, m_2 + 1\rangle^+. \end{aligned} \tag{6.99}$$

Returning the right-hand side of Equation 6.99 from Hermitian adjoint form back to the normal form gives:

$$\begin{aligned} &\langle j_1 m_1; j_2 m_2 |\mathbf{J}_-|\, jj\rangle \\ &= \left[(j_1 - m_1)(j_1 + m_1 + 1)\right]^{\frac{1}{2}} \langle j_1, m_1 + 1; j_2 m_2 |\, jj\rangle + \left[(j_2 - m_2)(j_2 + m_2 + 1)\right]^{\frac{1}{2}} \langle j_1 m_1; j_2, m_2 + 1 |\, jj\rangle. \end{aligned} \tag{6.100}$$

Substituting Equation 6.100 into Equation 6.98 produces a relation for the corresponding Clebsch–Gordan coefficient:

$$C(j_1 j_2 j; m_1 m_2 j-1) = \sqrt{\frac{(j_1-m_1)(j_1+m_1+1)}{2j}} \langle j_1, m_1+1; j_2 m_2 | jj\rangle + \sqrt{\frac{(j_2-m_2)(j_2+m_2+1)}{2j}} \langle j_1 m_1; j_2, m_2+1 | jj\rangle. \tag{6.101}$$

This equation may be examined for the specific case of $j = j_1 + j_2$. In that case, since $m = j - 1$ in Equation 6.101, $m_1 + m_2 = m = j_1 + j_2 - 1$. This, in turn, requires that either $m_1 = j_1$ or $m_2 = j_2$. For $m_1 = j_1$, the coefficient standing in front of the integral $\langle j_1 m_1 + 1; j_2 m_2 + 1 | jj\rangle$ annihilates the latter. With that, Equation 6.101 reduces to:

$$C(j_1 j_2 j;\ j_1 m_2 j-1) = \sqrt{\frac{(j_2-m_2)(j_2+m_2+1)}{2j}} \langle j_1 j_1; j_2, m_2+1 | jj\rangle. \tag{6.102}$$

The choice of $j = j_1 + j_2$ and $m_1 = j_1$ requires that $m_2 = j_2 - 1$ in order to fulfill the requirement that $m_1 + m_2 = j_1 + j_2 - 1$. This leads to:

$$C(j_1 j_2 j_1+j_2;\ j_1, j_2-1, j_1+j_2-1) = \sqrt{\frac{j_2}{j_1+j_2}} \langle j_1 j_1;\ j_2 j_2 | j_1+j_2, j_1+j_2\rangle = \sqrt{\frac{j_2}{j_1+j_2}}, \tag{6.103}$$

where $\langle j_1 j_1; j_2 j_2 | +j_1 + j_2, j_1 + j_2\rangle = 1$ according to Equation 6.39.

The second alternative case for Equation 6.101 to consider when $j = j_1 + j_2$ is $m_2 = j_2$. This condition zeroes the coefficient in front of $\langle j_1 m_1; j_2, m_2 + 1 | jj\rangle$, and that reduces Equation 6.101 to:

$$C(j_1 j_2 j;\ m_1 j_2 j-1) = \sqrt{\frac{(j_1-m_1)(j_1+m_1+1)}{2j}} \langle j_1, m_1+1;\ j_2 m_2 | jj\rangle. \tag{6.104}$$

The conditions of $m_2 = j_2$ and $m_1 + m_2 = j_1 + j_2 - 1$ require that $m_1 = j_1 - 1$. This resolves another Clebsch–Gordan coefficient:

$$C(j_1 j_2 j_1+j_2; j_1-1, j_2, j_1+j_2-1) = \sqrt{\frac{j_1}{j_1+j_2}} \langle j_1 j_1; j_2 m_2 | j_1+j_2, j_1+j_2\rangle = \sqrt{\frac{j_1}{j_1+j_2}}, \tag{6.105}$$

where, again, $\langle j_1 j_1; j_2 j_2 | j_1 + j_2, j_1 + j_2\rangle = 1$, due to Equation 6.39. Continuing this recurrence procedure generates Clebsch–Gordan coefficients for the remaining members of the j-multiplet.

This procedure may be generalized. The member ψ_{jm} and the corresponding Clebsch–Gordan coefficients $C(j_1 j_2 j;\ m_1 m_2 m)$ appear as the result of $j - m$ consecutive applications of the lowering operator:

$$C(j_1 j_2 j; m_1 m_2 m) = \left[\frac{(j+m)!}{(2j)!(j-m)!}\right]^{\frac{1}{2}} \langle j_1 m_1; j_2 m_2 | \mathbf{J}_-^{j-m} | jj\rangle \equiv \langle j_1 m_1; j_2 m_2 | jm\rangle. \tag{6.106}$$

The numerical coefficient standing in front of the integral in Equation 6.106 is the inverse of the normalization coefficient, which appears in Equation 6.93. These two equations are related because operator $\mathbf{J}_-^{j-m}$ turns into $\mathbf{J}_+^{j-m}$ when applied to the left side.

Since Clebsch–Gordan coefficients compose a unitary matrix, they may be viewed as a rotation of the system from the uncoupled to coupled state. Lowering and raising operator determine the transformation coefficients between those two states. The evaluation of Clebsch–Gordan coefficients proceeds from Equation 6.106. This equation remains true if the application of the shift operator is reversed on the left side of the integral when the operator acts on the uncoupled eigenfunctions $\varphi_{j_1m_1}\varphi_{j_2m_2}$. The task of redirecting the shift operator to the uncoupled functions is achieved by rewriting the integral in its Hermitian adjoint form:

$$C(j_1j_2j;m_1m_2m) = \langle j_1m_1;j_2m_2|jm\rangle = \left[\frac{(j+m)!}{(2j)!(j-m)!}\right]^{\frac{1}{2}} \langle jj|\mathbf{J}_+^{j-m}|j_1m_1;j_2m_2\rangle^+. \tag{6.107}$$

Finding the outcome of application of $\mathbf{J}_+^{j-m}$ to $|j_1m_1;j_2m_2\rangle$ requires expanding the raising operator using the binomial theorem. This gives:

$$\mathbf{J}_+^{j-m} = (\mathbf{J}_{1+} + \mathbf{J}_{2+})^{j-m} = \sum_{k=0}^{j-m} \frac{(j-m)!}{k!(j-m-k)!} \mathbf{J}_{1+}^k \mathbf{J}_{2+}^{j-m-k}. \tag{6.108}$$

Equation 6.108 replaces the coupled operator $\mathbf{J}_+^{j-m}$ with the product of uncoupled operators $\mathbf{J}_{1+}^k$ and $\mathbf{J}_{2+}^{j-m-k}$, greatly simplifying the task of resolving the operator equation. Substitution of Equation 6.108 into Equation 6.107 leads to:

$$C(j_1j_2j;m_1m_2m) = \left[\frac{(j+m)!}{(2j)!(j-m)!}\right]^{\frac{1}{2}} \sum_{k=0}^{j-m} \frac{(j-m)!}{k!(j-m-k)!} \langle jj|\mathbf{J}_{1+}^k \mathbf{J}_{2+}^{j-m-k}|j_1m_1;j_2m_2\rangle^+. \tag{6.109}$$

All that remains is to apply the operators $\mathbf{J}_{1+}^k$ and $\mathbf{J}_{2+}^{j-m-k}$ to their respective uncoupled functions. According to Equation 6.91, application of $\mathbf{J}_{1+}^k$ to $|j_1m_1\rangle$ leads to:

$$\mathbf{J}_{1+}^k|j_1m_1\rangle = \left[\frac{(j_1-m_1)!(j_1+m_1+k)!}{(j_1+m_1)!(j_1-m_1-k)!}\right]^{\frac{1}{2}} |j_1,m_1+k\rangle. \tag{6.110}$$

Similar to that, the application of $\mathbf{J}_{2+}^{j-m-k}$ to $|j_2m_2\rangle$ leads to:

$$\mathbf{J}_{2+}^{j-m-k}|j_2m_2\rangle = \left[\frac{(j_2-m_2)!(j_2+m_2+j-m-k)!}{(j_2+m_2)!(j_2-m_2-j+m+k)!}\right]^{\frac{1}{2}} |j_2,m_2+j-m-k\rangle. \tag{6.111}$$

Because $m_1 + m_2 = m$, replacing the simultaneous presence of m_2 and m with m_1 in the above equation provides a somewhat simpler expression:

$$\mathbf{J}_{2+}^{j-m-k}|j_2m_2\rangle = \left[\frac{(j_2-m_2)!(j_2+j-m_1-k)!}{(j_2+m_2)!(j_2-j+m_1+k)!}\right] |j_2,j-m_1-k\rangle. \tag{6.112}$$

Substituting these relations for the raising operators acting on the uncoupled eigenfunctions in Equation 6.109 gives:

$$C(j_1j_2j;m_1m_2m) = \left[\frac{(j+m)!}{(2j)!(j-m)!}\right]^{\frac{1}{2}} \sum_{k=0}^{j-m} \frac{(j-m)!}{k!(j-m-k)!} \left[\frac{(j_1-m_1)!(j_1+m_1+k)!}{(j_1+m_1)!(j_1-m_1-k)!}\right]^{\frac{1}{2}}$$
$$\times \left[\frac{(j_2-m_2)!(j_2+j-m_1-k)!}{(j_2+m_2)!(j_2-j+m_1+k)!}\right]^{\frac{1}{2}} \langle jj|j_1,m_1+k;j_2,j-m_1-k\rangle^+. \tag{6.113}$$

Moving the coefficients, which do not depend on index k, outside the sum, and returning the integral back to its normal form leads to:

$$C(j_1 j_2 j; m_1 m_2 m) = \left[\frac{(j_1-m_1)!(j_2-m_2)!}{(j_1+m_1)!(j_2+m_2)!}\frac{(j+m)!(j-m)!}{(2j)!}\right]^{\frac{1}{2}}$$
$$\sum_{k=0}^{j-m}\frac{1}{k!(j-m-k)!}\left[\frac{(j_1+m_1+k)!(j_2+j-m_1-k)!}{(j_1-m_1-k)!(j_2-j+m_1+k)!}\right]^{\frac{1}{2}}\langle j_1, m_1+k; j_2, j-m_1-k \,|\, jj\rangle. \tag{6.114}$$

Further progress in the derivation of coefficient $C(j_1 j_2 j;\, m_1 m_2 m)$ depends on resolving the integral $\langle j_1,\, m_1+k;\, j_2,\, j-m_1-k \,|\, jj\rangle$. This is where the previous solution for the Clebsch–Gordan coefficient for the upper-bound coupled eigenfunction determined in the step one becomes useful. The solution in Equation 6.83 defines a similar integral,

$$\langle j_1 m_1; j_2 m_2 \mid jj\rangle = \delta_{m_1+m_2,j}(-1)^{j_1-m_1}\left[\frac{(j_1+m_1)!(j_2+m_2)!}{(j_1-m_1)!(j_2-m_2)!}\frac{(j_2-j+j_1)!}{(j_2+j-j_1)!}\frac{(2j+1)!}{(j_1+j_2+j+1)!(j_1-j_2+j)!}\right]^{\frac{1}{2}}, \tag{6.115}$$

that can be transformed into the proper form required by Equation 6.114. Replacing the index m_1 by $m_1 + k$ and m_2 by $j - m_1 - k$ in Equation 6.115 leads to:

$$\langle j_1, m_1+k; j_2, j-m_1-k \,|\, jj\rangle$$
$$= \delta_{m_1+m_2,j}(-1)^{j_1-m_1-k}\left[\frac{(j_1+m_1+k)!(j_2+j-m_1-k)!}{(j_1-m_1-k)!(j_2-j+m_1+k)!}\frac{(j_2-j+j_1)!}{(j_2+j-j_1)!}\frac{(2j+1)!}{(j_1+j_2+j+1)!(j_1-j_2+j)!}\right]^{\frac{1}{2}}. \tag{6.116}$$

Note that the sum of two new azimuthal indices $m_1 + k$ and $j - m_1 - k$ is still equal to j after the substitution, as required. Substituting Equation 6.116 into Equation 6.114 gives:

$$C(j_1 j_2 j; m_1 m_2 m)$$
$$= \left[\frac{(j_1-m_1)!(j_2-m_2)!}{(j_1+m_1)!(j_2+m_2)!}\frac{(j+m)!(j-m)!}{(2j)!}\right]^{\frac{1}{2}}\sum_{k=0}^{j-m}\frac{1}{k!(j-m-k)!}\left[\frac{(j_1+m_1+k)!(j_2+j-m_1-k)!}{(j_1-m_1-k)!(j_2-j+m_1+k)!}\right]^{\frac{1}{2}}$$
$$\delta_{m_1+m_2,j}(-1)^{j_1-m_1-k} \tag{6.117}$$
$$\left[\frac{(j_1+m_1+k)!(j_2+j-m_1-k)!}{(j_1-m_1-k)!(j_2-j+m_1+k)!}\frac{(j_2-j+j_1)!}{(j_2+j-j_1)!}\frac{(2j+1)!}{(j_1+j_2+j+1)!(j_1-j_2+j)!}\right]^{\frac{1}{2}}.$$

Since, by definition, the index k in the sum assumes only positive values, it is convenient to replace the factor $(-1)^{j_1-m_1-k}$ by $(-1)^{j_1-m_1+k}$, so that the exponent always stays positive. Further on, canceling out the common term $(2j)!$, and rearranging the remaining terms leads to the general relation for Clebsch–Gordan coefficients:

$$
\begin{aligned}
&C(j_1 j_2 j; m_1 m_2 m) \\
&= \delta_{m_1+m_2,j}(-1)^{j_1-m_1+k}\left[\frac{(j_1-m_1)!(j_2-m_2)!(j+m)!(j-m)!(j_2-j+j_1)!(2j+1)}{(j_1+m_1)!(j_2+m_2)!(j_2+j-j_1)!(j_1+j_2+j+1)!(j_1-j_2+j)!}\right]^{\frac{1}{2}} \\
&\sum_{k=0}^{j-m}\frac{(j_1+m_1+k)!(j_2+j-m_1-k)!}{k!(j-m-k)!(j_1-m_1-k)!(j_2-j+m_1+k)!}.
\end{aligned}
\tag{6.118}
$$

This expression and its derivation gives a sense of meaning to the Clebsch–Gordan coefficients, but, other than that, and because of its complexity, the use of this equation in applied computations is impractical. In the Fast Multipole Method, only a portion of the Clebsch–Gordan theory that relates to efficient computation of the Wigner rotation matrix will be needed. The next section examines the actual matrix of Clebsch–Gordan coefficients that will later be used in the derivation of recurrence relations for the Wigner matrix.

6.3 Addition of Angular Momentum and Spin

Out of all possible combinations that lead to the total angular momentum $\mathbf{J} = \mathbf{J}_1 + \mathbf{J}_2$, the addition of angular momentum and spin is arguably the most frequently-encountered case that appears in physics applications. Although the Fast Multipole Method is considerably simpler than quantum theory, it can still benefit from the mathematical apparatus developed for physics problems that require the addition of angular and spin momenta. The following discussion aims at evaluating the matrix of Clebsch–Gordan coefficients that describes the outcome of addition of this type.

Addition of angular and spin momenta follows the same principles as that for the addition of any two general angular momenta that have integer or half-integer values. In order to keep the notation general, and to stress the addition of spin, it is convenient to denote the operator of the first angular momentum as $\mathbf{J}_1$, and that for spin as $\mathbf{S}$, so that they can be easily distinguished in the equations. With the above notation in place, the total angular momentum after the addition becomes $\mathbf{J} = \mathbf{J}_1 + \mathbf{S}$. With that, the operator of projection of the total angular momentum on quantization axis becomes:

$$\mathbf{J}_z = \mathbf{J}_{1z} + \mathbf{S}_z. \tag{6.119}$$

The square of the total angular momentum operator follows from Equation 6.46, and assumes the form:

$$\mathbf{J}^2 = \mathbf{J}_1^2 + \mathbf{S}^2 + 2\mathbf{J}_{1z}\mathbf{S}_z + \mathbf{J}_{+1}\mathbf{S}_- + \mathbf{J}_{-1}\mathbf{S}_+. \tag{6.120}$$

To define the angular momentum operator that represents the sum of $\mathbf{J}_1$ and $\mathbf{S}$, it is necessary to find eigenvalues and eigenfunctions of this operator. The eigenfunctions ψ of the operators $\mathbf{J}^2$ and $\mathbf{J}_z$ will be sought in the basis of functions of the uncoupled state $\varphi_{j_1 m_1}\varphi_{(1/2),\pm(1/2)}$. These basis functions hold the angular momentum j_1 and 1/2, and have the magnetic quantum numbers spanning in the range $-j_1 \le m_1 \le j_1$ and $m_2 = \pm 1/2$, respectively.

The eigenfunctions $\varphi_{j_1 m_1}$ and $\varphi_{(1/2),\pm(1/2)}$ of the angular and spin operators corresponding to individual states satisfy the eigenvalue equations:

$$
\begin{aligned}
\mathbf{J}_1^2\varphi_{j_1 m_1} &= j_1(j_1+1)\varphi_{j_1 m_1}, \quad \mathbf{J}_{1z}\varphi_{j_1 m_1} = m_1\varphi_{j_1 m_1}, \\
\mathbf{S}^2\varphi_{\frac{1}{2},\pm\frac{1}{2}} &= \tfrac{1}{2}\left(\tfrac{1}{2}+1\right)\varphi_{\frac{1}{2},\pm\frac{1}{2}}, \quad \mathbf{S}_z\varphi_{\frac{1}{2},\pm\frac{1}{2}} = \pm\tfrac{1}{2}\varphi_{\frac{1}{2},\pm\frac{1}{2}}.
\end{aligned}
\tag{6.121}
$$

The eigenfunctions ψ_{jm} of the total angular momentum are similar:

$$\mathbf{J}^2\psi_{jm} = j(j+1)\psi_{jm}, \quad \mathbf{J}_z\psi_{jm} = m\psi_{jm}; \tag{6.122}$$

these will be determined next. For the individual state $\varphi_{j_1 m_1}$ that has the quantum numbers j_1 and m_1, the admixture of spin generates two distinct angular quantum numbers $j = j_1 + (1/2)$ and $j = j_1 - (1/2)$, and two coupled eigenstates:

$$\psi_{j_1+\frac{1}{2},m_1+\frac{1}{2}} = C_{11}\varphi_{j_1,m_1}\varphi_{\frac{1}{2},\frac{1}{2}} + C_{12}\varphi_{j_1,m_1+1}\varphi_{\frac{1}{2},-\frac{1}{2}}, \tag{6.123}$$

$$\psi_{j_1-\frac{1}{2},m_1-\frac{1}{2}} = C_{21}\varphi_{j_1,m_1-1}\varphi_{\frac{1}{2},\frac{1}{2}} + C_{22}\varphi_{j_1,m_1}\varphi_{\frac{1}{2},-\frac{1}{2}}, \tag{6.124}$$

where C_{ij} are, at present, unknown Clebsch–Gordan coefficients.

The construction of the linear combination for coupled eigenstates $\psi_{j,m}$ in Equations 6.123 and 6.124 is guided by the rule that the azimuthal quantum numbers must obey the condition $m = m_1 + m_2$, where $m_2 = \pm(1/2)$ is the spin magnetic (azimuthal) number, so that the sum of azimuthal numbers inside each term $\varphi_{j_1,m_1}\varphi_{(1/2),m_2}$ is equal to m. This means that exactly two functions $\varphi_{j_1,m_1}\varphi_{(1/2),(1/2)}$ and $\varphi_{j_1,m_1+1}\varphi_{(1/2),-(1/2)}$ from the basis set $\varphi_{j_1,m_1}\varphi_{(1/2),m_2}$ can contribute to the coupled state $\psi_{j_1+(1/2),m_1+(1/2)}$, which has $m = m_1 + (1/2)$. Similar to that, the coupled state $\psi_{j_1-(1/2),m_1-(1/2)}$ is a linear combination of two basis functions $\varphi_{j_1,m_1-1}\varphi_{(1/2),(1/2)}$ and $\varphi_{j_1,m_1}\varphi_{(1/2),-(1/2)}$, which independently have the same combined azimuthal number $(m_1 - 1) + (1/2) = m_1 - (1/2)$ and $(m_1) - (1/2) = m_1 - (1/2)$, respectively, as that of the coupled state.

It can be shown that coupled states $\psi_{j_1+(1/2),m_1+(1/2)}$ and $\psi_{j_1-(1/2),m_1-(1/2)}$ fully determine all Clebsch–Gordan coefficients that appear in the addition of total angular momentum $\mathbf{J} = \mathbf{J}_1 + \mathbf{S}$. Other functions like $\psi_{j_1+(1/2),m_1-(1/2)}$ and $\psi_{j_1-(1/2),m_1+(1/2)}$ originate from Equations 6.123 and 6.124 *via* a substitution of m_1 by m_1-1 and by $m_1 + 1$, respectively. This corresponds to a trivial change in the value of m_1, which is not related to the addition of angular momenta. Therefore, coupled functions $\psi_{j_1+(1/2),m_1-(1/2)}$ and $\psi_{j_1-(1/2),m_1+(1/2)}$ use the same matrix of Clebsch–Gordan coefficients as that used by $\psi_{j_1+(1/2),m_1+(1/2)}$ and $\psi_{j_1-(1/2),m_1-(1/2)}$ in Equations 6.123 and 6.124.

The addition of angular and spin momenta leads to a 2 × 2 matrix of Clebsch–Gordan coefficients. The two rows in the matrix come from the two possible values of the angular quantum number j in the coupled eigenstate $\psi_{j,m}$: $j_1 + (1/2)$ and $j_1 - (1/2)$. The two columns in the Clebsch–Gordan matrix originate from the two possible values of index $m = m_1 \pm (1/2)$. Each initial state $\varphi_{j_1 m_1}$ can interact only with the two possible spin states $\varphi_{(1/2),\pm(1/2)}$ and that leads to Equations 6.123 and 6.124. Therefore, further analysis will focus on the coupled functions $\psi_{j_1+(1/2),m_1+(1/2)}$ and $\psi_{j_1-(1/2),m_1-(1/2)}$, which completely determine the matrix of Clebsch–Gordan coefficients for the case of addition of angular momenta $\mathbf{J} = \mathbf{J}_1 + \mathbf{S}$.

The unknown Clebsch–Gordan coefficients may be found from the application of the operator of the square of the angular momentum $\mathbf{J}^2$ to both sides of Equation 6.123. This leads to:

$$\mathbf{J}^2\psi_{j_1+\frac{1}{2},m_1+\frac{1}{2}} = \mathbf{J}^2\left(C_{11}\varphi_{j_1,m_1}\varphi_{\frac{1}{2},\frac{1}{2}} + C_{12}\varphi_{j_1,m_1+1}\varphi_{\frac{1}{2},-\frac{1}{2}}\right). \tag{6.125}$$

The action of the operator $\mathbf{J}^2$ in the left-hand side of Equation 6.125 has a known outcome:

$$\mathbf{J}^2\psi_{j_1+\frac{1}{2},m_1+\frac{1}{2}} = \left(j_1 + \frac{1}{2}\right)\left(j_1 + \frac{1}{2} + 1\right)\psi_{j_1+\frac{1}{2},m_1+\frac{1}{2}}. \tag{6.126}$$

Substituting $\psi_{j_1+(1/2),m_1+(1/2)}$ in the right-hand side of Equation 6.126 by its Clebsch–Gordan expansion from Equation 6.123 gives:

$$\mathbf{J}^2\psi_{j_1+\frac{1}{2},m_1+\frac{1}{2}} = \left(j_1 + \frac{1}{2}\right)\left(j_1 + \frac{3}{2}\right)C_{11}\varphi_{j_1,m_1}\varphi_{\frac{1}{2},-\frac{1}{2}} + \left(j_1 + \frac{1}{2}\right)\left(j_1 + \frac{3}{2}\right)C_{12}\varphi_{j_1,m_1+1}\varphi_{\frac{1}{2},-\frac{1}{2}}. \tag{6.127}$$

The right-hand side of Equation 6.125 resolves with help of Equation 6.120. This leads to:

$$\begin{aligned}
&\mathbf{J}^2\left(C_{11}\varphi_{j_1m_1}\varphi_{\frac{1}{2},\frac{1}{2}}+C_{12}\varphi_{j_1m_1+1}\varphi_{\frac{1}{2},-\frac{1}{2}}\right)\\
&=\left(\mathbf{J}_1^2+\mathbf{S}^2+2\mathbf{J}_{1z}\mathbf{S}_z+\mathbf{J}_{+1}\mathbf{S}_-+\mathbf{J}_{-1}\mathbf{S}_+\right)\left(C_{11}\varphi_{j_1m_1}\varphi_{\frac{1}{2},\frac{1}{2}}+C_{12}\varphi_{j_1m_1+1}\varphi_{\frac{1}{2},-\frac{1}{2}}\right)\\
&=\left[j_1(j_1+1)+\frac{3}{4}+2m_1\frac{1}{2}\right]C_{11}\varphi_{j_1m_1}\varphi_{\frac{1}{2},\frac{1}{2}}+\left[(j_1-m_1)(j_1+m_1+1)\right]^{\frac{1}{2}}C_{11}\varphi_{j_1m_1+1}\varphi_{\frac{1}{2},-\frac{1}{2}}\\
&+\left[j_1(j_1+1)+\frac{3}{4}+2(m_1+1)\left(-\frac{1}{2}\right)\right]C_{12}\varphi_{j_1m_1+1}\varphi_{\frac{1}{2},-\frac{1}{2}}+\left[(j_1+m_1+1)(j_1-m_1)\right]^{\frac{1}{2}}C_{12}\varphi_{j_1m_1}\varphi_{\frac{1}{2},\frac{1}{2}}.
\end{aligned}\tag{6.128}$$

Putting the left- and right-hand parts back together gives:

$$\begin{aligned}
&\left(j_1+\frac{1}{2}\right)\left(j_1+\frac{3}{2}\right)C_{11}\varphi_{j_1,m_1}\varphi_{\frac{1}{2},\frac{1}{2}}+\left(j_1+\frac{1}{2}\right)\left(j_1+\frac{3}{2}\right)C_{12}\varphi_{j_1,m_1+1}\varphi_{\frac{1}{2},-\frac{1}{2}}\\
&=\left[j_1(j_1+1)+\frac{3}{4}+2m_1\frac{1}{2}\right]C_{11}\varphi_{j_1m_1}\varphi_{\frac{1}{2},\frac{1}{2}}+\left[(j_1-m_1)(j_1+m_1+1)\right]^{\frac{1}{2}}C_{11}\varphi_{j_1m_1+1}\varphi_{\frac{1}{2},-\frac{1}{2}}\\
&+\left[j_1(j_1+1)+\frac{3}{4}+2(m_1+1)\left(-\frac{1}{2}\right)\right]C_{12}\varphi_{j_1m_1+1}\varphi_{\frac{1}{2},-\frac{1}{2}}+\left[(j_1+m_1+1)(j_1-m_1)\right]^{\frac{1}{2}}C_{12}\varphi_{j_1m_1}\varphi_{\frac{1}{2},\frac{1}{2}}.
\end{aligned}\tag{6.129}$$

Given that $\varphi_{j_1,m_1}\varphi_{(1/2),(1/2)}$ and $\varphi_{j_1,m_1+1}\varphi_{(1/2),-(1/2)}$ are orthogonal, both sides of Equation 6.129 can be multiplied by the complex conjugate form of either of the basis functions and integrated. Multiplying Equation 6.129 by $\varphi^*_{j_1,m_1}\varphi^*_{(1/2),(1/2)}$, and integrating leads to:

$$\left(j_1+\frac{1}{2}\right)\left(j_1+\frac{3}{2}\right)C_{11}=\left[j_1(j_1+1)+\frac{3}{4}+2m_1\frac{1}{2}\right]C_{11}+\left[(j_1+m_1+1)(j_1-m_1)\right]^{\frac{1}{2}}C_{12}.\tag{6.130}$$

Repeating the same procedure with $\varphi^*_{j_1,m_1+1}\varphi^*_{(1/2),-(1/2)}$ gives:

$$\left(j_1+\frac{1}{2}\right)\left(j_1+\frac{3}{2}\right)C_{12}=\left[(j_1-m_1)(j_1+m_1+1)\right]^{\frac{1}{2}}C_{11}+\left[j_1(j_1+1)+\frac{3}{4}+2(m_1+1)\left(-\frac{1}{2}\right)\right]C_{12}.\tag{6.131}$$

These are identical equations; therefore either one of them can be used in order to find the unknown coefficients. Simplifying the first equation leads to:

$$\left(j_1^2+2j_1+\frac{3}{4}\right)C_{11}=\left(j_1^2+j_1+\frac{3}{4}+m_1\right)C_{11}+\left[(j_1+m_1+1)(j_1-m_1)\right]^{\frac{1}{2}}C_{12},\tag{6.132}$$

which, after gathering similar terms, reduces to:

$$(j_1-m_1)C_{11}=\left[(j_1+m_1+1)(j_1-m_1)\right]^{\frac{1}{2}}C_{12}.\tag{6.133}$$

Therefore,

$$C_{11}=C_{12}\sqrt{\frac{j_1+m_1+1}{j_1-m_1}}.\tag{6.134}$$

The normalization property of the Clebsch–Gordan coefficients supplies the complementary equation:

$$C_{11}^2 + C_{12}^2 = 1. \tag{6.135}$$

Combining these two equations leads to:

$$C_{12}^2 \frac{j_1 + m_1 + 1}{j_1 - m_1} + C_{12}^2 = 1, \tag{6.136}$$

which, after simplification, yields the first coefficient:

$$C_{12} = \sqrt{\frac{j_1 - m_1}{2j_1 + 1}}. \tag{6.137}$$

Substituting this result back into Equation 6.134 gives the other coefficient:

$$C_{11} = \sqrt{\frac{j_1 + m_1 + 1}{2j_1 + 1}}. \tag{6.138}$$

Traditionally, literature defines Clebsch–Gordan coefficients in terms of variables j_1 and m. Since $m = m_1 + (1/2)$ and $m = m_1 - (1/2)$, substitution of m_1 by m leads to:

$$C_{11} = \sqrt{\frac{j_1 + m + \frac{1}{2}}{2j_1 + 1}}, \quad \text{and} \quad C_{12} = \sqrt{\frac{j_1 - m + \frac{1}{2}}{2j_1 + 1}}. \tag{6.139}$$

Using the same procedure produces the remaining two Clebsch–Gordan coefficients C_{21} and C_{22}. Application of the operator $\mathbf{J}^2$ to both side of Equation 6.124 then gives the equality:

$$\mathbf{J}^2 \psi_{j_1 - \frac{1}{2}, m_1 - \frac{1}{2}} = \mathbf{J}^2 \left(C_{21} \varphi_{j_1, m_1 - 1} \varphi_{\frac{1}{2}, \frac{1}{2}} + C_{22} \varphi_{j_1, m_1} \varphi_{\frac{1}{2}, -\frac{1}{2}} \right). \tag{6.140}$$

Its left-hand side resolves to:

$$\mathbf{J}^2 \psi_{j_1 - \frac{1}{2}, m_1 - \frac{1}{2}} = \left(j_1 - \frac{1}{2} \right)\left(j_1 - \frac{1}{2} + 1 \right) \psi_{j_1 - \frac{1}{2}, m_1 - \frac{1}{2}} = \left(j_1 - \frac{1}{2} \right)\left(j_1 + \frac{1}{2} \right)\left(C_{21} \varphi_{j_1, m_1 - 1} \varphi_{\frac{1}{2}, \frac{1}{2}} + C_{22} \varphi_{j_1, m_1} \varphi_{\frac{1}{2}, -\frac{1}{2}} \right), \tag{6.141}$$

which includes the replacement of $\psi_{j_1 - (1/2), m_1 - (1/2)}$ by its Clebsch–Gordan expansion from Equation 6.124. Action of the operator $\mathbf{J}^2$ on the right-hand side of Equation 6.140 leads to:

$$\begin{aligned}
&\mathbf{J}^2 \left(C_{21} \varphi_{j_1, m_1 - 1} \varphi_{\frac{1}{2}, \frac{1}{2}} + C_{22} \varphi_{j_1, m_1} \varphi_{\frac{1}{2}, -\frac{1}{2}} \right) \\
&= \left(\mathbf{J}_1^2 + \mathbf{S}^2 + 2\mathbf{J}_{1z}\mathbf{S}_z + \mathbf{J}_{+1}\mathbf{S}_- + \mathbf{J}_{-1}\mathbf{S}_+ \right)\left(C_{21} \varphi_{j_1, m_1 - 1} \varphi_{\frac{1}{2}, \frac{1}{2}} + C_{22} \varphi_{j_1, m_1} \varphi_{\frac{1}{2}, -\frac{1}{2}} \right) \\
&= \left[j_1(j_1 + 1) + \frac{3}{4} + 2(m_1 - 1)\frac{1}{2} \right] C_{21} \varphi_{j_1 m_1 - 1} \varphi_{\frac{1}{2}, \frac{1}{2}} + \left[(j_1 - m_1 + 1)(j_1 + m_1) \right]^{\frac{1}{2}} C_{21} \varphi_{j_1 m_1} \varphi_{\frac{1}{2}, -\frac{1}{2}} \\
&+ \left[j_1(j_1 + 1) + \frac{3}{4} + 2m_1 \left(-\frac{1}{2} \right) \right] C_{22} \varphi_{j_1 m_1} \varphi_{\frac{1}{2}, -\frac{1}{2}} + \left[(j_1 + m_1)(j_1 - m_1 + 1) \right]^{\frac{1}{2}} C_{22} \varphi_{j_1 m_1 - 1} \varphi_{\frac{1}{2}, \frac{1}{2}}.
\end{aligned} \tag{6.142}$$

Assembling the left- and right-hands together gives:

$$\left(j_1-\frac{1}{2}\right)\left(j_1+\frac{1}{2}\right)\left(c_{21}\varphi_{j_1,m_1-1}\varphi_{\frac{1}{2},\frac{1}{2}}+c_{22}\varphi_{j_1,m_1}\varphi_{\frac{1}{2},-\frac{1}{2}}\right)$$
$$=\left[j_1(j_1+1)+\frac{3}{4}+2(m_1-1)\frac{1}{2}\right]c_{21}\varphi_{j_1m_1-1}\varphi_{\frac{1}{2},\frac{1}{2}}+\left[(j_1-m_1+1)(j_1+m_1)\right]^{\frac{1}{2}}c_{21}\varphi_{j_1m_1}\varphi_{\frac{1}{2},-\frac{1}{2}}$$
$$+\left[j_1(j_1+1)+\frac{3}{4}+2m_1\left(-\frac{1}{2}\right)\right]c_{22}\varphi_{j_1m_1}\varphi_{\frac{1}{2},-\frac{1}{2}}+\left[(j_1+m_1)(j_1-m_1+1)\right]^{\frac{1}{2}}c_{22}\varphi_{j_1m_1-1}\varphi_{\frac{1}{2},\frac{1}{2}}. \tag{6.143}$$

Since the functions $\varphi_{j_1,m_1-1}\varphi_{(1/2),(1/2)}$ and $\varphi_{j_1,m_1}\varphi_{(1/2),-(1/2)}$ are orthogonal, both parts of this equation may be multiplied by the complex conjugate form of either one of those functions and integrated. Performing this procedure with $\varphi^*_{j_1,m_1-1}\varphi^*_{(1/2),(1/2)}$ produces the equation:

$$\left(j_1-\frac{1}{2}\right)\left(j_1+\frac{1}{2}\right)C_{21}=\left[j_1(j_1+1)+\frac{3}{4}+2(m_1-1)\frac{1}{2}\right]C_{21}+\left[(j_1+m_1)(j_1-m_1+1)\right]^{\frac{1}{2}}C_{22}. \tag{6.144}$$

Multiplying both sides of Equation 6.143 by $\varphi^*_{j_1,m_1}\varphi^*_{(1/2),-(1/2)}$, and integrating produces another equation:

$$\left(j_1-\frac{1}{2}\right)\left(j_1+\frac{1}{2}\right)C_{22}=\left[(j_1-m_1+1)(j_1+m_1)\right]^{\frac{1}{2}}C_{21}+\left[j_1(j_1+1)+\frac{3}{4}+2m_1\left(-\frac{1}{2}\right)\right]C_{22}. \tag{6.145}$$

Since these two equations are identical, either one of them can be selected for further analysis. Arbitrarily choosing the first equation, and simplifying leads to:

$$\left(j_1^2-\frac{1}{4}\right)C_{21}=\left(j_1^2+j_1+m_1-\frac{1}{4}\right)C_{21}+\left[(j_1+m_1)(j_1-m_1+1)\right]^{\frac{1}{2}}C_{22}. \tag{6.146}$$

Rearranging the terms in Equation 6.146 reduces it to:

$$-(j_1+m_1)C_{21}=\left[(j_1+m_1)(j_1-m_1+1)\right]^{\frac{1}{2}}C_{22}. \tag{6.147}$$

This simplifies to:

$$C_{21}=-\sqrt{\frac{j_1-m_1+1}{j_1+m_1}}C_{22}. \tag{6.148}$$

The normalization property of the Clebsch–Gordan coefficients provides the complementary relation that aids in evaluation of the individual coefficients:

$$C_{21}^2+C_{22}^2=1. \tag{6.149}$$

Substituting Equation 6.148 into Equation 6.149 leads to:

$$\frac{j_1-m_1+1}{j_1+m_1}C_{22}^2+C_{22}^2=1. \tag{6.150}$$

This resolves the first of two coefficients:

$$C_{22} = \sqrt{\frac{j_1 + m_1}{2j_1 + 1}}. \tag{6.151}$$

The other coefficient follows from substituting Equation 6.151 into Equation 6.148:

$$C_{21} = -\sqrt{\frac{j_1 - m_1 + 1}{2j_1 + 1}}. \tag{6.152}$$

Here, $m_1 = m + (1/2)$, so substitution of m_1 by m leads to:

$$C_{21} = -\sqrt{\frac{j_1 - m + \frac{1}{2}}{2j_1 + 1}}, \quad \text{and} \quad C_{22} = \sqrt{\frac{j_1 + m + \frac{1}{2}}{2j_1 + 1}}. \tag{6.153}$$

Finally, assembling the results of derivation produces the matrix of Clebsch–Gordan coefficients that describes the addition of an arbitrary angular momentum j_1 with spin 1/2:

$$\begin{array}{lccc} & & m_2 = \frac{1}{2} & m_2 = -\frac{1}{2} \\ j_1 + \frac{1}{2}, & m = m_1 + \frac{1}{2} & \sqrt{\frac{j_1 + m + \frac{1}{2}}{2j_1 + 1}} & \sqrt{\frac{j_1 - m + \frac{1}{2}}{2j_1 + 1}} \\ j_1 - \frac{1}{2}, & m = m_1 - \frac{1}{2} & -\sqrt{\frac{j_1 - m + \frac{1}{2}}{2j_1 + 1}} & \sqrt{\frac{j_1 + m + \frac{1}{2}}{2j_1 + 1}} \end{array} \tag{6.154}$$

This table will later be needed in the derivation of recurrence relation for the Wigner matrix elements.

6.4 Rotation of the Coupled Eigenstates of Angular Momentum

Section 6.1 explained that a two-component system, consisting of two particles carrying individual angular momenta j_1 and j_2, has a total angular momentum j that spans the range:

$$j = j_1 + j_2, j_1 + j_2 - 1, \ldots, j_1 - j_2, \tag{6.155}$$

where $j_1 \geq j_2$. The eigenfunction $\psi_m = |jm\rangle$ or, simply, $|jm\rangle$ that describes the coupled state of the system of two particles represents one of the many possible states of the system. Being an eigenfunction of the angular momentum operator, the function ψ_{jm} rotates according to the prescription of the Wigner matrix. Although rotation of the coupled eigenfunction follows the same principles as those governing the rotation of the eigenfunction of angular momentum of a single particle, analysis of rotation of the coupled eigenfunction reveals additional important relations that will be useful in computation of the Wigner matrix.

Rotation of the eigenfunction $|jm\rangle$ by means of a unitary transformation $\mathbf{U}$ is described by the standard equation:

$$\psi'_{jm} \equiv |jm\rangle' = \mathbf{U}\psi_{jm} = \sum_{s=-j}^{j} D^j_{sm} \,|\, js\rangle \equiv \sum_{s=-j}^{j} D^j_{sm}\psi_{js}, \tag{6.156}$$

which casts the rotated function $\psi'_{jm} \equiv |\, jm\rangle'$ into a linear combination of the states ψ_{js} belonging to the same j-multiplet as that of the original function ψ_{jm}, and where D^j is the Wigner matrix. Due to the orthonormality of the eigenfunctions, multiplying both sides of the equation by ψ^*_{ik}, and integrating leads to:

$$\langle \psi_{ik} \,|\, \mathbf{U} | \psi_{jm} \rangle = \sum_{s=-j}^{j} D^j_{sm} \langle \psi_{ik} | \psi_{js} \rangle = \delta_{ij} D^j_{km}. \tag{6.157}$$

In Equation 6.157, both i and j are the angular momenta of the coupled state arbitrarily picked from the space $(j_1 j_2) \equiv \{j_1 + j_2, j_1 + j_2 - 1, ..., j_1 - j_2\}$ constructed out of specific values of j_1 and j_2.

The expression obtained in Equation 6.157 looks similar to that for a regular non-composite eigenfunction of a single particle. However, since the angular momentum j of the coupled system is selected from the space defined by angular momenta j_1 and j_2, it implies that matrix D^j which rotates the coupled eigenfunction should depend on matrices D^{j_1} and D^{j_2}, which rotate eigenfunctions of the individual particles. This relationship is important, in that it permits an efficient recurrence computation of the Wigner matrix elements. The derivation starts from a recap of how the eigenfunction of the coupled system is constructed from the eigenfunctions of the individual particles.

Assembling the coupled system from the direct product of eigenfunctions $\varphi_{j_1, m_1} \equiv |\, j_1 m_1\rangle$ and $\varphi_{j_2, m_2} \equiv |\, j_2 m_2\rangle$ of the individual particles 1 and 2 produces an eigenfunction of the coupled system:

$$\psi_{j,m} \equiv |\, jm\rangle = \sum_{m_1 m_2} C(j_1 j_2 j; m_1 m_2 m) \,|\, j_1 m_1\rangle \otimes |\, j_2 m_2\rangle, \tag{6.158}$$

where $C(j_1 j_2 j;\, m_1 m_2 m)$ are Clebsch–Gordan coefficients. The symbol $\otimes$ indicates direct product. The meaning of direct product is most easily illustrated using an example of matrix multiplication:

$$\mathbf{A} \otimes \mathbf{B} = \begin{pmatrix} a_{11}\mathbf{B} & a_{12}\mathbf{B} \\ a_{21}\mathbf{B} & a_{22}\mathbf{B} \end{pmatrix} = \left(\begin{array}{cc|cc} a_{11}b_{11} & a_{11}b_{12} & a_{12}b_{11} & a_{12}b_{12} \\ a_{11}b_{21} & a_{11}b_{22} & a_{12}b_{21} & a_{12}b_{22} \\ \hline a_{21}b_{11} & a_{21}b_{12} & a_{22}b_{11} & a_{22}b_{12} \\ a_{21}b_{21} & a_{21}b_{22} & a_{22}b_{21} & a_{22}b_{22} \end{array}\right). \tag{6.159}$$

In the direct product, each element of matrix **A** gets multiplied to matrix **B**, as it is demonstrated in Equation 6.159.

The next task is to work out how the direct product of eigenfunctions transforms under rotation. Biedenharn and Louck provide a useful explanation of that topic. In the operator formalism, the unitary operator U that describes rotation in the composite space is:

$$U = e^{-i\theta \hat{\mathbf{n}} \cdot \mathbf{J}} = e^{-i\theta \hat{\mathbf{n}} \cdot \mathbf{J}_1} \otimes e^{-i\theta \hat{\mathbf{n}} \cdot \mathbf{J}_2}, \tag{6.160}$$

where θ is the rotation angle, $\hat{\mathbf{n}}$ is the unit vector of the rotation axis, and $\mathbf{J} = \mathbf{J}_1 + \mathbf{J}_2$ is the total angular momentum operator. Since operators $\mathbf{J}_1$ and $\mathbf{J}_2$ act on non-overlapping sub-spaces, the total rotation operator is a direct product of the individual rotation operators $e^{-i\theta \hat{\mathbf{n}} \cdot \mathbf{J}_1}$ and $e^{-i\theta \hat{\mathbf{n}} \cdot \mathbf{J}_2}$:

$$\psi'_{jm} = U\psi_{jm} = e^{i\theta \hat{\mathbf{n}} \cdot \mathbf{J}_1} \otimes e^{i\theta \hat{\mathbf{n}} \cdot \mathbf{J}_2} \sum_{m_1 m_2} C(j_1 j_2 j; m_1 m_2 m) \,|\, j_1 m_1\rangle \otimes |\, j_2 m_2\rangle. \tag{6.161}$$

Switching from the operator form to the Wigner matrix leads to:

$$\begin{aligned}\psi'_{jm} &= \sum_{m_1m_2} C(j_1j_2j;m_1m_2m)\left[e^{i\theta\hat{\mathbf{n}}\cdot\mathbf{J}_1}|j_1m_1\rangle\right]\otimes\left[e^{i\theta\hat{\mathbf{n}}\cdot\mathbf{J}_2}|j_2m_2\rangle\right] \\ &= \sum_{m_1m_2} C(j_1j_2j;m_1m_2m)\left[\sum_{s_1} D^{j_1}_{s_1m_1}|j_1s_1\rangle\right]\otimes\left[\sum_{s_2} D^{j_2}_{s_2m_2}|j_2s_2\rangle\right] \\ &= \sum_{m_1m_2}\sum_{s_1s_2} C(j_1j_2j;m_1m_2m)\left[D^{j_1}\otimes D^{j_2}\right]_{s_1s_2\,1m_2}|j_1s_1\rangle\otimes|j_2s_2\rangle,\end{aligned} \tag{6.162}$$

where $\left[D^{j_1}\otimes D^{j_2}\right]_{s_1s_2,m_1m_2}$ is a matrix element from the direct product of matrices $D^{j_1}\otimes D^{j_2}$. Both sides of Equation 6.162 may now be multiplied by

$$\psi^*_{ik} = \langle ik| = \sum_{k_1k_2} C(j_1j_2i;k_1k_2k)\,\langle j_1k_1|\otimes\langle j_2k_2|, \tag{6.163}$$

where i is one of the possible values of the angular momentum of the composite system (j_1j_2), from the left, and integrated. The value of the integral $\langle\psi_{ik}|\mathbf{U}|\psi_{jm}\rangle$ that appears on the left-hand side has already been calculated (Equation 6.157) and is equal to $\delta_{ij}D^j_{km}$; as a result, the right-hand side of Equation 6.162 can be transformed as follows:

$$\begin{aligned}&\sum_{k_1k_2}\sum_{m_1m_2}\sum_{s_1s_2} C(j_1j_2i;k_1k_2k)\,C(j_1j_2j;m_1m_2m)\left[D^{j_1}\otimes D^{j_2}\right]_{s_1s_2,m_1m_2}\langle j_1k_1|j_1s_1\rangle\otimes\langle j_2k_2|j_2s_2\rangle \\ &= \sum_{k_1k_2}\sum_{m_1m_2} C(j_1j_2i;k_1k_2k)\,C(j_1j_2j;m_1m_2m)\left[D^{j_1}\otimes D^{j_2}\right]_{k_1k_2\,m_1m_2}.\end{aligned} \tag{6.164}$$

Because the eigenfunctions are orthogonal, $s_1 = k_1$ and $s_2 = k_2$, as a result, the summations over indices s_1 and s_2 can be deleted.

Assembling the parts together gives:

$$\sum_{k_1k_2}\sum_{m_1m_2} C(j_1j_2i;k_1k_2k)\,C(j_1j_2j;m_1m_2m)\left[D^{j_1}\otimes D^{j_2}\right]_{k_1k_2\,m_1m_2} = \delta_{ij}D^j_{km}. \tag{6.165}$$

This can be simplified even more. In order to eliminate the dependence on index i, both sides of Equation 6.165 may be multiplied by $C(j_1j_2i;\ n_1n_2k)$, where $n_1 + n_2 = k$, and summed up over index i, giving:

$$\begin{aligned}&\sum_i C(j_1j_2i;n_1n_2k)\sum_{k_1k_2}\sum_{m_1m_2} C(j_1j_2i;k_1k_2k)\,C(j_1j_2j;m_1m_2m)\left[D^{j_1}\otimes D^{j_2}\right]_{k_1k_2\,m_1m_2} \\ &= \sum_i C(j_1j_2i;n_1n_2k)\,\delta_{ij}D^j_{km}.\end{aligned} \tag{6.166}$$

Due to the orthonormality of Clebsch–Gordan coefficients, described in Equation 6.35, the following relation exists:

$$\sum_{i=j_1-j_2}^{j_1+j_2} C(j_1j_2i;n_1n_2k)\,C(j_1j_2i;k_1k_2k) = \delta_{n_1k_1}\delta_{n_2k_2}. \tag{6.167}$$

Application of Equation 6.167 to Equation 6.166 gives:

$$\sum_{k_1k_2}\sum_{m_1m_2}\delta_{n_1k_1}\delta_{n_2k_2}C(j_1j_2j;m_1m_2m)\left[D^{j_1}\otimes D^{j_2}\right]_{k_1k_2\,m_1m_2}=\sum_i C(j_1j_2i;n_1n_2k)\,\delta_{ij}D^j_{km}. \tag{6.168}$$

Applying the conditions $n_1 = k_1$, $n_2 = k_2$, and $i = j$ imposed by the Kronecker deltas, and renaming n_1 to k_1, and n_2 to k_2 on both sides of the equation leads to:

$$\sum_{m_1m_2} C(j_1j_2j;m_1m_2m)\left[D^{j_1}\otimes D^{j_2}\right]_{k_1k_2,m_1m_2}=C(j_1j_2j;k_1k_2k)\,D^j_{k,m}, \tag{6.169}$$

or, in terms of individual matrix elements,

$$\sum_{m_1m_2} C(j_1j_2j;m_1m_2m)\,D^{j_1}_{k_1m_1}D^{j_2}_{k_2m_2}=C(j_1j_2j;k_1k_2k)\,D^j_{k,m}. \tag{6.170}$$

Due to the constraint of $m = m_1 + m_2$ coming from the Clebsch–Gordan coefficients, which become zero if this condition is not satisfied, for a fixed value of m, the sum in the left-hand side actually iterates over a single index, either m_1 or m_2, whichever is convenient. The value of the other index is then defined from the sum $m = m_1 + m_2$. Similarly, the Clebsch–Gordan coefficient in the right-hand side of the equation imposes the condition $k = k_1 + k_2$.

This equation establishes an important relation between the rotation matrices for the angular momenta j_1 and j_2 and the rotation matrix for the composite state created by the addition of the angular momenta, and serves as the basis for the recurrent computation of the Wigner matrix. In order to obtain the recurrence equations it is sufficient to consider the simplest case of $j_2 = ½$, and analyze a reduction of the direct product $D^{j_1}\otimes D^{(1/2)}$ into $D^{j_1+(1/2)}$ and $D^{j_2-(1/2)}$. The matrix of Clebsch–Gordan coefficients and the Wigner matrix needed at this point are:

$$C\left(j_1\tfrac{1}{2}j;m_1m_2m\right)\qquad\qquad d^{\frac{1}{2}}_{km}(\beta)$$

$$\begin{array}{c|cc|}
 & m_2=\frac{1}{2} & m_2=-\frac{1}{2}\\
j=j_1+\frac{1}{2} & \sqrt{\dfrac{j_1+m+\frac{1}{2}}{2j_1+1}} & \sqrt{\dfrac{j_1-m+\frac{1}{2}}{2j_1+1}}\\
j=j_1-\frac{1}{2} & -\sqrt{\dfrac{j_1-m+\frac{1}{2}}{2j_1+1}} & \sqrt{\dfrac{j_1+m+\frac{1}{2}}{2j_1+1}}
\end{array}\qquad
\begin{array}{c|cc|}
 & m=\frac{1}{2} & m=-\frac{1}{2}\\
k=\frac{1}{2} & \cos\dfrac{\beta}{2} & -\sin\dfrac{\beta}{2}\\
k=-\frac{1}{2} & \sin\dfrac{\beta}{2} & \cos\dfrac{\beta}{2}
\end{array}, \tag{6.171}$$

where index m_1 in the Clebsch–Gordan coefficients is determined from the condition $m_1 = m - m_2$. The complete Wigner matrix is:

$$D^{\frac{1}{2}}_{km}=e^{-ik\alpha}d^{(1/2)}_{km}(\beta)\,e^{-im\gamma}. \tag{6.172}$$

The left side of Equation 6.170 can now be written in terms of $d(\beta)$ by substituting the expression for complete Wigner matrix elements in Equation 6.172, and combining the exponential terms with the same angles α and γ, to give:

$$\sum_{m_1m_2} C(j_1j_2j;m_1m_2m)\,e^{-i(k_1+k_2)\alpha}d^{j_1}_{k_1m_1}(\beta)\,d^{j_2}_{k_2m_2}(\beta)\,e^{-i(m_1+m_2)\gamma}=C(j_1j_2j;k_1k_2k)\,e^{-ik\alpha}d^j_{km}(\beta)\,e^{-im\gamma}. \tag{6.173}$$

Due to the equalities $k = k_1 + k_2$ and $m = m_1 + m_2$ the exponential terms cancel out, so that only the dependence on angle β remains:

$$\sum_{m_1m_2} C(j_1j_2j;m_1m_2m)\,d^{j_1}_{k_1m_1}(\beta)\,d^{j_2}_{k_2m_2}(\beta)=C(j_1j_2j;k_1k_2k)\,d^j_{km}(\beta). \tag{6.174}$$

Hereinafter, for brevity of notation, symbol (β), which shows the dependence of the Wigner matrix elements on angle β, will be omitted from the equations.

In Equation 6.174, the value of the azimuthal index k is fixed, so that, for a given value of k_2, the value of k_1 is automatically determined from the relation $k_1 = k - k_2$. Also, because $j_2 = 1/2$ is a constant, it is convenient to run the sum in the left-hand side of Equation 6.174 through the related index m_2. For each value of m_2, the index m_1 follows from the condition $m_1 = m - m_2$. Implementing these changes into Equation 6.174 leads to:

$$\sum_{m_2} C(j_1,\tfrac{1}{2},j;m-m_2,m_2,m)\, d^{j_1}_{k-k_2,m-m_2}\, d^{\frac{1}{2}}_{k_2,m_2} = C(j_1,\tfrac{1}{2},j;k-k_2,k_2,k)\, d^{j}_{k,m}. \tag{6.175}$$

Explicitly writing the sum over index m_2 running through the values $m_2 = 1/2$ and $m_2 = -1/2$ gives:

$$\begin{aligned} &C(j_1,\tfrac{1}{2},j;m-\tfrac{1}{2},\tfrac{1}{2},m)\, d^{j_1}_{k-k_2,m-\frac{1}{2}}\, d^{\frac{1}{2}}_{k_2,\frac{1}{2}} + C(j_1,\tfrac{1}{2},j;m+\tfrac{1}{2},-\tfrac{1}{2},m)\, d^{j_1}_{k-k_2,m+\frac{1}{2}}\, d^{\frac{1}{2}}_{k_2,-\frac{1}{2}} \\ &= C(j_1,\tfrac{1}{2},j;k-k_2,k_2,k)\, d^{j}_{k,m}. \end{aligned} \tag{6.176}$$

From Equation 6.176, four different relations between matrix elements d follow, which correspond to all possible combinations of indices $j = j_1 \pm (1/2)$ and $k_2 = \pm (1/2)$.

Case #1: $j = j_1 \pm (1/2)$ and $k_2 = (1/2)$

Given that $m_1 = m - m_2$ and $k_1 = k - k_2$:

$$\begin{aligned} &C(j_1,\tfrac{1}{2},j_1+\tfrac{1}{2};m-\tfrac{1}{2},\tfrac{1}{2},m) d^{j_1}_{k-\frac{1}{2},m-\frac{1}{2}} d^{\frac{1}{2}}_{\frac{1}{2},\frac{1}{2}} + C(j_1,\tfrac{1}{2},j_1+\tfrac{1}{2};m+\tfrac{1}{2},-\tfrac{1}{2},m) d^{j_1}_{k-\frac{1}{2},m+\frac{1}{2}} d^{\frac{1}{2}}_{\frac{1}{2},-\frac{1}{2}} \\ &= C(j_1,\tfrac{1}{2},j_1+\tfrac{1}{2};k-\tfrac{1}{2},\tfrac{1}{2},k) d^{j}_{k,m}. \end{aligned} \tag{6.177}$$

Substituting the values of corresponding Clebsch–Gordan coefficients and the matrix elements of $d^{(1/2)}$ from Equation 6.171 into Equation 6.177 gives:

$$\sqrt{\frac{j_1+m+\frac{1}{2}}{2j_1+1}}\, d^{j_1}_{k-\frac{1}{2},m-\frac{1}{2}} \cos\frac{\beta}{2} - \sqrt{\frac{j_1-m+\frac{1}{2}}{2j_1+1}}\, d^{j_1}_{k-\frac{1}{2},m+\frac{1}{2}} \sin\frac{\beta}{2} = \sqrt{\frac{j_1+k+\frac{1}{2}}{2j_1+1}}\, d^{j}_{k,m}. \tag{6.178}$$

Canceling out the term $\sqrt{2j_1+1}$ in the denominator, and replacing j_1 by $j_1 = j - (1/2)$ derives the first relation that exists between matrix elements d:

$$\sqrt{j+m}\left(\cos\frac{\beta}{2}\right) d^{j-\frac{1}{2}}_{k-\frac{1}{2},m-\frac{1}{2}} - \sqrt{j-m}\left(\sin\frac{\beta}{2}\right) d^{j-\frac{1}{2}}_{k-\frac{1}{2},m+\frac{1}{2}} = \sqrt{j+k}\, d^{j}_{k,m}. \tag{6.179}$$

Case #2: $j = j_1 + (1/2)$ and $k_2 = -(1/2)$

Substituting the values of j and k_2 into Equation 6.176 leads to:

$$\begin{aligned} &C(j_1,\tfrac{1}{2},j_1+\tfrac{1}{2};m-\tfrac{1}{2},\tfrac{1}{2},m)\, d^{j_1}_{k+\frac{1}{2},m-\frac{1}{2}} d^{\frac{1}{2}}_{-\frac{1}{2},\frac{1}{2}} + C(j_1,\tfrac{1}{2},j_1+\tfrac{1}{2};m+\tfrac{1}{2},-\tfrac{1}{2},m)\, d^{j_1}_{k+\frac{1}{2},m+\frac{1}{2}} d^{\frac{1}{2}}_{-\frac{1}{2},-\frac{1}{2}} \\ &= C(j_1,\tfrac{1}{2},j_1+\tfrac{1}{2};k+\tfrac{1}{2},-\tfrac{1}{2},k)\, d^{j}_{k,m}. \end{aligned} \tag{6.180}$$

Inserting the values of the corresponding Clebsch–Gordan coefficients and the elements of the Wigner matrix from Equation 6.171 into this equation gives:

$$\sqrt{\frac{j_1+m+\frac{1}{2}}{2j_1+1}}\, d^{j_1}_{k+\frac{1}{2},m-\frac{1}{2}} \sin\frac{\beta}{2} + \sqrt{\frac{j_1-m+\frac{1}{2}}{2j_1+1}}\, d^{j_1}_{k+\frac{1}{2},m+\frac{1}{2}} \cos\frac{\beta}{2} = \sqrt{\frac{j_1-k+\frac{1}{2}}{2j_1+1}}\, d^{j}_{k,m}. \tag{6.181}$$

Multiplying both sides of the equation by $\sqrt{2j_1+1}$ and replacing j_1 by $j_1 = j - (1/2)$ produces the second relation between matrix elements d:

$$\sqrt{j+m}\left(\sin\frac{\beta}{2}\right) d^{j-\frac{1}{2}}_{k+\frac{1}{2},m-\frac{1}{2}} + \sqrt{j-m}\left(\cos\frac{\beta}{2}\right) d^{j-\frac{1}{2}}_{k+\frac{1}{2},m+\frac{1}{2}} = \sqrt{j-k}\, d^{j}_{k,m}. \tag{6.182}$$

Case #3: $j = j_1 - (1/2)$ and $k_2 = (1/2)$

Substituting the above values into Equation 6.176 leads to:

$$\begin{aligned} &C(j_1,\tfrac{1}{2},j_1-\tfrac{1}{2};m-\tfrac{1}{2},\tfrac{1}{2},m)\, d^{j_1}_{k-\frac{1}{2},m-\frac{1}{2}}\, d^{\frac{1}{2}}_{\frac{1}{2},\frac{1}{2}} + C(j_1,\tfrac{1}{2},j_1-\tfrac{1}{2};m+\tfrac{1}{2},-\tfrac{1}{2},m)\, d^{j_1}_{k-\frac{1}{2},m+\frac{1}{2}} d^{\frac{1}{2}}_{\frac{1}{2},-\frac{1}{2}} \\ &= C(j_1,\tfrac{1}{2},j_1-\tfrac{1}{2};k-\tfrac{1}{2},\tfrac{1}{2},k)\, d^{j}_{k,m}. \end{aligned} \tag{6.183}$$

Substituting the values of the Clebsch–Gordan coefficients and the elements of the Wigner matrix from Equation 6.171 into Equation 6.183 gives:

$$-\sqrt{\frac{j_1-m+\frac{1}{2}}{2j_1+1}}\, d^{j_1}_{k-\frac{1}{2},m-\frac{1}{2}} \cos\frac{\beta}{2} - \sqrt{\frac{j_1+m+\frac{1}{2}}{2j_1+1}}\, d^{j_1}_{k-\frac{1}{2},m+\frac{1}{2}} \sin\frac{\beta}{2} = -\sqrt{\frac{j_1-k+\frac{1}{2}}{2j_1+1}}\, d^{j}_{k,m}. \tag{6.184}$$

The term $\sqrt{2j_1+1}$ in the denominator and the negative sign uniformly cancel out. Since $j = j_1 - (1/2)$, this gives $j_1 = j + (1/2)$. Replacing j_1 by $j + (1/2)$ leads to the third relation between the d matrix elements:

$$\sqrt{j-m+1}\left(\cos\frac{\beta}{2}\right) d^{j+\frac{1}{2}}_{k-\frac{1}{2},m-\frac{1}{2}} + \sqrt{j+m+1}\left(\sin\frac{\beta}{2}\right) d^{j+\frac{1}{2}}_{k-\frac{1}{2},m+\frac{1}{2}} = \sqrt{j-k+1}\, d^{j}_{k,m}. \tag{6.185}$$

Case #4: $j = j_1 - (1/2)$ and $k_2 = -(1/2)$

Substituting these values of the indices into Equation 6.176 leads to:

$$\begin{aligned} &C(j_1,\tfrac{1}{2},j_1-\tfrac{1}{2};m-\tfrac{1}{2},\tfrac{1}{2},m)\, d^{j_1}_{k+\frac{1}{2},m-\frac{1}{2}} d^{\frac{1}{2}}_{-\frac{1}{2},\frac{1}{2}} + C(j_1,\tfrac{1}{2},j_1-\tfrac{1}{2};m+\tfrac{1}{2},-\tfrac{1}{2},m)\, d^{j_1}_{k+\frac{1}{2},m+\frac{1}{2}} d^{\frac{1}{2}}_{-\frac{1}{2},-\frac{1}{2}} \\ &= C(j_1,\tfrac{1}{2},j_1-\tfrac{1}{2};k+\tfrac{1}{2},-\tfrac{1}{2},k)\, d^{j}_{k,m}. \end{aligned} \tag{6.186}$$

Taking the values of the Clebsch–Gordan coefficients and the elements of the Wigner matrix from Equation 6.171 and substituting them into Equation 6.186 gives:

$$-\sqrt{\frac{j_1-m+\frac{1}{2}}{2j_1+1}}\, d^{j_1}_{k+\frac{1}{2},m-\frac{1}{2}} \sin\frac{\beta}{2} + \sqrt{\frac{j_1+m+\frac{1}{2}}{2j_1+1}}\, d^{j_1}_{k+\frac{1}{2},m+\frac{1}{2}} \cos\frac{\beta}{2} = \sqrt{\frac{j_1+k+\frac{1}{2}}{2j_1+1}}\, d^{j}_{k,m}. \tag{6.187}$$

Canceling out the term $\sqrt{2j_1+1}$ in the denominator, and replacing j_1 by $j+(1/2)$ leads to the fourth and final relation:

$$-\sqrt{j-m+1}\left(\sin\frac{\beta}{2}\right) d^{j+\frac{1}{2}}_{k+\frac{1}{2},m-\frac{1}{2}} + \sqrt{j+m+1}\left(\cos\frac{\beta}{2}\right) d^{j+\frac{1}{2}}_{k+\frac{1}{2},m+\frac{1}{2}} = \sqrt{j+k+1}\, d^{j}_{k,m}. \tag{6.188}$$

These four equations establish recurrence relations between the Wigner matrix element d^j on one hand, and, on the other hand, the matrix elements $d^{j-(1/2)}$ and $d^{+-(1/2)}$ corresponding to half-values of angular momentum. These relations serve as the starting point for the derivation of recurrence relations between the Wigner matrix elements corresponding to integer values of the angular momenta, which will be needed for rotation of spherical harmonic functions.

7

Recurrence Relations for Wigner Matrix

The purpose of this chapter is to deliver recurrence relations for computation of Wigner matrix elements.

7.1 Recurrence Relations with Increment in Index m

Practical applications of Wigner matrix require the knowledge of its matrix elements. For any arbitrary angular momentum j, the general form of the Wigner matrix element is:

$$D^j_{km}(\alpha\beta\gamma) = e^{-ik\alpha} d^j_{km}(\beta) e^{-im\gamma}, \tag{7.1}$$

where α, β, and γ are Euler angles. Computing the exponential terms in the Wigner matrix in Equation 7.1 that depend on angles α and γ is straightforward. The most time-consuming part is evaluating the matrix elements d^j_{km}, which describe rotation through angle β, so further analysis will be focused on the evaluation of those terms.

Direct evaluation of the Wigner matrix elements from Equation 5.230 is both computationally inefficient and can result in numerical instabilities. The most stable solutions rely on recurrence relations. A family of such relations comes as the result of reduction of the direct product $D^j \otimes D^{1/2}$ into $D^{j+(1/2)}$ and $D^{j-(1/2)}$, as pointed out by Gimbutas and Greengard. This reduction, explained in Chapter 6, produces:

$$(j+m)^{\frac{1}{2}}\left(\cos\frac{\beta}{2}\right) d^{j-\frac{1}{2}}_{k-\frac{1}{2},m-\frac{1}{2}} - (j-m)^{\frac{1}{2}}\left(\sin\frac{\beta}{2}\right) d^{j-\frac{1}{2}}_{k-\frac{1}{2},m+\frac{1}{2}} = (j+k)^{\frac{1}{2}} d^j_{k,m}, \tag{6.179}$$

and

$$(j+m)^{\frac{1}{2}}\left(\sin\frac{\beta}{2}\right) d^{j-\frac{1}{2}}_{k+\frac{1}{2},m-\frac{1}{2}} + (j-m)^{\frac{1}{2}}\left(\cos\frac{\beta}{2}\right) d^{j-\frac{1}{2}}_{k+\frac{1}{2},m+\frac{1}{2}} = (j-k)^{\frac{1}{2}} d^j_{k,m}. \tag{6.182}$$

Although these are already formally recurrence relations, using them for the computation of Wigner matrix elements for the purpose of rotation of spherical harmonics would still be impractical due to the presence of the terms corresponding to half-values of the angular momentum. These equations, however, open the path to other recurrence relations, which do not include the half-values in indices j and m, and are therefore better suited for rotation of spherical harmonics. In total, three new recurrence relations will be needed in order to cover all possible combinations of orbital and azimuthal indices. These equations can be derived as follows.

In Equation 6.179, performing index substitution $j \to j-(1/2)$, $k \to k-(1/2)$, and $m \to m-(1/2)$ gives:

$$(j+m-1)^{\frac{1}{2}}\left(\cos\frac{\beta}{2}\right) d^{j-1}_{k-1,m-1} - (j-m)^{\frac{1}{2}}\left(\sin\frac{\beta}{2}\right) d^{j-1}_{k-1,m} = (j+k-1)^{\frac{1}{2}} d^{j-\frac{1}{2}}_{k-\frac{1}{2},m-\frac{1}{2}}. \tag{7.2}$$

A similar but slightly different substitution, $j \to j-(1/2)$, $k \to k-(1/2)$, and $m \to m+(1/2)$, to Equation 6.179 produces:

$$(j+m)^{\frac{1}{2}}\left(\cos\frac{\beta}{2}\right) d^{j-1}_{k-1,m} - (j-m-1)^{\frac{1}{2}}\left(\sin\frac{\beta}{2}\right) d^{j-1}_{k-1,m+1} = (j+k-1)^{\frac{1}{2}} d^{j-\frac{1}{2}}_{k-\frac{1}{2},m+\frac{1}{2}}. \tag{7.3}$$

Note that these substitutions eliminate the half-values in the left side, but introduce them in the right side. The remaining presence of the half-value indices can be corrected in two steps. First, it is necessary to multiply Equation 6.179 by $(j + k - 1)^{1/2}$. This gives:

$$(j+k-1)^{\frac{1}{2}}(j+m)^{\frac{1}{2}}\left(\cos\frac{\beta}{2}\right)d^{j-\frac{1}{2}}_{k-\frac{1}{2},m-\frac{1}{2}}-(j+k-1)^{\frac{1}{2}}(j-m)^{\frac{1}{2}}\left(\sin\frac{\beta}{2}\right)d^{j-\frac{1}{2}}_{k-\frac{1}{2},m+\frac{1}{2}}=(j+k-1)^{\frac{1}{2}}(j+k)^{\frac{1}{2}}d^{j}_{k,m}. \tag{7.4}$$

In the second step, the expressions for $(j+k-1)^{1/2}d^{j-(1/2)}_{k-(1/2),m-(1/2)}$ from Equation 7.2 and $(j+k-1)^{1/2}d^{j-(1/2)}_{k-(1/2),m+(1/2)}$ from Equation 7.3 are used for replacing the corresponding terms in Equation 7.4. These substitutions lead to:

$$\begin{aligned}&(j+m)^{\frac{1}{2}}\left(\cos\frac{\beta}{2}\right)\left[(j+m-1)^{\frac{1}{2}}\left(\cos\frac{\beta}{2}\right)d^{j-1}_{k-1,m-1}-(j-m)^{\frac{1}{2}}\left(\sin\frac{\beta}{2}\right)d^{j-1}_{k-1,m}\right]\\&-(j-m)^{\frac{1}{2}}\left(\sin\frac{\beta}{2}\right)\left[(j+m)^{\frac{1}{2}}\left(\cos\frac{\beta}{2}\right)d^{j-1}_{k-1,m}-(j-m-1)^{\frac{1}{2}}\left(\sin\frac{\beta}{2}\right)d^{j-1}_{k-1,m+1}\right]\\&=(j+k-1)^{\frac{1}{2}}(j+k)^{\frac{1}{2}}d^{j}_{k,m}.\end{aligned} \tag{7.5}$$

Opening the square brackets and combining like terms produces the first recurrence relation that contains only integer values of the indices:

$$\begin{aligned}&(j+m)^{\frac{1}{2}}(j+m-1)^{\frac{1}{2}}\left(\cos\frac{\beta}{2}\right)^{2}d^{j-1}_{k-1,m-1}-2(j+m)^{\frac{1}{2}}(j-m)^{\frac{1}{2}}\left(\sin\frac{\beta}{2}\right)\left(\cos\frac{\beta}{2}\right)d^{j-1}_{k-1,m}\\&+(j-m)^{\frac{1}{2}}(j-m-1)^{\frac{1}{2}}\left(\sin\frac{\beta}{2}\right)^{2}d^{j-1}_{k-1,m+1}=(j+k-1)^{\frac{1}{2}}(j+k)^{\frac{1}{2}}d^{j}_{k,m}.\end{aligned} \tag{7.6}$$

Due to the presence of the coefficient $(j + k - 1)(j + k)$ on the right-hand side of Equation 7.6, this equation becomes undefined for $d^{j}_{k,m}$ when $k = -j + 1$ or $k = -j$. A different relation is necessary in that case, and that leads to the next recurrence relation.

Derivation of the second recurrence relation starts from Equation 6.182. As in the previous case, converting the left side to integral indices can be accomplished in many ways. The rationale of the substitution is to obtain such terms in the right-hand side of Equation 6.182, which are already present in the left-hand side of Equation 6.179. Applying index substitution $j \to j - (1/2)$, $k \to k - (1/2)$, and $m \to m - (1/2)$ to Equation 6.182 gives:

$$(j+m-1)^{\frac{1}{2}}\left(\sin\frac{\beta}{2}\right)d^{j-1}_{k,m-1}+(j-m)^{\frac{1}{2}}\left(\cos\frac{\beta}{2}\right)d^{j-1}_{k,m}=(j-k)^{\frac{1}{2}}d^{j-\frac{1}{2}}_{k-\frac{1}{2},m-\frac{1}{2}}. \tag{7.7}$$

This substitution offers an all-integer index equivalent for the term $d^{j-(1/2)}_{k-(1/2),m-(1/2)}$ present in Equation 6.179. The final term in Equation 6.179 which carries half-value indices, $d^{j-(1/2)}_{k-(1/2),m+(1/2)}$, can also be generated from Equation 6.182 by applying the following substitutions: $j \to j - (1/2)$, $k \to k - (1/2)$, and $m \to m + (1/2)$, leading to:

$$(j+m)^{\frac{1}{2}}\left(\sin\frac{\beta}{2}\right)d^{j-1}_{k,m}+(j-m-1)^{\frac{1}{2}}\left(\cos\frac{\beta}{2}\right)d^{j-1}_{k,m+1}=(j-k)^{\frac{1}{2}}d^{j-\frac{1}{2}}_{k-\frac{1}{2},m+\frac{1}{2}}. \tag{7.8}$$

Equations 7.7 and 7.8 offer a replacement of both half-value terms located in the left-hand side of Equation 6.179. An intermediate step before implementing the substitution to Equation 6.179 is to multiply both of its sides by $(j - k)^{1/2}$, which gives:

$$(j-k)^{\frac{1}{2}}(j+m)^{\frac{1}{2}}\left(\cos\frac{\beta}{2}\right)d^{j-\frac{1}{2}}_{k-\frac{1}{2},m-\frac{1}{2}}-(j-k)^{\frac{1}{2}}(j-m)^{\frac{1}{2}}\left(\sin\frac{\beta}{2}\right)d^{j-\frac{1}{2}}_{k-\frac{1}{2},m+\frac{1}{2}}=(j-k)^{\frac{1}{2}}(j+k)^{\frac{1}{2}}d^{j}_{k,m}. \tag{7.9}$$

Substituting Equations 7.7 and 7.8 into Equation 7.9 produces:

$$\begin{aligned}&(j+m)^{\frac{1}{2}}\left(\cos\frac{\beta}{2}\right)\left[(j+m-1)^{\frac{1}{2}}\left(\sin\frac{\beta}{2}\right)d_{k,m-1}^{j-1}+(j-m)^{\frac{1}{2}}\left(\cos\frac{\beta}{2}\right)d_{k,m}^{j-1}\right]\\&-(j-m)^{\frac{1}{2}}\left(\sin\frac{\beta}{2}\right)\left[(j+m)^{\frac{1}{2}}\left(\sin\frac{\beta}{2}\right)d_{k,m}^{j-1}+(j-m-1)^{\frac{1}{2}}\left(\cos\frac{\beta}{2}\right)d_{k,m+1}^{j-1}\right]\\&=(j-k)^{\frac{1}{2}}(j+k)^{\frac{1}{2}}d_{k,m}^{j}.\end{aligned}\tag{7.10}$$

Opening the square brackets gives:

$$\begin{aligned}&(j+m)^{\frac{1}{2}}(j+m-1)^{\frac{1}{2}}\left(\sin\frac{\beta}{2}\right)\left(\cos\frac{\beta}{2}\right)d_{k,m-1}^{j-1}+(j+m)^{\frac{1}{2}}(j-m)^{\frac{1}{2}}\left(\cos\frac{\beta}{2}\right)^{2}d_{k,m}^{j-1}\\&-(j+m)^{\frac{1}{2}}(j-m)^{\frac{1}{2}}\left(\sin\frac{\beta}{2}\right)^{2}d_{k,m}^{j-1}-(j-m)^{\frac{1}{2}}(j-m-1)^{\frac{1}{2}}\left(\sin\frac{\beta}{2}\right)\left(\cos\frac{\beta}{2}\right)d_{k,m+1}^{j-1}\\&=(j-k)^{\frac{1}{2}}(j+k)^{\frac{1}{2}}d_{k,m}^{j}.\end{aligned}\tag{7.11}$$

Finally, combining like terms leads to the second recurrence relation:

$$\begin{aligned}&(j+m)^{\frac{1}{2}}(j+m-1)^{\frac{1}{2}}\left(\sin\frac{\beta}{2}\right)\left(\cos\frac{\beta}{2}\right)d_{k,m-1}^{j-1}+(j+m)^{\frac{1}{2}}(j-m)^{\frac{1}{2}}\left[\left(\cos\frac{\beta}{2}\right)^{2}-\left(\sin\frac{\beta}{2}\right)^{2}\right]d_{k,m}^{j-1}\\&-(j-m)^{\frac{1}{2}}(j-m-1)^{\frac{1}{2}}\left(\sin\frac{\beta}{2}\right)\left(\cos\frac{\beta}{2}\right)d_{k,m+1}^{j-1}=(j-k)^{\frac{1}{2}}(j+k)^{\frac{1}{2}}d_{k,m}^{j}.\end{aligned}\tag{7.12}$$

In this equation, the coefficients $(j-k)(j+k)$ in front of $d_{k,m}^{j}$ make this relationship undefined when $k=j$ or $k=-j$. In order to address this and the previous restrictions on the value of indices in the derived recurrence relations one more equation is necessary.

The third recurrence relation starts by substituting the half-value index terms, $d_{k+(1/2),m-(1/2)}^{j-(1/2)}$ and $d_{k+(1/2),m+(1/2)}^{j-(1/2)}$, in Equation 6.182. The first of these may be converted to integer indices by applying a substitution $j\to j-(1/2)$, $k\to k+(1/2)$, and $m\to m-(1/2)$ to Equation 6.182, which gives:

$$(j+m-1)^{\frac{1}{2}}\left(\sin\frac{\beta}{2}\right)d_{k+1,m-1}^{j-1}+(j-m)^{\frac{1}{2}}\left(\cos\frac{\beta}{2}\right)d_{k+1,m}^{j-1}=(j-k-1)^{\frac{1}{2}}d_{k+\frac{1}{2},m-\frac{1}{2}}^{j-\frac{1}{2}}.\tag{7.13}$$

The second term, $d_{k+(1/2),m+(1/2)}^{j-(1/2)}$, is obtained by applying the substitutions $j\to j-(1/2)$, $k\to k+(1/2)$, and $m\to m+(1/2)$ to Equation 6.182. This leads to:

$$(j+m)^{\frac{1}{2}}\left(\sin\frac{\beta}{2}\right)d_{k+1,m}^{j-1}+(j-m-1)^{\frac{1}{2}}\left(\cos\frac{\beta}{2}\right)d_{k+1,m+1}^{j-1}=(j-k-1)^{\frac{1}{2}}d_{k+\frac{1}{2},m+\frac{1}{2}}^{j-\frac{1}{2}}.\tag{7.14}$$

Equations 7.13 and 7.14 complete the task of redefining terms $d_{k+(1/2),m-(1/2)}^{j-(1/2)}$ and $d_{k+(1/2),m+(1/2)}^{j-(1/2)}$ in terms that have integer indices. Since both latter equations carry a coefficient $(j-k-1)^{1/2}$ in their right-hand side, it is convenient to multiply Equation 6.182 by the same coefficient before performing the substitution:

$$\begin{aligned}&(j-k-1)^{\frac{1}{2}}(j+m)^{\frac{1}{2}}\left(\sin\frac{\beta}{2}\right)d_{k+\frac{1}{2},m-\frac{1}{2}}^{j-\frac{1}{2}}+(j-k-1)^{\frac{1}{2}}(j-m)^{\frac{1}{2}}\left(\cos\frac{\beta}{2}\right)d_{k+\frac{1}{2},m+\frac{1}{2}}^{j-\frac{1}{2}}\\&=(j-k-1)^{\frac{1}{2}}(j-k)^{\frac{1}{2}}d_{k,m}^{j}.\end{aligned}\tag{7.15}$$

Substituting the terms $(j-k-1)^{1/2} d^{j-(1/2)}_{k+(1/2),m-(1/2)}$ and $(j-k-1)^{1/2} d^{j-(1/2)}_{k+(1/2),m+(1/2)}$ in Equation 7.15 by the expressions in Equations 7.13 and 7.14 yields:

$$\begin{aligned}&(j+m)^{\frac{1}{2}}\left(\sin\frac{\beta}{2}\right)\left[(j+m-1)^{\frac{1}{2}}\left(\sin\frac{\beta}{2}\right)d^{j-1}_{k+1,m-1}+(j-m)^{\frac{1}{2}}\left(\cos\frac{\beta}{2}\right)d^{j-1}_{k+1,m}\right]\\&+(j-m)^{\frac{1}{2}}\left(\cos\frac{\beta}{2}\right)\left[(j+m)^{\frac{1}{2}}\left(\sin\frac{\beta}{2}\right)d^{j-1}_{k+1,m}+(j-m-1)^{\frac{1}{2}}\left(\cos\frac{\beta}{2}\right)d^{j-1}_{k+1,m+1}\right]\\&=(j-k-1)^{\frac{1}{2}}(j-k)^{\frac{1}{2}}d^{j}_{k,m}.\end{aligned} \tag{7.16}$$

Opening the square brackets and combining like terms produces the third recurrence relation:

$$\begin{aligned}&(j+m)^{\frac{1}{2}}(j+m-1)^{\frac{1}{2}}\left(\sin\frac{\beta}{2}\right)^{2}d^{j-1}_{k+1,m-1}+2(j+m)^{\frac{1}{2}}(j-m)^{\frac{1}{2}}\left(\sin\frac{\beta}{2}\right)\left(\cos\frac{\beta}{2}\right)d^{j-1}_{k+1,m}\\&+(j-m)^{\frac{1}{2}}(j-m-1)^{\frac{1}{2}}\left(\cos\frac{\beta}{2}\right)^{2}d^{j-1}_{k+1,m+1}=(j-k-1)^{\frac{1}{2}}(j-k)^{\frac{1}{2}}d^{j}_{k,m},\end{aligned} \tag{7.17}$$

which employs only integer values of angular and azimuthal indices. This equation still requires $k \neq j-1$ and $k \neq j$ due to the coefficient $(j-k-1)(j-k)$ on the right-hand side; for these particular values of index k one of the other two recurrence relations should be used. Combined together, Equations 7.6, 7.12, and 7.17 provide complete coverage of all possible values of the indices that are encountered during the calculation of the matrix elements $d^{j}_{k,m}$.

A few practical remarks need to be made regarding these derived recurrence relations. A close inspection of Equations 7.6, 7.12, and 7.17 reveals a few specific cases, which involve the index m. For instance, when $m = j$, the formulae for $d^{j}_{k,m}$ include various non-existent matrix elements, but in every case the coefficients in front of those terms vanish, and that zeroes out the corresponding terms. Therefore, for $m = j$, the recurrence relations reduce to:

$$(2j)^{\frac{1}{2}}(2j-1)^{\frac{1}{2}}\left(\cos\frac{\beta}{2}\right)^{2}d^{j-1}_{k-1,j-1}=(j+k-1)^{\frac{1}{2}}(j+k)^{\frac{1}{2}}d^{j}_{k,j}, \tag{7.18}$$

$$(2j)^{\frac{1}{2}}(2j-1)^{\frac{1}{2}}\left(\sin\frac{\beta}{2}\right)\left(\cos\frac{\beta}{2}\right)d^{j-1}_{k,j-1}=(j-k)^{\frac{1}{2}}(j+k)^{\frac{1}{2}}d^{j}_{k,j}, \tag{7.19}$$

$$(2j)^{\frac{1}{2}}(2j-1)^{\frac{1}{2}}\left(\sin\frac{\beta}{2}\right)^{2}d^{j-1}_{k+1,j-1}=(j-k-1)^{\frac{1}{2}}(j-k)^{\frac{1}{2}}d^{j}_{k,j}. \tag{7.20}$$

Similarly, for the case of $m = -j$, the original recurrence relations simplify to:

$$(2j)^{\frac{1}{2}}(2j-1)^{\frac{1}{2}}\left(\sin\frac{\beta}{2}\right)^{2}d^{j-1}_{k-1,-j+1}=(j+k-1)^{\frac{1}{2}}(j+k)^{\frac{1}{2}}d^{j}_{k,-j}, \tag{7.21}$$

$$-(2j)^{\frac{1}{2}}(2j-1)^{\frac{1}{2}}\left(\sin\frac{\beta}{2}\right)\left(\cos\frac{\beta}{2}\right)d^{j-1}_{k,-j+1}=(j-k)^{\frac{1}{2}}(j+k)^{\frac{1}{2}}d^{j}_{k,-j}, \tag{7.22}$$

$$(2j)^{\frac{1}{2}}(2j-1)^{\frac{1}{2}}\left(\cos\frac{\beta}{2}\right)^{2}d^{j-1}_{k+1,-j+1}=(j-k-1)^{\frac{1}{2}}(j-k)^{\frac{1}{2}}d^{j}_{k,-j}. \tag{7.23}$$

For the case of $m = j - 1$, the vanishing coefficients transform the original recurrence relations to:

$$(2j-1)^{\frac{1}{2}}(2j-2)^{\frac{1}{2}}\left(\cos\frac{\beta}{2}\right)^2 d^{j-1}_{k-1,j-2} - 2(2j-1)^{\frac{1}{2}}\left(\sin\frac{\beta}{2}\right)\left(\cos\frac{\beta}{2}\right) d^{j-1}_{k-1,j-1} = (j+k-1)^{\frac{1}{2}}(j+k)^{\frac{1}{2}} d^{j}_{k,j-1}, \quad (7.24)$$

$$(2j-1)^{\frac{1}{2}}(2j-2)^{\frac{1}{2}}\left(\sin\frac{\beta}{2}\right)\left(\cos\frac{\beta}{2}\right) d^{j-1}_{k,j-2} + (2j-1)^{\frac{1}{2}}\left[\left(\cos\frac{\beta}{2}\right)^2 - \left(\sin\frac{\beta}{2}\right)^2\right] d^{j-1}_{k,j-1} = (j-k)^{\frac{1}{2}}(j+k)^{\frac{1}{2}} d^{j}_{k,j-1}, \quad (7.25)$$

and

$$(2j-1)^{\frac{1}{2}}(2j-2)^{\frac{1}{2}}\left(\sin\frac{\beta}{2}\right)^2 d^{j-1}_{k+1,j-2} + 2(2j-1)^{\frac{1}{2}}\left(\sin\frac{\beta}{2}\right)\left(\cos\frac{\beta}{2}\right) d^{j-1}_{k+1,j-1} = (j-k-1)^{\frac{1}{2}}(j-k)^{\frac{1}{2}} d^{j}_{k,j-1}; \quad (7.26)$$

and for $m = -j + 1$, the original recurrence relations reduce to:

$$-2(2j-1)^{\frac{1}{2}}\left(\sin\frac{\beta}{2}\right)\left(\cos\frac{\beta}{2}\right) d^{j-1}_{k-1,-j+1} + (2j-1)^{\frac{1}{2}}(2j-2)^{\frac{1}{2}}\left(\sin\frac{\beta}{2}\right)^2 d^{j-1}_{k-1,-j+2} = (j+k-1)^{\frac{1}{2}}(j+k)^{\frac{1}{2}} d^{j}_{k,-j+1}, \quad (7.27)$$

$$(2j-1)^{\frac{1}{2}}\left[\left(\cos\frac{\beta}{2}\right)^2 - \left(\sin\frac{\beta}{2}\right)^2\right] d^{j-1}_{k,-j+1} - (2j-1)^{\frac{1}{2}}(2j-2)^{\frac{1}{2}}\left(\sin\frac{\beta}{2}\right)\left(\cos\frac{\beta}{2}\right) d^{j-1}_{k,-j+2} = (j-k)^{\frac{1}{2}}(j+k)^{\frac{1}{2}} d^{j}_{k,-j+1}, \quad (7.28)$$

$$2(2j-1)^{\frac{1}{2}}\left(\sin\frac{\beta}{2}\right)\left(\cos\frac{\beta}{2}\right) d^{j-1}_{k+1,-j+1} + (2j-1)^{\frac{1}{2}}(2j-2)^{\frac{1}{2}}\left(\cos\frac{\beta}{2}\right)^2 d^{j-1}_{k+1,-j+2} = (j-k-1)^{\frac{1}{2}}(j-k)^{\frac{1}{2}} d^{j}_{k,-j+1}. \quad (7.29)$$

Furthermore, for $k = j$ and $m = j$ the first recurrence relation in Equation 7.6 reduces to:

$$\left(\cos\frac{\beta}{2}\right)^2 d^{j-1}_{j-1,j-1} = d^{j}_{j,j}. \quad (7.30)$$

Similarly, for $k = j$ and $m = -j$ the vanishing coefficients turn Equation 7.6 into:

$$\left(\sin\frac{\beta}{2}\right)^2 d^{j-1}_{j-1,-j+1} = d^{j}_{j,-j}. \quad (7.31)$$

For $k = -j$, neither Equations 7.6 nor 7.12 are applicable, so the third recurrence relation, Equation 7.17, should be used. If $k = -j$ and $m = j$, Equation 7.17 simplifies to:

$$\left(\sin\frac{\beta}{2}\right)^2 d^{j-1}_{-j+1,j-1} = d^{j}_{-j,j}, \quad (7.32)$$

and, if $k = -j$ and $m = -j$, Equation 7.17 simplifies to:

$$\left(\cos\frac{\beta}{2}\right)^2 d^{j-1}_{-j+1,-j+1} = d^{j}_{-j,-j}. \quad (7.33)$$

Two useful symmetry properties of Wigner matrix elements can also be derived. If a rotation through angle β can be described by matrix $d^j(\beta)$, then rotation in the opposite direction, through angle $-\beta$, is the inverse operation, and is described by $d^j(-\beta) = [d^j(\beta)]^{-1}$. This allows $d^j(-\beta)$ to be quickly computed, given the known matrix $d^j(\beta)$. Due to the unitary property of rotation matrix $d^j(\beta)$, its inverse is its complex conjugate transpose, obtained by swapping rows and columns, and replacing each matrix element by its complex conjugate, but, because $d^j(\beta)$ is real, the step of taking the complex conjugate is unnecessary, so that:

$$d^j_{km}(-\beta) = d^j_{mk}(\beta). \tag{7.34}$$

Further analysis is possible. Substituting angle $-\beta$ instead of β into the Wigner equation

$$d^j_{km} = \sum_{s=0}^{j+m} \frac{(-1)^{k+s-m}\sqrt{(j+m)!(j-m)!(j+k)!(j-k)!}}{s!(j+m-s)!(j-k-s)!(k+s-m)!} \left(\cos\frac{\beta}{2}\right)^{2j+m-k-2s} \left(\sin\frac{\beta}{2}\right)^{k-m+2s}, \tag{7.35}$$

which follows from Equation 5.230, and using the parity property of trigonometric functions $\cos(-\beta/2) = \cos\beta/2$ and $\sin(-\beta/2) = -\sin\beta/2$, reveals that

$$d^j_{km}(-\beta) = (-1)^{k-m} d^j_{km}(\beta). \tag{7.36}$$

In the derivation of Equation 7.36, the equality $(-1)^{k-m+2s} = (-1)^{k-m}$ has been employed because $2s$ is, by definition, an even number. Combining Equations 7.34 and 7.36 leads to:

$$d^j_{mk}(\beta) = (-1)^{k-m} d^j_{km}(\beta). \tag{7.37}$$

Therefore, off-diagonal matrix elements located at symmetrical positions about the main diagonal are equal in absolute values, and swapping columns and rows in the Wigner matrix requires only a multiplication of the corresponding matrix element by the phase factor $(-1)^{k-m}$.

Inspection of the Wigner equation in Equation 7.35 reveals one more important relationship between indices m and k. Swapping the indices m and k, and simultaneously changing their signs, i.e. $m \to -m$ and $k \to -k$, restores the original value of the matrix element. This index substitution gives rise to the equality:

$$d^j_{-m,-k}(\beta) = d^j_{k,m}(\beta). \tag{7.38}$$

This symmetry relation significantly reduces the number of matrix elements, which need to be evaluated.

7.2 Recurrence Relations with Increment in Index *k*

Being a 2-dimensional array a Wigner matrix can be stored in computer memory in either a row-major or a column-major form. Choosing the right storage form for the array impacts the performance of computer code.

The typical use of a Wigner matrix involves rotation of eigenvectors of the angular momentum:

$$\psi'_{j,m} = \sum_{k=-j}^{j} d^j_{k,m}\psi_{j,k}, \tag{7.39}$$

where $\psi_{j,k}$ and $\psi'_{j,m}$ are initial and rotated eigenvectors, and $d^j_{k,m}$ is the Wigner matrix element for rotation about the y-axis with the orbital index j and the azimuthal indices k and m. Since the retrieval of a Wigner matrix from computer memory in Equation 7.39 happens sequentially by index k, the rotation operation favors the use of column-major format for the memory storage of a Wigner matrix.

An alternative, complementary, example of the use of the Wigner matrix elements involves their recurrence computation developed in Section 7.1. Equations 7.6, 7.12, and 7.17 describe recursion to the second index, m, in the matrix elements $d^j_{k,m}$. This operation requires consecutive data retrieval from memory storage of values for index m. In this type of calculation, row-major would be the preferred storage of the matrix.

Computation and consumption represent complete life cycle of the Wigner),matrix. As seen, computation prefers row-major storage for the Wigner matrix in computer memory whereas the consumption phase favors the column-major format. These contradictory requirements cannot be simultaneously satisfied. Choosing either storage format inevitably results in the performance of one or other of these operations being suboptimal. Since Equation 7.39 cannot be easily replaced, it would, therefore, be desirable to develop new recurrence relations for those computation steps that increment the first index k of the matrix elements $d^j_{k,m}$, so that both stages of the life cycle of the Wigner matrix could use the same storage format.

Derivation of the new recurrence relations relies on the reduction of the direct product $d^j \otimes d^{1/2}$ into $d^{j+(1/2)}$ and $d^{j-(1/2)}$, as described in Chapter 6, which produced the equations:

$$\sqrt{j+m}\left(\cos\frac{\beta}{2}\right)d^{j-\frac{1}{2}}_{k-\frac{1}{2},m-\frac{1}{2}} - \sqrt{j-m}\left(\sin\frac{\beta}{2}\right)d^{j-\frac{1}{2}}_{k-\frac{1}{2},m+\frac{1}{2}} = \sqrt{j+k}\; d^j_{k,m}, \tag{6.179}$$

$$\sqrt{j+m}\left(\sin\frac{\beta}{2}\right)d^{j-\frac{1}{2}}_{k+\frac{1}{2},m-\frac{1}{2}} + \sqrt{j-m}\left(\cos\frac{\beta}{2}\right)d^{j-\frac{1}{2}}_{k+\frac{1}{2},m+\frac{1}{2}} = \sqrt{j-k}\; d^j_{k,m}, \tag{6.182}$$

$$\sqrt{j-m+1}\left(\cos\frac{\beta}{2}\right)d^{j+\frac{1}{2}}_{k-\frac{1}{2},m-\frac{1}{2}} + \sqrt{j+m+1}\left(\sin\frac{\beta}{2}\right)d^{j+\frac{1}{2}}_{k-\frac{1}{2},m+\frac{1}{2}} = \sqrt{j-k+1}\; d^j_{k,m}, \tag{6.185}$$

$$-\sqrt{j-m+1}\left(\sin\frac{\beta}{2}\right)d^{j+\frac{1}{2}}_{k+\frac{1}{2},m-\frac{1}{2}} + \sqrt{j+m+1}\left(\cos\frac{\beta}{2}\right)d^{j+\frac{1}{2}}_{k+\frac{1}{2},m+\frac{1}{2}} = \sqrt{j+k+1}\; d^j_{k,m}. \tag{6.188}$$

In the current derivation, the indices j, k and m in Equations 6.185 and 6.188 will be manipulated by substituting $d^{j-(1/2)}_{k\mp(1/2),m-(1/2)}$ and $d^{j-(1/2)}_{k\mp(1/2),m+(1/2)}$ on the left-hand sides of Equations 6.179 and 6.182 with matrix elements of type $d^{j-1}_{k\mp1,m}$, so that they have integer azimuthal indices. This derivation produces two new recurrence relations in which index m is fixed and k becomes the moving index.

Derivation of the first recurrence relation starts with replacing $d^{j-(1/2)}_{k-(1/2),m-(1/2)}$ and $d^{j-(1/2)}_{k-(1/2),m+(1/2)}$ on the left-hand side of Equation 6.179 with $d^{j-1}_{k,m}$ and $d^{j-1}_{k-1,m}$, respectively, to obtain the increment in index k, and replace all half-value indices with those having integer values. Index substitution in Equations 6.185 and 6.188 then establishes the dependence between Wigner matrix elements with half-value indices with those having integer-value indices. The rationale for this index substitution will now be explained.

A visual comparison of Equation 6.179 with Equation 6.185 shows that the Wigner matrix elements have matching azimuthal indices and that, with the exception of the linear coefficients, the only difference between these two equations is the orbital index. Specifically, the matrix elements on the left-hand side of Equation 6.179 depend on index $j-(1/2)$ while in Equation 6.185 they depend on index $j+(1/2)$. This difference may be eliminated by subtracting unity from the orbital index in Equation 6.185. Once this is done, application of index substitution $j \to j-1$ in Equation 6.185 leads to:

$$\sqrt{j-m}\left(\cos\frac{\beta}{2}\right)d^{j-\frac{1}{2}}_{k-\frac{1}{2},m-\frac{1}{2}} + \sqrt{j+m}\left(\sin\frac{\beta}{2}\right)d^{j-\frac{1}{2}}_{k-\frac{1}{2},m+\frac{1}{2}} = \sqrt{j-k}\; d^{j-1}_{k,m}. \tag{7.40}$$

As the result of this substitution, Equation 7.40 now has the same matrix elements in the left-hand side as those in Equation 6.179. This allows the replacement of the matrix elements in Equation 6.179 that have half-value indices with the corresponding integer-value matrix element from the right-hand of Equation 7.40. But, before that can be done, the presence of the linear coefficients in front of the matrix elements must be addressed, so a companion equation to Equation 7.40 needs to be developed that will resolve the two unknowns $d^{j-(1/2)}_{k-(1/2),m-(1/2)}$ and $d^{j-(1/2)}_{k-(1/2),m+(1/2)}$ before they can be substituted into Equation 6.179.

The objective, then, is to come up with another independent linear combination of the same matrix elements $d^{j-(1/2)}_{k-(1/2),m-(1/2)}$ and $d^{j-(1/2)}_{k-(1/2),m+(1/2)}$ as those in Equation 7.40. This involves performing index manipulation on Equation 6.188 to make the matrix elements in it match those from Equation 6.179. Application of index substitution $j \to j-1$ and $k \to k-1$ to Equation 6.188 accomplishes this objective, leading to:

$$-\sqrt{j-m}\left(\sin\frac{\beta}{2}\right)d^{j-\frac{1}{2}}_{k-\frac{1}{2},m-\frac{1}{2}}+\sqrt{j+m}\left(\cos\frac{\beta}{2}\right)d^{j-\frac{1}{2}}_{k-\frac{1}{2},m+\frac{1}{2}}=\sqrt{j+k-1}\,d^{j-1}_{k-1,m}. \tag{7.41}$$

These two new independent linear combinations in Equations 7.40 and 7.41 provide the opportunity to resolve each of the of matrix elements $d^{j-(1/2)}_{k-(1/2),m-(1/2)}$ and $d^{j-(1/2)}_{k-(1/2),m+(1/2)}$ that have half-value indices into terms of matrix elements that have integer-value indices. Multiplying Equations 7.40 and 7.41 by sin (β/2) and cos (β/2), respectively, and adding them up as in the symbolic operation (7.40), sin (β/2) + (7.41) cos (β/2) then cancels out the term $d^{j-(1/2)}_{k-(1/2),m-(1/2)}$, giving:

$$\sqrt{j+m}\,d^{j-\frac{1}{2}}_{k-\frac{1}{2},m+\frac{1}{2}}=\sqrt{j-k}\left(\sin\frac{\beta}{2}\right)d^{j-1}_{k,m}+\sqrt{j+k-1}\left(\cos\frac{\beta}{2}\right)d^{j-1}_{k-1,m}. \tag{7.42}$$

Likewise, multiplying Equations 7.40 and 7.41 by cos (β/2) and sin (β/2), respectively, and subtracting the second equation from the first one as in symbolic operation (7.40), cos (β/2) − (7.41) sin (β/2) cancels out the other term $d^{j-(1/2)}_{k-(1/2),m+(1/2)}$, and gives:

$$\sqrt{j-m}\,d^{j-\frac{1}{2}}_{k-\frac{1}{2},m-\frac{1}{2}}=\sqrt{j-k}\left(\cos\frac{\beta}{2}\right)d^{j-1}_{k,m}-\sqrt{j+k-1}\left(\sin\frac{\beta}{2}\right)d^{j-1}_{k-1,m}. \tag{7.43}$$

Equations 7.42 and 7.43 express Wigner matrix elements having half-value indices over a linear combination of matrix elements $d^{j-1}_{k,m}$ and $d^{j-1}_{k-1,m}$ that have integer-value indices; in this form, they can now be used to substitute $d^{j-(1/2)}_{k-(1/2),m-(1/2)}$ and $d^{j-(1/2)}_{k-(1/2),m+(1/2)}$ in Equation 6.179. Multiplying Equation 6.179 by $\sqrt{(j-m)(j+m)}$ and substituting Equations 7.42 and 7.43 into it leads to:

$$\begin{aligned}&(j+m)\left(\cos\frac{\beta}{2}\right)\left\{\sqrt{j-k}\left(\cos\frac{\beta}{2}\right)d^{j-1}_{k,m}-\sqrt{j+k-1}\left(\sin\frac{\beta}{2}\right)d^{j-1}_{k-1,m}\right\}\\&-(j-m)\left(\sin\frac{\beta}{2}\right)\left\{\sqrt{j-k}\left(\sin\frac{\beta}{2}\right)d^{j-1}_{k,m}+\sqrt{j+k-1}\left(\cos\frac{\beta}{2}\right)d^{j-1}_{k-1,m}\right\}\\&=\sqrt{(j+k)(j-m)(j+m)}\,d^{j}_{k,m}.\end{aligned} \tag{7.44}$$

This equation can be simplified in a series of steps. Opening the curly brackets gives:

$$\begin{aligned}&(j+m)\sqrt{j-k}\left(\cos^2\frac{\beta}{2}\right)d^{j-1}_{k,m}-(j+m)\sqrt{j+k-1}\left(\sin\frac{\beta}{2}\cos\frac{\beta}{2}\right)d^{j-1}_{k-1,m}\\&-(j-m)\sqrt{j-k}\left(\sin^2\frac{\beta}{2}\right)d^{j-1}_{k,m}-(j-m)\sqrt{j+k-1}\left(\sin\frac{\beta}{2}\cos\frac{\beta}{2}\right)d^{j-1}_{k-1,m}\\&=\sqrt{(j+k)(j-m)(j+m)}\,d^{j}_{k,m}.\end{aligned} \tag{7.45}$$

Using the trigonometric identities

$$\sin^2\frac{\beta}{2}=\frac{1-\cos\beta}{2},\qquad \cos^2\frac{\beta}{2}=\frac{1+\cos\beta}{2},$$
$$2\sin\frac{\beta}{2}\cos\frac{\beta}{2}=\sin\beta,\quad \cos^2\frac{\beta}{2}-\sin^2\frac{\beta}{2}=\cos\beta, \tag{7.46}$$

and combining like matrix terms further reduces Equation 7.45 to:

$$\sqrt{j-k}\left\{(j+m)\frac{1+\cos\beta}{2}-(j-m)\frac{1-\cos\beta}{2}\right\}d_{k,m}^{j-1}$$
$$-\sqrt{j+k-1}\left\{(j+m)\frac{\sin\beta}{2}+(j-m)\frac{\sin\beta}{2}\right\}d_{k-1,m}^{j-1}=\sqrt{(j+k)(j-m)(j+m)}\;d_{k,m}^{j}. \tag{7.47}$$

Simplifying the expressions inside the curly brackets leads to:

$$\sqrt{j-k}\;(j\cos\beta+m)\,d_{k,m}^{j-1}-\sqrt{j+k-1}\;(j\sin\beta)\,d_{k-1,m}^{j-1}=\sqrt{(j+k)(j-m)(j+m)}\;d_{k,m}^{j}, \tag{7.48}$$

which is the first recurrence relation with increment in index k.

The applicability domain of Equation 7.48 is $-j+1\le k\le j$ and $-j+1\le m\le j-1$. Within either of the boundary values of index k one of the linear coefficients in the left-hand side of Equation 7.48 vanishes, thereby annihilating the corresponding term. Thus, for $k=j$, Equation 7.48 reduces to:

$$-\sqrt{2j-1}\;(j\sin\beta)\,d_{j-1,m}^{j-1}=\sqrt{(2j)(j-m)(j+m)}\;d_{j,m}^{j}. \tag{7.49}$$

Likewise, setting $k=-j+1$, simplifies Equation 7.48 to:

$$\sqrt{2j-1}\;(j\cos\beta+m)\,d_{-j+1,m}^{j-1}=\sqrt{(j-m)(j+m)}\;d_{-j+1,m}^{j}. \tag{7.50}$$

Note that the allowed range of index k in Equation 7.48 does not include $k=-j$ since that would cause the linear coefficient $\sqrt{(j+k)(j-m)(j+m)}$ to vanish. This means that one more similar recurrence relation is needed, one that would combine matrix elements $d_{k,m}^{j-1}$ and $d_{k+1,m}^{j-1}$, where index k goes up one notch rather than down, in comparison to Equation 7.48. The rationale behind the upward increment is that the term $d_{k+1,m}^{j-1}$ turns into the existing matrix element $d_{-j+1,m}^{j-1}$ when $k=-j$.

Before proceeding to the actual derivation, it is useful to reconstruct the overall picture of the derivation process that encompasses two recurrence relationships: one that has already been derived and the other that still has to be derived. Based on the value of the orbital indices in Equations 6.179, 6.182, 6.185, and 6.188, these equations split into two groups, Equations 6.179 and 6.182 on one hand, and Equations 6.185 and 6.188 on the other. These groups carry the indices $j-(1/2)$ and $j+(1/2)$ in their left-hand side, respectively. What makes this group-split special is that only the second group can be used to obtain index $j-1$, which is needed for recursion in the right-hand side, by applying the index substitution $j\rightarrow j-1$, whereas the first group should keep index j unchanged in its right-hand side.

A special feature introduced by the group split is that derivation of either recurrence relation allows those matrix elements in Equations 6.179 or 6.182 that have half-value indices to be replaced with matrix elements having integer-value indices. The use of Equation 6.179 in the derivation of Equation 7.48 leaves Equation 6.182 available for the derivation of a new recurrence relation. The purpose of the other two equations, Equations 6.185 and 6.188, is to establish a relationship between the matrix elements with half-value and integer-value indices. In the previous round of index substitution in Equations 6.185 and 6.188

matrix elements that had azimuthal indices k and $k-1$ on the right-hand side were generated, leading to the first recurrence relation. The next round of index substitution, to be made in Equations 6.185 and 6.188, will produce matrix elements with azimuthal indices k and $k+1$ in their right-hand side so that these indices can be used in forming the second recurrence relation.

Choosing between Equations 6.179 and 6.182 for the derivation of a recurrence relation with a specific change in index k, that is, k going to $k-1$ or $k+1$ going to k, depends on the presence of matrix elements $d^{j-(1/2)}_{k-(1/2),m-(1/2)}$ and $d^{j-(1/2)}_{k-(1/2),m+(1/2)}$ in Equation 6.179, which facilitates incorporating index $k-1$ into the recursion, and the presence of matrix elements $d^{j-(1/2)}_{k+(1/2),m-(1/2)}$ and $d^{j-(1/2)}_{k+(1/2),m+(1/2)}$ in Equation 6.182, which facilitates incorporating index $k+1$ into the recursion. This preference reveals itself when considering the options for index substitution in Equations 6.185 and 6.188. The presence of $d^{j+(1/2)}_{k-(1/2),m-(1/2)}$ and $d^{j+(1/2)}_{k-(1/2),m+(1/2)}$ in Equation 6.185 and $d^{j+(1/2)}_{k+(1/2),m-(1/2)}$ and $d^{j+(1/2)}_{k+(1/2),m+(1/2)}$ in Equation 6.188 justifies leaving index k in Equation 6.185 untouched while replacing index k by index $k-1$ in Equation 6.188, so that the resulting matrix elements will match those in Equation 6.179. Based on that match, the matrix elements from the right-hand side of Equations 6.185 and 6.188 replace the matrix elements on the left-hand side of Equation 6.179, leading to Equation 7.48, where the outcome of index substitution in the right-hand side of Equations 6.185 and 6.188 determines the actual resulting recursion in index j and the pattern of change in index k.

Similarly, observing the presence of matrix elements $d^{j-(1/2)}_{k+(1/2),m-(1/2)}$ and $d^{j-(1/2)}_{k+(1/2),m+(1/2)}$ in Equation 6.182 suggests replacing index k with $k+1$ in Equation 6.185, and leaving index k untouched in Equation 6.188 to give a positive increment in index k in the new recurrence relation. Following that plan, applying index substitution $j \to j-1$ and $k \to k+1$ to Equation 6.185 leads to:

$$\sqrt{j-m}\left(\cos\frac{\beta}{2}\right)d^{j-\frac{1}{2}}_{k+\frac{1}{2},m-\frac{1}{2}} + \sqrt{j+m}\left(\sin\frac{\beta}{2}\right)d^{j-\frac{1}{2}}_{k+\frac{1}{2},m+\frac{1}{2}} = \sqrt{j-k-1}\; d^{j-1}_{k+1,m}, \tag{7.51}$$

where the resulting matrix element $d^{j-1}_{k+1,m}$ in the right-hand side incorporates the indices expected in the new recurrence relation.

The companion Equation 6.188, which is paired with Equation 6.185, already has the correct value of index k in its matrix elements. With that, the only index that needs to be changed is j. Applying index substitution $j \to j-1$ to Equation 6.188 leads to:

$$-\sqrt{j-m}\left(\sin\frac{\beta}{2}\right)d^{j-\frac{1}{2}}_{k+\frac{1}{2},m-\frac{1}{2}} + \sqrt{j+m}\left(\cos\frac{\beta}{2}\right)d^{j-\frac{1}{2}}_{k+\frac{1}{2},m+\frac{1}{2}} = \sqrt{j+k}\; d^{j-1}_{k,m}. \tag{7.52}$$

Now that the index substitution has been completed, the resulting pair of equations, Equations 7.51 and 7.52, allows the matrix elements on the left-hand side having half-value indices to be expressed in terms of matrix elements from the right-hand side having integer-value indices. Performing a linear combination of those equations in the form (7.51) sin ($\beta/2$) + (7.52) cos ($\beta/2$) gives:

$$\sqrt{j+m}\; d^{j-\frac{1}{2}}_{k+\frac{1}{2},m+\frac{1}{2}} = \sqrt{j-k-1}\left(\sin\frac{\beta}{2}\right)d^{j-1}_{k+1,m} + \sqrt{j+k}\left(\cos\frac{\beta}{2}\right)d^{j-1}_{k,m}, \tag{7.53}$$

which expresses matrix element $d^{j-(1/2)}_{k+(1/2),m+(1/2)}$ in terms having integer-value indices. A similar linear combination in the form (7.51) cos ($\beta/2$) – (7.52) sin ($\beta/2$) provides:

$$\sqrt{j-m}\; d^{j-\frac{1}{2}}_{k+\frac{1}{2},m-\frac{1}{2}} = \sqrt{j-k-1}\left(\cos\frac{\beta}{2}\right)d^{j-1}_{k+1,m} - \sqrt{j+k}\left(\sin\frac{\beta}{2}\right)d^{j-1}_{k,m}, \tag{7.54}$$

which resolves the remaining matrix element $d^{j-(1/2)}_{k+(1/2),m-(1/2)}$, and serves as a complementary equation to Equation 7.53.

The new Equations 7.53 and 7.54 can now be used to substitute matrix elements $d^{j-(1/2)}_{k+(1/2),m+(1/2)}$ and $d^{j-(1/2)}_{k+(1/2),m-(1/2)}$ in Equation 6.182. Multiplying Equation 6.182 by $\sqrt{(j-m)(j+m)}$ and performing the substitution leads to:

$$\begin{aligned}&(j+m)\left(\sin\frac{\beta}{2}\right)\left\{\sqrt{j-k-1}\left(\cos\frac{\beta}{2}\right)d^{j-1}_{k+1,m}-\sqrt{j+k}\left(\sin\frac{\beta}{2}\right)d^{j-1}_{k,m}\right\}\\&+(j-m)\left(\cos\frac{\beta}{2}\right)\left\{\sqrt{j-k-1}\left(\sin\frac{\beta}{2}\right)d^{j-1}_{k+1,m}+\sqrt{j+k}\left(\cos\frac{\beta}{2}\right)d^{j-1}_{k,m}\right\}\\&=\sqrt{(j-k)(j-m)(j+m)}\;d^{j}_{k,m}.\end{aligned} \tag{7.55}$$

Opening the curly brackets, and applying trigonometric identities from Equation 7.46, gives:

$$\begin{aligned}&(j+m)\sqrt{j-k-1}\left(\frac{\sin\beta}{2}\right)d^{j-1}_{k+1,m}-(j+m)\sqrt{j+k}\left(\frac{1-\cos\beta}{2}\right)d^{j-1}_{k,m}\\&+(j-m)\sqrt{j-k-1}\left(\sin\frac{\beta}{2}\right)d^{j-1}_{k+1,m}+(j-m)\sqrt{j+k}\left(\frac{1+\cos\beta}{2}\right)d^{j-1}_{k,m}\\&=\sqrt{(j-k)(j-m)(j+m)}\;d^{j}_{k,m}.\end{aligned} \tag{7.56}$$

Combining like terms with matrix elements reduces this equation to:

$$\begin{aligned}&\sqrt{j-k-1}\,(j\sin\beta)\,d^{j-1}_{k+1,m}+\sqrt{j+k}\left\{(j-m)\left(\frac{1+\cos\beta}{2}\right)-(j+m)\left(\frac{1-\cos\beta}{2}\right)\right\}d^{j-1}_{k,m}\\&=\sqrt{(j-k)(j-m)(j+m)}\;d^{j}_{k,m}.\end{aligned} \tag{7.57}$$

Finally, simplifying the part in curly brackets leads to:

$$\sqrt{j-k-1}\,(j\sin\beta)\,d^{j-1}_{k+1,m}+\sqrt{j+k}\,(j\cos\beta-m)\,d^{j-1}_{k,m}=\sqrt{(j-k)(j-m)(j+m)}\;d^{j}_{k,m}. \tag{7.58}$$

which is the second recurrence relation in index j for those Wigner matrix elements that involve incrementing index k.

Equation 7.58 is applicable in the domain $-j\le k\le j-1$ and $-j+1\le m\le j-1$. For both boundary values of index k, the vanishing linear coefficient annihilates the corresponding term in the left-hand side of Equation 7.58. Because of this, when $k=j-1$, Equation 7.58 simplifies to:

$$\sqrt{2j-1}\,(j\cos\beta-m)\,d^{j-1}_{j-1,m}=\sqrt{(j-m)(j+m)}\;d^{j}_{j-1,m}. \tag{7.59}$$

Likewise, setting $k=-j$ reduces Equation 7.58 to:

$$\sqrt{2j-1}\,(j\sin\beta)\,d^{j-1}_{-j+1,m}=\sqrt{(2j)(j-m)(j+m)}\;d^{j}_{-j,m}. \tag{7.60}$$

The two recurrence relations Equations 7.48 and 7.58 combined together cover the full domain of indices j and k, and almost the entire range in index m, except $m=j$ and $m=-j$. Computation of the remaining matrix elements $d^{j}_{k,j}$ and $d^{j}_{k,-j}$ employs the former set of recurrence relations Equations 7.18 through 7.33 derived in Section 7.1.

8

Solid Harmonics

In mathematical physics, harmonic functions Φ are defined as solutions of the Laplace's equation:

$$\nabla^2\Phi = \frac{\partial^2\Phi}{\partial x^2} + \frac{\partial^2\Phi}{\partial y^2} + \frac{\partial^2\Phi}{\partial z^2} = 0. \tag{8.1}$$

Laplace's equation provides a description of the potential function in regions outside the location of the charge sources. The most common arrangement presumes placing the charge sources inside the bounding sphere, illustrated in Figure 8.1, and locating the points for the measurement of electrostatic potential, also called a field point, outside the sphere. This division of space between the source and field points leads to elegant equations to be developed in this chapter. When the location of a particle relative to the bounding sphere is important, the radius-vector **a** will be used to indicate the position of particle inside the bounding sphere and radius-vector **b** for the positions outside the sphere. In those situations where the particle's location is not specified, radius-vector **r** will be used.

8.1 Regular and Irregular Solid Harmonics

The solution to Laplace's equation in spherical polar coordinates r, θ, and ϕ, leads to a product of radial $R(r)$, angular $\Theta(\theta)$, and azimuthal $Z(\phi)$ terms as $\Phi = R(r)\Theta(\theta)Z(\phi)$. The product of angular and azimuthal components, $\Theta(\theta)Z(\phi)$ gives complex spherical harmonics, $Y_{lm}(\theta,\phi)$, of the type:

$$Y_{l,m}(\theta,\phi) = \sqrt{\frac{2l+1}{4\pi}\frac{(l-m)!}{(l+m)!}}\, P_l^m(\cos\theta)\, e^{im\phi}, \tag{8.2}$$

where the angular component is represented by the associated Legendre function $P_l^m(\cos\theta)$, and the azimuthal component by the $e^{im\phi}$ term. The degree l and order m are integer numbers that can assume values in the range $0 \le l < \infty$ and $-l \le m \le l$. The numerical coefficient in front of the product both normalizes the function and establishes the symmetry relationship between the positive and negative values of the index m,

$$Y_{l,-m} = (-1)^m Y^*_{l,m}, \tag{8.3}$$

where * indicates the complex conjugate.

Although spherical harmonics are flexible functions in their own right, they are inadequate for the series expansion of electrostatic potential, since the solution is typically required to be expressed in a 3D volume, and only rarely on the surface of a sphere. In order to make the final step to obtain the general solution of the Laplace equation in the form $R(r)\Theta(\theta)Z(\phi)$, it is necessary to combine the spherical harmonics $\Theta(\theta)Z(\phi)$ with a radial term $R(r)$. Since the radial term $R(r)$ exists in two forms, Ar^l and Br^{-l-1} where A and B are arbitrary linear coefficients, their combination with spherical harmonics leads to two different products:

$$\sqrt{\frac{4\pi}{2l+1}\frac{(l+m)!}{(l-m)!}}\, r^l\, Y_{lm}(\theta,\phi), \tag{8.4}$$

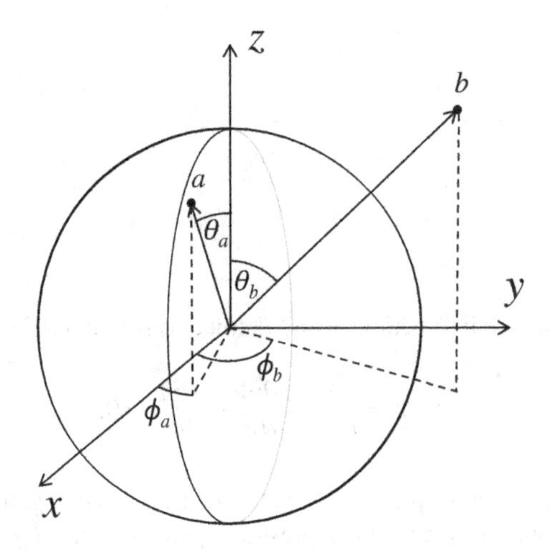

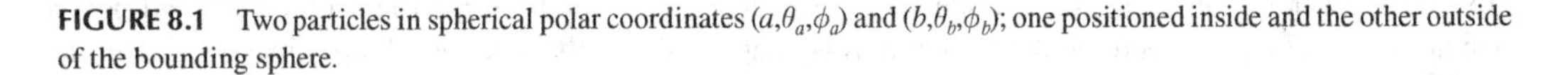

FIGURE 8.1 Two particles in spherical polar coordinates (a,θ_a,ϕ_a) and (b,θ_b,ϕ_b); one positioned inside and the other outside of the bounding sphere.

and

$$\sqrt{\frac{4\pi}{2l+1}\frac{(l+m)!}{(l-m)!}}\, r^{-l-1}\, Y_{lm}(\theta,\phi), \tag{8.5}$$

where

$$A = B = \sqrt{\frac{4\pi}{2l+1}\frac{(l+m)!}{(l-m)!}}. \tag{8.6}$$

The precursor linear coefficients are set to cancel out the normalization coefficient resulting from the spherical harmonics. Substituting the explicit form of the spherical harmonics, Equation 8.1, into Equations 8.5 and 8.6 and canceling out the normalization terms results in the solid harmonics:

$$R_{lm} = r^l P_l^m(\cos\theta)\, e^{im\phi} \qquad \text{regular solid harmonics} \tag{8.7}$$

$$S_{lm} = r^{-l-1} P_l^m(\cos\theta)\, e^{im\phi} \qquad \text{irregular solid harmonics} \tag{8.8}$$

The decision of having no linear coefficient in front of solid harmonics has the following rationale. Since Laplace's equation provides flexibility in choosing the linear coefficient, and because solid harmonics are regular mathematical functions to be used in the potential expansion, the actual choice of the coefficient should be made to best suit the purpose of generating the expansion series for electrostatic potential. Various sources have suggested different normalization constants for solid harmonics, each with its own advantages, but the most convenient approach is to start with the simple unscaled form in Equations 8.7 and 8.8, and then add whatever scaling coefficient is deemed necessary.

Classification of solid harmonics as regular R and irregular S harmonics comes from their different behavior at the origin. Regular harmonics vanish at the origin while irregular harmonics exhibit a singularity. The symmetry property of associated Legendre functions relating the terms with negative value $-m$ to those with positive m,

$$P_l^{-m} = (-1)^m \frac{(l-m)!}{(l+m)!} P_l^m, \tag{8.9}$$

leads to analogous relations for solid harmonics:

$$R_{l,-m} = r^l P_l^{-m} e^{-im\phi} = r^l (-1)^m \frac{(l-m)!}{(l+m)!} P_l^m e^{-im\phi} = (-1)^m \frac{(l-m)!}{(l+m)!} R_{lm}^*. \tag{8.10}$$

Similarly,

$$S_{l,-m} = (-1)^m \frac{(l-m)!}{(l+m)!} S_{lm}^*, \tag{8.11}$$

where the star in R^* and S^* indicates the complex-conjugate form.

An observant reader may recognize that solid harmonics have already been presented in previous chapters, although not explicitly defined as such. Indeed, the expansion of electrostatic potential $\Phi(\mathbf{r})$ was described as a series of spherical harmonics, as shown in Equation 8.12 for the problem illustrated in Figure 8.1:

$$\Phi = \frac{1}{|\mathbf{a}-\mathbf{b}|} = \sum_{l=0}^{\infty} \sum_{m=-l}^{l} \frac{4\pi}{2l+1} \frac{a^l}{b^{l+1}} Y_{lm}^*(\theta_a, \phi_a) Y_{lm}(\theta_b, \phi_b), \tag{8.12}$$

where $\mathbf{r} = \mathbf{a} - \mathbf{b}$ and $|\mathbf{a}| < |\mathbf{b}|$. Solid harmonics can be easily recognized in Equation 8.12 by grouping the radial part a^l with $Y_{lm}^*(\theta_a, \phi_a)$, and $1/(b^{l+1})$ with $Y_{lm}(\theta_b, \phi_b)$ so the expansion can be compactly rewritten in terms of solid harmonics:

$$\begin{aligned} \frac{1}{|\mathbf{a}-\mathbf{b}|} &= \sum_{l=0}^{\infty} \sum_{m=-l}^{l} \frac{(l-m)!}{(l+m)!} a^l P_l^m(\cos\theta_a) e^{-im\phi_a} b^{-l-1} P_l^m(\cos\theta_b) e^{im\phi_b} \\ &= \sum_{l=0}^{\infty} \sum_{m=-l}^{l} \frac{(l-m)!}{(l+m)!} R_{lm}^*(a,\theta_a,\phi_a) S_{lm}(b,\theta_b,\phi_b), \end{aligned} \tag{8.13}$$

where $|\mathbf{a}| < |\mathbf{b}|$.

The transition from Equation 8.12 to Equation 8.13 illustrates that carrying normalization coefficients from spherical harmonics into the potential expansion is indeed not justified because they will partially cancel out, which is why keeping those coefficients inside solid harmonics R and S in Equations 8.7 and 8.8 is unnecessary.

The purpose of introducing solid harmonics goes beyond the simplification of Equation 8.12. Due to the presence of a radial term, solid harmonics, unlike spherical harmonics, are no longer limited to the solutions on the unit sphere. Now they describe the potential in the continuum in 3D space, which opens up a whole array of interesting properties to be considered next.

8.2 Regular Multipole Moments

Placing a charge q in position of radius-vector $\mathbf{a}$ and measuring the potential $\Phi(\mathbf{b})$ in the position of radius-vector $\mathbf{b}$ in Figure 8.1 gives the following expression:

$$\Phi(\mathbf{b}) = \sum_{l=0}^{\infty} \sum_{m=-l}^{l} \frac{(l-m)!}{(l+m)!} q R_{l,m}^*(a,\theta_a,\phi_a) S_{l,m}(b,\theta_b,\phi_b). \tag{8.14}$$

It is customary to define regular multipole moments of charge q placed at the point **a** and having spherical polar coordinates (a,θ_a,ϕ_a) as:

$$M_{l,m}(q,a,\theta_a,\phi_a) = qa^l P_l^m(\cos\theta_a)e^{-im\phi_a} = qR_{l,m}^*(\mathbf{a}), \tag{8.15}$$

where the prefix regular in front of the name multipole moments implies combining the electrostatic charge with regular solid harmonics. With that, the expression for electrostatic potential in point **b**, Equation 8.14 can be rewritten as

$$\Phi(\mathbf{b}) = \sum_{l=0}^{\infty}\sum_{m=-l}^{l}\frac{(l-m)!}{(l+m)!}M_{l,m}(q,a,\theta_a,\phi_a)\,S_{l,m}(b,\theta_b,\phi_b), \tag{8.16}$$

where M_{lm} is the multipole moment. When referring to regular multipole moments, it is customary to drop the prefix "regular," and simply call them multipole moments.

In Equation 8.15, the value of the solid harmonics is scaled by the magnitude of the point charge; this is why the charge is sometimes called the strength of the multipole moment. The effect of omitting the charge in Equation 8.15, that is, of equating the multipole moment $M_{l,m}$ to regular solid harmonics $R_{l,m}^*$, is the same as using a multipole moment for a positive unit charge. Conversely, regular solid harmonics can be referred to as chargeless multipole moments.

Multipole moments with same l and m for a set of particles having a common origin can be added together:

$$M_{l,m}(A) = \sum_i M_{l,m}(\mathbf{a}_i) = \sum_i q_i R_{l,m}^*(\mathbf{a}_i). \tag{8.17}$$

The use of capital A as the argument of $M_{l,m}$ indicates the sum over all specified charges pointed to by radius-vector $\mathbf{a}_i$. The resulting multipole moments $M_{l,m}(A)$ of degree l and order m can then be used to compute the combined electrostatic potential at point **b**, located outside the sphere that encloses all charge sources $\mathbf{a}_i$ and with the center of the sphere coinciding with the origin of the coordinate system.

Now, the equation for electrostatic potential in point **b**, due to multiple charges q_i in position $\mathbf{a}_i$, where $|\mathbf{a}_i| < |\mathbf{b}|$, can be expressed in terms of multipole moments:

$$\Phi(\mathbf{b}) = \sum_{l=0}^{\infty}\sum_{m=-l}^{l}\frac{(l-m)!}{(l+m)!}\sum_i M_{l,m}(\mathbf{a}_i)S_{l,m}(\mathbf{b}). \tag{8.18}$$

For convenience, the sum over charge sources q_i can be taken outside the double sum and only needs to be computed once using Equation 8.17. A consequence of this is that:

$$\Phi(\mathbf{b}) = \sum_{l=0}^{\infty}\sum_{m=-l}^{l}\frac{(l-m)!}{(l+m)!}M_{l,m}(A)S_{l,m}(\mathbf{b}). \tag{8.19}$$

This equation implies that the aggregate potential due to a group of particles can be handled as a single entity *via* the multipole moments of the charge set.

Equation 8.19 has an important requirement in that both radius-vectors **a** and **b** have to be defined about the same origin. This aspect will be examined in detail later, when working on translation properties of multipole moments.

8.3 Irregular Multipole Moments

Similar to the multipole moments $M_{l,m}$ of a charge at point **a**, one can introduce irregular multipole moments $L_{l,m}$ by hooking up a charge q, which is placed in field point **b**, with irregular solid harmonics:

$$L_{l,m}(q,b,\theta_b,\phi_b) = qr^{-l-1}P_l^m(\cos\theta_b)\,e^{im\phi_b} = qS_{l,m}(\mathbf{b}), \tag{8.20}$$

where (b,θ_b,ϕ_b) are spherical polar coordinates of radius-vector **b**, and $|\mathbf{a}| < |\mathbf{b}|$. This arrangement is illustrated in Figure 8.1.

As the reader noticed, regular and irregular multipole moments describe the electrostatic potential created by different charge sources. An even more important difference is that these moments create electrostatic potential in different parts of the space.

From the viewpoint of the observer standing in the origin of the coordinate system, a regular multipole expansion of the charges located inside the bounding sphere creates potential in the points outside the sphere. Contrary to that, an irregular multipole expansion creates the potential due to the outside charges inside the bounding sphere.

Irregular multipole moments are a charge-scaled form of irregular solid harmonics. Using the irregular moments of the external charges, the electrostatic potential $\Phi(\mathbf{a})$ they create in the point **a** inside the bounding sphere is:

$$\Phi(\mathbf{a}) = \sum_{l=0}^{\infty}\sum_{m=-l}^{l}\frac{(l-m)!}{(l+m)!}R_{l,m}^{*}(\mathbf{a})\,L_{l,m}(\mathbf{b}). \tag{8.21}$$

Again, if there are several particles at positions $\mathbf{b}_j$, all defined about the same origin and carrying charges q_j, their moments can be summed up:

$$L_{l,m}(B) = \sum_{j} L_{l,m}(\mathbf{b}_j) = \sum_{i} q_j S_{l,m}(\mathbf{b}_j). \tag{8.22}$$

Using this property, the electrostatic potential of a set of point charges can be computed, working with them as if they were a single entity. The electrostatic potential in a point **a** inside the bounding sphere due to all charges q_j located in the position $\mathbf{b}_j$, where $|\mathbf{a}| < |\mathbf{b}|$, in terms of irregular moments becomes

$$\Phi(\mathbf{a}) = \sum_{l=0}^{\infty}\sum_{m=-l}^{l}\frac{(l-m)!}{(l+m)!}R_{l,m}^{*}(\mathbf{a})\,L_{l,m}(B). \tag{8.23}$$

Compare Equation 8.23 to Equation 8.19 to see how electrostatic potential inside and outside the bounding sphere is computed, respectively.

8.4 Computation of Electrostatic Energy via Multipole Moments

An important conclusion from the previous sections is that the computation of electrostatic potential between remote charge sets can be performed *via* their regular and irregular multipole moments. Computation of electrostatic interaction is similarly performed. Consider two remote charge sets A and B with their particle coordinates being defined about the same coordinate system A, as in Figure 8.2. Each charge set is placed inside their own bounding sphere. The bounding spheres are not allowed to overlap. The center of the sphere for charge set A is located at the center of the coordinate system A.

The electrostatic interaction energy, U, between charge sets A and B, is the product of the regular $M_{l,m}(A)$ and irregular $L_{l,m}(B)$ multipole moments of the corresponding charge sets, where the moments are computed according to Equations 8.17 and 8.22, respectively:

$$U = \sum_{l=0}^{\infty}\sum_{m=-l}^{l}\frac{(l-m)!}{(l+m)!}M_{l,m}(A)\,L_{l,m}(B). \tag{8.24}$$

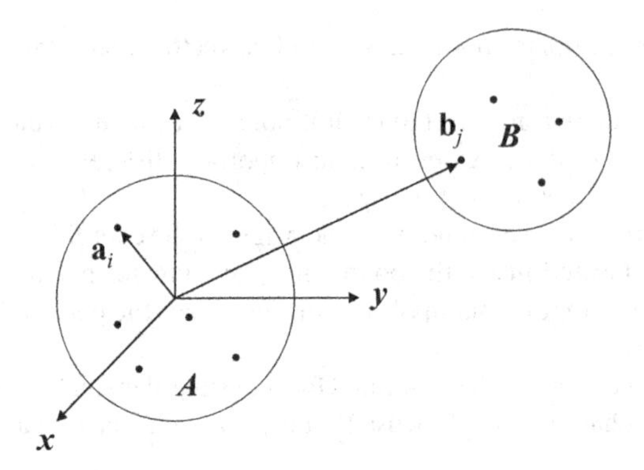

FIGURE 8.2 Two charge sets A and B defined about one coordinate system. The positions of charges are defined by radius-vectors $\mathbf{a}_i$ and $\mathbf{b}_j$, respectively. The spheres enclosing charge sets A and B are separated in space.

Equations 8.19, 8.23, and 8.24 represent three possible uses of regular and irregular multipole moments in computation of electrostatic interactions between groups of point charges. To put those equations into practical use one needs to compute the individual moments. Since, according to Equations 8.17 and 8.20, moments are a charge-scaled form of solid harmonics the computation of moments is practically equivalent to computation of solid harmonics. The next, step is to learn how to evaluate solid harmonics.

8.5 Recurrence Relations for Regular Solid Harmonics

Computation of the solid harmonics is a non-trivial procedure because they are defined in spherical polar coordinates while the particles are typically defined in Cartesian coordinates. In addition to that, the underlying associated Legendre functions, P_l^m, which are included in solid harmonics in Equations 8.7 and 8.8, introduce numerical instabilities due to the cancellation of large coefficients when computed from their analytic form,

$$P_l^m = \frac{(-1)^m}{2^l l!}(1-\mu^2)^{m/2}\frac{d^{l+m}}{d\mu^{l+m}}(\mu^2-1)^l, \tag{8.25}$$

where $\mu = \cos\theta$.

The problem of numerical instabilities can be resolved by using the previously-derived recurrence relations for associated Legendre functions. This requires incorporating those recurrence relations into the solid harmonics. To initiate the recursion, it is necessary to obtain the first few regular solid harmonics in their explicit form. Some modification to those equations will be required so they can accept particle coordinates in Cartesian form. After that the derivation of recursion relations to obtain the remaining regular solid harmonics will follow.

For a point P defined in Cartesian coordinates (x, y, z), the spherical polar representation (r, θ, ϕ) is defined in Figure 8.3, and the relationship between Cartesian and spherical polar coordinates established by means of Equation 8.26.

$$\begin{aligned}
r^2 &= x^2 + y^2 + z^2 \\
x &= r\sin\theta\cos\phi \\
y &= r\sin\theta\sin\phi \\
z &= r\cos\theta \\
x + iy &= r\sin\theta e^{i\phi}
\end{aligned} \tag{8.26}$$

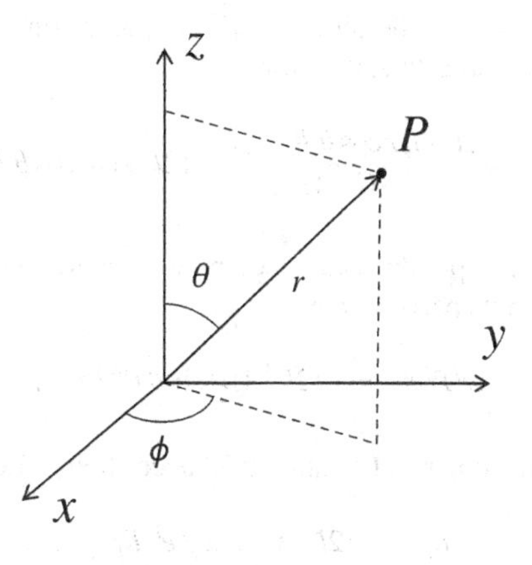

FIGURE 8.3 Representation of point P(x, y, z) in spherical polar coordinates.

After substituting the corresponding values of the associated Legendre functions, and applying Equation 8.26 to obtain the Cartesian representation, the first few regular solid harmonics become:

$$
\begin{aligned}
R_{0,0} &= P_0^0 = 1 \\
R_{1,0} &= r\,P_1^0 = r\cos\theta = z \\
R_{1,1} &= r\,P_1^1 e^{i\phi} = -r\sin\theta\, e^{i\phi} = -(x+iy) \\
R_{1,1}^* &= -(x-iy) \\
R_{2,0} &= r^2 P_2^0 = \frac{1}{2} r^2(3\cos^2\theta - 1) = \frac{3z^2-r^2}{2} = \frac{2z^2-x^2-y^2}{2} \\
R_{2,1} &= r^2 P_2^1 e^{i\phi} = -3r^2 \sin\theta\cos\theta\, e^{i\phi} = -3z(x+iy) \\
R_{2,1}^* &= -3z(x-iy) \\
R_{2,2} &= r^2 P_2^2 e^{2i\phi} = 3r^2\sin^2\theta\, e^{2i\phi} = 3(x+iy)^2 \\
R_{2,2}^* &= 3(x-iy)^2
\end{aligned}
\tag{8.27}
$$

From this point onward recurrence relations should be used for the remaining regular solid harmonics. The procedure for incorporating the recurrence relations starts from the previously derived equations for associated Legendre functions:

$$P_l^m = \frac{2l-1}{l-m}\cos\theta\, P_{l-1}^m - \frac{l+z-1}{l-m} P_{l-2}^m, \tag{8.28}$$

$$P_l^l = -(2l-1)\sin\theta\, P_{l-1}^{l-1}. \tag{8.29}$$

Using Equation 8.28 and multiplying both its sides by $r^l e^{im\phi}$, gives

$$r^l P_l^m e^{im\phi} = \frac{2l-1}{l-m}\cos\theta\, r^l\, P_{l-1}^m\, e^{im\phi} - \frac{l+m-1}{l-m} r^l\, P_{l-2}^m\, e^{im\phi}. \tag{8.30}$$

Replacing the product of the associated Legendre function, radial, and azimuthal terms with the corresponding regular solid harmonics, according to Equation 8.7, leads to a recurrence relationship:

$$R_{l,m} = \frac{(2l-1)r\cos\theta}{l-m} R_{l-1,m} - \frac{(l+m-1)r^2}{l-m} R_{l-2,m}. \tag{8.31}$$

This relation holds for the azimuthal values $0 \leq m \leq l-1$.

For the specific case of $m = l - 1$, the term $R_{l-2,l-1} = 0$, as a consequence of differentiation in the associated Legendre function zeroing P_{l-2}^{l-1}, therefore:

$$R_{l,l-1} = \frac{(2l-1)r\cos\theta\, R_{l-1,l-1}}{l-(l-1)} = (2l-1)r\cos\theta\, R_{l-1,l-1}. \tag{8.32}$$

For the condition of $m = l$, a separate recurrence relation is necessary. The derivation starts from Equation 8.29. Multiplying both its parts by $r^l e^{il\phi}$ gives:

$$r^l P_l^l e^{il\phi} = -(2l-1)r^l \sin\theta\, P_{l-1}^{l-1}\, e^{il\phi}. \tag{8.33}$$

Segregating regular solid harmonics in Equation 8.33 according to Equation 8.7 leads to:

$$R_{l,l} = -(2l-1)r\sin\theta\, e^{i\phi} R_{l-1,l-1}. \tag{8.34}$$

Utilizing the fact that $r\sin\theta\, e^{\pm i\phi} = x \pm iy$ produces:

$$R_{l,l} = -(2l-1)(x+iy)R_{l-1,l-1}. \tag{8.33}$$

Now one can gather the recurrence relations for regular complex harmonics in one place, and write them in the complex-conjugate form, which is necessary for multipole moments:

$$\begin{aligned} R^*_{l,m} &= \frac{(2l-1)r\cos\theta}{l-m} R^*_{l-1,m} - \frac{(l+m-1)r^2}{l-m} R^*_{l-2,m} \\ R^*_{l,l-1} &= (2l-1)r\cos\theta\, R^*_{l-1,l-1} \\ R^*_{l,l} &= -(2l-1)(x-iy)R^*_{l-1,l-1} \end{aligned} \tag{8.35}$$

Equation 8.35 provides the recurrence relations that allow the computation of regular solid harmonics of degree l based on a previous knowledge of terms of degree $l - 1$ and $l - 2$.

In concluding this section a comment can be made on the nature of the recurrent computation of solid harmonics. Unlike using the explicit equation Equation 8.7, where every harmonic (l, m) can be computed in parallel, independent of the other harmonics, recurrence computations are serial by nature, since the next term depends on the previous one and that prevents parallelization at the individual harmonics level. However, in practice, parallelization is performed over individual particles or the groups of particles. Particle harmonics depend on their respective particle coordinates only, and that allows treating each particle independently. Therefore, in a system consisting of a large number of particles, using recurrence relationships in the computation of solid harmonics does not introduce an obstacle to efficient parallelization.

8.6 Recurrence Relations for Irregular Solid Harmonics

The equations for the computation of irregular solid harmonics can be derived in a manner similar to the computation of regular solid harmonics. Using a system set up as illustrated in Figure 8.3, the first few terms involving irregular solid harmonics can be explicitly defined by using Equation 8.8, and recurrence relations subsequently derived to generate the higher-degree harmonics. As in the above, Equation 8.26 will help to relate the Cartesian coordinates of the particle P to spherical polar coordinates used in the definition of irregular solid harmonics. The use of Cartesian coordinates helps to minimize the use of trigonometric functions in the computations, since the latter are rather computationally expensive, and as such are inconvenient in computing.

Unlike the previous case of regular solid harmonics, there is no need to come up with the equations for complex-conjugate counterparts of irregular solid harmonics, since these are not present in the corresponding moments. Whenever explicit equations for complex-conjugate forms are needed, they can be obtained by changing the sign in front of the complex number i to its opposite. In the Fortran language this operation can be performed efficiently using an intrinsic function.

The analytic part of this section begins by directly obtaining first few terms of the irregular solid harmonics from Equation 8.8

$$\begin{aligned}
S_{0,0} &= r^{-1}\,P_0^0 = \frac{1}{r} \\
S_{1,0} &= r^{-2}\,P_1^0 = \frac{\cos\theta}{r^2} = \frac{z}{r^3} \\
S_{1,1} &= r^{-2}\,P_1^1\,e^{i\phi} = -\frac{\sin\theta\,e^{i\phi}}{r^2} = -\frac{x+iy}{r^3} \\
S_{2,0} &= r^{-3}\,P_2^0 = \frac{3\cos^2\theta-1}{2r^3} = \frac{2z^2-x^2-y^2}{2r^5} \\
S_{2,1} &= r^{-3}\,P_2^1\,e^{i\phi} = -\frac{3\sin\theta\cos\theta\,e^{i\phi}}{r^3} = -\frac{3z(x+iy)}{r^5} \\
S_{2,2} &= r^{-3}\,P_2^2\,e^{2i\phi} = \frac{3\sin^2\theta\,e^{2i\phi}}{r^3} = \frac{3(x+iy)^2}{r^5}
\end{aligned} \tag{8.36}$$

The remaining higher-degree irregular solid harmonics should be obtained *via* recurrence relations. The necessary equations can be derived using the corresponding recurrence relations for associated Legendre functions. Multiplying Equation 8.28 by $r^{-l-1}\,e^{im\phi}$ gives:

$$r^{-l-1}\,P_l^m\,e^{im\phi} = \frac{2l-1}{l-m}\cos\theta\,r^{-l-1}P_{l-1}^m\,e^{im\phi} - \frac{l+m-1}{l-m}r^{-l-1}P_{l-2}^m\,e^{im\phi}. \tag{8.37}$$

Applying Equation 8.8 to Equation 8.37 to segregate irregular solid harmonics yields:

$$S_{l,m} = \frac{(2l-1)\cos\theta}{(l-m)r}S_{l-1}^m - \frac{l+m-1}{(l-m)r^2}S_{l-2}^m. \tag{8.38}$$

Using the fact that $z = r\cos\theta$, a recurrence relation for irregular solid harmonics can be generated:

$$S_{l,m} = \frac{(2l-1)zS_{l-1,m} - (l+m-1)S_{l-2,m}}{(l-m)r^2}. \tag{8.39}$$

This equation restricts the azimuthal number m to the range $0 \leq m \leq l-1$. In a particular case of $m = l-1$, one can recognize that $S_{l-2,l-1} = 0$. Based on that, Equation 8.39 takes the form:

$$S_{l,l-1} = \frac{(2l-1)zS_{l-1,l-1}}{r^2}. \tag{8.40}$$

For $m = l$, a different recurrence relation is needed since the denominator and both terms in the numerator, $S_{l-1,m}$ and $S_{l-2,m}$, in Equation 8.39 are zero, and as a result $S_{l,m}$ is ill-defined. The sought equation can be derived from Equation 8.29, which is a recurrence relation for associated Legendre functions. Multiplying both sides in Equation 8.29 by $r^{-l-1}\,e^{il\phi}$ gives:

$$r^{-l-1}\,P_l^l\,e^{il\phi} = -(2l-1)\sin\theta\,r^{-l-1}\,P_{l-1}^{l-1}\,e^{il\phi}. \tag{8.41}$$

On the left-hand side is the irregular solid harmonic defined by Equation 8.8; the right-hand side can be rearranged to give:

$$S_{l,l} = -(2l-1)r^{-1}\sin\theta\,e^{i\phi}\,r^{-l}\,P_{l-1}^{l-1}e^{i(l-1)\phi}. \tag{8.42}$$

Substituting $r^{-l}P_{l-1}^{l-1}e^{i(l-1)\phi}$ by $S_{l-1,l-1}$ in the right-hand side and applying Equation 8.26 leads to the last recurrence relation for irregular solid harmonics:

$$S_{l,l} = -(2l-1)\frac{x+iy}{r^2}S_{l-1,l-1}. \tag{8.43}$$

To summarize, the derived recurrence relations for irregular solid harmonics can be collected in one place:

$$\begin{aligned} S_{l,m} &= \frac{(2l-1)\,z\,S_{l-1,m} - (l+m-1)\,S_{l-2,m}}{(l-m)r^2} \\ S_{l,l-1} &= \frac{(2l-1)\,z\,S_{l-1,l-1}}{r^2} \\ S_{l,l} &= -(2l-1)\frac{x+iy}{r^2}S_{l-1,l-1} \end{aligned} \tag{8.44}$$

These recurrence relations are numerically stable and allow reliable calculation of irregular solid harmonics of any degree.

8.7 Generating Functions for Solid Harmonics

Manipulation with the system of electrostatic charges often requires changing the reference frame of multipole expansion of charge sources. Translations of regular and irregular multipole moments from one coordinate system to another are handled by the addition theorem for solid harmonics. The standard path for deriving the addition theorem is through group theory. This route would require a strong mathematical background, and is unnecessarily complicated. Fortunately there is a simpler and a more visually appealing approach, which is suggested by Caola, to understanding the addition theorem that relies on generating functions for solid harmonics described by Hobson that will now be explained.

So far the solutions of Laplace's equation appeared only in spherical polar coordinates. The advantage of spherical polar over Cartesian coordinates is that spherical harmonics obtain a compact mathematical form in spherical polar coordinates, whereas the corresponding expression in Cartesian coordinates is very bulky. On the other hand, some operations are easier to handle in Cartesian coordinates. A brief review of Laplace's equation in Cartesian coordinates will demonstrate a useful connection between the Cartesian and spherical polar solutions.

Laplace's equation has the following form in Cartesian coordinates:

$$\left[\frac{\partial^2}{\partial x^2} + \frac{\partial^2}{\partial y^2} + \frac{\partial^2}{\partial z^2}\right]\Phi = 0. \tag{8.45}$$

Out of the infinite number of possible solutions, functions $(z \pm ix)^l$ and $(z \pm ix)^{-l-1}$ are of particular value. One can demonstrate that these functions satisfy Laplace's equation by using a direct substitution:

$$\begin{aligned} &\left[\frac{\partial^2}{\partial x^2} + \frac{\partial^2}{\partial y^2} + \frac{\partial^2}{\partial z^2}\right](z \pm ix)^l = \left[\pm il\frac{\partial}{\partial x} + l\frac{\partial}{\partial z}\right](z \pm ix)^{l-1} = \left[-l(l-1) + l(l-1)\right](z \pm ix)^{l-2} = 0 \\ &\left[\frac{\partial^2}{\partial x^2} + \frac{\partial^2}{\partial y^2} + \frac{\partial^2}{\partial z^2}\right](z \pm ix)^{-l-1} = \left[\mp i(l+1)\frac{\partial}{\partial x} - (l+1)\frac{\partial}{\partial z}\right](z \pm ix)^{-l-2} \\ &\qquad = \left[-(l+1)(l+2) + (l+1)(l+2)\right](z \pm ix)^{-l-3} = 0 \end{aligned} \tag{8.46}$$

A remarkable utility of these solutions is that they are generating functions for solid harmonics:

$$(z+ix)^l = \sum_{m=-l}^{l} i^{-m} \frac{l!}{(l+m)!} R_{l,m}(r,\theta,\phi) \tag{8.47}$$

$$(z+ix)^{-l-1} = \sum_{m=-l}^{l} i^{m} \frac{(l-m)!}{l!} S_{l,m}(r,\theta,\phi) + 2r^{-l-1} \sum_{m=l+1}^{\infty} i^{-m} \frac{(l+m)!}{l!} P_l^{-m}(\cos\theta)\cos(m\phi). \tag{8.48}$$

Note that the right-hand side in Equation 8.48 consists of two summands; the first is a conventional solid harmonic term and second is a residual term where $m > l$ with m going to infinity.

Seeing azimuthal index m going to infinity in Equation 8.48 might appear puzzling; discussing that range is indeed uncommon in the physics textbooks. Hence, a brief introduction to that matter might be useful. When referencing index m it is convenient to follow the previously-introduced convention of explicitly showing the negative sign outside parameter m while keeping the term itself positive. According to this convention, when a negative value of m is used, it will be represented as $-m$. For the positive range of index m, the associated Legendre function P_l^m becomes zero when $m > l$ due to differentiation. In the other direction, index $-m$ can theoretically go to negative infinity without zeroing the associated Legendre function, P_l^{-m}. The reason why the series P_l^{-m} does not terminate when index $-m$ goes to negative infinity can be understood from the fact that going from m to $m-1$, corresponds to integration. Because associated Legendre functions are integrable an infinite number of times, the value of m can go to negative infinity, and the corresponding associated Legendre functions are non-zero. Since associated Legendre functions with $|m| > l$ have no utility in physics, it makes no use going deeper into the subject.

The step-by-step derivation of equation Equation 8.48 for generating functions for irregular solid harmonics is provided in Appendix B.5. It is sufficient at this point to illustrate the procedure by deriving the generating function for regular solid harmonics (Equation 8.47). To start the derivation, first recall that in spherical polar coordinates

$$x = r\sin\theta\cos\phi \quad y = r\sin\theta\sin\phi \quad z = r\cos\theta, \tag{8.49}$$

$$e^{\pm i\phi} = \cos\phi \pm i\sin\phi, \tag{8.50}$$

$$(x \pm iy) = r\sin\theta\cos\phi \pm ir\sin\theta\sin\phi = r\sin\theta\, e^{\pm i\phi}. \tag{8.51}$$

Substituting z and x form Equation 8.49 into Equation 8.47 by their spherical polar counterparts gives:

$$(z+ix)^l = r^l(\cos\theta + i\sin\theta\cos\phi)^l. \tag{8.52}$$

Denoting $\mu = \cos\theta$ and $i\sin\theta = \sqrt{\mu^2-1}$ produces:

$$(z+ix)^l = r^l(\mu + \sqrt{\mu^2-1}\cos\phi)^l. \tag{8.53}$$

Since it follows from Equation 8.50 that $e^{i\phi} + e^{-i\phi} = 2\cos\phi$, the latter can be substituted into Equation 8.53, and the coefficient 1/2 can be moved outside the brackets:

$$(z+ix)^l = \frac{r^l}{2^l}\left[2\mu + \sqrt{\mu^2-1}(e^{i\phi}+e^{-i\phi})\right]^l. \tag{8.54}$$

Dividing this expression by $(\mu^2-1)^{\frac{l}{2}}$ and multiplying the terms in square brackets by the same quantity gives:

$$(z+ix)^l = \frac{r^l}{2^l(\mu^2-1)^{\frac{l}{2}}}\left[2\mu\sqrt{\mu^2-1} + (\mu^2-1)e^{i\phi} + (\mu^2-1)\,e^{-i\phi}\right]^l. \tag{8.55}$$

The common multiplier $e^{-il\phi}$ can be moved outside the square brackets and the terms inside rearranged, so that:

$$\begin{aligned}(z+ix)^l &= \frac{r^l e^{-il\phi}}{2^l(\mu^2-1)^{\frac{l}{2}}}\left[2\mu\sqrt{\mu^2-1}e^{i\phi}+(\mu^2-1)e^{2i\phi}+\mu^2-1\right]^l \\ (z+ix)^l &= \frac{r^l e^{-il\phi}}{2^l(\mu^2-1)^{\frac{l}{2}}}\left[\left(\mu+\sqrt{\mu^2-1}e^{i\phi}\right)^2-1\right]^l\end{aligned} \tag{8.56}$$

The expression in square brackets can be expanded in powers of $t=\sqrt{\mu^2-1}\,e^{i\phi}$ with help of the Taylor theorem,

$$f(a+t)=\sum_{k=0}^{\infty}\frac{1}{k!}t^k f^{(k)}(a), \tag{8.57}$$

where (k) indicates the k-th derivative. In its application to the present case, the Taylor theorem assumes the following form:

$$f(a+t)=[(\mu+t)^2-1]^l=\sum_{k=0}^{\infty}\frac{1}{k!}t^k([\mu^2-1]^l)^{(k)}, \tag{8.58}$$

where $[(\mu+t)^2-1]^l$ is the function to be expanded in a Taylor series. Since the expanded function is differentiable exactly $2l$ times, the sum terminates, giving

$$\begin{aligned}(z+ix)^l &= \frac{r^l e^{-il\phi}}{2^l(\mu^2-1)^{\frac{l}{2}}}\sum_{k=0}^{2l}\frac{1}{k!}(\mu^2-1)^{\frac{k}{2}}e^{ik\phi}\frac{d^k}{d\mu^k}(\mu^2-1)^l \\ &= r^l\sum_{k=0}^{2l}\frac{1}{2^l k!}e^{i(k-l)\phi}(\mu^2-1)^{\frac{k-l}{2}}\frac{d^k}{d\mu^k}(\mu^2-1)^l.\end{aligned} \tag{8.59}$$

Next, the variable substitution $k=l+m$ is made. Since index l is a fixed parameter, the variable index is m, so when index k goes from 0 to $2l$, index m will go from $-l$ to l.

$$(z+ix)^l=\sum_{m=-l}^{l}\frac{r^l e^{im\phi}}{2^l(l+m)!}(\mu^2-1)^{\frac{m}{2}}\frac{d^{l+m}}{d\mu^{l+m}}(\mu^2-1)^l. \tag{8.60}$$

Because $\mu=\cos\theta$ is a real number, and because $\mu^2<1$, the complex part can be separated off, $(\mu^2-1)^{\frac{m}{2}}=i^m(1-\mu^2)^{\frac{m}{2}}$. If the resulting expression is then multiplied and divided by $(-1)^m\,l!$ the associated Legendre function, $P_l^m=\frac{(-1)^m}{2^l l!}(1-\mu^2)^{\frac{m}{2}}\frac{d^{l+m}}{d\mu^{l+m}}(\mu^2-1)^l$, is obtained:

$$\begin{aligned}(z+ix)^l &= \sum_{m=-l}^{l}\frac{(-1)^m l!\ \ i^m r^l e^{im\phi}}{(-1)^m l!\ 2^l(l+m)!}(1-\mu^2)^{\frac{m}{2}}\frac{d^{l+m}}{d\mu^{l+m}}(\mu^2-1)^l \\ &= \sum_{m=-l}^{l}\frac{i^m l!}{(-1)^m(l+m)!}r^l e^{im\phi}P_l^m.\end{aligned} \tag{8.61}$$

Taking into account that $i^m/(-1)^m=i^{-m}$ and $R_{l,m}=r^l P_l^m e^{im\phi}$, the final result is:

$$(z+ix)^l=\sum_{m=-l}^{l}i^{-m}\frac{l!}{(l+m)!}R_{l,m}. \tag{8.62}$$

This proves Equation 8.47 and confirms that $(z + ix)^l$ is a generating function for regular solid harmonics, $R_{l,m}$. As a final touch, the complex-conjugate form of the generating function can be made by replacing i with $-i$ in all parts of the above equation:

$$(z - ix)^l = \sum_{m=-l}^{l} i^m \frac{l!}{(l+m)!} R^*_{l,m}. \tag{8.63}$$

The complex-conjugate form of regular solid harmonics is necessary for representing regular multipole moments. A remarkable property of Equations 8.47 and 8.63 is that while their left parts are in Cartesian coordinates, the right-hand sides are in spherical polar coordinates. This will be explored in the next section in the derivation of the addition theorem for regular solid harmonics.

8.8 Addition Theorem for Regular Solid Harmonics

The purpose of the addition theorem for regular solid harmonics is to translate the origin of a regular multipole expansion from one coordinate system to another. Initially, the relationship between the addition theorem and origin translation might not be obvious. To ease the learning curve, the abstract math of the addition theorem will be projected into the geometric language of vectors.

To begin with, there are a few apparent similarities between vectors and multipoles that can be favorably utilized. In linear algebra the vector addition operation is equivalent to a change in coordinate system. The same process can be applied to multipole moments. In addition to that, both vectors and odd moments reverse their sign upon coordinate inversion. Finally, a vector and a moment are uniquely identified for a point P, defined by coordinates x, y, z in Figure 8.3. Based on that analogy, a moment can be represented as a vector to provide the visual aid for an otherwise abstract mathematical description of multipole translations. This analogy should not be extended too far however: strictly speaking, moments are tensors and, unlike vectors, cannot be faithfully represented in 3D space.

Now the rules for vector diagram for multipole translations can be introduced. Vectors in the diagram will indicate either solid harmonics, $R^*_{l,m}$ and $S_{l,m}$, or moments, $M_{l,m}$ and $L_{l,m}$; no distinction will be made between them. Because solid harmonics are equivalent to moments when particles carry a unit positive charge, they will be referred to interchangeably in the vector diagrams. The vector origin should be placed in the origin of a coordinate system to indicate which of the present coordinate systems the moments are defined about. The direction of the vector will point toward the particle the moments and solid harmonics are defined for. When two or more particles are involved, as in the real case, the direction of the vector representing the sum of moments will point in the general direction of those particles without specifically locating any given one.

To understand the nature of multipole translation, first consider the case illustrated in Figure 8.4. It represents the systems of local charges A1, A2, A3, C1 and C2, each enclosed in their own bounding sphere. For each charge set there are regular solid harmonics, $R^*_{l,m}$, irregular solid harmonics, $S_{l,m}$, regular, $M_{l,m}$, and irregular $L_{l,m}$, multipole moments that can be computed by using Equations 8.27 and 8.35, 8.36 and 8.44, and 8.15 and 8.20, respectively, about the appropriate local coordinate system. The present goal is to compute the electrostatic potential created by charge sources located in A1, A2, and A3 in the position of particles located in C1 and C2.

The objective being sought is to sum up the multipoles from A1, A2, and A3 in order to obtain their collective effect on the external charges. This requires that a common coordinate system be used, so the next task is to define a common origin and then translate the multipoles to it. The center of sphere B provides a convenient common origin for the new coordinate system. Sphere B is defined such that it encloses the three charge sets A1, A2, and A3, and does not include any of the particles in the target charge sets C1 and C2.

The translation of regular solid harmonics $R^*(\mathbf{a})$ from the origin A1 to B will now be analyzed. Translations from A2 and A3 to B employ the same mechanism. The process of multipole translation goes through recreating the vector diagram in Figure 8.4 step-by-step. First, one draws a vector **a** that represents

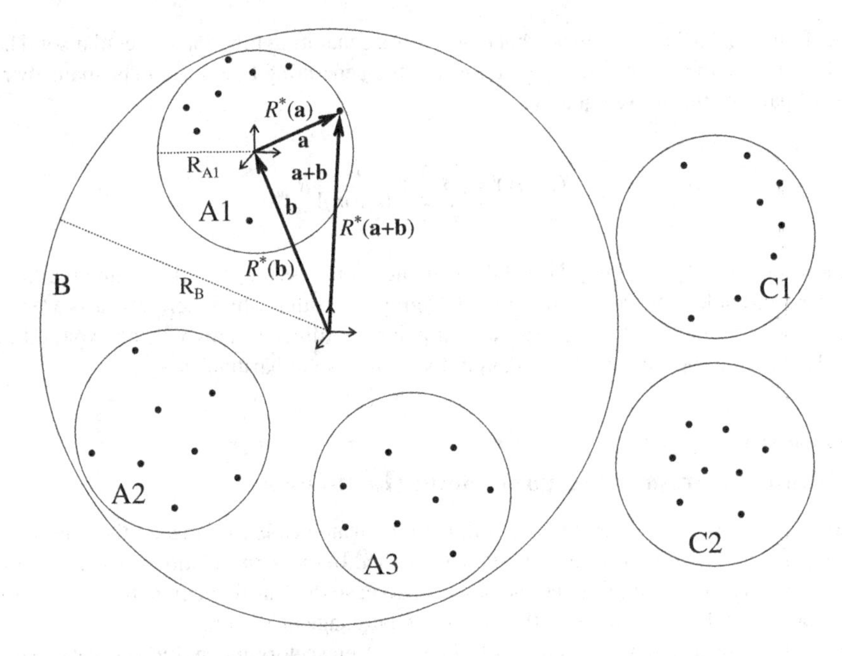

FIGURE 8.4 Translation of regular solid harmonics $R^*(\mathbf{a})$, defined about local coordinate system A1, to coordinate system B by means of helper solid harmonic $R^*(\mathbf{b})$ in order to create a translated moment $R^*(\mathbf{a}+\mathbf{b})$ about the origin of sphere B.

the initial expansion $R^*(\mathbf{a})$ using vector diagram rules, that is, the origin of the vector **a** gets placed in the origin of coordinate system A1 and the end of the vector points toward a charge source. A second vector, $\mathbf{a}+\mathbf{b}$, the outcome of translation representing $R^*(\mathbf{a}+\mathbf{b})$, is then added. Vector $\mathbf{a}+\mathbf{b}$ has its origin at the center of coordinate system B and terminates at the charge source. A third vector, **b**, connecting origins A1 and B is then to be added. A problem now appears in that, unlike vectors **a** and $\mathbf{a}+\mathbf{b}$, it is not immediately obvious where the origin of vector **b** should be placed. There are two options to choose from, with only one of these being correct. This is where the use of the vector diagram helps. The diagram rules establish that the remaining vector should be drawn in such a way as to complete the vector addition. This requirement places the origin of vector **b** in the center of coordinate system B, and has the vector tip pointing toward the center of coordinate system A1. The just defined radius-vector $\mathbf{b}=(x_{A1}-x_B, y_{A1}-y_B, z_{A1}-z_B)$ determines the remaining expansion $R^*(\mathbf{b})$, which is computed by using Equations 8.27 and 8.35.

To summarize, the expansion $R^*(\mathbf{a})$ is defined in the problem statement for the charge set A1. The expansion $R^*(\mathbf{b})$ has just been found. All that remains is to find $R^*(\mathbf{a}+\mathbf{b})$ by using the addition theorem for regular solid harmonics. The necessary mathematical tool will be developed next.

Having previously derived the generating function for regular solid harmonics (Equation 8.47),

$$(z_r+ix_r)^l=\sum_{m=-l}^{l} i^{-m}\frac{l!}{(l+m)!}R_{l,m}(\mathbf{r}), \tag{8.64}$$

it is possible now to take advantage of the simplicity with which the addition operation $\mathbf{r}=\mathbf{a}+\mathbf{b}$ can be introduced in the Cartesian portion of that equation in the left-hand side:

$$(z_r+ix_r)^l=[(z_a+ix_a)+(z_b+ix_b)]^l=\sum_{m=-l}^{l} i^{-m}\frac{l!}{(l+m)!}R_{l,m}(\mathbf{a}+\mathbf{b}). \tag{8.65}$$

Using the binomial expansion, the effect of the addition operation can be re-written as:

$$[(z_a+ix_a)+(z_b+ix_b)]^l=\sum_{j=0}^{l}\frac{l!}{j!(l-j)!}(z_a+ix_a)^j(z_b+ix_b)^{l-j}. \tag{8.66}$$

Substituting the generating functions $(z_a + ix_a)^j$ and $(z_b + ix_b)^{l-j}$ with the corresponding regular solid harmonics $R_{j,k}(\mathbf{a})$ and $R_{l-j,n}(\mathbf{b})$ from Equation 8.47 gives the following triple sum:

$$(z_r + ix_r)^l = \sum_{j=0}^{l} \frac{l!}{j!(l-j)!} \sum_{k=-j}^{j} i^{-k} \frac{j!}{(j+k)!} R_{j,k}(\mathbf{a}) \sum_{n=-l+j}^{l-j} i^{-n} \frac{(l-j)!}{(l-j+n)!} R_{l-j,n}(\mathbf{b}). \tag{8.67}$$

Rearranging the indices and simplifying produces:

$$(z_r + ix_r)^l = \sum_{j=0}^{l} \sum_{k=-j}^{j} \sum_{n=-l+j}^{l-j} i^{-k-n} \frac{l!}{(j+k)!(l-j+n)!} R_{j,k}(\mathbf{a}) R_{l-j,n}(\mathbf{b}). \tag{8.68}$$

Remembering that $\mathbf{r} = \mathbf{a} + \mathbf{b}$, and substituting the left-hand side in this equation with the right-hand side from Equation 8.65 leads to the equality:

$$\sum_{m=-l}^{l} i^{-m} \frac{l!}{(l+m)!} R_{l,m}(\mathbf{a}+\mathbf{b}) = \sum_{j=0}^{l} \sum_{k=-j}^{j} \sum_{n=-l+j}^{l-j} i^{-k-n} \frac{l!}{(j+k)!(l-j+n)!} R_{j,k}(\mathbf{a}) R_{l-j,n}(\mathbf{b}). \tag{8.69}$$

The sum of azimuthal numbers $k + n$ spans the range $-l$ to $+l$. This makes it possible to equate terms with the same value of azimuthal index m on both sides of Equation 8.69 by arranging the order of summation on the right-hand side to satisfy the condition $m = k + n$, that is, requiring that $n = m - k$.

$$i^{-m} \frac{l!}{(l+m)!} R_{l,m}(\mathbf{a}+\mathbf{b}) = \sum_{j=0}^{l} \sum_{k=-j}^{j} i^{-m} \frac{l!}{(j+k)!(l-j+m-k)!} R_{j,k}(\mathbf{a}) R_{l-j,m-k}(\mathbf{b}). \tag{8.70}$$

Terms $i^{-m}\, l!$ cancel out. Since l and m are constant parameters in this equation both sides can be multiplied by $(l + m)!$. These simplifications lead to the addition theorem for regular solid harmonics:

$$R_{l,m}(\mathbf{a}+\mathbf{b}) = \sum_{j=0}^{l} \sum_{k=-j}^{j} \frac{(l+m)!}{(j+k)!(l-j+m-k)!} R_{j,k}(\mathbf{a}) R_{l-j,m-k}(\mathbf{b}). \tag{8.71}$$

To find the complex-conjugate form of Equation 8.71, one can replace all instances of i by $-i$:

$$R^{*}_{l,m}(\mathbf{a}+\mathbf{b}) = \sum_{j=0}^{l} \sum_{k=-j}^{j} \frac{(l+m)!}{(j+k)!(l-j+m-k)!} R^{*}_{j,k}(\mathbf{a}) R^{*}_{l-j,m-k}(\mathbf{b}). \tag{8.72}$$

The next step is to transition from Equation 8.72 to a translation of multipole moments, $M_{l,m}$. Based on the system set up in Figure 8.4, $R^{*}(\mathbf{b})$ is a chargeless helper solid harmonic in Equation 8.72 that corresponds to vector $\mathbf{b} = (x_{A1} - x_B, y_{A1} - y_B, z_{A1} - z_B)$ connecting origins $O(x_{A1}, y_{A1}, z_{A1})$ and $O(x_B, y_B, z_B)$. The other two expansions $R^{*}(\mathbf{a})$ and $R^{*}(\mathbf{a} + \mathbf{b})$ are the initial and translated solid harmonics, respectively, corresponding to a system of positive unit charges. For a real system of electrostatic charges, the latter two expansions should be replaced by regular multipole moments $M_{l,m}$ as defined by Equation 8.17. Correspondingly, Equation 8.72 assumes the form:

$$M_{l,m}(\mathbf{a}+\mathbf{b}) = \sum_{j=0}^{l} \sum_{k=-j}^{j} \frac{(l+m)!}{(j+k)!(l-j+m-k)!} M_{j,k}(\mathbf{a}) R^{*}_{l-j,m-k}(\mathbf{b}). \tag{8.73}$$

The transition from Equation 8.72 to 8.73 can be verified by considering a system that consists only of a single particle carrying charge q. Due to the definition of regular multipole moments in Equation 8.15, both sides of Equation 8.72 can be multiplied by q, and the result matches Equation 8.73 exactly. For a system of multiple charges, each charge can be treated independently and the result summed. Alternatively, the multipoles of particles can be summed up and the combined multipole translated once. Both approaches give the same result due to the Distributive Law: (a + b)c = ac + bc that moments follow.

Equation 8.73 is conventionally abbreviated M2M that stands for regular multipole-to-multipole translation. M2M operation can be visualized as the addition of vectors representing multipoles while remembering that, strictly speaking, multipoles are tensors and not vectors. The directions of those vectors are shown in Figure 8.4. The M2M translation requires compliance with the condition $|R_B| > |R_{A1}| + |\mathbf{b}|$. This is a result of the solution of Laplace's equation in the form of an expansion of solid harmonics that requires that the points in C1 and C2 be located outside the boundaries of the sphere enclosing the charge sources.

A notable property of Equation 8.73 is the symmetry of the indices of multipole expansions. Specifically, the sum of indices in the product $M_{j,k}(\mathbf{a})R^*_{l-j,m-k}(\mathbf{b})$ is always equal to l and m in degree and order, respectively, and the indices in the product span the full permissible space. Because of that, the indices in the expansions can be swapped in their places leading to Equation 8.74, which is exactly equivalent to Equation 8.73 due to the sum being independent of the order of summation of its summands.

$$M_{l,m}(\mathbf{a}+\mathbf{b})=\sum_{j=0}^{l}\sum_{k=-j}^{j}\frac{(l+m)!}{(j+k)!(l-j+m-k)!}M_{l-j,m-k}(\mathbf{a})R^*_{j,k}(\mathbf{b}). \tag{8.74}$$

This property implies that the arguments to the M2M function can be freely interchanged in a computer program, that is, the order of the arguments would not be important. This feature will be used when discussing the computer implementation of M2M equation.

Finally, a translation in the opposite direction ($-\mathbf{b}$), that is moving from the origin B to origin A1, $\mathbf{r} = \mathbf{a} - \mathbf{b}$, can be obtained by repeating the above steps for the expression:

$$[(z_a - ix_a)-(z_b + ix_b)]^l = \sum_{m=-l}^{l} i^{-m}\frac{l!}{(l+m)!}R_{l,m}(\mathbf{a}-\mathbf{b}), \tag{8.75}$$

resulting in:

$$M_{l,m}(\mathbf{a}-\mathbf{b})=\sum_{j=0}^{l}\sum_{k=-j}^{j}\frac{(l+m)!}{(j+k)!(l-j+m-k)!}(-1)^{l-j}M_{j,k}(\mathbf{a})R^*_{l-j,m-k}(\mathbf{b}). \tag{8.76}$$

The only difference between Equation 8.76 and Equation 8.73 is the presence of an additional parity coefficient $(-1)^{l-j}$ generated as a result of the binomial expansion of the negative sign in the left-hand side of Equation 8.75. An alternative and shorter path to that equation can be developed by replacing $\mathbf{b}$ with $-\mathbf{b}$, that is, setting $\mathbf{r} = \mathbf{a} + (-\mathbf{b})$ and using $[(z_a + ix_a) + (-z_b - ix_b)]^l$ for which there is already a known solution. Using this technique, Equation 8.73 becomes:

$$M_{l,m}(\mathbf{a}+(-\mathbf{b}))=\sum_{j=0}^{l}\sum_{k=-j}^{j}\frac{(l+m)!}{(j+k)!(l-j+m-k)!}M_{j,k}(\mathbf{a})R^*_{l-j,m-k}(-\mathbf{b}). \tag{8.77}$$

Now one can inspect the term $R^*_{l-j,m-k}(-\mathbf{b})$ and see what can be done with the negative sign in $-\mathbf{b}$. According to the corresponding property of spherical harmonics, the replacement of $-\mathbf{b}$ by $\mathbf{b}$ in the argument of solid harmonics is the equivalent of performing a coordinate inversion operation:

$$R_{l,m}(-x,-y,-z)=(-1)^l R_{l,m}(x,y,z). \tag{8.78}$$

This property immediately leads to Equation 8.76 from Equation 8.77, and adds the parity coefficient $(-1)^{l-j}$ to $R^*_{l-j,m-k}(\mathbf{b})$. Obviously, the radial term in solid harmonics is not affected by coordinate inversion, so the property known for spherical harmonics directly applies to solid harmonics.

The physical meaning of Equation 8.76 is that the regular multipole representation of the charge distribution is translated from origin B to A1, that is, it is in reverse of the translation in Equation 8.73. If the expansion, $R^*(\mathbf{b})$ is inverted by means of Equation 8.78, then the same result as that of Equation 8.76 can be obtained using the original M2M operation, Equation 8.73. This means that a single implementation of M2M operation in the computer program can be used for performing regular multipole translation in either direction.

Now that the theoretical part of M2M translation has been completed, it is time to return to the physics problem of finding the potential due to charges A1, A2, and A3 in the points inside C1 and C2 that preceded this section. Given that the M2M operation translates the regular multipole moments $M_{l,m}$(A1), $M_{l,m}$(A2), $M_{l,m}$(A3), defined about their respective local coordinate systems, into $M_{l,m}$(A1,B), $M_{l,m}$(A2,B), $M_{l,m}$(A3,B) about the common origin B, they can now be combined to produce the collective regular moments $M_{l,m}$(A,B):

$$M_{l,m}(A,B) = M_{l,m}(A1,B) + M_{l,m}(A2,B) + M_{l,m}(A3,B). \tag{8.79}$$

Here, the arguments in $M_{l,m}$(A1,B) indicate that the regular multipole moments representing charge sources from A1 are defined about origin B. In those cases where multipole moments are defined about the same coordinate system as that of their original charge sources, the second index in the brackets will be omitted and the term simply referred to as $M_{l,m}$(A1). The collective moments $M_{l,m}$(A,B) now make it possible to compute the electrostatic potential due to charges A1, A2, and A3 anywhere outside the sphere B.

However, it is not yet time to compute the electrostatic potential inside C1 and C2. There are still two problems to overcome. First, the regular multipole moments $M_{l,m}$(A,B) are defined about origin B, whereas they should be defined about origins C1 and C2. Second, multipole moments $M_{l,m}$(A,B) are provided, whereas irregular moments $L_{l,m}$(A,C) are needed instead. These two requirements have to be solved in order for the potential from charge sets A to be visible in the points in C1 and C2. Both these problems are addressed in the addition theorem for irregular solid harmonics.

8.9 Addition Theorem for Irregular Solid Harmonics

Irregular solid harmonics are governed by a similar addition theorem as that for regular solid harmonics. To begin the derivation, consider the system of charges depicted in Figure 8.5. For simplicity let's assume that there is only one positive unit charge in the set B. As a starting condition, there is an expansion in regular solid harmonics $R^*(\mathbf{a})$ defined for that charge about origin B. According to vector diagram rules, the origin of vector **a** coincides with the center of coordinate system B, and the tip of the vector points toward the charge source. The expansion to obtain is $S(\mathbf{a} + \mathbf{b})$. Having the origin of vector $\mathbf{a} + \mathbf{b}$ placed in the center of coordinate system C1 and the tip of the vector pointing toward the charge source make the potential due to the charge source visible inside C1. One more vector is needed in order to complete the vector addition diagram. The vector addition rule is satisfied by a radius-vector $\mathbf{b} = (x_B - x_{CI}, y_B - y_{CI}, z_B - z_{CI})$ originating from the center of coordinate system C1 and pointing toward the center of coordinate system B. This vector corresponds to a helper irregular solid harmonic $S(\mathbf{b})$. Further details on which of those three vectors in the diagram corresponds to regular and which one to irregular harmonics will be provided later.

To find the expansion $S(\mathbf{a} + \mathbf{b})$ consider the generating function for irregular solid harmonics (Equation 8.48) written in the form of the sum $\mathbf{r} = \mathbf{a} + \mathbf{b}$,

$$(z_r + ix_r)^{-l-1} = [(z_a + ix_a) + (z_b + ix_b)]^{-l-1}, \tag{8.80}$$

and use the binomial theorem to expand the terms in square brackets. In this decomposition, the condition $|z_a + ix_a| < |z_b + ix_b|$ will be assumed. This condition determines how to treat the terms **a** and **b**. The distinction will become clearer in the course of the discussion.

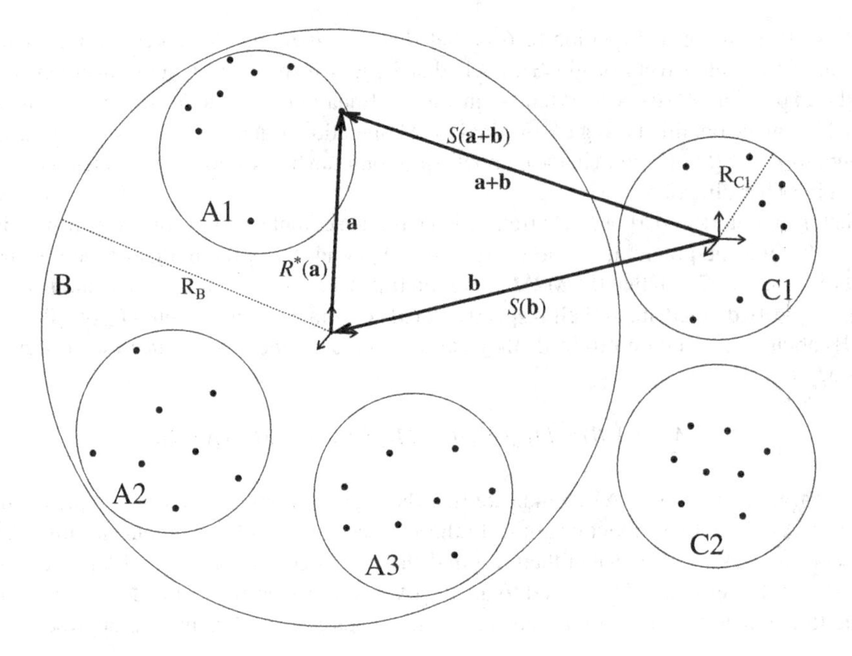

FIGURE 8.5 Translation of expansion $R^*(\mathbf{a})$ defined about coordinate system B into expansion $S(\mathbf{a}+\mathbf{b})$ centered about coordinate system C1 using $S(\mathbf{b})$ as a helper function.

Unlike the example of the generating function for regular solid harmonics where a binomial expansion produces a finite series, binomial expansion of the generating function for irregular solid harmonics produces an infinite series; this is because the power $-l-t-1$ can go to minus infinity.

$$[(z_a+ix_a)+(z_b+ix_b)]^{-l-1}=\sum_{t=0}^{\infty}\frac{(-l-1)!}{t!(-l-t-1)!}(z_a+ix_a)^t(z_b+ix_b)^{-l-t-1}. \tag{8.81}$$

This equation contains factorials of negative numbers, which individually go to infinity. The factorials can be simplified via the following reduction:

$$\frac{(-l-1)!}{(-l-1-t)!}=(-l-1)\ldots(-l-t)=(-1)^t(l+1)\ldots(l+t)=(-1)^t\frac{(l+t)!}{l!}. \tag{8.82}$$

Substituting this expression in Equation 8.81, one obtains

$$[(z_a+ix_a)+(z_b+ix_b)]^{-l-1}=\sum_{t=0}^{\infty}\frac{(-1)^t(l+t)!}{t!l!}(z_a+ix_a)^t(z_b+ix_b)^{-l-t-1}. \tag{8.83}$$

In order for this series to be convergent the condition $|z_a+ix_a|/|z_b+ix_b|<1$ is required, which is satisfied due to the previously defined condition that $|z_a+ix_a|<|z_b+ix_b|$. In the term $(z_a+ix_a)^t$ in Equation 8.83 one can recognize a generating function for regular spherical harmonics, Equation 8.47. Correspondingly, the term $(z_b+ix_b)^{-l-t-1}$ is a generating function for irregular solid harmonics, Equation 8.48. Substituting the appropriate expressions into Equation 8.83 yields:

$$(z_{a+b}+ix_{a+b})^{-l-1}=\sum_{t=0}^{\infty}(-1)^t\frac{(l+t)!}{t!l!}\sum_{n=-t}^{t}i^{-n}\frac{t!}{(t+n)!}R_{t,n}(\mathbf{a})\sum_{\substack{k=-l-t\\|k+n|\le l}}^{l+t}i^k\frac{(l+t-k)!}{(l+t)!}S_{l+t,k}(\mathbf{b})+\Delta_{RHS}. \tag{8.84}$$

In Equation 8.84, the product $R_{t,n}(\mathbf{a})\ S_{l+t,k}(\mathbf{b})$ has the sum of azimuthal indices $k + n$. Depending on the value of $|k + n|$, the term $R_{t,n}(\mathbf{a})\ S_{l+t,k}(\mathbf{b})$ can be identified either as essential or redundant. The condition of $|k + n| \leq l$ helps to gather only the essential components in Equation 8.84. The summands satisfying conditions $|k + n| > l$ or $k > l + t$ in their indices contribute to or directly represent the terms for which polynomial order exceeds the value of polynomial degree, that is, they go beyond the general requirement $-l \leq m \leq l$ on the bounds of azimuthal index while having non-zero value. Those summands are collected in the residual term, Δ_{RHS}:

$$\begin{aligned}\Delta_{RHS} = & \sum_{t=0}^{\infty}(-1)^t \frac{(l+t)!}{t!l!}\sum_{n=-t}^{t} i^{-n}\frac{t!}{(t+n)!}R_{t,n}(\mathbf{a})\sum_{\substack{k=-l-t \\ |k+n|>l}}^{l+t} i^k \frac{(l+t-k)!}{(l+t)!}S_{l+t,k}(\mathbf{b}) \\ & + 2\sum_{t=0}^{\infty}(-1)^t \frac{(l+t)!}{t!l!} r_b^{-l-t-1}\sum_{n=-t}^{t} i^{-n}\frac{t!}{(t+n)!}R_{t,n}(\mathbf{a})\sum_{k=l+t+1}^{\infty} i^{-k}\frac{(l+t+k)!}{(l+t)!}P_{l+t}^{-k}(\cos\theta_b)\cos(k\phi_b),\end{aligned} \tag{8.85}$$

where (r_b, θ_b, ϕ_b) are spherical polar coordinates of radius vector $\mathbf{b}$. Note that $S_{l+t,k}(\mathbf{b})$ is present in its normal form for $|k| \leq l + t$ in the first triple sum in the residual term, Δ_{RHS}, whereas it breaks down into product of associated Legendre function and cosine of azimuthal angle for particle $\mathbf{b}$ under the condition $k > l + 1$ in the second triple sum in Equation 8.85. Derivation of the generating function for irregular solid harmonics, which is presented in Appendix B.5, provides the necessary details explaining that matter.

Canceling out $(l + t)!$ and $t!$, and rearranging the parts gives the final expression for the residual term.

$$\begin{aligned}\Delta_{RHS} = & \sum_{t=0}^{\infty}\sum_{n=-t}^{t}\sum_{\substack{k=-l-t \\ |k+n|>l}}^{l+t} i^{k-n}(-1)^t \frac{(l+t-k)!}{l!(t+n)!}R_{t,n}(\mathbf{a})S_{l+t,k}(\mathbf{b}) \\ & + 2\sum_{t=0}^{\infty}\sum_{n=-t}^{t}\sum_{k=l+t+1}^{\infty} i^{-k-n}(-1)^t r_b^{-l-t-1}\frac{(l+t+k)!}{l!(t+n)!}R_{t,n}(\mathbf{a})P_{l+t}^{-k}(\cos\theta_b)\cos(k\phi_b).\end{aligned} \tag{8.86}$$

In like manner $(l + t)!$ and $t!$ get canceled out in Equation 8.84 to produce:

$$(z_{a+b} + ix_{a+b})^{-l-1} = \sum_{t=0}^{\infty}\sum_{n=-t}^{t}\sum_{\substack{k=-l-t \\ |k+n|\leq l}}^{l+t} i^{k-n}(-1)^t \frac{(l+t-k)!}{l!(t+n)!}R_{t,n}(\mathbf{a})S_{l+t,k}(\mathbf{b}) + \Delta_{RHS}. \tag{8.87}$$

Now one can replace the left-hand side in Equations 8.87 by the right-hand side from Equation 8.48 while canceling out the residual terms, so that we obtain:

$$\sum_{m=-l}^{l} i^m \frac{(l-m)!}{l!}S_{l,m}(\mathbf{a}+\mathbf{b}) = \sum_{t=0}^{\infty}\sum_{n=-t}^{t}\sum_{\substack{k=-l-t \\ |k+n|\leq l}}^{l+t} i^{k-n}(-1)^t \frac{(l+t-k)!}{l!(t+n)!}R_{t,n}(\mathbf{a})S_{l+t,k}(\mathbf{b}). \tag{8.88}$$

The equality of the residual terms can be verified numerically. The information in Appendix B.5 assists in the computation of the hypergeometric function used in the evaluation of the residual terms.

In Equation 8.88, one can equate $S_{l,m}(\mathbf{a} + \mathbf{b})$ and $R_{t,n}(\mathbf{a})\ S_{l+t,k}(\mathbf{b})$ on the left- and right-hand sides, respectively for which $m = k + n$. Now the reader may see a connection between this step and the condition $|k + n| \leq l$ that was previously introduces in Equation 8.84. For $m = k + n$, the requirement of $|k + n| \leq l$ in Equation 8.88 becomes $|m| \leq l$, which is satisfied by definition, so it no longer needs to be stated. The condition $m = k + n$ can be applied in two equivalent ways as $n = m - k$ and $k = m - n$. With those substitutions one obtains two equivalent forms of the same equation having different arrangements of the summation indices. Starting with $n = m - k$ and extracting the equated terms from Equation 8.88 leads to:

$$i^m \frac{(l-m)!}{l!} S_{l,m}(\mathbf{a}+\mathbf{b}) = \sum_{t=0}^{\infty} \sum_{k=-l-t}^{l+t} i^{2k-m}(-1)^t \frac{(l+t-k)!}{l!(t+m-k)!} R_{t,m-k}(\mathbf{a}) S_{l+t,k}(\mathbf{b}). \tag{8.89}$$

In it, multipliers $l!$ cancel out on both sides. Next, the regular solid harmonic R is converted into its complex-conjugate form, R^* by employing the relation:

$$R_{t,m-k} = (-1)^{k-m} \frac{(t+m-k)!}{(t-m+k)} R^*_{t,k-m}. \tag{8.90}$$

The purpose of this transformation is to obtain the final equation in terms of solid harmonics R^* and S, which customarily appear in the potential expansion. After replacing R by R^* in Equation 8.89, one obtains:

$$i^m (l-m)! S_{l,m}(\mathbf{a}+\mathbf{b}) = \sum_{t=0}^{\infty} \sum_{k=-l-t}^{l+t} i^{2k-m}(-1)^{t+k-m} \frac{(l+t-k)!}{(t-m+k)!} R^*_{t,k-m}(\mathbf{a}) S_{l+t,k}(\mathbf{b}). \tag{8.91}$$

Next, note that

$$i^{2k-m}(-1)^{k-m} = (-1)^k i^{-m}(-1)^{k-m} = i^{-m}(-1)^{-m} = i^m. \tag{8.92}$$

Canceling out i^m, and dividing both sides of Equation 8.91 by $(l - m)!$, gives:

$$S_{l,m}(\mathbf{a}+\mathbf{b}) = \sum_{t=0}^{\infty} \sum_{k=-l-t}^{l+t} (-1)^t \frac{(l+t-k)!}{(l-m)!(t+k-m)!} R^*_{t,k-m}(\mathbf{a}) S_{l+t,k}(\mathbf{b}). \tag{8.93}$$

Applying the substitution $j = l + t$ or $t = j - l$ leads to the addition theorem:

$$S_{l,m}(\mathbf{a}+\mathbf{b}) = \sum_{j=l}^{\infty} \sum_{k=-j}^{j} (-1)^{j-l} \frac{(j-k)!}{(l-m)!(j-l+k-m)!} R^*_{j-l,k-m}(\mathbf{a}) S_{j,k}(\mathbf{b}). \tag{8.94}$$

This is one of the possible forms of the addition theorem for irregular solid harmonics. The second form of this theorem is obtained by applying substitution $k = m - n$ to Equation 8.88 and equating the corresponding terms $S_{l,m}(\mathbf{a} + \mathbf{b})$ and $R_{t,n}(\mathbf{a}) S_{l+t,k}(\mathbf{b})$ in the left- and right-hand sides. This gives

$$i^m \frac{(l-m)!}{l!} S_{l,m}(\mathbf{a}+\mathbf{b}) = \sum_{t=0}^{\infty} \sum_{n=-t}^{t} i^{m-2n}(-1)^t \frac{(l+t-m+n)!}{l!(t+n)!} R_{t,n}(\mathbf{a}) S_{l+t,m-n}(\mathbf{b}). \tag{8.95}$$

Compare Equation 8.95 to Equation 8.89 to see the difference in the indices. The next step is to cancel out i^m and $l!$ in Equation 8.95, and divide both sides by $(l - m)!$ This leads to:

$$S_{l,m}(\mathbf{a}+\mathbf{b}) = \sum_{t=0}^{\infty} \sum_{n=-t}^{t} i^{-2n}(-1)^t \frac{(l+t-m+n)!}{(l-m)!(t+n)!} R_{t,n}(\mathbf{a}) S_{l+t,m-n}(\mathbf{b}). \tag{8.96}$$

Noticing that $i^{-2n} = (-1)^{-n}$, and substituting n by $-k$, and t by j, one obtains:

$$S_{l,m}(\mathbf{a}+\mathbf{b}) = \sum_{j=0}^{\infty} \sum_{k=-j}^{j} (-1)^{j+k} \frac{(l+j-m-k)!}{(l-m)!(j-k)!} R_{j,-k}(\mathbf{a}) S_{l+j,m+k}(\mathbf{b}). \tag{8.97}$$

As before, regular solid harmonics should be converted to their complex-conjugate form; the following expression provides the necessary conversion:

$$R_{j,-k} = (-1)^k \frac{(j-k)!}{(j+k)} R^*_{j,k}. \tag{8.98}$$

Substitution of Equation 8.98 into Equation 8.97 leads to the second form of the addition theorem for irregular solid harmonics:

$$S_{l,m}(\mathbf{a}+\mathbf{b}) = \sum_{j=0}^{\infty}\sum_{k=-j}^{j}(-1)^j \frac{(l+j-m-k)!}{(l-m)!(j+k)!} R^*_{j,k}(\mathbf{a}) S_{l+j,m+k}(\mathbf{b}). \tag{8.99}$$

Both equations Equation 8.94 and Equation 8.99 are entirely equivalent. Further derivations will be performed based on the second equation.

The next step is to transition from the translation of solid harmonics to the translation of multipole moments. The rationale behind the conversion is the same as in the derivation of the M2M translation. Since by assumption the system is composed of only a single charge in set B in the beginning, and because $R^*(\mathbf{a})$ and $S(\mathbf{a}+\mathbf{b})$ are initial and translated expansions in Figure 8.5, respectively, both sides of Equation 8.99 can be multiplied by a charge of magnitude q so that $R^*(\mathbf{a})$ and $S(\mathbf{a}+\mathbf{b})$ turn into corresponding moments. With that one obtains:

$$L_{l,m}(\mathbf{a}+\mathbf{b}) = \sum_{j=0}^{\infty}\sum_{k=-j}^{j}(-1)^j \frac{(l+j-m-k)!}{(l-m)!(j+k)!} M_{j,k}(\mathbf{a}) S_{l+j,m+k}(\mathbf{b}). \tag{8.100}$$

This equation remains the same if charge set B contains multiple charges due to the distributive property of multipole moments.

Equation 8.100 provides the missing link to the problem of obtaining the potential due to charge sources located in sets A1, A2, and A3 at the points inside C1 and C2. Specifically, the addition theorem changes the regular multipole moment $M(\mathbf{a},\mathrm{B})$, defined about the center of sphere B, into an irregular multipole moment $L(\mathbf{a}+\mathbf{b},\mathrm{C1})$, and defines the latter about origin C1. Because of these properties, Equation 8.100 is known as M2L, where letters M and L indicate regular and irregular multipole moments. This translation is only possible when $|\mathbf{a}| < |\mathrm{R_B}|$ and $|\mathbf{b}| > |\mathrm{R_B}| + |\mathrm{R_{C1}}|$ as depicted in Figure 8.5. These conditions are another way of saying that the spheres B and C1 should not overlap.

In the previous section, M2M translation has been used to change the origin of multipole moments for charge sets A1, A2, and A3 into the new origin B, where the translated moments were summed up into the collective multipole moments $M_{lm}(\mathrm{A,B})$ defined about origin B. The newly derived M2L translation converts $M_{lm}(\mathrm{A,B})$ into irregular moments $L_{lm}(\mathrm{A,C1})$ defined about origin C1. Based on those two consecutive translations the potential $\Phi(\mathbf{c})$ at a point $\mathbf{c}$ inside C1 can be computed *via* the following equation,

$$\Phi(\mathbf{c}) = \sum_{l=0}^{\infty}\sum_{m=-l}^{l}\frac{(l-m)!}{(l+m)!} R^*_{l,m}(\mathbf{c})\, L_{l,m}(A,C1), \tag{8.101}$$

where $R^*_{l,m}(\mathbf{c})$ is a chargeless regular solid harmonic for radius-vector $\mathbf{c}$ defined in local coordinate system C1. Repeating the application of M2L translation, the multipole moment $M_{lm}(\mathrm{A,B})$ can be turned into irregular moment $L_{lm}(\mathrm{A,C2})$ defined about origin C2 and the potential in the points inside C2 analogously computed.

Using M2M and M2L translations made possible computing the potential due to charge sources located in A1, A2, and A3 in the points inside C1 and C2. All that is left to complete the picture is to make the potential due to charge sources located in C1 and C2 available inside A1, A2, and A3. Since the potential generated by charge sources located in C1 and C2 is provided by their regular

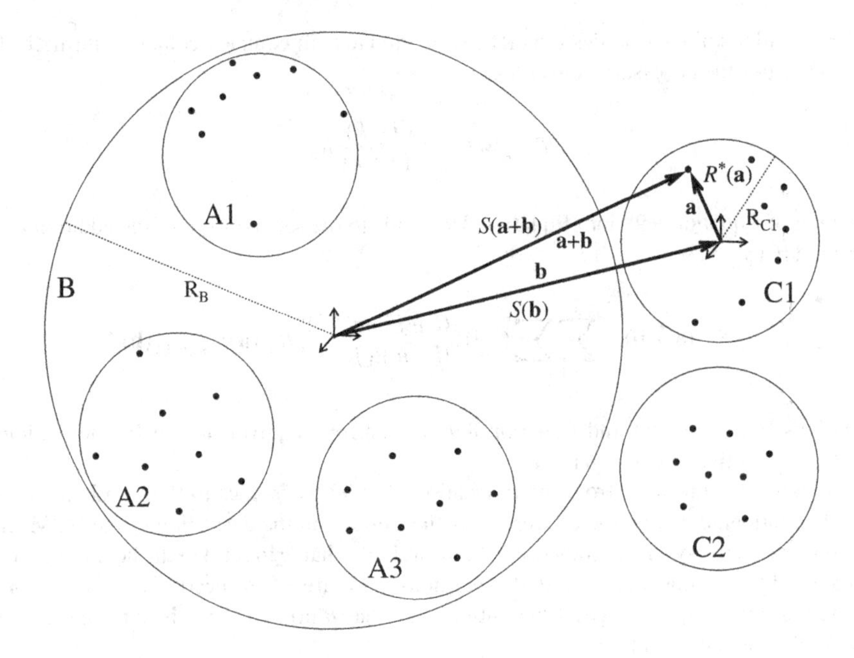

FIGURE 8.6 Translation of expansion $R^*(\mathbf{a})$ defined about coordinate system C1 into expansion $S(\mathbf{a}+\mathbf{b})$ centered about coordinate system B using $S(\mathbf{b})$ as a helper function.

multipole moments $M_{l,m}$(C1) and $M_{l,m}$(C2), one needs to use an M2L translation in order to convert them into irregular moments $L_{l,m}$(C1,B) and $L_{l,m}$(C2,B) about the new common origin, B. Again, as before, the common origin B as the destination for the M2L translation makes it possible to sum up the irregular multipole moments $L_{l,m}$(C1,B) and $L_{l,m}$(C2,B) to allow the subsequent transfer of the collective moments $L_{l,m}(\text{C,B}) = L_{l,m}(\text{C1,B}) + L_{l,m}(\text{C2,B})$ into A1, A2, and A3. This is preferable to moving the individual moments $L_{l,m}$(C1) and $L_{l,m}$(C2) as this technique helps to reduce the number of translations when dealing with large number of charge sets. The vector diagram for this translation is presented in Figure 8.6.

Drawing the vectors **a** and **a** + **b** corresponding to the initial $R^*(\mathbf{a})$ and translated $S(\mathbf{a}+\mathbf{b})$ expansions is trivial. Since expansion $R^*(\mathbf{a})$ is defined about origin C1, the origin of vector **a** is placed in the center of coordinate system C1. Using a similar argument, the origin of vector **a** + **b** is placed in the center of coordinate system B. The tip of both vectors points to the charge source. Determining the vector direction for the helper function could be confusing. However, the vector diagram makes that task unambiguous. The helper irregular solid harmonic $S(\mathbf{b})$ and its corresponding vector **b** are defined to complete the vector addition **a** + **b**. With all terms having been defined, Equation 8.100 now performs the M2L translation. The translation from C2 to B is performed in a similar manner. The results of M2L translations, $L_{l,m}$(C1,B) and $L_{l,m}$(C2,B) are then summed up into the cumulative irregular moments $L_{l,m}$(C,B), since both irregular multipole moments $L_{l,m}$(C1,B) and $L_{l,m}$(C2,B) are defined now about the same origin B.

The next step is to translate the origin of the irregular multipole moments $L_{l,m}$(C,B) into origins A1, A2, and A3 where they are needed. This is the topic for the next section.

8.10 Transformation of the Origin of Irregular Harmonics

The above derived addition theorems for regular and irregular solid harmonics constitute M2M and M2L translations. However, there is a room for one more translation. To see where it is coming from, a brief overview of the addition theorems might be helpful.

The addition theorem for regular solid harmonics (Equation 8.71) is a tensor product of two regular solid harmonics. Either of those can be chosen as the initial expansion and the other one would correspondingly become a helper regular solid harmonic that assists in the origin translation in an M2M operation. There is nothing more that can come out of this theorem due to its simplicity.

The addition theorem for irregular solid harmonics is a bit more complex. It is a tensor product of regular and irregular solid harmonics (Equation 8.99). In the derivation of the M2L translation the regular solid harmonics $R^*_{j,k}(\mathbf{a})$ served as an input expansion, and the irregular solid harmonics $S_{l+j,m+k}(\mathbf{b})$ functioned as a helper expansion assisting in the transformation. Now it is time to address the question that was deliberately omitted in the derivation of the M2L translation. How does one know which type of solid harmonics in the tensor product performs what role? Indeed, the addition theorem for irregular solid harmonics does not itself impose any specific role onto the individual solid harmonics in the product, therefore one is technically free to choose either solid harmonic as an input expansion, and assign the role of the helper expansion to the other. It means there is one more option in Equation 8.99 that remains unexplored, that is to use irregular solid harmonics $S_{l+j,m+k}(\mathbf{b})$ as the input expansion, and the regular solid harmonics $R^*_{j,k}(\mathbf{a})$ as a helper transformation. But what would be the physical meaning of such an assignment?

To find the answer, consider a system of electrostatic charges, as shown in Figure 8.7. This is the same system that was used in the derivation of the M2M and M2L translations.

The regular multipole moments $M_{l,m}$(C1) and $M_{l,m}$(C2) have previously been translated into irregular multipole moments $L_{l,m}$(C1,B) and $L_{l,m}$(C2,B) defined about origin B by using an M2L operation. The irregular moments from sets C1 and C2 have additionally been added up into the collective irregular moments $L_{l,m}$(C,B). Contracting the moments $M_{l,m}$(A,B) with $L_{l,m}$(C,B) by using Equation 8.24 gives the electrostatic interaction energy between the charge sets A and C. This contraction is possible because both moments are defined about the same coordinate system B. To obtain the electrostatic potential at the location of individual particles in A1, A2, and A3 due to the external charge sets C1 and C2 one more translation to redefine $L_{l,m}$(C,B) about each individual center A1, A2, and A3 is necessary.

The vector diagram for this translation is illustrated in Figure 8.7. Its appearance is quite different from the diagram in Figure 8.6; however, the difference is one of physics and not of math, as it will be clear later. The operation, Equation 8.99 that performs the required task for both diagrams has previously been derived. In the present case, the initial solid harmonic is $S(\mathbf{b})$. It can be replaced by the cumulative moment L_{lm}(C,B) that combines the effect of all charge sources in both sets C1 and C2. Recall that one can refer to $S(\mathbf{b})$ and $L(\mathbf{b})$ interchangeably.

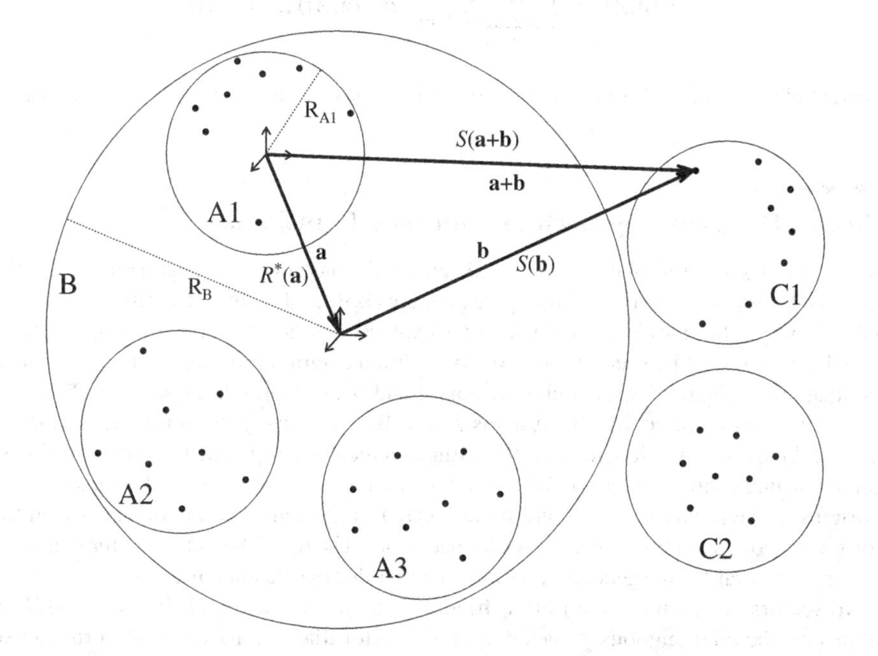

FIGURE 8.7 Translation of the origin of irregular solid harmonics $S(\mathbf{b})$ defined about coordinate system B to irregular solid harmonics $S(\mathbf{a} + \mathbf{b})$ defined about coordinate system A1. $R^*(\mathbf{a})$ is a chargeless helper regular solid harmonics.

The objective is to translate the irregular multipole moment L_{lm}(C,B) to a new origin A1. The resulting irregular moment is denoted as $L(\mathbf{a} + \mathbf{b})$. To perform the translation, one needs to compute a chargeless helper solid harmonic $R^*(\mathbf{a})$, which is defined for radius-vector $\mathbf{a} = (x_B - x_{A1}, y_B - y_{A1}, z_B - z_{A1})$. Note that the direction of this vector is determined by the vector diagram rules for completion of vector addition. An important requirement of this transformation is that sphere A1 is located inside the bounding sphere B, that is, $|R_B| > |\mathbf{a}| + |R_{A1}|$. Since $S(\mathbf{b})$ and $S(\mathbf{a} + \mathbf{b})$ are charge-scaled expansions, they can be replaced by irregular multipole moments in Equation 8.99. This gives:

$$L_{l,m}(\mathbf{a}+\mathbf{b}) = \sum_{j=0}^{\infty}\sum_{k=-j}^{j}(-1)^j \frac{(l+j-m-k)!}{(l-m)!(j+k)!} R^*_{j,k}(\mathbf{a}) L_{l+j,m+k}(\mathbf{b}). \tag{8.102}$$

This equation is known as an L2L operation because it translates the origin of irregular multipole moment L from one coordinate system to another. This is the physical meaning of this translation. Its relation to the sibling M2L translation is quite interesting, though. Although Equation 8.100 and 8.102 look different, the difference between them is only superficial. Multiplication of $R^*_{j,k}(\mathbf{a})S_{l+j,m+k}(\mathbf{b})$ in the right-hand side of Equation 8.99 by the value of charge q, gives either $M_{j,k}(\mathbf{a})\,S_{l+j,m+k}(\mathbf{b})$ or $R^*_{j,k}(\mathbf{a})L_{l+j,m+k}(\mathbf{b})$ depending on which one of the two terms the charge is assigned to. Whatever choice is made does not alter the product, and therefore the sum is the same. This proves that the math inside the M2L and L2L translations is exactly the same although their physical meaning is different. As the result, when implementing a computer program, both M2L and L2L transformations can be performed with the same subroutine in the program code. The switch is made by choosing $R^*_{j,k}(\mathbf{a})$ and $S_{l+j,m+k}(\mathbf{b})$ to be the input and helper solid harmonic, respectively, or choosing $R^*_{j,k}(\mathbf{a})$ and $S_{l+j,m+k}(\mathbf{b})$ to reverse their roles and assume the function of helper and input moments, respectively. The first choice gives the M2L translation, and the second option leads to the L2L operation.

Finally, contracting the collective moments L_{lm}(C,A1) with the chargeless multipoles $R^*_{l,m}(\mathbf{a}, A1)$ for a radius-vector **a** defined for a point inside A1 gives the desired potential $\Phi(\mathbf{a}, A1)$ inside A1 due to the external charges located in C1 and C2:

$$\Phi(\mathbf{a},A1) = \sum_{l=0}^{\infty}\sum_{m=-l}^{l}\frac{(l-m)!}{(l+m)!} R^*_{l,m}(\mathbf{a},A1)\, L_{l,m}(C,A1). \tag{8.103}$$

The potential inside A2 and A3 due to the external charges C1 and C2 is determined in a similar way.

8.11 Vector Diagram Approach to Multipole Translations

Concluding the description of the M2M, M2L and L2L operations one can summarize their basic elements. The translation of origin of multipole moments is described by the addition theorems for solid harmonics. However, the addition theorem does not explicitly say how to obtain the helper solid harmonic that is needed to perform the translation. A vector addition diagram provides the necessary answer.

In the diagram in Figure 8.8, the radius-vectors **a** and **b** can represent expansions R^*, S, M, or L for particle P in two different coordinate systems A and B, respectively. Because one vector is used for displaying all kinds of expansion, in order to distinguish them each specific type of expansion should be indicated by manually attaching a suitable label, for example, R^*, S, M, or L, to the vector.

The origins of the radius-vectors **a** and **b** are located in the center of coordinate system to indicate which origin the corresponding expansion is defined about. The tip of the radius-vector points toward the charge source or toward some general point in space that this expansion represents.

The two vectors representing the initial (**b**) and target (**a**) moments for the M2M, M2L, and L2L translations can be unambiguously placed onto the vector diagram based on the problem statement describing the starting condition and the result of the translation. The next step is to find the direction of the helper solid harmonic. The similarity of multipole translations to vector addition provides a

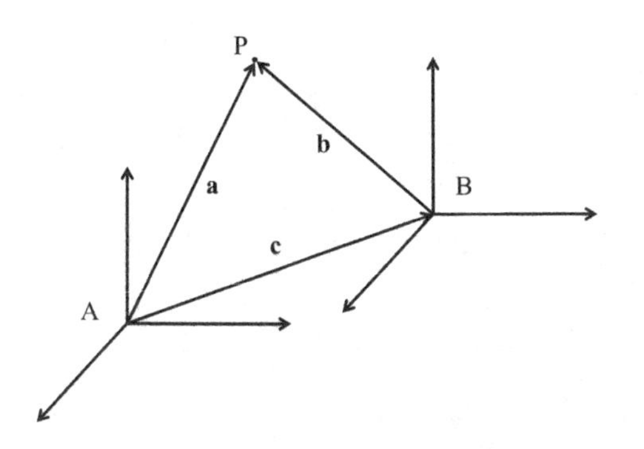

FIGURE 8.8 Translation of point P from coordinate system B to coordinate system A *via* vector addition $\mathbf{a} = \mathbf{b} + \mathbf{c}$.

useful aid. Recall that a coordinate translation of particle P from coordinate system B to coordinate system A in Figure 8.8 is described by a vector addition, $\mathbf{a} = \mathbf{b} + \mathbf{c}$. The addition diagram for multipole translations states that the virtual direction of the helper solid harmonic is the same as that of the underlying translation vector in the conventional vector addition operation. The direction of the vector **c** corresponding to the helper solid harmonic is determined by completing the vector addition diagram, $\mathbf{a} = \mathbf{b} + \mathbf{c}$.

Based on these definitions it is possible put together a short summary of translation operations as illustrated in Figure 8.9. The first operation is M2M translation. It is defined in Equation 8.73 and illustrated in Figure 8.9a. M2M operation translates a regular multipole moment $R^*(\mathbf{a})$ of a unit positive charge defined about origin O_2 into a new regular multipole moment $R^*(\mathbf{a} + \mathbf{b})$ centered about origin O_1. The centers of coordinate systems O_1 and O_2 have coordinates (x_1, y_1, z_1) and (x_2, y_2, z_2), respectively. The direction of vector $\mathbf{b} = (x_2 - x_1, y_2 - y_1, z_2 - z_1)$ for a chargeless helper solid harmonic $R^*(\mathbf{b})$ is determined by completing the vector addition diagram, $\mathbf{a} + \mathbf{b}$. The helper function $R^*(\mathbf{b})$ is computed by using Equations 8.27 and 8.35 for radius-vector **b**.

The next operation is the M2L translation. It is defined by Equation 8.100, and its diagram is illustrated in Figure 8.9b. The M2L operation transforms a multipole moment $R^*(\mathbf{a})$ of a unit positive charge, defined about origin O_2, into an irregular multipole moment $S(\mathbf{a} + \mathbf{b})$ defined about origin O_1. To perform this operation, it is necessary to generate a chargeless helper solid harmonic $S(\mathbf{b})$. The direction of vector $\mathbf{b} = (x_2 - x_1, y_2 - y_1, z_2 - z_1)$ is determined by completing the vector addition diagram, $\mathbf{a} + \mathbf{b}$. The helper function $S(\mathbf{b})$ is computed by using Equations 8.36 and 8.44 for radius-vector **b**.

The third operation, L2L, is defined by Equation 8.102. Its diagram is illustrated in Figure 8.9c. The L2L operation transforms an irregular multipole moment $S(\mathbf{b})$ defined for a unit positive charge about origin O_2 into another irregular multipole moment $S(\mathbf{a} + \mathbf{b})$ defined about origin O_1. This transformation is accomplished with help of a chargeless helper solid harmonic $R^*(\mathbf{a})$. The direction of the underlying vector $\mathbf{a} = (x_2 - x_1, y_2 - y_1, z_2 - z_1)$ is obtained by completing the vector addition diagram, $\mathbf{a} + \mathbf{b}$. The helper function $R^*(\mathbf{a})$ is computed by using Equations 8.27 and 8.35 for radius-vector **a**.

Together, the M2M, M2L, and L2L operations represent the core theoretical basis of the Fast Multipole Method.

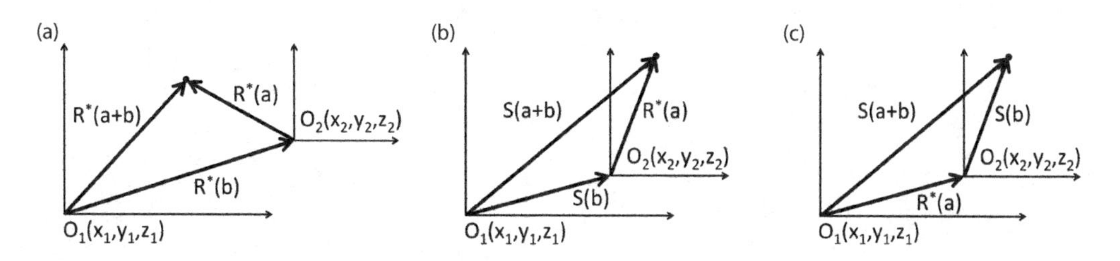

FIGURE 8.9 Vector addition diagram for the various FMM translations: M2M – panel (a); M2L – panel (b); L2L – panel (c).

9

Electrostatic Force

A charged particle exposed to other charge sources experiences a force trying to displace the particle from its current position. A knowledge of force is essential for molecular dynamics applications. Since force is equal to electric field scaled by the particle charge all that is needed to resolve force is to find the electric field created by all charge sources in the particle position. Help in solving this problem is provided by the superposition principle, which states that interaction between a pair of particles is unaffected by the presence of other particles, so that each pair can be treated independently. In conformance with this principle, if there is a system of charge sources q_i each located on distance $\boldsymbol{b}_i$ from the center of a coordinate system (Figure. 9.1), the electric field, $\mathbf{E}(\mathbf{r})$, created at a probe point P is the sum of contributions of individual charge sources:

$$\mathbf{E}(\mathbf{r}) = \sum_i \frac{q_i}{|\mathbf{r}-\mathbf{b}_i|^2} \frac{\mathbf{r}-\mathbf{b}_i}{|\mathbf{r}-\mathbf{b}_i|}. \tag{9.1}$$

If the particle that is placed at the field point carries charge Q, that particle will experience a force $\mathbf{F}(\mathbf{r})$ due to the electric field $\mathbf{E}(\mathbf{r})$ produced by the other charge sources:

$$\mathbf{F}(\mathbf{r}) = Q\mathbf{E}(\mathbf{r}). \tag{9.2}$$

Force is a directly observable property that can be measured experimentally. In simulations, the evaluation of force is essential for determining the magnitude and direction of displacement of each particle when searching for the energy minimum, and for generating particle dynamics trajectories. These applications require an efficient evaluation of the electric field.

Determining the electric field, which is a vector quantity, by Equation 9.1 is not an easy task for a large system of particles, due to quadratic scaling of the computational cost in terms of number of particles. Fortunately, this problem can be reduced to finding the electrostatic potential, which can be computed in a linear scaling fashion. Electrostatic potential $\Phi(\mathbf{r})$ represents potential energy of a unit positive charge placed in the point $\mathbf{r}$ in space. Since electrostatic potential satisfies Laplace's equation, $\nabla^2\Phi=0$, the potential is a continuous function everywhere in space within the specified boundaries, except at the position of the charge source, where it approaches infinity. The potential has no local maxima or minima, and its extreme values are always located at the boundaries of the physical system. Being a scalar function, it assumes a single real value for any given coordinate in space.

9.1 Gradient of Electrostatic Potential

For two particles whose positions are described by radius-vectors $\mathbf{r}$ and $\mathbf{b}$ (Figure 9.1), and are carrying charges Q and q_b, respectively, the electrostatic potential $\Phi(\mathbf{r})$ at the point $\mathbf{r}$, is:

$$\Phi(\mathbf{r}) = \frac{q_b}{|\mathbf{r}-\mathbf{b}|}. \tag{9.3}$$

Additional valuable information about the electrostatic system comes with differentiation of Equation 9.3. In order to conduct differentiation, the dependence of electrostatic potential in Equation 9.3 on particle positions $\mathbf{r}$ and $\mathbf{b}$ has to be clarified first.

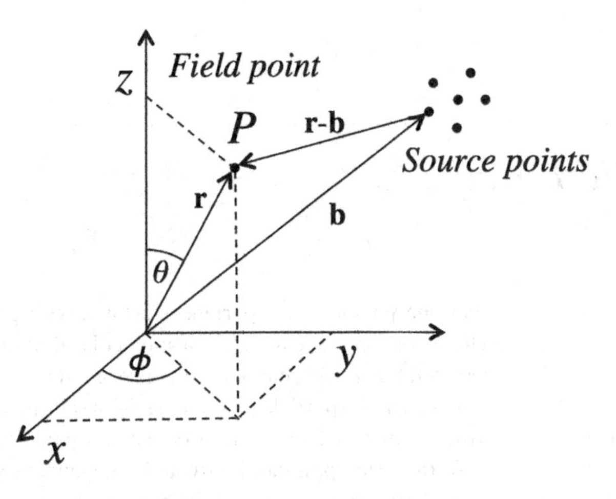

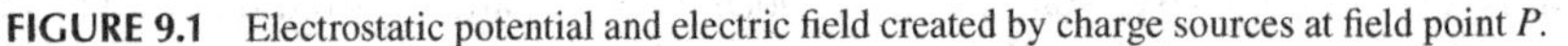

FIGURE 9.1 Electrostatic potential and electric field created by charge sources at field point *P*.

In a global coordinate system, as in the one shown in Figure 9.1, electrostatic potential at the point $\mathbf{r}$ depends on the position $\mathbf{b}$ of the charged particle. However, since electrostatics deals with a static charge distribution, the particle position $\mathbf{b} = b_x\,\hat{x} + b_y\,\hat{y} + b_z\,\hat{z}$ is a constant parameter, and the field point $\mathbf{r} = x\,\hat{x} + y\,\hat{y} + z\,\hat{z}$ is a variable of the function. This distinction helps finding a derivative of electrostatic potential function defined by Equation 9.3.

Numerical methods vividly explain the definition of derivative. In order to find a derivative in the position of particle $\mathbf{r}$, the particle $\mathbf{r}$ is moved by an infinitesimally small increment, Δ, and the potential, $\Phi(\mathbf{r})$ and $\Phi(\mathbf{r}+\Delta)$, in the initial, $\mathbf{r}$, and shifted, $\mathbf{r}+\Delta$, positions, respectively is measured. During this operation the particle (or particles, if there are more than one) at position $\mathbf{b}$ are kept fixed. This is reflected in the following equation representing the numerical derivative, $\Phi'(\mathbf{r})$:

$$\Phi'(\mathbf{r}) = \lim_{\Delta \to 0} \frac{\Phi(\mathbf{r}+\Delta) - \Phi(\mathbf{r})}{\Delta}. \tag{9.4}$$

For a small coordinate increment Δ, the derivative is the ratio of the difference in the function value over the coordinate increment.

Differentiation of electrostatic potential function leads to electric field, $\mathbf{E}(\mathbf{r})$, which is the negative gradient of the potential,

$$\mathbf{E}(\mathbf{r}) = -\nabla\Phi(\mathbf{r}). \tag{9.5}$$

If the field point $\mathbf{r}$ carries charge Q, multiplying the electric field by the magnitude of the charge gives force $\mathbf{F}(\mathbf{r})$ acting on that particle:

$$\mathbf{F}(\mathbf{r}) = Q\mathbf{E}(\mathbf{r}) = -Q\nabla\Phi = -Q\left(\frac{d}{dx}\hat{x} + \frac{d}{dy}\hat{y} + \frac{d}{dz}\hat{z}\right)\frac{q_b}{|\mathbf{r}-\mathbf{b}|}. \tag{9.6}$$

Here, the gradient ∇ is written in Cartesian coordinate form, and $\hat{x}$, $\hat{y}$, $\hat{z}$ are the unit Cartesian vectors. Using the Cartesian representation for vectors $\mathbf{r}$ and $\mathbf{b}$ in the form $\mathbf{r} = x\,\hat{x} + y\,\hat{y} + z\,\hat{z}$ and $\mathbf{b} = b_x\,\hat{x} + b_y\,\hat{y} + b_z\,\hat{z}$, the potential function in Equation 9.6 can be differentiated to produce an analytic expression for force $\mathbf{F}(\mathbf{r})$. For brevity, only one partial derivative over coordinate x is shown:

$$\frac{d}{dx}\hat{x}\left[(x-b_x)^2 + (y-b_y)^2 + (z-b_z)^2\right]^{-\frac{1}{2}} = \frac{-(x-b_x)\hat{x}}{\left[(x-b_x)^2 + (y-b_y)^2 + (z-b_z)^2\right]^{-\frac{3}{2}}}. \tag{9.7}$$

The other partial derivatives are obtained in a similar manner. Summing up the individual terms and substituting them into Equation 9.6 gives:

$$\mathbf{F}(\mathbf{r}) = Qq_b \frac{(x-b_x)\hat{x}+(y-b_y)\hat{y}+(z-b_z)\hat{z}}{\left[(x-b_x)^2+(y-b_y)^2+(z-b_z)^2\right]^{\frac{3}{2}}} = \frac{Qq_b}{|\mathbf{r}-\mathbf{b}|^2}\frac{\mathbf{r}-\mathbf{b}}{|\mathbf{r}-\mathbf{b}|}. \tag{9.8}$$

Due to the superposition principle, if there are multiple charge sources that contribute to the field in point **r**, the above equation turns into a sum:

$$\mathbf{F}(\mathbf{r}) = Q\sum_j \frac{q_j}{|\mathbf{r}-\mathbf{b}_j|^2}\frac{\mathbf{r}-\mathbf{b}_j}{|\mathbf{r}-\mathbf{b}_j|}. \tag{9.9}$$

Equation 9.9 represents Coulomb's law, a fundamental law of electrostatics, which provides the direct approach to the computation of forces and electric fields in a system of point charges. For a system of N particles, computing the force acting on each particle due to other $N-1$ particles requires $N(N-1)$ computations. Because this procedure scales as the square of the number of particles, the computational cost quickly becomes prohibitive. To avoid this bottleneck a more efficient ways of computing force and electric field is needed.

9.2 Differentiation of Multipole Expansion

The knowledge that electric field and force can be obtained by differentiation of the electrostatic potential, and that electrostatic potential can be computed with only a linear scaling computational cost, O(N), provides a guide on how to revise Equation 9.9 in order to avoid quadratic scaling. According to this, the electrostatic potential is formulated in a linear scaling form, and then the force and electric fields are obtained by differentiation of the electrostatic potential. Indeed, for a large set of charges, the potential $\Phi(\mathbf{r})$ in point **r** can be efficiently computed *via* multipole expansion,

$$\Phi(\mathbf{r}) = \sum_{l=0}^{\infty}\sum_{m=-l}^{l}\frac{(l-m)!}{(l+m)!}R^*_{lm}(\mathbf{r})L_{lm}(\mathbf{b}), \tag{9.10}$$

$$L_{lm}(\mathbf{b}) = \sum_i q_i S_{lm}(\mathbf{b}_i), \tag{9.11}$$

where $R^*_{l,m}(\mathbf{r})$ are regular solid harmonics, $L_{l,m}(\mathbf{b})$ are irregular multipole moments in terms of the irregular solid harmonics $S_{l,m}(\mathbf{b}_i)$, $\mathbf{b}_i$ is a particle coordinate vector, q_i is a particle charge, and l and m are polynomial degree and order, respectively. Note that Equation 9.10 requires $|\mathbf{r}| < |\mathbf{b}|$ and that both $R^*_{l,m}(\mathbf{r})$ and $L_{l,m}(\mathbf{b})$ are defined in spherical polar coordinates (r, θ, ϕ) about the same coordinate system. In subsequent discussions the cosine of the polar coordinate, $\mu = \cos\theta$ instead of angle θ, when referring to the angular degree of freedom in spherical polar coordinates will be used.

Having the analytic form of electrostatic potential function defined by Equation 9.10, it is necessary to differentiate it in order to obtain the electric field and force. In practice, in most applications, including molecular dynamics, particle coordinates are normally represented in Cartesian coordinates. This requires that the final derivatives be in Cartesian coordinates. Application of the chain rule accomplishes that objective, and leads to the following expression for the gradient of the potential:

$$\begin{aligned}\nabla\Phi = &\left(\frac{\partial\Phi}{\partial r}\frac{\partial r}{\partial x}+\frac{\partial\Phi}{\partial\mu}\frac{\partial\mu}{\partial x}+\frac{\partial\Phi}{\partial\phi}\frac{\partial\phi}{\partial x}\right)\hat{x}+\left(\frac{\partial\Phi}{\partial r}\frac{\partial r}{\partial y}+\frac{\partial\Phi}{\partial\mu}\frac{\partial\mu}{\partial y}+\frac{\partial\Phi}{\partial\phi}\frac{\partial\phi}{\partial y}\right)\hat{y}\\ &+\left(\frac{\partial\Phi}{\partial r}\frac{\partial r}{\partial z}+\frac{\partial\Phi}{\partial\mu}\frac{\partial\mu}{\partial z}+\frac{\partial\Phi}{\partial\phi}\frac{\partial\phi}{\partial z}\right)\hat{z}.\end{aligned} \tag{9.12}$$

Based on this, the Cartesian components of force acting on a particle at position **r** and carrying a charge Q become:

$$F_x = -Q\left(\frac{\partial \Phi}{\partial r}\frac{\partial r}{\partial x} + \frac{\partial \Phi}{\partial \mu}\frac{\partial \mu}{\partial x} + \frac{\partial \Phi}{\partial \phi}\frac{\partial \phi}{\partial x}\right),$$
$$F_y = -Q\left(\frac{\partial \Phi}{\partial r}\frac{\partial r}{\partial y} + \frac{\partial \Phi}{\partial \mu}\frac{\partial \mu}{\partial y} + \frac{\partial \Phi}{\partial \phi}\frac{\partial \phi}{\partial y}\right), \tag{9.13}$$
$$F_z = -Q\left(\frac{\partial \Phi}{\partial r}\frac{\partial r}{\partial z} + \frac{\partial \Phi}{\partial \mu}\frac{\partial \mu}{\partial z} + \frac{\partial \Phi}{\partial \phi}\frac{\partial \phi}{\partial z}\right).$$

This bulky equation can be simplified considerably. The first step in the differentiation is to find partial derivatives, $\partial\Phi/\partial r$, $\partial\Phi/\partial\mu$, $\partial\Phi/\partial\phi$, of the function given in the analytic form of Equation 9.10. Thus, it is helpful to rewrite Equations 9.10 and 9.5 in symbolic form:

$$\Phi(\mathbf{r}) = R^*(\mathbf{r})L(\mathbf{b}), \tag{9.14}$$

and

$$E(\mathbf{r}) = -\nabla\Phi(\mathbf{r}) = -\nabla[R^*(\mathbf{r})L(\mathbf{b})]. \tag{9.15}$$

Differentiation of the product in square brackets in Equation 9.14 follows the rule previously devised for Equation 9.5. According to it, the radius vector **r**, which defines the position of the observation point for electrostatic potential, serves as a differentiation coordinate. Likewise, the radius vector **b**, which expresses the coordinate of the charge source, is a constant parameter. This clarifies that the gradient in Equation 9.15 applies only to coordinate **r**. Correspondingly, coordinate **b** can be moved out of the differentiation clause. This simplifies Equation 9.15 to:

$$E(\mathbf{r}) = -\nabla\Phi(\mathbf{r}) = -\nabla[R^*(\mathbf{r})]L(\mathbf{b}). \tag{9.16}$$

This equation clarifies that using a multipole expansion to find the electric field at the point **r** requires only knowledge of the derivatives of the regular solid harmonics, $R^*_{lm}(\mathbf{r})$, and that no derivatives of the irregular solid harmonics $S_{lm}(\mathbf{b})$ appear in the equation.

With that, a more detailed equation for the calculation of force can be written:

$$F_x(\mathbf{r}) = -Q\sum_{l=0}^{\infty}\sum_{m=-l}^{l}\frac{(l-m)!}{(l+m)!}\left(\frac{\partial R^*_{l,m}(\mathbf{r})}{\partial r}\frac{\partial r}{\partial x} + \frac{\partial R^*_{l,m}(\mathbf{r})}{\partial \mu}\frac{\partial \mu}{\partial x} + \frac{\partial R^*_{l,m}(\mathbf{r})}{\partial \phi}\frac{\partial \phi}{\partial x}\right)L_{l,m}(\mathbf{b}),$$
$$F_y(\mathbf{r}) = -Q\sum_{l=0}^{\infty}\sum_{m=-l}^{l}\frac{(l-m)!}{(l+m)!}\left(\frac{\partial R^*_{l,m}(\mathbf{r})}{\partial r}\frac{\partial r}{\partial y} + \frac{\partial R^*_{l,m}(\mathbf{r})}{\partial \mu}\frac{\partial \mu}{\partial y} + \frac{\partial R^*_{l,m}(\mathbf{r})}{\partial \phi}\frac{\partial \phi}{\partial y}\right)L_{l,m}(\mathbf{b}), \tag{9.17}$$
$$F_z(\mathbf{r}) = -Q\sum_{l=0}^{\infty}\sum_{m=-l}^{l}\frac{(l-m)!}{(l+m)!}\left(\frac{\partial R^*_{l,m}(\mathbf{r})}{\partial r}\frac{\partial r}{\partial z} + \frac{\partial R^*_{l,m}(\mathbf{r})}{\partial \mu}\frac{\partial \mu}{\partial z} + \frac{\partial R^*_{l,m}(\mathbf{r})}{\partial \phi}\frac{\partial \phi}{\partial z}\right)L_{l,m}(\mathbf{b}).$$

At this point the individual terms appearing in Equation 9.17 can be worked out in the preparation of the assembly of the final result.

9.3 Differentiation of Regular Solid Harmonics in Spherical Polar Coordinates

In order to solve Equation 9.17, twelve different partial derivatives are needed. Three of these are for regular solid harmonics, and other nine are for spherical polar coordinates. The first step on that path will be to find partial derivatives of regular solid harmonics $R^*_{lm}(\mathbf{r})$ over spherical polar coordinates.

Regular solid harmonics are defined in spherical polar coordinates as:

$$R_{lm}^{*}(\mathbf{r}) = r^{l} P_{l}^{m}(\mu) e^{-im\phi}, \tag{9.18}$$

where r^{l} is a radial component, $P_{l}^{m}(\mu)$ is an associated Legendre function, $\mu = \cos\theta$ is the polar coordinate, and $e^{-im\phi}$ is the azimuthal component. As previously introduced, l is the polynomial degree, and m is the polynomial order. To keep the analysis compact, the associated Legendre function will be referred to simply as P_{l}^{m}. Similarly to that, all reference to radius-vector $\mathbf{r}$ from regular solid harmonics $R_{lm}^{*}(\mathbf{r})$ will be dropped, to reduce the unnecessary clutter in the equations. The dependence of R_{lm}^{*} on $\mathbf{r}$ is obvious, and will remain the same until the end of this chapter.

The partial derivative of R_{lm}^{*} over coordinate r is obtained by differentiating the radial term, r^{l}:

$$\frac{\partial}{\partial r} R_{lm}^{*} = l r^{l-1} P_{l}^{m} e^{-im\phi} = \frac{l}{r} R_{lm}^{*}. \tag{9.19}$$

Next to find is the derivative over polar coordinate. Since the final goal is to convert the derivative into Cartesian coordinates, performing differentiation directly over the polar coordinate θ can be avoided. Instead, finding the partial derivative over $\mu = \cos\theta$ is more convenient and less involved. Differentiation of associated Legendre function over argument μ leads to:

$$\frac{d}{d\mu} P_{l}^{m} = \frac{(l+m) P_{l-1}^{m} - l \mu P_{l}^{m}}{1-\mu^{2}}. \tag{9.20}$$

The detailed derivation of Equation 9.20 is given in Chapter 2. Based on Equations 9.18 and 9.20, the derivative of regular solid harmonics over μ is

$$\frac{d}{d\mu} R_{lm}^{*} = r^{l} \frac{d}{d\mu} P_{l}^{m} e^{-im\phi} = \frac{(l+m) r^{l} P_{l-1}^{m} e^{-im\phi} - l \mu r^{l} P_{l}^{m} e^{-im\phi}}{1-\mu^{2}} = \frac{(l+m) r R_{l-1,m}^{*} - l \mu R_{lm}^{*}}{1-\mu^{2}}. \tag{9.21}$$

In this equation, R_{lm}^{*} is first disassembled into its components based on Equation 9.18; differentiation is performed on the associated Legendre function; and the resulting expression is returned back to regular solid harmonics form by application of Equation 9.18 once again.

The remaining third partial derivative of regular solid harmonics over azimuthal angle ϕ is obtained by differentiation of the exponential term:

$$\frac{\partial}{\partial \phi} R_{lm}^{*} = r^{l} P_{l}^{m} \frac{\partial}{\partial \phi} e^{-im\phi} = -im\, r^{l} P_{l}^{m} e^{-im\phi} = -im\, R_{lm}^{*}. \tag{9.22}$$

This completes the derivation of partial derivatives of regular solid harmonics over spherical polar coordinates. To summarize this section, here are the partial derivatives of regular solid harmonics over spherical polar coordinates:

$$\begin{aligned} \frac{\partial}{\partial r} R_{lm}^{*} &= \frac{l}{r} R_{lm}^{*}, \\ \frac{\partial}{\partial \mu} R_{lm}^{*} &= \frac{(l+m) r R_{l-1,m}^{*} - l \mu R_{lm}^{*}}{1-\mu^{2}}, \\ \frac{\partial}{\partial \phi} R_{lm}^{*} &= -im\, R_{lm}^{*}. \end{aligned} \tag{9.23}$$

9.4 Differentiation of Spherical Polar Coordinates

Differentiation of spherical polar coordinates over Cartesian coordinates relies on the basic algebraic relationship that exists between Cartesian and spherical polar coordinates:

$$\begin{aligned} x &= r \sin\theta \cos\phi, \\ y &= r \sin\theta \sin\phi, \\ z &= r \cos\theta. \end{aligned} \tag{9.24}$$

The assignment of spherical polar coordinates is illustrated in Figure 9.1. Using Equation 9.24, a number of useful formulae can be generated:

$$\begin{aligned} r^2 &= x^2 + y^2 + z^2, \\ x^2 + y^2 &= r^2 \sin^2\theta, \\ \mu &= \cos\theta = \frac{z}{r}, \\ \frac{y}{x} &= \tan\phi. \end{aligned} \tag{9.25}$$

The differentiation of spherical polar coordinates (r,μ,ϕ) is straightforward and is independent of any other terms. The partial derivative of the radial component, r, over Cartesian coordinate x is obtained by expressing the distance r in Cartesian coordinates:

$$\frac{\partial r}{\partial x} = \frac{\partial}{\partial x}(x^2+y^2+z^2)^{\frac{1}{2}} = \frac{1}{2}2x(x^2+y^2+z^2)^{-\frac{1}{2}} = \frac{x}{\sqrt{x^2+y^2+z^2}} = \frac{x}{r}. \tag{9.26}$$

The derivatives of r over the other Cartesian coordinates y and z are similarly obtained due to symmetry in dependence of r on coordinates x, y, z. Thus, the partial derivatives of radial component r over Cartesian coordinates are:

$$\frac{\partial r}{\partial x} = \frac{x}{r}, \qquad \frac{\partial r}{\partial y} = \frac{y}{r}, \qquad \frac{\partial r}{\partial z} = \frac{z}{r}. \tag{9.27}$$

Next, $\mu = \cos\theta = z/r$ is differentiated. The partial derivative over coordinate x is

$$\frac{\partial \mu}{\partial x} = \frac{\partial}{\partial x}\cos\theta = \frac{\partial}{\partial x}\left(\frac{z}{r}\right) = z\frac{\partial}{\partial x}(x^2+y^2+z^2)^{-\frac{1}{2}} = -\frac{1}{2}2xz(x^2+y^2+z^2)^{-\frac{3}{2}} = -\frac{xz}{r^3}. \tag{9.28}$$

Since the angular coordinate is symmetric over coordinates x and y, it is sufficient to replace x by y in Equation 9.28 to obtain the partial derivative over coordinate y.

$$\frac{\partial \mu}{\partial y} = -\frac{yz}{r^3}. \tag{9.29}$$

The partial derivative of $\mu = z/r$ over coordinate z is obtained by differentiating it in the form of the product of two functions, z and $(x^2+y^2+z^2)^{-(1/2)}$.

$$\frac{\partial \mu}{\partial z} = \frac{\partial}{\partial z}\left(\frac{z}{r}\right) = \frac{\partial}{\partial z}z(x^2+y^2+z^2)^{-\frac{1}{2}} = (x^2+y^2+z^2)^{-\frac{1}{2}} - \frac{z^2}{r^3} = \frac{r^2-z^2}{r^3} = \frac{x^2+y^2}{r^3}. \tag{9.30}$$

The next step is to find partial derivatives of the azimuthal angle ϕ. Based on Equation 9.25, the azimuthal angle is expressed in the following form:

$$\begin{aligned} \phi &= \arctan\left(\frac{y}{x}\right) && \text{if } x > 0, \\ \phi &= \arctan\left(\frac{y}{x}\right) + \pi && \text{if } x < 0 \text{ and } y \geq 0, \\ \phi &= \arctan\left(\frac{y}{x}\right) - \pi && \text{if } x < 0 \text{ and } y < 0. \end{aligned} \tag{9.31}$$

Differentiating this function over coordinate x gives:

$$\frac{d}{dx}\phi = \frac{d}{dx}\left[\arctan\frac{y}{x}\right]. \tag{9.32}$$

Before going any further, the derivative of the arctan function is needed; this can be found in any mathematics handbook, but for convenience will be quickly re-derived here. The inverse of $f = \arctan(x)$ is $x = \tan(f)$, which can be easily differentiated using the well-known differentiation rules for the trigonometric functions sin and cos, *vis*:

$$dx = d\left[\tan(f)\right] = d\left[\frac{\sin(f)}{\cos(f)}\right] = \frac{\cos^2(f) + \sin^2(f)}{\cos^2(f)}\,df = \left[1 + \tan^2(f)\right]df = (1 + x^2)\,df. \tag{9.33}$$

Rearranging the terms in left- and right-hand sides leads to:

$$\frac{df}{dx} = \frac{1}{1 + x^2}. \tag{9.34}$$

Since $f = \arctan(x)$, the formula for differentiation becomes:

$$\frac{d}{dx}\arctan(x) = \frac{1}{1 + x^2}. \tag{9.35}$$

Returning to Equation 9.32 and resuming the work on finding the partial derivative over coordinate x, the chain rule can now be applied, yielding:

$$\frac{\partial\phi}{\partial x} = \frac{\partial\left[\arctan\left(\frac{y}{x}\right)\right]}{\partial\left(\frac{y}{x}\right)}\frac{\partial\left(\frac{y}{x}\right)}{\partial x} = \frac{1}{1 + \left(\frac{y^2}{x^2}\right)}\left(-\frac{y}{x^2}\right) = -\frac{x^2}{x^2 + y^2}\frac{y}{x^2} = -\frac{y}{x^2 + y^2}. \tag{9.36}$$

The partial derivative of azimuthal angle ϕ over Cartesian coordinate y is similarly obtained.

$$\frac{\partial\phi}{\partial y} = \frac{\partial\left[\arctan\left(\frac{y}{x}\right)\right]}{\partial\left(\frac{y}{x}\right)}\frac{\partial\left(\frac{y}{x}\right)}{\partial y} = \frac{1}{1 + \left(\frac{y^2}{x^2}\right)}\left(\frac{1}{x}\right) = \frac{x^2}{x^2 + y^2}\frac{1}{x} = \frac{x}{x^2 + y^2}. \tag{9.37}$$

The last partial derivative of the azimuthal angle over the z coordinate is the easiest to find. Since the angle ϕ does not depend on the z coordinate, the corresponding partial derivative is obviously zero.

$$\frac{\partial\phi}{\partial z} = 0. \tag{9.38}$$

Now the above-derived partial derivatives can be grouped together.

$$\begin{aligned} &\frac{\partial r}{\partial x}=\frac{x}{r}, && \frac{\partial r}{\partial y}=\frac{y}{r}, && \frac{\partial r}{\partial z}=\frac{z}{r}, \\ &\frac{\partial \mu}{\partial x}=-\frac{xz}{r^3}, && \frac{\partial \mu}{\partial y}=-\frac{yz}{r^3}, && \frac{\partial \mu}{\partial z}=\frac{x^2+y^2}{r^3}, \\ &\frac{\partial \phi}{\partial x}=-\frac{y}{x^2+y^2}, && \frac{\partial \phi}{\partial y}=\frac{x}{x^2+y^2}, && \frac{\partial \phi}{\partial z}=0. \end{aligned} \tag{9.39}$$

Equation 9.39 provides the means to convert the partial derivatives of a function taken in spherical polar coordinates into partial derivatives over Cartesian coordinates.

9.5 Differentiation of Regular Solid Harmonics in Cartesian Coordinates

Using the chain rule, the partial derivatives of regular solid harmonics over spherical polar coordinates can be combined with the partial derivatives of spherical polar coordinates over Cartesian coordinates to form partial derivatives of R^*_{lm} in Cartesian coordinates.

$$\begin{aligned} \frac{\partial}{\partial x}R^*_{l,m} &= \frac{\partial R^*_{l,m}}{\partial r}\frac{\partial r}{\partial x}+\frac{\partial R^*_{l,m}}{\partial \mu}\frac{\partial \mu}{\partial x}+\frac{\partial R^*_{l,m}}{\partial \phi}\frac{\partial \phi}{\partial x}, \\ \frac{\partial}{\partial y}R^*_{l,m} &= \frac{\partial R^*_{l,m}}{\partial r}\frac{\partial r}{\partial y}+\frac{\partial R^*_{l,m}}{\partial \mu}\frac{\partial \mu}{\partial y}+\frac{\partial R^*_{l,m}}{\partial \phi}\frac{\partial \phi}{\partial y}, \\ \frac{\partial}{\partial z}R^*_{l,m} &= \frac{\partial R^*_{l,m}}{\partial r}\frac{\partial r}{\partial z}+\frac{\partial R^*_{l,m}}{\partial \mu}\frac{\partial \mu}{\partial z}+\frac{\partial R^*_{l,m}}{\partial \phi}\frac{\partial \phi}{\partial z}. \end{aligned} \tag{9.40}$$

Inserting Equations 9.23 and 9.39 into Equation 9.40 produces Cartesian derivatives. The partial derivative over coordinate x then becomes:

$$\frac{\partial}{\partial x}R^*_{lm}=\frac{l}{r}R^*_{lm}\frac{x}{r}-\frac{(l+m)rR^*_{l-1,m}-l\mu R^*_{lm}}{1-\mu^2}\frac{xz}{r^3}+im\,R^*_{lm}\frac{y}{x^2+y^2}. \tag{9.41}$$

Separating terms with regular solid harmonics leads to:

$$\frac{\partial}{\partial x}R^*_{lm}=\frac{lx}{r^2}R^*_{lm}-\frac{(l+m)xz}{r^2(1-\mu^2)}R^*_{l-1,m}+\frac{l\mu xz}{r^3(1-\mu^2)}R^*_{lm}+\frac{imy}{x^2+y^2}R^*_{lm}. \tag{9.42}$$

Taking into account that $\mu=\cos\theta=z/r$ and $r^2(1-\mu^2)=r^2\sin^2\theta=x^2+y^2$, eliminating the trigonometric functions gives:

$$\frac{\partial}{\partial x}R^*_{lm}=\frac{lx}{r^2}R^*_{lm}-\frac{(l+m)xz}{x^2+y^2}R^*_{l-1,m}+\frac{lxz^2}{r^2(x^2+y^2)}R^*_{lm}+\frac{imy}{x^2+y^2}R^*_{lm}. \tag{9.43}$$

Next, grouping like terms together leads to:

$$\frac{\partial}{\partial x}R^*_{lm}=\left[\frac{lx}{r^2}+\frac{lxz^2}{r^2(x^2+y^2)}+\frac{imy}{x^2+y^2}\right]R^*_{lm}-\frac{(l+m)xz}{x^2+y^2}R^*_{l-1,m}. \tag{9.44}$$

The first two terms in the square brackets in Equation 9.44 can be combined:

$$\frac{\partial}{\partial x}R^*_{lm}=\left[\frac{lxx^2+lxy^2+lxz^2}{r^2(x^2+y^2)}+\frac{imy}{x^2+y^2}\right]R^*_{lm}-\frac{(l+m)xz}{x^2+y^2}R^*_{l-1,m}. \tag{9.45}$$

Canceling out $x^2+y^2+z^2$ in the denominator and r^2 in the numerator, and combining the parts in square brackets leads to the partial derivative of regular solid harmonics over coordinate x:

$$\frac{\partial}{\partial x}R^*_{lm} = \frac{lx+imy}{x^2+y^2}R^*_{lm} - \frac{(l+m)xz}{x^2+y^2}R^*_{l-1,m}. \tag{9.46}$$

The partial derivative of regular solid harmonics over coordinate y is obtained in the same manner. Substitution of Equations 9.23 and 9.39 into Equation 9.40 yields:

$$\frac{\partial}{\partial y}R^*_{lm} = \frac{l}{r}R^*_{lm}\frac{y}{r} - \frac{(l+m)rR^*_{l-1,m} - l\mu R^*_{lm}}{1-\mu^2}\frac{yz}{r^3} - imR^*_{lm}\frac{x}{x^2+y^2}. \tag{9.47}$$

Simplifying and segregating terms with regular solid harmonics gives:

$$\frac{\partial}{\partial y}R^*_{lm} = \frac{ly}{r^2}R^*_{lm} - \frac{(l+m)yz}{r^2(1-\mu^2)}R^*_{l-1,m} + \frac{l\mu yz}{r^3(1-\mu^2)}R^*_{lm} - \frac{imx}{x^2+y^2}R^*_{lm}. \tag{9.48}$$

Next, trigonometric functions in the multipliers in front of R^* are eliminated using the substitutions $\mu=\cos\theta=z/r$ and $r^2(1-\mu^2)=x^2+y^2$ to give:

$$\frac{\partial}{\partial y}R^*_{lm} = \frac{ly}{r^2}R^*_{lm} - \frac{(l+m)yz}{x^2+y^2}R^*_{l-1,m} + \frac{lyz^2}{r^2(x^2+y^2)}R^*_{lm} - \frac{imx}{x^2+y^2}R^*_{lm}. \tag{9.49}$$

Combining like terms leads to:

$$\frac{\partial}{\partial y}R^*_{lm} = \left[\frac{ly}{r^2} + \frac{lyz^2}{r^2(x^2+y^2)} - \frac{imx}{x^2+y^2}\right]R^*_{lm} - \frac{(l+m)yz}{x^2+y^2}R^*_{l-1,m}. \tag{9.50}$$

The first two terms in square brackets in Equation 9.50 can be added together:

$$\frac{\partial}{\partial y}R^*_{lm} = \left[\frac{lyx^2+lyy^2+lyz^2}{r^2(x^2+y^2)} - \frac{imx}{x^2+y^2}\right]R^*_{lm} - \frac{(l+m)yz}{x^2+y^2}R^*_{l-1,m}. \tag{9.51}$$

Canceling out $r^2=x^2+y^2+z^2$, and combining the parts in the square brackets leads to the final expression for the partial derivative of regular solid harmonics over coordinate y:

$$\frac{\partial}{\partial y}R^*_{l,m} = \frac{ly-imx}{x^2+y^2}R^*_{l,m} - \frac{(l+m)yz}{x^2+y^2}R^*_{l-1,m}. \tag{9.52}$$

The derivative of regular solid harmonics over coordinate z is obtained by substituting Equations 9.23 and 9.39 into Equation 9.40, and selecting the z component:

$$\frac{\partial}{\partial z}R^*_{lm} = \frac{l}{r}R^*_{lm}\frac{z}{r} + \frac{(l+m)rR^*_{l-1,m} - l\mu R^*_{lm}}{1-\mu^2}\frac{x^2+y^2}{r^3}. \tag{9.53}$$

Because $\partial\phi/\partial z=0$, the derivative of the azimuthal angle does not contribute to $(\partial/\partial z)R^*_{lm}$, so there are only two terms in the right-hand side of Equation 9.53. Segregating terms with solid harmonics leads to:

$$\frac{\partial}{\partial z}R^*_{lm} = \frac{lz}{r^2}R^*_{lm} + \frac{(l+m)(x^2+y^2)}{r^2(1-\mu^2)}R^*_{l-1,m} - \frac{l\mu(x^2+y^2)}{r^3(1-\mu^2)}R^*_{lm}. \tag{9.54}$$

Using the relation $r^2(1-\mu^2)=x^2+y^2$ and canceling out common terms, produces:

$$\frac{\partial}{\partial z}R^*_{lm}=\frac{lz}{r^2}R^*_{lm}+(l+m)R^*_{l-1,m}-\frac{l\mu}{r}R^*_{lm}. \tag{9.55}$$

Substituting $\mu=\cos\theta=z/r$ and canceling out equal terms gives the final expression for the partial derivative of regular solid harmonics over coordinate z:

$$\frac{\partial}{\partial z}R^*_{lm}=(l+m)R^*_{l-1,m}. \tag{9.56}$$

To summarize this analysis, the three derivatives of regular solid harmonics over Cartesian coordinates are:

$$\begin{aligned}\frac{\partial}{\partial x}R^*_{lm}&=\frac{lx+imy}{x^2+y^2}R^*_{lm}-\frac{(l+m)xz}{x^2+y^2}R^*_{l-1,m},\\ \frac{\partial}{\partial y}R^*_{lm}&=\frac{ly-imx}{x^2+y^2}R^*_{lm}-\frac{(l+m)yz}{x^2+y^2}R^*_{l-1,m},\\ \frac{\partial}{\partial z}R^*_{lm}&=(l+m)R^*_{l-1,m}.\end{aligned} \tag{9.57}$$

As demonstrated above in Equation 9.16, no derivatives of irregular solid harmonics are needed for the Fast Multipole Method. Therefore their derivation is unnecessary.

9.6 FMM Force in Cartesian Coordinates

Substituting Equation 9.40 into Equation 9.17 makes it possible to compactly rewrite the latter in the form of partial derivatives of regular solid harmonics over Cartesian coordinates:

$$\begin{aligned}F_x(\mathbf{r})&=-Q\sum_{l=0}^{\infty}\sum_{m=-l}^{l}\frac{(l-m)!}{(l+m)!}\left(\frac{\partial R^*_{lm}(\mathbf{r})}{\partial x}\right)L_{lm}(\mathbf{b}),\\ F_y(\mathbf{r})&=-Q\sum_{l=0}^{\infty}\sum_{m=-l}^{l}\frac{(l-m)!}{(l+m)!}\left(\frac{\partial R^*_{lm}(\mathbf{r})}{\partial y}\right)L_{lm}(\mathbf{b}),\\ F_z(\mathbf{r})&=-Q\sum_{l=0}^{\infty}\sum_{m=-l}^{l}\frac{(l-m)!}{(l+m)!}\left(\frac{\partial R^*_{lm}(\mathbf{r})}{\partial z}\right)L_{lm}(\mathbf{b}).\end{aligned} \tag{9.58}$$

Substituting Equation 9.57 into Equation 9.58, assembles the final equation for calculation of Cartesian components of force, which then becomes:

$$\begin{aligned}F_x(\mathbf{r})&=-Q\sum_{l=0}^{\infty}\sum_{m=-l}^{l}\frac{(l-m)!}{(l+m)!}\left[\frac{lx+imy}{x^2+y^2}R^*_{lm}(\mathbf{r})-\frac{(l+m)xz}{x^2+y^2}R^*_{l-1,m}(\mathbf{r})\right]L_{lm}(\mathbf{b}),\\ F_y(\mathbf{r})&=-Q\sum_{l=0}^{\infty}\sum_{m=-l}^{l}\frac{(l-m)!}{(l+m)!}\left[\frac{ly-imx}{x^2+y^2}R^*_{lm}(\mathbf{r})-\frac{(l+m)yz}{x^2+y^2}R^*_{l-1,m}(\mathbf{r})\right]L_{lm}(\mathbf{b}),\\ F_z(\mathbf{r})&=-Q\sum_{l=0}^{\infty}\sum_{m=-l}^{l}\frac{(l-m)!}{(l+m)!}(l+m)R^*_{l-1,m}(\mathbf{r})L_{lm}(\mathbf{b}).\end{aligned} \tag{9.59}$$

All that is necessary to compute the electric field $\mathbf{E}(\mathbf{r})$ instead of force $\mathbf{F}(\mathbf{r})$ by Equation 9.59 is to set the charge Q to unity. Discussion of calculation of force continues in Chapter 10 after an introduction of the scaled form of solid harmonics.

10

Scaling of Solid Harmonics

This chapter deals with optimization of the previously-derived mathematical equations used in the series expansion of the potential function for the purpose of their efficient computer implementation.

10.1 Optimization of Expansion of Inverse Distance Function

A series expansion of the inverse distance function representing electrostatic potential of a unit positive charge consists of a product of spherical harmonics:

$$\frac{1}{|\mathbf{r}|}=\sum_{l=0}^{\infty}\sum_{m=-l}^{l}\frac{a^{l}}{b^{l+1}}\frac{4\pi}{2l+1}Y_{lm}^{*}(A)Y_{lm}(B), \tag{10.1}$$

where $b > a$, and $Y_{lm}(A)$ and $Y_{lm}(B)$ are spherical harmonics of individual particles A and B, respectively. Figure 10.1 shows the relative orientation of the charge source at point A and that of the field point B, where $\boldsymbol{a}$ and $\boldsymbol{b}$ are their respective radius vectors. The present goal is to see what needs to be done to Equation 10.1 in order to ready it for computer implementation.

Although Equation 10.1 looks very simple, making the transition from the mathematically elegant form to one optimal for computer implementation requires a meticulous analysis of the equation.

Out of the available possibilities for optimization, the most straightforward one is to make use of the hidden symmetries in Equation 10.1. The symmetry of spherical harmonics in index m provides the opportunity to reduce the number of cycles in the loop that otherwise runs from $m = -l$ to $m = l$. Following that path, the sum over index m breaks down into two independent sums over negative and positive ranges of index m:

$$\sum_{m=-l}^{l}Y_{lm}^{*}(A)Y_{lm}(B)=Y_{00}(A)Y_{00}(B)+\sum_{m=1}^{l}Y_{lm}^{*}(A)Y_{lm}(B)+\sum_{m=-1}^{-l}Y_{lm}^{*}(A)Y_{lm}(B). \tag{10.2}$$

It is convenient to transform the right-most sum over negative values of index m that runs from –1 to $-l$ in Equation 10.2 into the sum over positive values of index m by explicitly moving the negative sign into the lower index of spherical harmonics. This leads to:

$$\sum_{m=-l}^{l}Y_{lm}^{*}(A)Y_{lm}(B)=Y_{00}(A)Y_{00}(B)+\sum_{m=1}^{l}Y_{lm}^{*}(A)Y_{lm}(B)+\sum_{m=1}^{l}Y_{l,-m}^{*}(A)Y_{l,-m}(B), \tag{10.3}$$

which is fully equivalent to Equation 10.2. The underlying symmetry is now expressed explicitly in Equation 10.3 in a form that allows the expansion to be simplified.

A symmetry rule of spherical harmonics states that:

$$Y_{l,-m}=(-1)^{m}Y_{l,m}^{*}. \tag{10.4}$$

Taking the complex conjugate of both sides of Equation 10.4 gives:

$$Y_{l,-m}^{*}=(-1)^{m}Y_{l,m}. \tag{10.5}$$

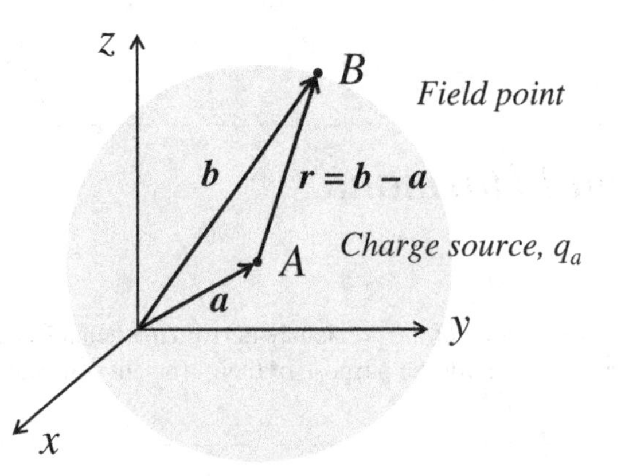

FIGURE 10.1 Electrostatic potential at point B due to a charge q_a placed at point A. Radius vectors $\boldsymbol{a}$ and $\boldsymbol{b}$ connect the center of the coordinate system with the points A and B, respectively. Radius vector $\boldsymbol{r}$ extends from the charge source toward the field point.

Substituting Equations 10.4 and 10.5 into Equation 10.3 and taking advantage of $(-1)^{2m} = 1$ eliminates the use of a negative sign in index m from that equation. Once that is done, combining the two sums on the right-hand side of Equation 10.3 into a single sum leads to:

$$\sum_{m=-l}^{l} Y_{lm}^{*}(A)Y_{lm}(B) = Y_{00}(A)Y_{00}(B) + \sum_{m=1}^{l}[Y_{lm}^{*}(A)Y_{lm}(B) + Y_{lm}(A)Y_{lm}^{*}(B)]. \tag{10.6}$$

By eliminating spherical harmonics with a negative value of index m, Equation 10.6 reduces the memory storage requirement by half. In addition, evaluation of the terms Y_{lm}^{*} may be performed on the fly from the already existing terms Y_{lm} without the need to store any intermediate complex conjugate quantities in memory.

Further progress toward improving the efficiency of evaluation of the series expansion of the inverse distance function requires a reduction in the number of floating-point operations over that of the naïve implementation. Complex conjugate sums of the type $Y_{lm}^{*}(A)Y_{lm}(B) + Y_{lm}(A)Y_{lm}^{*}(B)$ have a useful property in that the imaginary part vanishes and all that is left is the real part. Using a generic complex number $(x + iy)$ instead of Y_{lm} demonstrates how that property works:

$$\begin{aligned}
&(x_a + iy_a)^{*}(x_b + iy_b) + (x_a + iy_a)(x_b + iy_b)^{*} \\
&\quad = (x_a - iy_a)(x_b + iy_b) + (x_a + iy_a)(x_b - iy_b) \\
&\quad = x_a x_b + i x_a y_b - i y_a x_b + y_a y_b + x_a x_b - i x_a y_b + i y_a x_b + y_a y_b \\
&\quad = 2x_a x_b + 2y_a y_b.
\end{aligned} \tag{10.7}$$

Based on this property, the complex conjugate sum of a product of two complex numbers is numerically equivalent to twice the single product followed by discarding the imaginary part. This operation is denoted by a function Re():

$$(x_a + iy_a)^{*}(x_b + iy_b) + (x_a + iy_a)(x_b + iy_b)^{*} \equiv 2\mathrm{Re}\left[(x_a + iy_a)^{*}(x_b + iy_b)\right]. \tag{10.8}$$

This property of complex numbers eliminates half of the floating-point operations needed earlier. As a final remark, either of the summands from the left–hand–side of Equation 10.8 may be used as the argument to function Re() since both those products generate the same real component.

The ability of a complex conjugate sum to produce a real number simplifies Equation 10.6 to:

$$\sum_{m=-l}^{l} Y_{lm}^{*}(A)Y_{lm}(B) = Y_{00}(A)Y_{00}(B) + 2\sum_{m=1}^{l}\mathrm{Re}\left[Y_{lm}^{*}(A)Y_{lm}(B)\right]. \tag{10.9}$$

In this form, Equation 10.9 requires only half the floating-point operations of Equation 10.3. Based on this simplification, the series expansion of an inverse distance can be expressed as:

$$\frac{1}{|\mathbf{r}|}=\sum_{l=0}^{\infty}\frac{a^l}{b^{l+1}}\frac{4\pi}{2l+1}\left\{Y_{00}(A)Y_{00}(B)+2\sum_{m=1}^{l}\mathrm{Re}\left[Y_{lm}^{*}(A)Y_{lm}(B)\right]\right\}. \tag{10.10}$$

A continued analysis of the spherical harmonics functions allows an additional simplification of Equation 10.10 to improve its performance. Spherical harmonics break down to associated Legendre functions, P_l^m:

$$Y_{lm}(\theta,\phi)=\sqrt{\frac{2l+1}{4\pi}\frac{(l-m)!}{(l+m)!}}P_l^m(\theta)e^{im\phi}, \tag{10.11}$$

where θ and ϕ are polar and azimuthal angles of the particle radius vector. After noting that the associated Legendre function P_l^m reduces to the Legendre polynomial P_l when $m = 0$, substituting Equation 10.11 into Equation 10.10 leads to:

$$\frac{1}{|\mathbf{r}|}=\sum_{l=0}^{\infty}\frac{a^l}{b^{l+1}}\left\{P_l(\theta_a)P_l(\theta_b)+2\frac{(l-m)!}{(l+m)!}\sum_{m=1}^{l}\mathrm{Re}\left[P_l^m(\theta_a)e^{-im\phi_a}P_l^m(\theta_b)e^{im\phi_b}\right]\right\}, \tag{10.12}$$

where θ_a, ϕ_a and θ_b, ϕ_b are the polar and azimuthal angles of particles A and B, respectively. This equation shows that the normalization coefficient $(2l + 1)/4\pi$ originally used in calculating spherical harmonics is unnecessary, since it cancels out in the end.

An infinite summation over index l in Equation 10.12 is another place that requires attention since it is impractical for computer implementation. Fortunately, because $b > a$, the series quickly converges. Typically, when working with double precision, limiting the summation to $nL = 20 - 40$ provides sufficient numerical accuracy for the majority of molecular modeling applications. This leads to:

$$\frac{1}{|\mathbf{r}|}=\sum_{l=0}^{nL}\frac{a^l}{b^{l+1}}\left\{P_l(\theta_a)P_l(\theta_b)+2\frac{(l-m)!}{(l+m)!}\sum_{m=1}^{l}\mathrm{Re}\left[P_l^m(\theta_a)e^{-im\phi_a}P_l^m(\theta_b)e^{im\phi_b}\right]\right\}. \tag{10.13}$$

The derived equation completes the first round of optimization of the series expansion of inverse distance for numerical computation. Further fine-tuning steps will now be introduced in the following section.

10.2 Scaling of Associated Legendre Functions

Another improvement in the efficiency of calculation of Equation 10.13 can be made by incorporating the coefficient $((l-m)!)/(l+m)!)$, which appears in Equation 10.13 due to the orthonormality condition of spherical harmonics, into the associated Legendre functions, as suggested by White and Head-Gordon, such that

$$\bar{P}_l^m=\frac{1}{(l+m)!}P_l^m, \quad \tilde{P}_l^m=(l-m)!P_l^m. \tag{10.14}$$

The resulting scaled associated Legendre functions become identified as the bar- and tilde-scaled functions, respectively.

The benefit of scaling can immediately be seen in the symmetry relation, which in its conventional form states that:

$$P_l^{-m}=(-1)^m\frac{(l-m)!}{(l+m)!}P_l^m. \tag{2.49}$$

Replacing index m with $-m$ in Equation 10.14 for the bar-scaled form of associated Legendre function, and then substituting P_l^{-m} by its symmetric counterpart from Equation 2.49 leads to:

$$\bar{P}_l^{-m} = \frac{1}{(l-m)!} P_l^{-m} = \frac{1}{(l-m)!}(-1)^m \frac{(l-m)!}{(l+m)!} P_l^m = (-1)^m \frac{1}{(l+m)!} P_l^m = (-1)^m \bar{P}_l^m. \tag{10.15}$$

That is, switching to the bar-scaled form makes the inconvenient factorials vanish.

A similar symmetry relationship exists between pairs of tilde-scaled associated Legendre functions. Replacing index m with $-m$ in Equation 10.14 for the tilde-scaled function, and then substituting P_l^{-m} by its symmetric counterpart from Equation 2.49, leads to:

$$\tilde{P}_l^{-m} = (l+m)! P_l^{-m} = (l+m)!(-1)^m \frac{(l-m)!}{(l+m)!} P_l^m = (-1)^m (l-m)! P_l^m = (-1)^m \tilde{P}_l^m. \tag{10.16}$$

This result shows that the tilde-scaled associated Legendre functions have the same symmetry behavior as that of the bar-scaled functions: that is, the scaled functions with opposite signs of index m are equal in absolute value and are inverted in sign for odd values of m.

Regardless of these interesting symmetry relationships of associated Legendre functions, the actual purpose of scaling is to simplify manipulations with regular and irregular solid harmonics,

$$R_{lm} = a^l P_l^m e^{im\phi}, \quad S_{lm} = \frac{1}{b^{l+1}} P_l^m e^{im\phi}, \tag{10.17}$$

which appear in the series expansion of inverse distance function:

$$\begin{aligned} \frac{1}{|\mathbf{r}|} &= \sum_{l=0}^{\infty} \sum_{m=-l}^{l} \frac{(l-m)!}{(l+m)!} a^l P_l^m(\theta_a) e^{-im\phi_a} \frac{1}{b^{l+1}} P_l^m(\theta_b) e^{-im\phi_b} \\ &= \sum_{l=0}^{\infty} \sum_{m=-l}^{l} \frac{(l-m)!}{(l+m)!} R_{lm}^*(a) S_{lm}(b), \end{aligned} \tag{10.18}$$

where distances a and b migrate into the solid harmonics.

The benefits of scaling can now be demonstrated. Scaling of associated Legendre functions in Equation 10.18 according to Equation 10.14 changes the expansion into a compact expression:

$$\frac{1}{|\mathbf{r}|} = \sum_{l=0}^{\infty} \sum_{m=-l}^{l} \bar{R}_{lm}^* \tilde{S}_{lm}, \tag{10.19}$$

where

$$\bar{R}_{lm}^* = a^l \bar{P}_l^m(\theta_a) e^{-im\phi_a} \quad \text{and} \quad \tilde{S}_{lm} = \frac{1}{b^{l+1}} \tilde{P}_l^m(\theta_b) e^{im\phi_b} \tag{10.20}$$

are the bar- and tilde-scaled regular and irregular solid harmonics, respectively. Noticing that

$$\begin{aligned} \bar{R}_{l0}^* \tilde{S}_{l0} = \bar{R}_{l0} \tilde{S}_{l0} &= \frac{a^l}{b^{l+1}} \bar{P}_l(\theta_a) \tilde{P}_l(\theta_b) = \frac{a^l}{b^{l+1}} \frac{l!}{l!} P_l(\theta_a) P_l(\theta_b) \\ &= \frac{a^l}{b^{l+1}} P_l(\theta_a) P_l(\theta_b), \end{aligned} \tag{10.21}$$

followed by substituting the associated Legendre functions in Equation 10.13 by their scaled counterparts, and then switching to scaled solid harmonics, leads to:

$$\frac{1}{|\mathbf{r}|}=\sum_{l=0}^{\infty}\left\{\bar{R}_{l0}\,\tilde{S}_{l0}+2\sum_{m=1}^{l}\mathrm{Re}[\bar{R}^{*}_{lm}\,\tilde{S}_{lm}]\right\}. \tag{10.22}$$

This is the formulation of the inverse distance function optimized for numerical computations. The remaining step is to calculate the scaled solid harmonics.

10.3 Recurrence Relations for Scaled Regular Solid Harmonics

Recurrence relations provide the most efficient approach for the numerical computation of solid harmonics. These relations originate from the corresponding relations in associated Legendre functions developed in Chapter 2. All that is necessary in the derivation process is to convert the associated Legendre functions in recurrence relations to solid harmonics with the help of Equation 10.17. Three recurrence relations for each type of scaled solid harmonics need to be derived.

The first equation for bar-scaled regular solid harmonics originates from a recurrence relation between associated Legendre functions of different degrees in index l:

$$P_l^m=\frac{2l-1}{l-m}\cos\theta\,P_{l-1}^m-\frac{l+m-1}{l-m}P_{l-2}^m. \tag{2.71}$$

Conversion of Equation 2.71 into bar-scaled form makes use of the relation

$$P_l^m=(l+m)!\bar{P}_l^m, \tag{10.23}$$

which follows from Equation 10.14. Substituting Equation 10.23 into Equation 2.71 for P_l^m, P_{l-1}^m, and P_{l-2}^m leads to:

$$(l+m)!\bar{P}_l^m=\frac{2l-1}{l-m}\cos\theta\,(l+m-1)!\bar{P}_{l-1}^m-\frac{l+m-1}{l-m}(l+m-2)!\bar{P}_{l-2}^m. \tag{10.24}$$

Dividing by $(l+m-1)!$ gives:

$$(l+m)\,\bar{P}_l^m=\frac{2l-1}{l-m}\cos\theta\,\bar{P}_{l-1}^m-\frac{1}{l-m}\bar{P}_{l-2}^m. \tag{10.25}$$

Lastly, dividing both sides of Equation 10.25 by $(l+m)$ leads to the final expression of the recurrence relation for associated Legendre functions in bar-scaled form:

$$\bar{P}_l^m=\frac{(2l-1)\cos\theta\,\bar{P}_{l-1}^m-\bar{P}_{l-2}^m}{l^2-m^2}. \tag{10.26}$$

A comparison of Equation 2.71 with Equation 10.26 shows that the switch from conventional to the scaled form reduces the number of mathematical operations from 11 to 9, assuming that the cosine and associated Legendre functions have already been computed.

Obtaining a recurrence relation for bar-scaled regular solid harmonics from Equation 10.26 relies on Equation 10.20. Multiplying both sides of Equation 10.26 by $a^l e^{im\phi}$, and keeping in mind that θ and ϕ in the present context are the polar and azimuthal angles of radius vector $\boldsymbol{a}$, gives:

$$a^l\bar{P}_l^m e^{im\phi}=\frac{(2l-1)\cos\theta\,a^l\,\bar{P}_{l-1}^m e^{im\phi}-a^l\,\bar{P}_{l-2}^m\,e^{im\phi}}{l^2-m^2}. \tag{10.27}$$

Taking into account that $\bar{R}_{lm} = a^l \bar{P}_l^m e^{im\phi}$ accomplishes the switch to regular solid harmonics,

$$\bar{R}_{l,m} = \frac{(2l-1)a\cos\theta\,\bar{R}_{l-1,m} - a^2\bar{R}_{l-2,m}}{l^2-m^2}, \quad \text{and} \quad \bar{R}^*_{l,m} = \frac{(2l-1)a\cos\theta\,\bar{R}^*_{l-1,m} - a^2\bar{R}^*_{l-2,m}}{l^2-m^2}, \tag{10.28}$$

where the second equation is the complex conjugate of the first.

The recurrence relation for regular solid harmonics, Equation 10.28 operates in spherical polar coordinates, but the positions of particles in numerical computations are typically handled in Cartesian coordinates. This issue has a simple solution. The relationship between Cartesian and spherical polar coordinates follows the conversion rules:

$$\begin{aligned} x &= r\sin\theta\cos\phi, \\ y &= r\sin\theta\sin\phi, \\ z &= r\cos\theta, \end{aligned} \tag{10.29}$$

where $r \equiv a$ is the length of the radius vector of the coordinate point with Cartesian coordinates (x, y, z), and θ and ϕ are its polar and azimuthal angles, respectively. From Equation 10.29 it follows that $a\cos\theta = z$, and $a^2 = x^2 + y^2 + z^2$, therefore:

$$\bar{R}_{l,m} = \frac{(2l-1)\,z\,\bar{R}_{l-1,m} - (x^2+y^2+z^2)\,\bar{R}_{l-2,m}}{l^2-m^2}, \quad \text{and} \quad \bar{R}^*_{l,m} = \frac{(2l-1)\,z\bar{R}^*_{l-1,m} - (x^2+y^2+z^2)\bar{R}^*_{l-2,m}}{l^2-m^2}. \tag{10.30}$$

These equations represent the first recurrence relation for bar-scaled regular solid harmonics in Cartesian coordinates that works when the orbital and azimuthal indices satisfy the condition $l \geq 2$ and $|m| \leq l - 2$. Other recurrence relations are needed when the index m is greater than $l - 2$.

A quick inspection of Equation 10.30 reveals that for $m = l - 1$, the term $\bar{R}_{l-2,m}$ vanishes because the corresponding associated Legendre function becomes zero when the absolute value of the azimuthal index exceeds the value of the orbital index. Therefore the condition of $m = l - 1$ reduces Equation 10.30 to:

$$\bar{R}_{l,\,l-1} = z\,\bar{R}_{l-1,l-1}, \quad \text{and} \quad \bar{R}^*_{l,l-1} = z\,\bar{R}^*_{l-1,l-1}. \tag{10.31}$$

This is the second recurrence relation for numerical computation of bar-scaled regular solid harmonics that applies when $l \geq 1$ and $|m| = l - 1$.

The recurrence relations in Equations 10.30 and 10.31 together cover the range $|m| \leq l - 1$. This is nearly the full range of possible values in index m for the regular solid harmonics; all that remains is to find the last recurrence relation, $m = l$, that would allow calculation of the regular solid harmonics of type $\bar{R}_{ll}$. When computing $\bar{R}_{ll}$, it is not sufficient to have the recurrence relation that increments index l, since the terms having an azimuthal index $m = l$ do not exist in the already recursively computed pool of regular solid harmonics. Therefore, the new recurrence relation to be added should include the ability to simultaneously increment indices l and m. The necessary recurrence relation emerges from the relevant relation for associated Legendre functions:

$$P_l^l = -(2l-1)\sin\theta\,P_{l-1}^{l-1}. \tag{2.73}$$

The next step in the derivation of the target recurrence relation for bar-scaled regular solid harmonics is to convert Equation 2.73 into bar-scaled form. Replacing the canonical associated Legendre functions in Equation 2.73 with their bar-scaled counterpart defined in Equation 10.23 transforms this equation to:

$$\bar{P}_l^l = -(2l-1)\sin\theta\,\bar{P}_{l-1}^{l-1}. \tag{10.32}$$

Dividing both sides of Equation 10.32 by $(2l)!$ simplifies it to:

$$\bar{P}_l^l = -\frac{\sin\theta}{2l}\bar{P}_{l-1}^{l-1}. \tag{10.33}$$

This equation can now be converted into regular solid harmonics. Multiplying both sides of Equation 10.33 by $a^l e^{il\phi}$ gives:

$$a^l \bar{P}_l^l e^{il\phi} = -\frac{\sin\theta}{2l} a^l \bar{P}_{l-1}^{l-1} e^{il\phi}. \tag{10.34}$$

Applying Equations 10.20 through 10.34 in order to convert the latter into bar-scaled regular solid harmonics leads to:

$$\bar{R}_{l,l} = -\frac{a\sin\theta\, e^{i\phi}}{2l}\bar{R}_{l-1,l-1}, \quad \text{and} \quad \bar{R}_{l,l}^* = -\frac{a\sin\theta\, e^{-i\phi}}{2l}\bar{R}_{l-1,l-1}^*, \tag{10.35}$$

where the second equation is the complex conjugate of the first.

The recurrence relation in Equation 10.35 corresponds to the use of spherical polar coordinates. As before, for practical reasons, converting that equation to Cartesian coordinates would be most useful. Euler's formula $e^{\pm i\phi} = \cos\phi \pm i\sin\phi$ in combination with Equation 10.29 helps to accomplish this goal by providing the necessary intermediate equation:

$$r\sin\theta\, e^{\pm i\phi} = r\sin\theta\,(\cos\phi \pm i\sin\phi) = r\sin\theta\cos\phi \pm ir\sin\theta\sin\phi = x \pm iy. \tag{10.36}$$

Substituting Equation 10.36 into Equation 10.35 and using the fact that $r \equiv a$, produces the third and the final recurrence relation for the bar-scaled regular solid harmonics:

$$\bar{R}_{l,l} = -\frac{x+iy}{2l}\bar{R}_{l-1,l-1}, \quad \text{and} \quad \bar{R}_{l,l}^* = -\frac{x-iy}{2l}\bar{R}_{l-1,l-1}^*, \tag{10.37}$$

which applies to particle coordinates defined in Cartesian coordinates. This equation holds when $l > 0$, and $m = l$.

The obtained recurrence relations in Equations 10.30, 10.31, and 10.37 fully cover the range of possible non-negative values in index m. Although some of those equations are still valid for index $m < 0$, it is the easiest to obtain such harmonics from the symmetry relation

$$\bar{R}_{l,-m} = a^l \bar{P}_l^{-m} e^{-im\phi} = (-1)^m \bar{R}_{l,m}^*, \tag{10.38}$$

which follows from Equation 10.15.

From the practical standpoint, only a small number of regular solid harmonics need to be manually pre-computed in order for the recurrence relations to start working. Theoretically, only the value of the first term, $\bar{R}_{0,0} \equiv R_{0,0} = 1$, is required. All other regular solid harmonics emerge from the derived recurrence relations. For example, the term $\bar{R}_{1,0}$ originates from Equation 10.31 based on the known value of $\bar{R}_{0,0}$. Likewise, Equation 10.37 computes the term $\bar{R}_{1,1}$. Starting from orbital index $l = 2$, recursive computation employs Equation 10.30 for $0 \le m \le l-2$, then Equation 10.31 for $m = l-1$, and, finally, completes the cycle with the azimuthal index $m = l$ by employing Equation 10.37. Since Equation 10.22 eliminates the use of solid harmonics with negative value of index m, only non-negative values of index m appear in the computation.

The following analysis demonstrates how the first few bar-scaled regular solid harmonics can be computed manually. This relies on the equation:

$$\bar{R}_{lm} = \frac{a^l}{(l+m)!} P_l^m e^{im\phi}, \tag{10.39}$$

which follows from combining Equations 10.20 and 10.14.

Table 2.1 lists all the essential analytic expressions for the associated Legendre functions. Thus, for $l = 0$ and $m = 0$, $P_0^0 = 1$, so that

$$\bar{R}_{0,0} = \bar{R}^*_{0,0} = 1. \tag{10.40}$$

Similarly, for $l = 1$ and $m = 0$, substituting the value of $P_1^0 = \cos\theta$ into Equation 10.39 leads to:

$$\bar{R}_{1,0} = \bar{R}^*_{1,0} = a\cos\theta = z. \tag{10.41}$$

Finally, for $l = 1$ and $m = 1$, Table 2.1 gives $P_1^1 = -\sin\theta$. Combining the resulting expression with Equation 10.36 leads to:

$$\bar{R}_{1,1} = -\frac{a}{2}\sin\theta\, e^{i\phi} = -\frac{x+iy}{2}, \quad \text{and} \quad \bar{R}^*_{1,1} = -\frac{x-iy}{2}. \tag{10.42}$$

This completes the derivation of recurrence relations for scaled regular solid harmonics. The next step is to develop similar equations for irregular solid harmonics.

10.4 Recurrence Relations for Scaled Irregular Solid Harmonics

As with regular solid harmonics, the computation of irregular solid harmonics is most numerically stable when performed recursively. Irregular solid harmonics appear in Equation 10.22 in tilde-scaled form.

The first step in the derivation of the target equations is to convert the recurrence relations for associated Legendre functions to the tilde-scaled form. This conversion relies on Equation 10.14, which establishes the relationship between the canonical P_l^m and tilde-scaled $\tilde{P}_l^m$ associated Legendre functions. Using Equation 10.14, the tilde scaling determines that:

$$P_l^m = \frac{1}{(l-m)!}\tilde{P}_l^m. \tag{10.43}$$

With that, all that is necessary for the conversion is to express P_l^m in terms of $\tilde{P}_l^m$ in the recurrence relations for associated Legendre functions.

The first recurrence relation of interest, previously defined in Equation 2.71, establishes a relation between associated Legendre functions of different degree. That is, it computes the lth term based on the previously evaluated $(l-1)$th and $(l-2)$th terms:

$$P_l^m = \frac{2l-1}{l-m}\cos\theta\, P_{l-1}^m - \frac{l+m-1}{l-m}P_{l-2}^m. \tag{2.71}$$

Substituting Equation 10.43 into Equation 2.71 for P_l^m, P_{l-1}^m, and P_{l-2}^m leads to:

$$\frac{1}{(l-m)!}\tilde{P}_l^m = \frac{2l-1}{(l-m)}\cos\theta\frac{1}{(l-m-1)!}\tilde{P}_{l-1}^m - \frac{l+m-1}{(l-m)}\frac{1}{(l-m-2)!}\tilde{P}_{l-2}^m. \tag{10.44}$$

Afterwards, multiplying both sides of Equation 10.44 by $(l-m)!$ simplifies this relation to:

$$\tilde{P}_l^m = (2l-1)\cos\theta\,\tilde{P}_{l-1}^m - (l+m-1)(l-m-1)\,\tilde{P}_{l-2}^m. \tag{10.45}$$

Rearranging the linear coefficients in Equation 10.45 leads to the recurrence relation for tilde-scaled associated Legendre functions:

$$\tilde{P}_l^m = (2l-1)\cos\theta\,\tilde{P}_{l-1}^m - [(l-1)^2 - m^2]\,\tilde{P}_{l-2}^m. \tag{10.46}$$

The next step is to convert Equation 10.46 to a recurrence relation for irregular solid harmonics. Multiplying both sides of the equation by $b^{-l-1}\,e^{im\phi}$, and using Equation 10.20 in order to express the related terms as irregular solid harmonics, gives:

$$\tilde{S}_{l,m} = \frac{2l-1}{b}\cos\theta\,\tilde{S}_{l-1,m} - \frac{1}{b^2}[(l-1)^2 - m^2]\,\tilde{S}_{l-2,m}. \tag{10.47}$$

This equation uses spherical polar coordinates for the particle position, which are less convenient to deal with in practical applications than Cartesian coordinates. With that, it is necessary to convert Equation 10.46 into a Cartesian coordinate representation.

Conversion from spherical polar to Cartesian coordinates relies on Equation 10.29. Taking into account that in the present context $r \equiv b$, the following helpful relations emerge:

$$\frac{1}{b}\cos\theta = \frac{b\cos\theta}{b^2} = \frac{z}{b^2}, \quad \text{and} \quad b^2 = x^2 + y^2 + z^2. \tag{10.48}$$

Substituting these relations into Equation 10.47 leads to the first recurrence relation for tilde-scaled irregular solid harmonics in Cartesian coordinates:

$$\tilde{S}_{l,m} = \frac{z(2l-1)\,\tilde{S}_{l-1,m} - \left[(l-1)^2 - m^2\right]\tilde{S}_{l-2,m}}{x^2+y^2+z^2},$$
$$\text{and} \tag{10.49}$$
$$\tilde{S}^*_{l,m} = \frac{z(2l-1)\,\tilde{S}^*_{l-1,m} - \left[(l-1)^2 - m^2\right]\tilde{S}^*_{l-2,m}}{x^2+y^2+z^2}.$$

These equations only work in the domain $l \geq 2$ and $|m| \leq l-2$. The computation of irregular spherical harmonics for indices outside of these ranges requires introducing additional recurrence relations.

The first of these follows from Equation 10.49 by setting $m = l-1$. This condition zeroes the term $\tilde{S}_{l-2,m}$, since the coefficient in front of it vanishes. With that, Equation 10.49 transforms to a second recurrence relation:

$$\tilde{S}_{l,l-1} = \frac{z(2l-1)}{x^2+y^2+z^2}\tilde{S}_{l-1,l-1}, \quad \text{and} \quad \tilde{S}^*_{l,l-1} = \frac{z(2l-1)}{x^2+y^2+z^2}\tilde{S}^*_{l-1,l-1}. \tag{10.50}$$

A third recurrence relation should be derived for the case of $\tilde{S}_{ll}$, that is, where the azimuthal index m assumes its maximal value of $m = l$. The recurrence relation needed for the associated Legendre functions in which both orbital and azimuthal indices were simultaneously incremented was developed earlier:

$$P_l^l = -(2l-1)\sin\theta\,P_{l-1}^{l-1}. \tag{2.73}$$

The next step is to convert this equation into tilde-scaled form. When $m = l$, Equation 10.14 simplifies to $P_l^l = \tilde{P}_l^l$; and therefore Equation 2.73 becomes:

$$\tilde{P}_l^l = -(2l-1)\sin\theta\,\tilde{P}_{l-1}^{l-1}. \tag{10.51}$$

Expressing Equation 10.51 in irregular solid harmonics requires multiplying both sides of the equation by $b^{-l-1}\ e^{il\phi}$ and then applying the substitutions $\tilde{S}_{ll} = b^{-l-1}\tilde{P}_l^{\,l} e^{il\phi}$ and $\tilde{S}_{l-1,l-1} = b^{-l-2}\tilde{P}_{l-1}^{\,l-1} e^{i(l-1)\phi}$. This transforms Equation 10.51 to:

$$\tilde{S}_{l,l} = -(2l-1)\sin\theta \frac{e^{i\phi}}{b}\tilde{S}_{l-1,l-1}. \tag{10.52}$$

As with other recurrence equations, it is necessary to convert Equation 10.52 from spherical polar to Cartesian coordinates. This conversion relies on a useful intermediate relation:

$$\frac{\sin\theta\, e^{i\phi}}{r} = \frac{r\sin\theta\, e^{i\phi}}{r^2} = \frac{x+iy}{x^2+y^2+z^2}, \tag{10.53}$$

which follows from Equations 10.29 and 10.36. In the present context, $r \equiv b$, and x, y, z are Cartesian coordinates of radius vector **b**, so substituting Equation 10.53 into Equation 10.52 leads to the third and final recurrence relation for tilde-scaled irregular spherical harmonics:

$$\tilde{S}_{l,l} = -(2l-1)\frac{x+iy}{x^2+y^2+z^2}\tilde{S}_{l-1,l-1}, \quad \text{and} \quad \tilde{S}^*_{l,l} = -(2l-1)\frac{x-iy}{x^2+y^2+z^2}\tilde{S}^*_{l-1,l-1}. \tag{10.54}$$

Together, these three recurrence relations provide all the necessary machinery for computing the tilde-scaled irregular solid harmonics of a particle defined in Cartesian coordinates for a non-negative index m. The remaining harmonics for $m < 0$ emerge from the symmetry relation:

$$\tilde{S}_{l,-m} = \frac{1}{b^{l+1}}\tilde{P}_l^{\,-m} e^{-im\phi} = (-1)^m \tilde{S}^*_{l,m}, \tag{10.55}$$

due to Equation 10.16.

As with regular solid harmonics, an initial computation of the tilde-scaled irregular solid harmonics with orbital indices $l = 0$ and $l = 1$ in Cartesian coordinates must be done before the computer code that employs the recursion can be run. These initial functions follow from

$$\tilde{S}_{lm} = \frac{(l-m)!}{b^{l+1}} P_l^m e^{im\phi}, \tag{10.56}$$

which comes as a result of substituting Equation 10.14 into Equation 10.20.

Because $P_0^0 = 1$ for the lowest value of orbital and azimuthal indices, $l = 0$ and $m = 0$:

$$\tilde{S}_{0,0} = \frac{1}{b}P_0^{\,0} = \frac{1}{\sqrt{x^2+y^2+z^2}}, \tag{10.57}$$

where x, y, z are Cartesian coordinates of the radius vector **b**, and b is its length, as shown in Figure 10.1.

Derivation of tilde-scaled irregular solid harmonics for higher values of the orbital index l follows the same path. For $l = 1$ and $m = 0$, the value of $P_1^{\,0} = \cos\theta$ leads to:

$$\tilde{S}_{1,0} = \frac{1}{b^2}P_1^{\,0} = \frac{\cos\theta}{b^2} = \frac{b\cos\theta}{b^3} = \frac{z}{\left(\sqrt{x^2+y^2+z^2}\right)^3}, \tag{10.58}$$

where the application of $b\cos\theta = z$ finishes the conversion from spherical polar to Cartesian coordinate form.

For $l = 1$ and $m = 1$, the value of $P_1^{\,1} = -\sin\theta$ leads to:

$$\tilde{S}_{1,1} = \frac{1}{b^2}P_1^1 e^{i\phi} = -\frac{\sin\theta\, e^{i\phi}}{b^2} = -\frac{b\sin\theta\, e^{i\phi}}{b^3} = -\frac{x+iy}{\left(\sqrt{x^2+y^2+z^2}\right)^3}, \tag{10.59}$$

where $b \sin\theta e^{i\phi} = x + iy$, according to Equation 10.36. Together, the formulae in Equations 10.57 through 10.59 are sufficient to allow initiation of the recursive computation of all the tilde-scaled irregular solid harmonics.

10.5 First Few Terms of Scaled Solid Harmonics

Section 10.3 and 10.4 explained the derivation of bar-scaled regular solid harmonics and tilde-scaled irregular solid harmonics for the values of orbital indices $l = 0$ and $l = 1$, which serve as the starting point for recursive computation of other solid harmonics in scaled form. It is convenient to collect all these equations together so that they can be easily found when needed.

A brief outline of the derivation steps follows. This derivation relies on:

$$\begin{aligned} \bar{R}^*_{lm} &= r^l \bar{P}_l^m e^{-im\phi}, \quad \bar{P}_l^m = \frac{1}{(l+m)!} P_l^m, \\ \tilde{S}_{lm} &= \frac{1}{r^{l+1}} \tilde{P}_l^m e^{im\phi}, \quad \tilde{P}_l^m = (l-m)! P_l^m, \end{aligned} \tag{10.60}$$

and on the data from Table 2.1, which provides the analytic expressions for associated Legendre functions P_l^m:

$$P_0^0 = 1,\ P_1^0 = \cos\theta,\ P_1^1 = -\sin\theta,\ P_2^0 = \frac{1}{2}(3\cos^2\theta - 1),\ P_2^1 = -3\sin\theta\cos\theta,\ P_2^2 = 3\sin^2\theta. \tag{10.61}$$

Substituting the expressions from Equation 10.61 into Equation 10.60, and using Equations 10.29 and 10.36 to convert the resulting formulae from spherical polar to Cartesian coordinates results in the analytic expressions for scaled solid harmonic functions summarized in Table 10.1.

TABLE 10.1

Analytic Expressions for the First Few Scaled Solid Harmonics

$$\bar{R}^*_{0,0} = P_0^0 = 1$$

$$\bar{R}^*_{1,0} = rP_1^0 = r\cos\theta = z$$

$$\bar{R}^*_{1,1} = r\frac{1}{2}P_1^1 e^{-i\phi} = -\frac{r}{2}\sin\theta\,(\cos\phi - i\sin\phi) = -\frac{x-iy}{2}$$

$$\bar{R}^*_{2,0} = r^2\frac{1}{2}P_2^0 = \frac{r^2}{4}(3\cos^2\theta - 1) = \frac{3z^2 - r^2}{4} = \frac{2z^2 - x^2 - y^2}{4}$$

$$\bar{R}^*_{2,1} = r^2\frac{1}{6}P_2^1 e^{-i\phi} = -\frac{r^2}{2}\sin\theta\cos\theta\, e^{-i\phi} = -\frac{rz}{2}\sin\theta\, e^{-i\phi} = -z\frac{x-iy}{2}$$

$$\bar{R}^*_{2,2} = r^2\frac{1}{24}P_2^2 e^{-i2\phi} = \frac{r^2}{8}\sin^2\theta\, e^{-i2\phi} = \frac{(x-iy)^2}{8}$$

$$\tilde{S}_{0,0} = \frac{1}{r}P_0^0 = \frac{1}{r} = \frac{1}{\sqrt{x^2+y^2+z^2}}$$

$$\tilde{S}_{1,0} = \frac{1}{r^2}P_1^0 = \frac{r\cos\theta}{r^3} = \frac{z}{\left(\sqrt{x^2+y^2+z^2}\right)^3}$$

$$\tilde{S}_{1,1} = \frac{1}{r^2}P_1^1 e^{i\phi} = -\frac{r\sin\theta}{r^3}(\cos\phi + i\sin\phi) = -\frac{x+iy}{\left(\sqrt{x^2+y^2+z^2}\right)^3}$$

$$\tilde{S}_{2,0} = \frac{2}{r^3}P_2^0 = \frac{1}{r^3}(3\cos^2\theta - 1) = \frac{3z^2 - r^2}{r^5} = \frac{2z^2 - x^2 - y^2}{r^5}$$

$$\tilde{S}_{2,1} = \frac{1}{r^3}P_2^1 e^{i\phi} = -\frac{3}{r^3}\sin\theta\cos\theta\, e^{i\phi} = -\frac{3rz}{r^5}\sin\theta\, e^{i\phi} = -\frac{3z}{r^5}(x+iy)$$

$$\tilde{S}_{2,2} = \frac{1}{r^3}P_2^2 e^{i2\phi} = \frac{3}{r^3}\sin^2\theta\, e^{i2\phi} = \frac{3}{r^5}(x+iy)^2$$

Table 10.1 can be used in the derivation of analytic expressions for solid harmonics with higher values of azimuthal index l by employing the recurrence relations Equations 10.30, 10.31, and 10.37, for bar-scaled regular solid harmonics, and Equations 10.49, 10.50, 10.54 for tilde-scaled irregular solid harmonics.

10.6 Design of Computer Code for Computation of Solid Harmonics

Efficient computation of solid harmonics using recurrence relations requires careful design of the computer program. Writing an optimized computer code is a complex and error prone work that requires the equations be first fully understood before the fine-tuning can start. Therefore, it may sometimes be warranted to start the coding from implementing the equations in a simple naïve form to make sure that the program works and produces correct results, and only then start optimization of the program code. Figure 10.2 below provides an example of a naïve, easy to program implementation. Conditional constructs in the code in Figure 10.2 provide a simple mechanism to ensure that index m falls inside the range $0 \leq m \leq l-2$, or is equal to $l-1$ or to l in computation of solid harmonics so that proper recurrence relation may be invoked whenever needed.

Once the correctness of the simple implementation is verified against the expected results, the algorithm structure may be redesigned to improve its computational efficiency. Typical optimization work focuses on reviewing the loop structure, and eliminating conditional operators *if-then-else* and *select-case*. Branching disrupts the synchronization between the central processing unit and the cache memory, leading to computational performance degradation.

In order to avoid the use of conditional programming language constructs, the loop structure in Figure 10.2 needs to be revised. Fortunately, "*do*" loops provide sufficient programming language flexibility to reach the same outcome as that of conditional statements. Figure 10.3 provides schematic implementation.

Key to the performance solution is to set the external loop to start from the value of index $l = 2$, and the enclosed loop over index m going from 0 to $l-2$, while having harmonics with $l = 0$ and $l = 1$ pre-computed outside the loop. This arrangement automatically performs the task of conditional execution of the first recurrence relation given by Equations 10.30 and 10.49 inside the enclosed loop

```
pre-compute solid harmonics for l = 0
do l = 1, nL
  do m = 0, l
    if (m ≤ l – 2) then
      apply first recurrence relation
    end if
    if (m == l – 1) then
      apply second recurrence relation
    end if
    if (m == l) then
      apply third recurrence relation
    end if
  end do  ! index m
end do  ! index l
```

FIGURE 10.2 Naïve implementation of a computer program for numerical computation of solid harmonics. Parameter nL represents the upper boundary of index l in the truncated series expansion.

```
pre-compute solid harmonics for l = 0, 1
do l = 2, nL
  do m = 0, l – 2
    apply first recurrence relation
  end do   ! index m
  apply second recurrence relation  ! m = l – 1
  apply third recurrence relation  ! m = l
end do   ! index l
```

FIGURE 10.3 Optimized loop structure for recursive computation of solid harmonics.

without branching the code. Moreover, having the orbital index l in the external loop starting from the value of $l = 2$ makes it possible to have the second (Equations 10.31 and 10.50) and third (Equations 10.37 and 10.54) recurrence relations executed for each value of l. Therefore, none of these program blocks needs to be bracketed by conditional statements. Instead, they follow one after another in sequential order making the code cache-friendly. The next section turns the above algorithm into an actual computer code.

10.7 Program Code for Computation of Multipole Expansions

The completion of the derivation of the equations for solid harmonics opens a path for the computation of the multipole expansions. This computation uses the fact that in FMM theory multipoles can be expressed as a charge-scaled form of solid harmonics by computing the solid harmonics and then scaling them according to their atomic charges. In other words, for a system of N point charges, a scaled form of the regular and irregular solid harmonics,

$$M_{lm} = \sum_{i}^{N} q_i \bar{R}^*_{lm}, \quad L_{lm} = \sum_{i}^{N} q_i \tilde{S}_{lm}, \tag{10.62}$$

can be used to generate a scaled form of multipole expansions.

Before computing a multipole expansion for a set of point charges, a decision must be made regarding how it will be stored in the computer memory. Any storage method must allow for the retrieval of expansion elements M_{lm} and L_{lm}, given the values of the two indices l and m, where index l can assume any value from 0 to nL, the upper value of index l in the truncated expansion. For each l, index m varies in the range from 0 to l. There is no need to store expansion elements for negative value of index m since they can readily be obtained from the symmetry relations in Equations 10.38 and 10.55. This pattern of the variation of the indices l and m results in the expansion array assuming a triangular shape, as illustrated in Figure 10.4.

The dependence of the upper value of index m on the value of index l in the multipole expansion makes the usual square matrix storage wasteful because the upper triangle of that matrix will not be used. Since programming languages cannot handle triangular arrays, the multipole expansion is mapped onto a one-dimensional array, and the two-dimensional indices (l, m) are encoded into a sequential index, as shown in the right panel in Figure 10.4. Since indices l and m can have a value of zero, it is convenient to use a zero-base for the sequential index. If that is done, then index 0 holds the expansion element (0,0); index 1 holds element (1,0); and index $[(nL + 1)(nL + 2)/2] - 1$ holds element (nL, nL) in the one-dimensional array. Retrieving the expansion element (l, m) employs the conversion:

$$(l,m) = l(l+1)/2 + m, \tag{10.63}$$

0,0				0				
1,0	1,1			1	2			
2,0	2,1	2,2		3	4	5		
3,0	3,1	3,2	3,3	6	7	8	9	
...		...		...		...		
$nL,0$	$nL,1$	...	nL,nL	...		...		$(nL+1)(nL+2)/2-1$

FIGURE 10.4 Storage of the multipole expansion in computer memory. Row index l goes from 0 to nL, and the column index m goes from 0 to l. Parameter nL represents the upper value of index l in a truncated multipole expansion. The expansion elements are arranged by their indices (l, m) in the left panel, and the right panel shows the corresponding sequential enumeration of the same elements when stored in a one-dimensional array.

which returns the element index from the one-dimensional array. To reduce the cost of index computation, the index for elements $(l, 0)$ may be precomputed and stored in a one-dimensional array *tri*(0:*nL*), as illustrated in the following code snippet:

```
integer :: nL                    ! input data
integer :: nSzTri, tri(0:nL)     ! output data
integer :: L                     ! orbital index

! determine upper value nSzTri of a sequential index
! nSzTri is the index of expansion term (L,L)
nSzTri = (nL+1) * (nL+2) / 2 - 1

! precompute triangular pointers to expansion terms (L,0)
! so that expansion (L,0) can be accessed by zero-based index tri(L)
do L = 0, nL
  tri(L) = L*(L+1)/2
enddo
```

Evaluation of the regular multipole expansion can be expressed as a Fortran 90 subroutine called *RegularExpansion*(). On input, this subroutine receives an array of particle coordinates, *c*, an array of their point charges, *q*, the number of particles, *n*, and the coordinates of the origin *oxyz* of the coordinate system for the computation of multipole expansion. Next, the upper value of index $l = nL$ specifies how many terms need to be computed in the truncated multipole expansion, followed by a parameter, *nSzTri*, that represents the size of a triangular storage required for the multipole expansion, and an array, *tri*, of pointers to the expansion terms $(l, 0)$, both these entities having been computed earlier. Finally, the subroutine needs a scratch array, *tmpExpansion*, to store intermediate results of the computation. The array *regMultipole* concludes the list of formal arguments of the subroutine. On exit from the subroutine, *regMultipole* will contain the regular multipole expansion.

```
subroutine RegularExpansion(c, q, n, oxyz, nL, nSzTri, tri, tmpExpansion,
regMultipole)
!
! Compute regular multipole expansion for a set of point charges
!
implicit none
integer,           intent(in)  :: n                        ! number of particles
double precision,  intent(in)  :: c(3,n)                   ! particle coordinates
double precision,  intent(in)  :: q(n)                     ! particle charges
double precision,  intent(in)  :: oxyz(3)                  ! box origin
integer,           intent(in)  :: nL                       ! expansion truncation
integer,           intent(in)  :: nSzTri                   ! triangular size
integer,           intent(in)  :: tri(0:nL)                ! triangular pointers
complex(16),       intent(out) :: tmpExpansion(0:nSzTri)   ! scratch array
complex(16),       intent(out) :: regMultipole(0:nSzTri)   ! output
```

```
! Local variables
integer          :: L, m, iAtom, aL, aL1, aL2, LL
double precision :: x, y, z, r2
double precision :: z2lm1

regMultipole = dcmplx(0.0d0, 0.0d0)

!$omp parallel do reduction (+:regMultipole) &
!$omp default(none) shared(c, q, tri, n, nL, oxyz, nSzTri) &
!$omp private(x,y,z, r2, z2lm1,aL,aL1,aL2,LL, tmpExpansion)
do iAtom = 1, n

  ! Shift particle coordinates to the new coordinate system
  x  = c(1,iAtom) - oxyz(1)
  y  = c(2,iAtom) - oxyz(2)
  z  = c(3,iAtom) - oxyz(3)
  r2 = x*x + y*y + z*z

  ! Initialize first few regular harmonics
  ! (0,0) = (L,m)
  tmpExpansion(0) = dcmplx(1.0d0, 0.0d0)

  ! (1,0)
  tmpExpansion(1) = dcmplx(z, 0.0d0)
  ! (1,1)
  tmpExpansion(2) = dcmplx(-0.5d0*x, 0.5d0*y)

  ! Obtain higher-order harmonics from recurrence relations
  do L = 2, nL
    ! Initialize frequently used temporary variables
    z2lm1 = z * dble(2*L-1)
    aL    = tri(L)
    aL1   = tri(L-1)
    aL2   = tri(L-2)
    LL    = L*L

    ! (L,0) … (L,L-2)       use first recurrence relation
    do m = 0, L-2
      tmpExpansion(aL+m) = (z2lm1 * tmpExpansion(aL1+m) - r2 *
tmpExpansion(aL2+m)) / dble(LL-m*m)
    enddo

    ! (L,L-1)                  use second recurrence relation
    tmpExpansion(aL+L-1) = z * tmpExpansion(aL1+L-1)

    ! (L,L)                    use third recurrence relation
    tmpExpansion(aL+L) = tmpExpansion(aL1+L-1) * tmpExpansion(2) / dble(L)
  enddo

  ! Accumulate multipole expansion
  regMultipole = regMultipole + tmpExpansion * q(iAtom)
enddo

end subroutine RegularExpansion
```

The subroutine starts with a declaration of the formal arguments that are read-only having the attribute *intent(in)* to prevent them from being accidentally overwritten in the code. Likewise, declarations carrying the attribute *intent(out)* indicate those variables whose values will be set inside the subroutine.

The subroutine starts by initializing the complex array *regMultipole* to zero. This is necessary because multipole expansions computed for individual particles will be accumulated in this array. The ability of Fortran 90 to perform vector operations on array elements makes it possible to initialize the entire array using only a single line of code.

The multipole expansion is calculated using a loop that has index *iAtom* going over the list of charged particles in an FMM box. Array *oxyz* contains the Cartesian coordinates of the center of that box (it is presumed that the particle coordinates and the box center are defined in a global Cartesian coordinate system). Since *oxyz* is expected to be the origin of the coordinate system for the computed multipole expansion, the next few lines in the subroutine code shift each particle's coordinates to the new coordinate system. The particle's new coordinates are then stored in variables *x*, *y*, and *z*. The original particle coordinates are left intact.

The subroutine continues with the computation of the regular solid harmonics $R^*_{l,m}$ for the particle defined by radius vector (x, y, z), using the algorithm outlined in Figure 10.3. The first few regular harmonics for index $l = 0$ and $l = 1$ are initialized using Equations 10.40 through 10.42, and index l runs in a loop from 2 to *nL* computing the remaining harmonics using Equations 10.30, 10.31, and 10.37, which are then stored in a scratch array *tmpExpansion*.

At the end of the subroutine, *RegularExpansion*() converts the regular harmonics terms for a particle referred by index *iAtom* into a multipole expansion of that particle by scaling the computed *tmpExpansion* with the particle charge *q*(*iAtom*), and accumulating the result in array *regMultipole*. Once again, using the capability of Fortran 90 to perform vector operation on arrays allows this operation to be expressed in a single line of code.

Parallelization of the computation applies to the external loop over the number of particles. Placing the following OpenMP directives

```
!$omp parallel do reduction (+:regMultipole) &
!$omp default(none) shared(c, q, tri, n, nL, oxyz, nSzTri) &
!$omp private(x,y,z, r2, z2lm1,aL,aL1,aL2,LL, tmpExpansion)
```

in front of the *do*-loop over index *iAtom* accomplishes the parallelization. One nice aspect of the OpenMP parallelization model is that the code automatically compiles in a serial mode if the compiler turns off the recognition of OpenMP directives. In that case, everything on the right side of exclamation mark will be treated as a comment as defined in the Fortran 90 language.

The first OpenMP directive *!$omp parallel do reduction (+:regMultipole)* instructs a reduction operation to be applied to array *regMultipole*, which turns *regMultipole* into a private variable, replicating the array in the local memory of each OpenMPI thread. With that, after the *do*-loop over index *iAtom* is completed each instance of the array *regMultipole* holds the results generated by that thread. The reduction directive instructs OpenMP at runtime to merge these instances back into a single array *regMultipole* by summing up the corresponding elements of the array across the multiple copies held by OpenMP threads.

The second line of the OpenMP directives states that each OpenMP process has shared access to the variables *c, q, tri, n, nL, oxyz, nSzTri*, which are defined as read-only, and, therefore, are safe to share. The third line in the directives declares private variables *x, y, z, r2, z2lm1, aL, aL1, aL2, LL*, and *tmpExpansion*. Based on this latter instruction, each OpenMP thread creates a local instance of these variables during the parallel execution so that writing into these variables does not disrupt the work of other threads.

Complementary to the computation of regular multipole expansion is a subroutine *IrregularExpansion*() which computes the corresponding irregular multipole expansion:

```
subroutine IrregularExpansion(c, q, n, oxyz, nL, nSzTri, tri, tmpExpansion,
irrMultipole)
!
! Compute irregular multipole expansion for a set of point charges
!
implicit none
integer,            intent(in)  :: n                       ! number of particles
double precision,   intent(in)  :: c(3,n)                  ! particle coordinates
double precision,   intent(in   :: q(n)                    ! particle charges
double precision,   intent(in)  :: oxyz(3)                 ! box origin
```

```
integer,          intent(in)  :: nL                       ! expansion truncation
integer,          intent(in)  :: nSzTri                   ! triangular size
integer,          intent(in)  :: tri(0:nL)                ! triangular pointers
complex(16),      intent(out):: tmpExpansion(0:nSzTri)  ! scratch array
complex(16),      intent(out):: irrMultipole(0:nSzTri)  ! output

! Local variables
integer           :: L, m, iAtom, aL, aL1, aL2, LL1
double precision :: x, y, z, ir, ir2, ir3
double precision :: z2lm1
complex(16)       :: xpy

irrMultipole = dcmplx(0.0d0, 0.0d0)

!$omp parallel do reduction (+:irrMultipole) &
!$omp default(none) shared(c, q, tri, n, nL, oxyz, nSzTri) &
!$omp private(x,y,z, ir,ir2,ir3,xpy, z2lm1,aL,aL1,aL2,LL1, tmpExpansion)
do iAtom = 1, n

  x = c(1,iAtom) - oxyz(1)
  y = c(2,iAtom) - oxyz(2)
  z = c(3,iAtom) - oxyz(3)
  ir  = 1.0d0 / sqrt( dmax1(x*x + y*y + z*z, 1.0d-10) )
  ir2 = ir * ir
  ir3 = ir * ir2
  xpy = dcmplx(-x * ir2, -y * ir2)

  ! Initialize first few irregular harmonics
  ! (0,0)
  tmpExpansion(0) = dcmplx(ir, 0.0d0)

  ! (1,0)
  tmpExpansion(1) = dcmplx(z * ir3, 0.0d0)
  ! (1,1)
  tmpExpansion(2) = dcmplx(-x * ir3, -y * ir3)

  ! Obtain higher-order harmonics from recurrence relations
  do L = 2, nL
    z2lm1 = z * dble(2*L-1)
    aL  = tri(L)
    aL1 = tri(L-1)
    aL2 = tri(L-2)
    LL1 = (L-1)*(L-1)

    ! (L,0) … (L,L-2)   use first recurrence relation
    do m = 0, L-2
      tmpExpansion(aL+m) = (z2lm1 * tmpExpansion(aL1+m) - dble(LL1-m*m) *
tmpExpansion(aL2+m)) * ir2
    enddo

    ! (L,L-1)              use second recurrence relation
    tmpExpansion(aL+L-1) = z2lm1 * tmpExpansion(aL1+L-1) * ir2

    ! (L,L)                use third recurrence relation
    tmpExpansion(aL+L) = dble(2*L-1) * xpy * tmpExpansion(aL1+L-1)
  enddo

  ! Accumulate multipole expansion
  irrMultipole = irrMultipole + tmpExpansion * q(iAtom)
enddo
end subroutine IrregularExpansion
```

Because both subroutines for calculating the regular and irregular solid harmonics are very similar, much that has been said about the former code automatically applies here, so only a brief review of the second code is needed.

As in the previous case, the new subroutine takes an array of particle coordinates, *c*, and their point charges, *q*, on input. An array *irrMultipole* provides an output placeholder for the computed irregular multipole expansion.

The first executable line in the subroutine *IrregularExpansion*() initializes the array *irrMultipole*. This operation is followed by a group of OpenMP directives for parallelization of the loop over the list of particles. Within the loop, each particle is translated to the new origin *oxyz*, then the inverse distance quantities $1/r$, $1/r^2$, and $1/r^3$ are calculated (these will be used later on in the subroutine). To avoid the possibility of a floating–point exception due to division by zero, the double precision Fortran 90 intrinsic function *dmax1*() compares r^2 to the very small number 10^{-10}, and returns the larger of the two quantities.

Next, the first few irregular solid harmonics $\tilde{S}_{0,0}$, $\tilde{S}_{1,0}$, and $\tilde{S}_{1,1}$ are initialized by implementing Equations 10.57 through 10.59, respectively. Then the loop over index l computes the remaining irregular solid harmonics up to the upper value of index $l = nL$, using the recurrence relations described in Equations 10.49, 10.50, and 10.54. After the loop over index l ends, multiplication of the computed irregular harmonics *tmpExpansion* by the partial atomic charge *q*(*iAtom*) produces an irregular multipole expansion for that charged particle. The result of this multiplication goes into the array *irrMultipole* where it is summed up with the multipole expansion of the other particles.

10.8 Computation of Electrostatic Force Using Scaled Solid Harmonics

The utility of scaling the solid harmonics does not end with the improvement in efficiency of the calculation of electrostatic potential. Similar improvements can be made by the application of scaling to the equations of the electrostatic force.

The equation of electrostatic force **F**(**a**) in Cartesian coordinates acting on a charged particle located in point A, due to the presence of a charged particle positioned in point B is:

$$
\begin{aligned}
F_x(\mathbf{a}) &= -q_a q_b \sum_{l=0}^{\infty} \sum_{m=-l}^{l} \frac{(l-m)!}{(l+m)!} \left(\frac{\partial R_{lm}^*(\mathbf{a})}{\partial x} \right) S_{lm}(\mathbf{b}), \\
F_y(\mathbf{a}) &= -q_a q_b \sum_{l=0}^{\infty} \sum_{m=-l}^{l} \frac{(l-m)!}{(l+m)!} \left(\frac{\partial R_{lm}^*(\mathbf{a})}{\partial y} \right) S_{lm}(\mathbf{b}), \\
F_z(\mathbf{a}) &= -q_a q_b \sum_{l=0}^{\infty} \sum_{m=-l}^{l} \frac{(l-m)!}{(l+m)!} \left(\frac{\partial R_{lm}^*(\mathbf{a})}{\partial z} \right) S_{lm}(\mathbf{b}),
\end{aligned}
\tag{9.58}
$$

where ***a*** is a radius vector of local point A, ***b*** is a radius vector of distant point B, and q_a and q_b are their respective electrostatic charges; Figure 10.5 illustrates the relative positions of these two particles. Recall that, in the Fast Multipole Method, only the force on particle A, positioned in the local coordinate system, is evaluated, whereas the remote point B is typically the center of a multipole expansion of a large number of distant particles. Because A and B can be swapped, computation of the electrostatic force on the remote particles involves the same procedure as that for particle A in the respective local coordinate system for particles B.

The presence of the factor $((l - m)!)/((l + m)!)$ in Equation 9.58, which is the same as that in the equation for electrostatic potential in Equation 10.18, suggests a potential benefit from switching to the use of scaled solid harmonics. Scaling of irregular solid harmonics in Equation 9.58 proceeds according to:

$$(l-m)!S_{lm}(\mathbf{b}) = \tilde{S}_{lm}(\mathbf{b}). \tag{10.64}$$

The factor $(l - m)!$ that is needed to satisfy Equation 10.64 comes from the numerator portion of $((l - m)!)/((l + m)!)$ in Equation 9.58. With that, the remaining portion, $1/((l + m)!)$, goes to the scaling of regular solid harmonics.

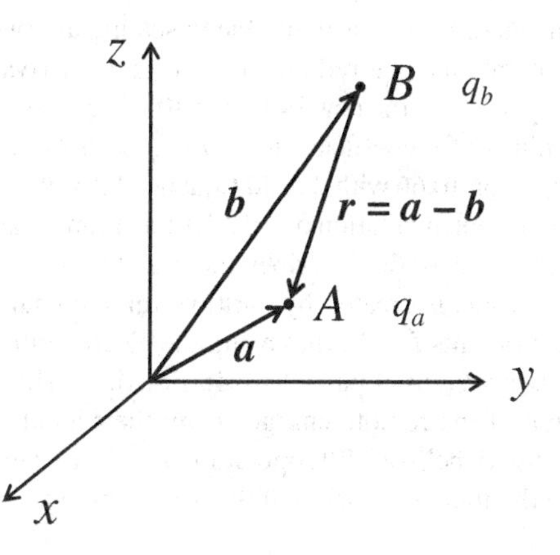

FIGURE 10.5 Exertion of electrostatic force on a particle A carrying charge q_a by the charge q_b centered at point B. Radius vectors $\boldsymbol{a}$ and $\boldsymbol{b}$ connect the center of coordinate system with the points A and B, respectively, and radius vector $\boldsymbol{r}$ extends from the point B toward the point A.

The dependence of Equation 9.58 on regular solid harmonics is hidden inside the derivatives of regular solid harmonics. These derivatives break down into functions of regular solid harmonics:

$$\begin{aligned}
\frac{\partial}{\partial x}R^*_{l,m} &= \frac{lx+imy}{x^2+y^2}R^*_{l,m} - \frac{(l+m)xz}{x^2+y^2}R^*_{l-1,m},\\
\frac{\partial}{\partial y}R^*_{l,m} &= \frac{ly-imx}{x^2+y^2}R^*_{l,m} - \frac{(l+m)yz}{x^2+y^2}R^*_{l-1,m},\\
\frac{\partial}{\partial z}R^*_{l,m} &= (l+m)R^*_{l-1,m}.
\end{aligned} \tag{9.57}$$

The task at hand is to convert these equations to bar-scaled form. Multiplying both sides of Equation 9.57 by the coefficient $1/((l+m)!)$ that comes from Equation 9.58, and taking into account that $\bar{R}^*_{l,m} = 1/((l+m)!)R^*_{l,m}$ and $\bar{R}^*_{l-1,m} = 1/((l+m-1)!)R^*_{l-1,m}$ leads to:

$$\begin{aligned}
\frac{\partial}{\partial x}\bar{R}^*_{l,m} &= \frac{lx+imy}{x^2+y^2}\bar{R}^*_{l,m} - \frac{xz}{x^2+y^2}\bar{R}^*_{l-1,m},\\
\frac{\partial}{\partial y}\bar{R}^*_{l,m} &= \frac{ly-imx}{x^2+y^2}\bar{R}^*_{l,m} - \frac{yz}{x^2+y^2}\bar{R}^*_{l-1,m},\\
\frac{\partial}{\partial z}\bar{R}^*_{l,m} &= \bar{R}^*_{l-1,m}.
\end{aligned} \tag{10.65}$$

Substituting the scaled forms of regular and irregular solid harmonics from Equations 10.65 and 10.64 into Equation 9.58 leads to the revised equations of force:

$$\begin{aligned}
F_x(\mathbf{a}) &= -q_a q_b \sum_{l=0}^{\infty}\sum_{m=-l}^{l}\left[\frac{lx+imy}{x^2+y^2}\bar{R}^*_{l,m}(\mathbf{a}) - \frac{xz}{x^2+y^2}\bar{R}^*_{l-1,m}(\mathbf{a})\right]\tilde{S}_{l,m}(\mathbf{b}),\\
F_y(\mathbf{a}) &= -q_a q_b \sum_{l=0}^{\infty}\sum_{m=-l}^{l}\left[\frac{ly-imx}{x^2+y^2}\bar{R}^*_{l,m}(\mathbf{a}) - \frac{yz}{x^2+y^2}\bar{R}^*_{l-1,m}(\mathbf{a})\right]\tilde{S}_{l,m}(\mathbf{b}),\\
F_z(\mathbf{a}) &= -q_a q_b \sum_{l=0}^{\infty}\sum_{m=-l}^{l}\bar{R}^*_{l-1,m}(\mathbf{a})\tilde{S}_{l,m}(\mathbf{b}),
\end{aligned} \tag{10.66}$$

which are now written in terms of scaled solid harmonics.

In summary, the improvements in Equation 10.66 due to scaling are two-fold. First, the scaling reduces the number of mathematical operations required for calculating the derivatives of regular solid harmonics, as is evident from a visual comparison of the new Equation 10.65 against the old Equation 9.57. Second, the scaling of solid harmonics eliminates the coefficient $((l-m)!)/((l+m)!)$ from the equation of force, as can be seen by comparing the new Equation 10.66 with the old Equation 9.58. Both these changes in Equation 10.66 lead to a more efficient computer implementation of calculation of force than that when using Equation 9.58.

The equation of force in Equation 10.66 describes the effect that a single charged particle at point B makes on the particle A. If the force is created by multiple remote charges, their combined effect sums up to the irregular multipole moments $L_{l,m}(\mathbf{b})$ that bring the electrostatic potential due to remote charge sources into the local coordinate system of particle A. In FMM, $L_{l,m}(\mathbf{b})$ is the result of origin translation of regular multipole expansion of the remote charges from their local coordinate system into the local coordinate system of particle A with help of M2L operation. Lastly, replacing irregular multipole moments $q_b\tilde{S}_{l,m}(\mathbf{b})$ with $L_{l,m}(\mathbf{b})$ leads to the final expression of the force acting on particle A due to remote charges:

$$
\begin{aligned}
F_x(\mathbf{a}) &= -q_a \sum_{l=0}^{\infty}\sum_{m=-l}^{l}\left[\frac{lx+imy}{x^2+y^2}\bar{R}^*_{l,m}(\mathbf{a}) - \frac{xz}{x^2+y^2}\bar{R}^*_{l-1,m}(\mathbf{a})\right]L_{l,m}(\mathbf{b}),\\
F_y(\mathbf{a}) &= -q_a \sum_{l=0}^{\infty}\sum_{m=-l}^{l}\left[\frac{ly-imx}{x^2+y^2}\bar{R}^*_{l,m}(\mathbf{a}) - \frac{yz}{x^2+y^2}\bar{R}^*_{l-1,m}(\mathbf{a})\right]L_{l,m}(\mathbf{b}),\\
F_z(\mathbf{a}) &= -q_a \sum_{l=0}^{\infty}\sum_{m=-l}^{l}\bar{R}^*_{l-1,m}(\mathbf{a})L_{l,m}(\mathbf{b}),
\end{aligned}
\tag{10.67}
$$

where multipole moments $\bar{R}^*_{l,m}(\mathbf{a})$ and $L_{l,m}(\mathbf{b})$ adhere to the local coordinate system of particle A. The next section explains additional modifications that need to be made to Equation 10.67 to adapt it for computer implementation.

10.9 Program Code for Computation of Force

Equation 10.67 provides a theoretical framework for computation of electrostatic field in the position of particle A having radius vector $\mathbf{a}$ and carrying charge q_a. Using it to write a computer code requires performing a few additional steps, though. Due to the symmetry of multipole expansions $\bar{R}^*(\mathbf{a})$ and $L(\mathbf{b})$ in azimuthal index m, the minimally required summation range over that index is only half of its theoretical size. Section 10.2 explains the underlying theory in details on the example of electrostatic potential. Application of similar steps to Equation 10.67 results in taking the element with $m = 0$ outside the sum, multiplying the remaining sum by factor of two while limiting the summation range to $1 \le m \le l$, and discarding the complex part in the summation element by means of function Re():

$$
\begin{aligned}
F_x(\mathbf{a}) = -q_a \sum_{l=0}^{\infty}\mathrm{Re}\Bigg\{&\left[\frac{lx}{x^2+y^2}\bar{R}^*_{l,0}(\mathbf{a}) - \frac{xz}{x^2+y^2}\bar{R}^*_{l-1,0}(\mathbf{a})\right]L_{l,0}(\mathbf{b})\\
&+2\sum_{m=1}^{l}\left[\frac{lx+imy}{x^2+y^2}\bar{R}^*_{l,m}(\mathbf{a}) - \frac{xz}{x^2+y^2}\bar{R}^*_{l-1,m}(\mathbf{a})\right]L_{l,m}(\mathbf{b})\Bigg\},\\
F_y(\mathbf{a}) = -q_a \sum_{l=0}^{\infty}\mathrm{Re}\Bigg\{&\left[\frac{ly}{x^2+y^2}\bar{R}^*_{l,0}(\mathbf{a}) - \frac{yz}{x^2+y^2}\bar{R}^*_{l-1,0}(\mathbf{a})\right]L_{l,0}(\mathbf{b})\\
&+2\sum_{m=1}^{l}\left[\frac{ly-imx}{x^2+y^2}\bar{R}^*_{l,m}(\mathbf{a}) - \frac{yz}{x^2+y^2}\bar{R}^*_{l-1,m}(\mathbf{a})\right]L_{l,m}(\mathbf{b})\Bigg\}\\
F_z(\mathbf{a}) = -q_a \sum_{l=0}^{\infty}\mathrm{Re}\Bigg\{&\bar{R}^*_{l-1,0}(\mathbf{a})\,L_{l,0}(\mathbf{b}) + 2\sum_{m=1}^{l}\bar{R}^*_{l-1,m}(\mathbf{a})L_{l,m}(\mathbf{b})\Bigg\}.
\end{aligned}
\tag{10.68}
$$

After that, the final step in transformation of Equation 10.68 into a computer-ready form is to identify all specific values of orbital and azimuthal indices l and m, which change the appearance of the equation, because each of those forms has to be coded separately. Since computation of force F(**a**) involves similar steps as those for calculation of electrostatic potential $\Phi(\mathbf{a})$, it is convenient to combine the computation of these two quantities. Equation 10.22 provides the starting point for electrostatic potential. Final equations for potential and force customized for specific ranges of orbital and azimuthal indices follow.

$l = 0$ and $m = 0$:

$$\Phi(\mathbf{a}) = \mathrm{Re}\left[\bar{R}^*_{0,0}(\mathbf{a})\,L_{0,0}(\mathbf{b})\right], \quad F_x(\mathbf{a}) = 0, \quad F_y(\mathbf{a}) = 0, \quad F_z(\mathbf{a}) = 0. \tag{10.69}$$

$l \geq 1$ and $m = 0$:

$$\begin{aligned}
\Phi(\mathbf{a}) &= \mathrm{Re}\left[\bar{R}^*_{l,0}(\mathbf{a})\,L_{l,0}(\mathbf{b})\right],\\
F_x(\mathbf{a}) &= -q_a \sum_{l=1}^{\infty}\left\{l\frac{x}{x^2+y^2}\mathrm{Re}\left[\bar{R}^*_{l,0}(\mathbf{a})\,L_{l,0}(\mathbf{b})\right] - z\frac{x}{x^2+y^2}\mathrm{Re}\left[\bar{R}^*_{l-1,0}(\mathbf{a})\,L_{l,0}(\mathbf{b})\right]\right\},\\
F_y(\mathbf{a}) &= -q_a \sum_{l=1}^{\infty}\left\{l\frac{y}{x^2+y^2}\mathrm{Re}\left[\bar{R}^*_{l,0}(\mathbf{a})\,L_{l,0}(\mathbf{b})\right] - z\frac{y}{x^2+y^2}\mathrm{Re}\left[\bar{R}^*_{l-1,0}(\mathbf{a})\,L_{l,0}(\mathbf{b})\right]\right\},\\
F_z(\mathbf{a}) &= -q_a \sum_{l=1}^{\infty}\mathrm{Re}\left[\bar{R}^*_{l-1,0}(\mathbf{a})\,L_{l,0}(\mathbf{b})\right].
\end{aligned} \tag{10.70}$$

$l \geq 1$ and $1 \leq m \leq l-1$:

$$\begin{aligned}
\Phi(\mathbf{a}) &= \mathrm{Re}\left[\bar{R}^*_{l,m}(\mathbf{a})\,L_{l,m}(\mathbf{b})\right],\\
F_x(\mathbf{a}) &= -2q_a \sum_{l=1}^{\infty}\left\{\mathrm{Re}\left[\left(l\frac{x}{x^2+y^2}, m\frac{x}{x^2+y^2}\right)\bar{R}^*_{l,m}(\mathbf{a})\,L_{l,m}(\mathbf{b})\right] - z\frac{x}{x^2+y^2}\mathrm{Re}\left[\bar{R}^*_{l-1,m}(\mathbf{a})\,L_{l,m}(\mathbf{b})\right]\right\},\\
F_y(\mathbf{a}) &= -2q_a \sum_{l=1}^{\infty}\left\{\mathrm{Re}\left[\left(l\frac{y}{x^2+y^2}, -m\frac{y}{x^2+y^2}\right)\bar{R}^*_{l,m}(\mathbf{a})\,L_{l,m}(\mathbf{b})\right] - z\frac{y}{x^2+y^2}\mathrm{Re}\left[\bar{R}^*_{l-1,m}(\mathbf{a})\,L_{l,m}(\mathbf{b})\right]\right\},\\
F_z(\mathbf{a}) &= -2q_a \sum_{l=1}^{\infty}\mathrm{Re}\left[\bar{R}^*_{l-1,m}(\mathbf{a})\,L_{l,m}(\mathbf{b})\right].
\end{aligned} \tag{10.71}$$

$l \geq 1$ and $m = l$:

$$\begin{aligned}
\Phi(\mathbf{a}) &= \mathrm{Re}\left[\bar{R}^*_{l,l}(\mathbf{a})\,L_{l,l}(\mathbf{b})\right],\\
F_x(\mathbf{a}) &= -2q_a \sum_{l=1}^{\infty}\left\{\mathrm{Re}\left[\left(l\frac{x}{x^2+y^2}, l\frac{x}{x^2+y^2}\right)\bar{R}^*_{l,l}(\mathbf{a})\,L_{l,l}(\mathbf{b})\right]\right\},\\
F_y(\mathbf{a}) &= -2q_a \sum_{l=1}^{\infty}\left\{\mathrm{Re}\left[\left(l\frac{y}{x^2+y^2}, -l\frac{y}{x^2+y^2}\right)\bar{R}^*_{l,l}(\mathbf{a})\,L_{l,l}(\mathbf{b})\right]\right\},\\
F_z(\mathbf{a}) &= 0.
\end{aligned} \tag{10.72}$$

In Equations 10.71 and 10.72, the expression in round brackets (x, y) represents a complex number $z = x + i\,y$. As it can be seen, each group of Equations 10.69 through 10.72 is considerably different one from another; and that justifies treating their index ranges separately in the computer program. These equations now directly translate into a computer code.

```
Subroutine ElectrostaticForceFMM(oxyz, coor, q, Lexpansion, energy, &
  potential, force, nL, nSzTri, tri, tmpExpansion, Rexpansion)
  ! The purpose of this subroutine is to compute potential and force due
  ! to far field in the position of particle (coor, q). It uses irregular
  ! multipole expansion (Lexpansion) of remote charges accumulated in the box
  ! holding that particle. Computed values are in Gauss units (charge in
  ! electrons, coordinates in Angstroms). Computation of energy includes
  ! correction against double counting.
  !
  implicit none
  !
  double precision, intent(in)    :: oxyz(3)   ! center of the box in global
coordinate system
  double precision, intent(in)    :: coor(3)   ! particle coordinates in
global coordinate system
  double precision, intent(in)    :: q         ! particle charge
  complex(16), intent(in)         :: Lexpansion(0:nSzTri) ! irregular
expansion of remote charges
  double precision, intent(inout) :: energy    ! energy due to far field
  double precision, intent(inout) :: potential ! potential on the particle
  double precision, intent(inout) :: force(3)  ! force on the particle
  integer,          intent(in)    :: nL        ! expansion truncation
  integer,          intent(in)    :: nSzTri    ! triangular size of expansion
array
  integer,          intent(in)    :: tri(0:nL) ! triangular pointers to
expansion elements
  complex(16), intent(inout)      :: tmpExpansion(0:nSzTri) ! scratch array
  complex(16), intent(inout)      :: Rexpansion(0:nSzTri)   ! scratch array
  !
  integer          :: i, L, aL1, m
  double precision :: x,y,z, x2y2, xDx2y2, yDx2y2, LxDx2y2, LyDx2y2, zxDx2y2,
zyDx2y2
  double precision :: dL, dm, sump, sumf(3), aprod, bprod
  double precision :: pot, potA, potB, r2, rij(3), forceA(3), forceB(3)
  complex(16)      :: cprod
  double precision, parameter :: zeroCoor = 1.0d-8, zero = 1.0d-14
  !
  ! determine particle coordinates in the local coordinate system of the box
  x = coor(1) - oxyz(1)
  y = coor(2) - oxyz(2)
  z = coor(3) - oxyz(3)

  x2y2 = x*x + y*y

  ! replace the particle coordinate by a very small increment to avoid
division by zero
  if(x2y2 < zero) then
    x = dsign(zeroCoor,x)
    x2y2 = x*x + y*y
  endif

  xDx2y2 = x / x2y2
  yDx2y2 = y / x2y2
  zxDx2y2 = z * xDx2y2
  zyDx2y2 = z * yDx2y2

  ! compute chargless multipole expansion for radius-vector of the particle
  call RegularExpansion(coor, 1.0d0, 1, oxyz, nL, nSzTri, tri, tmpExpansion,
Rexpansion)
```

```
  ! L = 0 and m = 0
  ! sump is sum for potential; sumf is sum for force
  ! divide by 2 so that the entire sum over m can be multiplied by 2 in the
end

  sump = dble(Rexpansion(0) * Lexpansion(0)) * 0.5d0
  sumf = 0.0d0

  ! L >= 1
  do L = 1, nL
    dL  = dble(L)
    aL1 = tri(L-1)
    i   = tri(L)

    LxDx2y2 = dL * xDx2y2
    LyDx2y2 = dL * yDx2y2

    ! m = 0
    ! potential
    sump  = sump + dble(Rexpansion(i) * Lexpansion(i)) * 0.5d0   ! Re[ R(l,0)

* L(l,0) ]

    ! force
    aprod = dble( Rexpansion(i)   * Lexpansion(i) ) * 0.5d0      ! Re[ R(l,0)
* L(l,0) ]
    bprod = dble( Rexpansion(aL1) * Lexpansion(i) ) * 0.5d0      ! Re[ R(l-
1,0) * L(l,0) ]
    sumf(1) = sumf(1) + LxDx2y2 * aprod - zxDx2y2 * bprod
    sumf(2) = sumf(2) + LyDx2y2 * aprod - zyDx2y2 * bprod
    sumf(3) = sumf(3) + bprod

    ! do half-range in index m (only positive values) and then multiply by 2
in the end
    ! 1 <= m <= L-1
    do m = 1, L-1
      i = i + 1

      ! potential
      sump = sump + dble(Rexpansion(i) * Lexpansion(i))  ! Re[ R(l,m) *
L(l,m) ]

      ! force
      dm = dble(m)
      bprod = dble( Rexpansion(aL1+m) * Lexpansion(i) )  ! Re[ R(l-1,m) *
L(l,m) ]
      cprod = Rexpansion(i) * Lexpansion(i)              ! complex R(l,m) *
L(l,m)
      sumf(1) = sumf(1) + dble( dcmplx(LxDx2y2,  dm*yDx2y2) * cprod ) - z *
xDx2y2 * bprod
      sumf(2) = sumf(2) + dble( dcmplx(LyDx2y2, -dm*xDx2y2) * cprod ) - z *
yDx2y2 * bprod
      sumf(3) = sumf(3) + bprod
    enddo

    ! m = L
    i = i + 1

    ! potential
```

```
    sump = sump + dble(Rexpansion(i) * Lexpansion(i))     ! Re[ R(1,1) *
L(1,1) ]

    ! force
    cprod = Rexpansion(i) * Lexpansion(i)                  ! complex R(1,1) *
L(1,1)
    sumf(1) = sumf(1) + dble( dcmplx(LxDx2y2,  LyDx2y2) * cprod )
    sumf(2) = sumf(2) + dble( dcmplx(LyDx2y2, -LxDx2y2) * cprod )
    !sumf(3) contribution is zero for m = 1
  enddo

  energy = energy + sump * q  ! skipping multiplication of sump by 2 corrects
against double counting
  potential = potential + sump * 2.0d0          ! multiply by 2
  force(1:3) = force(1:3) - sumf(1:3) * 2.0d0   ! to compensate for half-
summation in index m

end subroutine ElectrostaticForceFMM
```

Fortran subroutine *ElectrostaticForceFMM*() receives Cartesian coordinates of particle A in array *coor* on input. The center of the box, which holds that particle, has Cartesian coordinates stored in array *oxyz*. The coordinates of the particle and that of the box center come defined in the global coordinate system. Variable q carries the charge on particle A in units of electrons. The array *Lexpansion* carries irregular multipole moments $L(\mathbf{b})$ of the remote charges acting on particle A. Importantly, the origin of the coordinate system for expansion $L(\mathbf{b})$ is at the point *oxyz*. Variables *energy*, *potential*, and *force* serve as placeholders for the quantities to be computed. The discussed subroutine adds the computed value of each quantity on the top of the initial value coming with these variables. The initial value of energy, potential, and force could be those coming from the near-field contribution, and from the interaction of particle A with other particles located in the same box.

Computation begins from translating the particle coordinates into the local coordinate system of the box holding that particle. The next step is to check the condition of $x^2 + y^2 = 0$. When a point of measurement of force is exactly along the z axis, Equation 10.68 becomes ill-defined due to division by zero. Since there is always a probability of such a situation, a small corrective action is necessary.

One practical solution is to check on the particle's position relative to the z axis before calculating the electrostatic potential and force, and, if the condition is detected, add a small displacement to move the particle in the xy plane, resulting in the denominator becoming non-zero. The size of the displacement can be set small enough that no physical properties of the system will be affected. Such a procedure is illustrated in the following code snippet implemented in the Fortran language:

```
if(x*x + y*y < 1.0d-16) x = dsign(1.0d-8, x)
```

In the snippet, a test is made to detect whether $x^2 + y^2 < 10^{-16}$; if the pathologic condition is found to exist then the x coordinate is set to 10^{-8} Å retaining the sign of the original value. The criterion of 10^{-8} for coordinate increment comes as the result of numerical tests using double precision math.

Computation of potential and force on particle A employs the previously developed subroutine *RegularExpansion*(), which computes a chargeless ($q = 1$) regular multipole expansion $\bar{R}^*(\mathbf{a})$ of the radius vector of particle A with Cartesian coordinates *coor*. The computed expansion, *Rexpansion*, becomes defined about the origin *oxyz*. The remaining code in subroutine *ElectrostaticForceFMM*() represents a direct translation of Equations 10.69 through 10.72 along with their index conditions into Fortran program code.

To account for multiplication factor of 2, which appears in Equations 10.22, 10.71 and 10.72 when azimuthal index $m > 0$, the summation terms corresponding to $m = 0$ undergo multiplication by coefficient 0.5 so that the potential and force can be multiplied by factor of 2 only once after summation is completed thereby avoiding such operation in the inner-most *do-loop* over index m.

Computation of electrostatic energy from potential involves particle-particle interaction twice; once when particle A interacts with particle B being present in the expansion L, and second time when particle B in its turn interacts with the multipole expansion carrying the electrostatic contribution from particle *A*. To avoid arriving with the doubled value, computation of energy of particle *A* employs half-value of the potential.

Subroutine *ElectrostaticForceFMM*() computes energy, potential, and force for a single particle. However, in real life a box carries several particles in it. To account for that, subroutine *ElectrostaticForceFMM*() may be invoked from inside a *do-loop* over the particles in the box. Moreover, since a real system typically has its particles distributed over multiple boxes, an additional external *do-loop* over the boxes may be anticipated.

To test the correctness of the derived mathematical equations, and to check against possible bugs in the computer implementation, it is useful to compare the computed values of energy, potential, and force against those computed with help of direct sum method. The following Fortran program code illustrates the direct-sum computation of energy, potential, and force.

```
subroutine DirectSum(coor, q, nAtomsTotal, energy, potential, force)
  implicit none
  integer,              intent(in)  :: nAtomsTotal           ! number of particles
  double precision, intent(in)  :: coor(3,nAtomsTotal)   ! particle coordinates
  double precision, intent(in)  :: q(nAtomsTotal)        ! particle charge
  double precision, intent(out) :: energy                ! energy of the system
  double precision, intent(out) :: potential(nAtomsTotal),
force(3,nAtomsTotal)

  integer          :: iAtom, jAtom
  double precision :: r2, rij(3), pot, field(3)

  energy    = 0.0d0
  potential = 0.0d0
  force     = 0.0d0

  ! Compute electrostatic potential and force in the position of iAtom
  ! Opt for double summation to avoid writing into other processor memory
  !$omp parallel do reduction(+:energy) private(rij, r2, pot, field)
  do iAtom = 1, nAtomsTotal

    field = 0.0d0

    do jAtom = 1, iAtom-1
      rij(1:3)          = coor(1:3,iAtom) - coor(1:3,jAtom)
      r2                = rij(1)**2 + rij(2)**2 + rij(3)**2
      pot               = q(jAtom) / sqrt(r2)
      potential(iAtom)  = potential(iAtom) + pot
      field(1:3)        = field(1:3) + pot / r2 * rij(1:3)
    enddo

    do jAtom = iAtom+1, nAtomsTotal
      rij(1:3)          = coor(1:3,iAtom) - coor(1:3,jAtom)
      r2                = rij(1)**2 + rij(2)**2 + rij(3)**2
      pot               = q(jAtom) / sqrt(r2)
      potential(iAtom)  = potential(iAtom) + pot
      field(1:3)        = field(1:3) + pot / r2 * rij(1:3)
    enddo

    force(1:3,iAtom) = q(iAtom) * field(1:3)
    energy = energy + q(iAtom) * potential(iAtom)
  enddo

  energy = energy * 0.5d0    ! correct for double summation in energy
end subroutine DirectSum
```

Subroutine *DirectSum*() receives particle coordinates in array *coor* and particle charge in array q on input. Variable *nAtomsTotal* represents the number of particles in the system. Subroutine arguments *energy*, *potential*, and *force* hold the output of computation.

This example arranges computation in the form of a double sum where both the external and the internal *do-loops* iterate over the total number of particles. The requirement to avoid the condition $iAtom = jAtom$ breaks the internal *do-loop* in two parts. The double sum arrangement makes it easy to write a parallel computer code. Running the computation in parallel greatly reduces the time to solution.

In the above example, each parallel process deals with a separate particle, *iAtom*, and writes into the unique array elements, *potential*(*iAtom*) and *force*(1:3,*iAtom*) associated with that particle. Such localization of the write operations guarantees that each parallel process runs independently of other processes that maximizes computational efficiency. Unlike that, arranging the *do-loops* in a triangular shape by imposing condition $iAtom > jAtom$ will make multiple processes writing into the same array element. This complicates parallelization by the need to synchronize read/write operations to avoid a racing condition. The program code using a double sum in the *do-loop* arrangement avoids the synchronization issue, and that more than compensates for the time spent on computing same intermediary quantities twice.

Validation of the work of subroutine *ElectrostaticForceFMM*() presumes conducting one-to-one comparison of the computed values of potential and force in each particle position against the corresponding reference data obtained from subroutine *DirectSum*(). The compared values should agree up to ten or more significant digits when performing FMM computation with well separation parameter $ws = 2$ and with expansion truncation $nL = 30$. As an additional test of correctness of the implementation, the numerical difference must gradually decrease in absolute value when changing the expansion truncation from 10 to 50 in steps of 10.

11

Scaling of Multipole Translations

The multipole translation operations M2M, M2L, and L2L carry cumbersome factorial coefficients in their conventional formulation. Scaling of regular and irregular solid harmonics suggested by White and Head-Gordon eliminates those coefficients and greatly improves the efficiency of multipole operations. This, followed by meticulous analysis of continuous segments of azimuthal indices for the expansion terms in the multipole translation equations, leads to the development of highly optimized computer code.

11.1 Scaling of Multipole Translation Operations

Although the improvement in the computational efficiency of the electrostatic potential and force described in Chapter 10 has already established the usefulness of scaling solid harmonics, its most significant improvement results from the simplification of the multipole translation equations.

Chapter 8 defined the M2M, L2L, and M2L translations in canonical form in Equations 8.73, 8.94, and 8.100, respectively:

$$M_{lm}(B,A)=\sum_{j=0}^{l}\sum_{k=-j}^{j}\frac{(l+m)!}{(j+k)!\,(l-j+m-k)!}M_{jk}(B,B)R^{*}_{l-j,m-k}(B,A), \tag{8.73}$$

$$L_{lm}(B,A)=\sum_{j=l}^{\infty}\sum_{k=-j}^{j}(-1)^{j-l}\frac{(j-k)!}{(l-m)!\,(j-l+k-m)!}R^{*}_{j-l,k-m}(B,A)L_{jk}(B,B), \tag{8.94}$$

$$L_{lm}(B,A)=\sum_{j=0}^{\infty}\sum_{k=-j}^{j}(-1)^{j}\frac{(l+j-m-k)!}{(l-m)!\,(j+k)!}M_{jk}(B,B)S_{l+j,m+k}(B,A). \tag{8.100}$$

The L2L and M2L equations, Equations 8.94 and 8.100, also exist in the mathematically equivalent forms, depending on the designation of the helper expansions R^* and S:

$$L_{lm}(B,A)=\sum_{j=0}^{\infty}\sum_{k=-j}^{j}(-1)^{j}\frac{(l+j-m-k)!}{(l-m)!\,(j+k)!}R^{*}_{jk}(B,A)\,L_{l+j,m+k}(B,B), \tag{11.1}$$

$$L_{lm}(B,A)=\sum_{j=l}^{\infty}\sum_{k=-j}^{j}(-1)^{j-l}\frac{(j-k)!}{(l-m)!\,(j-l+k-m)!}M_{j-l,k-m}(B,B)S_{jk}(B,A). \tag{11.2}$$

To explain the notation, $M_{lm}(B, A)$ and $L_{lm}(B, A)$ are coefficients of a regular and irregular multipole expansion:

$$M_{lm}(B,A)=\sum_{i=1}^{N_B}q_i(B)R^{*}_{lm}(B,A),\quad\text{and}\quad L_{lm}(B,A)=\sum_{i=1}^{N_B}q_i(B)S_{lm}(B,A), \tag{11.3}$$

respectively. The first argument B in (B,A) points to the box, which holds the charge sources. The second argument A specifies the box, which provides the origin of the local coordinate system. In Equation 11.3, the sum runs over all point charges $q_i(B)$, where N_B is the number of particles in box B. Solid harmonics $R^*_{lm}(B,A)$ and $S_{lm}(B,A)$ represent expansion coefficients of a radius vector $\mathbf{r}_{AB}(i)$ of a unit charge i from box B in the coordinate system of box A. Similarly, helper expansions $R^*_{lm}(B,A)$ and $S_{lm}(B,A)$ in Equations 8.73, 8.94, and 8.100 represent expansion coefficients of a radius vector $\mathbf{r}_{AB}(B)$ that has its origin located in the center of box A and its tip pointing toward the center of box B.

Equations 8.73, 8.94, and 8.100 all have bulky factorial coefficients that are difficult to compute. An option to improve performance would be to use the scaled form of the solid harmonics:

$$\bar{R}^*_{lm} = \frac{1}{(l+m)!} R^*_{lm}, \quad \text{and} \quad \tilde{S}_{lm} = (l-m)!\, S_{lm}. \tag{11.4}$$

With that, rewriting Equation 8.73 to move the coefficient $(l + m)!$ to the left-hand side gives:

$$\frac{1}{(l+m)!} M_{lm}(B,A) = \sum_{j=0}^{l} \sum_{k=-j}^{j} \frac{1}{(j+k)!\,(l-j+m-k)!} M_{jk}(B,B) R^*_{l-j,m-k}(B,A), \tag{11.5}$$

and distributing the individual factorials to separate expansion terms such that

$$\frac{1}{(l+m)!} M_{lm}(B,A) = \sum_{i=1}^{N_B} q_i(B) \frac{1}{(l+m)!} R^*_{lm}(B,A) = \sum_{i=1}^{N_B} q_i(B) \bar{R}^*_{lm}(B,A) = \bar{M}_{lm}(B,A), \tag{11.6}$$

$$\frac{1}{(j+k)!} M_{jk}(B,B) = \bar{M}_{jk}(B,B), \tag{11.7}$$

$$\frac{1}{(l-j+m-k)!} R^*_{l-j,m-k}(B,A) = \bar{R}^*_{l-j,m-k}(B,A), \tag{11.8}$$

leads to the scaled form of M2M translation:

$$\bar{M}_{lm}(B,A) = \sum_{j=0}^{l} \sum_{k=-j}^{j} \bar{M}_{jk}(B,B) \bar{R}^*_{l-j,m-k}(B,A). \tag{11.9}$$

The simplicity of Equation 11.9 makes it computationally more efficient than its original form in Equation 8.73.

Rewriting the L2L translation given by Equation 8.94 proceeds in a similar way. Associating individual factorials with the matching expansion terms produces the following scaled expansions:

$$(l-m)!\, L_{lm}(B,A) = \tilde{L}_{lm}(B,A), \tag{11.10}$$

$$\frac{1}{(j-l+k-m)!} R^*_{j-l,k-m}(B,A) = \bar{R}^*_{j-l,k-m}(B,A), \tag{11.11}$$

$$(j-k)!\, L_{jk}(B,B) = \tilde{L}_{jk}(B,B), \tag{11.12}$$

and that leads to the scaled form of L2L equation:

$$\tilde{L}_{lm}(B,A) = \sum_{j=l}^{\infty}\sum_{k=-j}^{j}(-1)^{j-l}\bar{R}^{*}_{j-l,k-m}(B,A)\tilde{L}_{jk}(B,B). \tag{11.13}$$

Likewise pairing individual factorials with corresponding expansion terms in Equation 8.100 gives:

$$(l-m)!L_{lm}(B,A) = \tilde{L}_{lm}(B,A), \tag{11.14}$$

$$\frac{1}{(j+k)!}M_{jk}(B,B) = \bar{M}_{jk}(B,B), \tag{11.15}$$

$$(l+j-m-k)!S_{l+j,m+k}(B,A) = \tilde{S}_{l+j,m+k}(B,A), \tag{11.16}$$

and that leads to a scaled form of M2L translation:

$$\tilde{L}_{lm}(B,A) = \sum_{j=0}^{\infty}\sum_{k=-j}^{j}(-1)^{j}\bar{M}_{jk}(B,B)\tilde{S}_{l+j,m+k}(B,A). \tag{11.17}$$

Application of the same steps as above to the remaining two equations Equations 11.1 and 11.2 produces:

$$\tilde{L}_{lm}(B,A) = \sum_{j=0}^{\infty}\sum_{k=-j}^{j}(-1)^{j}\bar{R}^{*}_{jk}(B,A)\tilde{L}_{l+j,m+k}(B,B), \tag{11.18}$$

$$\tilde{L}_{lm}(B,A) = \sum_{j=l}^{\infty}\sum_{k=-j}^{j}(-1)^{j-l}\bar{M}_{j-l,k-m}(B,B)\tilde{S}_{jk}(B,A). \tag{11.19}$$

This completes the rewriting of the M2M, L2L, and L2M equations in a computationally efficient scaled form.

11.2 Program Code for M2M Translation

With the completion of the derivation of the equations for performing multipole translations, the next step is to convert these equations into computer code. Of the three, the M2M operation is the simplest and so its implementation will be done first.

The starting point for the computer code that performs an M2M operation on a scaled regular multipole expansion is Equation 11.9. As with the case of the computation of solid harmonics, it is beneficial to start the coding using an initially naïve implementation to ensure that all details of the equation are properly accounted for. Subroutine *M2M*() illustrates this implementation.

```
subroutine M2M(nL, nSzTri, tri, Ma, Mb, Mc)
!
! Shift the origin of a bar-scaled regular multipole expansion.
! Naive code.
!
implicit none
integer,      intent(in)   :: nL              ! expansion truncation
integer,      intent(in)   :: nSzTri          ! size of expansion array
integer,      intent(in)   :: tri(0:nL)       ! array of triangular pointers
complex(16),  intent(in)   :: Ma(0:nSzTri)    ! expansion to be translated
complex(16),  intent(in)   :: Mb(0:nSzTri)    ! helper expansion
```

```
complex(16), intent(out) :: Mc(0:nSzTri)    ! translated expansion
!
! Local variables
integer :: L, m, j, k, aJ, aLJ, aLm
!
! Zero-initialize the output array
Mc = dcmplx(0.0d0, 0.0d0)
!
do L = 0, nL
  do m = 0, L
    do j = 0, L

      aJ  = tri(j)
      aLJ = tri(L-j)
      aLm = tri(L)+m

      do k = -j, j
        if(iabs(m-k) <= L-j) then

          if(k < 0 .and. m-k < 0) then
            Mc(aLm) = Mc(aLm) + dconjg(Ma(aJ+iabs(k))) *
dconjg(Mb(aLJ+iabs(m-k))) * (-1.0d0)**m

          elseif(k < 0 .and. m-k >= 0) then
            Mc(aLm) = Mc(aLm) + dconjg(Ma(aJ+iabs(k))) * Mb(aLJ+m-k) * (-1.0d0)**k

          elseif(k >= 0 .and. m-k < 0) then
            Mc(aLm) = Mc(aLm) + Ma(aJ+k) * dconjg(Mb(aLJ+iabs(m-k))) * (-1.0d0)**(m-k)

          elseif(k >= 0 .and. m-k >= 0) then
            Mc(aLm) = Mc(aLm) + Ma(aJ+k) * Mb(aLJ+m-k)
          endif

        endif
      enddo  ! index k
    enddo  ! index j
  enddo  ! index m
enddo  ! index L
!
end subroutine M2M
```

Subroutine *M2M*() consists of four enclosed loops, reflecting the structure of Equation 11.9. The external loops over indices l and m determine the expansion element $\bar{M}_{l,m}$ to be computed. This element corresponds to variable *Mc*(*aLm*) in the program code. Using capital L in the program code for orbital index l helps to visually distinguish that letter from the numeric symbol of unity. The two inner loops, over j and k, sum up the contributions from the product $\bar{R}^*_{j,k}\bar{R}^*_{l-j,m-k}$, or $\bar{M}_{j,k}\bar{M}_{l-j,m-k}$: both notations of the expansion terms are equivalent, and will be used interchangeably. Since the values of index k depend on the values of index j, the loop over index k is inside the loop over j.

Before starting the loop over index k, the code initializes triangular pointers *aJ*, *aLJ*, and *aLm*, which represent the position of elements (j, 0), ($l-j$, 0), and (l, m), respectively, in a triangular-shaped array of the multipole expansion.

In the summation in Equation 11.9, index k varies from $-j$ to j. This requires using terms that have a negative azimuthal index, for example $\bar{R}^*_{j,-k}$, but, since only those terms that have a non-negative value of azimuthal index are stored, their negative counterparts have to be evaluated at run time. Because of the symmetry of scaled regular solid harmonics, the computer code for this computation is particularly simple:

$$\bar{R}^*_{j,-k} = (-1)^k \bar{R}_{j,k} = (-1)^k dconjg(\bar{R}^*_{j,k}), \tag{11.20}$$

where *dconjg*() is a Fortran 90 intrinsic function which performs complex conjugation.

The main part of M2M computation happens inside the loop over index k. Before the actual computation starts, two conditions that apply to azimuthal indices have to be tested. These requirements are that $-j \le k \le j$ and $-(l-j) \le m-k \le l-j$. The first of these is satisfied by setting the proper lower and upper values of index k in the corresponding *do*-loop. The second condition depends on two independently changing indices m and k, and, therefore, requires special attention. If the requirement of $-(l-j) \le m-k \le l-j$ cannot be satisfied then the expansion term $\bar{R}^*_{l-j,m-k}$ would be zero by definition. To keep the implementation simple, this condition is enforced in the code by placing the conditional *if (iabs(m – k) <= L – j)* inside the k loop. The Fortran 90 intrinsic function *iabs*() returns the absolute value of an argument of type integer.

The contribution of the product $\bar{M}_{j,k}\bar{M}_{l-j,m-k}$ breaks down into four conditions in the program code depending on whether or not the azimuthal indices k and $m-k$ are negative.

In the case where both indices are negative, that is $k < 0$ and $m-k < 0$, it is necessary to take the complex conjugate of both multipole terms, and multiply the product by the phase factor $(-1)^k(-1)^{m-k} = (-1)^m$, as indicated in Equation 11.20.

The second case of $k < 0$ and $m-k >= 0$ translates into $\bar{M}^*_{j,k}\bar{M}_{l-j,m-k}(-1)^k$, and at this point both of these multipole terms have already been computed; in addition, $(-1)^k = (-1)^{-k}$ so the simpler form of $(-1)^k$ can be used when $k < 0$.

Likewise, the third case of $k >= 0$ and $m-k < 0$ converts the original product into $\bar{M}_{j,k}\bar{M}^*_{l-j,|m-k|}(-1)^k$. Once again, $(-1)^{m-k} = (-1)^{k-m}$, regardless of whether $m-k$ is negative or non-negative.

In the fourth and final case, $k >= 0$ and $m-k >= 0$, that is, in the original product $\bar{M}_{j,k}\bar{M}_{l-j,m-k}$ both azimuthal indices are non-negative, so the elements of the expansion remain in their original form in the product.

To ensure that the complete subroutine *M2M*() works correctly, its results can be compared with known data. Computing the multipole expansions *M*1 and *M*2 for a charge set about two different origins, $O(x_1, y_1, z_1)$ and $O(x_2, y_2, z_2)$, respectively, then moving the expansion M1 from the origin $O(x_1, y_1, z_1)$ to $O(x_2, y_2, z_2)$, and comparing the result of translation against the previously computed expansion *M*2 provides the necessary validation mechanism. The following snippet illustrates this implementation.

```
implicit none
integer            :: n                  ! number of particles
double precision   :: c(3,n)             ! particle coordinates
double precision   :: q(n)               ! particle charges
integer            :: nL                 ! expansion truncation
integer            :: nSzTri             ! size of expansion array
integer            :: tri(0:nL)          ! triangular pointers
double precision   :: oxyz1(3)           ! origin1
double precision   :: oxyz2(3)           ! origin2
complex(16)        :: scr(0:nSzTri)      ! scratch array
complex(16)        :: M1(0:nSzTri)       ! regular expansion to be translated
complex(16)        :: M2(0:nSzTri)       ! reference regular expansion
complex(16)        :: Mb(0:nSzTri)       ! helper expansion
complex(16)        :: Mc(0:nSzTri)       ! translated regular expansion
logical            :: error

! set oxyz1 as the geometric center of particles
oxyz1 = 0.0d0
do i = 1, n
  oxyz1(1:3) = oxyz1(1:3) + c(1:3,i)
enddo
oxyz1 = oxyz1 / dble(n)

! set oxyz2 nearby of oxyz1
oxyz2 = oxyz1 + 1.0d0

! Get regular multipole expansion of a charge set about origin oxyz1
call RegularExpansion(c, q, n, oxyz1, nL, nSzTri, tri, scr, M1)

! Get regular multipole expansion of a charge set about origin oxyz2
call RegularExpansion(c, q, n, oxyz2, nL, nSzTri, tri, scr, M2)
```

```
! Get chargeless helper expansion for origin translation from oxyz1 to oxyz2
call RegularExpansion(oxyz1, 1.0d0, 1, oxyz2, nL, nSzTri, tri, scr, Mb)

! Perform M2M operation
call M2M(nL, nSzTri, tri, M1, Mb, Mc)

! Validate the result of transformation, Mc against the known data, M2
error = .FALSE.
do i = 0, nSzTri
    if(abs(dble( Mc(i) - M2(i))) > 1.0d-12) error = .TRUE.
    if(abs(aimag(Mc(i) - M2(i))) > 1.0d-12) error = .TRUE.
enddo
if(error) write(*,*) "Error in M2M computation"
```

In this validation, calculation of the multipole expansion about a specified origin employs the previously developed subroutine *RegularExpansion*(). A chargeless helper expansion, *Mb*, computed for the radius vector $(x_1 - x_2, y_1 - y_2, z_1 - z_2)$, assists in the origin translation, and subroutine *M2M*() performs the translation of the expansion *M*1 into an expansion *Mc*. The software implementation of subroutine *M2M*() is correct if the resulting expansion *Mc* is equivalent to the reference expansion *M*2, within the limits of double precision arithmetic.

This validation test assumes that the number of particles, *n*, particle coordinates, *c*, and their charges, *q*, have already been defined. The expansion truncation limit, *nL*, has a value in the range 30–50. Parameter *nSzTri* defines the size of expansion arrays, and array *tri* contains the triangular pointers. Details of their computation are given in Section 10.7.

Fortran 90 intrinsic functions *dble*() and *aimag*() employed in the code snippet return the real and imaginary parts of the complex number, respectively, so that these components can be individually validated. The tolerance error of 10^{-12} provides the lower boundary under double precision math.

Validation of the M2M code also provides a simple way to test a useful property of Equation 11.9. Using index substitution $l - j \to i$ and $m - k \to n$ that gives $j = l - i$ and $k = m - n$, respectively, followed by $i \to j$ and $n \to k$ leads to:

$$\bar{M}_{j,k}(B,B)\,\bar{R}^*_{l-j,m-k}(B,A) = \bar{M}_{l-i,m-n}(B,B)\,\bar{R}^*_{i,n}(B,A) = \bar{M}_{l-j,m-k}(B,B)\,\bar{R}^*_{j,k}(B,A). \tag{11.21}$$

Equation 11.21 indicates that the input expansions *Ma* and *Mb* in the call to subroutine *M2M*() are interchangeable. In other words, it does not matter whether the subroutine is invoked as *M2M(nL, nSzTri, tri, Ma, Mb, Mc)* or as *M2M(nL, nSzTri, tri, Mb, Ma, Mc)*: both generate the same result.

Once the naïve implementation of M2M operation is completed and its validity verified, the next step is to optimize the program code. This work requires a meticulous analysis of the different cases that apply to azimuthal indices, and finding a way to program the same outcome without using *if-then-else* operators, which slow down the program execution. As before, a rewriting of the *do*-loops helps achieve this goal. The result is the following computer code:

```
subroutine M2M(nL, nSzTri, tri, Ma, Mb, Mc)
!
! M2M operation:
! Shift the origin of a scaled regular multipole expansion.
! Optimized code.
! Mc(L,m) = sum(j=0,nL) sum(k=-j,+j) { Ma(j,k) * Mb(L-j,m-k) }
!
implicit none
integer,      intent(in)  :: nL               ! expansion truncation
integer,      intent(in)  :: nSzTri           ! size of expansion array
integer,      intent(in)  :: tri(0:nL)        ! array of triangular pointers
complex(16),  intent(in)  :: Ma(0:nSzTri)     ! expansion to be translated
complex(16),  intent(in)  :: Mb(0:nSzTri)     ! helper expansion
complex(16),  intent(out) :: Mc(0:nSzTri)     ! translated expansion
```

```
! Local variables
integer          :: L, m, j, k, aJ, aLJ, aLm

! (0,0)
Mc(0) = Ma(0) * Mb(0)

!$omp parallel do default(none) &
!$omp shared(nL,tri,nSzTri,Ma,Mb,Mc) &
!$omp private(L,m,j,k,aJ,aLJ,aLm)
do L = 1, nL
  do m = 0, L
    aLm = tri(L) + m

    Mc(aLm) = dcmplx(0.0d0, 0.0d0)

    ! j <= L-m
    do j = 0, L-m   !  0 <= m <= L-j
      aJ  = tri(j)
      aLJ = tri(L-j)

      ! k < 0;  m-k > 0;  m+|k| <= L-j
      do k = 1, min0(j,L-j-m)
        Mc(aLm) = Mc(aLm) + dconjg(Ma(aJ+k)) * Mb(aLJ+m+k) * dble((-1)**k)
      enddo

      ! k = 0
        Mc(aLm) = Mc(aLm) + Ma(aJ) * Mb(aLJ+m)

      ! 0 < k <= m;  m-k >= 0
      do k = 1, min0(j,m)
        Mc(aLm) = Mc(aLm) + Ma(aJ+k) * Mb(aLJ+m-k)
      enddo

      ! k > m;  m-k < 0
      do k = m+1, min0(j,L-j+m)
        Mc(aLm) = Mc(aLm) + Ma(aJ+k) * dconjg(Mb(aLJ+k-m)) * dble((-1)**(k-m))
      enddo

    enddo

    ! j > L-m
    do j = L-m+1, L   !  m > L-j
      aJ  = tri(j)
      aLJ = tri(L-j)

      ! k < 0;  m+|k| > L-j    (L-j,m-k) = 0

      ! k = 0 ... m-L+j-1    (L-j,m-k) = 0

      ! k = m-L+j ... m
      do k = m-L+j, min0(m,j)
        Mc(aLm) = Mc(aLm) + Ma(aJ+k) * Mb(aLJ+m-k)
      enddo

      ! k = m+1 ... L-j+m
      do k = m+1, min0(L-j+m,j)
```

```
        Mc(aLm) = Mc(aLm) + Ma(aJ+k) * dconjg(Mb(aLJ+k-m)) * dble((-1)**(k-m))
      enddo

    enddo

  enddo  ! index m
enddo  ! index L
!
end subroutine M2M
```

The first executable line computes the element $\bar{M}_{0,0}$ of the expansion. Since Equation 11.9 becomes trivial when $l = 0$ and $m = 0$ there is no need to have that element in the loop.

The loop structure for indices l and m remains the same as those in the naïve implementation of this subroutine, because the computation of each expansion element $\bar{M}_{l,m}$ is entirely independent of the computation of other elements for different value of indices l and m. With that, there is little to improve in the structure of these two *do*-loops, but their simplicity does provide a perfect opportunity for parallelization. Sorting out which variables are private and which are shared to each thread is all that is necessary for implementation of an elementary OpenMP parallelization. This analysis results in the following OpenMP directives:

```
!$omp parallel do default(none) &
!$omp shared(nL,tri,nSzTri,Ma,Mb,Mc) &
!$omp private(L,m,j,k,aJ,aLJ,aLm)
```

The directive "private" lists those variables that are going to be overwritten during the program execution. With this directive in place, each thread receives its own private copy of those variables. Shared variables are those that are typically used in read-only mode by the threads. An exception is the array *Mc*, which is simultaneously written to by multiple threads. Since each thread exclusively writes into a different array element, declaring array *Mc* as shared is safe.

Inside the two external *do*-loops over indices l and m, array element *Mc*(*aLm*), which represents expansion term $\bar{M}_{l,m}$, accumulates the contribution from the different summation elements defined by indices j and k. Because of the way it is used, this variable needs to be zero-initialized with the command *Mc*(*aLm*) = *dcmplx*(0.0*d*0, 0.0*d*0). The remaining part of the code represents a rearrangement of the enclosed *do*-loops over indices j and k to eliminate branching that had proliferated in the naïve implementation.

The reason the naïve implementation of the code was so entangled in *if-then-else* constructs is the appearance of term $\bar{M}_{l-j,m-k}$ in Equation 11.9, which has its orbital and azimuthal indices expressed in a non-trivial form. Having the index $m-k$ vary independently of the corresponding orbital index $l-j$ meant that the program code had to constantly check that the azimuthal index stayed within the boundaries of the corresponding orbital index.

Another complication comes from the decision to store only the expansion elements with non-negative value of the azimuthal index in the computer memory to save space. This choice requires tracking the sign of the azimuthal index, and expressing the expansion terms carrying a negative azimuthal index over those having a positive value of the index. Such tracking becomes difficult when indices m and k vary independently in their separate *do*-loops, and results in the use of *if-then-else* constructs in the naïve implementation of the program code.

In a first step to untangle the knot caused by conditional constructs, the *do*-loop over index j is broken into two loops that run to separate limits. One of those loops goes from 0 to $l-m$, and the other goes from $l-m+1$ to l, thereby jointly covering the required range for index j. The idea behind this decomposition is to make the value of index m fall into a predictable range depending on the value of indices l and j. For instance, the condition of $j \leq l-m$, which is imposed by the first *do*-loop, *do* $j = 0$, *L-m*, automatically translates to $m \leq l-j$. Likewise, the second *do*-loop, *do* $j = L-m+1$, L enforces the condition of $j \geq l-m+1$, which corresponds to $m > l-j$. The rearrangement of the *do*-loop over index j serves as a precondition for decomposition of the *do*-loop over index k which implements the tracking of the values of azimuthal index $m-k$ in expansion element $\bar{M}_{l-j,m-k}$.

Unlike index m, which is always non-negative, index k can assume negative values. Provided that the value of indices l, j, and m are already known, all that is left is to search for the possible values of index k such that the conditions $-j \le k \le j$ and $-l + j \le m - k \le l - j$ are met.

The presence of expansion term $\bar{M}_{j,k}$ in Equation 11.9 restricts the left-most contiguous range for negative values of index k by the condition $-j \le k \le -1$ or $|k| \le j$. For any $k < 0$, it automatically follows that index $m - k > 0$. With that, as the negative value of index k grows in absolute value, it becomes necessary to watch for the upper value of the azimuthal index $m - k$ to stay less or equal to the value of the orbital index $l - j$ in the second expansion term $\bar{M}_{l-j,m-k}$ in Equation 11.9. From the condition of $m - k \le l - j$, it follows that $-k \le l - j - m$. Since $k < 0$, which is equivalent to $-k > 0$, one can rewrite this condition as $|k| \le l - j - m$. Combining the two conditions $|k| \le j$ and $|k| \le l - j - m$ allows the negative values of index k to be replaced by the corresponding positive values in the *do*-loop, so that index k that starts from the value of 1, and goes all the way up to the value of j or $l - j - m$, whichever is smaller. Fortran 90 intrinsic function *min0()*, which works with integer arguments, performs the necessary testing of the upper value without branching the code. This leads to the following *do*-loop structure:

```
do k = 1, min0(j,L-j-m)
enddo
```

The case of negative values of index k in Equation 11.9 can be further simplified by explicitly moving the negative sign outside the symbol letter, and replacing the product $\bar{M}_{j,k}\bar{M}_{l-j,m-k}$ by $\bar{M}_{j,-k}\bar{M}_{l-j,m+k}$, where variable k is itself treated as positive number. Once this is done, symmetry relation Equation 10.38 can then be applied to the expansion terms that have a negative azimuthal index, turning the product of multipole terms in Equation 11.9 into:

$$\bar{M}_{j,-k}\,\bar{M}_{l-j,m+k} = \bar{M}^*_{j,k}\,\bar{M}_{l-j,m+k}(-1)^k = dconjg(\bar{M}_{j,k})\,\bar{M}_{l-j,m+k}(-1)^k. \tag{11.22}$$

This equation, together with the previously determined loop structure, translates into the code snippet:

```
! k < 0;  m-k > 0;  m+|k| <= L-j
do k = 1, min0(j,L-j-m)
   Mc(aLm) = Mc(aLm) + dconjg(Ma(aJ+k)) * Mb(aLJ+m+k) * dble((-1)**k)
enddo
```

In it, array indices $aJ + k$ and $aLJ + m + k$ represent addresses in memory that point to expansion elements $\bar{M}_{j,k}$ and $\bar{M}_{l-j,m+k}$, respectively. The expression *dble*((–1)**k) in the code snippet raises integer number -1 to the k-th power, and converts the result into a double precision number.

Choosing the right range for index k cannot by itself eliminate the use of a *if-then-else* construct in the *do*-loop over index k. Having index k exploring negative values brings forth an additional requirement that the value of $l - j - m$ in the inequality $|k| \le l - j - m$ must be non-negative for all values of variables l, j, and m. Having vertical bars around variable k in the inequality indicates that its left side must be non-negative. All these conditions require that the desired outcome of the expression $l - j - m$ must be programmatically enforced by making index j run from 0 to $l - m$ in the corresponding parent *do*-loop. Note that when $j = l - m$ the *do*-loop over index k becomes *do* $k = 1, 0$, which results in the loop being skipped.

The value of $k = 0$ represents a special case in which the product of the multipole terms in Equation 11.9 simplifies to $\bar{M}_{j,0}\,\bar{M}_{l-j,m}$. Including the case of $k = 0$ into the *do*-loop that deals with negative values of index k would redundantly apply a complex conjugate operation to $\bar{M}_{j,0}$, a real number, and would unnecessarily raise -1 to the zeroth power. Therefore, it is best from the performance standpoint to explicitly write the case of $k = 0$ as a single line of the program code

```
! k = 0
Mc(aLm) = Mc(aLm) + Ma(aJ) * Mb(aLJ+m)
```

that gets executed after the *do*-loop over negative values of index k is completed.

The case of positive values of index k is a bit more complicated than that for negative values. For any $k > 0$, the term $\bar{M}_{j,k}$ in Equation 11.9 does not require any special treatment, and remains in its original form. It is the term $\bar{M}_{l-j,m-k}$ that requires checking the azimuthal index $m - k$ to determine if it is negative or non-negative. With that, the next step is to address the case of $m - k \geq 0$, when $k > 0$.

Continuing the work from the case of $m - k \geq 0$ rather than from $m - k < 0$ is preferable because it preserves the continuity in the analysis of possible values of index k. The condition of $m - k \geq 0$ when $m \geq 0$ and $k > 0$ makes it possible for a *do*-loop to start from the value of $k = 1$. Correspondingly, rewriting $m - k \geq 0$ as $k \leq m$ gives a hint as to the upper value of index k. An additional constraint on the upper value of index k is that the expansion term $\bar{M}_{j,k}$ requires $k \leq j$. With that, for any given values of indices l, m, and j, the upper value for index k becomes either $k = m$ or $k = j$, whichever is smaller, so that both conditions $k \leq m$ and $k \leq j$ are satisfied simultaneously. This analysis translates into the following code snippet:

```
! 0 < k <= m implies m-k >= 0
do k = 1, min0(j,m)
   Mc(aLm) = Mc(aLm) + Ma(aJ+k) * Mb(aLJ+m-k)
enddo
```

Since both azimuthal indices in the expansion terms in the product from Equation 11.9 assume positive values, they appear in the *do*-loop in their original form. In the case of $j = 0$, the *do*-loop becomes *do* $k = 1, 0$, which makes the loop skip the computation since the case of $k > 0$ is not valid when $j = 0$ due to the requirement of $k \leq j$ that comes from $\bar{M}_{j,k}$. Likewise, the *do*-loop is skipped when $m = 0$ because of the failure to satisfy the requirement that $m - k \geq 0$ when $k > 0$.

The remaining continuous range of values of index k for $k > 0$ and $j \leq l -$ m comes with the condition of $m - k < 0$. This implies that the term $\bar{M}_{j,k}$ remains unmodified, and that the term $\bar{M}_{l-j,m-k}$ turns to $\bar{M}^{*}_{l-j,k-m}(-1)^{k-m}$, where $k - m > 0$. Determining the lower and upper values of the index k for the corresponding *do*-loop is a bit more complicated. In the previous case of $k > 0$ and $m - k \geq 0$, the loop stopped at $k = min0(j, m)$.

The next loop then starts immediately after the previous loop stopped, at $k = min0(j, m) + 1$. This expression can be simplified to $k = m + 1$ due to the constraint of $k > m$. The upper value of index k in this *do*-loop must satisfy two conditions. One is that $k \leq j$. The other is that the upper value of index $m - k$ must remain within the boundary of the corresponding orbital index $l - j$, that is $|m - k| \leq l - j$. Resolving the latter about variable k when $k > m$ gives $k - m \leq l - j$, leading to $k \leq l - j + m$. Using the Fortran 90 intrinsic function *min0* to determine the smallest value of k to simultaneously satisfy the conditions $k \leq j$ and $k \leq l - j + m$ leads to the following code snippet:

```
! k > m implies m-k < 0
do k = m+1, min0(j,L-j+m)
   Mc(aLm) = Mc(aLm) + Ma(aJ+k) * dconjg(Mb(aLJ+k-m)) * dble((-1)**(k-m))
enddo
```

Note that, as intended, this *do*-loop works only when $m < j$ and that when $m \geq j$, the loop `do k = j + 1, j` is skipped.

The second and final part of the computer program for M2M translation deals with the condition $j > l - m$, which represents the remaining range for index j. Since the upper limit for index j is fixed to $j = l$ in Equation 11.9, the *do*-loop over index j becomes:

```
! j > L-m
do j = L-m+1, L   !  m > L-j
```

Inside this loop, a number of different ranges for index k appear as separate *do*-loops to handle the various conditions. As before, the approach is to start from the left-most negative value of index k in the range, and continue the analysis to the right-most positive number. With that, the first domain to analyze is the negative range, $k < 0$.

A general rule for determining the allowed range for index k is to verify that all values within that range adhere to the requirement that the absolute value of the azimuthal indices k and $m-k$ in $\bar{M}_{j,k}$ and $\bar{M}_{l-j,m-k}$ expansion terms remain equal to or less than the corresponding orbital indices j and $l-j$, respectively.

Determining the allowed range of values of azimuthal index k in $\bar{M}_{j,k}$ is a straightforward procedure, which leads to $-j \le k \le j$. However, determining the acceptable range of values of index k in $\bar{M}_{l-j,m-k}$ is more complicated because the range boundaries depend on the running value of index m. This issue is managed by placing a condition $j > l - m$ onto the orbital index $l-j$, and using that to control the allowed values of the azimuthal index $m-k$. Having the parent *do*-loop over index j satisfying the condition $j > l - m$ helps to establish that $m > l - j$. Because of that, it is impossible for index k to have a negative value. Any negative value of k would only push $m-k$ to a greater positive number, which would zero out the contribution from $\bar{M}_{l-j,m-k}$. A similar analysis also rules out the possibility that index k could assume a zero value.

The next step is to find the proper positive range for index k that satisfies the condition $|m-k| \le l-j$. The starting value of index k immediately follows from resolving the equality $m-k = l-j$ about variable k; this results in $k = m - l + j$. As the positive value of index k increases, the index $m-k$ decreases to zero and then becomes negative. The change in sign of the azimuthal index $m-k$ requires splitting the positive range of index k into two segments, inside each of which the sign of the azimuthal index $m-k$ would remain constant.

The first segment in the positive range of index k keeps the azimuthal index $m-k$ non-negative. The upper value of index k in this segment becomes $k = m$ or $k = j$, whichever is smaller, ensuring that the index k stays in the right range for the expansion term $\bar{M}_{j,k}$. This portion of the computation is arranged by constructing the *do*-loop in the form

```
! k = m-L+j ... m
do k = m-L+j, min0(m,j)
  Mc(aLm) = Mc(aLm) + Ma(aJ+k) * Mb(aLJ+m-k)
enddo
```

Since the azimuthal indices of both expansion terms in this range are non-negative, the product $\bar{M}_{j,k}\,\bar{M}_{l-j,m-k}$ retains its original form in the summation element.

The second and final range of index k continues the previous range from the point where it ended. With that, the starting value for index k becomes $k = m + 1$. The end value comes from the relation $-(m-k) = l-j$, where the sign of index $m-k$ is inverted. Resolving that relation about variable k leads to $k = l - j + m$. Once again, the value of index k cannot exceed the value of the corresponding orbital index, j. This condition, combined with the determined lower and upper limits for index k, leads to the following *do*-loop structure:

```
! k = m+1 ... L-j+m
do k = m+1, min0(L-j+m,j)
  Mc(aLm) = Mc(aLm) + Ma(aJ+k) * dconjg(Mb(aLJ+k-m)) * dble((-1)**(k-m))
enddo
```

Since the value of azimuthal index $m-k$ is negative inside the *do*-loop, the program structure requires converting the expansion term $\bar{M}_{l-j,m-k}$ in Equation 11.9 to that with a positive value of the azimuthal index by using the relation $\bar{M}_{j,k}\,\bar{M}_{l-j,m-k} = \bar{M}_{j,k}\,\bar{M}^{*}_{l-j,k-m}\,(-1)^{k-m}$. This completes the conversion of Equation 11.9 into optimized computer code.

11.3 Program Code for M2L Translation

The M2L translation converts a regular multipole expansion M into an irregular multipole expansion L about a different origin with help of Equation 11.17. This translation requires a chargeless irregular solid harmonics expansion, S, to define the origin shift for the resulting irregular multipole expansion, L. To optimize computational performance, Equation 11.17 employs all expansion terms in their corresponding scaled forms:

$$\tilde{L}_{lm}(B,A) = \sum_{j=0}^{\infty}\sum_{k=-j}^{j}(-1)^j \bar{M}_{jk}(B,B)\,\tilde{S}_{l+j,m+k}(B,A). \tag{11.17}$$

As before, it is convenient to start the implementation of M2L program code by devising the simple naïve code to get a notion of the underlying equation. A straightforward translation of Equation 11.17 into program code requires little effort, and reduces the chance of errors due to potential misunderstanding of the intricate aspects of the equation. The following example provides a simple implementation of subroutine *M2L*():

```
subroutine M2L(nL, nSzTri, tri, Ma, Sb, Lc)
!
! Convert bar-scaled regular multipole expansion Ma into tilde-scaled
! irregular multipole expansion Lc defined about a different origin.
! Naive code.
!
implicit none
integer,      intent(in)  :: nL              ! expansion truncation
integer,      intent(in)  :: nSzTri          ! size of expansion array
integer,      intent(in)  :: tri(0:nL)       ! array of triangular pointers
complex(16),  intent(in)  :: Ma(0:nSzTri)    ! expansion to be translated
complex(16),  intent(in)  :: Sb(0:nSzTri)    ! helper expansion
complex(16),  intent(out) :: Lc(0:nSzTri)    ! translated expansion
!
! Local variables
integer :: L, m, j, k, aJ, aLJ, aLm
!
! Zero-initialize the output array
Lc = dcmplx(0.0d0, 0.0d0)
!
do L = 0, nL
  do m = 0, L
    do j = 0, nL
      if(L+j <= nL) then
        aJ  = tri(j)
        aLJ = tri(L+j)
        aLm = tri(L)+m
        do k = -j, j
          if(iabs(m+k) <= L+j) then
            if(k < 0 .and. m+k < 0) then
             Lc(aLm) = Lc(aLm) + dconjg(Ma(aJ+iabs(k))) * dconjg(Sb(aLJ+iabs(m+k)))
             * (-1.0d0)**(m+j)
            elseif(k < 0 .and. m+k >= 0) then
             Lc(aLm) = Lc(aLm) + dconjg(Ma(aJ+iabs(k))) * Sb(aLJ+m+k) * (-1.0d0)**(k+j)
            elseif(k >= 0 .and. m+k < 0) then
             Lc(aLm) = Lc(aLm) + Ma(aJ+k) * dconjg(Sb(aLJ+iabs(m+k))) * (-1.0d0)**(m+k+j)
            elseif(k >= 0 .and. m+k >= 0) then
              Lc(aLm) = Lc(aLm) + Ma(aJ+k) * Sb(aLJ+m+k) * (-1.0d0)**j
          endif
        enddo  ! index k
      endif
    enddo  ! index j
  enddo  ! index m
enddo  ! index L
!
end subroutine M2L
```

In this program code, expansion *Ma* represents the bar-scaled regular multipole expansion, $\bar{M}_{jk}$, to be shifted; expansion *Sb* is a tilde-scaled chargeless irregular helper expansion, $\tilde{S}_{l+j,m+k}$, which defines the origin shift; *Lc* is the resulting tilde-scaled irregular multipole expansion, $\tilde{L}_{lm}$.

Computation starts with initializing the array *Lc* to zero, so that the computed results can be accumulated in the array elements in the loop.

The main program code is arranged into four enclosed loops over indices *l*, *m*, *j*, and *k*, in that order. The outermost loop over index *l* spans the range $l = 0$ to $l = nL$, where parameter *nL* is the upper value of the orbital index in the truncated expansion.

In the next loop, index *m* spans all computed azimuthal indices in the output expansion *Lc*. The general range for index *m* is $-l \leq m \leq l$, but because of the symmetry of the azimuthal index only the terms with a non-negative value of index *m* need to be computed. Therefore the *do*-loop over index *m* is set to go from 0 to *l*.

The next *do*-loop over index *j* formally goes from zero to infinity in Equation 11.17. Since all expansions used here are truncated to the upper value, *nL*, of the orbital index, the loop over index *j* is limited to run from 0 to *nL*. Inside this loop an *if-then* clause checks to ensure that the orbital index $l + j$ does not exceed the value of *nL* in the similarly truncated expansion *Sb*, to avoid an out-of-bounds error when referring to array element $\tilde{S}_{l+j,m+k}$.

The fourth and last innermost loop over index *k* goes from $k = -j$ to $k = j$, according to Equation 11.17; this also automatically corresponds to the correct boundaries for the index *k* in expansion element $\bar{M}_{jk}$. For each specific expansion element $\tilde{S}_{l+j,m+k}$, the correct boundaries of azimuthal index $m + k$ need to be manually enforced on every cycle of the loop. This is achieved by the required condition $|m + k| \leq l + j$ in the *if-then* clause at the start of the loop over index *k*.

This loop structure takes care of all the allowed ranges of orbital and azimuthal indices in a simple manner. The last step is to properly express the expansion terms that have negative values of azimuthal indices in terms of their positive indices. The presence of two expansion terms $\bar{M}_{jk}$ and $\tilde{S}_{l+j,m+k}$ in Equation 11.17 gives rise to four cases for azimuthal indices *k* and $m + k$:

$$\begin{aligned}
k < 0 \quad &\text{and} \quad m + k < 0, \\
k < 0 \quad &\text{and} \quad m + k \geq 0, \\
k \geq 0 \quad &\text{and} \quad m + k < 0, \\
k \geq 0 \quad &\text{and} \quad m + k \geq 0.
\end{aligned}$$

The expansion terms that have a negative azimuthal index transform to those with a positive index using the standard relation $\bar{M}_{j,-k} = \bar{M}^*_{j,k}(-1)^k$ and $\tilde{S}_{l+j,-m-k} = \tilde{S}^*_{l+j,m+k}(-1)^{m+k}$. Fortran 90 intrinsic function *iabs*() returns the absolute value of its argument of integer type, so that the triangular index calculation incorporates the azimuthal offset as a non-negative number: *aJ* + *iabs*(*k*), in *Ma*(*aJ* + *iabs*(*k*)).

Validation of subroutine *M2L*() involves computing a regular multipole expansion, *M*1, of a charge set, converting it to an irregular multipole expansion *Lc* with the help of an M2L operation about a preset origin, and then comparing the resulting expansion against the reference expansion *L*2, which is computed directly about the same origin. To achieve that, the validation algorithm generates regular and irregular multipole expansions *M*1 and *L*2 about the origin $O(x_1, y_1, z_1)$ and $O(x_2, y_2, z_2)$, respectively, by calling subroutines *RegularExpansion*() and *IrregularExpansion*(). The origin $O(x_1, y_1, z_1)$ is the coordinate of the geometric center of the particles. The other origin $O(x_2, y_2, z_2)$ is the coordinate of the center of a distant box that encloses a set of charge sources. This fulfills the requirement that the virtual vector corresponding to origin translation should be longer than the virtual vector corresponding to the regular multipole expansion that is being transformed by the M2L operation.

In principle, calculation of the coordinate point $O(x_2, y_2, z_2)$ would be easy to perform by extending the radius vector of the center $O(x_1, y_1, z_1)$ for the system of charges in the direction away from the center of the global coordinate system by scaling up the coordinate component of the radius vector of point $O(x_1, y_1, z_1)$. However, it is possible that the point $O(x_1, y_1, z_1)$ is near to the center of the global coordinate system, so a simple scaling of the coordinate components of point $O(x_1, y_1, z_1)$ would be inadequate. In such a case, the use of a fixed multiplication factor might not result in a point to be outside the region occupied by the charge sources, and the choice of the scaling factor would then have to become system dependent. To avoid such a complication, the coordinate components of point $O(x_1, y_1, z_1)$ are first incremented by the value of the radius of a sphere that encloses the charge sources while preserving the sign of the components of point $O(x_1, y_1, z_1)$. Once that is done, the coordinate components can then be scaled by a constant factor to produce the new origin, $O(x_2, y_2, z_2)$. This procedure ensures that the generated point $O(x_2, y_2, z_2)$ is far away from the charge sources.

Computing a chargeless helper expansion *Sb* for the radius vector $(x_1 - x_2, y_1 - y_2, z_1 - z_2,)$ by using subroutine *IrregularExpansion*(), and substituting it together with the expansion *M*1 into subroutine *M2L*() produces the target expansion *Lc*, which would be equivalent to *L*2 if the program code for M2L translation is correct. The following code snippet implements the validation algorithm.

```
implicit none
integer           :: n                 ! number of particles
double precision :: c(3,n)             ! particle coordinates
double precision :: q(n)               ! particle charges
integer           :: nL                ! expansion truncation
integer           :: nSzTri            ! size of expansion array
integer           :: tri(0:nL)         ! triangular pointers
double precision :: oxyz1(3)           ! origin1
double precision :: oxyz2(3)           ! origin2
double precision :: radius             ! radius of enclosing sphere
complex(16)       :: scr(0:nSzTri)     ! scratch array
complex(16)       :: M1(0:nSzTri)      ! regular expansion to be translated
complex(16)       :: L2(0:nSzTri)      ! reference irregular expansion
complex(16)       :: Sb(0:nSzTri)      ! helper expansion
complex(16)       :: Lc(0:nSzTri)      ! translated irregular expansion
logical           :: error

! Determine oxyz1 as the geometric center of particles
oxyz1 = 0.0d0
do i = 1, n
  oxyz1(1:3) = oxyz1(1:3) + c(1:3,i)
enddo
oxyz1 = oxyz1 / dble(n)

! Place oxyz2 far away of the charge set
oxyz2  = maxval(c, dim = 2) - oxyz1
radius = dsqrt(oxyz2(1)**2 + oxyz2(2)**2 + oxyz2(3)**2)
oxyz2  = (oxyz1 + dsign(radius, oxyz1)) * 3.0d0

! Get regular multipole expansion of a charge set about origin oxyz1
call RegularExpansion(c, q, n, oxyz1, nL, nSzTri, tri, scr, M1)

! Get irregular multipole expansion of a charge set about origin oxyz2
call IrregularExpansion(c, q, n, oxyz2, nL, nSzTri, tri, scr, L2)

! Get chargeless helper expansion for origin translation from oxyz1 to oxyz2
call IrregularExpansion(oxyz1, 1.0d0, 1, oxyz2, nL, nSzTri, tri, scr, Sb)

! Perform M2L operation
call M2L(nL, nSzTri, tri, M1, Sb, Lc)

! Validate the result of transformation, Lc against the known data, L2
error = .FALSE.
do i = 0, nSzTri
   if(abs(dble( Lc(i) - L2(i))) > 1.0d-12) error = .TRUE.
```

```
    if(abs(aimag(Lc(i) - L2(i))) > 1.0d-12) error = .TRUE.
enddo
if(error) write(*,*) "Error in M2L computation"
```

In this example, the number of particles, *n*, particle coordinates, *c*, and their charges, *q*, have already been defined. The expansion truncation, *nL*, is a number in the range from 30 to 50. Parameter *nSzTri* provides the address of the last element in the multipole expansion, and array *tri* holds precomputed triangular pointers. Section 10.7 explains the details of the computation of parameter *nSzTri* and the content of array *tri*. Fortran 90 intrinsic functions *dble*() and *aimag*() return the real and imaginary parts of the complex number, respectively so that each of those components may be independently evaluated against the reference value. The criterion for an error, 10^{-12}, represents the lower boundary of a double precision computation. However, truncation of the infinite summation in Equation 11.17 can cause additional numerical errors, so, during testing, setting the value of parameter *nL* to 30 or higher will help minimize the truncation error.

No optimization of the computer code for M2L is needed at this moment since the L2L computer code to be discussed next performs both types of operation and supersedes the implementation.

11.4 Program Code for L2L Translation

The purpose of the L2L operation is to translate the origin of an irregular multipole expansion from one coordinate system to another. Equation 11.13 describes the math of this translation for the scaled form of multipole expansions:

$$\tilde{L}_{lm}(B,A)=\sum_{j=l}^{\infty}\sum_{k=-j}^{j}(-1)^{j-l}\bar{R}^{*}_{j-l,k-m}(B,A)\,\tilde{L}_{jk}(B,B). \tag{11.13}$$

The naïve coding of this equation follows the same rules previously outlined for the M2M and M2L translations: setting up the four enclosed loops, checking that the orbital and azimuthal indices remain within the allowed limits, and converting the expansion elements with negative value of the azimuthal index to those with the positive value. Implementing these basic principles leads to the following L2L program code:

```
subroutine L2L(nL, nSzTri, tri, Ra, Lb, Lc)
!
! Translate tilde-scaled irregular multipole expansion
! from one coordinate system to another. Naive implementation.
!
implicit none
integer,       intent(in)  :: nL               ! expansion truncation
integer,       intent(in)  :: nSzTri           ! size of expansion array
integer,       intent(in)  :: tri(0:nL)        ! array of triangular pointers
complex(16),   intent(in)  :: Ra(0:nSzTri)     ! helper expansion
complex(16),   intent(in)  :: Lb(0:nSzTri)     ! expansion to be translated
complex(16),   intent(out) :: Lc(0:nSzTri)     ! translated expansion
!
! Local variables
integer :: L, m, j, k, aJ, aJL, aLm
!
! Zero-initialize the output array
Lc = dcmplx(0.0d0, 0.0d0)
!
do L = 0, nL
  do m = 0, L
    do j = L, nL
```

```
      aJ  = tri(j)
      aJL = tri(j-L)
      aLm = tri(L)+m
      do k = -j, j
        if(iabs(k-m) <= j-L) then
          if(k-m < 0 .and. k < 0) then
            Lc(aLm) = Lc(aLm) + dconjg(Ra(aJL+iabs(k-m))) *
             dconjg(Lb(aJ+iabs(k))) * (-1.0d0)**(j-L-m)
          elseif(k-m <  0 .and. k >= 0) then
            Lc(aLm) = Lc(aLm) + dconjg(Ra(aJL+iabs(k-m))) * Lb(aJ+iabs(k)) *
             (-1.0d0)**(j-L+k-m)
          elseif(k-m >= 0 .and. k < 0) then
            Lc(aLm) = Lc(aLm) + Ra(aJL+iabs(k-m)) * dconjg(Lb(aJ+iabs(k))) *
             (-1.0d0)**(j-L+k)
          elseif(k-m >= 0 .and. k >= 0) then
           Lc(aLm) = Lc(aLm) + Ra(aJL+iabs(k-m)) * Lb(aJ+iabs(k)) * (-1.0d0)**(j-L)
          endif
        endif
      enddo  ! index k
    enddo  ! index j
  enddo  ! index m
enddo  ! index L
!
end subroutine L2L
```

In this code, the results are accumulated in the array *Lc*, so *Lc* needs to be zero-initialized before the loops start.

```
! Zero-initialize the output array
Lc = dcmplx(0.0d0, 0.0d0)
```

To keep the implementation simple, the *do*-loop structure for indices l, m, j, and k directly corresponds to that in Equation 11.13. Since all expansions are truncated, the *do*-loop over index l goes from 0 to nL, the *do*-loop over index m adheres to the requirement of $0 \leq m \leq l$, and the *do*-loop over index j runs from l to nL, in accordance with Equation 11.13, and accounts for the truncation of the expansion. The innermost *do*-loop over index k runs from $-j$ to j, as specified in Equation 11.13.

The first requirement to check inside the *do*-loop over index k is that the azimuthal index $k - m$ remains within the boundaries of orbital index $j - l$. The following code snippet illustrates this implementation:

```
do k = -j, j
  if(iabs(k-m) <= j-L) then
    ...
  endif
enddo  ! index k
```

After that, the usual conditionals will be those that represent all possible combinations of the sign of the azimuthal indices:

$$k-m<0 \quad \text{and} \quad k<0,$$
$$k-m<0 \quad \text{and} \quad k\geq 0,$$
$$k-m\geq 0 \quad \text{and} \quad k<0,$$
$$k-m\geq 0 \quad \text{and} \quad k\geq 0.$$

When both azimuthal indices have negative values, that is, $k - m < 0$ and $k < 0$, the product of the expansion terms $(-1)^{j-l}\bar{R}^*_{j-l,-k+m}\tilde{L}_{j,-k}$ in Equation 11.13 is replaced by $(-1)^{j-l-m}\, dconjg(\bar{R}^*_{j-l,k-m})\, dconjg(\tilde{L}_{j,k})$. Likewise, the condition of $k-m<0$ and $k\geq 0$ leads to the replacement of $(-1)^{j-l}\bar{R}^*_{j-l,-k+m}\tilde{L}_{j,k}$ by

$(-1)^{j-l+k-m} dconjg(\bar{R}^*_{j-l,k-m}) \tilde{L}_{j,k}$, where the Fortran 90 intrinsic function *dconjg*() yields the complex conjugate of the variable specified in its argument. In the third case of $k - m \geq 0$ and $k < 0$ the product $(-1)^{j-l} \bar{R}^*_{j-l,k-m} \tilde{L}_{j,-k}$ is replaced by $(-1)^{j-l+k} \bar{R}^*_{j-l,k-m} dconjg(\tilde{L}_{j,k})$, and in the fourth and the final case of $k - m \geq 0$ and $k \geq 0$ the product, $(-1)^{j-l} \bar{R}^*_{j-l,k-m} \tilde{L}_{j,k}$, is unmodified.

The use of Fortran 90 intrinsic function *iabs*() in specifying the triangular index in *Ra*(*aJL* + *iabs*(*k* − *m*)) and *Lb*(*aJ* + *iabs*(*k*)) ensures that both azimuthal indices $k - m$ and k result in non-negative offsets.

Validation of the subroutine *L2L*() can be performed in two different ways. Since L2L and M2L translations are equivalent in a spherical harmonics representation, one option is to replace the subroutine *M2L*() by *L2L*() in the code that performs the validation of M2L translation, while switching the position of expansions *Ra* and *Lb* in the subroutine arguments. This is a simple but indirect approach that does not require additional coding. Alternatively, using a more direct technique for validating the L2L program code may be a better way to illustrate the nature of the L2L operation. A simple code snippet that implements this is:

```
implicit none
integer          :: n              ! number of particles
double precision :: c(3,n)         ! particle coordinates
double precision :: q(n)           ! particle charges
integer          :: nL             ! expansion truncation
integer          :: nSzTri         ! size of expansion array
integer          :: tri(0:nL)      ! triangular pointers
double precision :: oxyz1(3)       ! origin1
double precision :: oxyz2(3)       ! origin2
double precision :: radius         ! radius of enclosing sphere
complex(16)      :: scr(0:nSzTri)  ! scratch array
complex(16)      :: Ra(0:nSzTri)   ! helper expansion
complex(16)      :: L1(0:nSzTri)   ! irregular expansion to be translated
complex(16)      :: L2(0:nSzTri)   ! reference irregular expansion
complex(16)      :: Lc(0:nSzTri)   ! translated irregular expansion
logical          :: error

! Temporarily define oxyz1 as the geometric center of particles
oxyz1 = 0.0d0
do i = 1, n
  oxyz1(1:3) = oxyz1(1:3) + c(1:3,i)
enddo
oxyz1 = oxyz1 / dble(n)

! Place oxyz2 far away of the charge set
oxyz2  = maxval(c, dim = 2) - oxyz1
radius = dsqrt(oxyz2(1)**2 + oxyz2(2)**2 + oxyz2(3)**2)
oxyz2  = (oxyz1 + dsign(radius, oxyz1)) * 2.0d0

! Redefine oxyz1 as a point in relative vicinity of oxyz2
oxyz1 = oxyz2 + 1.0d0

! Get irregular multipole expansion of a charge set about origin oxyz1
call IrregularExpansion(c, q, n, oxyz1, nL, nSzTri, tri, scr, L1)

! Get irregular multipole expansion of a charge set about origin oxyz2
call IrregularExpansion(c, q, n, oxyz2, nL, nSzTri, tri, scr, L2)

! Get chargeless helper expansion for origin translation from oxyz1 to oxyz2
call RegularExpansion(oxyz1, 1.0d0, 1, oxyz2, nL, nSzTri, tri, scr, Ra)

! Perform L2L operation
call L2L(nL, nSzTri, tri, Ra, L1, Lc)
```

```
! Validate the result of transformation, Lc against the known data, L2
error = .FALSE.
do i = 0, nSzTri
    if(abs(dble( Lc(i)  - L2(i))) > 1.0d-12) error = .TRUE.
    if(abs(aimag(Lc(i)  - L2(i))) > 1.0d-12) error = .TRUE.
enddo
if(error) write(*,*) "Error in L2L computation"
```

Of its nature, the code for validation of subroutine *L2L*() is very similar to that used in the validation of subroutine *M2L*(), so only the key aspects of the validation algorithm, which make it different from that for M2L translation, need to be reviewed.

A unique feature of the L2L translation is that both origins *oxyz*1 and *oxyz*2 for the initial and transformed irregular multipole expansions have to be placed outside the sphere that encircles the charge sources. This is a consequence of the definition of the irregular multipole expansion. In the present validation code, the position of point *oxyz*2 remained the same as that in the validation code for M2L operation, which placed that point outside the sphere encircling the charge sources. However, the point *oxyz*1, which had previously been the geometric center of a charge set, must now also to be placed outside the encircling sphere. In addition to that, the length of the virtual vector of the helper expansion connecting *oxyz*1 to *oxyz*2 must be less than that for the irregular multipole expansion that has its origin at point *oxyz*2. The easiest way of satisfying these requirements is to place *oxyz*1 close to *oxyz*2 by adding a small increment to the latter point, which the Fortran 90 array operation $oxyz1 = oxyz2 + 1.0d0$ accomplishes using only a single line of program code.

In this validation code, the irregular multipole expansion *L*1 of a charge set is calculated using the origin *oxyz*1; later on in the program code this expansion will be shifted to the origin *oxyz*2. The irregular multipole expansion *L*2, computed about the origin *oxyz*2, is used as reference data in the validation. A helper chargeless regular multipole expansion *Ra* assists in origin translation from *oxyz*1 to *oxyz*2. Further down in the code, subroutine *L2L*() transforms the expansion *L*1 into *Lc*. Finally, each array element of expansion *Lc* is compared with its corresponding element in *L*2. If all differences are smaller than 10^{-12}, then the match is confirmed. Truncation of the infinite summation in Equation 11.13 by a finite upper value of the orbital index, *nL*, contributes some numerical error in *Lc*, but setting the parameter *nL* to 30 or more helps minimize this error.

After checking that the direct implementation of subroutine *L2L*() works as expected, the program code can be optimized to improve its performance. The following code illustrates the optimized implementation.

```
subroutine L2L(nL, nSzTri, tri, Ra, Sb, Sc)
!
! Sc(L,m) = sum(j=L,nL) sum(k=-j,+j) { (-1)**(j-L) * Ra(j-L,k-m) * Sb(j,k) }
! Optimized implementation.
!
! L2L operation:
! Transforms irregular expansion from one coordinate system to another
! Ra - bar-scaled chargeless regular helper expansion
! Sb - tilde-scaled irregular multipole expansion to be shifted
! Sc - resulting tilde-scaled irregular multipole expansion
!
! M2L operation:
! Transforms regular expansion into irregular one about different origin
! Ra - bar-scaled regular multipole expansion to be transformed
! Sb - tilde-scaled chargeless irregular helper expansion
! Sc - resulting tilde-scaled irregular multipole expansion
!
implicit none
!
! Formal arguments
integer,       intent(in)      :: nL              ! expansion truncation
integer,       intent(in)      :: nSzTri          ! size of expansion array
```

```
integer,      intent(in)    :: tri(0:nL)       ! array of triangular pointers
complex(16), intent(in)     :: Ra(0:nSzTri)
complex(16), intent(in)     :: Sb(0:nSzTri)
complex(16), intent(inout) :: Sc(0:nSzTri)
!
! Local variables
integer :: L, m, j, k, aL, aJ, aJL, aLm
double precision :: rsum
!
! Zero-initialize the output array
Sc = dcmplx(0.0d0, 0.0d0)
!
!$omp parallel do default(none) &
!$omp shared(nL,tri,nSzTri,Ra,Sb,Sc) &
!$omp private(L,m,j,k,aL,aJ,aJL,aLm,rsum)
do L = 0, nL
  aL = tri(L)
  do j = L, nL
    aJ  = tri(j)
    aJL = tri(j-L)

    ! m = 0
    ! k = 0
    rsum = dble( Ra(aJL) * Sb(aJ) ) * 0.5d0

    ! k > 0;  k <= j-L;   includes k > 0 and k < 0 for m = 0
    do k = 1, j-L
      rsum = rsum + dble( Ra(aJL+k) * Sb(aJ+k) )
    enddo
    Sc(aL) = Sc(aL) + dcmplx(rsum * 2.0d0, 0.0d0) * dble((-1)**(j-L))

    do m = 1, min0(j-L-1,L)                ! m < j - L
      aLm = aL+m

      do k = 1, j-L-m                       ! k < 0
        Sc(aLm) = Sc(aLm) + dconjg(Ra(aJL+k+m)) * dconjg(Sb(aJ+k)) *
        dble((-1)**(j-L-m))
      enddo

      do k = 0, min0(m-1,j-L+m)             ! 0 <= k < m
       Sc(aLm) = Sc(aLm) + dconjg(Ra(aJL-k+m)) * Sb(aJ+k) * dble((-1)**(j-L+k-m))
      enddo

      do k = m, j-L+m                       ! k >= 0; k - m >= 0
        Sc(aLm) = Sc(aLm) + Ra(aJL+k-m) * Sb(aJ+k) * dble((-1)**(j-L))
      enddo  ! index k
    enddo

    do m = max0(1,j-L), L                  ! m >= j - L
      aLm = aL+m

      do k = L-j+m, min0(m-1,j-L+m)  ! 0 <= k < m
       Sc(aLm) = Sc(aLm) + dconjg(Ra(aJL-k+m)) * Sb(aJ+k) * dble((-1)**(j-L+k-m))
      enddo

      do k = m, j-L+m                       ! k >= 0; k - m >= 0
        Sc(aLm) = Sc(aLm) + Ra(aJL+k-m) * Sb(aJ+k) * dble((-1)**(j-L))
      enddo
    enddo
```

```
  enddo   ! index j
enddo   ! index L
!
end subroutine L2L
```

Optimization of the naïve implementation of subroutine *L2L*() involves changing the loop structure to get rid of conditional *if-then-else* operators, which slow down the execution. The first such construct to work on in the naïve implementation is the one that encloses everything inside the *do*-loop over index k:

```
do k = -j, j
  if(iabs(k-m) <= j-L) then
    ...
  endif
enddo
```

In this code snippet, the *do*-loop maintains the requirement that $-j \leq k \leq j$ and the *if-then* operator adds another condition, that $|k - m| \leq j - l$. Both these conditions come from the standard requirement that the azimuthal index has to stay within the boundaries of the corresponding orbital index in the two expansion terms in Equation 11.13; both these logical conditions may be implemented by redefining the starting and ending value of the *do*-loop index.

The first step to developing a solution is to rewrite the condition $|k - m| \leq j - l$ as $l - j \leq k - m \leq j - l$, and then solve the latter inequality about variable k, to give $l - j + m \leq k \leq j - l + m$. Next, this inequality is combined with the inequality $-j \leq k \leq j$ to produce a starting value of index k in the revised *do*-loop of either $l - j + m$ or $-j$, whichever is the smaller in absolute value. Thus, it is straightforward to show that $|l - j + m| \leq |-j|$. Rewriting the latter as $|l - j + m| \leq j$, and then as $|j - l - m| \leq j$ provides the proof that this inequality is true, based on $l \geq 0$, $m \geq 0$, and $j \geq l$. With that, the starting value of index k in the new *do*-loop becomes $k = l - j + m$.

Likewise, combining the condition $l - j + m \leq k \leq j - l + m$ with $-j \leq k \leq j$ makes the ending value for index k in the new *do*-loop to be either $j - l + m$ or j, whichever is smaller. Once again, $j - l + m \leq j$, since $l \geq m$, and each variable in this inequality is non-negative. This development gives the upper value of index k in the new *do*-loop as $j - l + m$. Finally, having the former *if-then* construct from the naïve implementation fused into the *do*-loop leads to an improved piece of code:

```
do k = L-j+m, j-L+m
  ...
enddo
```

which accomplishes the same task as the combination of *do*-loop and *if-then* operator in the naïve implementation.

The new *do*-loop over index k inherits four different cases for azimuthal indices from the naïve code that need to be considered inside the loop. One way to sort out these cases is to split the present *do*-loop into two separate *do*-loops, each exploring its own individual range of index k. The two such segments are $l - j + m \leq k < m$ and $m \leq k \leq j - l + m$, which translate to the following code snippet:

```
do k = L-j+m, min0(m-1,j-L+m)
  ...
enddo
do k = m, j-L+m
  ...
enddo
```

Since indices m and j vary independently, it is necessary to make sure that the value of $m - 1$ does not become greater than $j - l + m$ in the first of these *do*-loops. Fortran 90 intrinsic function *min0*($m - 1$, $j - l + m$) performs this test, and returns the lesser out of the two values. The second *do*-loop will only be executed when $j \geq l$.

A convenient feature of the above split is that the second *do*-loop always represents the condition when $k \geq 0$ and $k - m \geq 0$. This maps Equation 11.13 into the straightforward piece of code:

```
do k = m, j-L+m                          ! k >= 0; k - m >= 0
   Sc(aLm) = Sc(aLm) + Ra(aJL+k-m) * Sb(aJ+k) * dble((-1)**(j-L))
enddo
```

Further work on the allowed ranges of index k depends on the value of index m. Therefore, the next step in optimizing the code is to decide where to put the *do*-loop over index m in the overall loop structure of the subroutine *L2L*(). Index l remains as the outermost *do*-loop in the optimized code, but the next *do*-loop could be either over index m or j. Ideally, longer *do*-loops should go inside the shorter ones, but in this case the length of the *do*-loop over index m is practically the same as that for *do*-loop over index j, which makes deciding which loop should be external and which one internal largely a matter of personal preference. Since indices k and m together form the index $k - m$, it makes sense to arrange *do*-loops over indices m and k next to each other. Based on this reasoning, the *do*-loops were arranged in the order l, j, m, and then k.

Considering the structure of *do*-loop over index m, it makes sense to treat $m = 0$ as a separate case, and to handle it outside the corresponding *do*-loop. When $m = 0$, Equation 10.85 simplifies to:

$$\tilde{L}_{l,0} = \sum_{j=l}^{\infty}\sum_{k=-j}^{j}(-1)^{j-l}\bar{R}^{*}_{j-l,k}\,\tilde{L}_{j,k} = \sum_{j=l}^{\infty}(-1)^{j-l}\left\{\bar{R}^{*}_{j-l,0}\,\tilde{L}_{j,0} + 2\sum_{k=1}^{j-l}\bar{R}^{*}_{j-l,k}\,\tilde{L}_{j,k}\right\}, \tag{11.23}$$

where $\tilde{L}_{l,0}$ is a real number. Since index k in $\bar{R}^{*}_{j-l,k}$ cannot exceed the value of $j - l$, the upper value of index k in the sum becomes $j - l$. Because of the simplicity of Equation 11.23 it readily maps into the following code snippet:

```
! m = 0
! k = 0
rsum = dble( Ra(aJL) * Sb(aJ) ) * 0.5d0

! k > 0;  k <= j-L;    includes k > 0 and k < 0 for m = 0
do k = 1, j-L
   rsum = rsum + dble( Ra(aJL+k) * Sb(aJ+k) )
enddo
Sc(aL) = Sc(aL) + dcmplx(rsum * 2.0d0, 0.0d0) * dble((-1)**(j-L))
```

Continuation of the work of sorting out the individual cases corresponding to a specific sign of azimuthal indices of expansion terms $\bar{R}^{*}_{j-l,k-m}$ and $\tilde{L}_{jk}$, present in Equation 11.13, requires breaking down the *do*-loop over index m in two different segments: $1 \leq m < j - l$ and $j - l \leq m \leq l$. The rationale behind this division is to aid in the work on ranges of index k. For instance, when $m \geq j - l$, only non-negative values of index k contribute in Equation 11.13, whereas negative values of that variable push the azimuthal index $k - m$ outside the allowed range for the expansion term $\bar{R}^{*}_{j-l,k-m}$, thereby zeroing its contribution. With that, for $m \geq j - l$, only two previously introduced cases $0 \leq k < m$ and $k \geq m$ remain valid. This translates to the following implementation:

```
do m = max0(1,j-L), L                    ! m >= j - L
   aLm = aL+m

   do k = L-j+m, min0(m-1,j-L+m)    ! 0 <= k < m
      Sc(aLm) = Sc(aLm) + dconjg(Ra(aJL-k+m)) * Sb(aJ+k) * dble((-1)**(j-L+k-m))
   enddo

   do k = m, j-L+m                      ! k >= 0; k - m >= 0
      Sc(aLm) = Sc(aLm) + Ra(aJL+k-m) * Sb(aJ+k) * dble((-1)**(j-L))
   enddo
enddo
```

Since the case of $m = 0$ is handled separately, it is necessary to make sure that the starting value of index m in the *do*-loop is unity when $j - l = 0$. The Fortran 90 intrinsic function $max0(1, j - l)$ performs the necessary test.

The remaining case of $1 \leq m < j - l$ for index m allows the range for index k to be split in three segments, $k < 0$, $0 \leq k < m$, and $k \geq m$, as illustrated in the following code snippet:

```
do m = 1, min0(j-L-1,L)                 ! m < j - L
  aLm = aL+m

  do k = 1, j-L-m                       ! k < 0
    Sc(aLm) = Sc(aLm) + dconjg(Ra(aJL+k+m)) * dconjg(Sb(aJ+k)) * dble((-1)**
(j-L-m))
  enddo

  do k = 0, min0(m-1,j-L+m)             ! 0 <= k < m
    Sc(aLm) = Sc(aLm) + dconjg(Ra(aJL-k+m)) * Sb(aJ+k) * dble((-1)**(j-L+k-m))
  enddo

  do k = m, j-L+m                       ! k >= 0; k - m >= 0
    Sc(aLm) = Sc(aLm) + Ra(aJL+k-m) * Sb(aJ+k) * dble((-1)**(j-L))
  enddo  ! index k
enddo
```

In this code, the second and third *do*-loops for index k are exact replicas of similar *do*-loops from the previous implementation for $m \geq j - l$. The *do*-loop for the negative range of index k, $k < 0$ is unique, however. If written directly, the *do*-loop would look like:

```
do k = L-j+m, -1                ! k < 0
  Lc(aLm) = Lc(aLm) + dconjg(Ra(aJL+iabs(k-m))) * dconjg(Lb(aJ+iabs(k))) *
  dble((-1)**(j-L-m))
enddo
```

where variable k is actually a negative number. Replacing the variable k by positive numbers, and switching the starting and ending values of the index gives:

```
do k = 1, j-L-m                 ! k < 0
  Lc(aLm) = Lc(aLm) + dconjg(Ra(aJL+iabs(-k-m))) * dconjg(Lb(aJ+iabs(-k))) *
dble((-1)**(j-L-m))
enddo
```

Finally, the intrinsic function *iabs*() can be deleted if $|-k - m|$ and $|-k|$ are replaced by $k + m$ and k, respectively, leading to the final optimized form of the *do*-loop:

```
do k = 1, j-L-m                 ! k < 0
Lc(aLm) = Lc(aLm) + dconjg(Ra(aJL+k+m)) * dconjg(Lb(aJ+k)) * dble((-1)**(j-L-m))
enddo
```

This completes the rewriting of *do*-loops in the optimized version of subroutine *L2L*() to eliminate the use of time-consuming *if-then* operators.

Some additional speedup can be obtained by parallelizing the code, and for this the OpenMP model provides the simplest and least invasive approach. In the presence of multiple enclosed loops, parallelization typically applies to the outermost *do*-loop to minimize the cost of spawning OpenMP processes. Variables that are written to by the parallel processes are declared as private, and those that are read-only or written to exclusively by a single process can be defined as shared. For subroutine *L2L*(), these directives are:

```
!$omp parallel do default(none) &
!$omp shared(nL,tri,nSzTri,Ra,Sb,Sc) &
!$omp private(L,m,j,k,aL,aJ,aJL,aLm,rsum)
```

Because the arrays *Ra* and *Sb* are used in a read-only mode throughout the program code they are declared as shared variables. Array *Sc* works in read-write mode while accumulating results, but, since each OpenMP process exclusively writes into its own array element determined by index l, the array *Sc* can also be included in the list of shared variables. This saves time and memory by avoiding creating multiple private copies out of the array and then merging them back.

At first glance it might be puzzling that the expansion elements are labeled *Ra*, *Sb*, and *Sc*, reminiscent of solid harmonics labels R and S, in subroutine *L2L*(), instead of the multipole expansion labels M and L. The driving force for this is to emphasize the notion that this subroutine is a computer implementation of the L2L and M2L operations simultaneously, where the expansions of a and b may be interchanged. This duality comes from Equations 11.13 and 11.17 being mathematically equivalent. Because of this, by using *Ra* as a chargeless regular helper expansion, and *Sb* and *Sc* as input and output irregular multipole expansions, this subroutine can perform L2L translations. Likewise, supplying a regular multipole expansion in the place of array *Ra*, and using *Sb* and *Sc* as input and output irregular expansions allows the same subroutine to perform M2L translations. Due to the dual roles that arrays *Ra* and *Sb* play in the program code, denoting them by their underlying solid harmonics labels emphasizes their general nature more clearly than if multipole labels had been used instead of these variables.

Finally the choice of Equation 11.13 over that of Equation 11.17 for the optimized implementation of M2L and L2L translations in subroutine *L2L*() may be explained by the difference in the sum over index j when comparing these equations side by side. In Equation 11.13, index j goes from l to nL, which makes the sum visibly shorter than that in Equation 11.17 where index j goes from 0 to nL. Since both equations are mathematically equivalent, the difference in the starting values in index j suggests that the sum in Equation 11.17 has many summation elements that make no contribution, all of which need to be properly sorted out during the implementation of computer code. Knowing beforehand that $j \geq l$ enhances the utility of Equation 11.13 when determining continuous segments of azimuthal indices for the expansion terms based on the interplay of the relative value of indices l, j, m, and k.

12

Fast Multipole Method

When dealing with functions, which fall off rapidly with increasing distance, a mechanism that uses cut-offs can efficiently mitigate the high computational cost of treating the pairwise interactions in large systems. Unfortunately, this solution cannot be applied to the electrostatic potential because the inverse distance function that describes electrostatic potential falls off too slowly to allow its tail to be truncated without introducing significant numerical errors. Dealing with such functions requires accounting for all pairwise interactions in the system, even those separated by very large distances. This leads to a quadratically rising cost of the evaluation of electrostatic interactions, resulting in the direct sum method quickly becoming the bottleneck of the computation.

Seeking a solution to the general potential energy function for large systems, Greengard and Rokhlin proposed space decomposition into the set of boxes, generating a multipole expansion for each box, and then assembling the potential energy of the whole system from the contribution of the individual boxes. They named this technique the Fast Multipole Method, dubbed FMM. The present chapter will provide an overview of this approach.

12.1 Near and Far Fields: Prerequisites for the Use of the Fast Multipole Method

Using the FMM for calculating a system of point charges begins with decomposition of the space occupied by the particles into a collection of tightly fitted boxes, so that any part of the space where a particle resides or may happen to occur at some moment in time becomes assigned to a unique box. The use of cubic boxes makes it easy to avoid introducing gaps during the division of space and also simplifies accounting.

Figure 12.1 provides a 2D example of the decomposition of a system of point charges into a set of cubic boxes. After the decomposition, each newly-created box is given a unique identifier based on the side indices, which start from zero in each dimension of the system. The choice of starting with zero makes the index computation particularly easy.

Once the division of the system into boxes is complete, each charged particle is assigned to its respective box. Then, the FMM algorithm computes the multipole expansion $\bar{M}_{lm}(A,A)$ of the charge set confined in each box A (indicated by first argument) about the local coordinate system of the box A (specified by second argument). For each box, this gives:

$$\bar{M}_{lm}(A,A) = \sum_{i=1}^{N_A} q_i(A)\bar{R}^*_{lm}(\mathbf{r}_{iA}), \tag{12.1}$$

where i is an index of the charged particle inside the box, and $\bar{R}^*_{lm}(\mathbf{r}_{iA})$ is a bar-scaled regular solid harmonic. The radius vector $\mathbf{r}_{iA}$ connects the center of the box with the position of the particle i. This center serves as the origin of a local coordinate system, and is obtained by translating the origin of the global coordinate system to the center of the box. With that, each particle gets redefined in the local coordinate system of its respective box.

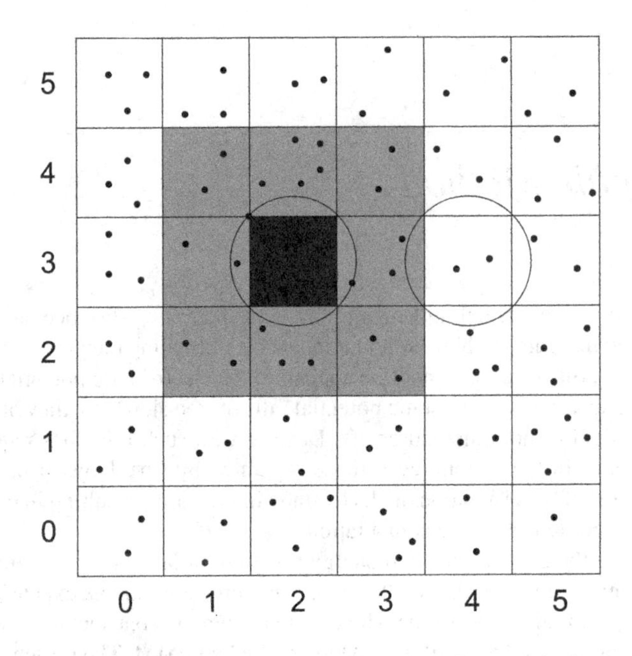

FIGURE 12.1 Decomposition of a 2D system of point charges into a set of rectangular boxes. A box with index 2-3 is shown in black, boxes composing the near field of box 2-3 are shown in gray, and boxes composing the far field of the box 2-3 are shown in white. The circle around boxes 2-3 and 4-3 represents the boundary that separates the enclosed charge sources from the external points in the evaluation of the electrostatic potential using FMM.

Multipole moments computed in Equation 12.1 offer the possibility of evaluating the electrostatic interaction energy U_{AB} between a pair of boxes A and B by means of the multipole expansion:

$$U_{AB} = \sum_{l=0}^{\infty}\sum_{m=-l}^{l} \overline{M}_{lm}(A,A)\tilde{L}_{lm}(B,A), \tag{12.2}$$

where $\tilde{L}_{lm}(B,A)$ is an irregular multipole expansion of point charges from box B about the origin of box A. The irregular moments $\tilde{L}_{lm}(B,A)$ can be computed from the regular multipole moments $\overline{M}_{lm}(B,B)$ of box B by application of Equation 11.17, as described in Chapter 11.

Using Equation 12.2, all suitable pair interactions between the boxes in the system can be evaluated. Not every pair of boxes can be treated in that way, though. First of all, there is a general restriction on the eligibility of individual charge sets confined in the boxes for their interaction with other charge sets *via* multipole expansion. This eligibility condition originates from the definition of multipole expansion. According to that definition, a multipole expansion of a charge set can only be used for computing electrostatic potentials at points located outside a sphere that encompasses the set. Therefore, two interacting charge groups must be sufficiently separated spatially from each other to ensure that their enclosing spheres do not overlap. This condition is achieved as follows.

In any rectangular-shaped container, the encircling sphere must encompass all eight vertices of the box since a particle, which must be inside the sphere, may potentially reside in any part of the box including the vertices. Given this, the encircling sphere inevitably extends into the space of adjacent boxes, and therefore the encircling spheres of the adjacent boxes always overlap. The overlap, albeit in a single point, also takes place when the boxes reside in a main diagonal next to each other. With that, any pair of boxes that touch each other by a side, edge, or vertex, becomes automatically ineligible for interaction *via* multipole expansion. To account for this condition, FMM views each box in the system as being surrounded by a layer of boxes called the *near field* of that box, abbreviated *NF*. In a computer code, each box has to keep a list of the boxes in its near field.

The name "near field" literally means that the space occupied by these boxes is too close to the central box for the multipole expansion to be used in evaluating the electrostatic potential. Instead, the electrostatic interaction of a box with any other box in its near field has to be computed conventionally by means of the direct sum. A box and its near field are illustrated in Figure 12.1.

In addition to the notion of near field, FMM introduces the concept of *far field*, abbreviated *FF*, which encompasses boxes located outside of the near field, and therefore capable of interacting with the central box *via* a multipole expansion. These boxes are shown in white in Figure 12.1. The far field is an external layer of boxes that is further away from the central box than the near field. As with the near field, each box in the system has to keep track of the list of boxes belonging to its far field.

These definitions of near and far fields, which are based on the conditions of overlap of the spheres encircling the boxes, represent a minimum requirement in the definition of near and far fields. FMM adds an additional criterion to determine the extent of near and far fields. This will be addressed in the following section.

12.2 Series Convergence and Truncation of Multipole Expansion

The infinite series expansion of the electrostatic interaction energy in Equation 12.2 is always guaranteed to converge when charge sets in two interacting boxes are in each other's far field. The series has to be truncated at some point to make the computation feasible. Keeping too many terms in the expansion in order to produce the converged result is obviously an impractical option since the computational cost is proportional to the size of the expansion. To avoid this, it is necessary to keep the expansion as small as possible while paying close attention to reliability of the numerical computation.

Before proceeding, a brief clarification of the employed terminology might be helpful. It is convenient to refer to the convergence of a series in terms of its speed, as a measure of how many expansion terms need to be summed up in order to have the result converge within an acceptable numerical threshold. A *faster*-converging series requires fewer terms to sum up until reaching the convergent state. Similarly, the series that requires more terms to sum up will be called a *slower*-converging one. Establishing how many terms a truncated multipole expansion needs to include, and determining other factors affecting the convergence, is the objective of this section.

Factors affecting convergence of multipole series are best seen on a pair of interacting particles *A* and *B* that carry charges q_A and q_B, respectively. Rewriting Equation 12.2 in terms of underlying components leads to:

$$U_{AB} = \sum_{l=0}^{\infty}\sum_{m=-l}^{l} q_A a^l \bar{P}_l^m(\theta_A) e^{-im\phi_A} \frac{q_B}{b^{l+1}} \tilde{P}_l^m(\theta_B) e^{im\phi_B}, \quad (12.3)$$

where $\bar{P}_l^m$ and $\tilde{P}_l^m$ are scaled associated Legendre functions, and (a,θ_A,ϕ_A) and (b,θ_B,ϕ_B) are spherical polar coordinates of points *A* and *B*, respectively, in a global coordinate system.

The primary factor that determines the series convergence in Equation 12.3 is the ratio of radial distances (a/b), where distances a and b connect the points *A* and *B* with the center of the coordinate system, respectively. Under the condition of $a < b$ the series in Equation 12.3 is absolutely convergent, and the speed of convergence depends on the magnitude of the ratio. The closer the ratio is to unity, the slower the convergence. Likewise, making the ratio smaller makes the series converge faster.

In a two-particle system, moving the origin of the coordinate system in order to minimize the distance a and simultaneously maximize the distance b achieves the stated goal of optimizing the ratio (a/b), and thus improves the speed of convergence of the multipole expansion.

A similar approach applies to large multi-particle systems where there are often a huge number of pairwise interactions. Placing the local coordinate system in the center of the box carrying charge sources minimizes the distance a from the origin of the coordinate system to the particles. Further on, choosing the field point b far from the center of charge sources favorably reduces the ratio (a/b).

To control the distance that separates a pair of interacting boxes, the FMM algorithm introduces a *well-separation* parameter, which is abbreviated *ws*. This parameter indicates how many insulating layers of boxes exist between a box and its far field. In other words, *ws* represents the thickness of the *NF* layer

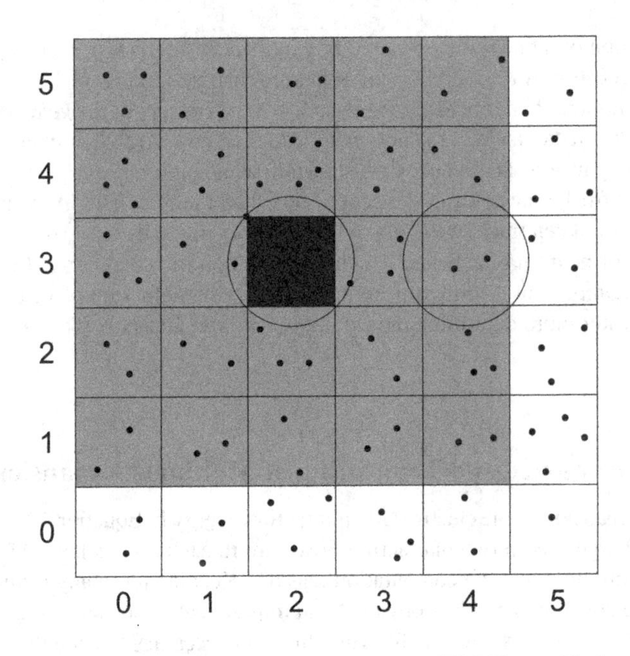

FIGURE 12.2 Decomposition of a 2D system of point charges into a set of FMM boxes. The box with index 2-3 is shown in black. Boxes in the near field of the box 2-3 under $ws = 2$ are shown in gray and the boxes in the far field of 2-3 are shown in white. Circles enclose the inner boundary of multipole expansion of the corresponding boxes. Due to $ws = 2$, the electrostatic interaction between boxes 2-3 and 4-3 has to be computed by using a direct sum method because box 4-3 is formally in the near field of box 2-3 despite their encircling circles not overlapping.

that separates any given box from its *FF* region. For example, Figure 12.1 illustrates the case of $ws = 1$, whereas Figure 12.2 corresponds to the case of $ws = 2$.

Saying that any two boxes are *well separated* indicates that the boxes are in far field relative to each other. Similarly, saying that the boxes are *not well separated* indicates that they are in the position of near field one to another so that their mutual electrostatic interaction should be treated by the application of the direct sum method.

Changing the value of parameter *ws* provides a mechanism to control the rate of convergence of, and therefore the size of, the multipole expansion. A two-particle system in Figure 12.3 provides a convenient example for demonstration. The objective of the analysis is to see how the ratio (a/b) changes with the increase in the distance between boxes on the example of the least favorable case of (a_{max}/b_{min}) when the distance a is maximized, and the distance b is minimized.

In FMM, each box sees the other boxes from the point of view of its own local coordinate system, with its origin in the center of that box. For a pair of interacting boxes, this origin may be assigned to either box.

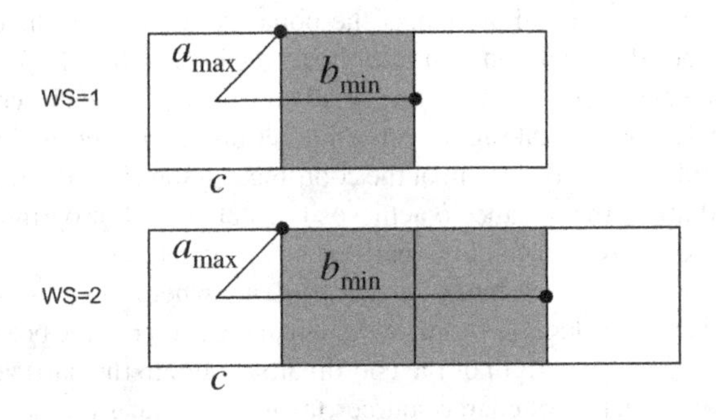

FIGURE 12.3 Examples of a_{max} and b_{min} resulting from different value of parameter *ws*. Interacting boxes, which are in *FF* position one to another, are shown in white. Boxes in gray represent *NF* region. The origin of the local coordinate system resides in the center of the left-most box. Parameter c represents the box side length.

Figure 12.3 has the origin arbitrarily placed into the left-most box. Under this choice of the coordinate system, the largest possible value of distance a_{max} from the origin to a particle occurs when that particle resides in a vertex of the box. This position caps the maximum value of distance a to half the value of the main diagonal of the box. For a box in 2D with equal side length c the maximum value of distance a is $a_{max} = (\sqrt{c^2 + c^2})/2 = c/\sqrt{2}$.

The distance b_{min} from the origin to the field point, which is located in the right-most box, can be similarly evaluated. Using the setup in Figure 12.3 and the case of $ws = 1$, the smallest possible distance b from the origin to the nearest point in the right-most box occurs when the particle resides in the middle of the side of that box facing the origin of the coordinate system, and thus $b_{min} = (c/2) + c = (3/2)c$. With the extreme value of distances a and b identified, the worst possible ratio (a/b) for a pair of interacting boxes separated by a single layer $ws = 1$ becomes $(a_{max}/b_{min}) = (\sqrt{2}/3) \approx 0.5$.

What happens to the ratio (a_{max}/b_{min}) when the distance between interacting boxes increases? Placing two near-field boxes instead of one between the interacting boxes increases the separation; this corresponds to the case of $ws = 2$ in Figure 12.3. In the new configuration, the value of a_{max} remains unchanged, since it is defined only by the box size. The minimum value of distance b from the origin to the field point incorporates the distance increase, and becomes $b_{min} = (c/2) + 2c = (5/2)c$. With the distances to particles being determined, their ratio becomes $(a_{max}/b_{min}) = (\sqrt{2}/5) \approx 0.3$.

The change in convergence of multipole series between the cases of $ws = 1$ and $ws = 2$ that corresponds to the increase in distance between interacting boxes in Figure 12.3 may roughly be estimated as a ratio of their radial factors, $(\sqrt{2}/3)/(\sqrt{2}/5) = (5/3) \approx 1.7$. This shows that the case of $ws = 2$ has a better convergence profile by factor of 1.7 than that of the case of $ws = 1$, and therefore fewer terms are needed in the multipole expansion. A more rigorous formula for the required length of the multipole expansion will be derived later on in this section.

Once the role of ws in the rate of convergence of multipole expansion is established, it is necessary to come up with a reliable estimate of how many terms in the multipole expansion would be necessary in order to converge the series within a preset limit. This requires determining a numerical criterion for convergence and an analytic procedure for evaluating the terms in the expansion.

Modern computing architectures offer single and double precision real numbers. Single precision mathematical operations guarantees have 6 or 7 digits significant figures. Double precision increases the number of significant figures to 16 or 17. Taking into account the complexity of FMM computations, it is unlikely that single precision calculations using the FMM would ever be useful because of the rapid accumulation of numerical errors. When double precision math is used a convergence criterion of about 10^{-12} and 10^{-14} can readily be achieved. Reaching this numerical threshold would then unambiguously indicate that the series sum had been fully converged, and that the remaining terms in the series may be safely discarded. The next step is to find the way to sum up the multipole series.

A two-particle system of the type described by Equation 12.3, and illustrated in 3D in Figure 13.4, provides a suitable model for the analytic evaluation of the multipole expansion series. Charged particles in this system reside in separate boxes located in the far field one to another, with parameter ws determining the distance between the interacting boxes. As in the previous example of 2D system, the positions of the particles are chosen so as to maximize the ratio (a_{max}/b_{min}) in the multipole expansion and thus make it harder for the series to converge. This would result in the largest possible error in the computation. In addition to that, from all possible positions of particle A, which satisfy a_{max}, the one being closest to particle B is selected because the convergence is more difficult to achieve for the nearest particles. Together, these arrangements guarantee obtaining the worst convergence properties for the series.

The convenience of a two-particle system is its ability to simplify the multipole expansion to the state when the numerical value of each member of the expansion can be easily evaluated. The addition theorem for spherical harmonics, given by Equation 5.251, is useful for that simplification. Rewriting that equation in terms of scaled associated Legendre functions $\bar{P}_l^m = (1/(l+m!))\,P_l^m$ and $\tilde{P}_l^m = (l-m)!\,P_l^m$ leads to:

$$P_l(\gamma) = \frac{4\pi}{2l+1}\sum_{m=-l}^{l} Y_{lm}^*(\theta_A,\phi_A)\,Y_{lm}(\theta_B,\phi_B) = \sum_{m=-l}^{l} \bar{P}_l^m(\theta_A)\,e^{-im\phi_A}\,\tilde{P}_l^m(\theta_B)\,e^{im\phi_B}, \tag{12.4}$$

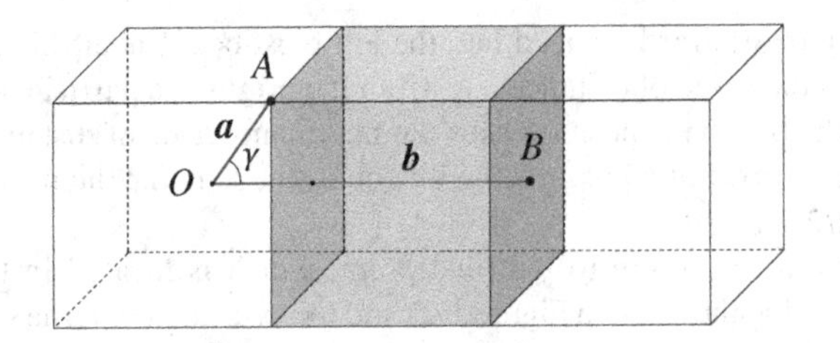

FIGURE 12.4 Electrostatic interaction of two particles A and B placed in well-separated boxes with $ws = 1$. A box in gray represents near field. Boxes in white represent far field. Particle A resides in a vertex of the left-most box on the side toward particle B. Particle B resides in the middle of the side of the right-most box facing particle A. Radius vectors $\boldsymbol{a}$ and $\boldsymbol{b}$ connect the particles A and B with the origin O of the coordinate system, and have angle γ subtended between them.

where γ is the angle subtended between radius vectors $\boldsymbol{a}$ and $\boldsymbol{b}$, and (a,θ_A,ϕ_A) and (a,θ_B,ϕ_B) are spherical polar coordinates of points A and B, respectively, illustrated in Figure 12.4. Substituting Equation 12.4 into Equation 12.3 leads to a simplified form of the multipole expansion:

$$U_{AB} = q_A q_B \sum_{l=0}^{\infty} \frac{a^l}{b^{l+1}} P_l(\gamma), \tag{12.5}$$

which is written in terms of angle γ; and where $P_l(\gamma)$ are the Legendre polynomials.

Since convergence of the expansion in Equation 12.5 does not depend on the value of charges q_A and q_B it is convenient to assume their value is unity. This reduces Equation 12.5 to:

$$U_{AB} = \sum_{l=0}^{\infty} \frac{a^l}{b^{l+1}} P_l(\gamma). \tag{12.6}$$

Truncating this expansion on the last term $l = nL$, which still needs to be determined, leads to:

$$U_{AB} = \sum_{l=0}^{nL} \frac{a^l}{b^{l+1}} P_l(\gamma) + Q(Z), \tag{12.7}$$

where $Q(Z)$ is the remainder of the series, and Z is the number of significant figures in the convergence of the multipole series.

Due to Equation 12.6 the analytic form of $Q(Z)$ is known, and is:

$$Q(Z) = \sum_{l=nL+1}^{\infty} \frac{a^l}{b^{l+1}} P_l(\gamma) \leq 10^{-Z}. \tag{12.8}$$

The next task is to determine the value of parameter nL such that the remainder of the series $Q(Z) \leq 10^{-Z}$ is small enough that it can be neglected. Solving Equation 12.8 establishes a link between the value of ws, the box side length c, the polynomial degree nL, and the numerical precision Z. Parameters nL, and Z are already in Equation 12.8, and ws and c will be added shortly.

Because $P_l(\gamma) < 1$, deleting the Legendre polynomial $P_l(\gamma)$ from Equation 12.8 leads to a stronger inequality:

$$Q(Z) = \frac{1}{b} \sum_{l=nL+1}^{\infty} \left(\frac{a}{b}\right)^l \leq 10^{-Z}, \tag{12.9}$$

where $(1/b)$ is put in front of the sum to reshape it in a form of geometric series.

The series in Equation 12.9 can now be evaluated. Appendix A.5 shows how this is done, and provides the sum S for geometric series of the type:

$$S = \sum_{k=A}^{N} x^k = \frac{x^A - x^{N+1}}{1-x}. \tag{12.10}$$

Substituting $A = nL + 1$ and $x = a/b$ under the condition of $(a/b) < 1$ leads to $\lim_{N\to\infty} (a/b)^{N+1} = 0$. Therefore,

$$Q(Z) = \frac{1}{b}\frac{(a/b)^{nL+1}}{1-(a/b)} = \frac{(a/b)^{nL+1}}{b-a} \leq 10^{-Z}. \tag{12.11}$$

The inequality in Equation 12.11 is easily satisfied for any small value of parameter Z. Since the goal is to find the largest possible Z, it is appropriate to replace the inequality with an equals sign. Doing this, and multiplying both sides of Equation 12.11 by $b - a$ leads to:

$$\left(\frac{a}{b}\right)^{nL+1} = 10^{-Z}(b-a). \tag{12.12}$$

Further work on this equation involves replacing the distances a and b by box side length, c and the well-separation parameter, ws, that they incorporate. From the system geometry and particles arrangement in Figure 12.4, it follows that:

$$a = \frac{\sqrt{3}}{2}c, \quad \text{and} \quad b = \frac{c}{2} + c\,ws = c\frac{(1+2\,ws)}{2}. \tag{12.13}$$

Therefore,

$$\frac{a}{b} = \frac{\sqrt{3}}{1+2ws}, \quad \text{and} \quad b-a = c\left(\frac{1-\sqrt{3}}{2} + ws\right). \tag{12.14}$$

Substituting Equation 12.14 into Equation 12.12 leads to:

$$\left(\frac{\sqrt{3}}{1+2ws}\right)^{nL+1} = 10^{-Z}c\left(\frac{1-\sqrt{3}}{2} + ws\right). \tag{12.15}$$

Equation 12.15 establishes the relationship between parameters ws, c, nL, and Z, so that given the values of any three parameters, the value of the forth parameter can be easily calculated. This allows c, the box side length, to be defined in terms of the values of ws, nL, and Z as:

$$c = 10^{Z}\left(\frac{\sqrt{3}}{1+2ws}\right)^{nL+1} \Bigg/ \left(\frac{1-\sqrt{3}}{2} + ws\right). \tag{12.16}$$

Resolving parameters Z and nL requires taking the log of both sides of Equation 12.15, this gives:

$$(nL+1)\log\left(\frac{\sqrt{3}}{1+2ws}\right) = \log c + \log\left(\frac{1-\sqrt{3}}{2} + ws\right) - Z. \tag{12.17}$$

Rearranging this equation to solve for parameter Z gives:

$$Z = \log c + \log\left(\frac{1-\sqrt{3}}{2} + ws\right) - (nL+1)\log\left(\frac{\sqrt{3}}{1+2ws}\right). \tag{12.18}$$

Equation 12.18 is useful in practical applications of FMM, in that, for given specific values of *ws*, *c*, and *nL*, it can predict the precision of the resulting multipole expansion. Note that Fortran 90 intrinsic function *log*() corresponds to a natural logarithm, whereas the intrinsic function for base ten logarithm is *log10*().

Another useful equation follows from Equation 12.17 by solving it for parameter *nL*. This transformation leads to:

$$nL = \left[\log c + \log\left(\frac{1-\sqrt{3}}{2} + ws\right) - Z\right] \Big/ \log\left(\frac{\sqrt{3}}{1+2ws}\right) - 1. \tag{12.19}$$

This provides an answer to the important question of what should the highest value of polynomial degree $l = nL$ be in the truncated multipole expansion for the given values of parameters *ws* and *c* so that the precision *Z* is satisfied.

Determining the remaining parameter *ws* from Equation 12.15 is more difficult because of the nonlinear dependence on *ws*. Therefore, it is necessary to go back to Equation 12.12.

Because $b - a > 1$ for typical values of parameters *c* and *ws*, dropping the term $b - a$ from Equation 12.12 leads to a stronger relation:

$$\left(\frac{a}{b}\right)^{nL+1} = 10^{-Z}, \tag{12.20}$$

which now allows the determination of parameter *ws*.

Replacing (a/b) in Equation 12.20 by its value from Equation 12.14 leads to:

$$\left(\frac{\sqrt{3}}{1+2ws}\right)^{nL+1} = 10^{-Z}. \tag{12.21}$$

Taking the log of both sides of Equation 12.21 gives:

$$(nL+1)\log\left(\frac{\sqrt{3}}{1+2ws}\right) = -Z. \tag{12.22}$$

Then dividing both sides of Equation 12.22 by $nL + 1$, and rewriting the expression in exponential form produces:

$$\frac{\sqrt{3}}{1+2ws} = 10^{-\frac{Z}{nL+1}}. \tag{12.23}$$

Or:

$$\sqrt{3} = (1+2ws)10^{-\frac{Z}{nL+1}}. \tag{12.24}$$

Resolving this expression about parameter *ws* leads to the final equation:

$$ws = \frac{\sqrt{3}\,10^{\frac{Z}{nL+1}} - 1}{2}. \tag{12.25}$$

Together, Equations 12.16, 12.18, 12.19, and 12.25 offer the possibility of computing the value of parameters *c*, *Z*, *nL*, and *ws* from the known value of other three parameters.

Although each of the four parameters may be computed from the value of the other parameters, in practice, three of them need to be guessed. Typically, the value of *ws* is set to 1, 2, or 3, based on personal preferences. Similarly, *c*, is guessed to be around 5–10 Å. One parameter, *Z*, is easy to define; this is set to the number of significant digits expected for the precision of the calculation. Finally, the remaining parameter *nL* is computed from Equation 11.19. Typical sample values of parameters *ws*, *c*, *Z*, and *nL* are listed in Table 12.1.

TABLE 12.1
Computed Values of nL as a Function of Parameters ws, c, and Z

ws	c	Z	nL
1	5	12	47
2	5	12	23
3	5	12	17
1	10	12	46
2	10	12	22
3	10	12	16
1	5	14	56
2	5	14	27
3	5	14	20
1	10	14	54
2	10	14	27
3	10	14	20

Analysis of data in Table 12.1 provides several useful insights. For example, the change in numerical precision Z of convergence has the most profound influence on parameter nL, and through that on the size of the truncated multipole expansion. Although it is tempting to push the value of parameter Z down so that fewer terms in the multipole expansion will be needed to achieve the specified convergence threshold, making it smaller than 10 faces an increasing risk of numerical instability. As expected, making nL larger increases the value of parameter Z, that is, it improves the accuracy of convergence of the multipole expansion.

According to the data from Table 12.1, another parameter that has a strong influence on the value of nL is the well-separation parameter, ws. For a given value of Z, using larger values of ws makes it possible to use fewer terms in multipole expansion in order to achieve the specified convergence threshold. Once the value of ws is decided, the value of nL is uniquely known from Equation 12.19.

Although larger values of parameter ws makes multipole expansions converge faster, other parts of the FMM algorithm depend differently on parameter ws, and that places a practical limit on the size of ws. Increasing the value of ws lowers the computational cost in the *FF* region though its connection with parameter nL, and, at the same time, increases the computational cost in *NF* region due to the increase in its size. Since the cost of computation in the *NF* region grows quadratically in terms of number of particles whereas the change in the *FF* region changes linearly, it quickly offsets the performance gain in the *FF* region. Finding a sweet spot in the value of ws typically requires experimentation. Popular values for well separation are $ws = 1$ and $ws = 2$.

The interdependence of parameter nL and ws helps to better understand the meaning of NF region. Employing larger values of ws and, as the result of that, using smaller nL and shorter multipole expansion, makes the multipole methods inapplicable to the boxes in the NF region even if they are formally eligible for this type of interaction based on the criterion of non-overlapping of their encircling spheres, as illustrated in Figure 12.2. If a truncated multipole expansion optimized for FF region were applied to NF boxes, it would produce an insufficiently converged series sum, and that would lead to a numerically wrong result.

With that, principles that govern the use of multipole expansion in the computation of electrostatic interactions may now be fully formalized. The first rule requires that the bounding spheres of the interacting boxes must not overlap. According to this rule, there should be at least a single layer of boxes, which corresponds to $ws = 1$, placed between the pair of boxes in order for them to be eligible for interaction *via* multipole expansion. The second principle builds upon the objective of minimizing the cost of computation by using fewer terms in the multipole expansion. This requires increasing the distance between the interacting boxes by placing additional layers of *NF* boxes in between the box and its *FF* region. With that, the *NF* region becomes fully defined as a segment of boxes where multipole expansion either cannot be used due to overlap of the encircling spheres or should not be used due to inability of the specifically truncated expansion to provide a converged result. That completes the definition of NF region.

One remaining task is to complete the specification of the *FF* region. This involves the logical partitioning of the system of electrostatic charges into a hierarchical set of boxes.

12.3 Hierarchical Division of Boxes in the Fast Multipole Method

The division of the system into a set of boxes and then using the multipole expansion for evaluation of electrostatic interactions between each pair of remote boxes does not yet make the method linear scaling since the number of interacting box pairs in this arrangement still grows quadratically with the system size. The FMM algorithm resolves this problem by dividing the system of point charges into a hierarchical set of boxes in the form of binary tree and sorting the box-box interactions out to various hierarchy levels in order to keep the number of interacting pairs growing linearly with the increase in the system size. This division is performed in a sequence of steps.

The first step places the whole system into a single large cubic box as depicted in Figure 12.5. The top box serves as a root of the binary tree. After that, each side of the box is branched in two by dividing the side length in half. Depending on the system dimension, branching of a parent box creates 2, 4, and 8 smaller child boxes in 1D, 2D, and 3D, respectively. The branching of the boxes continues until the desired depth of division is reached.

Placement of an asymmetric system into a cubic box proceeds in two steps. In the first step, a best-fit rectangular cuboid is placed around the system. In the second step, the cuboid is expanded into a cube. Figure 12.6 illustrates the outcome of this procedure. Implementation of these steps in a computer code will now be described.

The task of determining the best-fit cuboid that encloses a system of charges is the easiest to accomplish by employing Fortran 90 intrinsic functions *minval*() and *maxval*(). These functions take an array of Cartesian coordinates of the particles on input, and return the coordinates of the vertices *vertexMin*(3) and *vertexMax*(3), located at the opposite ends of the main diagonal of the cuboid. These are the lower left and the upper right vertices of the cuboid shown in Figure 12.6 that encloses the system of particles.

Since the generated cuboid has unequal side lengths, it needs to be converted to a regular cube before the binary tree construction can proceed. This is done in the second step. This task is most easily accomplished by expanding the cuboid into a cube. The algorithm picks the longest side of the cuboid to serve as the side length *cubeEdge* of the final cube. Then it extends the other sides of the cuboid to match the longest side. In that process, the position of the lowest left vertex *vertexMin*(3) remains unchanged, and the coordinate of the highest right vertex *vertexMax*(3) gets computed from the vector operation `vertexMax = vertexMin + cubeEdge`. Fortran 90 automatically vectorizes array manipulations so no explicit reference to array indices in the computer program is required.

The following code snippet provides a sample implementation of the box placement algorithm:

```
double precision                :: vertexMin(3), vertexMax(3)
double precision, allocatable :: coor(3,:)
double precision                :: cubeEdge

! the code to allocate array coor goes here
! the code to read coordinates into array coor goes here

! step 1
vertexMin = minval(coor, dim = 2)
vertexMax = maxval(coor, dim = 2)

! step 2
cubeEdge  = dmax1(vertexMax(1)-vertexMin(1), vertexMax(2)-vertexMin(2),
vertexMax(3)-vertexMin(3))
vertexMax = vertexMin + cubeEdge
```

Since *coor* is a two-dimensional array that holds particle Cartesian coordinates, functions *minval*() and *maxval*() requires "*dim* = 2" as the second argument. Those parts of the code that deal with allocation and initialization of the array *coor* have been omitted for the sake of brevity.

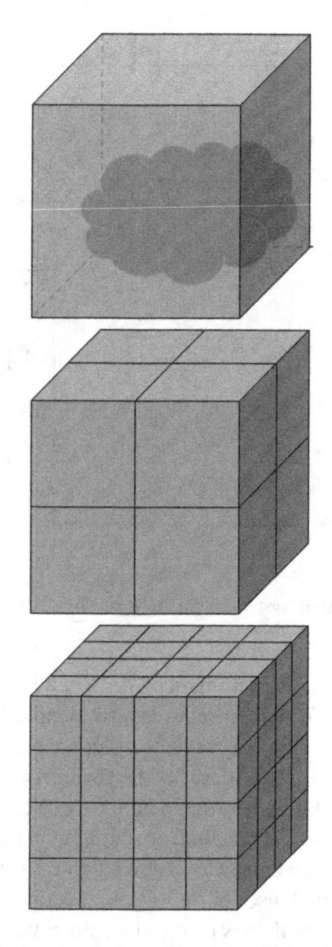

FIGURE 12.5 Division of the system of point charges into a hierarchy of boxes in the fast multipole method. The hierarchy levels are visually disentangled from each other in order to better illustrate the character of fragmentation and parent-child relationship.

The ability of this implementation to fit in a few lines of Fortran 90 code adds an extra value to this algorithm. Intrinsic functions *minval*() and *maxval*() that determine the vertex points do their job very efficiently. Handcoding their substitutes to iterate through the particle coordinates, and using conditional operators to locate the extreme values of coordinate components would most likely produce a less efficient code.

In the code, points *vertexMin*(3) and *vertexMax*(3) should not be confused with coordinates of the real particles from array *coor*. Because of the way functions *minval*() and *maxval*() work, the Cartesian components x, y, and z of the vertex point usually obtain their values from Cartesian coordinates of different particles. Therefore, the vertices of the encompassing cuboid should be treated as new points, that is, points that do not exist in the array *coor*.

The task of enclosing a non-uniform system of charges into a cubic box may apparently have many possible implementations. A specific aspect of the above code is that the sides of the generated best-fit cuboid are aligned along the axes of coordinate system. This is automatically achieved due to the use of functions *minval*() and *maxval*(). Apparently, for a system of point charges of arbitrary shape, the generated cuboid will be somewhat larger than the ideal best-fit cuboid. This does not represent a problem, though. In any general system with a non-uniform geometry, the enclosing cube will always include some empty space that is not occupied by charge sources. A standard approach to deal with it is to purge empty boxes from the box list, thereby removing them from the bookkeeping procedure. Therefore having the top cube created slightly larger in size than the best-fitting cube makes no significant difference.

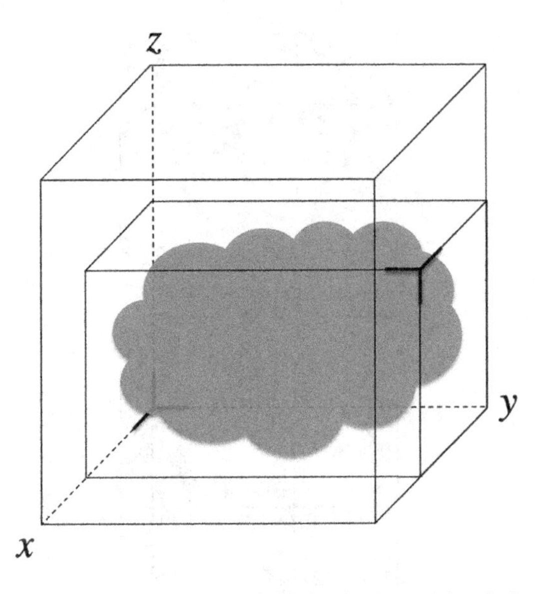

FIGURE 12.6 Placement of an asymmetric system into a cubic box by determining the best-fit cuboid and expanding it to the cube.

Having the sides of the top cube being automatically aligned with the coordinate axes makes it easy to compute coordinates of each box from their relative position to other boxes after the system is divided into boxes; and that also greatly simplifies the bookkeeping procedure.

There may be rare circumstances when the described algorithm would be less optimal than usually. The examples are systems of rod- or disk-like geometry in 3D space that exhibit a large ratio of the longest side to the shortest one. Such systems may benefit from employing a more complex algorithm that will be only briefly outlined here. First, it is necessary to assign a unit mass to individual particles in the system, and to compute the moment of inertia tensor of the system. Second, the 3 by 3 tensor matrix needs to be diagonalized, to determine the principal axes of rotation. Eigenvalues of the tensor compose a rotation matrix. Rotating the coordinate system to align it with the principal axes of rotation produces a new set of coordinates for the system of point charges. After that, application of the previously discussed algorithm, which relies on functions *minval*() and *maxval*(), generates the ideal best-fit cuboid. Alternatively, using the standard approach followed by purging the unoccupied cubes typically takes care of non-uniformity in the system geometry, rendering the complex implementation unnecessary.

After the system of point charges is successfully enclosed in a single cube, the FMM algorithm starts the division process by dividing each side of the cube in half. Each cube to be divided is called a parent cube and the ones created are called child cubes. A cube may simultaneously be a child to one cube and a parent to another. Parents and their child cubes occupy the same space.

All the cubes created at the same fragmentation depth share the same hierarchy level. Each hierarchy level is associated with an index that represents the number of division steps that need to be applied to descend from the top cube to that level. Zero-based indexing provides the most convenient approach for the numbering scheme as it makes counting the number of cubes *nCubes* in each hierarchy level k particularly easy:

$$nCubes = 2^{3h}, \tag{12.26}$$

where 2 is the base of the binary tree division, and 3 is the number of dimensions of the space. How this counting works will now be described.

The top hierarchy level has the 0th index. Raising any number to the zero-th power produces unity. This is consistent with the 0th level being represented by a single cube, $nCubes = 1$. This is the only cube that has no parents. Division of the top cube generates eight smaller cubes, $2^3 = 8$ which compose the 1st level of hierarchy, $h = 1$. Each of the eight cubes further divide into eight smaller grand-child cubes leading to the 2nd hierarchy level, $h = 2$ that consists of 64 cubes, $2^6 = 64$. The simplicity of determining the number of cubes in each hierarchy level h makes the zero-based numbering scheme the method of choice in the binary tree algorithm.

This example provides an insight into the origin of Equation 12.26: a binary tree division cuts each side of the top box on 2^h pieces. In 3D space, division of the top cube creates $2^h 2^h 2^h = 2^{3h}$ smaller cubes, and that leads to Equation 12.26.

The division of a top box into smaller ones may be instructed by specifying either the hierarchy depth, H, of the tree ($h = 0, 1, 2, \ldots H$) or by choosing the side length, c, of the smallest box to be produced, whichever is convenient. Each of those parameters can independently be used to control the division process. Due to their interdependence, only one of them is necessary to initiate the division. The other one will appear as the result of division. The most common strategy is to specify the target side length, c, and let H to be automatically determined. The rationale behind this choice is that the value of the lowest level box may be visually related to the system of study, whereas the hierarchy depth is a more abstract quantity, and is therefore more difficult to guess. Choosing the value of side length of the smallest box in the range of 5–10 Å is a common choice for molecular systems. With that, the goal is to determine the value of parameter H, based on the user-provided value of parameter c.

A relation between parameters H and c originates from the use of the binary tree algorithm to perform the division that cuts a single side of the top cube into 2^H pieces. Recall that *cubeEdge* is the side length of the top cube, so multiplying the number of generated cubes 2^H along the side onto side length c of the smallest created boxes gives the formula for determining the hierarchy depth H:

$$c\,2^H = cubeEdge. \tag{12.27}$$

A matter that perceivably complicates the use of Equation 12.27 is that the product $c\,2^H$ may not exactly be equal to the value of *cubeEdge* for a requested value of side length c. Indeed, since the value of *cubeEdge* is system dependent, and 2^H is always an integer, choosing the right value of parameter c to satisfy the equality in Equation 12.27 can be difficult. Fortunately, specifying a precise value for parameter c is unnecessary. Since purging takes care of empty boxes, letting $c\,2^H$ become somewhat larger than *cubeEdge* retains the applicability of Equation 12.27 for any given value of c. Fortran function *ceiling*() takes care of the uncertainty that exists around the choice of parameter c. Here is how the proposed solution works.

The rounding operation accomplishes a virtual expansion of the value of *cubeEdge* through a series of transformations of Equation 12.27. In the first step, it is necessary to ensure that *cubeEdge* is divisible by the parameter c. Rounding the ratio (*cubeEdge*/*c*) to the greatest whole number leads to:

$$2^H = ceiling\left(\frac{cubeEdge}{c}\right). \tag{12.28}$$

Finding the unknown H from Equation 12.28 requires taking a logarithm from both sides of the equation. This assumes that *ceiling*(*cubeEdge*/*c*) is an exact power of 2 due to H being a whole number. Once again application of function *ceiling*() enforces that condition, and leads to:

$$H = ceiling\left(\frac{\log\left[ceiling\left[\frac{cubeEdge}{c}\right]\right]}{\log 2}\right). \tag{12.29}$$

Equation 12.29 is the transformed form of Equation 12.27 that accepts any value of parameter c, and computes the hierarchy depth H.

Before the system is divided into boxes, it is necessary to figure out how many boxes will be created during the division process so that the memory storage for each box can be allocated dynamically. The use of a binary tree algorithm in the division process, and the choice of boxes with equal side length makes this straightforward. The number of boxes along a single dimension in each hierarchy level h is 2^h, where

$h = 0, \ldots, H$. In 3D, the number of boxes becomes 2^{3h}. Therefore, the total number of boxes N_b becomes a sum along all hierarchy levels:

$$N_b = \sum_{h=0}^{H} 2^{3h} = \sum_{h=0}^{H} 8^h. \tag{12.30}$$

According to Appendix A.5, the sum of the geometric power series $\sum_{k=0}^{n} x^k$, where $x > 1.0$ is:

$$N_b = \frac{x^{n+1} - 1}{x - 1}. \tag{12.31}$$

Substituting x by 8, and n by H in Equation 12.31 leads to:

$$N_b = \frac{8^{H+1} - 1}{7}. \tag{12.32}$$

Since the value of H is already known, based on the known values of *cubeEdge* and c, the side lengths of the zeroth and the lowest hierarchy level boxes, respectively, the total number of boxes N_b across all hierarchy levels can immediately be determined. For a non-periodic system, the number of empty boxes, determined separately, should be subtracted from N_b before performing the memory allocation. For a periodic system that has cubic symmetry the total number of cubes across all hierarchy levels is exactly equal to N_b.

This completes the definition of hierarchy levels. This definition will be essential in the following section for developing a comprehensive description of the far field.

12.4 Far Field

Near and far fields are the parts of the bookkeeping mechanism that the FMM algorithm uses to track how the electrostatic interaction between a pair of boxes should be computed. In Section 12.1 the introduction of *NF* and *FF* concepts was made, based on the criterion of overlap of the encircling spheres. According to that definition, the boxes that have their encircling spheres mutually overlapping are in *NF* position one to another. Such boxes cannot interact *via* the multipole expansion, so all electrostatic interactions of a box with the boxes from its *NF* region require using direct sum methods. The *FF* region of a box then becomes automatically defined as all parts of the system that are outside of *NF* region. The electrostatic interaction between a box and any box in its far field employs the multipole expansion method.

With the introduction of the well-separation parameter, *ws*, in Section 12.2, the definitions of *NF* and *FF* regions are further refined. The *NF* region is expanded as *ws* layers placed between a box and its *FF* region in order to allow a truncated multipole expansion to be used in the description of the electrostatic interaction between the box and its *FF*. The value chosen for parameter *ws* determines the degree of truncation of the multipole expansion.

The introduction of hierarchy levels in the FMM algorithm adds additional important details into the interplay of the *NF* and *FF* regions. Its purpose is to address a current limitation in the definition of the *FF* region. In its present formulation, letting the *FF* region expand to the outer boundaries of the system would result in the electrostatic computation scaling quadratically with the size of the system, which would defeat the goal of making the method linear scaling.

One solution to that problem is to split the system into a hierarchy of boxes using larger boxes for more distant multipole interactions and smaller boxes for closer interactions. It includes redefining the *FF* region in terms of parent-child relationship while avoiding any overlap of their *FF* regions. In other words, any part of the system that is defined as an *FF* region at the child level must not simultaneously be defined at the parent level. This requirement can only be fulfilled by constraining the *FF* region at the child level to fit into the boundary of the *NF* region at the parent level. The meaning behind this division

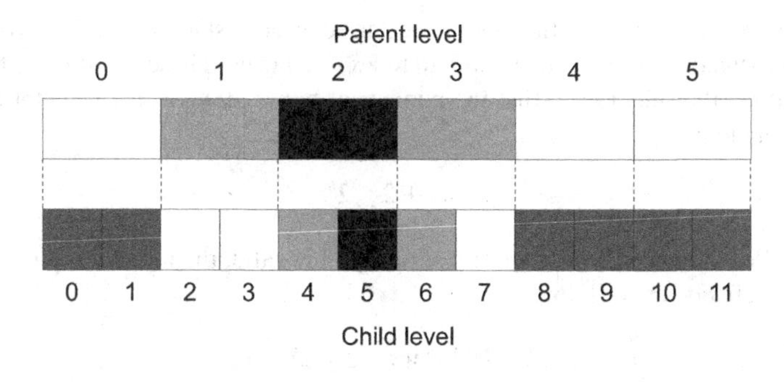

FIGURE 12.7 Assignment of the *NF* and *FF* regions for a box shown in black at parent and child hierarchy levels, in a 1D system with $ws = 1$. The *NF* boxes are shown in light gray and the *FF* boxes in white. Boxes in dark gray are ignored at the child level.

is that any part of the global *FF* region, which is not handled at the child level, will be addressed at its parent level or, if necessary, at the grandparent level, and so on.

Figure 12.7 illustrates the working of this methodology, using the example of a 1D system. The *NF* region includes a *ws* number of layers surrounding the box. With that, for a box number 5 at the child level and $ws = 1$, the *NF* region includes boxes 4 and 6. Likewise, the *NF* region of box number 2 at the parent level consists of boxes 1 and 3. Since the *NF* region at the parent level defines the boundary of the *FF* region at the child level, child boxes 2, 3, and 7 become the *FF* region of child box number 5. Similar to that, parent boxes 0, 4, and 5 constitute the *FF* region at the parent level based on the requirement to fit into the *NF* boundary of the grandparent level (not shown). Boxes 0, 1, 8, 9, 10, and 11, which are left unassigned at the child level, will be ignored since they are not contributing at this hierarchy level. The particles assigned into these boxes will be taken into account in boxes 0, 4, and 5 at the parent level; the working of that mechanism will be explained later.

With the assignment of the *FF* region of the child boxes within the boundary of the *NF* region of the parent box, the definition of the *FF* region becomes complete. Now that the principles governing the system division into *NF* and *FF* regions have been fully described, the next step is to analytically evaluate the number of boxes that need to be tracked in each region to determine the required dynamic memory storage they will need.

12.5 Near Field and Far Field Pair Counts

In the FMM, each box needs to carry a list of its *NF* and *FF* boxes so that it knows which other boxes it interacts with, and whether to use direct sum or the multipole expansion method when calculating the electrostatic interaction. Dynamic memory allocation for these arrays requires a knowledge of how many boxes of each type appear in each box's neighborhood. This requires counting the boxes in the *NF* and *FF* regions, minus the number of empty boxes. The number of empty boxes is empirically determined based on the system shape at the time the system is decomposed into boxes, and the assignment of particles into those boxes. Determining the number of boxes in the *NF* and *FF* regions will now be addressed.

The number and location of *NF* and *FF* boxes around a box depend on the choice of the well-separation parameter and hierarchy level. When computing the box count, it is most convenient to traverse the hierarchy tree from the top to bottom. This begins by establishing the hierarchy level at which the *NF* and *FF* boxes will both have a non-zero count. Having only the *NF* count non-zero while the *FF* count is zero is unimportant, since no FMM computation would actually be performed at such a hierarchy level. Therefore, the question translates to finding the highest hierarchy level, h, at which the *FF* count of well-separated boxes is non-zero.

To resolve the stated problem, it is sufficient to consider a one-dimensional case; the results that follow will be fully applicable to the 3D case. At this point, the objective is to derive an equation to relate the variables *ws* and h.

For two boxes to be well-separated they should be parted by at least ws boxes. The two interacting boxes and the ws boxes located in between them add up to $ws + 2$ boxes. These should all fit in the number of available boxes along the side. Given that the number of boxes along a side is 2^h for hierarchy level h, this leads to the relation:

$$ws + 2 \leq 2^h. \tag{12.33}$$

To find the smallest number h for which Equation 12.33 is valid, the number $ws + 2$ must be formally bounded from the left side as well, thus:

$$2^{h-1} < ws + 2 \leq 2^h, \tag{12.34}$$

where $h - 1$ is a preceding hierarchy level that has a smaller capacity than is necessary to accommodate $ws + 2$ boxes. Simply replacing the inequality sign in Equation 12.33 by an equality sign is sufficient to allow the unknown h to be defined. Taking the logarithm of both sides of Equation 12.33, and using the Fortran function *ceiling*() to round the result of division up to the nearest whole number leads to:

$$h = \text{ceiling}[\log(ws + 2)/\log(2)]. \tag{12.35}$$

Using cubic boxes makes Equation 12.35 applicable to both 2D and 3D systems.

The next task is to determine how many *NF* and *FF* pairs exist at each hierarchy level, starting at level h. Although, for all practical applications, the FMM will be applied to 3D systems, it is easier to derive the key rules to obtain the *NF* and *FF* box counts based on the analysis of 2D systems, and then generalize the conclusions to 3D systems.

Due to the symmetry that comes with the use of cubic boxes, and the application of the binary tree algorithm, it is only necessary to count the number of pairs in one quadrant in 2D or in one octant in 3D, and then multiply the result by the number of equivalent units.

Figure 12.8 shows a series of snapshots of a box in black traversing the lower-left quadrant of a 2D system. In it, the assignment of *NF* and *FF* regions corresponds to the choice of parameter $ws =$ 1. The unit-base indexing scheme adopted in Figure 12.8 facilitates the counting process. Counting the number of *NF* boxes is straightforward. The gray and black boxes together form a rectangle, so the number of *NF* boxes is the product of the side-lengths of the gray rectangle, minus one for the box in black.

Two integer numbers x and y are necessary to track the position of boxes in 2D. Using these indices, the number of boxes along each side of the gray rectangle is equal to the larger index minus the smaller index plus one. The larger index is given by the addition of ws to the index of the box in black along a side, e.g. $x + ws$, while the smaller index is given by subtracting ws from the index of the box in black, $x - ws$.

Care must be taken to ensure that the addition and subtraction of ws boxes does not violate the allowed boundaries for the index. Fortran functions *min*0() and *max*0() provide the necessary assistance with that by determining the smaller and larger indices as $max0(x - ws, 1)$ and $min0(x + ws, 2^h)$, respectively, where h is the depth of the hierarchy, and 2^h is the number of boxes along a side for that hierarchy level. The functions *min*0() and *max*0() take whole numbers as their arguments, and return the least and the greatest whole number from the list of arguments, respectively. With that, the number of boxes along each side of the gray rectangle is:

$$min0(x + ws, 2^h) - max0(x - ws, 1) + 1, \tag{12.36}$$

and

$$min0(y + ws, 2^h) - max0(y - ws, 1) + 1. \tag{12.37}$$

In these expressions, function $min0(i + ws, 2^h)$ ensures that the sum $i + ws$ does not exceed the maximum box index 2^h along the side, and the function $max0(i - ws, 1)$ ensures that the result of $i - ws$ does not become less than 1, where symbol i designates index x or y.

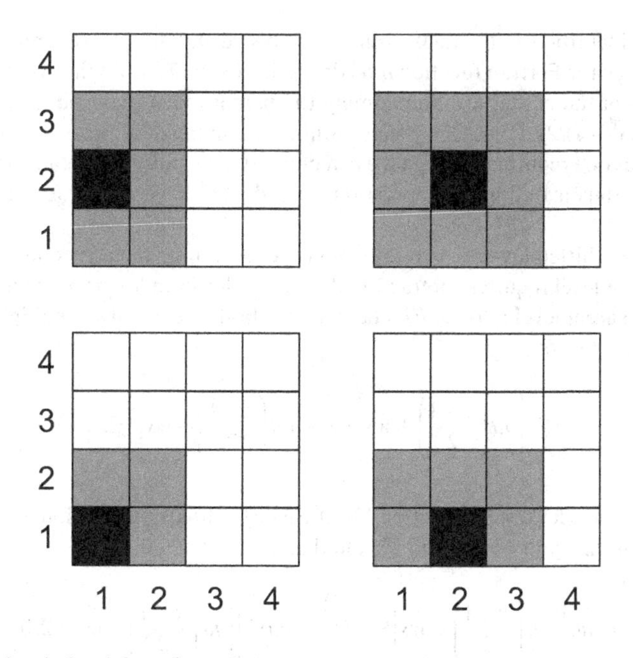

FIGURE 12.8 Traversing the low-left quadrant of a 2D system by a box in black with $ws = 1$. *NF* boxes are shown in gray and *FF* boxes in white. Boxes are numbered using a unit-based indexing scheme to aid in the counting process.

The count $N_{NF}(x, y)$ of near field boxes for a box specified by indices x and y in 2D is a product of Equations 12.36 and 12.37, minus one:

$$N_{NF}(x,y) = [min0(x+ws,2^h) - max0(x-ws,1)+1] \times [min0(y+ws,2^h) - max0(y-ws,1)+1] - 1. \quad (12.38)$$

This result can be readily extended to 3D, which differs from 2D only by the presence of an extra dimension tracked by index z:

$$N_{NF}(x,y,z) = [min0(x+ws,2^h) - max0(x-ws,1)+1] \times [min0(y+ws,2^h) - max0(y-ws,1)+1] \times [min0(z+ws,2^h) - max0(z-ws,1)+1] - 1. \quad (12.39)$$

The number of boxes in the *FF* region is determined in a similar manner. The *FF* region as illustrated in Figure 12.8 has a highly irregular shape that makes direct counting of the constituent boxes difficult. The fact that *FF* boxes, together with *NF* boxes, and including the box in black, compose a *FF* superbox of rectangular shape can be used in simplifying this problem. The total number of boxes in this superbox is a product of its side lengths in the units of boxes, so the number of *FF* boxes is simply the number of boxes in the *FF* superbox minus those in the already determined *NF* region minus 1 for the box in black.

Calculating the number of boxes in the *FF* superbox starts with an analysis of snapshots presented in Figure 12.8. Once again the box in black will be traversing the lower-left quadrant of the system in 2D. Determining the side-lengths of the *FF* superbox uses a procedure similar to that used in the construction of the *FF* region. The *FF* region for the box in black, identified by the indices x and y, extends to the outer boundaries of the *NF* region of its parents. Figure 12.7 illustrates the construction of a child *FF* region based on the span of the parent *NF* region in 1D. Since the size of the child *FF* superbox is equal to the size of the parent *NF* superbox, all that needs to be done is to obtain the parent index from the child index, and then convert the size of the parent *NF* superbox into the units of child boxes. Similar to the case of seeking the *NF* count, the unit-based index proves to be the most efficient for the purpose of counting the number of *FF* boxes.

For a child box identified with index x in a unit-based numbering scheme, the parent index is $int((x+1)/2)$, in which the Fortran function *int*() drops the decimal part of the argument. The boundaries of the *NF* superbox at the parent level are simply the parent index plus and minus *ws*. With that, the upper index is $int((x+1)/2)+ws$. Since determining the number of boxes by subtracting the lower index from the upper one requires adding an extra unity to the result, it is better to shift the lower index one box to the left, thus avoiding that operation. The shifted lower index for parent *NF* superbox is $int((x+1)/2)-ws$.

Once the upper and shifted lower indices are known, determining the box count along a side of the *NF* superbox at the parent level requires subtracting the shifted lower index from the upper one. Converting the *NF* count at the parent level into the *FF* count at the child level involves multiplying everything by a factor of two, which leads to:

$$\left[int\left(\frac{x+1}{2}\right)+ws\right]*2-\left[int\left(\frac{x-1}{2}\right)-ws\right]*2. \tag{12.40}$$

Next is to add a boundaries test to Equation 12.40. Since the final computation is performed at the child level, the limits to check against are 0 and 2^h; and that leads to:

$$min0\left(\left[int\left(\frac{x+1}{2}\right)+ws\right]*2,2^h\right)-max0\left(\left[int\left(\frac{x-1}{2}\right)-ws\right]*2,0\right). \tag{12.41}$$

The box count along the other sides of a 3D system is obtained by simply replacing the index x by indices y and z. The final count $N_{FF}(x, y, z)$ of *FF* boxes at the child level is a product of the number of boxes along each side of *FF* super-box minus the size of *NF* super-box:

$$\begin{aligned} N_{FF}(x,y,z)=&\left\{min0\left(\left[int\left(\frac{x+1}{2}\right)+ws\right]*2,2^h\right)-max0\left(\left[int\left(\frac{x-1}{2}\right)-ws\right]*2,0\right)\right\}\\ &\times\left\{min0\left(\left[int\left(\frac{y+1}{2}\right)+ws\right]*2,2^h\right)-max0\left(\left[int\left(\frac{y-1}{2}\right)-ws\right]*2,0\right)\right\}\\ &\times\left\{min0\left(\left[int\left(\frac{z+1}{2}\right)+ws\right]*2,2^h\right)-max0\left(\left[int\left(\frac{z-1}{2}\right)-ws\right]*2,0\right)\right\}-N_{NF}-1, \end{aligned} \tag{12.42}$$

where the value of N_{NF} is determined in Equation 12.39.

The following code snippet illustrates the calculation of the number of *NF* and *FF* boxes in each hierarchy level for a 3D system.

```
program neighbor_count
implicit none

! input data
integer :: ws = 1                  ! Well separation parameter
integer :: nLevels = 9             ! Total number of hierarchy levels in the system

! local variables
integer :: hLevel, x, y, z   ! current hierarchy level and coordinate indices
integer (kind=8) :: superBoxNF ! child NF superbox, which includes box (x,y,z)
integer (kind=8) :: superBoxFF ! child FF superbox for a box (x,y,z)
integer :: nBoxes                ! number of boxes along a side in hLevel
integer :: nScan                 ! number of boxes to scan
```

```
! output data
integer (kind=8) :: nNFlevel    ! number of NF boxes in a hierarchy level
integer (kind=8) :: nFFlevel    ! number of FF boxes in a hierarchy level
integer (kind=8) :: nNFtotal    ! total NF box count across the system
integer (kind=8) :: nFFtotal    ! total FF box count across the system

! initialize counters
nNFtotal = 0
nFFtotal = 0

write(*,'(a,i2)') "ws = ", ws
write(*,'(a)') &
    "hLevel    nNFlevel        nNFtotal        nFFlevel        nFFtotal"
!
! loop over hierarchy levels
do hLevel = 1, nLevels
  !
  nBoxes = 2 ** hLevel        ! number of boxes along a side
  nScan  = nBoxes / 2         ! scan half of nBoxes for each box side
  !
  ! loop over each cube x,y,z in the hierarchy
  ! work on a single octant and then multiply the total count by 8
  !
  nNFlevel = 0
  nFFlevel = 0
  do x = 1, nScan
    do y = 1, nScan
      do z = 1, nScan
        !
        ! NF count
        !
        superBoxNF = (min0(x+ws,nBoxes) - max0(x-ws,1) + 1)* &
                     (min0(y+ws,nBoxes) - max0(y-ws,1) + 1)* &
                     (min0(z+ws,nBoxes) - max0(z-ws,1) + 1)
        nNFlevel = nNFlevel + superBoxNF - 1 ! subtract box (x,y,z)
        !
        ! FF count
        !
        superBoxFF = &
        (min0((int((x+1)/2)+ws)*2, nBoxes) - max0((int((x-1)/2)-ws)*2, 0)) *&
        (min0((int((y+1)/2)+ws)*2, nBoxes) - max0((int((y-1)/2)-ws)*2, 0)) *&
        (min0((int((z+1)/2)+ws)*2, nBoxes) - max0((int((z-1)/2)-ws)*2, 0))
        nFFlevel = nFFlevel + superBoxFF - superBoxNF

      enddo
    enddo
  enddo

  nNFtotal = nNFtotal + nNFlevel
  nFFtotal = nFFtotal + nFFlevel

  ! print NF and FF counts for each hierarchy level
  write(*,'(i2,4i16)') hLevel, nNFlevel*8, nNFtotal*8, nFFlevel*8, nFFtotal*8
enddo

end program neighbor_count
```

Input for this program consists of the well-separation parameter, *ws*, and the number of layers, *nLevels*. The total number of *NF* and *FF* boxes is stored in variables *nNFtotal* and *nFFtotal*, respectively, and the number of boxes for each of the hierarchy levels is stored in variables *nNFlevel* and *nFFlevel*. The computation is arranged in a loop over hierarchy levels. There are no neighbors at the top 0th hierarchy level, which consists of a single cube, so the *do*-loop starts from *hLevel* = 1.

For each hierarchy level, *hLevel*, variable *nBoxes* holds the number of boxes along a side. Because of symmetry only half of those boxes need to be scanned. The scan is performed using a triple loop over side-indices *x*, *y*, and *z*, and the starting unit value of these indices indicates the use of the unit-based numbering scheme. Variable *superBoxNF* represents the number of boxes in the *NF* superbox, and includes the box (*x*, *y*, *z*) in the count. Variable *nNFlevel* accumulates the number of *NF* box pairs.

Variable *superBoxFF* holds the number of boxes for the *FF* superbox. Therefore, when counting the number of *FF* boxes in a hierarchy level, variable *superBoxFF* is added to the variable *nFFlevel*, and the variable *superBoxNF* is subtracted. At the end, the program multiplies the computed box counts by a factor of eight to convert them from an octant to the whole cube, and prints their values.

Although computation of the box counts takes a fraction of a second, since the values of *NF* and *FF* boxes depend only on the hierarchy level and on the value of parameter *ws*, it is possible to compute them once and store the results for the future use, so that they do not have to be calculated every time the FMM computation starts.

The values of the *NF* and *FF* counts show how much memory should be allocated for the corresponding lists. These strongly depend on the choice of hierarchy depth and the well-separation parameter, and increase rapidly with the values of the input parameters. The large memory requirement vividly demonstrates that performing electrostatic computations in large systems requires the use of distributed memory programming models.

12.6 FMM Algorithm

Decomposition of the system of point charges into a hierarchy of boxes and the generation of *NF* and *FF* lists completes the initialization step of the FMM computation. At this point, the charges are assigned to individual boxes at the bottom of the hierarchy tree; and the *FF* lists are created so that each box knows which other boxes it should interact with using the multipole expansion method.

Every box, with the exception of a few at the top of the hierarchy tree, needs to know the multipole expansion for the set of charges they contain. Computing the multipole expansion for a set of point charges is straightforward, but doing that for each box in the system would be computationally inefficient. For that reason only the lowest hierarchy boxes keep a record of the charges contained within their boundaries. Parent and grandparent boxes get their multipole expansion assembled from that of their child boxes, as will be explained shortly.

The first step in the FMM algorithm after initialization is to generate the multipole expansion for the boxes located at the lowest level of the hierarchy. This is accomplished by looping over the particles assigned to a box using Equation 12.1, and accumulating the resulting multipole expansion in that box. Multipole expansions of different particles with the same indices l and m are summed together. This operation is conducted in a local coordinate system whose origin is located in the center of the box. Since each particle is accessed only once, the computational effort scales linearly with the number of particles in the system.

Using the multipole expansions for the boxes located at the lowest hierarchy, the FMM generates the multipole expansions for parent, grandparent, and other higher placed boxes. This step is called the *upward translation of multipole expansions*, and is performed with the help of an M2M operation. In this translation, the coordinate origin for the multipole expansion of each of the child boxes, defined to be at the center of the child box, is translated to the center of the parent box. This allows the multipole expansions of children to be summed up in the parent box.

Figure 12.9 illustrates the work of upward translations using the example of a 1D system. This procedure takes a multipole expansion from the boxes at the lowest hierarchy level $h = 5$, and accumulates the contributions in the respective parent boxes at the level $h = 4$. Then the computation repeats by taking the multipole expansions at the level $h = 4$, and accumulates the expansions in respective parent boxes at the level $h = 3$. This upward translation continues until the highest level that participates in multipole interactions is reached. Translation of the multipole expansions to the boxes in hierarchy levels 0 and 1 is unnecessary, unless periodic boundary conditions are present. In that case, multipole expansions are needed for all boxes, including the top box. In a non-periodic case, the boxes at level $h = 2$ are eligible for multipole interactions only when $ws = 1$ or $ws = 2$.

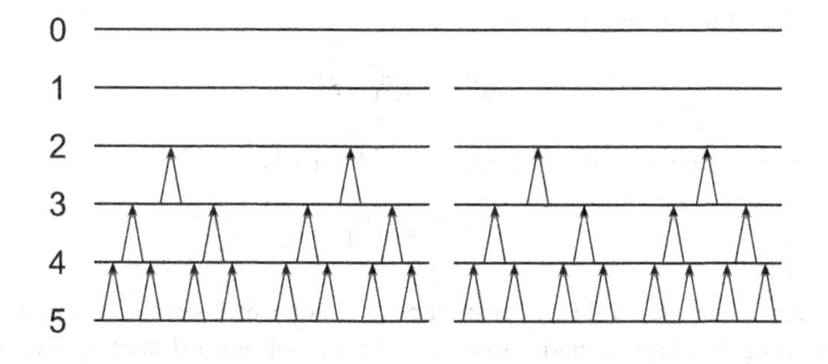

FIGURE 12.9 Upward translation of regular multipole expansions for different hierarchy levels in a non-periodic 1D system where $ws = 1$.

The 1D system illustrated in Figure 12.9 provides a convenient platform to see how well the upward translation scales with the change in the system size. If the system is made small enough, so that its decomposition into a hierarchy of boxes ends up with the lowest hierarchy level being $h = 3$, such a system would have only $s_3^{1D} = 2^3$ upward translations. Doubling the size of this system while maintaining the side-length of the lowest hierarchy box at the same value as in the smaller system produces a system in which the lowest hierarchy level has the index $h = 4$. In this larger system there would be a total of $s_4^{1D} = 2^3 + 2^4$ upward translations. Expanding the size of the system by another factor of two, while maintaining the side-length of the lowest hierarchy box constant across the series of studied systems, produces a lowest hierarchy level with the index $h = 5$, and such a system would have $s_5^{1D} = 2^3 + 2^4 + 2^5$ upward translations. Doubling the system size one more time leads to $h = 6$ and $s_6^{1D} = 2^3 + 2^4 + 2^5 + 2^6$ translations, and so on.

The number of upward translations can now be generalized. Arranging the created systems in a sequence from the smallest to the largest, in the order of their lowest hierarchy level indices: $h = 3$, 4, 5, 6, leads to the sum s_k^{1D} of upward translations forming a series:

$$\begin{aligned} s_3^{1D} &= 2^3, \\ s_4^{1D} &= 2^4 + 2^3 = 2(2^3) + 2^3 = 2s_3^{1D} + 2^3, \\ s_5^{1D} &= 2^5 + 2^4 + 2^3 = 2(2^4 + 2^3) + 2^3 = 2s_4^{1D} + 2^3, \\ s_6^{1D} &= 2^6 + 2^5 + 2^4 + 2^3 = 2(2^5 + 2^4 + 2^3) + 2^3 = 2s_5^{1D} + 2^3. \end{aligned} \tag{12.43}$$

Examination of Equation 12.43 leads to a simple, general, recursion formula for any arbitrary index h:

$$s_h^{1D} = 2s_{h-1}^{1D} + 2^3. \tag{12.44}$$

Equation 12.44 establishes the relationship between the sums of upward translations of two 1D systems that differ in size by factor of two, and have the same box side length in the lowest hierarchy level.

For sufficiently large h this relation becomes:

$$s_h^{1D} \approx 2s_{h-1}^{1D}, \tag{12.45}$$

which shows that the upward translation step of the multipole expansions is linear scaling, that is, doubling the size of a 1D system leads to a two-fold increase in the number of upward translations.

A similar analysis can be extended to 3D systems. In this case, the relative size of the systems would increase by a factor of 8 instead of 2 because each side of a 3D system has to be doubled in size when generating a larger system from the smaller one. With that, the number of upward translations s_h^{3D} of multipole expansion in 3D forms the series:

$$\begin{aligned} s_3^{3D} &= 8^3, \\ s_4^{3D} &= 8^4 + 8^3 = 8(8^3) + 8^3 = 8s_3^{3D} + 8^3, \\ s_5^{3D} &= 8^5 + 8^4 + 8^3 = 8(8^4 + 8^3) + 8^3 = 8s_4^{3D} + 8^3, \\ s_6^{3D} &= 8^6 + 8^5 + 8^4 + 8^3 = 8(8^5 + 8^4 + 8^3) + 8^3 = 8s_5^{3D} + 8^3, \end{aligned} \tag{12.46}$$

which generalizes into a recurrence relation:

$$s_h^{3D} = 8s_{h-1}^{3D} + 8^3. \tag{12.47}$$

For sufficiently large values of index h, Equation 12.47 turns to:

$$s_h^{3D} \approx 8s_{h-1}^{3D}. \tag{12.48}$$

This shows that increasing the total number of cubes by factor of 8 leads to the equivalent increase in the number of upward translations, and proves that the step of upward multipole translations is linear scaling in 3D.

As the result of the upward translation of the multipole expansions, each box in the system that participates in multipole interactions now carries a multipole expansion $\bar{M}$ of the charges located inside the box. Evaluation of electrostatic interaction between a pair of boxes by means of Equation 12.2 requires an irregular expansion, $\tilde{L}$, to be defined. The expansion $\tilde{L}$ brings the electrostatic contribution from the far field region into the interior space of the recipient box. This expansion is not available yet. Therefore, the next step in FMM algorithm is the *computation of the irregular multipole expansion*, $\tilde{L}$.

The computation of expansion $\tilde{L}$ for a recipient box requires accumulating the electrostatic contribution that comes from the far field region. To aid in that task, each box that can participate in a multipole interaction stores a list of its far field boxes in memory. For each recipient box, the computation iterates through the list of *FF* boxes, and employs a M2L transformation to transform a regular multipole expansion $\bar{M}$ of a *FF* box, defined about the center of that box, into an irregular multipole expansion $\tilde{L}$ defined about the center of the recipient box. The expansion terms $\tilde{L}_{lm}$ obtained from different *FF* boxes sum up for the same values of the pair of indices l and m; and the result accumulates in the recipient box.

At this point it would be helpful to know how the computational effort of computing expansion $\tilde{L}$ scales with the size of the system. The relative computational cost of this step is equal to the number of times a M2L transformation is invoked. That is, to the total number of boxes in the *FF* lists summed over all the recipient boxes across the entire system. Therefore, scaling is given by how the total number of boxes in the *FF* lists changes in response to a change in the size of the system.

The total numbers of *FF* boxes are listed in Table 12.2. Calculating the size dependency uses the technique of doubling the length of each side of the top (the enclosing) box to create a larger system from the reference system. Starting with a reference system of arbitrary size, the replication operation in 3D creates a comparison system, which is larger than the initial one in terms of number of particles by a factor of eight.

Keeping the side length of the lowest level hierarchy box in the reference and comparison systems the same makes it possible to attribute a record with hierarchy level h in Table 12.2 to the reference system, and the record with hierarchy level $h+1$ to the comparison system. For a fixed value of parameter ws, the scaling is given by the ratio of *nFFtotal* for the hierarchy levels $h+1$ and h, which, for a sufficiently large h, converges on the number 8. From this, the conclusion can be made that the computational effort of the expansion $\tilde{L}$ scales linearly with the size of the system.

Now, when each box that participates in a multipole expansion is supplied with irregular multipole expansion, $\tilde{L}$, which comes from its *FF* environment at a given hierarchy level, that information needs to be communicated down to the boxes at the lowest hierarchy level. This operation is performed in the step of *downward translation of irregular multipole expansion*, $\tilde{L}$, from a parent to child boxes.

The purpose of the downward translation is to give the child box access to the far field contribution, which is calculated at the parent level, and is beyond the child's immediate reach. All that is needed for that is to translate the origin of the parent expansion $\tilde{L}$ from the coordinate frame of the parent box to the origin of the coordinate system of the child box. This is accomplished with help of an L2L translation. The parent irregular multipole expansion terms $\tilde{L}_{lm}$ that arrive in the child box are then added to the corresponding expansion terms created in the child box from its immediate *FF* neighborhood in the preceding M2L step.

The downward translation starts from the highest hierarchy level where any boxes contain an expansion $\tilde{L}$. This situation is illustrated in Figure 12.10 for the example of a 1D system for $ws = 1$.

TABLE 12.2

Number of Near Field and Far Field Boxes (*nNFlevel* and *nFFlevel*) for a Single Hierarchy Level, *hLevel*, Together with Their Cumulative Number (*nNFtotal* and *nFFtotal*) Summed Up over Preceding Hierarchy Levels, Depending on the Choice of Parameter *ws* for a System of Point Charges Placed in a Cubic Box

ws = 1: *hLevel*	*nNFlevel*	*nNFtotal*	*nFFlevel*	*nFFtotal*
1	56	56	0	0
2	936	992	3096	3096
3	10,136	11,128	53,352	56,448
4	93,240	104,368	584,136	640,584
5	797,816	902,184	5,398,920	6,039,504
6	6,596,856	7,499,040	46,298,376	52,337,880
7	53,645,816	61,144,856	383,233,032	435,570,912
8	432,677,880	493,822,736	3,118,094,856	3,553,665,768
9	3475,523,576	3,969,346,312	25,155,384,840	287,090,506,08
ws* = 2: *hLevel	***nNFlevel***	***nNFtotal***	***nFFlevel***	***nFFtotal***
1	56	56	0	0
2	2680	2736	1352	1352
3	38,792	41,528	136,312	137,664
4	401,128	442,656	2,110,232	2,247,896
5	3,619,496	4,062,152	22,282,072	24,529,968
6	30,697,000	34,759,152	202,785,752	227,315,720
7	252,742,952	287,502,104	1,726,545,112	1,953,860,832
8	2,051,021,608	2,338,523,712	14,241,967,832	16,195,828,664
9	16,525,309,736	18,863,833,448	115,679,597,272	131,875,425,936
ws* = 3: *hLevel	***nNFlevel***	***nNFtotal***	***nFFlevel***	***nFFtotal***
1	56	56	0	0
2	4032	4088	0	0
3	84,672	88,760	176,960	176,960
4	995,904	1,084,664	4,451,776	4,628,736
5	9,495,360	10,580,024	54,471,872	59,100,608
6	82,619,712	93,199,736	526,918,336	586,018,944
7	688,709,952	781,909,688	4,613,631,680	5,199,650,624
8	5,622,974,784	6,404,884,472	38,571,902,656	43,771,553,280
9	45,441,587,520	51,846,471,992	3,153,68,322,752	359,139,876,032

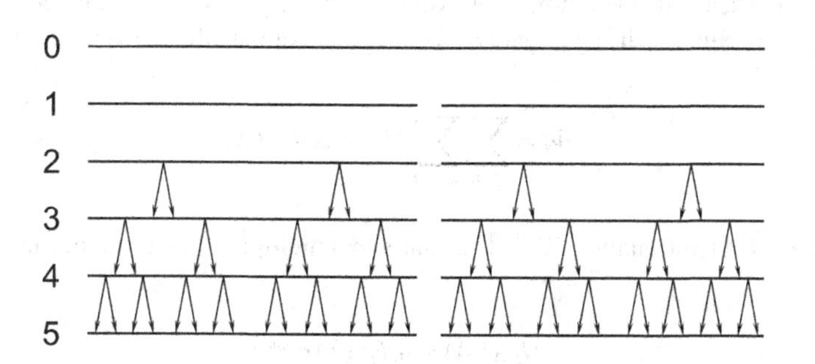

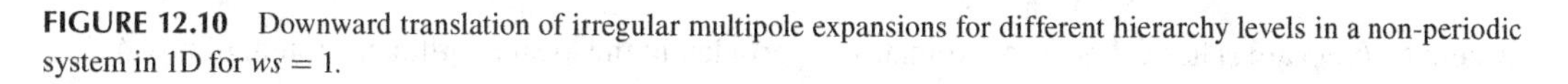

FIGURE 12.10 Downward translation of irregular multipole expansions for different hierarchy levels in a non-periodic system in 1D for $ws = 1$.

First, the child list for each box in this level (level $h = 2$) is retrieved; then the parent irregular multipole expansion is pushed down into each child box. Upon completing the transfer from the parent boxes located at the hierarchy level h to their child boxes, residing at the hierarchy level $h+1$, the child boxes become ready to transfer the accumulated irregular multipole expansion to their own children. With that, the action moves to the boxes located at the next lower hierarchy level, and the operation repeats until the lowest hierarchy level is reached.

There is an apparent similarity between the downward translation of the irregular multipole expansion, $\tilde{L}$, shown in Figure 12.10 and the upward translation of the regular multipole expansion, $\bar{M}$, shown in Figure 12.9. The topological equivalence of the upward and downward translations implies that both steps have an identical scaling profile. Since the previous analysis determined that the step of upward multipole translation is linear scaling, it follows that the downward translation step is also linear scaling.

At the end of the downward translation of the irregular multipole expansion, each box at the lowest hierarchy level acquires the expansion $\tilde{L}$ that brings the global *FF* contribution into that particular box. With that, the regular and irregular multipole expansions in the lowest hierarchy boxes are now ready to be contracted to compute the *FF* portion of electrostatic interactions in the *energy computation step*. This involves determining the *FF* portion of the potential energy of the entire system, and finding the *FF* part of the electrostatic potential at the position of each particle. Computing the energy is straightforward, and consists of obtaining the energy component U_A^{FF} for each lowest-level hierarchy box, A, using Equation 12.49,

$$U_A^{FF} = \sum_{l=0}^{\infty}\sum_{m=-l}^{l} \bar{M}_{lm}(A)\tilde{L}_{lm}(A), \tag{12.49}$$

followed by summing up the energy contribution from all such boxes. The multipole expansions $\bar{M}(A)$ and $\tilde{L}(A)$ are those that were previously computed for each box A, and are defined about the local coordinate system of that box.

The scalability of the energy computation step depends on the way the number of boxes in the lowest hierarchy level varies with the size of the system. Determining the scalability involves the procedure that was developed earlier for this purpose. Doubling each side length of the top-most box, and filling up the newly created boxes with the replicated image of the original box, increases the size of the system in terms of number of particles by factor of 8. Keeping the side length of the lowest hierarchy box at the fixed value and decomposing the reference and comparison systems into a hierarchy of boxes leads to their lowest hierarchy levels having the indices h and $h+1$, respectively. In these circumstances, the binary tree algorithm predicts that the ratio of the number of boxes in the lowest hierarchy level between the reference and comparison systems is $2^{3(h+1)}/2^{3h} = 8$. Since the size of the system and the number of boxes involved in the computation both increase with exactly the same factor, it follows that the computational effort for evaluating the *FF* contribution to the potential energy of the system scales linearly.

The next step in the *FF* contribution is to account for the electrostatic potential Φ_i induced by the remote charges on every particle i inside a box A. Making use of the previously computed irregular multipole expansion $\tilde{L}(A)$ to account for the aggregate *FF* contribution of the other particles in the system, leads to:

$$\Phi_i = \sum_{l=0}^{\infty}\sum_{m=-l}^{l} \bar{M}_{lm}(i,A)\tilde{L}_{lm}(A). \tag{12.50}$$

The expansion $\bar{M}_{lm}(i,A)$ in Equation 12.50 is a chargeless multipole expansion of the particle i about the center of box A:

$$\bar{M}_{lm}(i,A) = a_i^l \bar{P}_l^m(\theta_i)e^{-im\phi_i}, \tag{12.51}$$

where (a_i,θ_i,ϕ_i) are spherical polar coordinates of particle i in the local coordinate system of box A.

Evaluating the electrostatic potential involves looping over all boxes in the lowest hierarchy level in a manner similar to that used in the energy calculation. The higher prefactor in the computation of the electrostatic potential over that for energy comes from the additional need to loop over particles that are inside each box. Since each particle is accessed only once during the electrostatic potential computation this step is also linear scaling with the number of particles in the system.

The next piece in the energy computation step is accounting for *NF* contribution to energy and electrostatic potential, which must be done at the lowest hierarchy level. The computation loops over the lowest-level hierarchy boxes, and computes the energy and electrostatic potential contribution from the particles enclosed in *NF* boxes. To assist with this operation, each box carries a list of its *NF* boxes so that those boxes can readily be accessed.

Obtaining a scaling profile of this computational step requires assessing the way in which the number of boxes in the *NF* lists varies with the size of the system. An assessment of this number's behavior follows from the data listed in Table 12.2 by taking the ratio of the number of boxes *nNFlevel* for individual hierarchy levels $h+1$ and h. This ratio corresponds to the change in system size by a factor of 8 in a model that doubles the side lengths of the top-enclosing box, replicates it to fill the newly created boxes, and decomposes the reference and comparison systems into a hierarchy of boxes, all while keeping the side length of the lowest-level hierarchy box constant. A result of the match of the ratio of box counts *nNFlevel* between the comparison and reference systems is that the number of boxes in the *NF* lists, and therefore the amount of computational work due to *NF* contributions, scales linearly.

Finally, the last step in the FMM algorithm is computing the local contribution to energy and electrostatic potential from the particles enclosed into the lowest-level box. This accounts for the electrostatic interaction of each particle with all other particles located in the same box. The computation loops over the boxes in the lowest hierarchy level. Since the number of boxes at this level changes linearly with the change in system size, this step is linear scaling.

For each box in the lowest-level hierarchy, the calculation employs a direct sum method that implements a double loop over the number of enclosed particles. Therefore, increasing the side-length of the box that places more particles into it increases the computational cost quadratically.

In summary, FMM computation involves the following basic steps:

1. Decomposition of the system of point charges into a hierarchy of boxes and the generation of *NF* and *FF* lists.
2. Computation of the regular multipole expansion, $\bar{M}$.
3. Upward translation of the regular multipole expansion, $\bar{M}$, using an M2M operation.
4. Computation of the irregular multipole expansion, $\tilde{L}$, using an M2L operation.
5. Downward translation of the irregular multipole expansion, $\tilde{L}$, from a parent to its child boxes, using an L2L operation.
6. Computation of the *FF* portion of electrostatic interactions.
7. Accounting for *NF* contributions to the energy and electrostatic potential.
8. Computation of the local contribution to the energy and electrostatic potential from the particles contained in each lowest-level box.

These steps constitute the backbone of the FMM method. Now that the basic steps have been described, the entire FMM algorithm may be schematically represented in a single picture that shows the relationship between the multipole operations. The dotted arrow lines in Figure 12.11 represent the upward translation of regular multipole expansions across the hierarchy levels, and the accumulation of the expansions in parent boxes. The horizontal arrow lines represent the computation of irregular multipole expansions at hierarchy level 2. The dashed arrow lines indicate the downward translation of the irregular multipole expansions that makes the cumulative electrostatic effect of the charge sources from the entire right half of the system visible to the left-side boxes.

The purpose of this chain of translations is to make the number of box-box interactions scale linearly. For simplicity, the schematic drawing of the various FMM operations presented in Figure 12.11 is intentionally left incomplete by omitting other arrows that would be present in the full picture.

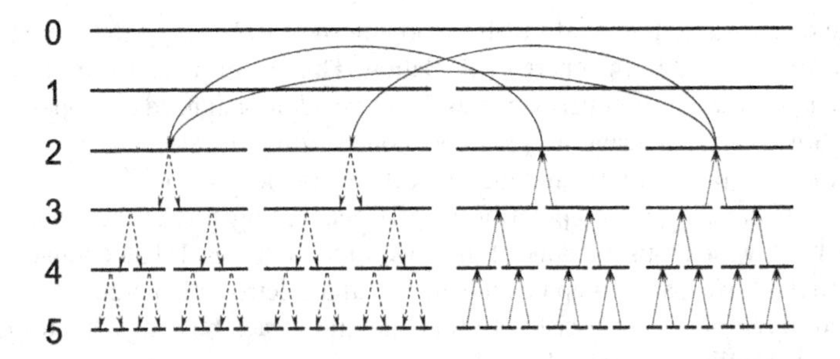

FIGURE 12.11 Simplified schematic representation of the FMM algorithm for a 1D system, for $ws = 1$. Dotted arrow lines represent the upward translations of regular multipole expansions; horizontal arrow lines describe the computation of irregular multipole expansions due to far-field charges; and the dashed lines denote the downward translation of irregular multipole expansions.

12.7 Accuracy Assessment of Multipole Operations

Section 12.2 established factors determining the accuracy of electrostatic interactions calculated using the FMM method. Although parameters like box-side length and well separation are among the contributing factors, the root cause of numerical error is truncation of the multipole expansion. Truncating the multipole expansion of the electrostatic interaction energy between distant boxes A and B, which originally includes infinite sum, on orbital index $l = nL$ changes Equation 12.2 into:

$$U_{AB} = \sum_{l=0}^{nL} \sum_{m=-l}^{l} \bar{M}(A,A)\tilde{L}(B,A) + Q(Z), \tag{12.52}$$

where $Q(Z)$ is the remainder of the series representing a numerical error due to truncation. The secondary accuracy factors, such as box side length and well-separation parameter, contribute to the error through the choice of radial factors a and b in multipole expansion terms $\bar{M}_{lm}$ and $\tilde{L}_{lm}$, respectively. As the ratio (a/b) moves closer to unity the series becomes harder and harder to converge, resulting in errors due to the use of a finite sum.

Additional secondary factors that influence numerical precision are multipole translation operations. Among those, the upward translation of multipole expansion, the M2M operation, is the simplest to analyze since it works in a finite space of summation elements:

$$\bar{M}_{lm}(B,A) = \sum_{j=0}^{l} \sum_{k=-j}^{j} \bar{M}_{jk}(B,B)\,\bar{R}^{*}_{l-j,m-k}(B,A). \tag{11.9}$$

This operation transforms the multipole expansion $\bar{M}(B,B)$ of the charge set of box B into a multipole expansion $\bar{M}(B,A)$ about the coordinate system of box A. The charge-free helper expansion $\bar{R}^{*}_{l-j,m-k}(B,A)$, which corresponds to a radius vector of the center point of box B in coordinate system of box A, assists in this transformation. No approximations are involved in Equation 11.9, so the M2M operation is analytically accurate.

Unlike the M2M operation, the remaining two M2L and L2L translations require special attention because of the presence of an infinite sum. The equation for these transformations exists in two equivalent forms:

$$\tilde{L}_{lm}(B,A) = \sum_{j=0}^{\infty} \sum_{k=-j}^{j} (-1)^{j} \bar{M}_{jk}(B,B)\,\tilde{S}_{l+j,m+k}(B,A), \tag{11.17}$$

and

$$\tilde{L}_{lm}(B,A) = \sum_{j=l}^{\infty}\sum_{k=-j}^{j}(-1)^{j-l}\bar{R}^{*}_{j-l,k-m}(B,A)\,\tilde{L}_{jk}(B,B). \tag{11.13}$$

Although these equations are equivalent, it is convenient to use Equation 11.17 in the error analysis of the M2L operation because of the simplicity of working with the orbital index j in the source expansion of $\bar{M}_{jk}(B,B)$, and Equation 11.13 for the error analysis of the L2L operation for the source expansion of $\tilde{L}_{jk}(B,B)$ for a similar reason.

In the M2L translation, the effect of a multipole translation is to convert the regular multipole expansion $\bar{M}_{jk}(B,B)$ of the remote charge distribution defined about box B into an irregular multipole expansion $\tilde{L}_{lm}(B,A)$ defined about box A. This uses the charge-free irregular multipole expansion $\tilde{S}_{l+j,m+k}(B,A) \equiv \tilde{L}_{l+j,m+k}(B,A)$, which corresponds to a radius vector that starts at the origin of the coordinate system of box A, that is, at the center of box A, and terminates at the center of box B.

Errors in the M2L computation are caused by two effects. The first of those relates to the truncation of the infinite summation in the orbital index j in Equation 11.17 at the upper value, nL. Therefore, box B carries the expansion $\bar{M}(B,B)$ only up to $j = nL$, and that makes the sum over index j incomplete.

Secondly, Equation 11.17 requires terms $\tilde{S}_{l+j,m\pm k}(B,A)$ to be available for each $\bar{M}_{jk}(B,B)$. This makes $2nL$ the upper value of the orbital index $l+j$ in $\tilde{S}_{l+j,m+k}(B,A)$. However, if, like $\bar{M}(B,B)$, the expansion $\tilde{S}(B,A)$ is also truncated because of the condition $l+j \leq nL$, this would result in an additional source of error in the M2L operation. In this case, some of the source expansion terms $\bar{M}_{jk}(B,B)$ would be left out of the summation in the M2L equation. The missing terms would thus introduce errors into the translation operation.

To minimize the error in the M2L translation, it is possible to let the index $l+j$ run up to $2nL$ in value. This is beneficial because $\tilde{S}(B,A)$ is a helper expansion that is easily calculated directly from the radius vector connecting box centers, and is fully accurate, unlike the multiply translated irregular multipole expansions of charge distribution. In addition, computing $\tilde{S}(B,A)$ up to orbital index $2nL$, and using its full range in M2L operations is far less expensive computationally than raising the value of nL across all expansions.

The last multipole translation to analyze is the L2L operation. This operation, described by Equation 11.13, translates the irregular multipole expansion $\tilde{L}_{jk}(B,B)$, defined about the center of box B, into an irregular multipole expansion $\tilde{L}_{lm}(B,A)$ defined about the center of box A. A charge-free helper expansion $\bar{R}^{*}_{j-l,k-m}(B,A) \equiv \bar{M}_{j-l,k-m}(B,A)$, which corresponds to a radius vector originating from the center of box A and pointing toward the center of box B, assists in this transformation.

The numerical error in the L2L operation is a result of the truncation of the infinite series in Equation 11.13 caused by box B carrying a truncated expansion $\tilde{L}_{jk}(B,B)$, where $j \leq nL$. Unlike the M2L translation, the L2L operation has only one source of error: the helper expansion in the L2L operation does not contribute to the error. In $\bar{R}^{*}_{j-l,k-m}(B,A)$, index $j-l$ is already less than or equal to nL, so that no special effort is required to keep the L2L transformation at its best accuracy level.

Truncation of the infinite sum in the M2L and L2L operations makes the resulting expansion $\tilde{L}_{lm}(B,A)$ become a little less accurate than the source expansion. This error adds to the error caused by truncation of the multipole expansions in energy and the electrostatic potential computation. Unfortunately, a fully analytic estimation of the error of M2L and L2L operations is difficult. Considering that the error estimates derived for the multipole series truncation in Section 12.2 represent a worst-case scenario, it roughly covers the error generated by the M2L and L2L operations. Therefore, for all practical purposes, the estimate of the truncation error derived in Section 12.2 can be used as the overall error predictor in the FMM calculation.

13

Multipole Translations along the z-Axis

Typical uses of FMM involve systems containing millions or billions of particles. For systems of this size, time spent in evaluating particle interactions becomes considerable. As the number of particles in the systems of interest grows, so does the cost of computation of electrostatic interactions. Despite FMM making such cost growing linearly with the system size, the time to solution still remains substantial. Attempts to speed up FMM drive the search for new techniques, which could accomplish the required computations in shorter time.

FMM spends most of its time in performing multipole translation operations. Because of that, the choice of which particular form of the M2M, M2L, and L2L operations to use is important. Chapter 11 introduced a scaled form of M2M, L2L, and M2L operations:

$$\bar{M}_{l,m}(B,A) = \sum_{j=0}^{l} \sum_{k=-j}^{j} \bar{M}_{j,k}(B,B)\, \bar{R}^{*}_{l-j,m-k}(B,A), \tag{11.9}$$

$$\tilde{L}_{l,m}(B,A) = \sum_{j=l}^{\infty} \sum_{k=-j}^{j} (-1)^{j-l}\, \bar{R}^{*}_{j-l,k-m}(B,A)\, \tilde{L}_{j,k}(B,B), \tag{11.13}$$

$$\tilde{L}_{l,m}(B,A) = \sum_{j=0}^{\infty} \sum_{k=-j}^{j} (-1)^{j}\, \bar{M}_{j,k}(B,B)\, \tilde{S}_{l+j,m+k}(B,A). \tag{11.17}$$

The dependence of the multipole translation equations on four indices, l, m, j, and k, results in their computational cost scaling as $O(N^4)$ of the size, N, of the truncated multipole expansion. Although the computation still obeys linear scaling in terms of the number of particles in the system, the rapidly rising penalty incurred by using a larger multipole expansion, which may be necessary to keep the potential energy computation accurate, determines the high computational cost of the FMM method in comparison to that of its rival, Particle Mesh Ewald method.

Remarkably, the cost of multipole translation operations significantly drops if the translation happens along the z-axis: that is, along the quantization axis of the angular momentum operator. The following sections develop the necessary theory to illustrate that advantage.

13.1 M2M Translation along the *z*-Axis

The M2M operation defined by Equation 11.9 translates a regular multipole expansion, $\bar{M}(B,B) \equiv \boldsymbol{b}$ that corresponds to converting a set of particles confined to box B from a coordinate system of box B into the coordinate system of box A, producing $\bar{M}(B,A) \equiv \boldsymbol{a} + \boldsymbol{b}$. The left panel in Figure 13.1 illustrates this translation in vector diagram form. In this notation, the first argument, B, in $\bar{M}(B,A)$ defines the box containing the charge sources, and the second argument, A, defines the center of box A that will serve as the origin of the coordinate system for the multipole expansion.

To accomplish this, the M2M translation requires a chargeless helper bar-scaled regular expansion $\bar{R}^{*}(B,A) \equiv \boldsymbol{a}$. This corresponds to a virtual radius vector, $\boldsymbol{a}$, originating from the center of the coordinate system of box A, that points toward the center of the coordinate system of box B.

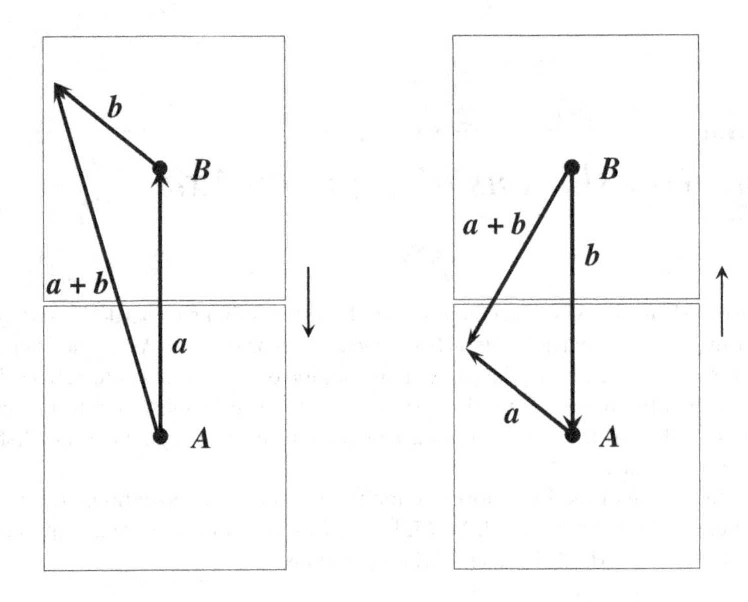

FIGURE 13.1 Vector diagram of an M2M translation of a regular multipole expansion. Left panel: translation of the origin of the multipole expansion from box B to box A. Right panel: translation from box A to box B.

The vector diagram displayed in Figure 13.1 conforms to the following rules. Initially, there are two local coordinate systems A and B that have origins at the center of their respective boxes. Each vector in the diagram starts from the origin of the local coordinate system it is defined in. The left panel in Figure 13.1 corresponds to Equation 11.9. In it, the virtual vector $\boldsymbol{b}$, with its origin at the center of box B, indicates the multipole expansion $\bar{M}(B,B)$ to be translated.

The vector, $\boldsymbol{a}$, in the left panel in Figure 13.1, determines the translation, and, therefore, is defined about the origin of the local coordinate system of the box A, which is the recipient of the translation. Because the purpose of vector $\boldsymbol{a}$ is to describe the center of box B using the local coordinate system of box A, the tip of vector $\boldsymbol{a}$ points toward the center of box B. It is also worthwhile to mention that the direction of vector $\boldsymbol{a}$ corresponds to the shift of point A into point B upon coordinate translation, whereas the origin translation happens in the opposite direction, that is, from origin B to origin A.

The last vector $\boldsymbol{a}+\boldsymbol{b}$ in the left panel in Figure 13.1 represents the result of vector addition. This vector simulates the translated multipole expansion defined about the local coordinate system of box A. Upon performing the vector addition, the direction vectors $\boldsymbol{a}$ and $\boldsymbol{b}$ uniquely determine the direction of the resulting vector, such that whatever point in space is selected for the tip of vector $\boldsymbol{b}$, the tip of vector $\boldsymbol{a}+\boldsymbol{b}$ would follow that point in accordance with the rules of classical vector addition.

Thus far, the analysis of an M2M translation has considered only the case of translation from box B to box A. For completeness, it is necessary to consider a translation in the opposite direction, that is, from box A to box B. This is illustrated in the right panel in Figure 13.1. The corresponding mathematical equation can be derived by interchanging arguments A and B in Equation 11.9, leading to:

$$\bar{M}_{l,m}(A,B)=\sum_{j=0}^{l}\sum_{k=-j}^{j}\bar{M}_{j,k}(A,A)\,\bar{R}^{*}_{l-j,m-k}(A,B). \tag{13.1}$$

A side-by-side comparison of the left and right diagrams in Figure 13.1 shows that the translation vector $\boldsymbol{a}$ from the left panel is equal in absolute value to the translation vector $\boldsymbol{b}$ from the right panel but has its sign reversed. This illustrates that the helper expansions $\bar{R}^{*}(A,B)$ and $\bar{R}^{*}(B,A)$, which correspond to these vectors, are related to each other through a coordinate inversion:

$$\bar{R}^{*}_{l-j,m-k}(A,B) = (-1)^{l-j}\bar{R}^{*}_{l-j,m-k}(B,A). \tag{13.2}$$

An immediate outcome of this relation is that, once computed, the expansion $\bar{R}^{*}(B,A)$ may be used to perform an M2M translation from box B to A and then backwards with minimal modification.

In the general picture, choosing the donor box and the recipient box uniquely determines the math of the M2M computations and automatically establishes the translation direction, thereby eliminating the need to keep track of it. With that, choosing either of the equations Equation 11.9 or Equation 13.1, and the corresponding panel in the vector diagram in Figure 13.1 to help see the right direction of vector corresponding to helper expansion, provides everything needed to perform M2M translations throughout the entire FMM computation.

The visual representation of multipole translation is also useful when taking advantage of the simplification of the M2M equation that occurs after aligning the translation axis with the z-axis of the local coordinate system of the boxes. Using the convention established in Equation 11.9, that box B is the donor of multipole expansion and that box A is the recipient of the translated multipole expansion, and aligning the centers of these two boxes with the z-axis, creates two possible cases when the multipole translation is performed, either in the negative or the positive direction of the z-axis. The vector diagrams in Figure 13.2 illustrate these two dispositions.

Aligning the translation axis with the z-axis of the coordinate system results in a simplification of the helper expansion and thus in a simplification in the multipole translation equation. In this orientation, the angular dependence terms in the expansion vanish, leaving only the radial component. This can be demonstrated as follows: If the radius vector $\boldsymbol{a} = (r,\theta,\phi)$ of a solid harmonics expansion $\bar{R}^{*}_{l,m}(r,\theta,\phi)$ is aligned along the positive direction of the z-axis, then the polar and azimuthal angles θ and ϕ are both zero. Noting that $P_l^{\,m}(0) = \delta_{m,0}$, this alignment of vector $\boldsymbol{a}$ leads to:

$$\bar{R}^{*}_{l,m}(r,\theta,\phi) \equiv \frac{r^l}{(l+m)!}P_l^m(\theta)e^{-im\phi} \rightarrow \bar{R}^{*}_{l,m}(r,0,0) = \frac{r^l}{l!}\delta_{m,0}. \tag{13.3}$$

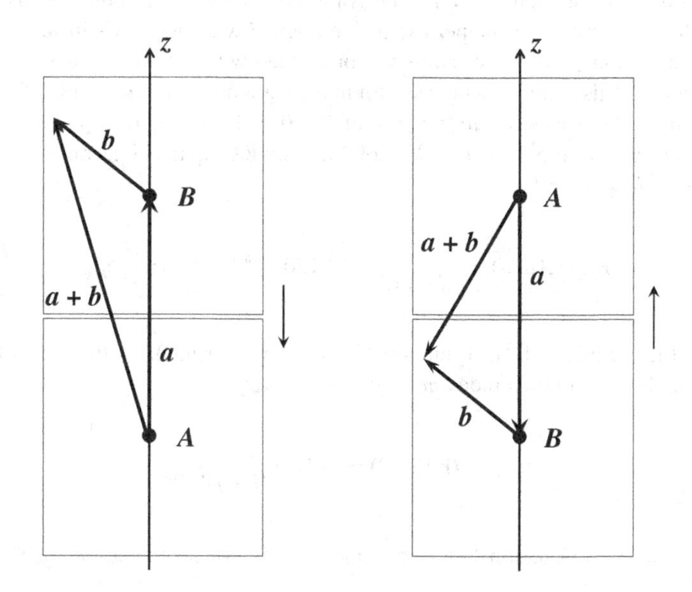

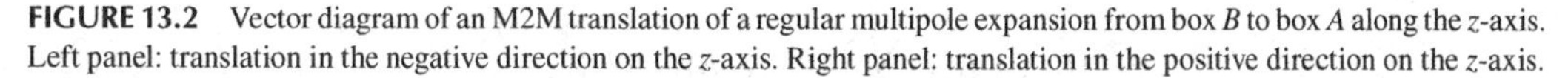
FIGURE 13.2 Vector diagram of an M2M translation of a regular multipole expansion from box B to box A along the z-axis. Left panel: translation in the negative direction on the z-axis. Right panel: translation in the positive direction on the z-axis.

where $\delta_{m,0}$ is Kronecker delta:

$$\delta_{m,0} = \begin{cases} 0 & \text{if } m \neq 0, \\ 1 & \text{if } m = 0. \end{cases} \tag{13.4}$$

The presence of the Kronecker delta in the right-hand side of Equation 13.3 causes all expansion terms that have a non-zero azimuthal index to vanish. Likewise, the alignment of the radius vector $\boldsymbol{a}$ of the actual helper expansion $\bar{R}^*_{l-j,m-k}(r,\theta,\phi)$ along the positive direction of the z-axis leads to:

$$\bar{R}^*_{l-j,m-k}(r,0,0) = \frac{r^{l-j}}{(l-j)!}\delta_{m,k}. \tag{13.5}$$

The result in Equation 13.5 can now be used to rewrite Equation 11.9 to produce a customized M2M equation for translation along the z-axis. This replacement zeroes all summation elements in the sum over index k except those that satisfy the condition $m - k = 0$, which is equivalent to $k = m$, and reduces Equation 11.9 to:

$$\bar{M}_{l,m}(B,A) = \sum_{j=0}^{l} \bar{M}_{j,k}(B,B)\frac{r^{l-j}}{(l-j)!}\delta_{m,k} = \sum_{j=0}^{l} \bar{M}_{j,m}(B,B)\frac{r^{l-j}}{(l-j)!}. \tag{13.6}$$

One problem still remains. Since in spherical polar coordinates the range of values of radial component r is from zero to positive infinity Equation 13.6 can only be used for the case displayed in the left panel in Figure 12.2, where the vector $\boldsymbol{a}$ points in the same direction as the z-axis, and the translation of the origin of the multipole expansion is in the opposite direction to the z-axis. The other possible alignment of the vector of helper expansion with the z-axis is that displayed in the right panel of Figure 13.2, where vector $\boldsymbol{a}$ and the z-axis point in opposite directions, and the origin of multipole expansion moves toward the positive direction on the z-axis. This latter case is outside of the domain of applicability of Equation 13.6, and requires performing an independent derivation of its own M2M equation, as follows.

Having the radius vector $\boldsymbol{a}$ of the helper expansion aligned with the z-axis in such way that their tips point in opposite directions makes the radius vector defined with spherical polar coordinates $(r,180,0)$, where $\theta = 180$ degrees; this arrangement is shown in the right panel in Figure 13.2. For comparison, the left panel in Figure 13.2 corresponds to the case of $\theta = 0$. Substituting the spherical polar coordinates $(r,180,0)$ of the radius vector into the bar-scaled solid harmonics expansion term, and taking into account that $P_l^m(180) = (-1)^l\,\delta_{m,0}$, leads to:

$$\bar{R}^*_{l,m}(r,180,0) = \frac{1}{(l+m)!} r^l P_l^m(180) e^{-im0} = (-1)^l \frac{r^l}{l!}\delta_{m,0}. \tag{13.7}$$

Now, applying the technique from Equation 13.7 to $\bar{R}^*_{l-j,m-k}(r,\theta,\phi)$, so that the orbital index l gets replaced by $l - j$ and the azimuthal index m becomes $m - k$, gives:

$$\bar{R}^*_{l-j,m-k}(r,180,0) = (-1)^{l-j}\frac{r^{l-j}}{(l-j)!}\delta_{m,k}. \tag{13.8}$$

Substituting the result from Equation 13.8 into Equation 11.9 reduces the latter equation to:

$$\bar{M}_{l,m}(A,B) = \sum_{j=0}^{l} \bar{M}_{j,m}(B,B)(-1)^{l-j}\frac{r^{l-j}}{(l-j)!}. \tag{13.9}$$

This is the M2M equation for the multipole translation that corresponds to the right panel in Figure 13.2, where the origin of the multipole expansion moves in the positive direction along the z-axis. Note the presence of the extra factor $(-1)^{l-j}$ when comparing Equations 13.9 to 13.6.

The existence of two slightly different forms of the M2M equation for translation along the z-axis makes it necessary to pay extra attention to the specific alignment of the axis connecting the centers of boxes A and B with the direction of the z-axis. This complication can be avoided by a simple workaround that merges Equations 13.6 and 13.9 into a single equation. If the centers of boxes A and B are aligned with the z-axis of the coordinate system, and the Cartesian coordinates of those centers are $A = (0,0,z_A)$ and $B = (0,0,z_B)$, respectively, then replacing the radial component r of the radius vector $\boldsymbol{a}$ with the relation $r = z_B - z_A$ in Equation 13.6 would produce a unified M2M equation for translation along the z-axis:

$$\bar{M}_{l,m}(B,A) = \sum_{j=m}^{l} \frac{(z_B - z_A)^{l-j}}{(l-j)!} \bar{M}_{j,m}(B,B), \tag{13.10}$$

which is now applicable to both possible orientations of the boxes A and B as displayed in the left and right panels in Figure 13.2. Note the change of the starting value of index j in the sum from the previous value of $j = 0$ to $j = m$; this is dictated by the presence of expansion term $\bar{M}_{j,m}$ in the sum, which requires $j \geq m$.

Now that the derivation has been completed, the next step is to develop a computer code to implement Equation 13.10. As usual, starting the coding from the naïve straightforward implementation is the best way to ensure the correctness of the code before embarking on performance optimization. This starts with the following sample code:

```
subroutine M2Mz(nL, nSzTri, tri, Mb, Tz, Ma)
!
! M2M translation from box B to box A along the z-axis. Naïve code.
! Ma(L,m) = sum(j=m,L) { Mb(j,m) * (Tz)^(L-j) / (L-j)! }
!
! Box centers have Cartesian coordinates A(0,0,zA) and B(0,0,zB)
! Tz = zB - zA is translation distance along the z-axis
! Negative sign of Tz implies translation in direction of z-axis
! Positive sign of Tz performs translation in opposite direction
!
implicit none
!
integer,      intent(in)          :: nL            ! expansion truncation
integer,      intent(in)          :: nSzTri        ! size of expansion array
integer,      intent(in)          :: tri(0:nL)     ! array of triangular pointers
complex(16), intent(in)           :: Mb(0:nSzTri) ! expansion to be shifted
double precision, intent(in)      :: Tz            ! translation distance
complex(16), intent(inout)        :: Ma(0:nSzTri) ! translated expansion
!
! Local variables
integer           :: L, m, j, LmJ, aJ, aL
double precision :: factor

do L = 0, nL
  aL = tri(L)
  do m = 0, L
    Ma(aL+m) = dcmplx(0.0d0, 0.0d0)
    do j = m, L
      factor = 1.0d0
      do LmJ = 1, L-j
        factor = factor * Tz / dble(LmJ)
```

```
        enddo
        aJ = tri(j)
        Ma(aL+m) = Ma(aL+m) + Mb(aJ+m) * factor
      enddo
    enddo
  enddo

end subroutine M2Mz
```

In this program code, parameters *nL*, *nSzTri*, and array *tri*() are defined elsewhere, as in previous implementations. The main body of the program code consists of a series of enclosed *do*-loops where the two most external ones are over indices l and m:

```
do L = 0, nL
  aL = tri(L)
  do m = 0, L
    Ma(aL+m) = dcmplx(0.0d0, 0.0d0)
    ...
  enddo
enddo
```

In this code snippet, index l varies from 0 to *nL*, where *nL* is the last element in the truncated expansion. Values of azimuthal index m trail the value of orbital index l, such that $0 \leq m \leq l$. Variable $aL = tri(L)$ is the address of expansion element $\bar{M}_{l,0}(B,A)$ in the triangular array *Ma*. The enclosed *do*-loop over index m starts with initializing the array element $Ma(aL + m)$, which corresponds to $\bar{M}_{l,m}(B,A)$.

The next *do*-loop encountered in the program code is that over index j. The starting and ending values of this index directly follow from Equation 13.10. Inside the *do*-loop over index j, there is an additional *do*-loop over index *LmJ* that runs from 1 to $l - j$, and which carries out the computation of the multiplication factors $(z_B - z_A)^{l-j}/(l-j)!$. The following code snippet illustrates this arrangement:

```
do j = m, L
  factor = 1.0d0
  do LmJ = 1, L-j
    factor = factor * Tz / dble(LmJ)
  enddo
  aJ = tri(j)
  Ma(aL+m) = Ma(aL+m) + Mb(aJ+m) * factor
enddo
```

The final two lines of the code inside the j *do*-loop accumulate the contribution to $\bar{M}_{l,m}(B,A)$ from each cycle of the loop. This completes the naïve implementation of Equation 12.10.

Checking the correctness of subroutine *M2Mz*() uses the same code that was previously developed for validation of subroutine *M2M*() in Section 11.2, the only change being that the origins *oxyz*1 and *oxyz*2 for the multipole computation are now placed on a line parallel to the z-axis. An example of that is shown in the following code snippet:

```
oxyz2(1) = oxyz1(1)
oxyz2(2) = oxyz1(2)
oxyz2(3) = oxyz1(3) + 1.0d0
```

Now that the correctness of the naïve implementation of subroutine *M2Mz*() has been verified, a few rearrangements can be made that will improve its performance. Due to the simplicity of Equation 13.10, its naïve computer implementation does not require the use of conditional statements. The suboptimal part in the naïve code is the position of the multiplication factor $(z_B - z_A)^{l-j}/(l-j)!$ inside the *do*-loop over index m. This quantity has to be calculated repeatedly, but the repetition can be avoided if variable *factor* is moved outside the m loop:

```
subroutine M2Mz(nL, nSzTri, tri, Mb, Tz, Ma)
!
! M2M translation from box B to box A along the z-axis. Optimized code.
! Ma(L,m) = sum(j=m,L) { Mb(j,m) * (Tz)^(L-j) / (L-j)! }
!
! Box centers have Cartesian coordinates A(0,0,zA) and B(0,0,zB)
! Tz = zB - zA is translation distance along the z-axis
! Negative sign of Tz implies translation in direction of z-axis
! Positive sign of Tz performs translation in opposite direction
!
implicit none
!
integer,          intent(in)      :: nL           ! expansion truncation
integer,          intent(in)      :: nSzTri       ! size of expansion array
integer,          intent(in)      :: tri(0:nL)    ! array of triangular pointers
complex(16),      intent(in)      :: Mb(0:nSzTri) ! expansion to be shifted
double precision, intent(in)      :: Tz           ! translation distance
complex(16),      intent(inout)   :: Ma(0:nSzTri) ! translated expansion
!
! Local variables
integer              :: L, m, j, LmJ, aJ, aL
double precision :: factor
! Each Ma(L,m) includes Mb(L,m) when j = L
Ma = Mb

! (L,m)=(0,0), j=L=0 has already been accounted above
! Loop over LmJ = L - j virtually represents the loop over j

factor = 1.0d0
do  LmJ = 1, nL                            ! loop over (L-j)
  factor = factor * Tz / dble(LmJ)        ! appears in Ma(L,m) for each L >= LmJ
  do L = LmJ, nL                          ! enforces j < L
    j = L - LmJ                           ! obtains j from LmJ when L is known
    aL = tri(L)
    aJ = tri(j)
    do m = 0, j
      Ma(aL+m) = Ma(aL+m) + Mb(aJ+m) * factor
    enddo
  enddo
enddo

end subroutine M2Mz
```

When considering ways to optimize the naïve program code, it helps to notice that, in Equation 13.10, the same multiplication factor, $(z_B - z_A)^{l-j}/(l-j)!$, is used in multiple expansion elements $\bar{M}_{l,m}(B,A)$ because the indices l and j are simultaneously incremented, resulting in the difference $l-j$ remaining fixed. In the specific case of $l = j$, the factor that enters into each element $\bar{M}_{l,m}(B,A)$ of the translated expansion is unity. This fact allows the summation in *Ma* to start at 1 rather than 0, provided the entire array *Mb* is initially copied into *Ma*:

```
! Each Ma(L,m) includes Mb(L,m) when j = L
Ma = Mb
```

A useful byproduct of this assignment is a restatement of the equality $\bar{M}_{0,0}(B,A) = \bar{M}_{0,0}(B,B)$, which is the final result for $l = 0$ and $m = 0$. A brief inspection of Equation 13.10 confirms that the regular multipole expansion element $\bar{M}_{0,0}$ does indeed not change upon the translation of coordinate origin along the z-axis.

The presence of factorial $(l-j)!$ in the expression for the multiplication factor requires the latter be iteratively computed. Such a computation may be accomplished in a *do*-loop, with the difference $l-j$ serving as the loop index. The allowed values of index $l-j$ can be inferred from Equation 13.10, which imposes the condition $l \geq j$, more conveniently expressed as $l-j \geq 0$. With the array *Ma* initialized to *Mb*, the rest of the multiplication factors involve the condition that $l-j>0$. With that, and representing the difference $l-j$ by variable *LmJ*, the starting value for index *LmJ* in the *do*-loop becomes $LmJ=1$. The upper value of this *do*-loop relies on the fact that the orbital indices l and j are independently defined in the range from zero to *nL* in the truncated expansion, so that the maximum value of $l-j$ is *nL* when $l=nL$ and $j=0$ under the condition $l>j$. With that, the upper value of the *do*-loop index becomes $LmJ=nL$, as shown in the following code snippet:

```
factor = 1.0d0
do  LmJ = 1, nL
   factor = factor * Tz / dble(LmJ)
   ...
enddo
```

Optimizing each cycle of the *do*-loop that runs over index $LmJ \equiv l-j$ involves determining all values of indices l and j that satisfy the running value of $l-j$ and using the latest value of the multiplication factor with all suitable multipole moments $\bar{M}_{l,m}$ and $\bar{M}_{j,m}$ in Equation 13.10 before the value of the factor becomes modified in the next cycle of the *do*-loop. Under the condition that the value of $l-j$ is fixed, the only available freedom to vary indices l and j is to increment them simultaneously. This task may be solved by making one of the variables l or j an index of an enclosed *do*-loop, and by determining the other variable from the two known values.

As presented, the *do*-loop is set to run over index l, with the starting and ending values being $l=LmJ$ and $l=nL$, respectively. To understand the choice of the starting value of index l, note that, since $l \geq j$ in Equation 13.10, it follows that $l \geq l-j$ and therefore the ending value of the index l reflects the size of the truncated multipole expansion.

Once the values of *LmJ* and l are set in the corresponding *do*-loops, the value of index j immediately follows from the identity $j=l-LmJ$, that is, $j=l-(l-j)$. This is illustrated in the following code snippet:

```
do LmJ = 1, nL       ! L - j
   ...
   do L = LmJ, nL
      j = L - LmJ    ! j = L - (L - j)
      ...
   enddo
enddo
```

Finally, in the third and innermost *do*-loop over index m, the math described in Equation 13.10 is performed. In the program code, index m goes from 0 to j. While the starting value of $m=0$ comes directly from the definition of indices l and m, the less obvious ending value of index m is a result of the presence of the expansion term $\bar{M}_{j,m}(B,B)$ in Equation 13.10, which gives rise to the condition $m \leq j \leq l$. This completes the analysis of the optimized version of subroutine *M2Mz*().

To summarize this section, the presence of only three indices l, m, and j in Equation 13.10 makes the scaling $O(N^3)$. This compares favorably with the $O(N^4)$ scaling of Equation 11.9, where four indices, l, m, j, and k, are used. This improved performance suggests that if it were possible to perform all multipole translations along the z-axis only, then the FMM computations would run much faster.

13.2 L2L Translation along the z-Axis

The purpose of the L2L operation is to shift the origin of the coordinate system for an irregular multipole expansion to a new location. There is a great degree of overlap in the rationale of multipole translation between L2L and M2M operations; much that has been said in Section 13.1 also applies here, and, therefore, the discussion proceeds by employing the concepts explained or developed earlier in this chapter.

Translation of an irregular solid harmonics expansion, $\tilde{L}(B,B)$, defined about the origin in the center of box B, into an irregular solid harmonics expansion, $\tilde{L}(B,A)$, defined about the center of the box A, is performed with the assistance of a bar-scaled regular expansion $\bar{R}^*(B,A)$, according to Equation 11.13,

$$\tilde{L}_{l,m}(B,A)=\sum_{j=l}^{\infty}\sum_{k=-j}^{j}(-1)^{j-l}\bar{R}^*_{j-l,k-m}(B,A)\tilde{L}_{j,k}(B,B), \tag{11.13}$$

In its conventional form, the L2L operation defined by Equation 11.13 has a $O(N^4)$ scaling profile, where N is the size of the truncated multipole expansion. The vector diagram in Figure 13.3 illustrates the arrangement of boxes and virtual vectors representing multipole expansions. There are two possible ways to align the vector $\boldsymbol{a}$ of a helper expansion $\bar{R}^*(B,A)$ with the z-axis of coordinate system: the co-directional alignment of the vector and the z-axis, shown in the left panel in Figure 13.3, and the contra-directional alignment, shown in the right panel.

The left panel alignment will be analyzed first. Similar to the derivation of Equation 13.5, the alignment of the radius vector $\boldsymbol{a} = (r,\theta,\phi)$ of the helper expansion $\bar{R}^*_{j-l,k-m}(r,\theta,\phi)$ with the positive direction of the z-axis zeroes the polar and azimuthal angles $\theta = 0$ and $\phi = 0$, respectively, and leads to:

$$\bar{R}^*_{j-l,k-m}(r,0,0)=\frac{r^{j-l}}{(j-l)!}\delta_{m,k}. \tag{13.11}$$

Substituting the obtained Equation 13.11 into Equation 11.13 produces:

$$\tilde{L}_{l,m}(B,A)=\sum_{j=l}^{\infty}\sum_{k=-j}^{j}(-1)^{j-l}\frac{r^{j-l}}{(j-l)!}\delta_{m,k}\tilde{L}_{j,k}(B,B)=\sum_{j=l}^{\infty}(-1)^{j-l}\frac{r^{j-l}}{(j-l)!}\tilde{L}_{j,m}(B,B), \tag{13.12}$$

which corresponds to the left panel in Figure 13.3.

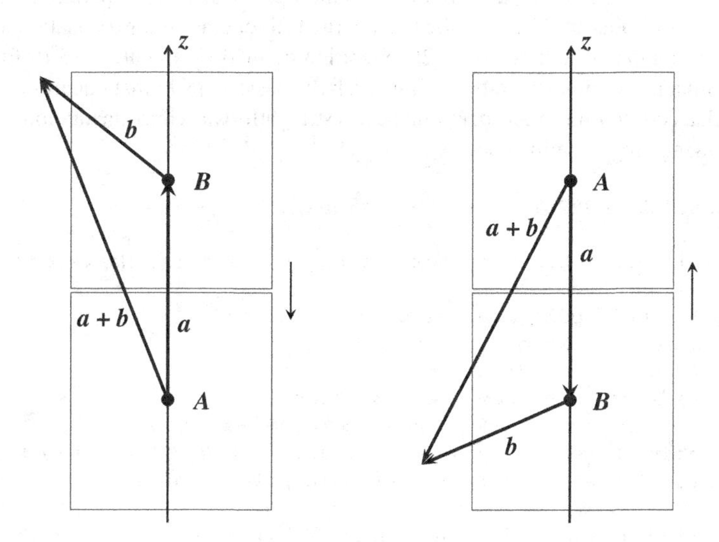

FIGURE 13.3 Vector diagram of the L2L translation of an irregular multipole expansion from box B to box A along the z-axis. Left panel: translation in the negative direction to the z-axis. Right panel: translation in the positive direction to the z-axis. The tip of vector $\boldsymbol{b}$ is positioned outside the boundary of box B to emphasize the remote location of the charge sources.

Proceeding to the case illustrated in the right panel of Figure 13.3, the alignment of radius vector $\boldsymbol{a} = (r,\theta,\phi)$ is in the direction opposite to the positive direction of the z-axis, which makes the polar and azimuthal angles $\theta = 180$ and $\phi = 0$, respectively, and leads to:

$$\bar{R}^*_{j-l,k-m}(r,180,0) = (-1)^{j-l}\frac{r^{j-l}}{(j-l)!}\delta_{m,k}. \tag{13.13}$$

Substituting this equation into Equation 11.13 produces:

$$\tilde{L}_{l,m}(B,A) = \sum_{j=l}^{\infty}\sum_{k=-j}^{j}(-1)^{j-l}(-1)^{j-l}\frac{r^{j-l}}{(j-l)!}\delta_{m,k}\tilde{L}_{j,k}(B,B) = \sum_{j=l}^{\infty}\frac{r^{j-l}}{(j-l)!}\tilde{L}_{j,m}(B,B), \tag{13.14}$$

which corresponds to the case when the origin of irregular multipole expansion is shifted in the direction of positive values on the z-axis.

The existence of two slightly different formulations, Equations 13.12 and 13.14, for the L2L translation along the z-axis reflects the dependence of the equation on the sign of the alignment of the translation vector with the z-axis of coordinate system. Given that $A = (0,0,z_A)$ and $B = (0,0,z_B)$ are the Cartesian coordinates of the centers of boxes A and B, respectively, the radial coordinate component r, which assumes only positive values, can, in Equations 13.12 and 13.14, be replaced by the signed difference of the Cartesian components, and, as a result, vector $\boldsymbol{a}$ can be replaced by $r = z_B - z_A$ in Figure 13.3. With that, equations Equations 12.12 and 12.14 converge to give a single expression in Cartesian representation:

$$\tilde{L}_{l,m}(B,A) = \sum_{j=l}^{\infty}\frac{(z_A - z_B)^{j-l}}{(j-l)!}\tilde{L}_{j,m}(B,B) \equiv \sum_{j=l}^{\infty}(-1)^{j-l}\frac{(z_B - z_A)^{j-l}}{(j-l)!}\tilde{L}_{j,m}(B,B), \tag{13.15}$$

which is the final L2L equation for translation of an irregular multipole expansion from box B into box A along the z-axis.

Of the two possible forms in Equation 13.15, the left-hand is clearly the more compact and would appear to have a clear advantage. However, the right-hand form benefits from the same radius-vector helper expansion in the form $z_B - z_A$ being used in all three multipole translation operations. Because of that, the right-hand form of Equation 13.15 will be used in the L2L operation for translation along the z-axis.

Development of computer code for the L2L operation defined by Equation 13.15 follows the usual route. It starts with a naïve implementation to establish the basic steps in that equation, and then applies optimization to that code to improve its performance. A straightforward implementation of Equation 13.15 leads to the following programming code:

```
subroutine L2Lz(nL, nSzTri, tri, Lb, Tz, La)
!
! L2L translation from box B to box A along the z-axis. Naïve code.
!
! Lb - irregular multipole expansion to be shifted
! Tz - translation distance along the z-axis
!      negative sign - in direction of z-axis
!      positive sign - in opposite direction
!      Tz = zB - zA for translation from box B to box A
!      box centers have Cartesian coordinates A(0,0,zA) and B(0,0,zB)
! La - the resulting shifted irregular multipole expansion
!
! La(L,m) = sum(j=L,nL) { Lb(j,m) * (-Tz)^(j-L) / (j-L)! }
!
implicit none
!
```

```
integer,       intent(in)        :: nL              ! expansion truncation
integer,       intent(in)        :: nSzTri          ! size of expansion array
integer,       intent(in)        :: tri(0:nL)       ! array of triangular pointers
complex(16),   intent(in)        :: Lb(0:nSzTri)    ! expansion to be shifted
double precision, intent(in)     :: Tz              ! translation distance
complex(16),   intent(inout)     :: La(0:nSzTri)    ! translated expansion
!
! Local variables
integer          :: L, m, j, LmJ, aJ, aL
double precision :: factor, zT

zT = -Tz
do L = 0, nL
  aL = tri(L)
  do m = 0, L
    La(aL+m) = dcmplx(0.0d0, 0.0d0)
    do j = L, nL
      factor = 1.0d0
      do JmL = 1, j-L
        factor = factor * zT / dble(JmL)
      enddo
      aJ = tri(j)
      La(aL+m) = La(aL+m) + Lb(aJ+m) * factor
    enddo
  enddo
enddo

end subroutine L2Lz
```

Due to the close similarity between Equations 13.10 and 13.15 for the M2M and L2L translations along the z-axis, respectively, their computer implementations should also be similar. However, in addition to replacing the regular multipole expansions by irregular ones, a few other minor differences exist. One of these is the sign inversion in translation distance Tz,

```
zT = -Tz,
```

that accounts for the relation $(-1)^{j-l}r^{j-l} = (-r)^{j-l}$, implicit in Equation 13.15, and using variable zT instead of the original Tz in the program code. Another detail of subroutine *L2Lz*() is the choice of l and nL for the starting and ending values of index j in the respective *do*-loop, which is also determined by Equation 13.15. Finally, the dependence of the multiplication factor in Equation 13.15 on $l - j$ results in the innermost *do*-loop running from 1 to $j - l$.

A simple modification of the validation technique developed earlier for subroutine *L2L*() in Section 11.4, that involved placing the origin points $oxyz1 = (x_1, y_1, z_1)$ and $oxyz2 = (x_2, y_2, z_2)$ on a line parallel to the z-axis, was used in testing the computer implementation of subroutine *L2Lz*(). The following code snippet implements this validation process:

```
implicit none
integer          :: n                 ! number of particles
double precision :: c(3,n)            ! particle coordinates
double precision :: q(n)              ! particle charges
integer          :: nL                ! expansion truncation
integer          :: nSzTri            ! size of expansion array
integer          :: tri(0:nL)         ! triangular pointers
double precision :: oxyz1(3)          ! origin1
double precision :: oxyz2(3)          ! origin2
double precision :: radius            ! radius of enclosing sphere
complex(16)      :: scr(0:nSzTri)     ! scratch array
complex(16)      :: L1(0:nSzTri)      ! irregular expansion to be translated
```

```
complex(16)        :: L2(0:nSzTri)      ! reference irregular expansion
complex(16)        :: La(0:nSzTri)      ! translated irregular expansion
logical            :: error

! Temporarily define oxyz1 as the geometric center of particles
oxyz1 = 0.0d0
do i = 1, n
  oxyz1(1:3) = oxyz1(1:3) + c(1:3,i)
enddo
oxyz1 = oxyz1 / dble(n)

! Place oxyz2 far away of the charge set
oxyz2  = maxval(c, dim = 2) - oxyz1
radius = dsqrt(oxyz2(1)**2 + oxyz2(2)**2 + oxyz2(3)**2)
oxyz2  = (oxyz1 + dsign(radius, oxyz1)) * 3.0d0

! Redefine oxyz1 as a point in relative vicinity of oxyz2
! Make x1=x2 and y1=y2 while having z1 and z2 different
oxyz1(1) = oxyz2(1)
oxyz1(2) = oxyz2(2)
oxyz1(3) = oxyz2(3) + 1.0d0

! Get irregular multipole expansion of a charge set about origin oxyz1
call IrregularExpansion(c, q, n, oxyz1, nL, nSzTri, tri, scr, L1)

! Get irregular multipole expansion of a charge set about origin oxyz2
call IrregularExpansion(c, q, n, oxyz2, nL, nSzTri, tri, scr, L2)

! Perform L2Lz operation by moving the origin from oxyz1 to oxyz2
call L2Lz(nL, nSzTri, tri, L1, oxyz1(3)-oxyz2(3), La)

! Validate the result of transformation, La against the known data, L2
error = .FALSE.
do i = 0, nSzTri
    if(abs(dble( La(i) - L2(i))) > 1.0d-12) error = .TRUE.
    if(abs(aimag(La(i) - L2(i))) > 1.0d-12) error = .TRUE.
enddo
if(error) write(*,*) "Error in L2Lz computation"
```

Having shown that the program code works correctly, the next step is to optimize it. This involves rearranging all the *do*-loops so as to avoid the repetitive calculation of the multiplication factor $(-1)^{j-l}((z_B - z_A)^{j-l}/(j-l)!)$. Since this factor includes factorial $(j - l)!$, it is necessary to conduct its computation iteratively in a separate *do*-loop that uses $j - l$ as an index of the loop. At this point, the goal is to have the running value of the factor used before updating it in the next cycle of the loop. This is illustrated in the following program code:

```
subroutine L2Lz(nL, nSzTri, tri, Lb, Tz, La)
!
! L2L translation from box B to box A along the z-axis. Optimized code.
!
! Lb - irregular multipole expansion to be shifted
! Tz - translation distance along the z-axis
!      negative sign - in direction of z-axis
!      positive sign - in opposite direction
!      Tz = zB - zA for translation from box B to box A
!      box centers have Cartesian coordinates A(0,0,zA) and B(0,0,zB)
! La - the resulting shifted irregular multipole expansion
!
! La(L,m) = sum(j=L,nL) { Lb(j,m) * (-Tz)^(j-L) / (j-L)! }
```

```
!
implicit none
!
integer,      intent(in)       :: nL            ! expansion truncation
integer,      intent(in)       :: nSzTri        ! size of expansion array
integer,      intent(in)       :: tri(0:nL)     ! array of triangular pointers
complex(16),  intent(in)       :: Lb(0:nSzTri)  ! expansion to be shifted
double precision, intent(in)   :: Tz            ! translation distance
complex(16),  intent(inout)    :: La(0:nSzTri)  ! translated expansion
!
! Local variables
integer            :: L, m, j, LmJ, aJ, aL
double precision :: factor, zT

zT = -Tz  ! invert the sign of translation distance to account for (-1)^(j-L)

La = Lb   ! handle the case of j-L=0 or j=L

factor = 1.0d0
do JmL = 1, nL
  factor = factor * zT / dble(JmL)
  do j = JmL, nL    ! to ensure j > L
    L  = j - JmL
    aJ = tri(j)
    aL = tri(L)
    do m = 0, L
      La(aL+m) = La(aL+m) + Lb(aJ+m) * factor
    enddo
  enddo
enddo

end subroutine L2Lz
```

Because of the similarity of the M2M and L2L equations for translation along the z-axis, their optimized program codes are also very similar. A few minor differences exist, however. The line of the code:

```
La = Lb
```

copies the initial values of the entire array *Lb* into array *La*. This operation handles the case of $j = l$ or $j - l = 0$.

The remainder of the code deals with the case $j - l > 0$ by casting it in a *do*-loop over index $j - l$ that runs from 1 to *nL*. Using variable *JmL* to denote $j - l$ gives the following loop structure:

```
factor = 1.0d0
do JmL = 1, nL
  factor = factor * zT / dble(JmL)
  ...
enddo
```

The first line inside this *do*-loop computes the running value of the multiplication factor,

```
factor = factor * zT / dble(JmL),
```

and the following lines deal with its being used. This requires generating all suitable values of indices j and l, which satisfy the condition $JmL = j - l$, and is accomplished in an enclosed *do*–loop that runs over the index j, and computes the value of index l from the known values of *JmL* and j:

```
do j = JmL, nL
  L  = j - JmL
  ...
enddo
```

An alternative would be to run the *do*-loop over index l, and compute the value of index j in an equivalent piece of code:

```
do L = 0, nL-JmL
  j  = L + JmL
  ...
enddo
```

The third, and final, enclosed *do*-loop over index m uses the running value of the multiplication factor, and accumulates the resulting contribution in the expansion term $\tilde{L}_{lm}$.

```
do m = 0, L
  La(aL+m)  = La(aL+m) + Lb(aJ+m)  * factor
enddo
```

After the work inside the *do*-loops over j and m is complete, control returns to the external loop over *JmL*, and the computation continues using the updated value of the multiplication factor corresponding to the running value of variable $j - l$. As expected, because the optimized program code of *L2Lz* now clearly includes only three *do*-loops that have their respective indices running from 1 to *nL*, the computation now scales as $O(N^3)$. The same cannot be said of the naïve version of subroutine *L2Lz*(), which includes four enclosed *do*-loops, and, therefore, scales as $O(N^4)$.

13.3 M2L Translation along the z-Axis

The third, and last, translation in the series of multipole translation operations is the M2L translation, which is defined by Equation 11.17. This transforms a regular multipole moment $\bar{M}_{j,k}(B,B)$ into an irregular multipole moment $\tilde{L}_{l,m}(B,A)$ about a new origin. During this operation, and unlike the other two multipole translations, the M2L operation performs the multipole transformation and the origin translation at the same time:

$$\tilde{L}_{l,m}(B,A) = \sum_{j=0}^{\infty} \sum_{k=-j}^{j} (-1)^j \bar{M}_{j,k}(B,B)\, \tilde{S}_{l+j,m+k}(B,A). \tag{11.17}$$

A close analog of the M2L operation is the L2L translation described by Equation 11.13. Due to the mathematical equivalence of Equations 11.13 and 11.17, proved in Section 8.9, a single software implementation of either equation is sufficient for both of the M2L and L2L operations. However, that flexibility no longer exists in the reduced form of these equations, which performs translation along the z-axis; and, the previously developed Equation 13.15 is good only for L2L operation. With that, it is necessary to develop a separate reduced form for M2L translation.

Development of an equation for an M2L translation along the z-axis begins with a vector diagram, presented in Figure 13.4, which closely resembles the equivalent diagram used in the M2M translation. In this diagram, vector $\boldsymbol{a}$ corresponds to a helper expansion, vector $\boldsymbol{b}$ represents a regular multipole expansion defined about the local coordinate system of box B, and vector $\boldsymbol{a} + \boldsymbol{b}$ denotes the resulting translated irregular multipole expansion defined about the local coordinate system of box A.

The derivation starts with the evaluation of the quantities $\tilde{S}_{l+j,m+k}(r,0,0)$ and $\tilde{S}_{l+j,m+k}(r,180,0)$ corresponding to the virtual vector $\boldsymbol{a}$ shown in the left and right panels in Figure 13.4, respectively. Given $\tilde{P}_l^m(0) = (l-m)!\,P_l^m(0) = l!\,\delta_{m0}$ and $\tilde{P}_l^m(180) = (l-m)!\,P_l^m(180) = (-1)^l l!\,\delta_{m0}$ it follows that:

$$\tilde{S}_{l+j,m+k}(r,0,0) = \frac{(l+j)!}{r^{l+j+1}}\,\delta_{-m,k}, \tag{13.16}$$

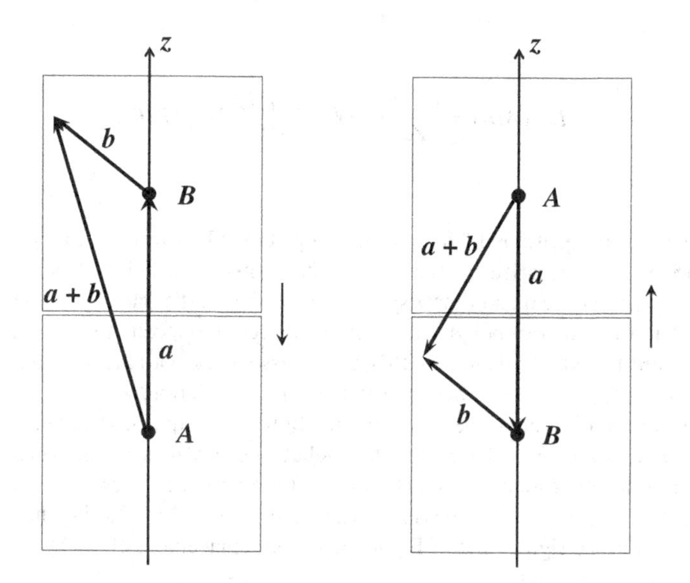

FIGURE 13.4 Vector diagram of the M2L translation of a regular multipole expansion defined about the coordinate system of box B into an irregular multipole expansion defined about box A along the z-axis. Left panel: origin translation in the negative direction to the z-axis. Right panel: origin translation in the positive direction to the z-axis.

and

$$\tilde{S}_{l+j,m+k}(r,180,0) = (-1)^{l+j}\frac{(l+j)!}{r^{l+j+1}}\delta_{-m,k}. \tag{13.17}$$

Now that the simplified forms of the helper expansion aligned along the z-axis have been derived, the next step is to obtain the respective simplified forms of the M2L equation. Substituting Equation 13.16 into Equation 11.17, and recalling that $\bar{M}_{j,-m} = (-1)^m \bar{M}^*_{j,m}$, gives:

$$\tilde{L}_{l,m}(B,A) = \sum_{j=m}^{\infty}(-1)^{j+m}\frac{(l+j)!}{r^{l+j+1}}\bar{M}^*_{j,m}(B,B), \tag{13.18}$$

which represents the M2L equation for the left panel in Figure 13.4, where $\bar{M}_{j,-m}$ comes from $\bar{M}_{j,k}$ due to $\delta_{-m,k}$, and the starting value of index $j = m$ replaces the former value $j = 0$ in the sum due to the requirement of $j \geq m$ coming from $\bar{M}_{j,m}$.

Likewise, substituting Equation 13.17 into Equation 11.17 leads to:

$$\tilde{L}_{l,m}(B,A) = \sum_{j=m}^{\infty}(-1)^{l+m}\frac{(l+j)!}{r^{l+j+1}}\bar{M}^*_{j,m}(B,B), \tag{13.19}$$

which represents the M2L equation for the right panel in Figure 13.4.

Having two different equations for translation along the z–axis is undesirable. To combine Equations 13.18 and 13.19 into a single equation, it is helpful by move $1/r$ outside of the sum. This gives:

$$\tilde{L}_{l,m}(A,B) = \frac{1}{r}\sum_{j=m}^{\infty}(-1)^{j+m}\frac{(l+j)!}{r^{l+j}}\bar{M}^*_{j,m}(B,B), \tag{13.20}$$

and

$$\tilde{L}_{l,m}(A,B) = \frac{1}{r}\sum_{j=m}^{\infty}(-1)^{l+m}\frac{(l+j)!}{r^{l+j}}\bar{M}^*_{j,m}(B,B), \tag{13.21}$$

respectively.

Continuing the work on Equations 13.20 and 13.21 depicted in Figure 13.4, it is necessary to designate one of the translations as the negative and the other as the positive translation, depending on the direction of motion of the origin of the local coordinate system. Since, during the translation, the origin moves from the center of box B to the center of box A, that is, in the direction opposite to the direction of the z-axis, the left panel in Figure 13.4 and Equation 13.20 becomes associated with the negative sign of translation, and the right panel and Equation 13.21 becomes the positive translation.

Once the centers of the local coordinate system are aligned with the z-axis, their origins get assigned coordinates $A = (x,y,z_A)$ and $B = (x,y,z_B)$ in the global coordinate system, where $x_A = x_B = x$ and $y_A = y_B = y$ are arbitrary coordinates. This shows that the relation $z_B - z_A > 0$ takes place in the left panel in Figure 13.4, and $z_B - z_A < 0$ in the right one, where $r = |z_B - z_A|$. Taking into account that $r > 0$, and since $z_B - z_A < 0$ in the right panel of Figure 13.4, the replacement of r^{l+j} by $(-1)^{l+j}(z_B - z_A)^{l+j}$ in Equation 13.21 turns that equation to:

$$\tilde{L}_{l,m}(B,A) = \frac{1}{|z_B - z_A|}\sum_{j=m}^{\infty}(-1)^{l+m}(-1)^{l+j}\frac{(l+j)!}{(z_B - z_A)^{l+j}}\bar{M}^*_{j,m}(B,B). \tag{13.22}$$

Simplification in the phase factors reduces this equation to:

$$\tilde{L}_{l,m}(B,A) = \frac{1}{|z_B - z_A|}\sum_{j=m}^{\infty}(-1)^{j+m}\frac{(l+j)!}{(z_B - z_A)^{l+j}}\bar{M}^*_{j,m}(B,B). \tag{13.23}$$

A comparison of Equations 13.20 and 13.23 shows that the latter is equivalent to Equation 13.20 when $z_B - z_A > 0$. Likewise, having $z_B - z_A < 0$ reduces Equations 13.23 to 13.21. This confirms that Equation 13.23 is a valid final unified form of the original Equations 13.18 and 13.19 that describes both negative and positive origin translations in M2L operation along the z-axis.

The next task is to convert the unified equation into program code. Adopting a straightforward coding style while temporarily ignoring performance concerns reduces the potential for errors, and gives rise to the following program code:

```
subroutine M2Lz(nL, nSzTri, tri, Mb, Tz, La)
!
! M2L translation from box B to box A along the z-axis. Naïve code.
!
! Mb - regular multipole expansion to be shifted
! Tz - translation distance along the z-axis
!      negative sign - in direction of z-axis
!      positive sign - in opposite direction
!      Tz = zB - zA for translation from box B to box A
!      box centers have Cartesian coordinates A(0,0,zA) and B(0,0,zB)
! La - the resulting shifted irregular multipole expansion
!
! La(L,m) = 1/|Tz| * sum(j=m,nL) { (-1)^(j+m) (L+j)! (1/Tz)^(L+j) Mb*(j,m) }
!
implicit none
!
integer,     intent(in)      :: nL          ! expansion truncation
integer,     intent(in)      :: nSzTri      ! size of expansion array
```

```
integer,       intent(in)        :: tri(0:nL)      ! array of triangular pointers
complex(16),   intent(in)        :: Mb(0:nSzTri)   ! expansion to be shifted
double precision, intent(in)     :: Tz             ! translation distance
complex(16),   intent(inout)     :: La(0:nSzTri)   ! translated expansion
!
! Local variables
integer          :: L, m, j, LpJ, aJ, aL
double precision :: factor, Tzinv

Tzinv = 1.0d0 / Tz

do L = 0, nL
  aL = tri(L)
  do m = 0, L
    La(aL+m) = dcmplx(0.0d0, 0.0d0)
    do j = m, nL
      aJ = tri(j)
      factor = dabs(Tzinv)    ! l+j=0
      do LpJ = 1, L+j
        factor = factor * dble(LpJ) * Tzinv
      enddo
      La(aL+m) = La(aL+m) + dconjg(Mb(aJ+m)) * factor * (-1.0d0)**(j+m)
    enddo
  enddo
enddo

end subroutine M2Lz
```

Subroutine *M2Lz*() starts by computing the inverse distance, $Tzinv = 1/(z_B - z_A)$, in the first executable line of the program code:

```
Tzinv = 1.0d0 / Tz
```

where translation distance *Tz* is assigned the value of $z_B - z_A$ on input, under the assumption that the centers of boxes *A* and *B* are aligned with the *z*-axis and have Cartesian coordinates $A = (x,y,z_A)$ and $B = (x,y,z_B)$ in the global coordinate system. In principle, a test should be performed to ensure that the value of variable *Tz* is finite before dividing by it, to avoid the error of division by zero. However, in the present formulation of the FMM algorithm, no pair of boxes has overlapping or closely positioned box centers. This rules out the possibility of division by zero, making such a test unnecessary.

The main body of computation is arranged in a series of four enclosed *do*–loops, three of which go over indices *l*, *m*, and *j*. The *do*-loops over indices *l* and *m* have the task of computing all expansion terms $L_{l,m}(B,A)$ for the allowed range of indices *l* and *m*. The starting value of index *l*, including the starting and ending values of index *m*, follows from the standard definition of spherical harmonics, while, at the same time, restricting the computation of expansion terms only to non-negative values of index *m* saves on memory storage. Truncation of the multipole expansion results in the upper bounds of orbital indices *l* and *j* being *nL*. The lower bound of index *j* follows from Equation 13.13.

The fourth innermost *do*-loop calculates the multiplication factor

$$\frac{1}{|z_B - z_A|} \frac{(l+j)!}{(z_B - z_A)^{l+j}}.$$

This computation needs to be done in a *do*-loop because of the presence of factorial $(l+j)!$.

```
factor = dabs(Tzinv)    ! l+j=0
do LpJ = 1, L+j
  factor = factor * dble(LpJ) * Tzinv
enddo
```

Computation of the contribution of $\bar{M}^*_{j,m}(B,B)$ to $L_{l,m}(B,A)$ proceeds in the line

```
La(aL+m) = La(aL+m) + dconjg(Mb(aJ+m)) * factor * (-1.0d0)**(j+m)
```

which employs Fortran function *dconjg*() to obtain $\bar{M}^*_{j,m}$ from the value of $\bar{M}_{j,m}$ provided on input.

Validation of the new subroutine *M2Lz*() employs the program code previously developed in Section 11.3 for the general M2L translation. The only change that needs to be made in it is to place the local coordinate origin centers *oxyz*1 and *oxyz*2 on a line parallel to the z-axis. An easy way to accomplish this is to assign the x and y components of point *oxyz*1 to the respective components of point *oxyz*2:

```
oxyz2(1) = oxyz1(1)
oxyz2(2) = oxyz1(2)
```

while leaving the z-components at their original value. The following code snippet gives the resulting validation code:

```
implicit none
integer          :: n                ! number of particles
double precision :: c(3,n)           ! particle coordinates
double precision :: q(n)             ! particle charges
integer          :: nL               ! expansion truncation
integer          :: nSzTri           ! size of expansion array
integer          :: tri(0:nL)        ! triangular pointers
double precision :: oxyz1(3)         ! origin1
double precision :: oxyz2(3)         ! origin2
double precision :: radius           ! radius of enclosing sphere
complex(16)      :: scr(0:nSzTri)    ! scratch array
complex(16)      :: M1(0:nSzTri)     ! regular expansion to be translated
complex(16)      :: L2(0:nSzTri)     ! reference irregular expansion
complex(16)      :: La(0:nSzTri)     ! translated irregular expansion
logical          :: error

! Determine oxyz1 as the geometric center of particles
oxyz1 = 0.0d0
do i = 1, n
  oxyz1(1:3) = oxyz1(1:3) + c(1:3,i)
enddo
oxyz1 = oxyz1 / dble(n)

! Place oxyz2 far away of the charge set
oxyz2  = maxval(c, dim = 2) - oxyz1
radius = dsqrt(oxyz2(1)**2 + oxyz2(2)**2 + oxyz2(3)**2)
oxyz2  = (oxyz1 + dsign(radius, oxyz1)) * 3.0d0
oxyz2(1) = oxyz1(1)
oxyz2(2) = oxyz1(2)

! Get regular multipole expansion of a charge set about origin oxyz1
call RegularExpansion(c, q, n, oxyz1, nL, nSzTri, tri, scr, M1)

! Get irregular multipole expansion of a charge set about origin oxyz2
call IrregularExpansion(c, q, n, oxyz2, nL, nSzTri, tri, scr, L2)

! Perform M2L operation
call M2Lz(nL, nSzTri, tri, M1, oxyz1(3)-oxyz2(3), La)

! Validate the result of transformation, Lc against the known data, L2
error = .FALSE.
do i = 0, nSzTri
```

```
    if(abs(dble( La(i) - L2(i))) > 1.0d-12) error = .TRUE.
    if(abs(aimag(La(i) - L2(i))) > 1.0d-12) error = .TRUE.
enddo
if(error) write(*,*) "Error in M2Lz computation"
```

Once the correctness of the naïve implementation is verified, the next step is to critically analyze the program code, and locate and eliminate the performance limiting constructs. An obvious inefficiency in the current implementation of subroutine *M2Lz*() is the repetitive evaluation of the multiplication factor,

$$\frac{1}{|z_B - z_A|}\frac{(l+j)!}{(z_B - z_A)^{l+j}}.$$

In the naïve solution, a fourth enclosed *do*-loop causes the code to scale unnecessarily as $O(N^4)$; Equation 13.13 implies that it should theoretically scale as $O(N^3)$, based on the presence of only three indices l, m, and j.

The solution is to rearrange the *do*-loops so that the factorial is calculated in the outermost one. The indices l and j are then defined in a second *do*-loop over index either l or j, while using the provided value of $l+j$ to determine the complementary index. This is illustrated in the following program code:

```
subroutine M2Lz(nL, nSzTri, tri, Mb, Tz, La)
!
! M2L translation from box B to box A along the z-axis. Optimized code.
!
! Mb - regular multipole expansion to be shifted
! Tz - translation distance along the z-axis
!      negative sign - in direction of z-axis
!      positive sign - in opposite direction
!      Tz = zB - zA for translation from box B to box A
!      box centers have Cartesian coordinates A(0,0,zA) and B(0,0,zB)
! La - the resulting shifted irregular multipole expansion
!
! La(L,m) = 1/|Tz| * sum(j=m,nL) { (-1)^(j+m) (L+j)! (1/Tz)^(L+j) Mb*(j,m) }
!
implicit none
!
integer,          intent(in)    :: nL             ! expansion truncation
integer,          intent(in)    :: nSzTri         ! size of expansion array
integer,          intent(in)    :: tri(0:nL)      ! array of triangular pointers
complex(16),      intent(in)    :: Mb(0:nSzTri)   ! expansion to be shifted
double precision, intent(in)    :: Tz             ! translation distance
complex(16),      intent(inout) :: La(0:nSzTri)   ! translated expansion
!
! Local variables
integer            :: L, m, j, LpJ, aJ, aL
double precision :: factor, Tzinv

Tzinv = 1.0d0 / Tz
factor = dabs(Tzinv)   ! l+j=0

La = dcmplx(0.0d0, 0.0d0)

! L+j=0
La(0) = Mb(0) * factor

do LpJ = 1, nL+nL
  factor = factor * dble(LpJ) * Tzinv
  do L = max0(0,LpJ-nL), min0(LpJ,nL)
    j = LpJ - L
```

```
    aJ = tri(j)
    aL = tri(L)
    ! m = 0
    La(aL) = La(aL) + Mb(aJ) * factor * dble((-1)**j)
    do m = 1, min0(j,L)
      La(aL+m) = La(aL+m) + dconjg(Mb(aJ+m)) * factor * dble((-1)**(j+m))
    enddo
  enddo
enddo

end subroutine M2Lz
```

The first executable line in this program code computes the inverse distance, *Tzinv*, and its absolute value then gives the starting value of the multiplication factor,

```
factor = dabs(Tzinv)   ! l+j=0
```

which corresponds to the case of $l+j=0$.

The next line initializes the *La*:

```
La = dcmplx(0.0d0, 0.0d0)
```

so that it is ready to accumulate the results of the computation.

The starting value of the multiplication factor is used in the computation of the expansion term $\tilde{L}_{0,0}(B,A)$ in the following line of the program code:

```
! L+j=0
La(0) = Mb(0) * factor
```

This arrangement corresponds to a special case of $l=j=m=0$, which reduces Equation 13.23 to:

$$\tilde{L}_{0,0}(B,A)=\frac{1}{|z_B-z_A|}\bar{M}_{0,0}(B,B). \tag{13.24}$$

This outcome highlights the fact that the transformation of the term $\bar{M}_{0,0}(B,B)$ into $\tilde{L}_{0,0}(B,A)$ does not depend on the sign of the translation.

The rest of the subroutine deals with the case of $l+j>0$. Computation of the scale factor that includes factorial $(l+j)!$ is best accomplished in a *do*-loop over index $l+j$. This index has a starting value of 1 and is denoted by variable *LpJ*. Since each index l and j can independently run up to *nL*, the upper value for their sum becomes $nL+nL$, as illustrated in the following code snippet:

```
do LpJ = 1, nL+nL
  ...
enddo
```

Since this *do*-loop establishes the value of the sum $l+j$, it is necessary to add only one extra *do*-loop inside it, over either index l or j, and compute the other index from their sum and from the paired index. An example of a *do*-loop of this type follows:

```
do LpJ = 1, nL+nL
  ...
  do L = max0(0,LpJ-nL), min0(LpJ,nL)
    j = LpJ - L
    ...
  enddo
enddo
```

The use of Fortran 90 function *max*0(0, *LpJ* – *nL*) guarantees that the starting value of index l is zero when $(l+j) \leq nL$, and is $l+j-nL$ when $(l+j) > nL$. Likewise, the upper value of index l becomes $l+j$ when $(l+j) \leq nL$, and nL when $(l+j) > nL$. Within the *do*-loop, the value of the paired index j is trivially determined as $j = (l+j) - l$.

Obviously this same *do*-loop could equally well be made using index j, in which case the value of index l would be determined inside the *do*-loop.

```
do j = max0(0,LpJ-nL), min0(LpJ,nL)
  L = LpJ - L
  ...
enddo
```

In the third, the innermost, *do*-loop, m runs from 1 to the smaller of j or l. The value of $m = 0$ represents a special case in M2L computation since it does not require taking the complex conjugate of $\bar{M}_{j,0}$, and as such can be more efficiently computed outside of the *do*-loop via:

```
! m = 0
La(aL) = La(aL) + Mb(aJ) * factor * dble((-1)**j)
```

With this brief review of the *do*-loop structure completed it is helpful to do an additional analysis of the outermost *do*-loop over index *LpJ*. This index runs from 1 to $nL + nL$ because each index l and j independently has a maximum value of nL. However, it also might be observed that the combined index $l+j$ in $\tilde{S}_{l+j,m+k}$ in the original Equation 11.17 runs only up to nL due to the decision to uniformly truncate all multipole expansions on the upper value of orbital index set to nL; this is how the general M2L translation is implemented in Section 11.3. If subroutine *M2Lz*() were to be used to produce the same result as the outcome of subroutine *M2L*(), this can be achieved by setting the upper value of index *LpJ* to nL, resulting in the following revision of the M2Lz code:

```
do LpJ = 1, nL
  factor = factor * dble(LpJ) * Tzinv
  do L = 0, LpJ
    j  = LpJ - L
    aJ = tri(j)
    aL = tri(L)
    ! m = 0
    La(aL) = La(aL) + Mb(aJ) * factor * dble((-1)**j)
    do m = 1, min0(j,L)
      La(aL+m) = La(aL+m) + dconjg(Mb(aJ+m)) * factor * dble((-1)**(j+m))
    enddo
  enddo
enddo
```

Although this revised code would have a slightly lower accuracy than the previous implementation, it would be faster by a factor of 2. Provided the value of nL was already chosen sufficiently high to guarantee the required accuracy of the M2M and L2L translations, the preferred solution for subroutine *M2Lz*() would be to go with the revised loop structure that limits the upper value of index *LpJ* to nL in favor of better performance.

Validation of the correctness of the optimized subroutine *M2Lz*() relies on the same procedure developed for the naïve implementation of the code, and the presence of only three *do*–loops in the program structure confirms that it scales as $O(N^3)$, as predicted by theory.

14

Rotation of Coordinate System

Multipole operations, M2M, M2L, and L2L are most computationally efficient when the origin translation happens along the z-axis. However, this condition only occurs in a few box-pairs in a real 3D system. Most box-pairs have their translation axis arbitrarily oriented relative to the z-axis, and that makes the multipole translation along the z-axis inapplicable.

Flexibility in the choice of coordinate system provides the mechanism to achieve the required parallel alignment of the z-axis with the translation vector. Implementation of this idea requires developing a coordinate system rotation procedure to align the z-axis with the axis of the multipole translation, which connects the box centers. The present chapter addresses the problem of developing a computationally efficient procedure for rotation of coordinate system of multipole expansion.

14.1 Rotation of Coordinate System to Align the z-axis with the Axis of Translation

Before it is possible to speed up FMM calculations by carrying out multipole translation operations along the z-axis, it is necessary to develop a coordinate system rotation procedure to align the z-axis with the axis of the multipole translation, $\boldsymbol{t}$, which connects the box centers in Figure 14.1.

A convenient approach for describing rotations begins with the parametric form of the Euler angles discussed in Section 5.1. Figure 14.1 illustrates its application to the present problem. For a multipole translation going from box B to box A, the translation vector $\boldsymbol{t}$ originates at the center of box B and points toward the center of box A. Each box has its own local coordinate system that has its origin placed in the center of the box. Coordinate systems of these boxes can be transformed from one to the other *via* a parallel shift operation.

For a translation vector $\boldsymbol{t}$ arbitrarily positioned relative to the coordinate system in Figure 14.1, only two coordinate rotations are required in order to align the z-axis with the translation vector. In the first step, rotation through angle α about the z-axis places the x'-axis into the plane formed by the z-axis and vector $\boldsymbol{t}$. This gives a new coordinate system $x'y'z$ in which the y'-axis is perpendicular to the plane formed by the z-axis and vector $\boldsymbol{t}$. A second rotation through angle β about the y'-axis applied to the coordinate system $x'y'z$ results in vector $\boldsymbol{t}$ becoming aligned with the z-axis.

Although this operation looks straightforward, the change in the rotation frame that happens between the first and second rotations makes deriving a rotation matrix impractical. Fortunately, there is a simple way to recast the transformation steps into rotations about the coordinate axes of the initial coordinate system so that the individual rotation matrices required for each of the rotations can easily be obtained. All that is needed is to reverse the order of rotations about the original coordinate system as described in Section 5.1. With that, the first rotation through angle β is made around the y-axis. This transforms the z-axis into the z^*-axis in Figure 14.1. A second rotation through angle α about the original z-axis then aligns the z^*-axis with vector $\boldsymbol{t}$. The rotation matrix $\mathbf{R}$ of this transformation is the product of the two independently-determined rotation matrices $\mathbf{R}_y(\beta)$ and $\mathbf{R}_z(\alpha)$ around the initial fixed y- and z-axes, respectively, so that

$$\mathbf{R} = \mathbf{R}_z(\alpha)\mathbf{R}_y(\beta). \tag{14.1}$$

The next task is to determine the rotation angles α and β based on the known coordinates of vector $\boldsymbol{t}$.

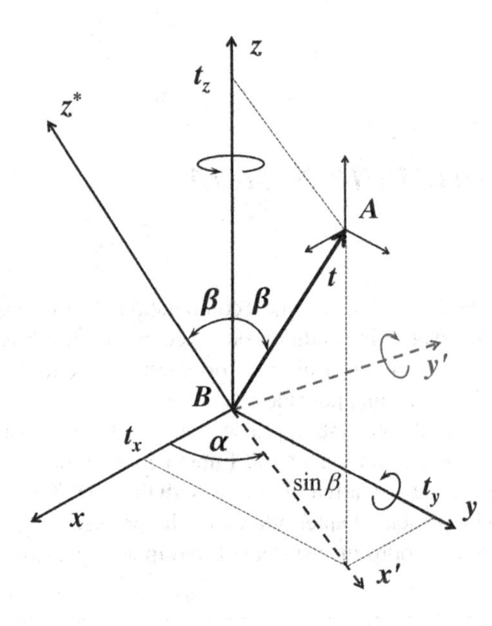

FIGURE 14.1 Rotation of coordinate system to align the z-axis with the vector $\boldsymbol{t}$, which corresponds to origin translation from box B to box A. The rotation through angle alpha takes place around the z-axis and puts the x'-axis into the plane formed by the vector $\boldsymbol{t}$ and the z-axis; this is followed by rotating the new coordinate system through angle beta about the y'-axis. Alternatively, the original coordinate system can be rotated around the y-axis through angle beta, and then the new coordinate system is rotated around the original z-axis on angle alpha to finally align the z^* axis with the vector $\boldsymbol{t}$.

Of the two, angle β is the easiest to determine. It is simply

$$\beta = \arccos(t_z / |t|), \tag{14.2}$$

where t_z is the z-component of the vector $\boldsymbol{t}$, and $|t|$ is the vector norm. This equation supports the entire domain of angle β.

The angle α has values in the range 0 to 2π radians, including the boundaries. Using the trigonometric function arccos() to resolve angle α, however, encounters a minor difficulty in that the range of this function is limited to the interval $[0, \pi]$, which is insufficient to allow the angle α to be defined. Because of this, it is necessary to split the domain for angle α into two segments, $[0, \pi]$ and $]\pi, 2\pi[$, where outward-looking brackets indicate that the limits are not included, and test the location of angle α in the particular interval. Checking the sign of the component t_y provides the necessary mechanism to fully determine the angle α:

$$\alpha = \begin{cases} \arccos(\cos\alpha) & \text{if } t_y \geq 0, \\ 2\pi - \arccos(\cos\alpha) & \text{if } t_y < 0, \end{cases} \tag{14.3}$$

where $\cos\alpha = t_x \big/ \sqrt{t_x^2 + t_y^2}$. The condition of $t_y \geq 0$ in Equation 14.3 constrains the angle α to the interval $[0, \pi]$; and the condition of $t_y < 0$ associates angle α with the second interval $[\pi, 2\pi]$. This is illustrated in the following sample computer implementation of Equations 14.2 and 14.3:

```
subroutine EulerAngles(b, a, alpha, beta)
!
! Determine Euler angles. See Fig. 14.1.
! B - donor box
! A - recipient box
!
! Function arguments
```

```
double precision, intent(in)  :: a(3), b(3)    ! origin of boxes A and B
double precision, intent(out) :: alpha, beta   ! Euler angles
!
! Local variables
double precision :: t(3)                       ! translation axis
double precision, parameter :: pi = 3.14159265358979323846d0
!
! Determine translation axis
t = a - b

! Check whether z-axis and t-vector are parallel
if(abs(t(1)) + abs(t(2)) > 1.0d0-8) then
   ! t and z-axis are not parallel
   ! Determine angle alpha subtended between x-axis and rotated x'-axis
   ! cosalp = tx / sqrt(tx^2 + ty^2)
   ! if ty >= 0 then 0 <= alpha <= 180; therefore, alpha = acos(cosalp)
   ! if ty <  0 then 180 < alpha < 360; therefore, alpha = 2pi - acos(cosalp)
   ! this leads to: alpha = [1 - sign(1,ty)] * pi + sign(1,ty)  * acos(cosalp)

   alpha  = (1.0d0 - dsign(1.0d0,t(2))) * pi + &
            dsign(1.0d0,t(2)) * dacos( t(1) / sqrt(t(1)*t(1) + t(2)*t(2)) )

   ! Determine angle beta subtended between z-axis and vector t
   ! cos(bet) = tz / sqrt(tx^2 + ty^2 + tz^2)

   beta = dacos( t(3) / sqrt(t(1)*t(1) + t(2)*t(2) +t(3)*t(3)) )
else
   ! t and z-axis are parallel
   alpha = 0.0d0
   beta  = pi * (0.5d0 - dsign(0.5d0,t(3)))  ! beta = 0 for positive; beta =
180 for negative sign of alignment
endif

end subroutine EulerAngles
```

As shown, the input to this computer code consists of the Cartesian coordinates of points *a* and *b*, which correspond to the centers of boxes *A* and *B*, respectively. Box *B* is the donor, and box *A* is the recipient of the multipole translation. The output variables *alpha* and *beta* represent the angles, in radians, to be computed.

The first executable line in the program code computes the Cartesian coordinates of radius vector *t*:

```
t = a - b
```

In this expression, Fortran 90 array handling takes care of the array indices so that they don't have to be programmed individually.

Before proceeding to the determination of angles α and β, the program needs to check whether the vector ***t*** and the *z*-axis are already parallel. This is a special case, which requires separate handling. If the vectors ***t*** and the *z*-axis are parallel then both components t_x and t_y would be zero. The program code determines the condition of alignment of vector ***t*** and the *z*-axis by summing up the absolute value of components t_x and t_y and checking the sum against a very small number representing zero, to allow for rounding error. This test uses a conditional statement that branches the computation to separately handle the non-parallel and parallel orientations.

If the vector ***t*** and the *z*-axis are not parallel, which happens in the vast majority of cases, the program code evaluates the angles α and β. Note than an elegant way to avoid the inefficient conditional statement in Equation 14.3 would be to integrate the sign test of component t_y into the computation of angle α:

$$\alpha = [1 - \mathrm{sign}(1,t_y)]\,\pi + \mathrm{sign}(1,t_y)\arccos\left(t_x / \sqrt{t_x^2 + t_y^2}\right), \tag{14.4}$$

which turns the two formulae in Equation 14.3 into a single expression.

Equation 14.4 uses the Fortran intrinsic function sign() to transfer the sign of component t_y to its first argument; unity in the present case. With that, when the sign of t_y is positive, the expression in square brackets becomes zero, converting Equation 14.4 into the first line of Equation 14.3. When the sign of t_y is negative, the expression in square brackets becomes equal to two, and the entire expression in Equation 14.4 becomes equivalent to the second line of Equation 14.3. Computation of angle β based on Equation 14.2 finalizes the computation in the first branch.

The second branch in the program code handles the case where the alignment of the z-axis is parallel with the vector $\boldsymbol{t}$ in the *else* portion of the conditional statement. Parallel alignment implies that angle α is zero, and leaves the choice for angle β to be either zero or π, depending on the sign of the component t_z. A positive sign of this component implies that the vector $\boldsymbol{t}$ and the z-axis point in the same direction: in this case the angle β is zero, while a negative sign of t_z indicates that the vector $\boldsymbol{t}$ and the z-axis point in opposite directions: this makes the angle β, which is subtended between these two vectors, equal to π. Integrating the sign-test of the component t_z into the equation for angle β leads to:

$$\beta = [0.5 - \text{sign}(0.5, t_z)]\,\pi. \tag{14.5}$$

The expression in square brackets in Equation 14.5 becomes zero for $t_z \geq 0$, and unity for $t_z < 0$. Equation 14.5 ends the second branch of computation of angles α and β in subroutine *EulerAngles*().

The resulting Euler angles will be used in building the rotation matrix.

14.2 Rotation Matrix

The present task is to determine the rotation matrices for multipole expansions. Because they are constructed from spherical harmonic functions, multipole expansions rotate in the same way as the spherical harmonics. In turn, being eigenvectors of the angular momentum operator, spherical harmonics $Y_{l,m}$ of degree l rotate according to Equation 5.63, described in Section 5.2, in the basis of their azimuthal components $-l, -l+1, \dots m \dots, l-1, l$:

$$Y'_{l,m} = \sum_{k=-l}^{l} D^l_{k,m}(\alpha, \beta, \gamma)\, Y_{l,k}, \tag{14.6}$$

where the prime symbol indicates a rotated function, while $D^l_{k,m}(\alpha,\beta)$ is a matrix element with indices k and m of rotation matrix D^l, and α, β, and γ are Euler angles. Since angle γ is equal to zero for the purpose of aligning the multipole translation axis with the z-axis of the coordinate system, it reduces Equation 14.6 to:

$$Y'_{l,m} = \sum_{k=-l}^{l} D^l_{k,m}(\alpha, \beta)\, Y_{l,k}. \tag{14.7}$$

The rotation matrix elements to be found are

$$D^l_{k,m}(\alpha, \beta) = e^{-i\alpha k} d^l_{k,m}(\beta), \tag{14.8}$$

which relates to Equation 14.1, and where $d^l_{k,m}(\beta)$ is the Wigner matrix element for the rotation around the y-axis. Rotation through angle α about the z-axis corresponds to multiplication by a complex factor $e^{-i\alpha k} = \cos(\alpha k) - i\sin(\alpha k)$.

Deriving the matrix elements $d^l_{k,m}(\beta)$ for rotation on angle β around the y-axis requires a major effort. Because of that, it is convenient to separately consider the rotation on angle β starting with:

$$Y'_{l,m} = \sum_{k=-l}^{l} d^l_{k,m} Y_{l,k}, \tag{14.9}$$

where the dependence on angle β is omitted for brevity.

The slight difference that exists between the equation for rotation of spherical harmonics and that for multipole expansions, which needs to be addressed, is the difference in the normalization or scaling coefficient that these functions employ. With that, the first step in transitioning from the rotation of spherical harmonics $Y_{l,m}$ to the rotation of bar- and tilde-scaled multipole expansions is to rewrite Equation 14.9 in bare angular terms. Substituting spherical harmonics in Equation 14.9 by their canonical form gives:

$$\sqrt{\frac{(2l+1)}{4\pi}\frac{(l-m)!}{(l+m)!}}\,P_l^m(\cos(\theta+\beta))\,e^{im\phi}=\sum_{k=-l}^{l}d_{k,m}^l\sqrt{\frac{(2l+1)}{4\pi}\frac{(l-k)!}{(l+k)!}}\,P_l^k(\cos\theta)\,e^{ik\phi}. \tag{14.10}$$

Cancelling out the term $(2l+1)/4\pi$ in Equation 14.10 leads to:

$$\sqrt{\frac{(l-m)!}{(l+m)!}}\,P_l^m(\cos(\theta+\beta))\,e^{im\phi}=\sum_{k=-l}^{l}d_{k,m}^l\sqrt{\frac{(l-k)!}{(l+k)!}}\,P_l^k(\cos\theta)\,e^{ik\phi}. \tag{14.11}$$

Multiplying both sides of Equation 14.11 by a scalar radial component r^l or $1/r^{l+1}$ and rearranging the coefficients inside the square root gives:

$$\begin{aligned}R_{l,m}(r,\theta+\beta,\phi)&=\sum_{k=-l}^{l}\sqrt{\frac{(l-k)!(l+m)!}{(l-m)!(l+k)!}}\,d_{k,m}^l R_{l,k}(r,\theta,\phi),\\S_{l,m}(r,\theta+\beta,\phi)&=\sum_{k=-l}^{l}\sqrt{\frac{(l-k)!(l+m)!}{(l-m)!(l+k)!}}\,d_{k,m}^l S_{l,k}(r,\theta,\phi),\end{aligned} \tag{14.12}$$

which are the equations for rotation of unscaled solid harmonics $R_{l,m}$ and $S_{l,m}$, respectively.

The next step is to convert Equation 14.12 into the bar- and tilde-scaled form of multipole expansions. Taking into account that

$$\bar{R}_{l,m}=\frac{1}{(l+m)!}R_{l,m}\quad\text{and}\quad\tilde{S}_{l,m}=(l-m)!S_{l,m}, \tag{14.13}$$

multiplying both sides of the first line in Equation 14.12 by $\frac{1}{(l+m)!}$ and those of the second line by $(l-m)!$ gives:

$$\begin{aligned}\bar{R}_{l,m}(r,\theta+\beta,\phi)&=\sum_{k=-l}^{l}\frac{1}{(l+m)!}\sqrt{\frac{(l-k)!(l+m)!}{(l-m)!(l+k)!}}\,\frac{(l+k)!}{(l+k)!}\,d_{k,m}^l R_{l,k}(r,\theta,\phi),\\\tilde{S}_{l,m}(r,\theta+\beta,\phi)&=\sum_{k=-l}^{l}(l-m)!\sqrt{\frac{(l-k)!(l+m)!}{(l-m)!(l+k)!}}\,\frac{(l-k)!}{(l-k)!}\,d_{k,m}^l S_{l,k}(r,\theta,\phi).\end{aligned} \tag{14.14}$$

Simplifying the right-hand side of Equation 14.14 leads to:

$$\begin{aligned}\bar{R}_{l,m}(r,\theta+\beta,\phi)&=\sum_{k=-l}^{l}\sqrt{\frac{(l-k)!(l+k)!}{(l-m)!(l+m)!}}\,d_{k,m}^l\bar{R}_{l,k}(r,\theta,\phi),\\\tilde{S}_{l,m}(r,\theta+\beta,\phi)&=\sum_{k=-l}^{l}\sqrt{\frac{(l-m)!(l+m)!}{(l-k)!(l+k)!}}\,d_{k,m}^l\tilde{S}_{l,k}(r,\theta,\phi),\end{aligned} \tag{14.15}$$

which are the equations for rotation of scaled solid harmonics. Because of their identical rotation properties, transitioning from a rotation of scaled solid harmonics to a rotation of scaled multipole expansions is a simple matter of replacing symbols R and S with M and L, respectively. This transition is allowed because the additional particle charge factors present in the multipole expansions are scalar numbers.

The appearance of a bulky square-root factor in Equation 14.15 makes this equation less than optimal for numerical computation. If coded naïvely, the computer program may run into overflow or precision loss, especially for large values of orbital index l, not mentioning the performance penalty associated with computing large factorials. These problems can be avoided by switching from the canonical form of the Wigner matrix elements to their scaled form. An examination of Equation 14.15 suggests two scaling forms for the Wigner matrix element:

$$\begin{aligned} \bar{d}^l_{k,m} &= \sqrt{\frac{(l-k)!(l+k)!}{(l-m)!(l+m)!}}\, d^l_{k,m}, \\ \tilde{d}^l_{k,m} &= \sqrt{\frac{(l-m)!(l+m)!}{(l-k)!(l+k)!}}\, d^l_{k,m}, \end{aligned} \tag{14.16}$$

which lead to:

$$\begin{aligned} \bar{R}_{l,m}(r,\theta+\beta,\phi) &= \sum_{k=-l}^{l} \bar{d}^l_{k,m} \bar{R}_{l,k}(r,\theta,\phi), \\ \tilde{S}_{l,m}(r,\theta+\beta,\phi) &= \sum_{k=-l}^{l} \tilde{d}^l_{k,m} \tilde{S}_{l,k}(r,\theta,\phi). \end{aligned} \tag{14.17}$$

Before Equation 14.17 can be used, it is necessary to compute matrix elements $\bar{d}^l_{k,m}$ and $\tilde{d}^l_{k,m}$ efficiently. Sections 7.1 and 7.2 describe two different sets of recurrence relations for the Wigner matrix elements that provide the starting point for development of recurrence relations for scaled matrix elements. Derivation of these equations will now be addressed.

14.3 Computation of Scaled Wigner Matrix Elements with Increment in Index *m*

Sections 7.1 establishes the first set of conventional recurrence relations for unscaled Wigner matrix elements $d^l_{k,m}$:

$$\begin{aligned} &\sqrt{(l+m-1)(l+m)}\left(\frac{1+\cos\beta}{2}\right) d^{l-1}_{k-1,m-1} - \sqrt{(l-m)(l+m)}(\sin\beta)\, d^{l-1}_{k-1,m} \\ &\quad + \sqrt{(l-m-1)(l-m)}\left(\frac{1-\cos\beta}{2}\right) d^{l-1}_{k-1,m+1} = \sqrt{(l+k-1)(l+k)}\, d^l_{k,m}, \end{aligned} \tag{7.6}$$

$$\begin{aligned} &\sqrt{(l+m-1)(l+m)}\left(\frac{\sin\beta}{2}\right) d^{l-1}_{k,m-1} + \sqrt{(l-m)(l+m)}(\cos\beta)\, d^{l-1}_{k,m} \\ &\quad - \sqrt{(l-m-1)(l-m)}\left(\frac{\sin\beta}{2}\right) d^{l-1}_{k,m+1} = \sqrt{(l-k)(l+k)}\, d^l_{k,m}, \end{aligned} \tag{7.12}$$

$$\begin{aligned} &\sqrt{(l+m-1)(l+m)}\left(\frac{1-\cos\beta}{2}\right) d^{l-1}_{k+1,m-1} + \sqrt{(l-m)(l+m)}(\sin\beta)\, d^{l-1}_{k+1,m} \\ &\quad + \sqrt{(l-m-1)(l-m)}\left(\frac{1+\cos\beta}{2}\right) d^{l-1}_{k+1,m+1} = \sqrt{(l-k-1)(l-k)}\, d^l_{k,m}, \end{aligned} \tag{7.17}$$

which incorporate the increment in index m, and allow the evaluation of all elements of the matrix $\mathbf{d}^l$ based on the known matrix $\mathbf{d}^{l-1}$. From these relations, a few useful specific cases that deal with $m = \pm l$ follow:

$m = l$ and $k = l$,

$$\frac{1+\cos\beta}{2} d^{l-1}_{l-1,l-1} = d^l_{l,l}, \tag{7.30}$$

$m = l$ and $l-1 \geq k \geq -l+1$,

$$\sqrt{(2l)(2l-1)}\left(\frac{\sin\beta}{2}\right) d^{l-1}_{k,l-1} = \sqrt{(l-k)(l+k)}\, d^l_{k,l}, \tag{7.19}$$

$m = l$ and $k = -l$,

$$\frac{1-\cos\beta}{2} d^{l-1}_{-l+1,l-1} = d^l_{-l,l}, \tag{7.32}$$

$m = -l$ and $k = l$,

$$\frac{1-\cos\beta}{2} d^{l-1}_{l-1,-l+1} = d^l_{l,-l}, \tag{7.31}$$

$m = -l$ and $l-1 \geq k \geq -l+1$,

$$-\sqrt{(2l)(2l-1)}\left(\frac{\sin\beta}{2}\right) d^{l-1}_{k,-l+1} = \sqrt{(l-k)(l+k)}\, d^l_{k,-l}, \tag{7.22}$$

$m = -l$ and $k = -l$,

$$\frac{1+\cos\beta}{2} d^{l-1}_{-l+1,-l+1} = d^l_{-l,-l}. \tag{7.33}$$

Equations involving matrix elements with indices $|m| = |k|$ are identical in scaled and unscaled forms, because this condition turns the scale coefficient to unity. For all other equations, the first step in transitioning to the scaled form defined by Equation 14.16 is to determine the scale coefficient for the individual matrix elements present in the recurrence relations. For the bar-scaled form, these matrix elements are:

$$\begin{aligned}
&\bar{d}^{l-1}_{k-1,m-1} = \sqrt{\frac{(l-k)!(l+k-2)!}{(l-m)!(l+m-2)!}}\, d^{l-1}_{k-1,m-1}, \quad \bar{d}^{l-1}_{k-1,m} = \sqrt{\frac{(l-k)!(l+k-2)!}{(l-m-1)!(l+m-1)!}}\, d^{l-1}_{k-1,m},\\
&\bar{d}^{l-1}_{k-1,m+1} = \sqrt{\frac{(l-k)!(l+k-2)!}{(l-m-2)!(l+m)!}}\, d^{l-1}_{k-1,m+1},\\
&\bar{d}^{l-1}_{k,m-1} = \sqrt{\frac{(l-k-1)!(l+k-1)!}{(l-m)!(l+m-2)!}}\, d^{l-1}_{k,m-1}, \quad \bar{d}^{l-1}_{k,m} = \sqrt{\frac{(l-k-1)!(l+k-1)!}{(l-m-1)!(l+m-1)!}}\, d^{l-1}_{k,m},\\
&\bar{d}^{l-1}_{k,m+1} = \sqrt{\frac{(l-k-1)!(l+k-1)!}{(l-m-2)!(l+m)!}}\, d^{l-1}_{k,m+1},\\
&\bar{d}^{l-1}_{k+1,m-1} = \sqrt{\frac{(l-k-2)!(l+k)!}{(l-m)!(l+m-2)!}}\, d^{l-1}_{k+1,m-1}, \quad \bar{d}^{l-1}_{k+1,m} = \sqrt{\frac{(l-k-2)!(l+k)!}{(l-m-1)!(l+m-1)!}}\, d^{l-1}_{k+1,m},\\
&\bar{d}^{l-1}_{k+1,m+1} = \sqrt{\frac{(l-k-2)!(l+k)!}{(l-m-2)!(l+m)!}}\, d^{l-1}_{k+1,m+1},\\
&\bar{d}^{l-1}_{k,l-1} = \sqrt{\frac{(l-k-1)!(l+k-1)!}{(2l-2)!}}\, d^l_{k,l-1}, \quad \bar{d}^{l-1}_{k,-l+1} = \sqrt{\frac{(l-k-1)!(l+k-1)!}{(2l-2)!}}\, d^l_{k,-l+1},\\
&\bar{d}^l_{k,l} = \sqrt{\frac{(l-k)!(l+k)!}{(2l)!}}\, d^l_{k,l}, \quad \bar{d}^l_{k,-l} = \sqrt{\frac{(l-k)!(l+k)!}{(2l)!}}\, d^l_{k,-l},
\end{aligned} \tag{14.18}$$

$$\bar{d}^{\,l}_{l,m} = \sqrt{\frac{(2l)!}{(l-m)!(l+m)!}}\, d^{\,l}_{l,m}, \qquad \bar{d}^{\,l}_{-l,m} = \sqrt{\frac{(2l)!}{(l-m)!(l+m)!}}\, d^{\,l}_{-l,m},$$

$$\bar{d}^{\,l}_{l-1,m} = \sqrt{\frac{(2l-1)!}{(l-m)!(l+m)!}}\, d^{\,l}_{l-1,m}, \qquad \bar{d}^{\,l}_{-l+1,m} = \sqrt{\frac{(2l-1)!}{(l-m)!(l+m)!}}\, d^{\,l}_{-l+1,m},$$

$$\bar{d}^{\,l-1}_{l-1,m} = \sqrt{\frac{(2l-2)!}{(l-m-1)!(l+m-1)!}}\, d^{\,l-1}_{l-1,m}, \quad \bar{d}^{\,l-1}_{-l+1,m} = \sqrt{\frac{(2l-2)!}{(l-m-1)!(l+m-1)!}}\, d^{\,l-1}_{-l+1,m}.$$

Derivation of recurrence relations for bar-scaled matrix elements starts from Equations 7.6, 7.12, 7.17, and Equation 7.19. Another equation that may also be employed is Equation 7.22. However, because it is similar to Equation 7.19, only one of them is to be included in the derivation; the other one will follow from the former equation by analogy, since they share the same set of scaling coefficients. Derivation proceeds by multiplying Equations 7.6, 7.12, and 7.17 by the factor $\sqrt{((l-k)!(l+k)!)/((l-m)!(l+m)!)}$ and the Equation 7.19 by $\sqrt{((l-k)!(l+k)!)/(2l)!}$. After that, making the substitution defined in Equation 14.18 and dividing both sides of the equation by the factor positioned in the right-hand side gives:

$$\begin{aligned}
&\sqrt{\frac{(l+m-1)(l+m)}{(l+k-1)(l+k)}}\sqrt{\frac{(l-k)!(l+k)!}{(l-m)!(l+m)!}}\left(\frac{1+\cos\beta}{2}\right) d^{\,l-1}_{k-1,m-1} \\
&\quad -\sqrt{\frac{(l-m)(l+m)}{(l+k-1)(l+k)}}\sqrt{\frac{(l-k)!(l+k)!}{(l-m)!(l+m)!}}(\sin\beta)\, d^{\,l-1}_{k-1,m} \\
&\quad +\sqrt{\frac{(l-m-1)(l-m)}{(l+k-1)(l+k)}}\sqrt{\frac{(l-k)!(l+k)!}{(l-m)!(l+m)!}}\left(\frac{1-\cos\beta}{2}\right) d^{\,l-1}_{k-1,m+1} = \bar{d}^{\,l}_{k,m},
\end{aligned} \tag{14.19}$$

$$\begin{aligned}
&\sqrt{\frac{(l+m-1)(l+m)}{(l-k)(l+k)}}\sqrt{\frac{(l-k)!(l+k)!}{(l-m)!(l+m)!}}\left(\frac{\sin\beta}{2}\right) d^{\,l-1}_{k,m-1} \\
&\quad +\sqrt{\frac{(l-m)(l+m)}{(l-k)(l+k)}}\sqrt{\frac{(l-k)!(l+k)!}{(l-m)!(l+m)!}}(\cos\beta)\, d^{\,l-1}_{k,m} \\
&\quad -\sqrt{\frac{(l-m-1)(l-m)}{(l-k)(l+k)}}\sqrt{\frac{(l-k)!(l+k)!}{(l-m)!(l+m)!}}\left(\frac{\sin\beta}{2}\right) d^{\,l-1}_{k,m+1} = \bar{d}^{\,l}_{k,m},
\end{aligned} \tag{14.20}$$

$$\begin{aligned}
&\sqrt{\frac{(l+m-1)(l+m)}{(l-k-1)(l-k)}}\sqrt{\frac{(l-k)!(l+k)!}{(l-m)!(l+m)!}}\left(\frac{1-\cos\beta}{2}\right) d^{\,l-1}_{k+1,m-1} \\
&\quad +\sqrt{\frac{(l-m)(l+m)}{(l-k-1)(l-k)}}\sqrt{\frac{(l-k)!(l+k)!}{(l-m)!(l+m)!}}(\sin\beta)\, d^{\,l-1}_{k+1,m} \\
&\quad +\sqrt{\frac{(l-m-1)(l-m)}{(l-k-1)(l-k)}}\sqrt{\frac{(l-k)!(l+k)!}{(l-m)!(l+m)!}}\left(\frac{1+\cos\beta}{2}\right) d^{\,l-1}_{k+1,m+1} = \bar{d}^{\,l}_{k,m},
\end{aligned} \tag{14.21}$$

and

$$\sqrt{\frac{(2l)(2l-1)}{(l-k)(l+k)}}\sqrt{\frac{(l-k)!(l+k)!}{(2l)!}}\left(\frac{\sin\beta}{2}\right) d^{\,l-1}_{k,l-1} = \bar{d}^{\,l}_{k,l}. \tag{14.22}$$

This operation results in the replacement of matrix element $d^{\,l}_{k,m}$ with $\bar{d}^{\,l}_{k,m}$, and $d^{\,l}_{k,l}$ with $\bar{d}^{\,l}_{k,l}$ on the right-hand side of the equations.

Continuing the transformation, the left-hand side is simplified as follows:

$$
\begin{aligned}
&\sqrt{\frac{(l-k)!(l+k-2)!}{(l-m)!(l+m-2)!}}\left(\frac{1+\cos\beta}{2}\right)d^{l-1}_{k-1,m-1}-\sqrt{\frac{(l-k)!(l+k-2)!}{(l-m-1)!(l+m-1)!}}(\sin\beta)\,d^{l-1}_{k-1,m}\\
&\quad+\sqrt{\frac{(l-k)!(l+k-2)!}{(l-m-2)!(l+m)!}}\left(\frac{1-\cos\beta}{2}\right)d^{l-1}_{k-1,m+1}=\bar{d}^{l}_{k,m},
\end{aligned}
\tag{14.23}
$$

$$
\begin{aligned}
&\sqrt{\frac{(l-k-1)!(l+k-1)!}{(l-m)!(l+m-2)!}}\left(\frac{\sin\beta}{2}\right)d^{l-1}_{k,m-1}+\sqrt{\frac{(l-k-1)!(l+k-1)!}{(l-m-1)!(l+m-1)!}}(\cos\beta)\,d^{l-1}_{k,m}\\
&\quad-\sqrt{\frac{(l-k-1)!(l+k-1)!}{(l-m-2)!(l+m)!}}\left(\frac{\sin\beta}{2}\right)d^{l-1}_{k,m+1}=\bar{d}^{l}_{k,m},
\end{aligned}
\tag{14.24}
$$

$$
\begin{aligned}
&\sqrt{\frac{(l-k-2)!(l+k)!}{(l-m)!(l+m-2)!}}\left(\frac{1-\cos\beta}{2}\right)d^{l-1}_{k+1,m-1}+\sqrt{\frac{(l-k-2)!(l+k)!}{(l-m-1)!(l+m-1)!}}(\sin\beta)\,d^{l-1}_{k+1,m}\\
&\quad+\sqrt{\frac{(l-k-2)!(l+k)!}{(l-m-2)!(l+m)!}}\left(\frac{1+\cos\beta}{2}\right)d^{l-1}_{k+1,m+1}=\bar{d}^{l}_{k,m},
\end{aligned}
\tag{14.25}
$$

$$
\sqrt{\frac{(l-k-1)!(l+k-1)!}{(2l-2)!}}\left(\frac{\sin\beta}{2}\right)d^{l-1}_{k,l-1}=\bar{d}^{l}_{k,l}.
\tag{14.26}
$$

The next step in the transformation is to convert the matrix elements on the left-hand side into their scaled form. Matching the square-root coefficients in the above recurrence relations to those in Equation 14.18 finishes the conversion. Combining these results with those for $m=\pm l$ leads to:

$$
\left(\frac{1+\cos\beta}{2}\right)\bar{d}^{l-1}_{k-1,m-1}-(\sin\beta)\bar{d}^{l-1}_{k-1,m}+\left(\frac{1-\cos\beta}{2}\right)\bar{d}^{l-1}_{k-1,m+1}=\bar{d}^{l}_{k,m},
\tag{14.27}
$$

$$
\left(\frac{\sin\beta}{2}\right)\bar{d}^{l-1}_{k,m-1}+(\cos\beta)\,\bar{d}^{l-1}_{k,m}-\left(\frac{\sin\beta}{2}\right)\bar{d}^{l-1}_{k,m+1}=\bar{d}^{l}_{k,m},
\tag{14.28}
$$

$$
\left(\frac{1-\cos\beta}{2}\right)\bar{d}^{l-1}_{k+1,m-1}+(\sin\beta)\,\bar{d}^{l-1}_{k+1,m}+\left(\frac{1+\cos\beta}{2}\right)\bar{d}^{l-1}_{k+1,m+1}=\bar{d}^{l}_{k,m},
\tag{14.29}
$$

$m=l$ and $k=l$:

$$
\frac{1+\cos\beta}{2}\,\bar{d}^{l-1}_{l-1,l-1}=\bar{d}^{l}_{l,l},
\tag{14.30}
$$

$m=l$ and $l-1\geq k\geq -l+1$:

$$
\left(\frac{\sin\beta}{2}\right)\bar{d}^{l-1}_{k,l-1}=\bar{d}^{l}_{k,l},
\tag{14.31}
$$

$m=l$ and $k=-l$:

$$
\frac{1-\cos\beta}{2}\,\bar{d}^{l-1}_{-l+1,l-1}=\bar{d}^{l}_{-l,l},
\tag{14.32}
$$

$m = -l$ and $k = l$:

$$\frac{1-\cos\beta}{2}\,\bar{d}^{l-1}_{l-1,-l+1} = \bar{d}^{l}_{l,-l}, \tag{14.33}$$

$m = -l$ and $l - 1 \geq k \geq -l + 1$:

$$-\left(\frac{\sin\beta}{2}\right)\bar{d}^{l-1}_{k,-l+1} = \bar{d}^{l}_{k,-l}, \tag{14.34}$$

$m = -l$ and $k = -l$:

$$\frac{1+\cos\beta}{2}\,\bar{d}^{l-1}_{-l+1,-l+1} = \bar{d}^{l}_{-l,-l}, \tag{14.35}$$

where Equation 14.31 follows from Equation 14.26. Equation 14.34 is obtained using the same steps as Equation 14.31 because its scale factor remains unchanged for $\pm m$. The other relations for $|m| = |k|$ follow directly because their scale factors are all unity.

These recurrence relationships for bar-scaled Wigner matrix elements are remarkably compact, contain no square roots and do not involve division of indices, and, therefore, are more easily computed than the original recurrence relations for unscaled matrix elements.

The complementary recurrence relations that remain to be derived are those for tilde-scaled Wigner matrix elements. The first step in the derivation process is to obtain tilde-scaled expressions for individual matrix elements present in Equations 7.6, 7.12, 7.17, and 7.19. Converting all matrix elements that are present in those equations into the tilde scaled form by applying the tilde-substitution from Equation 14.16 leads to the following transformations:

$$\begin{aligned}
\tilde{d}^{l-1}_{k-1,m-1} &= \sqrt{\frac{(l-m)!(l+m-2)!}{(l-k)!(l+k-2)!}}\,d^{l-1}_{k-1,m-1}, & \tilde{d}^{l-1}_{k-1,m} &= \sqrt{\frac{(l-m-1)!(l+m-1)!}{(l-k)!(l+k-2)!}}\,d^{l-1}_{k-1,m}, \\
& & \tilde{d}^{l-1}_{k-1,m+1} &= \sqrt{\frac{(l-m-2)!(l+m)!}{(l-k)!(l+k-2)!}}\,d^{l-1}_{k-1,m+1}, \\
\tilde{d}^{l-1}_{k,m-1} &= \sqrt{\frac{(l-m)!(l+m-2)!}{(l-k-1)!(l+k-1)!}}\,d^{l-1}_{k,m-1}, & \tilde{d}^{l-1}_{k,m} &= \sqrt{\frac{(l-m-1)!(l+m-1)!}{(l-k-1)!(l+k-1)!}}\,d^{l-1}_{k,m}, \\
& & \tilde{d}^{l-1}_{k,m+1} &= \sqrt{\frac{(l-m-2)!(l+m)!}{(l-k-1)!(l+k-1)!}}\,d^{l-1}_{k,m+1}, \\
\tilde{d}^{l-1}_{k+1,m-1} &= \sqrt{\frac{(l-m)!(l+m-2)!}{(l-k-2)!(l+k)!}}\,d^{l-1}_{k+1,m-1}, & \tilde{d}^{l-1}_{k+1,m} &= \sqrt{\frac{(l-m-1)!(l+m-1)!}{(l-k-2)!(l+k)!}}\,d^{l-1}_{k+1,m}, \\
& & \tilde{d}^{l-1}_{k+1,m+1} &= \sqrt{\frac{(l-m-2)!(l+m)!}{(l-k-2)!(l+k)!}}\,d^{l-1}_{k+1,m+1}, \\
\tilde{d}^{l-1}_{k,l-1} &= \sqrt{\frac{(2l-2)!}{(l-k-1)!(l+k-1)!}}\,d^{l}_{k,l-1}, & \tilde{d}^{l-1}_{k,-l+1} &= \sqrt{\frac{(2l-2)!}{(l-k-1)!(l+k-1)!}}\,d^{l}_{k,-l+1}, \\
\tilde{d}^{l}_{k,l} &= \sqrt{\frac{(2l)!}{(l-k)!(l+k)!}}\,d^{l}_{k,l}, & \tilde{d}^{l}_{k,-l} &= \sqrt{\frac{(2l)!}{(l-k)!(l+k)!}}\,d^{l}_{k,-l}, \\
\tilde{d}^{l}_{l,m} &= \sqrt{\frac{(l-m)!(l+m)!}{(2l)!}}\,d^{l}_{l,m}, & \tilde{d}^{l}_{-l,m} &= \sqrt{\frac{(l-m)!(l+m)!}{(2l)!}}\,d^{l}_{-l,m}, \\
\tilde{d}^{l}_{l-1,m} &= \sqrt{\frac{(l-m)!(l+m)!}{(2l-1)!}}\,d^{l}_{l-1,m}, & \tilde{d}^{l}_{-l+1,m} &= \sqrt{\frac{(l-m)!(l+m)!}{(2l-1)!}}\,d^{l}_{-l+1,m}, \\
\tilde{d}^{l-1}_{l-1,m} &= \sqrt{\frac{(l-m-1)!(l+m-1)!}{(2l-2)!}}\,d^{l-1}_{l-1,m}, & \tilde{d}^{l-1}_{-l+1,m} &= \sqrt{\frac{(l-m-1)!(l+m-1)!}{(2l-2)!}}\,d^{l-1}_{-l+1,m}.
\end{aligned} \tag{14.36}$$

Conversion of the original recurrence relation to the tilde-scaled form involves multiplying Equation 7.6, 7.12, and 7.17 by $\sqrt{((l-m)!(l+m)!)\,/\,((l-k)!(l+k)!)}$, and Equation 7.19 by $\sqrt{(2l)!/((l-k)!(l+k)!)}$. This, together with dividing both sides of the equation with the coefficient from the right-hand side, leads to:

$$\begin{aligned}&\sqrt{\frac{(l+m-1)(l+m)}{(l+k-1)(l+k)}}\sqrt{\frac{(l-m)!(l+m)!}{(l-k)!(l+k)!}}\left(\frac{1+\cos\beta}{2}\right)d^{l-1}_{k-1,m-1}\\&\quad-\sqrt{\frac{(l-m)(l+m)}{(l+k-1)(l+k)}}\sqrt{\frac{(l-m)!(l+m)!}{(l-k)!(l+k)!}}(\sin\beta)\,d^{l-1}_{k-1,m}\\&\quad+\sqrt{\frac{(l-m-1)(l-m)}{(l+k-1)(l+k)}}\sqrt{\frac{(l-m)!(l+m)!}{(l-k)!(l+k)!}}\left(\frac{1-\cos\beta}{2}\right)d^{l-1}_{k-1,m+1}=\tilde{d}^{l}_{k,m},\end{aligned}\tag{14.37}$$

$$\begin{aligned}&\sqrt{\frac{(l+m-1)(l+m)}{(l-k)(l+k)}}\sqrt{\frac{(l-m)!(l+m)!}{(l-k)!(l+k)!}}\left(\frac{\sin\beta}{2}\right)d^{l-1}_{k,m-1}\\&\quad+\sqrt{\frac{(l-m)(l+m)}{(l-k)(l+k)}}\sqrt{\frac{(l-m)!(l+m)!}{(l-k)!(l+k)!}}(\cos\beta)\,d^{l-1}_{k,m}\\&\quad-\sqrt{\frac{(l-m-1)(l-m)}{(l-k)(l+k)}}\sqrt{\frac{(l-m)!(l+m)!}{(l-k)!(l+k)!}}\left(\frac{\sin\beta}{2}\right)d^{l-1}_{k,m+1}=\tilde{d}^{l}_{k,m},\end{aligned}\tag{14.38}$$

$$\begin{aligned}&\sqrt{\frac{(l+m-1)(l+m)}{(l-k-1)(l-k)}}\sqrt{\frac{(l-m)!(l+m)!}{(l-k)!(l+k)!}}\left(\frac{1-\cos\beta}{2}\right)d^{l-1}_{k+1,m-1}\\&\quad+\sqrt{\frac{(l-m)(l+m)}{(l-k-1)(l-k)}}\sqrt{\frac{(l-m)!(l+m)!}{(l-k)!(l+k)!}}(\sin\beta)\,d^{l-1}_{k+1,m}\\&\quad+\sqrt{\frac{(l-m-1)(l-m)}{(l-k-1)(l-k)}}\sqrt{\frac{(l-m)!(l+m)!}{(l-k)!(l+k)!}}\left(\frac{1+\cos\beta}{2}\right)d^{l-1}_{k+1,m+1}=\tilde{d}^{l}_{k,m},\end{aligned}\tag{14.39}$$

$$\sqrt{\frac{(2l)(2l-1)}{(l-k)(l+k)}}\sqrt{\frac{(2l)!}{(l-k)!(l+k)!}}\left(\frac{\sin\beta}{2}\right)d^{l-1}_{k,l-1}=\tilde{d}^{l}_{k,l}.\tag{14.40}$$

This multiplication turns the matrix elements $d^{l}_{k,m}$ and $d^{l}_{k,l}$ on the right-hand side to $\tilde{d}^{l}_{k,m}$ and $\tilde{d}^{l}_{k,l}$, respectively. Next, simplifying the left-hand side of the equations produces:

$$\begin{aligned}&\frac{(l+m-1)(l+m)}{(l+k-1)(l+k)}\sqrt{\frac{(l-m)!(l+m-2)!}{(l-k)!(l+k-2)!}}\left(\frac{1+\cos\beta}{2}\right)d^{l-1}_{k-1,m-1}\\&\quad-\frac{(l-m)(l+m)}{(l+k-1)(l+k)}\sqrt{\frac{(l-m-1)!(l+m-1)!}{(l-k)!(l+k-2)!}}(\sin\beta)\,d^{l-1}_{k-1,m}\\&\quad+\frac{(l-m-1)(l-m)}{(l+k-1)(l+k)}\sqrt{\frac{(l-m-2)!(l+m)!}{(l-k)!(l+k)!}}\left(\frac{1-\cos\beta}{2}\right)d^{l-1}_{k-1,m+1}=\tilde{d}^{l}_{k,m},\end{aligned}\tag{14.41}$$

$$\begin{aligned}&\frac{(l+m-1)(l+m)}{(l-k)(l+k)}\sqrt{\frac{(l-m)!(l+m-2)!}{(l-k-1)!(l+k-1)!}}\left(\frac{\sin\beta}{2}\right)d^{l-1}_{k,m-1}\\&\quad+\frac{(l-m)(l+m)}{(l-k)(l+k)}\sqrt{\frac{(l-m-1)!(l+m-1)!}{(l-k-1)!(l+k-1)!}}(\cos\beta)\,d^{l-1}_{k,m}\\&\quad-\frac{(l-m-1)(l-m)}{(l-k)(l+k)}\sqrt{\frac{(l-m-2)!(l+m)!}{(l-k-1)!(l+k-1)!}}\left(\frac{\sin\beta}{2}\right)d^{l-1}_{k,m+1}=\tilde{d}^{l}_{k,m},\end{aligned}\tag{14.42}$$

$$\frac{(l+m-1)(l+m)}{(l-k-1)(l-k)}\sqrt{\frac{(l-m)!(l+m-2)!}{(l-k-2)!(l+k)!}}\left(\frac{1-\cos\beta}{2}\right)d^{l-1}_{k+1,m-1}$$
$$+\frac{(l-m)(l+m)}{(l-k-1)(l-k)}\sqrt{\frac{(l-m-1)!(l+m-1)!}{(l-k-2)!(l+k)!}}(\sin\beta)\,d^{l-1}_{k+1,m} \tag{14.43}$$
$$+\frac{(l-m-1)(l-m)}{(l-k-1)(l-k)}\sqrt{\frac{(l-m-2)!(l+m)!}{(l-k-2)!(l+k)!}}\left(\frac{1+\cos\beta}{2}\right)d^{l-1}_{k+1,m+1}=\tilde{d}^{l}_{k,m},$$

$$(2l)(2l-1)\sqrt{\frac{(2l-2)!}{(l-k-1)!(l+k-1)!}}\left(\frac{\sin\beta}{2}\right)d^{l-1}_{k,l-1}=\tilde{d}^{l}_{k,l}. \tag{14.44}$$

Finally, converting the unscaled matrix elements in the above equations into a tilde-scaled form based on Equation 14.36, and including the cases for $m=\pm l$ gives the desired recurrence relations for tilde-scaled Wigner matrix elements:

$$\frac{(l+m-1)(l+m)}{(l+k-1)(l+k)}\left(\frac{1+\cos\beta}{2}\right)\tilde{d}^{l-1}_{k-1,m-1}-\frac{(l-m)(l+m)}{(l+k-1)(l+k)}(\sin\beta)\,\tilde{d}^{l-1}_{k-1,m}$$
$$+\frac{(l-m-1)(l-m)}{(l+k-1)(l+k)}\left(\frac{1-\cos\beta}{2}\right)\tilde{d}^{l-1}_{k-1,m+1}=\tilde{d}^{l}_{k,m}, \tag{14.45}$$

$$\frac{(l+m-1)(l+m)}{(l-k)(l+k)}\left(\frac{\sin\beta}{2}\right)\tilde{d}^{l-1}_{k,m-1}+\frac{(l-m)(l+m)}{(l-k)(l+k)}(\cos\beta)\,\tilde{d}^{l-1}_{k,m}$$
$$-\frac{(l-m-1)(l-m)}{(l-k)(l+k)}\left(\frac{\sin\beta}{2}\right)\tilde{d}^{l-1}_{k,m+1}=\tilde{d}^{l}_{k,m}, \tag{14.46}$$

$$\frac{(l+m-1)(l+m)}{(l-k-1)(l-k)}\left(\frac{1-\cos\beta}{2}\right)\tilde{d}^{l-1}_{k+1,m-1}+\frac{(l-m)(l+m)}{(l-k-1)(l-k)}(\sin\beta)\,\tilde{d}^{l-1}_{k+1,m}$$
$$+\frac{(l-m-1)(l-m)}{(l-k-1)(l-k)}\left(\frac{1+\cos\beta}{2}\right)\tilde{d}^{l-1}_{k+1,m+1}=\tilde{d}^{l}_{k,m}, \tag{14.47}$$

$m=l$ and $k=l$:

$$\frac{1+\cos\beta}{2}\,\tilde{d}^{l-1}_{l-1,l-1}=\tilde{d}^{l}_{l,l}, \tag{14.48}$$

$m=l$ and $l-1\geq k\geq -l+1$:

$$\frac{(2l)(2l-1)}{(l-k)(l+k)}\left(\frac{\sin\beta}{2}\right)\tilde{d}^{l-1}_{k,l-1}=\tilde{d}^{l}_{k,l}, \tag{14.49}$$

$m=l$ and $k=-l$:

$$\frac{1-\cos\beta}{2}\,\tilde{d}^{l-1}_{-l+1,l-1}=\tilde{d}^{l}_{-l,l}, \tag{14.50}$$

$m=-l$ and $k=l$:

$$\frac{1-\cos\beta}{2}\,\tilde{d}^{l-1}_{l-1,-l+1}=\tilde{d}^{l}_{l,-l}, \tag{14.51}$$

$m = -l$ and $l - 1 \geq k \geq -l + 1$:

$$-\frac{(2l)(2l-1)}{(l-k)(l+k)}\left(\frac{\sin\beta}{2}\right)\tilde{d}^{l-1}_{k,-l+1} = \tilde{d}^{l}_{k,-l}, \tag{14.52}$$

$m = -l$ and $k = -l$:

$$\frac{1+\cos\beta}{2}\tilde{d}^{l-1}_{-l+1,-l+1} = \tilde{d}^{l}_{-l,-l}. \tag{14.53}$$

Equation 14.49 follows from Equation 14.44, completing the derivation for $m = l$ and $l - 1 \geq k \geq -l + 1$, and, because the scale coefficients are identical for $\pm m$, the same derivation path also leads to Equation 14.52 for $m = -l$. Equations satisfying the condition of $|m| = |k|$ directly follow from the unscaled equations, because the scale coefficient is unity.

Comparing these equations with the original Equation 7.6, 7.12, and Equation 7.17 shows that tilde-scaling eliminates the square root operation in the left-hand side. Although the new equations are relatively bulky they require fewer floating-point operations than the original equations and therefore they represent a valuable improvement.

To summarize this section, scaling applied to Wigner matrix elements reduces the cost of recursive computation of the matrix elements especially in the bar-scaled form given by Equations 14.27–14.35, with a somewhat lesser but still noticeable improvement for tilde-scaled matrix elements defined by Equations 14.45–14.53. The improvement due to scaling is actually two-fold since in addition it allows the use of the efficient Equation 14.17 for rotation of scaled solid harmonics.

14.4 Computation of Scaled Wigner Matrix Elements with Increment in Index *k*

Sections 7.2 introduced a set of alternative recurrence relations for unscaled Wigner matrix elements $d^{l}_{k,m}$:

$$\sqrt{l-k}(l\cos\beta + m)d^{l-1}_{k,m} - \sqrt{l+k-1}(l\sin\beta)d^{l-1}_{k-1,m} = \sqrt{(l+k)(l-m)(l+m)}\,d^{l}_{k,m}, \tag{7.48}$$

$$\sqrt{l-k-1}(l\sin\beta)d^{l-1}_{k+1,m} + \sqrt{l+k}(l\cos\beta - m)d^{l-1}_{k,m} = \sqrt{(l-k)(l-m)(l+m)}\,d^{l}_{k,m}, \tag{7.58}$$

which employ increment in index k. These equations lead to a few others, which will be handy in computer code implementation:

$k = l$:

$$-\sqrt{2l-1}(l\sin\beta)\,d^{l-1}_{l-1,m} = \sqrt{(2l)(l-m)(l+m)}\,d^{l}_{l,m}, \tag{7.49}$$

$k = l - 1$:

$$\sqrt{2l-1}(l\cos\beta - m)\,d^{l-1}_{l-1,m} = \sqrt{(l-m)(l+m)}\,d^{l}_{l-1,m}, \tag{7.59}$$

$k = -l + 1$:

$$\sqrt{2l-1}(l\cos\beta + m)\,d^{l-1}_{-l+1,m} = \sqrt{(l-m)(l+m)}\,d^{l}_{-l+1,m}, \tag{7.50}$$

$k = -l$:

$$\sqrt{2l-1}(l\sin\beta)\,d^{l-1}_{-l+1,m} = \sqrt{(2l)(l-m)(l+m)}\,d^{l}_{-l,m}. \tag{7.60}$$

The objective at this point is to convert Equations 7.48, 7.58, 7.715, and 7.59 to bar- and tilde-scaled forms so that they can be used for performing rotation of the scaled multipole expansions according to Equation 14.17. The remaining recurrence relations Equation 7.50 and 7.60 follow from the transformed forms of Equation 7.59 and 7.49, respectively, since they share similar scaling coefficients.

Conversion to the bar-scaled form starts with help of Equations 14.16 and 14.18. Multiplying the recurrence relations by $\sqrt{((l-k)!(l+k)!)/((l-m)!(l+m)!)}$ and simultaneously dividing each line by the square root coefficient on the right-hand side gives:

$$\begin{aligned}&\sqrt{\frac{l-k}{(l+k)(l-m)(l+m)}}\sqrt{\frac{(l-k)!(l+k)!}{(l-m)!(l+m)!}}(l\cos\beta+m)d^{l-1}_{k,m}\\&-\sqrt{\frac{l+k-1}{(l+k)(l-m)(l+m)}}\sqrt{\frac{(l-k)!(l+k)!}{(l-m)!(l+m)!}}(l\sin\beta)d^{l-1}_{k-1,m}=\bar{d}^{l}_{k,m},\end{aligned}\tag{14.54}$$

$$\begin{aligned}&\sqrt{\frac{l-k-1}{(l-k)(l-m)(l+m)}}\sqrt{\frac{(l-k)!(l+k)!}{(l-m)!(l+m)!}}(l\sin\beta)d^{l-1}_{k+1,m}\\&+\sqrt{\frac{l+k}{(l-k)(l-m)(l+m)}}\sqrt{\frac{(l-k)!(l+k)!}{(l-m)!(l+m)!}}(l\cos\beta-m)d^{l-1}_{k,m}=\bar{d}^{l}_{k,m},\end{aligned}\tag{14.55}$$

$k=l$:

$$-\sqrt{\frac{2l-1}{(2l)(l-m)(l+m)}}\sqrt{\frac{(2l)!}{(l-m)!(l+m)!}}(l\sin\beta)d^{l-1}_{l-1,m}=\bar{d}^{l}_{l,m},\tag{14.56}$$

$k=l-1$:

$$\sqrt{\frac{2l-1}{(l-m)(l+m)}}\sqrt{\frac{(2l-1)!}{(l-m)!(l+m)!}}(l\cos\beta-m)d^{l-1}_{l-1,m}=\bar{d}^{l}_{l-1,m}.\tag{14.57}$$

This operation turns the matrix element $d^{l}_{k,m}$ into $\bar{d}^{l}_{k,m}$ in the right-hand side of the equations.

Next, the square root coefficients in the left-hand side are simplified to give:

$$\begin{aligned}&\frac{(l-k)}{(l-m)(l+m)}\sqrt{\frac{(l-k-1)!(l+k-1)!}{(l-m-1)!(l+m-1)!}}(l\cos\beta+m)d^{l-1}_{k,m}\\&-\frac{(l+k-1)}{(l-m)(l+m)}\sqrt{\frac{(l-k)!(l+k-2)!}{(l-m-1)!(l+m-1)!}}(l\sin\beta)d^{l-1}_{k-1,m}=\bar{d}^{l}_{k,m},\end{aligned}\tag{14.58}$$

$$\begin{aligned}&\frac{(l-k-1)}{(l-m)(l+m)}\sqrt{\frac{(l-k-2)!(l+k)!}{(l-m-1)!(l+m-1)!}}(l\sin\beta)d^{l-1}_{k+1,m}\\&+\frac{(l+k)}{(l-m)(l+m)}\sqrt{\frac{(l-k-1)!(l+k-1)!}{(l-m-1)!(l+m-1)!}}(l\cos\beta-m)d^{l-1}_{k,m}=\bar{d}^{l}_{k,m},\end{aligned}\tag{14.59}$$

$k=l$:

$$-\frac{2l-1}{(l-m)(l+m)}\sqrt{\frac{(2l-2)!}{(l-m-1)!(l+m-1)!}}(l\sin\beta)d^{l-1}_{l-1,m}=\bar{d}^{l}_{l,m},\tag{14.60}$$

$k=l-1$:

$$\frac{2l-1}{(l-m)(l+m)}\sqrt{\frac{(2l-2)!}{(l-m-1)!(l+m-1)!}}(l\cos\beta-m)d^{l-1}_{l-1,m}=\bar{d}^{l}_{l-1,m}.\tag{14.61}$$

Finally, matching the square root coefficients in the above equations to those in Equations 14.16 and 14.18 converts the equations into the recurrence relation for bar-scaled Wigner matrix elements with increment in index k:

$$\frac{(l-k)}{(l-m)(l+m)}(l\cos\beta+m)\bar{d}^{l-1}_{k,m}-\frac{(l+k-1)}{(l-m)(l+m)}(l\sin\beta)\bar{d}^{l-1}_{k-1,m}=\bar{d}^{l}_{k,m}, \tag{14.62}$$

$$\frac{(l-k-1)}{(l-m)(l+m)}(l\sin\beta)\bar{d}^{l-1}_{k+1,m}+\frac{(l+k)}{(l-m)(l+m)}(l\cos\beta-m)\bar{d}^{l-1}_{k,m}=\bar{d}^{l}_{k,n} \tag{14.63}$$

$k = l$:

$$-\frac{2l-1}{(l-m)(l+m)}(l\sin\beta)\bar{d}^{l-1}_{l-1,m}=\bar{d}^{l}_{l,m}, \tag{14.64}$$

$k = l - 1$:

$$\frac{2l-1}{(l-m)(l+m)}(l\cos\beta-m)\bar{d}^{l-1}_{l-1,m}=\bar{d}^{l}_{l-1,m}, \tag{14.65}$$

$k = -l + 1$:

$$\frac{2l-1}{(l-m)(l+m)}(l\cos\beta+m)\bar{d}^{l-1}_{-l+1,m}=\bar{d}^{l}_{-l+1,m}, \tag{14.66}$$

$k = -l$:

$$\frac{2l-1}{(l-m)(l+m)}(l\sin\beta)\bar{d}^{l-1}_{-l+1,m}=\bar{d}^{l}_{-l,m}, \tag{14.67}$$

where the last two recurrence relations follow from Equations 7.50 and 7.60 by analogy with Equations 14.65 and 14.64, respectively. Conveniently, application of bar-scaling to matrix elements eliminates the use of the square-root operations in the recurrence relations.

The next task is to convert the original recurrence relations into a tilde-scaled form suitable for rotating the tilde-scaled irregular multipole expansions. This conversion involves multiplying Equations 7.48, 7.58, 7.49, and 7.59 by $\sqrt{((l-m)!(l+m)!)/((l-k)!(l+k)!)}$. Simultaneously dividing each line by the coefficient from its right-hand side leads to:

$$\begin{aligned}&\sqrt{\frac{l-k}{(l+k)(l-m)(l+m)}}\sqrt{\frac{(l-m)!(l+m)!}{(l-k)!(l+k)!}}(l\cos\beta+m)d^{l-1}_{k,m}\\&-\sqrt{\frac{l+k-1}{(l+k)(l-m)(l+m)}}\sqrt{\frac{(l-m)!(l+m)!}{(l-k)!(l+k)!}}(l\sin\beta)d^{l-1}_{k-1,m}=\tilde{d}^{l}_{k,m},\end{aligned} \tag{14.68}$$

$$\begin{aligned}&\sqrt{\frac{l-k-1}{(l-k)(l-m)(l+m)}}\sqrt{\frac{(l-m)!(l+m)!}{(l-k)!(l+k)!}}(l\sin\beta)d^{l-1}_{k+1,m}\\&+\sqrt{\frac{l+k}{(l-k)(l-m)(l+m)}}\sqrt{\frac{(l-m)!(l+m)!}{(l-k)!(l+k)!}}(l\cos\beta-m)d^{l-1}_{k,m}=\tilde{d}^{l}_{k,m},\end{aligned} \tag{14.69}$$

$k = l$:

$$-\sqrt{\frac{2l-1}{(2l)(l-m)(l+m)}}\sqrt{\frac{(l-m)!(l+m)!}{(2l)!}}(l\sin\beta)d^{l-1}_{l-1,m} = \tilde{d}^{l}_{l,m}, \tag{14.70}$$

$k = l - 1$:

$$\sqrt{\frac{2l-1}{(l-m)(l+m)}}\sqrt{\frac{(l-m)!(l+m)!}{(2l-1)!}}(l\cos\beta - m)d^{l-1}_{l-1,m} = \tilde{d}^{l}_{l-1,m}. \tag{14.71}$$

This operation turns the matrix element $d^l_{k,m}$ into $\tilde{d}^l_{k,m}$ in the right-hand side of the equations.

Simplifying the left-hand side in the above recurrence relations produces:

$$\begin{aligned}&\frac{1}{(l+k)}\sqrt{\frac{(l-m-1)!(l+m-1)!}{(l-k-1)!(l+k-1)!}}(l\cos\beta + m)d^{l-1}_{k,m}\\&-\frac{1}{(l+k)}\sqrt{\frac{(l-m-1)!(l+m-1)!}{(l-k)!(l+k-2)!}}(l\sin\beta)d^{l-1}_{k-1,m} = \tilde{d}^{l}_{k,m},\end{aligned} \tag{14.72}$$

$$\begin{aligned}&\frac{1}{(l-k)}\sqrt{\frac{(l-m-1)!(l+m-1)!}{(l-k-2)!(l+k)!}}(l\sin\beta)d^{l-1}_{k+1,m}\\&+\frac{1}{(l-k)}\sqrt{\frac{(l-m-1)!(l+m-1)!}{(l-k-1)!(l+k-1)!}}(l\cos\beta - m)d^{l-1}_{k,m} = \tilde{d}^{l}_{k,m},\end{aligned} \tag{14.73}$$

$k = l$:

$$-\sqrt{\frac{(l-m-1)!(l+m-1)!}{(2l-2)!}}\left(\frac{\sin\beta}{2}\right)d^{l-1}_{l-1,m} = \tilde{d}^{l}_{l,m}, \tag{14.74}$$

$k = l - 1$:

$$\sqrt{\frac{(l-m-1)!(l+m-1)!}{(2l-2)!}}(l\cos\beta - m)d^{l-1}_{l-1,m} = \tilde{d}^{l}_{l-1,m}. \tag{14.75}$$

Matching the square root coefficients in the above expressions to those in Equation 14.36 finally delivers the recurrence relations for tilde-scaled Wigner matrix elements:

$$\frac{1}{(l+k)}(l\cos\beta + m)\tilde{d}^{l-1}_{k,m} - \frac{1}{(l+k)}(l\sin\beta)\tilde{d}^{l-1}_{k-1,m} = \tilde{d}^{l}_{k,m}, \tag{14.76}$$

$$\frac{1}{(l-k)}(l\sin\beta)\tilde{d}^{l-1}_{k+1,m} + \frac{1}{(l-k)}(l\cos\beta - m)\tilde{d}^{l-1}_{k,m} = \tilde{d}^{l}_{k,m}, \tag{14.77}$$

$k = l$:

$$-\left(\frac{\sin\beta}{2}\right)\tilde{d}^{l-1}_{l-1,m} = \tilde{d}^{l}_{l,m}, \tag{14.78}$$

$k = l - 1$:

$$(l\cos\beta - m)\tilde{d}^{l-1}_{l-1,m} = \tilde{d}^{l}_{l-1,m} \tag{14.79}$$

$k = -l + 1$:

$$(l \cos\beta + m)\tilde{d}^{l-1}_{-l+1,m} = \tilde{d}^{l}_{-l+1,m}, \tag{14.80}$$

$k = -l$:

$$\left(\frac{\sin\beta}{2}\right)\tilde{d}^{l-1}_{-l+1,m} = \tilde{d}^{l}_{-l,m}. \tag{14.81}$$

In these equations the square root operations are eliminated, providing a more computationally efficient formulation than the original unscaled equations Equations 7.48 and 7.58.

Having obtained two alternative sets of recurrence relations for scaled Wigner matrix elements with increment in different indices provides the opportunity to compare them and determine their relative efficiency for rotation of scaled multipole expansions. To refer to these two sets without reciting each time their full name, it is useful to devise a simple naming convention. Accordingly, the first set derived in Section 14.3 that employs increment in index m receives the name m-set. The second set derived in this Section 14.4 that applies increment to index k will likewise be called a k-set. The primary advantage of scaling is the simplification of the rotation equation that results from the replacement of the cumbersome Equation 14.15 by the more efficient Equation 14.17. Both sets of recurrence relations benefit from this scaling, so the next step is to compare them, and determine which one is the more efficient.

A direct approach to compare computational efficiency of two equations is to count the number of floating-point operations that they involve. In its simplest form, this consists of counting the number of addition, subtraction, multiplication, and division operations between the variables of floating-point type in the equation. A manipulation between an integral number and a floating-point number also counts as a floating-point operation since the result of such a process is a floating-point number. The computational cost of multiplication, addition, and subtraction between two variables of integer type can be ignored since such operations are considerably faster than the operation involving floating-point variables. Conversely, division between two integral numbers should be regarded as a floating-point operation in computer code, because if programmed differently it may result in loss of precision. Finally, the scaled Wigner matrices $\bar{\mathbf{d}}^{l-1}$ and $\tilde{\mathbf{d}}^{l-1}$ and trigonometric functions $\sin\beta$ and $\cos\beta$ that are present in the recurrence relations come precomputed, so their retrieval from computer memory is not counted toward the number of floating-point operations.

Applying the counting procedure to Equations 14.27, 14.28, 14.29, 14.45, 14.46, and 14.47, which are the most general equations in the m-set of recurrence relations, gives 9, 7, 9, 15, 11, and 15 floating-point operations, respectively. Likewise, applying the same procedure to Equations 14.62, 14.63, 14.76, and 14.77 in the k-set produces 9, 9, 8, and 8 floating-point operations, respectively. To obtain a more detailed picture, it is convenient to compare the cost in the subsets of equations. For the bar-scaled subset, the computational cost is 9, 7, 9 *vs.* 9, 9 floating-point operations in the m- and k-sets, respectively. Likewise, the computational cost in the tilde-scaled subset is 15, 11, 15 *vs.* 9, 9 for the m- and k-sets, respectively. According to these data, the computational cost in the bar-scaled subset of m- and k-sets is approximately the same. However, the tilde-scaled subset in the k-set needs significantly fewer floating-point operations than that in the m-set.

Another important factor in determining computational performance of an equation is the pattern of data retrieval of array elements from computer memory defined by the structure of that equation. Data retrieval cost depends on the stride between array elements. Retrieving array elements with unit stride fully utilizes the cache buffers, and, therefore, incurs the least cost. Equations that use a unit stride will be executed faster than the other equations that read array elements using large or variable stride. Therefore, maximizing the performance of an equation requires minimizing the stride in the data arrangement in computer memory.

Application of that strategy to computation of the Wigner matrix elements by using either the k- or m-sets, which employ multiple array elements with the increment along the k and m indices, means using column-major and row-major array storage formats, respectively, to place the increment along the faster running index. The other use of the Wigner matrix elements is in Equation 14.17 to perform rotations of the scaled multipole expansions. As this latter equation iterates over index k, it establishes the preferred storage format for the Wigner matrix elements to be column-major. Out of the two recurrence sets, the k-set for the recursive computation of the Wigner matrix elements naturally satisfies the preference of Equation 14.17 for the column-major format of the Wigner matrix, whereas the m-set may incur a small penalty, the extent of which depends on processor architecture and the size of cache memory.

Now that the optimal set of recurrence relations has been determined, a final remark needs to be made about the nature in which scaling affects the properties of the Wigner matrix. One needs to be aware that scaling breaks the normalization and symmetry property between the elements of the Wigner matrix, so that the matrix is no longer unitary. Without scaling, only the symmetry-independent portion of the matrix would need to be computed, with the rest of the matrix elements being generated by using symmetry conditions, but switching to the scaled form would require explicit evaluation of all the elements of the matrix. For instance, it is no longer possible to obtain the matrix $\mathbf{d}^l(-\beta)$ from $\mathbf{d}^l(\beta)$ by exchanging the column and row indices in the scaled matrix; rather the entire matrix needs to be computed from scratch. However, the gain in computational performance due to scaling, and the improved efficiency of data retrieval from computer memory outweighs the loss of symmetry.

14.5 Program Code for Computation of Scaled Wigner Matrix Elements Based on the *k*-set

Completion of the derivation of recurrence relations for computing the scaled Wigner matrix elements opens the path for writing a computer code. Before doing it, it is necessary to resolve a possible ambiguity in terminology. The name *Wigner matrix* collectively represents a set of regular matrices, one matrix per orbital index l, each of those also being called a Wigner matrix. To distinguish the reference to a single matrix from the reference to the entire set of matrices they will be called a Wigner matrix and a Wigner set, respectively. With the exception of the first matrix $\mathbf{d}^0$, which is a single number, all matrices $\mathbf{d}^l$ are square tables of the size $(2l+1) \times (2l+1)$, where $l = 0, 1, 2, \ldots, nL$; and $nL + 1$ is the number of the matrices. Each matrix in the Wigner set has an individual row and a column index. The row index k starts from the value $k = l$ in the top row of the matrix all the way down to $k = -l$, which marks the bottom row of the matrix. Likewise, column index m labels the left-most column in the table with the index value $m = l$, and descends to the value $m = -l$ on the right-most side of the matrix. These two index rules create the following table of matrix elements:

$$\mathbf{d}^l = \begin{array}{c|cccc} & m=l & l-1 & \ldots & -l \\ \hline k=l & d^l_{l,l} & d^l_{l,l-1} & \cdots & d^l_{l,-l} \\ l-1 & d^l_{l-1,l} & d^l_{l-1,l-1} & \cdots & d^l_{l-1,-l} \\ \ldots & \ldots & \ldots & \ldots & \ldots \\ -l & d^l_{-l,l} & d^l_{-l,l-1} & \cdots & d^l_{-l,-l} \end{array}. \tag{14.82}$$

Storing the Wigner set in a computer memory implies dealing with $nL + 1$ individual matrices $\mathbf{d}^l$ starting from the first matrix $\mathbf{d}^0$ for $l = 0$ all the way up to the last matrix $\mathbf{d}^{nL}$ for $l = nL$, where nL is the upper value of the orbital index in the truncated multipole expansion. The preferred way to store each individual matrix $\mathbf{d}^l$ in computer memory is the column-major format so that index k runs faster than index m.

Before starting computation of Wigner matrices it is necessary to determine the number of matrix elements that need to be stored in the computer memory so that the required space may be allocated. The number of elements *nSzWset*, which constitute the Wigner set, is the sum of sizes of individual matrices:

$$nSzWset = 1 + 3^2 + 5^2 + \cdots + (2l+1)^2 \ldots (2nL+1)^2 = \sum_{l=0}^{nL} (2l+1)^2. \tag{14.83}$$

The resulting series sums up to:

$$nSzWset = \sum_{l=0}^{nL} (2l+1)^2 = (nL+1)(2nL+1)(2nL+3)/3. \tag{14.84}$$

Allocating a one-dimensional array $d(1{:}\ nSzWset)$, which includes $nSzWset$ array elements, provides the necessary space to hold the Wigner set of matrices in computer memory.

Because the Wigner set consists of square matrices of varying size, using a programming language mechanism for multi-dimensional arrays to access the individual array elements would appear to be impractical. An alternative would be to manually map the Wigner set onto a one-dimensional array $d(1{:}\ nSzWset)$. If this is done, then a custom mechanism for retrieving a matrix element $d^l_{k,m}$ from the array must be designed. This is where the equation to determine the sum of elements in the Wigner set additionally proves to be useful. Since computer memory is a linear storage, where each consecutive element goes one after another in a linear order, the first step to find a pointer to the matrix element $d^l_{k,m}$ is to sum up the number of elements in the preceding matrices $\mathbf{d}^0,\mathbf{d}^1,\ldots,\mathbf{d}^{l-1}$. Replacing parameter nL with $l-1$ in Equation 14.84 sums up the power series to $l(2l-1)(2l+1)/3$, which is the cumulative number of stored matrix elements in all preceding matrices up to and including matrix $\mathbf{d}^{l-1}$ in the Wigner set. Adding a unity to that sum gives the number $l(2l-1)(2l+1)/3+1$, and that serves as a pointer to the first element $d^l_{l,l}$ in matrix $\mathbf{d}^l$ according to the ordering scheme explained in Equation 14.82. This step accounts for index l in $d^l_{k,m}$.

Computing the remaining shift due to indices k and m involves determining the number of array elements between $d^l_{l,l}$ and $d^l_{k,m}$. Since these indices start from the value of l and end at $-l$, before proceeding any further it is convenient to convert them to zero-base. Using the values $(l-k)$ and $(l-m)$ instead of k and m, respectively in the computational formula accomplishes this conversion. After that, taking into account that each column holds $(2l+1)$ matrix elements, the shift due to indices k and m becomes $(l-m)(2l+1)+(l-k)$.

This analysis of the index computation can now be converted into an integer function $fptr(l, k, m)$ that takes the value of indices l, k, and m on input, and returns the position of the matrix element $d^l_{k,m}$ in array $d(1{:}\ nSzWset)$:

```
integer function fptr(L,k,m)
! compute pointer to element d(L,k,m) of Wigner d-matrix(k,m) of degree L
! k is row index:    L, L-1, ..., 0, ..., -L+1, -L
! m is column index: L, L-1, ..., 0, ..., -L+1, -L
! d-matrix assumes column-major order, where index k runs faster than index m
implicit none
integer, intent(in) :: L, k, m

! determine the pointer to d(L,L,L) as a sum of preceding elements plus 1
fptr = L * (2*L - 1) * (2*L + 1) / 3 + 1

! add shift due to index m
fptr = fptr + (L - m) * (2*L + 1)

! add shift due to index k
fptr = fptr + (L - k)
end function fptr
```

Recursive computation of the Wigner matrix elements is initiated using the unscaled, bar-scaled, and tilde-scaled forms of the first two matrices $\mathbf{d}^0$ and $\mathbf{d}^1$, that is, for $l=0$ and $l=1$. In principle, only the first matrix, for $l=0$, is necessary to obtain all succeeding matrices from the recurrence relations, however, using the matrix $\mathbf{d}^1$ to initiate the series leads to a more efficient computer code.

For $l=0$, matrix element $d^0_{0,0}$ may be directly obtained from the Wigner equation, Equation 5.230. Setting all indices in that equation to zero leads to $d^0_{0,0}=1$. This value holds for any value of the rotation angle, which implies that a multipole expansion with orbital index $l=0$ is rotationally invariant, that is, it rotates into itself. Because both the orbital and azimuthal indices are zero, bar and tilde scaling does not affect the value of this matrix. With that:

$$d^0_{0,0} = \bar{d}^0_{0,0} = \tilde{d}^0_{0,0} = 1. \tag{14.85}$$

The unscaled form of the Wigner matrix $\mathbf{d}^1$ has already been derived in Section 5.2 and has the structure:

$$\mathbf{d}^1(\beta) = \begin{pmatrix} \frac{1+\cos\beta}{2} & -\frac{\sin\beta}{\sqrt{2}} & \frac{1-\cos\beta}{2} \\ \frac{\sin\beta}{\sqrt{2}} & \cos\beta & -\frac{\sin\beta}{\sqrt{2}} \\ \frac{1-\cos\beta}{2} & \frac{\sin\beta}{\sqrt{2}} & \frac{1+\cos\beta}{2} \end{pmatrix}. \tag{5.64}$$

Transitioning from Equation 5.64 to the scaled form requires multiplying each matrix element $d^l_{k,m}$ with the appropriate coefficient from Equation 14.16. To assist in that process, it is useful to compose a table of scaling coefficients for each pair of indices k and m. Substituting the value of indices k and m into the scaling coefficients in Equation 14.16 gives the matrices of bar and tilde scaling coefficients:

	$m=1$	0	-1
$k=1$	1	$\sqrt{2}$	1
0	$\frac{1}{\sqrt{2}}$	1	$\frac{1}{\sqrt{2}}$
-1	1	$\sqrt{2}$	1

and

	$m=1$	0	-1
$k=1$	1	$\frac{1}{\sqrt{2}}$	1
0	$\sqrt{2}$	1	$\sqrt{2}$,
-1	1	$\frac{1}{\sqrt{2}}$	1

(14.86)

respectively.

Multiplying each element of the matrix in Equation 5.64 by a respective element from Equation 14.86 leads to:

$$\bar{\mathbf{d}}^1(\beta) = \begin{pmatrix} \frac{1+\cos\beta}{2} & -\sin\beta & \frac{1-\cos\beta}{2} \\ \frac{\sin\beta}{2} & \cos\beta & -\frac{\sin\beta}{2} \\ \frac{1-\cos\beta}{2} & \sin\beta & \frac{1+\cos\beta}{2} \end{pmatrix} \quad \tilde{\mathbf{d}}^1(\beta) = \begin{pmatrix} \frac{1+\cos\beta}{2} & -\frac{\sin\beta}{2} & \frac{1-\cos\beta}{2} \\ \sin\beta & \cos\beta & -\sin\beta \\ \frac{1-\cos\beta}{2} & \frac{\sin\beta}{2} & \frac{1+\cos\beta}{2} \end{pmatrix}. \tag{14.87}$$

Comparing these matrix elements to those in the original unscaled matrix in Equation 5.64 shows that scaling has eliminated the square root operation, making the scaled matrix elements easier to evaluate.

Although the program code for evaluating the bar-scaled Wigner matrix elements involves a straightforward coding of the recurrence relations, breaking down the computation into different segments, and deciding which recurrence relation to use where and when, is quite complicated. For that reason, it is more convenient to reconstruct the program design based on an analysis of a working code. The following code insert illustrates implementation of subroutine *dmatkbar*().

```
subroutine dmatkbar(nL, nSzWset, beta, d)
  ! Compute bar-scaled Wigner d-matrix using k-set of recurrence
  !   relations that involve increment in index k.
  ! Use a few equations from the m-set to augment the k-set.
  ! Use column-major storage format for d-matrix in computer memory,
```

```
!   where index k runs faster than index m.
!
implicit none
!
! Subroutine arguments
integer, intent(in)               :: nL          ! max value of index L
integer, intent(in)               :: nSzWset     ! size of the Wigner set
double precision, intent(in)      :: beta        ! Euler angle
double precision, intent(out)     :: d(1:nSzWset) ! Wigner set of d-matrices
!
! Local variables
integer          :: L, m, k
double precision :: cosb, sinb, sinb2, opcosb2, omcosb2
!
! Function declarations
integer          :: fptr

! initialize trigonometric functions
cosb   = dcos(beta)
sinb   = dsin(beta)
sinb2  = sinb * 0.5d0
opcosb2 = (1.0d0 + cosb) * 0.5d0
omcosb2 = (1.0d0 - cosb) * 0.5d0

! zero all elements
d = 0.0d0

! L = 0    Eq. 14.85
d(1)  = 1.0d0

! L = 1    Eq. 14.87
d(2)  = opcosb2                ! k= 1  m= 1
d(3)  = sinb2                  ! k= 0  m= 1
d(4)  = omcosb2                ! k=-1  m= 1
d(5)  = -sinb                  ! k= 1  m= 0
d(6)  = cosb                   ! k= 0  m= 0
d(7)  = sinb                   ! k=-1  m= 0
d(8)  = omcosb2                ! k= 1  m=-1
d(9)  = -sinb2                 ! k= 0  m=-1
d(10)= opcosb2                 ! k=-1  m=-1

! recursively compute d-matrix
do L = 2, nL

  ! m = L;  k = L                 Eq. 14.30 m-set
  d(fptr(L,L,L)) = opcosb2 * d(fptr(L-1,L-1,L-1))

  ! m = L;  L-1 >= k >= -L+1     Eq. 14.31 m-set
  do k = L-1, -L+1, -1
    d(fptr(L,k,L)) = sinb2 * d(fptr(L-1,k,L-1))
  enddo

  ! m = L;  k = -L                Eq. 14.32 m-set
  d(fptr(L,-L,L)) = omcosb2 * d(fptr(L-1,-L+1,L-1))

  ! L-1 >= m >= -L+1
  do m = L-1, -L+1, -1
    ! k = L       Eq. 14.64 k-set
```

```
      d(fptr(L,L,m)) = -dble(2*L-1) / dble((l-m)*(L+m)) &
                     * dble(L) * sinb * d(fptr(L-1,L-1,m))

      ! k = L-1    Eq. 14.65 k-set
      d(fptr(L,L-1,m)) = dble(2*L-1) / dble((L-m)*(L+m)) &
                     * (dble(L)*cosb - dble(m)) * d(fptr(L-1,L-1,m))

      ! L-2 >= k >= -L+2
      do k = L-2, -L+2, -1
        !          Eq. 14.63 k-set
        d(fptr(L,k,m)) = ( dble((L-k-1) * L) * sinb * d(fptr(L-1,k+1,m)) &
                 + dble(L+k) * (dble(L)*cosb - dble(m)) &
                 * d(fptr(L-1,k,m)) ) / dble((L-m)*(L+m))
      enddo

      ! k = -L+1   Eq. 14.66 k-set
      d(fptr(L,-L+1,m)) = dble(2*L-1) / dble((L-m)*(L+m)) &
                     * (dble(L)*cosb + dble(m)) * d(fptr(L-1,-L+1,m))

      ! k = -L     Eq. 14.67 k-set
      d(fptr(L,-L,m)) = dble(2*L-1) / dble((L-m)*(L+m)) &
                     * dble(L) * sinb * d(fptr(L-1,-L+1,m))
    enddo

    ! m = -L;  k = L               Eq. 14.33 m-set
    d(fptr(L,L,-L)) = omcosb2 * d(fptr(L-1,L-1,-L+1))

    ! m = -L;  L-1 >= k >= -L+1    Eq. 14.34 m-set
    do k = L-1, -L+1, -1
      d(fptr(L,k,-L)) = -sinb2 * d(fptr(L-1,k,-L+1))
    enddo

    ! m = -L;  k = -L              Eq. 14.35 m-set
    d(fptr(L,-L,-L)) = opcosb2 * d(fptr(L-1,-L+1,-L+1))

  enddo
  !
  end subroutine dmatkbar
```

Subroutine *dmatkbar*() takes parameters *nL*, *nSzWset*, and *beta* on input. These are, in order, the maximum value of orbital index l in the truncated multipole expansion, the size of the Wigner set of *d*-matrices, and the rotation angle β. On exit, the subroutine returns array *d*, which holds all the *d*-matrices in the Wigner set. The memory required to hold the array *d* is presumed dynamically allocated outside subroutine *dmatkbar*().

Computation begins with initialization of the various trigonometric functions used, and zeroing the array *d*. Initialization begins with matrix $\bar{\mathbf{d}}^0$, which holds a single matrix element, by setting it to unity according to Equation 14.85. After that, matrix $\bar{\mathbf{d}}^1$ is filled using the values taken from Equation 14.87. With that, recursive computation of the remaining *d*-matrices starts with matrix $\bar{\mathbf{d}}^2$ and continues to matrix $\bar{\mathbf{d}}^{nL}$; this can be seen in the starting, 2, and ending, *nL*, values of index l in the outermost *do-loop*.

Since each matrix element $\bar{d}^l_{k,m}$ involves three indices l, k, and m, each index needs to be iterated within its own *do-loop*. With that, and the external *do-loop* over index l, the computer code has two additional enclosed *do-loops*. The outer *do-loop* goes over index m, and the inner *do-loop* iterates over index k, so that index k always runs faster than index m to take advantage of the column-major arrangement of the *d*-matrix in computer memory.

Computation of matrix elements inside the *do-loop* over index l is broken down into several segments, each of which handles specific ranges of indices m and k. For each set of indices there are three segments: a beginning segment, then a continuous segment, which runs inside its own *do-loop*, and finally an ending segment.

Analysis of the example of the *do-loop* over index m illustrates the use of these computational segments, which arise from the need to handle specific values of index m in different ways. The use of the k-set of recurrence relations limits the range of index m by the condition $l-1 \geq m \geq -l+1$, where the inequality reads from left to right in the same order as the index m changes in the d-matrix, that is, from the largest positive to the smallest negative value. The index limitation of the k-set necessitates considering the values of $m = \pm l$ as a special case to be handled outside the *do-loop* over index m. This condition breaks the computation down into the beginning ($m = l$), continuous ($l-1 \geq m \geq -l+1$), and ending ($m = -l$) segments.

The first segment inside the *do-loop* over index l in the program code deals with the condition $m = l$. This breaks down into three sub-segments depending on the value of index k. These sub-segments use three different equations from the m-set of recurrence relations. The beginning sub-segment for $m = l$ and $k = l$ employs Equation 14.30, the continuous sub-segment for $m = l$ and $l-1 \geq k \geq -l+1$ uses Equation 14.31, and the ending sub-segment for $m = l$ and $k = -l$ uses Equation 14.32. Within this block of three sub-segments, the value of index m remains fixed to the value of $m = l$, whereas index k changes from the value of $k = l$ down to $k = -l$ thereby optimally utilizing the column-major storage format of d-matrix in the computer memory.

Most of the work is done in the next block, the *do-loop* over index m. This block employs the k-set of recurrence relations. Once again it has to be broken down into several segments to account for the different ranges in index k, each of which has to be handled by specific equations. Two beginning segments treat the conditions of $k = l$ and $k = l-1$ by employing Equations 14.64 and 14.65, respectively. Each of these equations refers to a single matrix element from the preceding d-matrix, and as such they have to be coded separately from the general case, which uses two matrix elements. The continuous segment that follows handles the condition of $l-2 \geq k \geq -l+2$ as defined by the general equation Equation 14.63 for the k-set of recurrence relations. After the *do-loop* ends, the ending segment handles the remaining conditions of $k = -l+1$ and $k = -l$ by using Equations 14.66 and 14.67, respectively. Once again, these equations refer to a single matrix element from the preceding d-matrix, and as such do not conform with the general case handled in the preceding *do-loop* over index k. Together, these three segments hold the value of index m fixed, and have index k changing from the top-most positive value of $k = l$, going down step-by-step to the most negative value of $k = -l$, in accord with the column-major storage format of the d-matrix to minimize the data retrieval time.

The completion of the *do-loop* over index m marks the end of the corresponding continuous segment.

The final segment that follows handles the case of $m = -l$, and once again this breaks down into three sub-segments that process different ranges of index k, using several special-case equations from the m-set of recurrence relations. The first sub-segment addresses the condition of $k = l$ by using Equation 14.33, while the second sub-segment deals with the continuous range of $l-1 \geq k \geq -l+1$ by applying Equation 14.34, and the third sub-segment handles the condition of $k = -l$ by employing Equation 14.35. In summary, this part, comprising three sub-segments, deals with the case of $m = -l$, and, as with the other segments, optimizes the data retrieval of matrix elements from computer memory.

Once the breakdown of the recursive computation of the Wigner d-matrix into individual blocks is complete, the coding of the subroutine *dmatkbar*() represents a straightforward implementation of the respective equations. Using an integer function *fptr*() to handle indices l, k, and m in the d-matrix simplifies mapping the multidimensional d-matrix onto a one-dimensional array d. Together, these steps accomplish the goal of keeping the computer code easy to read and implement.

A simple validation test of the correctness of implementation of the produced computer code relies on a check of unitary and symmetry properties of each computed d-matrix in the Wigner set. This requires the matrices to be unscaled, but, since efficiency is not important at this stage, simply multiplying the matrix elements by their inverse scaling coefficient is sufficient to obtain the unscaled values.

Earlier it was explained that direct computation of the coefficient defined by Equation 14.16 was prone to numerical instabilities, so an alternative, more stable, approach is needed. Noticing that the indices k and m in Equation 14.16 are simultaneously used in addition and subtraction operations leads to the equality:

$$\sqrt{\frac{(l-k)!(l+k)!}{(l-m)!(l+m)!}} \equiv \sqrt{\frac{(l-|k|)!(l+|k|)!}{(l-|m|)!(l+|m|)!}}. \tag{14.88}$$

Equation 14.88 greatly simplifies the analysis in that the signs of indices k and m have no effect on the final value of the coefficient. This allows like terms in the numerator and denominator to be canceled, thereby avoiding the risk of overflow. Pairing the smallest factorial in the numerator with the smallest factorial in the denominator, followed by application of the same procedure to the pair of the largest factorials, and canceling out like terms leads to:

$$\begin{aligned} |k|>|m|: \quad & \frac{(l-|k|)!}{(l-|m|)!} = \frac{1}{(l-|k|+1)\ldots(l-|m|)}, \qquad \frac{(l+|k|)!}{(l+|m|)!} = \frac{(l+|m|+1)\ldots(l+|k|)}{1}, \\ |k|<|m|: \quad & \frac{(l-|k|)!}{(l-|m|)!} = \frac{(l-|m|+1)\ldots(l-|k|)}{1}, \qquad \frac{(l+|k|)!}{(l+|m|)!} = \frac{1}{(l+|k|+1)\ldots(l+|m|)}. \end{aligned} \tag{14.89}$$

Merging the results from Equation 14.89 back into the square root form simplifies Equation 14.16 to:

$$\sqrt{\frac{(l-k)!(l+k)!}{(l-m)!(l+m)!}} = \begin{cases} \sqrt{\dfrac{(l+|m|+1)\ldots(l+|k|)}{(l-|k|+1)\ldots(l-|m|)}} & \text{for } |k|>|m|, \\ \sqrt{\dfrac{(l-|m|+1)\ldots(l-|k|)}{(l+|k|+1)\ldots(l+|m|)}} & \text{for } |k|<|m|. \end{cases} \tag{14.90}$$

Equation 14.90 has two useful properties. First, the number of terms in the numerator is the same as in the denominator. Second, the terms go from the smallest to the largest value from left to right in both the numerator and denominator. With that, instead of separately multiplying the terms in the numerator and in the denominator that may lead to overflow, and then dividing a very large number by another very large number that may potentially lead to the loss of precision, a better approach is to divide the smallest term in the numerator by the smallest term in the denominator, and repeat that operation with the next larger pair of terms until reaching the last one. Consecutive multiplication of fractions can satisfactorily handle values of orbital index l on the order of several hundreds in double precision math. Finally, to give an example of hidden symmetry in the scaling coefficient, setting $k=6$ and $m=3$ leads to $\sqrt{(l+4)(l+5)(l+6)/(l-5)(l-4)(l-3)}$. Similarly, the scaling coefficient assumes the form $\sqrt{(l-5)(l-4)(l-3)/(l+4)(l+5)(l+6)}$ for $k=3$ and $m=6$. This symmetry allows avoiding the use of conditional statement when implementing Equation 14.90 in a computer program. This is illustrated by the following program code for computation of the scaling coefficient:

```
double precision function Coeff(L,k,m)
  !
  ! k = |k| and m = |m|
  ! For k > m:
  ! SQRT[(L-k)!(L+k)!/{(L-m)!(L+m)!}] = SQRT[(L+m+1)...(L+k)/(L-k+1)...
  (L-m)]
  ! For k < m:
  ! SQRT[(L-k)!(L+k)!/{(L-m)!(L+m)!}] = SQRT[(L-m+1)...(L-k)/(L+k+1)...
  (L+m)]
  !
  implicit none
  integer, intent(in) :: L, k, m
  !
  ! Local variables
  integer :: i, ka, ma, ks, ms
  !
  ka = iabs(k)                 ! k = |k|                       absolute value
  ma = iabs(m)                 ! m = |m|
  ks = isign(1, ma-ka) * ka    ! k = -|k|   if |k| > |m|       signed value
```

```
   ms = isign(1, ka-ma) * ma      ! m = -|m|    if |m| > |k|
   !
   Coeff = 1.0d0
   do i = 1, max0(ka,ma) - min0(ka,ma)    ! if |k| == |m|, the do-loop skips
     Coeff = Coeff * dble(L+ms+i) / dble(L+ks+i)
   enddo
   Coeff = dsqrt(Coeff)
   !
end function Coeff
```

Function *Coeff(l, k, m)* takes the parameters *l*, *k*, and *m* as input, and returns the value of the scaling coefficient defined by Equation 14.90. Inside the function, variables *ka* and *ma* represent the absolute value of the parameters *k* and *m*, respectively. Signed parameters *ks* and *ms* get their value based on the conditions $if(|k|>|m|)\ then\ ks=-|k|\ otherwise\ ks=|k|$ and $if(|m|>|k|)\ then\ ms=-|m|\ otherwise\ ms=|m|$, respectively. This provides the opportunity for replacing the two parts in Equation 14.90 with a single *do-loop*. The expression *max0(ka, ma) – min0(ka, ma)* returns the number of terms in the product in Equation 14.90.

The use of function *Coeff(l, k, m)* makes it easy to scale and unscale the *d*-matrix for test purposes. To invoke function *Coeff()* from inside subroutine *dmatkbar()*, only requires a declaration of this function in the appropriate section of the code of subroutine *dmatkbar()*.

```
! Function declarations
integer          :: fptr
double precision :: Coeff
```

Unscaling, validation, and back scaling of *d*-matrices is most conveniently done at the end of subroutine *dmatkbar()*, as shown in the following code:

```
! Validation test for bar-scaled Wigner d-matrix
! loop over d-matrices
do L = 1, nL
  ! unscale the d-matrix
  do m = L, -L, -1
    do k = L, -L, -1
      d(fptr(L,k,m)) = d(fptr(L,k,m)) * Coeff(L,m,k)  ! unscale
    enddo
  enddo

  ! run validation test
  call checkdmat(d(fptr(L,L,L)), L, 2*L+1)

  ! scale back the d-matrix
  do m = L, -L, -1
    do k = L, -L, -1
      d(fptr(L,k,m)) = d(fptr(L,k,m)) * Coeff(L,k,m)  ! scale back
    enddo
  enddo
enddo
```

A useful trick for computing the inverse value of the scaling coefficient is to swap the position of indices *k* and *m* in the argument list of the function *Coeff(l, k, m)*.

Standard unitary and symmetry tests of *d*-matrix can be performed using the validation subroutine *checkdmat()*, as illustrated in the following implementation:

```
subroutine checkdmat(d, L, n)
  !
  ! Validate Wigner d-matrix
```

```
! Map the linear array d() to two-dimensional matrix d(k,m)
!
implicit none
integer,            intent(in) :: L,n
double precision, intent(in) :: d(n,n)
integer                        :: m, k, i, j, a, b
double precision, parameter  :: zero = 1.0d-10
double precision               :: sum
!
! row normalization
do i = 1, n
  k = L - i + 1
  sum = 0.0d0
  do j = 1, n
    sum = sum + d(i,j)**2
  enddo
  if(dabs(sum - 1.0d0) > zero) then
    write(*,'(a,2i3,f16.8)') &
    "Normalization test failed for L, k = ", L, k, sum
    stop
  endif
enddo
!
! column normalization
do j = 1, n
  m = L - j + 1
  sum = 0.0d0
  do i = 1, n
    sum = sum + d(i,j)**2
  enddo
  if(dabs(sum - 1.0d0) > zero) then
    write(*,'(a,2i3,f16.8)') &
    "Normalization test failed for L, m = ", L, m, sum
    stop
  endif
enddo
!
! row orthogonalization
do i = 1, n-1
  k = L - i + 1
  sum = 0.0d0
  do a = i+1, n
    b = L - a + 1
    do j = 1, n
      sum = sum + d(i,j)*d(a,j)
    enddo
    if(dabs(sum) > zero) then
      write(*,'(a,3i3,f16.8)') &
      "Orthogonalization test failed for L, k1, k2 = ", L, k, b, sum
      stop
    endif
  enddo
enddo
!
! column orthogonalization
do j = 1, n-1
  m = L - j + 1
  sum = 0.0d0
  do b = j+1, n
```

```
      a = L - b + 1
      do i = 1, n
        sum = sum + d(i,j)*d(i,b)
      enddo
      if(dabs(sum) > zero) then
        write(*,'(a,3i3,f16.8)') &
        "Orthogonalization test failed for L, m1, m2 = ", L, m, a, sum
        stop
      endif
    enddo
  enddo
  !
  ! symmetry across the main diagonal: d(k,m) = d(m,k) *(-1)^(m-k)
  do j = 2, n
    m = L - j + 1
    do i = 1, j-1
      k = L - i + 1
      if(dabs(d(i,j) - d(j,i) * (-1.0d0)**(j-i)) > zero) then
        write(*,'(a,2f8.2,4x,2i3)') &
        "Symmetry test failed for d(k,m) d(m,k) k m =", d(i,j), d(j,i), k, m
        stop
      endif
    enddo
  enddo
  !
  ! symmetry in the index sign: d(k,m) = d(-m,-k)
  do m = L, 0, -1
    j = L - m + 1    ! index of +m in 2D array d(i,j)
    b = L + m + 1    ! index of -m in 2D array d(i,b); m self is always +
    do k = L, -L, -1
      i = L - k + 1 ! index of  k in 2D array d(i,j)
      a = 2*L+2 - i ! index of -k in 2D array d(a,j); k self may be + or -
      if(dabs(d(i,j) - d(b,a)) > zero) then
        write(*,'(a,2f8.2,4x,2i3,4x,2i3)') &
        "Symmetry test failed for d(k,m) d(-m,-k)  k m -m -k", &
        d(i,j), d(b,a), k, m, -m, -k
        stop
      endif
    enddo
  enddo
  !
  write(*,'(a,i3)') "Validation test passed for d(L), L = ", L
  !
end subroutine checkdmat
```

Within subroutine *checkdmat*(), a one-dimensional array d is mapped onto a two-dimensional d-matrix by supplying a pointer $d(fptr(l, l, l))$ to the first element $d^l_{l,l}$ of matrix $\mathbf{d}^l$ in the subroutine argument list:

```
! run validation test
call checkdmat(d(fptr(L,L,L)), L, 2*L+1)
```

The second argument in this call is the orbital index l; and the third argument is the number of elements in a column or row. This completes the description of the validation test for the algorithm for computing bar-scaled d-matrices. Obviously, once testing is complete, the validation code should be removed to prevent slowing down the computation.

The next step is to compute the tilde-scaled d-matrix used in rotating the tilde-scaled irregular multipole expansions.

The code for computing the tilde-scaled Wigner d-matrix is presented in the following subroutine *dmatktilde*():

```
subroutine dmatktilde(nL, nSzWset, beta, d)
  ! Compute tilde-scaled Wigner d-matrix using k-set of recurrence
  !   relations that involve increment in index k.
  ! Use a few equations from the m-set to augment the k-set.
  ! Use column-major storage format for d-matrix in computer memory,
  !   where index k runs faster than index m.
  !
  implicit none
  !
  ! Subroutine arguments
  integer, intent(in)             :: nL            ! max value of index L
  integer, intent(in)             :: nSzWset       ! size of the Wigner set
  double precision, intent(in)    :: beta          ! Euler angle
  double precision, intent(out)   :: d(1:nSzWset)  ! Wigner set of d-matrices
  !
  ! Local variables
  integer          :: L, m, k
  double precision :: cosb, sinb, sinb2, opcosb2, omcosb2
  !
  ! Function declarations
  integer          :: fptr
  !
  ! initialize d-matrix
  cosb  = dcos(beta)
  sinb  = dsin(beta)
  sinb2 = sinb * 0.5d0
  opcosb2 = (1.0d0 + cosb) * 0.5d0
  omcosb2 = (1.0d0 - cosb) * 0.5d0

  ! zero all elements
  d = 0.0d0

  ! L = 0       Eq. 14.85
  d(1) = 1.0d0

  ! L = 1       Eq. 14.87
  d(2)  = opcosb2                     ! k= 1  m= 1
  d(3)  = sinb                        ! k= 0  m= 1
  d(4)  = omcosb2                     ! k=-1  m= 1
  d(5)  = -sinb2                      ! k= 1  m= 0
  d(6)  = cosb                        ! k= 0  m= 0
  d(7)  = sinb2                       ! k=-1  m= 0
  d(8)  = omcosb2                     ! k= 1  m=-1
  d(9)  = -sinb                       ! k= 0  m=-1
  d(10)= opcosb2                      ! k=-1  m=-1

  ! recursively compute d-matrix
  do L = 2, nL

    ! m = L;  k = L                   Eq. 14.48  m-set
    d(fptr(L,L,L)) = opcosb2 * d(fptr(L-1,L-1,L-1))

    ! m = L;  L-1 >= k >= -L+1     Eq. 14.49  m-set
    do k = L-1, -L+1, -1
      d(fptr(L,k,L)) = dble(L*(2*L-1)) / dble((L-k)*(L+k)) &
```

```
                             * sinb * d(fptr(L-1,k,L-1))
    enddo

    ! m = L;   k = -L                    Eq. 14.50  m-set
    d(fptr(L,-L,L)) = omcosb2 * d(fptr(L-1,-L+1,L-1))

    ! L-1 >= m >= -L+1
    do m = L-1, -L+1, -1
      ! k = L                            Eq. 14.78 k-set
      d(fptr(L,L,m)) = -sinb2 * d(fptr(L-1,L-1,m))

      ! k = L-1                          Eq. 14.79 k-set
      d(fptr(L,L-1,m)) = (dble(L)*cosb - dble(m)) * d(fptr(L-1,L-1,m))

      ! L-2 >= k >= -L+2                 Eq. 14.77 k-set
      do k = L-2, -L+2, -1
        d(fptr(L,k,m)) = ( dble(L)*sinb * d(fptr(L-1,k+1,m)) &
           + (dble(L)*cosb - dble(m)) * d(fptr(L-1,k,m)) ) / dble(L-k)
      enddo

      ! k = -L+1                         Eq. 14.80 k-set
      d(fptr(L,-L+1,m)) = (dble(L)*cosb + dble(m)) * d(fptr(L-1,-L+1,m))

      ! k = -L                           Eq. 14.81 k-set
      d(fptr(L,-L,m)) = sinb2 * d(fptr(L-1,-L+1,m))
    enddo

    ! m = -L;  k = L                     Eq. 14.51  m-set
    d(fptr(L,L,-L)) = omcosb2 * d(fptr(L-1,L-1,-L+1))

    ! m = -L;  L-1 >= k >= -L+1          Eq. 14.52  m-set
    do k = L-1, -L+1, -1
      d(fptr(L,k,-L)) = -dble(L*(2*L-1)) / dble((L-k)*(L+k)) &
                        * sinb * d(fptr(L-1,k,-L+1))
    enddo

    ! m = -L;  k = -L                    Eq. 14.53  m-set
    d(fptr(L,-L,-L)) = opcosb2 * d(fptr(L-1,-L+1,-L+1))

  enddo
  !
end subroutine dmatktilde
```

In this code, the input parameter *nL* represents the size of the multipole expansion; *nSzWset* determines the memory size occupied by the Wigner *d*-matrix; *beta* is the angle of rotation around the *y*-axis; and *d* is the array to hold the set of Wigner *d*-matrices allocated in computer memory before entering subroutine *dmatktilde*().

The structure of subroutine *dmatktilde*() is the same as that of the previously-described subroutine *dmatkbar*(); the only difference between them is that they work with different sets of equations. Subroutine *dmatktilde*() uses the tilde-scaled *k*-set of recurrence relations for the range $l-1 \geq m \geq -l+1$ in index *m* and at the limits, $m = \pm l$, it uses the tilde-scaled *m*-set of recurrence relations.

The validation procedure for subroutine *dmatktilde*() is also similar in structure to that of subroutine *dmatkbar*(). Once unscaling is performed by multiplication of the tilde-scaled *d*-matrix element by function *Coeff*(*l*, *k*, *m*), subroutine *checkdmat*() can check the unitary and symmetry properties of the *d*-matrix. After the test is completed, multiplication of the unscaled *d*-matrix element by function *Coeff*(*l*, *m*, *k*) returns the matrix element to its tilde-scaled form.

As with subroutine *dmatkbar*(), validation of subroutine *dmatktilde*() requires inserting a declaration of function *Coeff*() into the appropriate section near the start of the subroutine:

```
! Function declarations
integer          :: fptr
double precision :: Coeff
```

This allows the test of subroutine *dmatktilde*() to be performed using the following code insert placed at the end of the subroutine:

```
! Validation test for tilde-scaled Wigner d-matrix
! loop over d-matrices
do L = 1, nL
  ! unscale the d-matrix
  do m = L, -L, -1
    do k = L, -L, -1
      d(fptr(L,k,m)) = d(fptr(L,k,m)) * Coeff(L,k,m)   ! unscale
    enddo
  enddo

  ! run validation test
  call checkdmat(d(fptr(L,L,L)), L, 2*L+1)

  ! scale back the d-matrix
  do m = L, -L, -1
    do k = L, -L, -1
      d(fptr(L,k,m)) = d(fptr(L,k,m)) * Coeff(L,m,k)   ! scale back
    enddo
  enddo
enddo
```

Once the test is successfully completed, disabling the portion of the code performing the validation test returns the subroutine *dmatktilde*() to its production state.

14.6 Program Code for Computation of Scaled Wigner Matrix Elements Based on the *m*-set

The bar scaling of the *m*-set of recurrence relations produces a compact set of equations that makes them computationally attractive. In the present section, the computer implementation of the *m*-set for the recursive computation of bar-scaled Wigner matrix elements will be described. The implementation that follows stores the scaled *d*-matrix elements in column-major format with index k running faster than index m to facilitate the computation in Equation 14.17

Two sets of recurrence relations, Equations 7.6 and 7.17, together span the entire range of index k:

$l \geq k \geq -l+2$:

$$\sqrt{(l+m-1)(l+m)}\left(\frac{1+\cos\beta}{2}\right)d^{l-1}_{k-1,m-1} - \sqrt{(l-m)(l+m)}(\sin\beta)d^{l-1}_{k-1,m} + \sqrt{(l-m-1)(l-m)}\left(\frac{1-\cos\beta}{2}\right)d^{l-1}_{k-1,m+1} = \sqrt{(l+k-1)(l+k)}\,d^{l}_{k,m}, \tag{7.6}$$

$l-2 \geq k \geq -l$:

$$\sqrt{(l+m-1)(l+m)}\left(\frac{1-\cos\beta}{2}\right)d^{l-1}_{k+1,m-1} + \sqrt{(l-m)(l+m)}(\sin\beta)d^{l-1}_{k+1,m} + \sqrt{(l-m-1)(l-m)}\left(\frac{1+\cos\beta}{2}\right)d^{l-1}_{k+1,m+1} = \sqrt{(l-k-1)(l-k)}\,d^{l}_{k,m}, \tag{7.17}$$

Section 14.5 describes the technique for conversion of recurrence relations to bar-scaled form. Application of that procedure to Equation 7.6 and Equation 7.17 produces:

$l \geq k \geq -l + 2$:

$$\left(\frac{1+\cos\beta}{2}\right)\bar{d}^{\,l-1}_{k-1,m-1} - (\sin\beta)\bar{d}^{\,l-1}_{k-1,m} + \left(\frac{1-\cos\beta}{2}\right)\bar{d}^{\,l-1}_{k-1,m+1} = \bar{d}^{\,l}_{k,m}, \tag{14.91}$$

$l - 2 \geq k \geq -l$:

$$\left(\frac{1-\cos\beta}{2}\right)\bar{d}^{\,l-1}_{k+1,m-1} + (\sin\beta)\bar{d}^{\,l-1}_{k+1,m} + \left(\frac{1+\cos\beta}{2}\right)\bar{d}^{\,l-1}_{k+1,m+1} = \bar{d}^{\,l}_{k,m}. \tag{14.92}$$

In addition to these equations, a few more equations covering the specific values of $m = \pm l$ and $m = \pm(l - 1)$ are needed.

For $m = l$, the vanishing coefficients in Equation 7.6 and Equation 7.17 reduce these equations to:

$$\sqrt{(2l)(2l-1)}\left(\frac{1+\cos\beta}{2}\right)d^{l-1}_{k-1,l-1} = \sqrt{(l+k-1)(l+k)}\,d^{l}_{k,l}. \tag{7.18}$$

$$\sqrt{(2l)(2l-1)}\left(\frac{1-\cos\beta}{2}\right)d^{l-1}_{k+1,l-1} = \sqrt{(l-k-1)(l-k)}\,d^{l}_{k,l}. \tag{7.20}$$

For $m = l - 1$, the original recurrence relations transform to:

$$\sqrt{(2l-1)(2l-2)}\left(\frac{1+\cos\beta}{2}\right)d^{l-1}_{k-1,l-2} - \sqrt{2l-1}(\sin\beta)d^{l-1}_{k-1,l-1} = \sqrt{(l+k-1)(l+k)}\,d^{l}_{k,l-1} \tag{7.24}$$

$$\sqrt{(2l-1)(2l-2)}\left(\frac{1-\cos\beta}{2}\right)d^{l-1}_{k+1,l-2} + \sqrt{2l-1}(\sin\beta)d^{l-1}_{k+1,l-1} = \sqrt{(l-k-1)(l-k)}\,d^{l}_{k,l-1}. \tag{7.26}$$

Likewise, for $m = -l + 1$:

$$-\sqrt{2l-1}(\sin\beta)d^{l-1}_{k-1,-l+1} + \sqrt{(2l-1)(2l-2)}\left(\frac{1-\cos\beta}{2}\right)d^{l-1}_{k-1,-l+2} = \sqrt{(l+k-1)(l+k)}\,d^{l}_{k,-l+1}, \tag{7.27}$$

$$\sqrt{2l-1}(\sin\beta)d^{l-1}_{k+1,-l+1} + \sqrt{(2l-1)(2l-2)}\left(\frac{1+\cos\beta}{2}\right)d^{l-1}_{k+1,-l+2} = \sqrt{(l-k-1)(l-k)}\,d^{l}_{k,-l+1}. \tag{7.29}$$

For $m = -l$:

$$\sqrt{(2l)(2l-1)}\left(\frac{1-\cos\beta}{2}\right)d^{l-1}_{k-1,-l+1} = \sqrt{(l+k-1)(l+k)}\,d^{l}_{k,-l}, \tag{7.21}$$

$$\sqrt{(2l)(2l-1)}\left(\frac{1+\cos\beta}{2}\right)d^{l-1}_{k+1,-l+1} = \sqrt{(l-k-1)(l-k)}\,d^{l}_{k,-l}. \tag{7.23}$$

Converting these equations into the bar-scaled form proceeds as follows:

Multiplying the above equations by the proper scale coefficient and simultaneously dividing both sides of each equation by the square root coefficient on the right-hand side gives:

$m = l$:

$$\sqrt{\frac{(2l)(2l-1)}{(l+k-1)(l+k)}}\sqrt{\frac{(l-k)!(l+k)!}{(2l)!}}\left(\frac{1+\cos\beta}{2}\right)d^{l-1}_{k-1,l-1} = \bar{d}^{l}_{k,l}, \tag{14.93}$$

$$\sqrt{\frac{(2l)(2l-1)}{(l-k-1)(l-k)}}\sqrt{\frac{(l-k)!(l+k)!}{(2l)!}}\left(\frac{1-\cos\beta}{2}\right)d^{l-1}_{k+1,l-1} = \bar{d}^{l}_{k,l}, \tag{14.94}$$

$m = l - 1$:

$$\begin{aligned}&\sqrt{\frac{(2l-1)(2l-2)}{(l+k-1)(l+k)}}\sqrt{\frac{(l-k)!(l+k)!}{(2l-1)!}}\left(\frac{1+\cos\beta}{2}\right)d^{l-1}_{k-1,l-2}\\ &-\sqrt{\frac{2l-1}{(l+k-1)(l+k)}}\sqrt{\frac{(l-k)!(l+k)!}{(2l-1)!}}(\sin\beta)d^{l-1}_{k-1,l-1} = \bar{d}^{l}_{k,l-1},\end{aligned} \tag{14.95}$$

$$\begin{aligned}&\sqrt{\frac{(2l-1)(2l-2)}{(l-k-1)(l-k)}}\sqrt{\frac{(l-k)!(l+k)!}{(2l-1)!}}\left(\frac{1-\cos\beta}{2}\right)d^{l-1}_{k+1,l-2}\\ &+\sqrt{\frac{2l-1}{(l-k-1)(l-k)}}\sqrt{\frac{(l-k)!(l+k)!}{(2l-1)!}}(\sin\beta)d^{l-1}_{k+1,l-1} = \bar{d}^{l}_{k,l-1},\end{aligned} \tag{14.96}$$

$m = -l + 1$:

$$\begin{aligned}&-\sqrt{\frac{2l-1}{(l+k-1)(l+k)}}\sqrt{\frac{(l-k)!(l+k)!}{(2l-1)!}}(\sin\beta)d^{l-1}_{k-1,-l+1}\\ &+\sqrt{\frac{(2l-1)(2l-2)}{(l+k-1)(l+k)}}\sqrt{\frac{(l-k)!(l+k)!}{(2l-1)!}}\left(\frac{1-\cos\beta}{2}\right)d^{l-1}_{k-1,-l+2} = \bar{d}^{l}_{k,-l+1},\end{aligned} \tag{14.97}$$

$$\begin{aligned}&\sqrt{\frac{2l-1}{(l-k-1)(l-k)}}\sqrt{\frac{(l-k)!(l+k)!}{(2l-1)!}}(\sin\beta)d^{l-1}_{k+1,-l+1}\\ &+\sqrt{\frac{(2l-1)(2l-2)}{(l-k-1)(l-k)}}\sqrt{\frac{(l-k)!(l+k)!}{(2l-1)!}}\left(\frac{1+\cos\beta}{2}\right)d^{l-1}_{k+1,-l+2} = \bar{d}^{l}_{k,-l+1},\end{aligned} \tag{14.98}$$

$m = -l$:

$$\sqrt{\frac{(2l)(2l-1)}{(l+k-1)(l+k)}}\sqrt{\frac{(l-k)!(l+k)!}{(2l)!}}\left(\frac{1-\cos\beta}{2}\right)d^{l-1}_{k-1,-l+1} = \bar{d}^{l}_{k,-l}, \tag{14.99}$$

$$\sqrt{\frac{(2l)(2l-1)}{(l-k-1)(l-k)}}\sqrt{\frac{(l-k)!(l+k)!}{(2l)!}}\left(\frac{1+\cos\beta}{2}\right)d^{l-1}_{k+1,-l+1} = \bar{d}^{l}_{k,-l}. \tag{14.100}$$

Canceling like terms under the square roots leads to:

$m = l$:

$$\left(\frac{1+\cos\beta}{2}\right)\sqrt{\frac{(l-k)!(l+k-2)!}{(2l-2)!}}\,d^{l-1}_{k-1,l-1} = \bar{d}^{l}_{k,l}, \tag{14.101}$$

$$\left(\frac{1-\cos\beta}{2}\right)\sqrt{\frac{(l-k-2)!(l+k)!}{(2l-2)!}}\, d_{k+1,l-1}^{l-1} = \bar{d}_{k,l}^{l}, \tag{14.102}$$

$m = l - 1$:

$$\left(\frac{1+\cos\beta}{2}\right)\sqrt{\frac{(l-k)!(l+k-2)!}{(2l-3)!}}\, d_{k-1,l-2}^{l-1} - (\sin\beta)\sqrt{\frac{(l-k)!(l+k-2)!}{(2l-2)!}}\, d_{k-1,l-1}^{l-1} = \bar{d}_{k,l-1}^{l}, \tag{14.103}$$

$$\left(\frac{1-\cos\beta}{2}\right)\sqrt{\frac{(l-k-2)!(l+k)!}{(2l-3)!}}\, d_{k+1,l-2}^{l-1} + (\sin\beta)\sqrt{\frac{(l-k-2)!(l+k)!}{(2l-2)!}}\, d_{k+1,l-1}^{l-1} = \bar{d}_{k,l-1}^{l}, \tag{14.104}$$

$m = -l + 1$:

$$-(\sin\beta)\sqrt{\frac{(l-k)!(l+k-2)!}{(2l-2)!}}\, d_{k-1,-l+1}^{l-1} + \left(\frac{1-\cos\beta}{2}\right)\sqrt{\frac{(l-k)!(l+k-2)!}{(2l-3)!}}\, d_{k-1,-l+2}^{l-1} = \bar{d}_{k,-l+1}^{l}, \tag{14.105}$$

$$(\sin\beta)\sqrt{\frac{(l-k-2)!(l+k)!}{(2l-2)!}}\, d_{k+1,-l+1}^{l-1} + \left(\frac{1+\cos\beta}{2}\right)\sqrt{\frac{(l-k-2)!(l+k)!}{(2l-3)!}}\, d_{k+1,-l+2}^{l-1} = \bar{d}_{k,-l+1}^{l}, \tag{14.106}$$

$m = -l$:

$$\left(\frac{1-\cos\beta}{2}\right)\sqrt{\frac{(l-k)!(l+k-2)!}{(2l-2)!}}\, d_{k-1,-l+1}^{l-1} = \bar{d}_{k,-l}^{l}, \tag{14.107}$$

$$\left(\frac{1+\cos\beta}{2}\right)\sqrt{\frac{(l-k-2)!(l+k)!}{(2l-2)!}}\, d_{k+1,-l+1}^{l-1} = \bar{d}_{k,-l}^{l}. \tag{14.108}$$

Converting the unscaled d-matrix elements in the left-hand side into their scaled form, then combining these equations with Equations 14.91 and 14.92, leads to the final computer-implementation ready m-set of bar-scaled recurrence relations covering the entire range in index m:

$m = l$ and $l \geq k \geq -l + 2$:

$$\left(\frac{1+\cos\beta}{2}\right)\bar{d}_{k-1,l-1}^{l-1} = \bar{d}_{k,l}^{l}, \tag{14.109}$$

$m = l$ and $l - 2 \geq k \geq -l$:

$$\left(\frac{1-\cos\beta}{2}\right)\bar{d}_{k+1,l-1}^{l-1} = \bar{d}_{k,l}^{l}, \tag{14.110}$$

$m = l - 1$ and $l \geq k \geq -l + 2$:

$$\left(\frac{1+\cos\beta}{2}\right)\bar{d}_{k-1,l-2}^{l-1} - (\sin\beta)\bar{d}_{k-1,l-1}^{l-1} = \bar{d}_{k,l-1}^{l}, \tag{14.111}$$

$m = l - 1$ and $l - 2 \geq k \geq -l$:

$$\left(\frac{1-\cos\beta}{2}\right)\bar{d}_{k+1,l-2}^{l-1} + (\sin\beta)\bar{d}_{k+1,l-1}^{l-1} = \bar{d}_{k,l-1}^{l}, \tag{14.112}$$

$l-2 \geq m \geq -l+2$ and $l \geq k \geq -l+2$:

$$\left(\frac{1+\cos\beta}{2}\right)\bar{d}^{l-1}_{k-1,m-1} - (\sin\beta)\bar{d}^{l-1}_{k-1,m} + \left(\frac{1-\cos\beta}{2}\right)\bar{d}^{l-1}_{k-1,m+1} = \bar{d}^{l}_{k,m}, \tag{14.113}$$

$l-2 \geq m \geq -l+2$ and $l-2 \geq k \geq -l$:

$$\left(\frac{1-\cos\beta}{2}\right)\bar{d}^{l-1}_{k+1,m-1} + (\sin\beta)\bar{d}^{l-1}_{k+1,m} + \left(\frac{1+\cos\beta}{2}\right)\bar{d}^{l-1}_{k+1,m+1} = \bar{d}^{l}_{k,m}, \tag{14.114}$$

$m = -l+1$ and $l \geq k \geq -l+2$:

$$-(\sin\beta)\bar{d}^{l-1}_{k-1,-l+1} + \left(\frac{1-\cos\beta}{2}\right)\bar{d}^{l-1}_{k-1,-l+2} = \bar{d}^{l}_{k,-l+1}, \tag{14.115}$$

$m = -l+1$ and $l-2 \geq k \geq -l$:

$$(\sin\beta)\bar{d}^{l-1}_{k+1,-l+1} + \left(\frac{1+\cos\beta}{2}\right)\bar{d}^{l-1}_{k+1,-l+2} = \bar{d}^{l}_{k,-l+1}, \tag{14.116}$$

$m = -l$ and $l \geq k \geq -l+2$:

$$\left(\frac{1-\cos\beta}{2}\right)\bar{d}^{l-1}_{k-1,-l+1} = \bar{d}^{l}_{k,-l}, \tag{14.117}$$

$m = -l$ and $l-2 \geq k \geq -l$:

$$\left(\frac{1+\cos\beta}{2}\right)\bar{d}^{l-1}_{k+1,-l+1} = \bar{d}^{l}_{k,-l}. \tag{14.118}$$

Finally, the range in index k between Equations 14.91 and 14.92 contains an overlapping region, so a decision must be made on how to split this range. The simplest solution is to use the first equation for $l \geq k \geq l-1$, and the other for $l-2 \geq k \geq -l$. This is implemented in the following computer code:

```
subroutine dmatmbar(nL, nSzWset, beta, d)
  ! Compute bar-scaled Wigner d-matrix using m-set of recurrence
  !   relations that involves the increment in index m.
  ! Use column-major storage format for d-matrix in computer memory,
  !   where index k runs faster than index m.
  !
  implicit none
  !
  ! Subroutine arguments
  integer, intent(in)               :: nL            ! max value of index L
  integer, intent(in)               :: nSzWset       ! size of the Wigner set
  double precision, intent(in)      :: beta          ! Euler angle
  double precision, intent(out)     :: d(1:nSzWset)  ! Wigner set of d-matrices
  !
  ! Local variables
  integer            :: L, m, k
  double precision   :: cosb, sinb, sinb2, opcosb2, omcosb2
  !
  ! Function declarations
  integer            :: fptr
```

```
double precision :: Coeff
!
! initialize d-matrix
cosb   = dcos(beta)
sinb   = dsin(beta)
sinb2 = sinb * 0.50d0
opcosb2 = (1.0d0 + cosb) * 0.5d0
omcosb2 = (1.0d0 - cosb) * 0.5d0

! zero all elements
d = 0.0d0

! L = 0          Eq. 14.85
d(1) = 1.0d0

! L = 1          Eq. 14.87
d(2)  = opcosb2                      ! k= 1  m= 1
d(3)  = sinb2                        ! k= 0  m= 1
d(4)  = omcosb2                      ! k=-1  m= 1
d(5)  = -sinb                        ! k= 1  m= 0
d(6)  = cosb                         ! k= 0  m= 0
d(7)  = sinb                         ! k=-1  m= 0
d(8)  = omcosb2                      ! k= 1  m=-1
d(9)  = -sinb2                       ! k= 0  m=-1
d(10)= opcosb2                       ! k=-1  m=-1

! recursively compute d-matrix
do L = 2, nL
  !
  do k = L, L-1, -1
    ! m=L                  Eq. 14.109  m-set
    d(fptr(L,k,L)) = opcosb2 * d(fptr(L-1,k-1,L-1))
    ! m=L-1                Eq. 14.111  m-set
    d(fptr(L,k,L-1)) = opcosb2 * d(fptr(L-1,k-1,L-2)) &

           - sinb * d(fptr(L-1,k-1,L-1))

    ! m= L-2 ... -L+2  Eq. 14.91    m-set
    do m = L-2, -L+2, -1
      d(fptr(L,k,m)) = opcosb2 * d(fptr(L-1,k-1,m-1)) &
           - sinb * d(fptr(L-1,k-1,m)) + omcosb2 * d(fptr(L-1,k-1,m+1))
    enddo

    ! m= -L+1              Eq. 14.115  m-set
    d(fptr(L,k,-L+1)) = -sinb * d(fptr(L-1,k-1,-L+1)) &
           + omcosb2 * d(fptr(L-1,k-1,-L+2))

    ! m= -L                Eq. 14.117  m-set
    d(fptr(L,k,-L)) = omcosb2 * d(fptr(L-1,k-1,-L+1))
  enddo

  do k = L-2, -L, -1
    ! m=L                  Eq. 14.110  m-set
    d(fptr(L,k,L)) = omcosb2 * d(fptr(L-1,k+1,L-1))

    ! m=L-1                Eq. 14.112  m-set
    d(fptr(L,k,L-1)) = omcosb2 * d(fptr(L-1,k+1,L-2)) &
           + sinb * d(fptr(L-1,k+1,L-1))
```

```
      ! m= L-2 ... -L+2  Eq. 14.92    m-set
      do m = L-2, -L+2, -1
        d(fptr(L,k,m)) = omcosb2 * d(fptr(L-1,k+1,m-1)) &
             + sinb * d(fptr(L-1,k+1,m)) + opcosb2 * d(fptr(L-1,k+1,m+1))
      enddo

      ! m= -L+1             Eq. 14.116  m-set
      d(fptr(L,k,-L+1)) = sinb * d(fptr(L-1,k+1,-L+1)) &
             + opcosb2 * d(fptr(L-1,k+1,-L+2))
      ! m= -L               Eq. 14.118
      d(fptr(L,k,-L)) = opcosb2 * d(fptr(L-1,k+1,-L+1))
    enddo
    !
  enddo

  ! Validation test
  do L = 1, nL
    ! unscale the d-matrix
    do m = L, -L, -1
      do k = L, -L, -1
        d(fptr(L,k,m)) = d(fptr(L,k,m)) * Coeff(L,m,k)   ! unscale
      enddo
    enddo

    ! run validation test
    call checkdmat(d(fptr(L,L,L)), L, 2*L+1, .false., .true.)

    ! scale back the d-matrix
    do m = L, -L, -1
      do k = L, -L, -1
        d(fptr(L,k,m)) = d(fptr(L,k,m)) * Coeff(L,k,m)   ! scale back
      enddo
    enddo
  enddo

end subroutine dmatmbar
```

Computer implementation of subroutine *dmatmbar*() involves cycling over index l in the outermost *do-loop* that initiates the computation of the bar-scaled Wigner d-matrix of degree l. This is followed by a *do-loop* that runs over index k, and an innermost *do-loop* cycles over index m. This arrangement of loops follows from the structure of the m-set of recurrence relations that applies the increment to index m, thereby requiring the *do-loop* over index m to be the innermost one even though the d-matrix is stored in column-major format with index k running faster than index m to facilitate the computation in Equation 14.17.

Within the *do-loop* over index l, the computer program divides the work on the allowed values of index k into two *do-loops*. The first one processes a short segment $l \geq k \geq l-1$, and the second *do-loop* handles the remaining $l-2 \geq k \geq -l$. Both *do-loops* then use the appropriate equation for each specific set of values of index m.

Subroutine *dmatmbar*() ends with validation of the computed bar-scaled Wigner d-matrices by first unscaling them, running the validation test, and finally returning the matrix elements back to the bar-scaled form. Once the validation test is satisfactorily completed, the corresponding block should be removed from the production code to prevent any unnecessary slowdown caused by running the test.

15

Rotation-Based Multipole Translations

Conventional multipole operations, M2M, M2L, and L2L exhibit $O(N^4)$ scaling with respect to the size, N, of the multipole expansion. The scaling profile improves to $O(N^3)$ upon performing translation the origin of the multipole expansion along the z-axis. Generally, only a fraction of boxes in FMM happen to be aligned along the z-axis. Flexibility in the choice of coordinate system provides a mechanism to achieve the required parallel alignment of the z-axis with the translation vector. Implementation of this idea turns the general multipole translation into a three-step operation known as rotation-based multipole translation.

In the first step, the local coordinate system of the multipole expansion to be translated is rotated to align its z-axis with the line connecting the centers of the donor and recipient boxes. The second step performs a multipole translation along the z-axis from the donor to the recipient box. Finally, in the third step, the local coordinate system of the translated expansion is rotated to match that of the recipient box. These steps now need to be formalized.

15.1 Assembly of Rotation Matrix

Rotation of the coordinate system of multipole expansion involves computation of the Wigner matrix. Both the rotation operation and matrix computation have an $O(N^3)$ scaling profile in respect to size of the multipole expansion. A significant reduction in the cost of rotation occurs when choosing the bar-scaled m-set and the tilde scaled k-set of recurrence relations for computation of Wigner d-matrix.

Having the elements of Wigner d-matrix computed allows the assembly of the complete rotation D-matrix, which is the product of three non-commutative rotations around the coordinate axes. Before attempting its computer implementation it is helpful to briefly overview the basic properties of the D-matrix.

The notation of matrix element $D^l_{k,m}(\alpha,\beta,\gamma)$ establishes a strict relationship between the lower indices k and m, and the angles α and γ. The association rule connects the first lower index, k in this case, to the first angle α in the argument list, and establishes a similar relation between the second index m and the last angle γ in the list:

$$D^l_{k,m}(\alpha,\beta,\gamma) = e^{-ik\alpha}\, d^l_{k,m}(\beta)\, e^{-im\gamma}, \tag{15.1}$$

where angle β uniquely associates with matrix element $d^l_{k,m}(\beta)$. The product of the lower index and the rotation angle forms a complex exponent in Equation 15.1. Because exponents are simple multiplication factors, they can commute freely from one side to the other of the matrix element $d^l_{k,m}(\beta)$, as long as the association of the specific lower index with the corresponding angle remains unchanged. Unlike that, changing the order of the lower indices or that of the rotation angles on the left-hand side of Equation 15.1 leads to a different matrix element, requiring the reconstruction of the right-hand side of Equation 15.1 to reflect the new arrangement. Finally, the order of the lower indices k and m in $D^l_{k,m}$ directly translates to that in the matrix element $d^l_{k,m}$.

Making the complex conjugate of an element of the rotation matrix $D^l_{k,m}(\alpha,\beta,\gamma)$ involves only the simple procedure of reversing the sign in front of the complex number i. The d-matrix is unaffected since all of its elements are real. With that, taking the complex conjugate of the original Equation 15.1 leads to:

$$D^{l*}_{k,m}(\alpha,\beta,\gamma) = e^{ik\alpha}\, d^l_{k,m}(\beta)\, e^{im\gamma}. \tag{15.2}$$

For every forward rotation $\mathbf{D}^l(\alpha,\beta,\gamma)$ of the coordinate system there exists a corresponding backward rotation $[\mathbf{D}^l(\alpha,\beta,\gamma)]^{-1}$, its matrix inverse, which would restore the coordinate system to its original position. Finding the inverse of a rotation matrix relies on the unitary nature of the rotation operator, R, which states that the matrix element $\left[D^l\right]_{k,m}^{-1} = \langle lk|R^{-1}|lm\rangle$ is equal to that of its Hermitian adjoint $\left[D^l\right]_{m,k}^{*} \equiv D_{m,k}^{l*} = \langle lm|R|lk\rangle^{*}$:

$$\langle lk|R^{-1}|lm\rangle = \langle lm|R|lk\rangle^{*}. \tag{15.3}$$

The result of $[D^l]_{k,m}^{-1} = D_{m,k}^{l*}$ corresponds to the matrix transpose that swaps columns with rows followed by taking the complex conjugate of each matrix element.

Another way to arrive at the same conclusion is to use the fact that a backward rotation corresponds to undoing a forward rotation by applying the rotation steps in the reverse order with the angles also reversed. If the forward rotation corresponds to coordinate system rotation around the z-axis on angle γ, followed by rotation around the original y-axis on angle β, and finally rotating the coordinate system around the original z-axis on angle α, the backward rotation would correspond to the rotation of the coordinate system around the z-axis on angle $-\alpha$, followed by rotation around the original y-axis on angle $-\beta$, and concluding the rotation operation by rotating the transformed coordinate system around the original z-axis on angle $-\gamma$. Combining these three latter steps into a single equation leads to the rotation matrix $\mathbf{D}^l(-\gamma,-\beta,-\alpha)$ that has the matrix element:

$$D_{k,m}^{l}(-\gamma,-\beta,-\alpha) = e^{ik\gamma}\, d_{k,m}^{l}(-\beta)\, e^{im\alpha}. \tag{15.4}$$

The right-hand side of Equation 15.4 follows from the association between the lower indices and the rotation angles indicated on the left-hand side.

The remaining negative sign in Equation 15.4 can be eliminated by using a consequence of the unitary property of the d-matrix:

$$d_{k,m}^{l}(-\beta) = d_{m,k}^{l}(\beta), \tag{15.5}$$

which corresponds to a matrix transpose. It is unnecessary to take the complex conjugate in this case as all the elements of the d-matrix are real. By substituting Equation 15.5 into Equation 15.4, and interchanging the position of exponents, leads to:

$$D_{k,m}^{l}(-\gamma,-\beta,-\alpha) = e^{ik\gamma}\, d_{k,m}^{l}(-\beta)\, e^{im\alpha} = e^{im\alpha}\, d_{m,k}^{l}(\beta)\, e^{ik\gamma} = D_{m,k}^{l*}(\alpha,\beta,\gamma), \tag{15.6}$$

where the right-most matrix element follows from the established relation between the order of the lower indices in the matrix element and that of the rotation angles in the argument list. Equation 15.6 represents a proof of the previous result, that the inverse of a rotation matrix is the complex conjugate of the transpose of the matrix.

The purpose of the D-matrix in FMM is to rotate the coordinate system of the multipole expansion to align the z-axis with the vector of origin translation of multipole operations, as explained in Chapter 14. Since achieving the target alignment requires only two rotations around the coordinate axes, it makes angle γ equal to zero. With that, angle γ can be dropped from the list of arguments of D-matrix in the discussion that follows.

The use of spherical harmonics provides a starting point for readying the rotation equations for computer implementation. Rewriting the portion of the sum that deals with the negative azimuthal index, and applying the symmetry relation, $Y_{l,-k} = Y_{l,k}^{*}(-1)^k$ leads to:

$$\begin{aligned} Y_{l,m}' = \sum_{k=-l}^{l} D_{k,m}^{l}(\alpha,\beta)\, Y_{l,k} &= D_{0,m}^{l}(\alpha,\beta)\, Y_{l,0} \\ &+ \sum_{k=1}^{l}\left[D_{k,m}^{l}(\alpha,\beta)\, Y_{l,k} + D_{-k,m}^{l}(\alpha,\beta)\, Y_{l,k}^{*}(-1)^k\right], \end{aligned} \tag{15.7}$$

where the prime symbol indicates a rotated function. Explicitly writing out the matrix element $D_{k,m}^{l}(\alpha,\beta)$ gives the detailed equation for a forward rotation of spherical harmonics:

$$\sum_{k=-l}^{l} D^{l}_{k,m}(\alpha,\beta)\,Y_{l,k} = d^{l}_{0,m}(\beta)\,Y_{l,0} + \sum_{k=1}^{l}\left[e^{-ik\alpha}\,d^{l}_{k,m}(\beta)\,Y_{l,k} + e^{ik\alpha}\,d^{l}_{-k,m}(\beta)\,Y^{*}_{l,k}(-1)^{k}\right]. \tag{15.8}$$

Another implementation-ready equation that needs to be derived, for the purpose of multipole translations, is that for backward rotation. This transformation corresponds to the use of the inverse rotation matrix:

$$\sum_{k=-l}^{l} D^{l}_{k,m}(-\beta,-\alpha)\,Y_{l,k} = \sum_{k=-l}^{l} e^{im\alpha}\,d^{l}_{k,m}(-\beta)\,Y_{l,k}. \tag{15.9}$$

Rewriting the sum in Equation 15.9 to have index k going over only positive values leads to the equation for a backward rotation:

$$\sum_{k=-l}^{l} D^{l}_{k,m}(-\beta,-\alpha)\,Y_{l,k} = e^{im\alpha}\left\{d^{l}_{0,m}(-\beta)\,Y_{l,0} + \sum_{k=1}^{l}\left[d^{l}_{k,m}(-\beta)\,Y_{l,k} + d^{l}_{-k,m}(-\beta)\,Y^{*}_{l,k}(-1)^{k}\right]\right\}. \tag{15.10}$$

It is convenient to keep the negative rotation angle $-\beta$ in the d-matrix elements in Equation 15.10 unchanged for now, since replacing it with the matrix element over the positive angle β would prevent its use in the scaled form of the d-matrix.

The final pair of forward and backward rotation equations to be derived using spherical harmonics is that for complex-conjugated functions, which appear in a regular multipole expansion. The complex conjugate of Equation 15.8 leads to the equation for a forward rotation of the corresponding complex-conjugated functions:

$$\sum_{k=-l}^{l} D^{l*}_{k,m}(\alpha,\beta)\,Y^{*}_{l,k} = d^{l}_{0,m}(\beta)\,Y_{l,0} + \sum_{k=1}^{l}\left[e^{ik\alpha}\,d^{l}_{k,m}(\beta)\,Y^{*}_{l,k} + e^{-ik\alpha}\,d^{l}_{-k,m}(\beta)\,Y_{l,k}(-1)^{k}\right]. \tag{15.11}$$

The elements of the d-matrix and $Y_{l,0}$ are real, so complex conjugation leaves them unchanged. Likewise, the complex conjugate of Equation 15.10 gives the equation for a backward rotation of the complex-conjugated functions:

$$\sum_{k=-l}^{l} D^{l*}_{k,m}(-\beta,-\alpha)\,Y^{*}_{l,k} = e^{-im\alpha}\left\{d^{l}_{0,m}(-\beta)\,Y_{l,0} + \sum_{k=1}^{l}\left[d^{l}_{k,m}(-\beta)\,Y^{*}_{l,k} + d^{l}_{-k,m}(-\beta)\,Y_{l,k}(-1)^{k}\right]\right\}. \tag{15.12}$$

Equations for rotation of the coordinate system of bar-scaled regular and tilde-scaled irregular multipole expansions follow directly from these derived relations. Performing the substitutions $D^{l}_{k,m} \to \bar{D}^{l}_{k,m}$, $d^{l}_{k,m} \to \bar{d}^{l}_{k,m}$, $Y^{*}_{l,m} \to \bar{M}_{l,m}$, and $Y_{l,m} \to \bar{M}^{*}_{l,m}$ in Equation 15.11, and remembering that the functions with zero azimuthal index are real, automatically leads to the equation for a forward rotation of bar-scaled regular multipole expansions:

$$\sum_{k=-l}^{l} \bar{D}^{l*}_{k,m}(\alpha,\beta)\,\bar{M}_{l,k} = \bar{d}^{l}_{0,m}(\beta)\,\bar{M}_{l,0} + \sum_{k=1}^{l}\left[e^{ik\alpha}\,\bar{d}^{l}_{k,m}(\beta)\,\bar{M}_{l,k} + e^{-ik\alpha}\,\bar{d}^{l}_{-k,m}(\beta)\,\bar{M}^{*}_{l,k}(-1)^{k}\right]. \tag{15.13}$$

This simple transition becomes possible due to the isomorphism that exists between Equations 14.6 and 14.17 from Chapter 14. Applying the same set of substitutions to Equation 15.12 leads to the equation for a backward rotation of bar-scaled regular multipole expansions:

$$\sum_{k=-l}^{l} \bar{D}^{l*}_{k,m}(-\beta,-\alpha)\,\bar{M}_{l,k} = e^{-im\alpha}\left\{\bar{d}^{l}_{0,m}(-\beta)\,\bar{M}_{l,0} + \sum_{k=1}^{l}\left[\bar{d}^{l}_{k,m}(-\beta)\,\bar{M}_{l,k} + \bar{d}^{l}_{-k,m}(-\beta)\,\bar{M}^{*}_{l,k}(-1)^{k}\right]\right\}. \tag{15.14}$$

The derivation of the equations for rotation of a coordinate system of the tilde-scaled irregular multipole expansions employs the substitution $D^l_{k,m} \rightarrow \tilde{D}^l_{k,m}$, $d^l_{k,m} \rightarrow \tilde{d}^l_{k,m}$, $Y_{l,m} \rightarrow \tilde{L}_{l,m}$, and $Y^*_{l,m} \rightarrow \tilde{L}^*_{l,m}$. Application of these changes to Equation 15.8 leads to:

$$\sum_{k=-l}^{l} \tilde{D}^l_{k,m}(\alpha,\beta)\tilde{L}_{l,k} = \tilde{d}^l_{0,m}(\beta)\tilde{L}_{l,0} + \sum_{k=1}^{l}\left[e^{-ik\alpha}\tilde{d}^l_{k,m}(\beta)\tilde{L}_{l,k} + e^{ik\alpha}\tilde{d}^l_{-k,m}(\beta)\tilde{L}^*_{l,k}(-1)^k\right], \tag{15.15}$$

which is the equation for forward rotation of tilde-scaled irregular multipole expansions. Application of the same substitutions to Equation 15.10 provides:

$$\sum_{k=-l}^{l} \tilde{D}^l_{k,m}(-\beta,-\alpha)\tilde{L}_{l,k} = e^{im\alpha}\left\{\tilde{d}^l_{0,m}(-\beta)\tilde{L}_{l,0} + \sum_{k=1}^{l}\left[\tilde{d}^l_{k,m}(-\beta)\tilde{L}_{l,k} + \tilde{d}^l_{-k,m}(-\beta)\tilde{L}^*_{l,k}(-1)^k\right]\right\}, \tag{15.16}$$

which describes the backward rotation of tilde-scaled irregular multipole expansions. As with the case of bar-scaled regular multipole expansions, the similarity of Equations 14.6 to 14.17 from Chapter 14 makes obtaining Equations 15.15 and 15.16 a straightforward procedure. The resulting equations (Equations 15.13 through 15.16) are computer-ready equations for the rotation of scaled multipole expansions that rely on the use of scaled Wigner matrix elements.

Another set of implementation-ready rotation equations, this time using the unscaled Wigner matrix elements, follows from the conversion of Equation 14.15 to multipole expansions. In it, replacing the symbols of scaled solid harmonics with those for scaled multipole expansions gives:

$$\begin{aligned}\bar{M}_{l,m} &= \sum_{k=-l}^{l}\sqrt{\frac{(l-k)!(l+k)!}{(l-m)!(l+m)!}}D^{l*}_{k,m}(\alpha,\beta)\bar{M}_{l,k},\\ \tilde{L}_{l,m} &= \sum_{k=-l}^{l}\sqrt{\frac{(l-m)!(l+m)!}{(l-k)!(l+k)!}}D^{l}_{k,m}(\alpha,\beta)\tilde{L}_{l,k},\end{aligned} \tag{15.17}$$

where the conversion to the regular multipole expansion $\bar{M}_{l,m} = \sum_i q_i \bar{R}^*_{l,m}(\mathbf{r}_i)$ requires taking the complex conjugate of both sides. Rewriting the sum in the manner just described, the forward and backward rotations become:

$$\begin{aligned}\sum_{k=-l}^{l} D^{l*}_{k,m}(\alpha,\beta)\bar{M}_{l,k} &= \sqrt{\frac{(l)!(l)!}{(l-m)!(l+m)!}}d^l_{0,m}(\beta)\bar{M}_{l,0}\\ &\quad + \sqrt{\frac{(l-k)!(l+k)!}{(l-m)!(l+m)!}}\sum_{k=1}^{l}\left[e^{ik\alpha}d^l_{k,m}(\beta)\bar{M}_{l,k} + e^{-ik\alpha}d^l_{-k,m}(\beta)\bar{M}^*_{l,k}(-1)^k\right],\end{aligned} \tag{15.18}$$

$$\begin{aligned}\sum_{k=-l}^{l} D^{l*}_{k,m}(-\beta,-\alpha)\bar{M}_{l,k} &= e^{-im\alpha}\left\{\sqrt{\frac{(l)!(l)!}{(l-m)!(l+m)!}}d^l_{0,m}(-\beta)\bar{M}_{l,0}\right.\\ &\quad \left. + \sqrt{\frac{(l-k)!\,(l+k)!}{(l-m)!\,(l+m)!}}\sum_{k=1}^{l}\left[d^l_{k,m}(-\beta)\bar{M}_{l,k} + d^l_{-k,m}(-\beta)\bar{M}^*_{l,k}(-1)^k\right]\right\}\\ &= e^{-im\alpha}\left\{\sqrt{\frac{(l)!(l)!}{(l-m)!(l+m)!}}d^l_{m,0}(\beta)\bar{M}_{l,0}\right.\\ &\quad \left. + \sqrt{\frac{(l-k)!(l+k)!}{(l-m)!(l+m)!}}\sum_{k=1}^{l}\left[d^l_{m,k}(\beta)\bar{M}_{l,k} + d^l_{m,-k}(\beta)\bar{M}^*_{l,k}(-1)^k\right]\right\},\end{aligned} \tag{15.19}$$

$$\sum_{k=-l}^{l} D^{l}_{k,m}(\alpha,\beta)\tilde{L}_{l,k} = \sqrt{\frac{(l-m)!(l+m)!}{(l)!(l)!}}\, d^{l}_{0,m}(\beta)\,\tilde{L}_{l,0} + \sqrt{\frac{(l-m)!(l+m)!}{(l-k)!(l+k)!}} \sum_{k=1}^{l}\left[e^{-ik\alpha} d^{l}_{k,m}(\beta)\,\tilde{L}_{l,k} + e^{ik\alpha} d^{l}_{-k,m}(\beta)\,\tilde{L}^{*}_{l,k}(-1)^{k}\right], \tag{15.20}$$

$$\begin{aligned}\sum_{k=-l}^{l} D^{l}_{k,m}(-\beta,-\alpha)\tilde{L}_{l,k} &= e^{im\alpha}\left\{\sqrt{\frac{(l-m)!(l+m)!}{(l-k)!(l+k)!}}\, d^{l}_{0,m}(-\beta)\,\tilde{L}_{l,0} + \sqrt{\frac{(l-m)!(l+m)!}{(l-k)!(l+k)!}} \sum_{k=1}^{l}\left[d^{l}_{k,m}(-\beta)\,\tilde{L}_{l,k} + d^{l}_{-k,m}(-\beta)\,\tilde{L}^{*}_{l,k}(-1)^{k}\right]\right\} \\ &= e^{im\alpha}\left\{\sqrt{\frac{(l-m)!(l+m)!}{(l-k)!(l+k)!}}\, d^{l}_{m,0}(\beta)\,\tilde{L}_{l,0} + \sqrt{\frac{(l-m)!(l+m)!}{(l-k)!(l+k)!}} \sum_{k=1}^{l}\left[d^{l}_{m,k}(\beta)\,\tilde{L}_{l,k} + d^{l}_{m,-k}(\beta)\,\tilde{L}^{*}_{l,k}(-1)^{k}\right]\right\}.\end{aligned} \tag{15.21}$$

These derived equations can be used for debugging the rotation equations that use scaled Wigner matrix elements, or even in the actual applied computations. In that case, to maintain reasonable computational performance, precomputing and tabulating the values of the square root coefficients will be necessary, even though it requires additional memory storage.

Completed characterization of the rotation matrix together with previous work conducted in Chapters 13 and 14 laid the ground for substituting the conventional multipole translation operations, M2M, M2L, and L2L, which have $O(N^4)$ scaling with respect to the size N of multipole expansion, with a better scaling $O(N^3)$ algorithm, which consists of three major steps. In the first step, the program rotates the coordinate system of the bar-scaled regular multipole expansion to align its z-axis with the translation vector, which connects the origin of the coordinate system of the donor box with the center of the coordinate system of the recipient box. In the second step, the program translates the origin of the rotated bar-scaled regular multipole expansion from the donor box to the origin of the recipient box. In the final third step, the computer code rotates the coordinate system of the translated bar-scaled regular multipole expansion to align its z-axis with the z-axis of the recipient box. Because the alignment of the z-axis of the local coordinate system is the same in all boxes, the second rotation technically corresponds to a straightforward backward rotation by the same angles used in the forward rotation, but with the signs reversed.

The remainder of this chapter explains computer implementation of the rotation-based multipole operations.

15.2 Rotation-Based M2M Operation

Program code for the rotation-based M2M operation shifts the origin of the bar-scaled regular multipole expansion along the translation vector. The program code, which reuses a number of previously developed subroutines, and illustrates the implementation of the rotation equations, follows:

```
subroutine M2Mrotate(nL, tri, nSzTri, nSzWset, d, tvector, expalp,
inpExpansion, tmpExpansion, outExpansion)
! Rotate the coordinate system of bar-scaled regular multipole expansion
! Translate the coordinate system of the rotated expansion along the z-axis
! Rotate back the coordinate system of the translated expansion
  !
  implicit none
  !
```

```
! Subroutine arguments
integer, intent(in)               :: nL             ! max value of index L
integer, intent(in)               :: tri(0:nL)      ! triangular pointers
integer, intent(in)               :: nSzTri         ! size of expansion array
integer, intent(in)               :: nSzWset        ! size of the Wigner set
double precision, intent(in)      :: beta           ! Euler angle
double precision, intent(out)     :: d(1:nSzWset)   ! Wigner set of d-matrices
double precision, intent(in)      :: tvector(3)     ! translation vector
complex(16), intent(inout)        :: expalp(0:nL)   ! complex exponents
complex(16), intent(in)         :: inpExpansion(0:nSzTri) ! to be rotated
complex(16), intent(inout)      :: tmpExpansion(0:nSzTri) ! temporary expansion
complex(16), intent(out)        :: outExpansion(0:nSzTri) ! output expansion
!
! Local variables
double precision, parameter :: phase(0:1) = (/ 1.0d0, -1.0d0 /)
double precision            :: alpha, beta   ! Euler angles
integer                     :: L, k, m       ! indices of d-matrix d(L,k,m)
integer            :: aL
complex(16)        :: csum
double precision :: kalp, distance
!
! Function declarations
integer            :: fptr

! Step 1: rotate forward the bar-scaled regular multipole expansion

! Determine Euler angles
call EulerAngles(tvector, alpha, beta)

! Compute bar-scaled Wigner d-matrix for angle beta
call dmatmbar(nL, nSzWset, beta, d)

! compute complex exponents, exp(i k alpha)
expalp(0) = dcmplx(1.0d0, 0.0d0)
do k = 1, nL
  kalp = alpha * dble(k)
  expalp(k) = dcmplx(dcos(kalp), dsin(kalp))
enddo

! rotate forward using bar-scaled d(beta)
do L = 0, nL
  aL = tri(L)
  !
  do m = 0, L
    ! k = 0
    csum = d(fptr(L,0,m)) * inpExpansion(aL)
    do k = 1, L
      csum = csum + expalp(k) * inpExpansion(aL+k)  * d(fptr(L, k, m)) + &
            dconjg(expalp(k) * inpExpansion(aL+k)) * d(fptr(L,-k, m)) * &
            phase(mod(k,2))
    enddo
    outExpansion(aL+m) = csum
  enddo
enddo

! Step 2: shift the coordinate system along z-axis

! Compute translation distance
distance = dsqrt( tvector(1)*tvector(1) + &
                tvector(2)*tvector(2) + tvector(3)*tvector(3) )
```

```
    ! Move the origin in positive direction defined by -distance
    call M2Mz(nL, nSzTri, tri, outExpansion, -distance, tmpExpansion)

    ! Step 3: rotate backward the bar-scaled regular multipole expansion

    ! Compute bar-scaled Wigner d-matrix for angle -beta
    call dmatmbar(nL, nSzWset, -beta, d)

    ! rotate backward using bar-scaled d(-beta)
    do L = 0, nL
      aL = tri(L)
      !
      do m = 0, L
        ! k = 0
        csum = d(fptr(L,0,m)) * tmpExpansion(aL)
        do k = 1, L
          csum = csum + tmpExpansion(aL+k)   * d(fptr(L, k, m)) + &
                 dconjg(tmpExpansion(aL+k)) * d(fptr(L,-k, m)) *phase(mod(k,2))
        enddo
        outExpansion(aL+m) = csum * dconjg(expalp(m))
      enddo
    enddo
end subroutine M2Mrotate
```

Subroutine M2Mrotate() receives the size of expansion, *nL*, the array of triangular pointers, *tri*, the size of triangular array, *nSzTri*, and the size of elements in the entire set of Wigner *d*-matrices, *nSzWset*, on input, treating these variables as read-only parameters. Array *d* is the placeholder for all the Wigner *d*-matrices to be computed. Parameter *tvector* is a vector that originates at the center of the local coordinate system of the donor box and points to the center of the local coordinate system of the recipient box. The donor box is the holder of the regular multipole expansion to be translated, and the recipient box is the place the regular multipole expansion is shifted to. Array *expalp* is the placeholder for the complex exponent part of the rotation matrix, and array *inpExpansion* is the bar-scaled regular multipole expansion to be translated. Array *tmpExpansion* is a scratch space. Subroutine *M2Mrotate*() returns the translated bar-scaled regular multipole expansion in array *outExpansion*.

Subroutine *M2Mrotate*() begins by calling subroutine *EulerAngles*(), previously developed in Section 14.1, to evaluate the Euler angles *alpha* and *beta* for the translation vector *tvector*. After that, the action proceeds to the computation of the bar-scaled Wigner *d*-matrices of degree *l* for the range 0 through *nL*. Section 14.6 describes how this is done in subroutine *dmatmbar*(). At this point, either of the *k*-set or *m*-set of recurrence relations can be used in computing the bar-scaled Wigner *d*-matrices, but because the *m*-set requires the least number of floating point operations when cast in the bar-scaled form, the code shown uses subroutine *dmatmbar*(). On most computational platforms this should be the faster procedure.

Next, the complex exponents, $e^{ik\alpha}$, for the elements of array *expalp* are evaluated. Here the array index *k* spans all values from zero to *nL*. Conventionally, one would need to compute the exponent with a negative sign in front of each complex number when constructing the rotation matrix, but initializing the array *expalp* with the positive power reduces the number of times the operation of complex conjugation would need to be performed in the process of applying the rotation to the bar-scaled regular multipole expansion.

Step one in the multipole-based M2M operation ends by applying the forward rotation to the coordinate system of the bar-scaled regular multipole expansion to be translated. The rotated bar-scaled regular multipole expansion temporarily occupies the array *outExpansion*. The result of computing each element of the *do-loop* over index *k* accumulates in the complex variable *csum*, and the program code moves the result to array element *outExpansion(aL+m)* only after the *do-loop* finishes. This avoids computing the array index in each cycle of the *do-loop*, although the Fortran compiler optimizes such operations automatically.

The lines *tmpExpansion(aL+k)* * *d(fptr(L,k,m))* and *dconjg(expalp(k)* * *inpExpansion(aL+k))* * *d(fptr(L,−k,m))* * *phase(mod(k,2))* of the code correspond to the parts $e^{ik\alpha}\,\bar{d}^{\,l}_{k,m}(\beta)\,\bar{M}_{l,k}$ and $e^{-ik\alpha}\,\bar{d}^{\,l}_{-k,m}(\beta)\,\bar{M}^{*}_{l,k}(-1)^k$ of Equation 15.13, respectively. Initialization of array *expalp* with the values of $e^{ik\alpha}$ provides the opportunity to avoid calling Fortran intrinsic function *dconjg*(), which performs complex conjugation, in the first line, and to apply it to the product *expalp*(*k*) * *inpExpansion*(*aL*+*k*) in the second line so that this function is invoked only once in each cycle of the *do-loop* over index *k*. The multiplication factor *phase*(*mod*(*k*, 2)) returns 1 or −1 in decimal form from the two-value array *phase* for even and odd values of index *k*, respectively.

Step two in the rotation-based M2M operation computes the length of the translation vector, *tvector*, and stores the result in variable *direction*. Then the origin of the forward-rotated expansion, temporarily held in the array *outExpansion*, is translated along the positive direction of the *z*-axis by using subroutine *M2Mz*(), described earlier in Section 13.1. Within that implementation, a negative value of the vector length implies translation in the positive direction. On exit, subroutine *M2Mz*() returns the expansion, which has its origin translated, in the array *tmpExpansion*.

The final step in the rotation-based M2M operation starts with calculating the bar-scaled Wigner *d*-matrix for the backward rotation angle −*beta*. This operation is necessary, since bar–scaling makes the *d*-matrix non-unitary and therefore the usual way to reuse the *d*-matrix computed for the positive angle *beta* no longer works. Invocation of subroutine *dmatmbar*() performs the computation. On the other hand, there is no need to compute the complex exponents for angle *alpha* because these can readily be obtained from the already computed exponents for the positive angle stored in the array *expalp* by using intrinsic function *dconjg*().

Computing the backward rotation uses Equation 15.14. This rotation applies to the shifted bar-scaled regular multipole expansion held in array *tmpExpansion*. The result inside the *do-loop* over index *k* accumulates in the temporary complex variable *csum*, and then moves to the output array *outExpansion*. Fewer floating-point operations are needed in the backward rotation because the complex exponent appears only in the *do-loop* over index *m*.

Subroutine *M2Mrotate*() returns the transformed bar-scaled regular multipole expansion in array *outExpansion*. Validation consists of comparing each element of that array against the result of conventional computation in subroutine *M2M*(). The compared values should agree within the machine precision since no additional approximation is made in the rotation-based M2M operation over that in the conventional method.

15.3 Rotation-Based M2L Operation

The rotation-based M2L operation transforms a bar-scaled regular multipole expansion defined about the center of a donor box into a tilde-scaled irregular multipole expansion associated with the local coordinate system of the recipient box. This M2L computation breaks down into three major steps. Step one rotates the coordinate system of a bar-scaled regular multipole expansion to align its *z*-axis with the translation vector, which originates from the center of the donor box, and points toward the center of the local coordinate system of the recipient box. Step two transforms the rotated bar-scaled regular multipole expansion into an intermediate tilde-scaled irregular multipole expansion. Step three finally rotates the coordinate system of the generated tilde-scaled irregular expansion backwards to align its *z*-axis with that of the local coordinate system of the recipient box. These steps can be seen in the following code for subroutine *M2Lrotate*():

```
subroutine M2Lrotate(nL, tri, nSzTri, nSzWset, d, tvector, expalp,
inpExpansion, tmpExpansion, outExpansion)
! Rotate the coordinate system of bar-scaled regular multipole expansion
! Translate the coordinate system of the rotated expansion along the z-axis
! Rotate back the coordinate system of the translated expansion
  !
  implicit none
  !
  ! Subroutine arguments
  integer, intent(in)          :: nL          ! max value of index L
  integer, intent(in)          :: tri(0:nL)   ! triangular pointers
  integer, intent(in)          :: nSzTri      ! size of expansion array
```

```
integer, intent(in)              :: nSzWset        ! size of the Wigner set
double precision, intent(in)     :: beta           ! Euler angle
double precision, intent(out)    :: d(1:nSzWset)   ! Wigner set of d-matrices
double precision, intent(in)     :: tvector(3)     ! translation vector
complex(16), intent(inout)       :: expalp(0:nL)   ! complex exponents
complex(16), intent(in)       :: inpExpansion(0:nSzTri) ! to be rotated
complex(16), intent(inout)    :: tmpExpansion(0:nSzTri) ! temporary expansion
complex(16), intent(out)      :: outExpansion(0:nSzTri) ! output expansion
!
! Local variables
double precision, parameter :: phase(0:1) = (/ 1.0d0, -1.0d0 /)
double precision            :: alpha, beta   ! Euler angles
integer                     :: L, k, m       ! indices of d-matrix d(L,k,m)
integer          :: aL
complex(16)      :: csum
double precision :: kalp, distance
!
! Function declarations
integer          :: fptr

! Step 1: rotate forward the bar-scaled regular multipole expansion

! Determine Euler angles
call EulerAngles(tvector, alpha, beta)

! Compute bar-scaled Wigner d-matrix for angle beta
call dmatmbar(nL, nSzWset, beta, d)

! compute complex exponents, exp(i k alpha)
expalp(0) = dcmplx(1.0d0, 0.0d0)
do k = 1, nL
  kalp = alpha * dble(k)
  expalp(k) = dcmplx(dcos(kalp), dsin(kalp))
enddo

! rotate forward using bar-scaled d(beta)
do L = 0, nL
  aL = tri(L)
  !
  do m = 0, L
    ! k = 0
    csum = d(fptr(L,0,m)) * inpExpansion(aL)
    do k = 1, L
      csum = csum + expalp(k) * inpExpansion(aL+k)  * d(fptr(L, k, m)) + &
             dconjg(expalp(k) * inpExpansion(aL+k)) * d(fptr(L,-k, m)) * &
             phase(mod(k,2))
    enddo
    outExpansion(aL + m) = csum
  enddo
enddo

! Step 2: shift the coordinate system along z-axis

! Compute translation distance
distance = dsqrt( tvector(1)*tvector(1) + &
                  tvector(2)*tvector(2) + tvector(3)*tvector(3) )

! Move the origin in positive direction defined by -distance
call M2Lz(nL, nSzTri, tri, outExpansion, -distance, tmpExpansion)
```

```
  ! Step 3: rotate backward the tilde-scaled irregular multipole expansion

  ! Compute tilde-scaled Wigner d-matrix for angle -beta
  call dmatktilde(nL, nSzWset, -beta, d)

  ! rotate backward using tilde-scaled d(-beta)
  do L = 0, nL
    aL = tri(L)
    !
    do m = 0, L
      ! k = 0
      csum = d(fptr(L,0,m)) * tmpExpansion(aL)
      do k = 1, L
        csum = csum + tmpExpansion(aL + k)   * d(fptr(L, k, m)) + &
               dconjg(tmpExpansion(aL+k)) * d(fptr(L,-k, m)) * &
               phase(mod(k,2))
      enddo
      outExpansion(aL + m) = csum * expalp(m)
    enddo
  enddo
end subroutine M2Lrotate
```

Due to their algorithmic similarity, the program code for the rotation-based M2L transformation reuses a significant portion of the program code developed for the rotation-based M2M operation. In particular, the interface to subroutine *M2Lrotate*() and the portion of the code performing Step 1 replicate those of subroutine *M2Mrotate*().

Step 2, in addition to computing the translation distance between the centers of the donor and recipient boxes, translates the coordinate system of the rotated bar-scaled regular multipole expansion stored in array *outExpansion* along the translation vector using subroutine *M2Lz*() developed in Section 13.3. This translation produces a tilde-scaled irregular multipole expansion stored in array *tmpExpansion*.

Step 3 begins by computing the complete set of tilde-scaled Wigner *d*-matrices for the backward rotation through angle −*beta*. This uses subroutine *dmatktilde*() developed in Section 14.5 and the *k*-set of recurrence relations for computation of tilde-scaled *d*-matrices to produce the most efficient code. After that, the program code uses Equation 15.16 to perform a backward rotation of the coordinate system of the tilde-scaled irregular multipole expansion stored in array *tmpExpansion*. The final tilde-scaled irregular multipole expansion defined about the local coordinate system of the recipient box is accumulated in the array *outExpansion*.

Validation of subroutine M2Lrotate() consists of comparing each element of the array *outExpansion* with the result of conventional computation in subroutine *M2L*() and verifying that the data agree for ten or more digits after the decimal point.

15.4 Rotation-Based L2L Operation

The rotation-based L2L operation recasts a tilde-scaled irregular multipole expansion from the local coordinate system of a donor box into the local coordinate system of a recipient box. This involves three major steps. First, the local coordinate system of the tilde-scaled irregular multipole expansion is rotated to align its z-axis with the vector of translation from the center of the donor box to the center of the recipient box, this resulting in a rotated tilde-scaled irregular multipole expansion. Second, the origin of the rotated tilde-scaled irregular expansion is translated along the z-axis in the positive direction defined by the translation vector. The third step rotates the coordinate system of the translated tilde-scaled irregular expansion back into its original orientation. This is illustrated in the following implementation:

```
subroutine L2Lrotate(nL, tri, nSzTri, nSzWset, d, tvector, expalp,
inpExpansion, tmpExpansion, outExpansion)
! Rotate the coordinate system of tilde-scaled irregular multipole expansion
```

```
! Translate the coordinate system of the rotated expansion along the z-axis
! Rotate back the coordinate system of the translated expansion
  !
  implicit none
  !
  ! Subroutine arguments
  integer, intent(in)                  :: nL           ! max value of index L
  integer, intent(in)                  :: tri(0:nL)    ! triangular pointers
  integer, intent(in)                  :: nSzTri       ! size of expansion array
  integer, intent(in)                  :: nSzWset      ! size of the Wigner set
  double precision, intent(in)         :: beta         ! Euler angle
  double precision, intent(out)        :: d(1:nSzWset) ! Wigner set of d-matrices
  double precision, intent(in)         :: tvector(3)   ! translation vector
  complex(16), intent(inout)           :: expalp(0:nL) ! complex exponents
  complex(16), intent(in)        :: inpExpansion(0:nSzTri) ! to be rotated
  complex(16), intent(inout)     :: tmpExpansion(0:nSzTri) ! temporary expansion
  complex(16), intent(out)       :: outExpansion(0:nSzTri) ! output expansion
  !
  ! Local variables
  double precision, parameter :: phase(0:1) = (/ 1.0d0, -1.0d0 /)
  double precision            :: alpha, beta  ! Euler angles
  integer                     :: L, k, m      ! indices of d-matrix d(L,k,m)
  integer          :: aL
  complex(16)      :: csum
  double precision :: kalp, distance
  !
  ! Function declarations
  integer          :: fptr

  ! Step 1: rotate forward the tilde-scaled irregular multipole expansion

  ! Determine Euler angles
  call EulerAngles(tvector, alpha, beta)

  ! Compute tilde-scaled Wigner d-matrix for angle beta
  call dmatktilde(nL, nSzWset, beta, d)

  ! compute complex exponents, exp(-i k alpha)
  expalp(0) = dcmplx(1.0d0, 0.0d0)
  do k = 1, nL
    kalp = alpha * dble(k)
    expalp(k) = dcmplx(dcos(kalp), -dsin(kalp))
  enddo

  ! rotate forward using tilde-scaled d(beta)
  do L = 0, nL
    aL = tri(L)
    !
    do m = 0, L
      ! k = 0
      csum = d(fptr(L,0,m)) * inpExpansion(aL)
      do k = 1, L
        csum = csum + expalp(k) * inpExpansion(aL + k)   * d(fptr(L, k, m)) + &
               dconjg(expalp(k) * inpExpansion(aL + k)) * d(fptr(L,-k, m)) * &
               phase(mod(k,2))
      enddo
      outExpansion(aL + m) = csum
    enddo
  enddo
```

```
  ! Step 2: shift the coordinate system along z-axis

  ! Compute translation distance
  distance = dsqrt( tvector(1)*tvector(1) + &
                    tvector(2)*tvector(2) + tvector(3)*tvector(3) )

  ! Move the origin in positive direction defined by -distance
  call L2Lz(nL, nSzTri, tri, outExpansion, -distance, tmpExpansion)

  ! Step 3: rotate backward the tilde-scaled irregular multipole expansion

  ! Compute tilde-scaled Wigner d-matrix for angle -beta
  call dmatktilde(nL, nSzWset, -beta, d)

  ! rotate backward using tilde-scaled d(-beta)
  do L = 0, nL
    aL = tri(L)
    !
    do m = 0, L
      ! k = 0
      csum = d(fptr(L,0,m)) * tmpExpansion(aL)
      do k = 1, L
        csum = csum + tmpExpansion(aL + k)  * d(fptr(L, k, m)) + &
               dconjg(tmpExpansion(aL + k)) * d(fptr(L,-k, m)) * &
               phase(mod(k,2))
      enddo
      outExpansion(aL + m) = csum * dconjg(expalp(m))
    enddo
  enddo
end subroutine L2Lrotate
```

Subroutine *L2Lrotate*() uses the argument *tvector* to establish the direction of translation of the local coordinate system of the tilde-scaled irregular multipole expansion. This vector originates from the center of the coordinate system of the donor box and terminates at the center of the coordinate system of the recipient box. Computation begins with determining Euler angles *alpha* and *beta* by calling subroutine *EulerAngles*(), developed in Section 14.1. Evaluation of the tilde-scaled Wigner *d*-matrices for angle *beta* uses the *k*-set of recurrence relations implemented in subroutine *dmatktilde*(), as described in Section 14.5. The array *expalp* is then initialized with the complex exponents $e^{-ik\alpha}$ for rotation angle *alpha*. In *expalp*, index *k* varies from zero to the last orbital index, *nL*, of the multipole expansion. Step 1 finishes with the forward rotation that implements Equation 15.15 and temporarily saves the rotated bar-scaled irregular multipole expansion in array *outExpansion*.

Step 2 determines the distance between the origins of the donor and recipient boxes by computing the length of *tvector*, and performs the coordinate system translation of the rotated bar-scaled irregular multipole expansion along the *z*-axis by invoking subroutine *L2Lz*(), developed in Section 13.2. The program code temporarily saves the translated tilde-scaled irregular multipole expansion in array *tmpExpansion*.

Step 3 proceeds by using subroutine *dmatktilde*() to compute the tilde-scaled Wigner *d*-matrices for the *k*-set of recurrence relations for angle *beta*, followed by a backward rotation using Equation 15.16. In this operation, phase factor *phase(mod(k*, 2)) returns 1 and −1 in decimal form for even and odd values of index *k*, respectively. The resulting tilde-scaled irregular multipole expansion is then saved in the array *outExpansion*.

To verify the correctness of the computation, the simplest approach is to compare each element of the array *outExpansion* with the result of conventional computations performed using subroutine *L2L*(). The compared data should agree for ten or more digits after decimal point.

Subroutines *M2Mrotate*(), *M2Lrotate*(), and *L2Lrotate*() each have a maximum of three enclosed *do-loops* in the Wigner *d*-matrix computation and in the rotation operation, with each *do-loop* being bound by the size of the multipole expansion. A consequence of this is that the resulting codes scale as $O(N^3)$ versus the $O(N^4)$ scaling exhibited by the conventional algorithms, and therefore the computational cost of rotation-base multipole translation operations grows more slowly than that of the conventional multipole operations as the size of the multipole expansion increases. However, due to the considerable cost of calculating the Wigner d-matrix, the crossover point between the $O(N^3)$ and $O(N^4)$ algorithms lies in the region of $nL = 30$. For multipole expansions smaller than that, conventional $O(N^4)$ algorithms might be faster than the $O(N^3)$ one, so any actual FMM computer program should have both the $O(N^3)$ and $O(N^4)$ algorithms for multipole operations implemented. Depending on the size of the multipole expansion selected on the input, the program could choose the use of $O(N^3)$ or $O(N^4)$ algorithm, whichever would be faster.

16

Periodic Boundary Conditions

Condensed-phase phenomena play important role in chemistry and biology; therefore, these disciplines qualify as material sciences.

A characteristic feature of any material is that each particle in it interacts with an infinite number of other particles. A straightforward approach for constructing a computationally manageable atomic-level model of a material would be to cut a representative cluster of molecules out of the condensed matter continuum to limit the number of interacting particles and the accompanying computational cost. However, the resulting surface effects in such system easily reach the reactive center deviating the model from the true condensed phase. Increasing the thickness of the cluster shell around the reactive center is of only limited use since the number of interacting particles increases as the cube of the distance from the center to the surface.

One common solution to the problem of material simulation is to use periodic boundary conditions (PBC), as illustrated in Figure 16.1 in two dimensions, where image cells are periodic replicas of the central unit cell (CUC). A system modeled with PBC is infinite in size, having neither walls nor surfaces. In this model, mobile molecules can travel freely from one unit cell of the dynamic system to another. The contribution of all image cells into the energy of the CUC is known as the *lattice sum*. An undesirable effect of PBC in application to liquids is the emergence of a crystal-like structure when a small CUC is used, but, even so, the ability of PBC to efficiently represent condensed phase outweigh its limitations.

16.1 Principles of Periodic Boundary Conditions

The notion of PBC comes from crystallography, from the study of atomic structure of condensed matter. Crystallography employs a reciprocal space in which the CUC obtains the unit side length (multiplied by the factor 2π). This is where the word *unit* in the name *central unit cell* comes from. Although PBC operate in direct space, and do not require unit side lengths, referring to the central cell as an elemental CUC remains a useful way of distinguishing it from composite cells (or super-cells), which will appear in periodic FMM.

The CUC consists of a collection of particles. Replicating those in the global coordinate system makes the rest of the periodic system. To that end, the lowest left front corner of the CUC defines the origin of the global coordinate system; the rest of the periodic system is then symmetrically constructed around the central cell. The word *central* thus emphasizes the unique position that the *central unit cell* has in the global coordinate system. Note that the global coordinate system is only used for the purpose of replicating the central cell, and should not be confused with the local coordinate system associated with the center of each cell that will be used in FMM work.

A cell needs three vectors, $\mathbf{t}^a$, $\mathbf{t}^b$, and $\mathbf{t}^c$, placed along the cell edges, to describe its shape. In the general case of an arbitrarily-oriented global coordinate system and the CUC of arbitrary crystallographic type, these vectors have coordinates $\left(t_x^a, t_y^a, t_z^a\right)$, $\left(t_x^b, t_y^b, t_z^b\right)$, and $\left(t_x^c, t_y^c, t_z^c\right)$, respectively. Of all possible cell types, the simplest and most common case is that of a rectangular-shaped cell. Aligning the edges of the CUC with the axes of the global coordinate system allows the cell vectors to be simplified to $\left(t_x^a, 0, 0\right)$, $\left(0, t_y^b, 0\right)$, and $\left(0, 0, t_z^c\right)$. Since FMM works with cubic boxes, this additionally simplifies the construction of the periodic lattice in that, for a cubic central cell, the cell vectors $\mathbf{t}^a$, $\mathbf{t}^b$, and $\mathbf{t}^c$ reduce to $(t, 0, 0)$, $(0, t, 0)$, and $(0, 0, t)$, respectively, where t is the length of a cell side.

Replication of the central cell along the cell vectors in both positive and negative directions generates an infinite number of image cells. To track the locations of the image cells, each of them receives a 3-digit

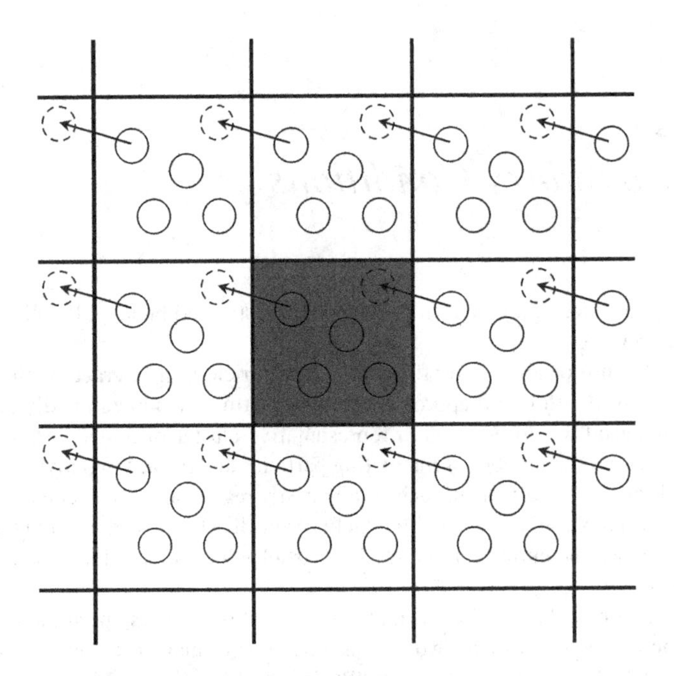

FIGURE 16.1 Two-dimensional model of a liquid system under periodic boundary conditions. The shaded box is the CUC, and the other boxes are its periodic replicas. The arrows indicate an object leaving the cell on one side and simultaneously entering the same cell from the opposite side in the process of molecular dynamics simulation.

cell identifier $\{a, b, c\}$ that specifies the distances from the center of the global coordinate system to the lowest left corner of the cell in integer multiples, a, b, c of the side-length of the central cell along the cell vectors $\mathbf{t}^a$, $\mathbf{t}^b$, and $\mathbf{t}^c$, respectively, as illustrated in Figure 16.2. Using this convention, the identifier {0, 0, 0} defines the location of the central unit cell.

A system under PBC consists of infinite number of interacting particles. Out of those, only the particles from the CUC and those in the adjacent unit cells explicitly interact one with another. The more distant cells only contribute to the CUC *via* the lattice sum, as will be explained next.

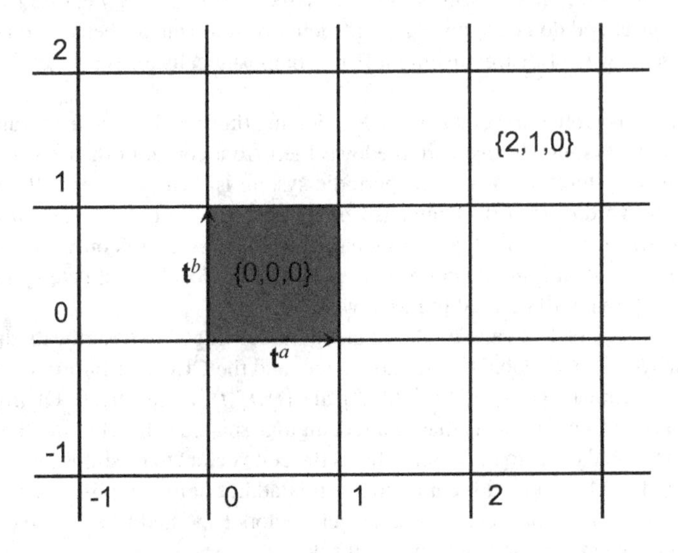

FIGURE 16.2 A two-dimensional slice of a 3D periodic system at coordinate $z = 0$ with the z-axis pointing toward the reader. Cell vectors $\mathbf{t}^i$ having coordinates $\left(t_x^i, t_y^i, t_z^i\right)$ define the geometry of the cell where $i = \{a, b, c\}$. Integer numbers along cell vectors in units of the respective vector length indicate the location of a cell in the lattice.

16.2 Lattice Sum for Energy in Periodic FMM

For a set of particles located in a cubic CUC of side length t, and having Cartesian coordinates $\left(x_i^0, y_i^0, z_i^0\right)$ in the global coordinate system, where i is the index of a particle, the position of each particle in an image cell is given by:

$$x_i^a = x_i^0 + at, \quad y_i^b = y_i^0 + bt, \quad z_i^c = z_i^0 + ct, \tag{16.1}$$

where the superscript index indicates the appropriate cell along the respective axis. Positive integer values of the image index span in the positive direction of the coordinate axis, and negative integer values of the index go in the negative direction of the coordinate axis.

The total electrostatic energy U_{total} of the CUC under PBC is

$$U_{total} = \frac{1}{2}\sum_{n=-\infty}^{\infty}\sum_{i=1}^{N}\sum_{j=1}^{N}\frac{q_i^0 q_j^n}{\left|\mathbf{r}_i^0 - \mathbf{r}_j^n\right|}, \tag{16.2}$$

where index n represents a triple sum over indices a, b, c, and goes through all the unit cells in the periodic system. The index i iterates over all particles in the central cell, and index j iterates over all particles in cell n. When $n = 0$, which is equivalent to $n = \{0, 0, 0\}$, index j represents particles in the CUC. In that case, the summation requires that $i \neq j$.

The sum in Equation 16.2 is conditionally convergent; that makes it numerically sensitive to the order of summation of terms. It would diverge to infinity if summed along the positive or negative direction of the indices. Even when using precaution to avoid the divergence, the slow rate of convergence of Equation 16.2 makes the direct summation impractical. Using periodic FMM provides a mechanism that delivers a rapidly converging lattice sum. The computation relies on a non-periodic FMM to generate a starting multipole expansion of the charge sources located in the CUC. PBC replicates the multipole expansion of the CUC to each image cell, where the replicated expansion becomes defined about the center of the respective image cell. Finally, periodic FMM recasts the problem of particle summation into the task of accounting for multipole contribution from the image cells into the central cell.

The renormalization approach developed by Kudin and Scuseria provides a particularly elegant formulation of periodic FMM. This method creates an upside-down hierarchy of super-cells constructed on the top of the CUC like nesting dolls as illustrated in Figure 16.3 for 1D. As usual, the parent boxes encompass child boxes. However, unlike the case of a non-periodic FMM, the hierarchy level numbering begins from the child side of the tree, because the parent end of the tree goes off to infinity. This results in the tree standing upside down.

The tree includes an infinite number of hierarchy levels, each represented by *central*, *near-field*, and *far-field* super-cells. The principal concept of the tree construction algorithm involves the arranging of the group of smaller cells into a larger super-cell. In the beginning, at $h = 0$, the central super-cell matches the CUC. The first actual super-cell, of size $(2ws + 1)$ appears at $h = 1$. It is created by padding the CUC by ws unit cells on each side, as shown in black on Figure 16.3. At the hth step, the 1D central super-cell has the size $(2ws + 1)^h$. Switching from 1D to 3D increases the size of the central super-cell to $(2ws + 1)^{3h}$.

Starting from $h = 1$ and up, the central super-cell at each hierarchy level serves as a building block for construction of the near- and far-field regions at that level. Padding each side of the central super-cell with ws replicas creates the near-field zone shown in dark gray in Figure 16.3. The created near-field zone in turn appears as a *near-field* super-cell that incorporates the *central* super-cell inside it. The size of the near-field super-cell is $(2ws + 1)$ in 1D, or $(2ws + 1)^3$ in 3D, in units of the central super-cell at the respective hierarchy level. This number remains constant along the tree. Recasting the size of the near-field super-cell in units of the elementary unit cell gives $(2ws + 1)^{h+1}$ in 1D, or $(2ws + 1)^{3(h+1)}$ in 3D. A comparison of the assembly of the near-field super-cell with that of the central super-cell shows that both mechanisms are identical. With that, the near-field super-cell created at the child hierarchy level, h, becomes the central super-cell when viewed at the parent level, $h + 1$.

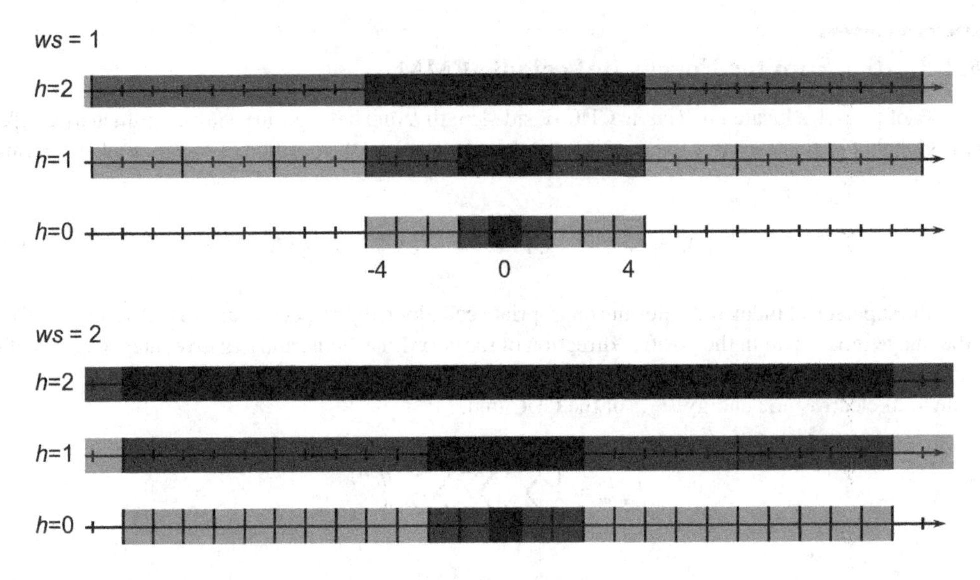

FIGURE 16.3 1D representation of the hierarchy of super-cells in PBC for well separation parameters $ws = 1$ and $ws = 2$. The section in black is the central super-cell at the hierarchy level h. For $h = 0$, the central super-cell matches the CUC. The region in dark gray adjacent to the cells in black represents the near-field zone. The next section, in light gray, constitutes the far-field zone. Short vertical bars outline the boundary of the unit cells, while long vertical bars represent the boundary of super-cells serving the role of building blocks in the construction of near- and far-field regions at the respective hierarchy level.

The space decomposition in periodic FMM continues with the definition of a far-field super-cell; this uses the previously introduced notion of a near-field super-cell. Once again, the 1D system is a good template for practicing the construction of the hierarchy tree. Padding each side of the near-field super-cell with the *ws* number of its replicas creates the far-field zone, shown in light gray in Figure 16.3. With that, the size of the far-field super-cell, which incorporates everything inside it, becomes $(2ws + 1)$ in units of the near-field super-cell. The far-field super-cell can also be expressed in units of the central super cell. Given that the near-field super-cell consists of $(2ws + 1)$ central super-cells, the size, in units of the central super-cell, of the far-field super-cell is $(2ws + 1)^2$ in 1D. Switching from 1D to 3D simply involves raising the obtained value to the third power, which gives $(2ws + 1)^6$.

Conversion of the size of the far-field super-cell all the way down to units of the CUC relies on the previously determined cell count for the central super-cell as a function of h, which is $(2ws + 1)^h$ and $(2ws + 1)^{3h}$ in 1D and 3D, respectively. Substituting that into the expression for the cell count of the far-field super-cell, previously determined in units of the central super-cell, gives $(2ws + 1)^{h+2}$ and $(2ws + 1)^{3h+6}$, respectively, which is the size of the far-field super-cell expressed in units of the CUC.

This mechanism for the construction of the hierarchical tree contains a useful property. On going up one step in the hierarchy level, the far-field super-cell established on a child hierarchy level, h, becomes the near-field super-cell on the parent hierarchy level, $h + 1$. This feature of periodic FMM implies that the near-field super-cell from the parent level carries the far-field contribution accounted for at the child level, so that no actual evaluation of near-field contribution is required anywhere in the hierarchy tree, except at the level $h = 0$.

A formal removal of the near-field interactions from the periodic FMM reduces the task of computing the lattice sum $U_{lattice\ sum}$ to a simple summation of the far-field contribution U_{FF}^h from each hierarchy level, h:

$$U_{total} = U_{unit\ cell} + U_{lattice\ sum} = U_{unit\ cell} + U_{NF}^0 + \sum_{h=0}^{\infty} U_{FF}^h, \tag{16.3}$$

where $U_{unit\ cell}$ and U_{NF}^0 are the internal electrostatic energy of the CUC and the interaction energy of the CUC with its images from the near-field zone, respectively, computed at the hierarchy level $h = 0$ using non-periodic FMM.

The use of super-cells as the building blocks for the construction of near- and far-field zones at each hierarchy level, instead of employing individual unit cells for that same purpose, has a number of important advantages. Primarily, it minimizes the number of multipole operations that are required to collect the effect of the far-field zone into the central cell on each hierarchy level in comparison to the exponentially increasing number of such operations that would be required when dealing with the elementary unit cells. In fact, when the size of the central super-cell increases on going from a child to the parent hierarchy level, the number of replicas of the central super-cell used in the construction of the near- and far-field zones does not change. As the result of that, the number of multipole operations required on each hierarchy level remains constant across the hierarchy tree.

A second important benefit of using the super-cell mechanism is the ease with which the far-field contribution from an arbitrary hierarchy level, h, can be accumulated in the CUC. At each hierarchy level, the effect of the far-field zone normally goes into the respective central super-cell at that level. Since the origin of the central super-cell coincides with the origin of the CUC, any irregular multipole expansion of the far-field zone translated to the center of the central super-cell simultaneously becomes defined about the center of the CUC, and carries the electrostatic potential of the far-field zone into that cell. This outcome allows the summing up of the far-field contribution from each hierarchy level directly into the CUC, without requiring any additional transformation.

An introduction to the development of the mathematical theory of periodic FMM will now be presented.

16.3 Multipole Moments of the Central Super-Cell

Formulation of the periodic FMM in terms of super-cells requires developing a technique for determining multipole moments of the central super-cell at an arbitrary hierarchy level, h, since those moments will be necessary for the generation of the moments of the image super-cells composing the far-field zone. Hereinafter, symbol $\bar{M}^h_{l,m}$ will be used to indicate the multipole moments of a central super-cell where the upper index indicates the hierarchy level, and the letters l and m indicate the orbital and azimuthal indices, respectively. For $h = 0$, the functions $\bar{M}^0_{l,m}$ are simply the multipole moments of the CUC computed by using the non-periodic FMM.

Evaluating the multipole moments for $\bar{M}^1_{l,m}$ relies on the equivalence of the central super-cell at the hierarchy level $h = 1$ to the near-field super-cell from $h = 0$. The near-field super-cell consists of the CUC and the elementary image cells. Multipole moments of the CUC have already been derived from a non-periodic FMM computation, and moments of the elementary image cells follow from PBC, where each elementary image cell carries a replica, that is, $\bar{M}^0_{j,k}$, of the multipole moments of the CUC defined about the center of the respective image cell. Translation of the origin of the multipole expansion of each image cell located in the near-field zone, with the help of the M2M operation, into the origin of the CUC, and the summing up of the translated multipoles, produces the multipole moments of the near-field zone. After that, addition of the regular multipole moments $\bar{M}^0_{l,m}$ of the CUC to the sum yields the multipole moments of the near-field super-cell:

$$\bar{M}^1_{l,m} = \sum_{n \in NF^0}^{(2ws+1)^3-1} \sum_{j=0}^{l} \sum_{k=-j}^{j} \bar{R}^*_{l-j,m-k}\left(\mathbf{a}^0_n\right) \bar{M}^0_{j,k} + \bar{M}^0_{l,m}. \tag{16.4}$$

In Equation 16.4, $\bar{R}^*_{l-j,m-k}\left(\mathbf{a}^0_n\right)$ is a chargeless expansion of translation vector $\mathbf{a}^0_n$, which has its origin in the center of the CUC, and has its tip pointing toward the center of the image cell n in the near-field zone, NF, at the hierarchy level $h = 0$, which is denoted in the upper index of vector $\mathbf{a}^0_n$. The number of unit cells composing the near-field zone is $(2ws + 1)^3 - 1$ in 3D, where the "−1" reflects the removal of the CUC from the cell count of the near-field super-cell.

Equation 16.4 explicitly describes all the steps needed to compute the multipole moments of the central super-cell at the hierarchy level $h = 1$. A separate treatment of the moments of the CUC from those of the image cells underlines the fact that, unlike the moments of the image cells, the moments of the CUC are already defined about the correct origin and therefore do not require translation. However, in the mathematical operations that follow, it is convenient to treat the moments of the CUC on equal footing with those of the image cells; that is, that they be placed inside the sum. Extending the range of index n in the summation from the near-field zone, NF, to the near-field super-cell, NFB, to include the CUC $n = 0$, brings in an extra summation term $\bar{R}^*_{l-j,m-k}(\mathbf{a}^0_0)M^0_{l,m}$. This term corresponds to an identity translation of the origin of the CUC onto itself, and, therefore, has as its translation vector $\mathbf{a}^0_0 = 0$. Under this condition, the radial component $|\mathbf{a}|^{l-j}$ in the regular solid harmonics $\bar{R}^*_{l-j,m-k}(\mathbf{a}^0_0)$ becomes unity for $l - j = 0$ due to $0^0 = 1$, and zero for other values of $l - j$. Therefore, $\bar{R}^*_{0,0}(0) = 1$ for $l - j = 0$, and is equal to zero otherwise. Because of that, each moment $\bar{M}^0_{l,m}$ directly sums up into $\bar{M}^1_{l,m}$ as prescribed by Equation 16.4; and has zero contribution when $j \neq l$ or $k \neq m$. This outcome is a simple restatement of the fact that an M2M translation of the origin of a multipole expansion onto itself is mathematically legitimate, and that it does not change the multipole expansion. Taking this into account simplifies Equation 16.4 to:

$$\bar{M}^1_{l,m} = \sum_{n \in NFB^0}^{(2ws+1)^3} \sum_{j=0}^{l} \sum_{k=-j}^{j} \bar{R}^*_{l-j,m-k}(\mathbf{a}^0_n)\bar{M}^0_{j,k}. \tag{16.5}$$

This expression for $\bar{M}^1_{l,m}$ holds the potential for additional simplification. In it, only the translation moments $\bar{R}^*_{l-j,m-k}(\mathbf{a}^0_n)$ depend on cell index n. With that, they may be separately summed up to produce the cumulative translation moments $\bar{R}^{*0}_{l-j,m-k}$:

$$\bar{R}^{*0}_{l-j,m-k} = \sum_{n \in NFB^0}^{(2ws+1)^3} \bar{R}^*_{l-j,m-k}(\mathbf{a}^0_n). \tag{16.6}$$

Incorporating this step into Equation 16.5 transforms the latter to:

$$\bar{M}^1_{l,m} = \sum_{j=0}^{l} \sum_{k=-j}^{j} \bar{R}^{*0}_{l-j,m-k}\bar{M}^0_{j,k}. \tag{16.7}$$

Finally, replacing the double sum in Equation 16.7, that represents the M2M operation, with a symbolic operator $\oplus$, gives the compact expression:

$$\bar{M}^1 = \bar{R}^{*0} \oplus \bar{M}^0, \tag{16.8}$$

where $\bar{M}^0$ are multipole moments of the CUC, $\bar{M}^1$ are the moments of the central super-cell at the hierarchy level $h = 1$, and $\bar{R}^{*0}$ are the cumulative regular translation moments in the near-field zone at $h = 0$.

Starting with $h = 2$, all multipole operations are performed on super-cells. Since super-cells have the same structure as the elementary cells, but are larger, the math of multipole operations on super-cells remains exactly the same as that on elementary cells. Once again, the multipole moments of the central super-cell at $h = 2$ are those of the near-field super-cell from $h = 1$. This turns the task of finding the multipole moments of the central super-cell at $h = 2$ into the task of finding the multipole moments of the near-field super-cell at $h = 1$. The previous determination of multipole moments $\bar{M}^1$ of the central super-cell at $h = 1$ provides the necessary condition to accomplish this. Details of the operation follow.

The recipe to accomplish multipole operations on each following hierarchy level is to newly apply the periodic boundary conditions to the super-cells created on that level so that to avoid directly dealing with the increased number of elementary cells. According to PBC and the described construction of the hierarchy tree, each image of the central super-cell at $h = 1$ carries a replica of the multipole moments $\bar{M}^1_{j,k}$ of the central super-cell at that hierarchy level. Translation of the origin of the multipole moments of the image super-cells composing the near-field zone at the level $h = 1$ to the center of the central super-cell with help of M2M operation leads to:

$$\bar{M}^2_{l,m} = \sum_{n \in NFB^1}^{(2ws+1)^3} \sum_{j=0}^{l} \sum_{k=-j}^{j} \bar{R}^*_{l-j,m-k}(\mathbf{a}^1_n)\bar{M}^1_{j,k} = \sum_{j=0}^{l} \sum_{k=-j}^{j} \bar{R}^{*1}_{l-j,m-k}\bar{M}^1_{j,k}, \tag{16.9}$$

where $\bar{M}^2_{l,m}$ are the desired multipole moments of the central super-cell at the hierarchy level $h = 2$, $\bar{M}^1_{j,k}$ are multipoles of the central super-cell at the hierarchy level $h = 1$ obtained from Equation 16.5, and $\bar{R}^*_{l-j,m-k}(\mathbf{a}^1_n)$ are multipole moments of the translation vector $\mathbf{a}^1_n$, which connects the center of the central super-cell at the hierarchy level $h = 1$ with the center of the image super-cell n at the same level. Summation of the translation moments $\bar{R}^*_{l-j,m-k}(\mathbf{a}^1_n)$ over index n gives the cumulative moments $\bar{R}^{*1}_{l-j,m-k}$. Note that the sum over index n in Equation 16.9 has the same structure as that in Equation 16.5.

Applying the symbolic form of M2M operation to Equation 16.9 simplifies this equation to:

$$\bar{M}^2 = \bar{R}^{*1} \oplus \bar{M}^1, \tag{16.10}$$

where $\bar{M}^2$ are the moments of the central super-cell at $h = 2$, $\bar{M}^1$ are the moments of the central super-cell at $h = 1$, and $\bar{R}^{*1}$ are the cumulative translation moments at $h = 1$. In this form, Equation 16.10 looks similar to Equation 16.8.

Extending the line of reasoning, which was used to obtain multipole moments $\bar{M}^1_{l,m}$ and $\bar{M}^2_{l,m}$, to any arbitrary hierarchy level h, leads to:

$$\bar{M}^h = \bar{R}^{*h-1} \oplus \bar{M}^{h-1}, \tag{16.11}$$

where $\bar{M}^h$ and $\bar{M}^{h-1}$ are multipole moments of the central super-cell at hierarchy level h and $h - 1$, respectively, and $\bar{R}^{*h-1}$ are the cumulative translation moments at hierarchy level $h - 1$.

The multipole moments of the central super-cell obtained in Equation 16.11 includes the as-yet unknown multipoles $\bar{M}^{h-1}$ on the right-hand side of the equation. Resolving this part formally requires computing the moments of the central super-cell of each preceding hierarchy level. Implementing that solution while employing the recursion explicitly present in Equation 16.11, and substituting the expression for each intermediate multipole moment back into Equation 16.11, leads to:

$$\bar{M}^h = \bar{R}^{*h-1} \oplus \left(\bar{R}^{*h-2} \oplus \left(\bar{R}^{*h-3} \oplus \left(\ldots\left(\bar{R}^{*0} \oplus \bar{M}^0\right)\right)\right)\right), \tag{16.12}$$

where each bracket encloses a multipole moment of the central super-cell of the respective child hierarchy level.

The presence of brackets in Equation 16.12 forces this equation to be evaluated from right to left, starting with the multipole moments $\bar{M}^0$, but, since the translation moments $\bar{R}^*$ are easy to compute, it makes sense to investigate the possibility of performing the M2M operations in Equation 16.12 from left to right. The operation of computing the multipole moments of the central super-cell for $h = 2$ in Equation 16.9 provides a suitable test bed. The term $\bar{M}^1_{j,k}$ present in Equation 16.9 has already been defined in Equation 16.7, although with different subscripts. Applying the index substitution $j \to p$ and $k \to q$ followed by $l \to j$ and $m \to k$ to Equation 16.7 replaces $\bar{M}^1_{l,m}$ with $\bar{M}^1_{j,k}$ on the left-hand side of that equation. Substituting that term into Equation 16.9 and placing the result of substitution in square brackets gives:

$$\bar{M}^2_{l,m} = \sum_{j=0}^{l} \sum_{k=-j}^{j} \bar{R}^{*1}_{l-j,m-k} \left[\sum_{p=0}^{j} \sum_{q=-p}^{p} \bar{R}^{*0}_{j-p,k-q} \bar{M}^0_{p,q} \right]. \tag{16.13}$$

Equation 16.13 consists of two consecutive M2M operations performed from right to left. The first M2M operation takes place inside the square brackets; the result becomes the subject of the second M2M operation. Since $\bar{M}^0_{p,q}$ does not depend on the indices *j or k*, it should be possible to perform the summation in a different order, in which indices *j and k* run faster than indices *p and q*. Before trying that out, it is helpful to note that, although the pairs of summation indices (*j*, *k*) and (*p*, *q*) are formally independent, they are connected *via* the multipole moments $\bar{R}^{*1}_{l-j,m-k}$, $\bar{R}^{*0}_{j-p,k-q}$, and $\bar{M}^0_{p,q}$ leading to the conditions $l \geq j \geq p \geq 0$. For all other values of the orbital indices, the summation elements are zero. With that, for a given value of indices (*p*, *q*), the sum over index *j* would actually go over index $j - p$ to ensure that $j - p \geq 0$. Subtracting index *p* from the lower and upper index of the sum over index *j* accomplishes the required modification. Likewise, the sum over index k needs to be performed over index $k-q$ to satisfy the condition $|k-q| \leq j-p$. The lower and upper boundaries of index $k-q$ follow from the term $\bar{R}^{*0}_{j-p,k-q}$. Incorporating all this into the summation over indices (j, k) for a given value of indices (p, q) in Equation 16.13, and writing that summation out separately, gives:

$$\bar{A}_{l-p,m-q} = \sum_{j-p=0}^{l-p} \sum_{k-q=-j+p}^{j-p} \bar{R}^{*1}_{l-j,m-k} \bar{R}^{*0}_{j-p,k-q}. \tag{16.14}$$

This represents an M2M operation. Comparison of the right-hand side of Equation 16.14 to that of Equation 16.7 confirms that assertion. This conclusion helps to determine the orbital and azimuthal indices in the left-hand side of Equation 16.14. According to the index pattern of the M2M operation, the index $l-p$ in $\bar{A}_{l-p,m-q}$ is a result of the addition $(l-j)+(j-p)$ of the two orbital indices from the multipole moments $\bar{R}^{*1}_{l-j,m-k}$ and $\bar{R}^{*0}_{j-p,k-q}$, and the index $m-q$ resolves from the addition $(m-k)+(k-q)$ of the respective azimuthal indices of the moments.

Substituting Equation 16.14 back into Equation 16.13 leads to:

$$\bar{M}^2_{l,m} = \sum_{p=0}^{j} \sum_{q=-p}^{p} \bar{A}_{l-p,m-q} \bar{M}^0_{p,q}. \tag{16.15}$$

This equation once again represents a regular M2M operation. Comparison of the indices in Equation 16.15 to those in Equation 16.7 provides the necessary proof.

Charting the path from Equations 16.13 to 16.15 demonstrates that the M2M operations in Equation 16.13 can be accomplished equally well from either right-to-left or from left-to-right, whichever is more convenient for any specific purpose, that is, that:

$$\bar{M}^2 = \bar{R}^{*1} \oplus (\bar{R}^{*0} \oplus \bar{M}^0) = (\bar{R}^{*1} \oplus \bar{R}^{*0}) \oplus \bar{M}^0. \tag{16.16}$$

Extending this conclusion to the case of arbitrary hierarchy level, h changes Equation 16.12 to:

$$\bar{M}^h = \left(\left(\left(\bar{R}^{*h-1} \oplus \bar{R}^{*h-2}\right) \oplus \bar{R}^{*h-3}\right) \oplus \ldots \bar{R}^{*0}\right) \oplus \bar{M}^0, \tag{16.17}$$

which combines all of the easily obtainable multipoles $\bar{R}^*$ together, and places the multipole moments $\bar{M}^0$ of the central-unit cell in the right-most position. The ability to rearrange the order of multipole operations in this way will later be instrumental in the derivation of the equation for the lattice sum.

Development of the mathematical theory of periodic FMM continues by exploring one more useful feature of the hierarchical tree construction algorithm presented as a uniform distance stretching. The systematic increase in the side length of the central super-cell by the factor $(2ws+1)$, on going from a child level to its parent level, makes the parent set of super-cells appearing as a blown out replica of the child-level super-cells. This feature maintains the number of boxes in each level constant, and, which is important, preserves the orientation of the translation vector $\mathbf{a}_n$, which connects the center of the central super-cell with the center of its periodic image n at each level. Because of that, for each vector $\mathbf{a}^{h-1}_n$ at the child level $h-1$ there exists a similarly oriented vector $\mathbf{a}^h_n$ at the parent level h that assists in a similar M2M operation, and is stretched by a constant factor of $(2ws+1)$, so that $\mathbf{a}^h_n = (2ws+1)\mathbf{a}^{h-1}_n$. Figure 16.4 illustrates this relation for $h=1$ and $h=0$, and parameter $ws=1$.

The significance of vectors $\mathbf{a}^h_n$ and $\mathbf{a}^{h-1}_n$ being aligned is that the uniform distance stretching introduces a distance scaling relation between the translation moments of the parent and child levels, $\bar{R}^*_{l-j,m-k}\left(\mathbf{a}^h_n\right)$ and $\bar{R}^*_{l-j,m-k}\left(\mathbf{a}^{h-1}_n\right)$. Since the radial term $\left|\mathbf{a}^h_n\right|^{l-j}$ is the only distance-dependent part in the function $\bar{R}^*_{l-j,m-k}\left(\mathbf{a}^h_n\right)$, the parent moment, $\bar{R}^*_{l-j,m-k}\left(\mathbf{a}^h_n\right)$, resolves from the child moment, $\bar{R}^*_{l-j,m-k}\left(\mathbf{a}^{h-1}_n\right)$ based on the distance

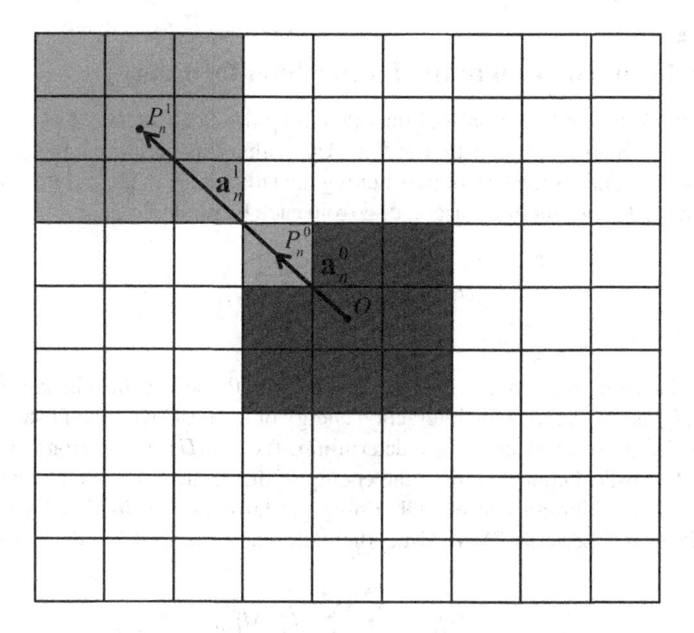

FIGURE 16.4 A matching pair of translation vectors $\mathbf{a}_n^0$ and $\mathbf{a}_n^1$ from the hierarchy levels $h = 0$ and $h = 1$, respectively. Both vectors have the origin in point O, and direct toward the center P_n^0 of cell n^0 and the center P_n^1 of cell n^1, respectively. The uniform distance stretching by a factor of $2ws + 1$, where $ws = 1$, translates the point P_n^0 into the point P_n^1 and stretches out a unit image cell n^0 into an image super-cell n^1, where both cells are shown in light gray. The rest of the central super-cell at $h = 1$ is in dark gray.

stretching relationship $\left|\mathbf{a}_n^h\right|^{l-j} \equiv \left|(2ws+1)\mathbf{a}_n^{h-1}\right|^{l-j}$. Writing that relation between the translation multipole moments of the parent and child hierarchy levels in operator form leads to:

$$\bar{R}^*\left(\mathbf{a}_n^h\right) = T\,\bar{R}^*\left(\mathbf{a}_n^{h-1}\right), \quad \text{or, explicitly:} \quad \bar{R}^*_{l-j,m-k}\left(\mathbf{a}_n^h\right) = T_{l-j}\bar{R}^*_{l-j,m-k}\left(\mathbf{a}_n^{h-1}\right), \tag{16.18}$$

where T is a stretching operator, which substitutes the radial term $\left|\mathbf{a}_n^{h-1}\right|^{l-j}$ in $\bar{R}^*_{l-j,m-k}\left(\mathbf{a}_n^{h-1}\right)$ by $\left|(2ws+1)\mathbf{a}_n^{h-1}\right|^{l-j}$, and where $T_{l-j} = (2ws + 1)^{l-j}$ serves as the multiplicative factor.

The use of the operator T extends to the cumulative translation moment $\bar{R}^*_{l-j,m-k}$, which is the sum of individual translation moments over index n. Because of this, the translation operator acting on the individual translation moments can be moved outside the sum giving:

$$\begin{aligned} \bar{R}^{*h}_{l-j,m-k} &= \sum_{n\in NFB^h}^{(2ws+1)^3} \bar{R}^*_{l-j,m-k}\left(\mathbf{a}_n^h\right) = \sum_{n\in NFB^{h-1}}^{(2ws+1)^3} T_{l-j}\bar{R}^*_{l-j,m-k}\left(\mathbf{a}_n^{h-1}\right) \\ &= T_{l-j}\sum_{n\in NFB^{h-1}}^{(2ws+1)^3} \bar{R}^*_{l-j,m-k}\left(\mathbf{a}_n^{h-1}\right) = T_{l-j}\bar{R}^{*h-1}_{l-j,m-k}. \end{aligned} \tag{16.19}$$

Application of the stretching operator to the unit moment $\bar{R}^*_{l-j,m-k}\left(\mathbf{a}_0^{h-1}\right)$, which corresponds to the translation of the origin onto itself, in the sum for the cumulative moment, $\bar{R}^{*h-1}_{l-j,m-k}$ introduced in Equation 16.5, does not cause any problem because $(2ws+1)^{l-j}\bar{R}^*_{l-j,m-k}(0) = (2ws+1)^{l-j}0^{l-j} = 0^{l-j} = \bar{R}^*_{l-j,m-k}(0)$ is equal to unity for $l - j = 0$ and is zero for $l - j > 0$. This is consistent with the notion that a zero-value translation vector $\mathbf{a}_0^{h-1}$ does not stretch.

The mathematical apparatus developed in this section provides a starting point for the evaluation of the lattice sum in a periodic FMM. The next step in determining the lattice sum is the evaluation of the far-field contribution.

16.4 Far-Field Contribution to the Lattice Sum for Energy

The procedure established in Equation 16.3 for calculating the electrostatic energy of the CUC defines the lattice sum as the energy of interaction of the CUC with all its periodic images. This is an infinite sum. Periodic FMM organizes the task of partitioning the lattice sum $U_{lattice\ sum}$ into a near-field term and a summation of far-field contributions originating from each hierarchy level, h:

$$U_{lattice\ sum} = U_{NF}^{0} + \sum_{h=0}^{\infty} U_{FF}^{h}, \tag{16.20}$$

where U_{NF}^{0} is the electrostatic interaction energy of the CUC with the near-field images from the hierarchy level $h = 0$, and U_{FF}^{h} is the electrostatic interaction energy of the CUC with the far-field images from the hierarchy level h. The present task consists of determining the term U_{FF}^{h} for an arbitrary hierarchy level, h.

Derivation of the far-field contribution to the energy of the central super-cell begins with an explicit analysis of the first few hierarchy levels. Obtaining the far-field contribution for $h = 0$ employs the familiar principles of non-periodic FMM. Using these, the energy of the CUC due to the far-field zone is:

$$U_{FF}^{h=0} = \sum_{l=0}^{N} \sum_{m=-l}^{l} \tilde{L}_{l,m}^{0} \bar{M}_{l,m}^{0}, \tag{16.21}$$

where $\bar{M}_{l,m}^{0}$ are the known regular multipole moments of the CUC, and $\tilde{L}_{l,m}^{0}$ are the cumulative irregular multipole moments of the far-field zone defined about the center of the CUC. The moments $\tilde{L}_{l,m}^{0}$ translate the electrostatic potential of the image cells from the far-field zone into the CUC.

In periodic FMM, each unit image cell at $h = 0$ carries a replica of the regular multipole moments $\bar{M}_{j,k}^{0}$ of the CUC defined about the center of the image cell. Determining the electrostatic potential that the image cell creates inside the CUC requires converting the regular multipole moments of the former into irregular multipole moments about the center of the CUC with the help of an M2L operation. Finally, summing up the irregular multipole moments of the individual image cells produces the cumulative moment:

$$\tilde{L}_{l,m}^{0} = \sum_{n \in FF}^{(2ws+1)^6-(2ws+1)^3} \sum_{j=0}^{N} \sum_{k=-j}^{j} (-1)^j \tilde{S}_{l+j,m+k}\left(\mathbf{r}_n^0\right) \bar{M}_{j,k}^{0}, \tag{16.22}$$

where $\tilde{S}_{l+j,m+k}\left(\mathbf{r}_n^0\right)$ is a tilde scaled irregular translation moment of translation vector $\mathbf{r}_n^0$, which has its origin in the center of the CUC, and its tip pointing toward the center of the image cell n in the far field zone of the hierarchy level $h = 0$. The number of image boxes in the far-field zone FF is $(2ws + 1)^6 - (2ws + 1)^3$ in 3D. That is, the difference in size of the far-field super-cell and that of the enclosed near-field super-cell.

Equation 16.22 can be simplified slightly. Since the individual irregular translation moments $\tilde{S}_{l+j,m+k}\left(\mathbf{r}_n^0\right)$ have the same origin, they can be summed up. Summation of the individual translational moments over all image cells n belonging to the far-field zone at the hierarchy level $h = 0$ then produces the cumulative irregular translation moments $\tilde{S}_{l+j,m+k}^{0}$:

$$\tilde{S}_{l+j,m+k}^{0} = \sum_{n \in FF}^{(2ws+1)^6-(2ws+1)^3} \tilde{S}_{l+j,m+k}\left(\mathbf{r}_n^0\right). \tag{16.23}$$

Taking that summation into account reduces Equation 16.22 to:

$$\tilde{L}_{l,m}^{0} = \sum_{j=0}^{N} \sum_{k=-j}^{j} (-1)^j \tilde{S}_{l+j,m+k}^{0} \bar{M}_{j,k}^{0}. \tag{16.24}$$

This equation, when written in symbolic form using symbol $\otimes$ to indicate the M2L operation, becomes:

$$\tilde{L}^0 = \tilde{S}^0 \otimes \bar{M}^0. \tag{16.25}$$

The next step in working out the pattern for computing the far-field contributions in the lattice sum is to review the computation of the irregular moments of the far-field zone for the hierarchy level $h = 1$. This level accounts for contributions from the more distant layer of image cells, as illustrated in Figure 16.3. As before, the electrostatic energy $U_{FF}^{h=1}$ of the CUC due to those image cells is:

$$U_{FF}^{h=1} = \sum_{l=0}^{N} \sum_{m=-l}^{l} \tilde{L}_{l,m}^1 \bar{M}_{l,m}^0, \tag{16.26}$$

where the $\tilde{L}_{l,m}^1$ are the cumulative irregular translational moments of the far-field zone. Determining these moments from the contribution of individual image cells, as was done in Equation 16.22, is no longer practical because the number of elementary image cells in the far-field zone grows exponentially as the index h increases. Instead, a convenient solution would involve recasting the problem in terms of super-cells.

Key elements of the super-cell structure of the hierarchy tree are the constant number of super-cells composing the near- and far-field zones in every hierarchy level and the constant orientation of the translation vectors that connect the origin of the central super-cell with the center of the image super-cell. These properties of the hierarchy tree make the problem of finding the far-field contribution at any hierarchy level similar to that for $h = 0$.

Having established that the far-field zone at $h = 1$ consists of image super-cells, which carry a replica of the multipole moments of the central super-cell $\bar{M}_{l,m}^1$, defined in Equation 16.5, the next step is to convert these regular multipole moments into irregular multipole moments, $\tilde{L}_{l,m}^1$, defined about the center of the CUC. The familiar M2L operation accomplishes this transition:

$$\tilde{L}_{l,m}^1 = \sum_{n \in FF^1}^{(2ws+1)^6-(2ws+1)^3} \sum_{j=0}^{N} \sum_{k=-j}^{j} (-1)^j \tilde{S}_{l+j,m+k}\left(\mathbf{r}_n^1\right) \bar{M}_{j,k}^1 = \sum_{j=0}^{N} \sum_{k=-j}^{j} (-1)^j \tilde{S}_{l+j,m+k}^1 \bar{M}_{j,k}^1, \tag{16.27}$$

where $\tilde{S}_{l+j,m+k}\left(\mathbf{r}_n^1\right)$ are tilde-scaled irregular translation moments of the translation vector $\mathbf{r}_n^1$, which has its origin at the center of the CUC and its tip pointing toward the center of the image super-cell n in the far-field zone of the hierarchy level $h = 1$, and $\tilde{S}_{l+j,m+k}^0$ are cumulative translation moments obtained from summing up the individual moments $\tilde{S}_{l+j,m+k}\left(\mathbf{r}_n^1\right)$ over all image cells n belonging to the far-field zone FF at the hierarchy level $h = 1$.

Once again, it is convenient to rewrite Equation 16.27 in symbolic form using symbol $\otimes$ to indicate the M2L operation:

$$\tilde{L}^1 = \tilde{S}^1 \otimes \bar{M}^1. \tag{16.28}$$

In it, substituting $\bar{M}^1$ from Equation 16.8 into Equation 16.28 leads to:

$$\tilde{L}^1 = \tilde{S}^1 \otimes (\bar{R}^{*0} \oplus \bar{M}^0), \tag{16.29}$$

which simplifies construction of the irregular multipole moments $\tilde{L}^1$ of the far-field zone by using components already developed.

In Equation 16.29 the cumulative irregular translation moment, $\tilde{S}^1$, which is on the right-hand side of the equation, carries the attribute that belongs to the parent hierarchy level $h = 1$, whereas the remaining terms, $\bar{R}^{*0}$ and $\bar{M}^0$, come from the child hierarchy level $h = 0$. To eliminate the mix up of different hierarchy levels in the right-hand side of the equation, it is necessary to convert the term $\tilde{S}^1$ to those from the child level. Implementing this arrangement is straightforward. Since each hierarchy level has a constant number of super-cells and a constant orientation of the translation vectors, for every translation vector $\mathbf{r}_n^h$ at the hierarchy level h there exists a similarly oriented vector $\mathbf{r}_n^{h-1}$ at the hierarchy level $h-1$, which satisfies the condition $\mathbf{r}_n^h = (2ws+1)\mathbf{r}_n^{h-1}$. Building on the derivation of the stretching operator T for regular multipole expansions in Equation 16.18, a similar stretching operator X can now be introduced for $\tilde{S}^1$ to express the latter using the corresponding term $\tilde{S}^0$ from the child hierarchy level. The operator X replaces the radial term $|\mathbf{r}|^{-l-j-1}$ with $|(2ws+1)\mathbf{r}|^{-l-j-1}$, when it acts on the irregular moment $\tilde{S}^h_{l+j,m+k}$, so that:

$$\tilde{S}^{h+1} = X\tilde{S}^h, \quad \text{and} \quad \tilde{S}^{h+1}_{l+j,m+k} = X_{l+j,m+k}\tilde{S}^h_{l+j,m+k}, \tag{16.30}$$

where $X_{l+j,\,m+k} = (2ws+1)^{-l-j-1}$ is a multiplication factor. Substituting Equation 16.30 for $h = 0$ into Equation 16.29 leads to:

$$\tilde{L}^1 = (X\tilde{S}^0) \otimes (\bar{R}^{*0} \oplus \bar{M}^0), \tag{16.31}$$

which now produces the irregular moments $\tilde{L}^1$ of the far-field zone at $h = 1$ entirely from the moments available at $h = 0$.

The final step to be performed on Equation 16.31 is to rearrange the sequence of multipole operations. The presence of brackets in this equation establishes that the sequence is from right to left. First, the regular cumulative translation moments, $\bar{R}^{*0}$, are used in an M2M operation on the multipole moments, $\bar{M}^0$ of the CUC. The resulting multipole moment is then subjected to an M2L operation using scaled irregular cumulative translation moments $X\tilde{S}^0$ to finally produce the irregular moments L^1 that carry the electrostatic effect of the far-field zone into the CUC. Since translational moments $\tilde{S}^0$ and $\bar{R}^{*0}$ are simple constructs, depending only on the size of the CUC, and are independent of the charge distribution inside the cell, it is useful to take the opportunity to perform the multipole operations on them first, and leaving the operation on the multipole moments of the CUC until later.

To explore the opportunity for rewriting Equation 16.31 for an alternative sequence of multipole operations, it is necessary to first convert the equation into its explicit form, showing all the indices. The orbital and azimuthal indices of $\tilde{S}^0$ follow from Equation 16.27 and Equation 16.30. $\bar{R}^{*0} \oplus \bar{M}^0$ resolves from Equation 16.7. However, since Equation 16.27 requires $\bar{M}^1_{j,k}$ as the outcome of an M2M operation, it is necessary to perform an index substitution in Equation 16.7 to produce $\bar{M}^1_{j,k}$ instead of $\bar{M}^1_{l,m}$ on the left-hand side. Replacing indices (l, m) and (j, k) in Equation 16.7 with (j, k) and (p, q), respectively, and substituting the obtained explicit form of $\bar{R}^{*0} \oplus \bar{M}^0$ into Equation 16.31 leads to:

$$\tilde{L}^1_{l,m} = \sum_{j=0}^{N}\sum_{k=-j}^{j}(-1)^j\left(X_{l+j,m+k}\tilde{S}^0_{l+j,m+k}\right)\sum_{p=0}^{j}\sum_{q=-p}^{p}\bar{R}^{*0}_{j-p,k-q}\bar{M}^0_{p,q}, \tag{16.32}$$

which is the expanded form of Equation 16.31.

The resulting equation includes two groups of summation indices (j, k) and (p, q), which can be run independently. This makes it possible to sum up the product $(-1)^{j-p}\left(X_{l+j,m+k}\tilde{S}^0_{l+j,m+k}\right)\bar{R}^{*0}_{j-p,k-q}$ over indices (j, k) for a fixed value of indices (p, q). The phase factor $(-1)^{j-p}$ associated with this product comes from splitting the original factor $(-1)^j$ into two parts, $(-1)^{j-p}$ and $(-1)^p$, with the latter portion going to the second summation over indices (p, q).

The presence of terms $\tilde{S}^0_{l+j,m+k}$ and $\bar{R}^{*0}_{j-p,k-q}$ in the first sum over indices (j, k) imposes certain limits on the range of value of those indices. For example, term $\bar{R}^{*0}_{j-p,k-q}$ requires that $j \geq p$. Therefore, for any given value of p, the value of index j in the sum can be replaced by $j-p$. In addition to that, for any given value of index q, the value of index k is bounded by the condition $|k-q| \leq j-p$. With that, the original sum over index k is transformed to a sum over index $k-q$ with respective boundaries. All other values of the indices (j, k) outside these boundaries do not contribute to the sum.

The next step in rearranging Equation 16.32 is to determine the result of summation over indices (j, k). The phase factor $(-1)^{j-p}$ and the product of irregular $\tilde{S}^0_{l+j,m+k}$ and regular $\bar{R}^{*0}_{j-p,k-q}$ multipole moments in the sum indicate that it must be an M2L operation. That in turn implies that the result of summation over indices (j, k) is an irregular multipole moment $\tilde{S}$. The orbital and azimuthal indices of the latter resolve from the structure of the M2L operation, therefore the resultant orbital and azimuthal indices are those of the irregular moment $\tilde{S}^0_{l+j,m+k}$ minus the indices of the regular moment $\bar{R}^{*0}_{j-p,k-q}$, that is $l+j-j+p=l+p$ and $m+k-k+q=m+q$. This rule establishes that the current M2L operation produces the multipole moment $\tilde{S}_{l+p,m+q}$. Taking all the above factors into consideration turns the summation over indices (j, k) from Equation 16.32 to:

$$\tilde{S}_{l+p,m+q} = \sum_{j-p=0}^{N} \sum_{k-q=-j+p}^{j-p} (-1)^{j-p} \left(X_{l+j,m+k} \tilde{S}^0_{l+j,m+k} \right) \bar{R}^{*0}_{j-p,k-q}. \tag{16.33}$$

The presence of the multiplier $X_{l+j,\,m+k}$ alongside $\tilde{S}^0_{l+j,m+k}$ does not change the structure of this equation since $X_{l+j,m+k}\tilde{S}^0_{l+j,m+k}$ is a usual irregular moment, therefore Equation 16.33 qualifies as an M2L operation.

The second, and final step, in the analysis of Equation 16.32 consists of substituting Equation 16.33 back into Equation 16.32; this produces the following expression:

$$\tilde{L}^1_{l,m} = \sum_{p=0}^{N} \sum_{q=-p}^{p} (-1)^p \tilde{S}_{l+p,m+q} \bar{M}^0_{p,q}. \tag{16.34}$$

Recall that the phase factor $(-1)^p$ comes from splitting the original term $(-1)^j$ from Equation 16.32 into $(-1)^{j-p}$ and $(-1)^p$. The factor $(-1)^{j-p}$ was used in Equation 16.33, and the factor $(-1)^p$ goes into Equation 16.34. In this equation, the upper boundary of index p in the sum becomes N, the size of the truncated multipole expansion, instead of the former value j, which has been integrated out. Technically, the upper boundary should be $N-l$, but setting it to N does not change the outcome because the extra summation terms are all zero, and using the simpler upper boundary makes the equation look more conventional.

A visual inspection of Equation 16.34 reveals that the orbital and azimuthal indices of the resulting irregular multipoles $\tilde{L}^1_{l,m}$ come from the indices of the irregular multipoles $\tilde{S}_{l+p,m+q}$ minus the respective indices of the regular multipoles $\bar{M}^0_{p,q}$. Also, the power of the factor $(-1)^p$ matches the orbital index of the multipole $\bar{M}^0_{p,q}$. These aspects qualify Equation 16.34 as an M2L operation.

Combining the outcome of the M2L operation in Equations 16.33 and 16.34, and using symbolic notation, allows Equation 16.31 to be rewritten as:

$$\tilde{L}^1 = (X\tilde{S}^0) \otimes (\bar{R}^{*0} \oplus \bar{M}^0) = \left((X\tilde{S}^0) \otimes \bar{R}^{*0} \right) \otimes \bar{M}^0. \tag{16.35}$$

This underscores the possibility of processing multipole terms from right to left and from left to right in the computation of the irregular multipole moments $\tilde{L}^1$.

A generalization of Equation 16.35 and using symbols $\tilde{L}$ and $\bar{M}$ for the irregular and regular multipole moments, respectively, summarizes an important property of multipole operations, that:

$$\tilde{L}_3 \otimes (\bar{M}_2 \oplus \bar{M}_1) = (\tilde{L}_3 \otimes \bar{M}_2) \otimes \bar{M}_1, \tag{16.36}$$

where the lower indices distinguish one multipole moment from another. This relationship lays the groundwork for the following steps.

Continuing on to the hierarchy level $h=2$, and based on Equation 16.28, the far-field electrostatic potential expansion $\tilde{L}^2$ inside the CUC emerges as the result of the M2L operation:

$$\tilde{L}^2 = \tilde{S}^2 \otimes \bar{M}^2, \tag{16.37}$$

where $\bar{M}^2$ are the multipole moments of the central super-cell, and $\tilde{S}^2$ are cumulative irregular moments of the translation vectors which connect the origin of the central super-cell with the center of its respective far-field image at $h=2$.

Equation 16.37 needs to be reduced to the form that uses the multipoles from the lowest hierarchy level in the right hand side while maintaining the sequence of multipole operations from left to right as needed in the lattice sum. Substitution of $\bar{M}^2$ from Equation 16.10 into Equation 16.37 leads to:

$$\tilde{L}^2 = \tilde{S}^2 \otimes (\bar{R}^{*1} \oplus \bar{M}^1). \tag{16.38}$$

Application of the equality from Equation 16.36 to Equation 16.38 transforms the latter equation to a sequence of operations from left to right:

$$\tilde{L}^2 = (\tilde{S}^2 \otimes \bar{R}^{*1}) \otimes \bar{M}^1. \tag{16.39}$$

Reduction in the hierarchy level order of $\bar{M}^1$ continues with the help of Equation 16.8, leading to:

$$\tilde{L}^2 = (\tilde{S}^2 \otimes \bar{R}^{*1}) \otimes (\bar{R}^{*0} \oplus \bar{M}^0). \tag{16.40}$$

The next step is to rearrange the sequence of multipole operations. Since the term $\left(\tilde{S}^2 \otimes \bar{R}^{*1}\right)$ is an irregular multipole moment, while $\bar{R}^{*0}$ and $\bar{M}^0$ are regular multipole moments, application of Equation 16.36 to Equation 16.40 changes the sequence of multipole operations in the latter equation to the desired direction of left to right:

$$\tilde{L}^2 = \left((\tilde{S}^2 \otimes \bar{R}^{*1}) \otimes \bar{R}^{*0}\right) \otimes \bar{M}^0. \tag{16.41}$$

In Equation 16.41, only the translation moments $\tilde{S}^2$ and $\bar{R}^{*1}$ still belong to a non-zero hierarchy level on the right hand side. Stretching operators T and X can help with the transition of these translation moments to the zeroth-level. Using Equation 16.18 for $\bar{R}^{*1} = T\bar{R}^{*0}$, and Equation 16.30 for $\tilde{S}^2 = XX\tilde{S}^0$ transforms Equation 16.41 to:

$$\tilde{L}^2 = \left(\left((XX\tilde{S}^0) \otimes (T\bar{R}^{*0})\right) \otimes \bar{R}^{*0}\right) \otimes \bar{M}^0. \tag{16.42}$$

In Equation 16.42, all the components are defined at the zeroth hierarchy level, and there is no need to invoke any attributes from a higher hierarchy level. The final step in Equation 16.42 is to simplify the term $(XX\tilde{S}^0) \otimes (T\bar{R}^{*0})$.

To keep the derivation work general, it is convenient to analyze first the case of $(X\tilde{S}) \otimes (T\bar{R}^*)$, where each translation moment in the product is the subject of a separate stretching operation. It also helps to drop the upper index in the analyzed multipole moments, since reference to a hierarchy level does not add anything useful to the present analysis. Because it is a regular M2L operation, the above product expands out to:

$$\tilde{L}_{l,m} = \sum_{j=0}^{N} \sum_{k=-j}^{j} (-1)^j \left[X_{l+j,m+k}\tilde{S}_{l+j,m+k}(\mathbf{r})\right]\left[T_j \bar{R}^*_{j,k}(\mathbf{a})\right], \tag{16.43}$$

where operator X stretches the vector $\mathbf{r}$, and the operator T applies to vector $\mathbf{a}$. The irregular multipole moment $\tilde{L}_{l,m}$ is the scaled result of an M2L operation.

The effect that operators X and T produce in Equation 16.43 is:

$$\frac{1}{(2ws+1)^{l+j+1}} \tilde{S}_{l+j,m+k}(\mathbf{r})(2ws+1)^j \bar{R}^*_{j,k}(\mathbf{a}) = \frac{1}{(2ws+1)^{l+1}} \tilde{S}_{l+j,m+k}(\mathbf{r}) \bar{R}^*_{j,k}(\mathbf{a}), \tag{16.44}$$

where $(2ws + 1)$ is a stretching factor. After canceling out the common terms in the numerator and denominator in the left hand side, the remaining term $(2ws + 1)^{-l-1}$ corresponds to the factor X_l, which goes outside the sum and scales out the result of the M2L operation:

$$\tilde{L}_{l,m} = X_l \sum_{j=0}^{N} \sum_{k=-j}^{j} (-1)^j \tilde{S}_{l+j,m+k}(\mathbf{r}) \bar{R}^*_{j,k}(\mathbf{a}). \tag{16.45}$$

This result leads to the formulation of an important property of multipole operations, that:

$$(X\tilde{S}) \otimes (T\bar{R}^*) = X(\tilde{S} \otimes \bar{R}^*), \tag{16.46}$$

which states that the M2L product of scaled multipoles is a scaled M2L product of unscaled multipoles. Application of this rule to Equation 16.42 transforms the latter to:

$$\tilde{L}^2 = \left(X \left((X\tilde{S}^0) \otimes \bar{R}^{*0} \right) \otimes \bar{R}^{*0} \right) \otimes \bar{M}^0. \tag{16.47}$$

This is the final form of the irregular multipole expansion of the far-field zone from hierarchy level $h = 2$ needed for the derivation of the lattice sum equation.

The irregular multipole moments $\tilde{L}^h$ of the far-field zone of the remaining hierarchy levels h that are needed for completion of the lattice sum emerge from an analysis of the previously derived equations for irregular moments. Lining up Equations 16.25, 16.35, and 16.47 together:

$$\tilde{L}^0 = \tilde{S}^0 \otimes \bar{M}^0, \quad \tilde{L}^1 = \left((X\tilde{S}^0) \otimes \bar{R}^{*0} \right) \otimes \bar{M}^0, \quad \tilde{L}^2 = \left(X \left((X\tilde{S}^0) \otimes \bar{R}^{*0} \right) \otimes \bar{R}^{*0} \right) \otimes \bar{M}^0, \tag{16.48}$$

draws attention to the fact that they all have the same component $\otimes \bar{M}^0$ in their right-most part. The portion to the left of $\bar{M}^0$ represents the irregular partial moments, $\tilde{B}^h$:

$$\tilde{B}^0 = \tilde{S}^0, \quad \tilde{B}^1 = (X\tilde{S}^0) \otimes \bar{R}^{*0}, \quad \tilde{B}^2 = X\left((X\tilde{S}^0) \otimes \bar{R}^{*0} \right) \otimes \bar{R}^{*0}, \tag{16.49}$$

so that

$$\tilde{L}^h = \tilde{B}^h \otimes \bar{M}^0. \tag{16.50}$$

A valuable feature of partial moments is that they obey the recurrence relation:

$$\tilde{B}^{h+1} = (X\tilde{B}^h) \otimes \bar{R}^{*0}. \tag{16.51}$$

This relation, when substituted into Equation 16.50, produces the irregular far-field moments $\tilde{L}^h$ for an arbitrary hierarchy level, h.

Determination of the irregular multipole moments $\tilde{L}^h$ of the far-field zone completes the preparation needed for the derivation of the lattice sum. All that remains is to sum up their contribution to the energy of the CUC:

$$U_{FF} = \left(\sum_{h=0}^{\infty} \tilde{L}^h \right) \otimes \bar{M}^0, \tag{16.52}$$

where each irregular moment is defined about the same coordinate system. Since Equation 16.52 uses the sum of the moments $\tilde{L}^h$, it is natural to sum up the partial moments $\tilde{B}^h$ into cumulative irregular lattice moments, $\tilde{C}^t$:

$$\tilde{C}^t = \sum_{h=0}^{t} \tilde{B}^h, \quad \text{so that} \quad \tilde{C}^t \otimes \bar{M}^0 = \sum_{h=0}^{t} \tilde{L}^h, \tag{16.53}$$

which accumulate the far-field contributions up to the hierarchy tree level $h = t$. The advantage of lattice moments is that they satisfy the recurrence relation:

$$\tilde{C}^{t+1} = (X\tilde{C}^{t}) \otimes \bar{R}^{*0} + \tilde{S}^{0}, \quad (16.54)$$

which sums up the irregular far-field contribution starting from $\tilde{C}^{0} = \tilde{S}^{0}$, thereby avoiding the need to compute the individual moments $\tilde{B}^{h}$.

Letting index t in Equation 16.54 go to infinity delivers the cumulative moments $\tilde{C}^{t\to\infty}$ of the infinite lattice, which carry the electrostatic potential of all periodic images into the CUC. Finally, the energy of the CUC due to the far field under PBC becomes:

$$U_{FF} = (\tilde{C}^{t\to\infty} \otimes \bar{M}^{0}) \otimes \bar{M}^{0}. \quad (16.55)$$

In practice, the lattice moments quickly converge, and setting the upper bound of t in Equation 16.54 to 20 is sufficient for double precision work.

16.5 Contribution of the Near-Field Zone into the Central Unit Cell

The application of PBC to a system of electrostatic charges arranged in a cubic box, which represents the CUC, divides the periodic continuum around the CUC into a finite near-field zone, *NF* with a size depending on the choice of the parameter *ws*, and an infinite far-field zone, *FF* both constructed out of elementary images of the CUC. Periodic FMM, described in Section 16.4, manages the *FF* portion of the lattice sum by a renormalization approach using super-cells. All that remains is the derivation of the contribution U_{NF}^{0} of the periodic *NF* to the lattice sum in Equation 16.20.

As with non-periodic FMM, the name *NF* suggests using a direct sum to compute the electrostatic interaction between particles of the periodic *NF* zone and those of the CUC. But, because the periodic *NF* zone is so large that a direct sum approach is impractical, a non-periodic FMM approach is used in deriving this interaction.

The *NF* zone emerges in PBC as the result of padding the CUC with *ws* images of itself on each side of the box. In 3D, the use of $ws = 1$ places 26 images of the CUC in the immediate vicinity of the CUC, as shown in Figure 16.5. Most CUCs are so large that splitting them into a hierarchy of fine-grain boxes is justified when applying non-periodic FMM to the multipole expansion of electrostatic particles in the CUC. In addition, the same fine-grain structure of the CUC automatically translates to the periodic NF, as illustrated in Figure 16.5.

The resulting fine-grain structure of the *NF* zone surrounding the CUC and the application of non-periodic FMM to the electrostatic interaction of the *NF* zone with the CUC break down the *NF* contribution, U_{NF}^{0} in Equation 16.20 into three terms:

$$U_{NF}^{0} = \sum_{p}\left(\sum_{t}\sum_{n\in p}\sum_{k\in t}\frac{q_n q_k}{|\mathbf{r}_k - \mathbf{r}_n|} + \sum_{s}\tilde{B}_p \otimes \bar{A}_s\right) + \tilde{C}_{\neq p} \otimes \bar{M}^{0}, \quad (16.56)$$

where the direct sum accounts for the particle-particle interactions between fine-grain boxes p in the *NF* zone and t in the CUC that satisfy the *NF*-condition for a non-periodic FMM, and the multipole terms $\tilde{B}_p \otimes \bar{A}_s$ and $\tilde{C}_{\neq p} \otimes \bar{M}^{0}$ which describe the electrostatic interaction between the distant fine-grain boxes. Continuing the explanation of Equation 16.56: A fine-grain box, p, in the *NF* zone, which participates in particle-particle interactions, might also be involved with fine-grain box, s, in the CUC in a *FF* position relative to each other. Those boxes interact through the multipole term $\tilde{B}_p \otimes \bar{A}_s$, where $\tilde{B}_p$ is an irregular multipole expansion and $\bar{A}_s$ is a regular multipole expansion of the respective fine-grain boxes, both expansions being defined about the same origin. The final third term, $\tilde{C}_{\neq p} \otimes \bar{M}^{0}$ in Equation 16.56, accounts for the electrostatic interaction of the remaining fine-grain boxes of the *NF* zone, which are in

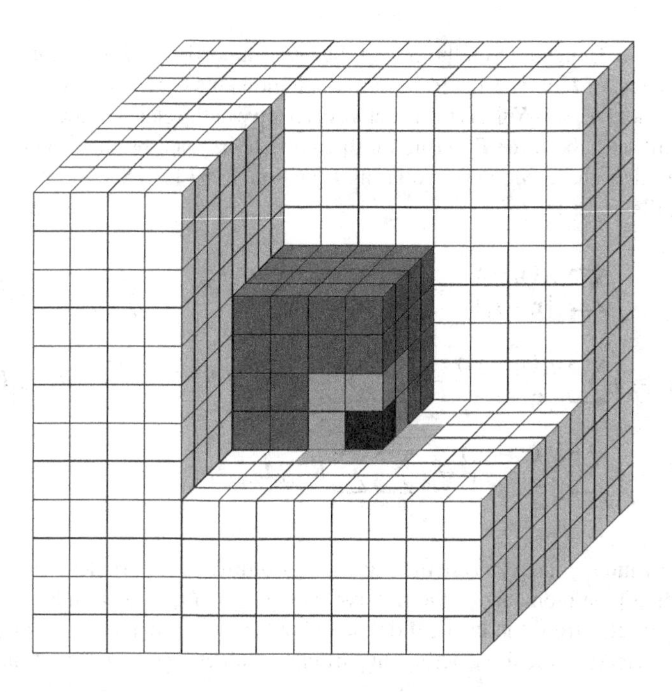

FIGURE 16.5 A periodic *NF* super-box, shown in white, envelops the CUC, shown in gray. The division of the CUC into a set of fine-grain boxes translates the same fine-grain structure into the periodic *NF* zone. For an individual fine-grain box in the CUC, shown in black, the immediate near-field zone for $ws = 1$ consists of both a set of fine-grain boxes, shown in light gray, belonging to the CUC, and a set of the fine-grain boxes, shown in light gray, associated with the periodic near-field zone.

a relative *FF*-position to every fine-grain box from the CUC. The expansion $\tilde{C}_{\neq p}$ is a sum of irregular multipole expansions of every fine-grain box from the *NF* zone defined about the center of the CUC, which does not participate in particle-particle interactions, and, is therefore not included in the range of index p. The regular multipole expansion $\bar{M}^0$ is that of the entire CUC.

Together, Sections 16.4 and 16.5 complete the description of the electrostatic interaction energy of CUC with the periodic continuum.

16.6 Derivative of Electrostatic Energy on Particles in the Central Unit Cell

Resolving the energy due to electrostatic charges in systems that have PBC opens a path to finding the electrostatic field acting on particles, that is, the derivative of the electrostatic potential about each particle. Since particles in the image cells are exact replicas of those of the CUC, only the derivative for particles located inside the CUC needs to be evaluated.

The electrostatic potential, Φ_i, measured at the position of particle i located in the CUC, consists of two distinct terms:

$$\Phi_i = \sum_j \frac{q_j}{|r_i - r_j|} + \bar{R}^*(\mathbf{r}_i) \bullet \tilde{L}. \tag{16.57}$$

Here, the direct-sum part describes the electrostatic contribution arising from the particles j, which is within the near-field boundary in respect to particle i for particle-particle interactions, and are located in the CUC or in the *NF* zone of the periodic system.

The second term, $\bar{R}^*(\mathbf{r}_i) \bullet \tilde{L} \equiv \sum_{l=0}^{\infty} \sum_{m=-l}^{l} \bar{R}^*_{l,m} \tilde{L}_{l,m}$ represents the *FF*-contribution coming from the particles positioned in the *FF* zone relative to particle i, where $\bar{R}^*(\mathbf{r}_i)$ is a regular multipole expansion of the radius vector $\mathbf{r}_i$, and $\tilde{L}$ is an irregular multipole expansion of the potential generated by particles located in the *FF* zone. Depending on the size of the CUC, a portion of $\tilde{L}$ may come from the interior of the CUC while the rest of $\tilde{L}$ carries the irregular lattice moments of the energy.

The electrostatic field, $\mathbf{E}_i = -\nabla\Phi_i$, is the negative derivative of the electrostatic potential, Φ_i with respect to particle positions, and the force $\mathbf{F}_i$ acting on that particle is given by the product of the electrostatic field with the particle's charge q_i. Differentiating Equation 16.57 (see Section 9.1 and Section 10.8 for mathematical details) leads to:

$$F_{i,x} = q_i \sum_j \frac{q_k(x_i - x_j)}{|\mathbf{r}_i - \mathbf{r}_j|^3} - q_i \sum_{l=0}^{\infty}\sum_{m=-l}^{l}\left[\frac{lx_i + imy_i}{x_i^2 + y_i^2}\bar{R}^*_{l,m} - \frac{xz}{x_i^2 + y_i^2}\bar{R}^*_{l-1,m}\right]\tilde{L}_{l,m},$$

$$F_{i,y} = q_i \sum_j \frac{q_k(y_i - y_j)}{|\mathbf{r}_i - \mathbf{r}_j|^3} - q_i \sum_{l=0}^{\infty}\sum_{m=-l}^{l}\left[\frac{ly_i - imx_i}{x_i^2 + y_i^2}\bar{R}^*_{l,m} - \frac{yz}{x_i^2 + y_i^2}\bar{R}^*_{l-1,m}\right]\tilde{L}_{l,m}, \tag{16.58}$$

$$F_{i,z} = q_i \sum_j \frac{q_k(z_i - z_j)}{|\mathbf{r}_i - \mathbf{r}_j|^3} - q_i \sum_{l=0}^{\infty}\sum_{m=-l}^{l}\bar{R}^*_{l-1,m}\tilde{L}_{l,m},$$

where $r_i = (x_i, y_i, z_i)$ and $r_j = (x_j, y_j, z_j)$ are the Cartesian coordinates of particles *i* and *j*, respectively, while $\bar{R}^*_{l,m}$ are regular multipole moments of the radius-vector r_i, and $\tilde{L}_{l,m}$ are cumulative irregular multipole moments of the particles from the far field defined about the coordinate system of particle *i*. Equation 16.58 describes the electrostatic force acting on particle *i*, with charge q_i, arising from all charge sources in the system.

Taking into account that the initial Equation 16.57, which led to the equation of force, applies equally to both non-periodic and periodic systems, the same conclusion also applies to Equation 16.58. The only difference in the application of Equations 16.57 and 16.58 to non-periodic and periodic systems is that the presence of PBS adds additional particle-particle interactions from the nearest boxes surrounding the CUC into the direct sum, and augments the term $\tilde{L}$ by a lattice sum.

16.7 Stress Tensor

Simulating a system that has PBC requires a knowledge of the position of all particles located both inside and outside the CUC. While all the particles inside the CUC have independent degrees of freedom, particle coordinates of the image cells depend on the cell vectors, **t**, that determine the geometry of the CUC as well as on the coordinates of the real particles. Given that, and a knowledge of the electric forces acting on the particles, a complete characterization of the system also requires knowing the forces acting on the cell walls so that the cell geometry can respond appropriately.

Unit cell boundaries are observable physical properties. In order to draw a physical boundary around a cell, it needs to be placed into a periodic continuum and then have the entire system equilibrated. An application of an external force, **F**, to the cell faces by the material surrounding it (see Figure 16.6) induces *stress* in the body. Since overall force is proportional to the area of the surface, stress, σ, is the force, *F*, applied to the body per unit area:

$$\sigma = \frac{F}{A}, \tag{16.59}$$

where *A* is the cross-section area. Although force is a vector that has three components, a cubic solid unit cell has three perpendicularly positioned cross-sections, so a complete specification of stress needs nine components. Each face of the unit cell has three stress components, which define the force acting on that face. When the external force and the body response balance, the unit cell is in equilibrium and the components of stress on parallel faces are of equal magnitude but opposite in sign. Stress components σ_{ij} form a tensor, which is a 3×3 matrix. The diagonal elements σ_{11}, σ_{22}, and σ_{33} are the *normal* components of stress, which are aligned with the coordinate axes. The off-diagonal elements σ_{ij}, where $i \neq j$, are the

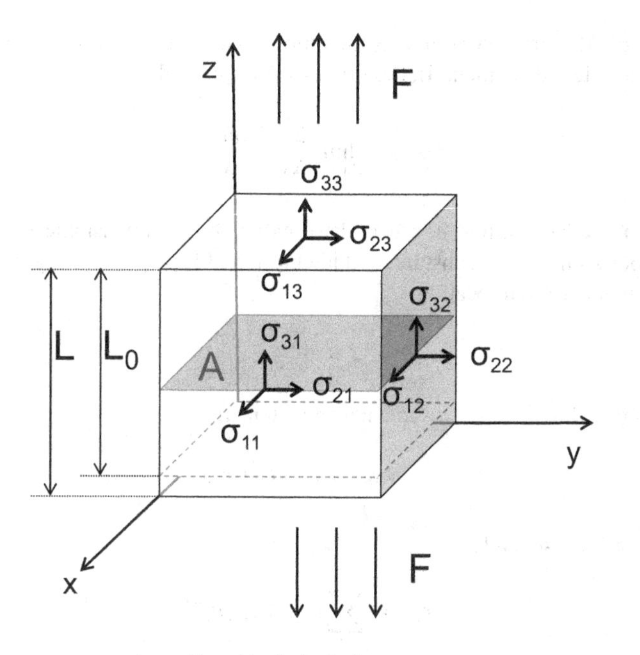

FIGURE 16.6 Stress tensor components σ_{ij} aligned to the body faces.

shear components. If there is no torque, the relationship $\sigma_{ij} = \sigma_{ji}$ holds. If the applied force is uniform, that is, isotropic, and the material is homogeneous, as in the present case, then the stress will be homogeneous, and its value will be the same at every point on the cross-section.

Stress is a second-rank tensor. A tensor is a matrix that represents a physical quantity and the rank of the tensor refers to the number of indices, or dimensions, in that matrix. Since stress has two indices that makes it a tensor of the second rank. An electric field, being a vector, has one index and that makes it a tensor of the first rank. Finally, electrostatic potential, as a scalar number, has no indices, and is, therefore, a tensor of the zeroth rank.

Quantification of stress requires defining one more related physical property of the material called strain, which is the body's response to the stress. Stress induces *strain* in the body. Strain, ε is deformation per unit length:

$$\varepsilon = \frac{L - L_0}{L_0}, \tag{16.60}$$

where L_0 is the initial length and L is the final length, as shown in Figure 16.6. If the deformation is elastic, as in the present case, the body returns to its original state after removal of the stress.

Strain deals with deformation of a continuous material. The one-dimensional case provides a convenient tool for developing the basic math of strain. Figure 16.7 shows an elastic string subjected to stress. A small segment, $\Delta \boldsymbol{x}$, beginning at point $\boldsymbol{P}$ and having coordinate $\boldsymbol{x}$, extends under the pulling force to become $\Delta \boldsymbol{x} + \Delta \boldsymbol{u}$. At the same time, the part of the string to the left of the point $\boldsymbol{P}$ extends to become $\boldsymbol{x} + \boldsymbol{u}$.

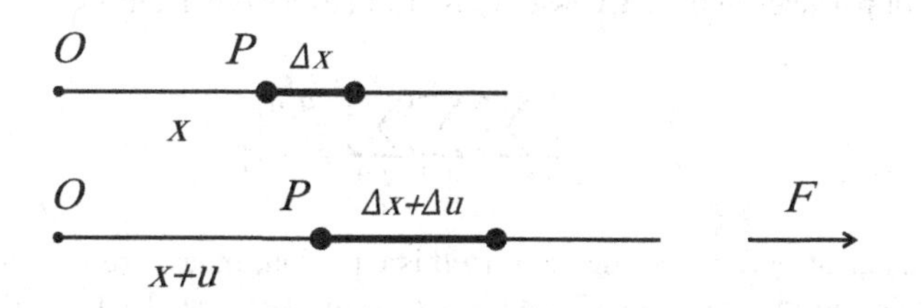

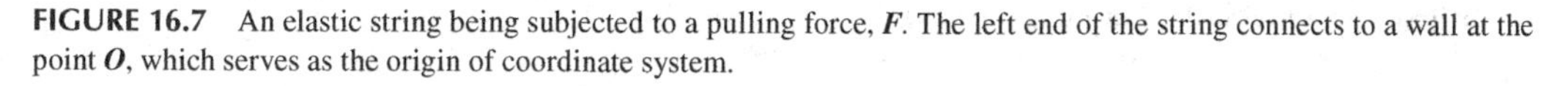
FIGURE 16.7 An elastic string being subjected to a pulling force, $\boldsymbol{F}$. The left end of the string connects to a wall at the point $\boldsymbol{O}$, which serves as the origin of coordinate system.

Using Equation 16.60, the strain at point $\boldsymbol{P}$ is the ratio of the increase Δu in the stretched segment over the length Δx of the original segment. In the limit, as Δx approaches zero:

$$\varepsilon = \lim_{\Delta x \to 0} \frac{\Delta u}{\Delta x} = \frac{du}{dx}. \tag{16.61}$$

Equation 13.61 shows that strain is the rate of deformation. It is a dimensionless quantity. In a particular case of homogeneous strain, ε is a constant, and integration of Equation 13.61 gives the displacement, u, of point $\boldsymbol{P}$ relative to its original location:

$$u = \varepsilon x. \tag{16.62}$$

Point $\boldsymbol{P}$, initially at coordinate x_0, moves to a new position, x:

$$x = x_0 + \varepsilon x_0 = (1+\varepsilon)x_0. \tag{16.63}$$

Going from that to a 3D case leads to:

$$r_\alpha^{new} = \sum_\beta (\delta_{\alpha\beta} + \varepsilon_{\alpha\beta}) r_\beta^{old}, \tag{16.64}$$

where r^{old} and r^{new} are the old and new radius-vector coordinates of the point, respectively; Greek symbols α and β each represent the Cartesian components x, y, and z; $\delta_{\alpha\beta}$ is the Kronecker delta; $\varepsilon_{\alpha\beta}$ is a matrix element of the stain tensor, and is proportional to the corresponding element of the stress tensor. This model of strain theory is general, so it is applicable to the CUC in a non-equilibrium condition as well.

Stress and strain are interrelated physical quantities. Recall that stress is an external force applied to the body, whereas strain is an internal property of the material in response to stress. These phenomena hinge on the reaction of the total energy to the change in unit cell geometry:

$$\sigma_{\alpha\beta} = \frac{\partial U_{total}}{\partial \varepsilon_{\alpha\beta}}, \tag{16.65}$$

that is, stress is a derivative of energy, U_{total}, over strain components. This relationship allows the stress tensor to be obtained analytically, as suggested by Kudin and Scuseria, by conducting the differentiation shown in Equation 16.65.

16.8 Analytic Expression for Stress Tensor

The total energy of a CUC includes all the particle-particle interactions taking place inside the CUC as well as interaction of particles in the CUC with those from every other cell:

$$U_{total} = \frac{1}{2} \sum_{n=-\infty}^{\infty} \sum_{i=1}^{N} \sum_{j=1}^{N} \frac{q_i^0 q_j^n}{\left|\mathbf{r}_i^0 - \mathbf{r}_j^n\right|}, \tag{16.2}$$

where N is the number of particles in the cell. In this equation, index i runs over particles located in the CUC, that is reflected in coordinate $\mathbf{r}_i^0$; and index j runs over particles located in cell n and having coordinate $\mathbf{r}_j^n$. Obviously, $i \neq j$ when $n = 0$.

Due to the indirect dependence of energy on strain components, $\varepsilon_{\alpha\beta}$, differentiation of U_{total} over the strain components requires using a chain rule that includes summations over all particles and all cells in the system:

$$\frac{\partial U_{total}}{\partial \varepsilon_{\alpha\beta}} = \sum_{p=-\infty}^{\infty} \sum_{k=1}^{N} \sum_{\mu} \frac{\partial r_{k\mu}^{p}}{\partial \varepsilon_{\alpha\beta}} \frac{\partial U_{total}}{\partial r_{k\mu}^{p}}, \tag{16.66}$$

where $r_{k\mu}^{p}$ is a radius-vector element of particle k in each cell p, and μ is the Cartesian component of the vector $\mathbf{r}_k^p = \left(r_{kx}^p \hat{\mathbf{x}} + r_{ky}^p \hat{\mathbf{y}} + r_{kz}^p \hat{\mathbf{z}}\right)$. The new cell index p is introduced to distinguish it from the cell index n, which runs inside the expression for U_{total} in Equation 16.2.

Using the dependence of each particle coordinate on the cell strain already derived in Equation 16.64, the first partial derivative of Equation 16.66 can be resolved as:

$$\sum_{\mu} \frac{\partial r_{k\mu}^{p}}{\partial \varepsilon_{\alpha\beta}} \frac{\partial U_{total}}{\partial r_{k\mu}^{p}} = \sum_{\mu} \frac{\partial}{\partial \varepsilon_{\alpha\beta}} \left[\sum_{\nu} \left(\delta_{\mu\nu} + \varepsilon_{\mu\nu}\right) r_{j\nu}^{p} \right] \frac{\partial U_{total}}{\partial r_{k\mu}^{p}} = r_{k\beta}^{p} \frac{\partial U_{total}}{\partial r_{k\alpha}^{p}}. \tag{16.67}$$

In this expression, only terms having indices $\mu = \alpha$ and $\nu = \beta$ survive the differentiation (remember that indices α and β each represent the Cartesian components x, y, and z). When these substitutions are made in Equation 16.66, it simplifies to:

$$\frac{\partial U_{total}}{\partial \varepsilon_{\alpha\beta}} = \sum_{p=-\infty}^{\infty} \sum_{k=1}^{N} r_{k\beta}^{p} \frac{\partial U_{total}}{\partial r_{k\alpha}^{p}}. \tag{16.68}$$

With that, all that remains is to differentiate the electrostatic energy U_{total} of the CUC over each coordinate component $r_{k\alpha}^{p}$ of particle k in cell p. But before moving on to that goal, it is useful to revisit the differentiation of the energy over particle coordinate $\mathbf{r}_k$.

The derivative of the energy U_{total}, defined for PBC in Equation 16.2, over the position of particle k in the CUC gives the force, $\mathbf{F}_k$, acting on that particle:

$$\mathbf{F}_k = \frac{dU_{total}}{d\mathbf{r}_k} = \frac{1}{2} \sum_{n=-\infty}^{\infty} \sum_{i=1}^{N} \sum_{j=1}^{N} \frac{d}{d\mathbf{r}_k} \frac{q_i^0 q_j^n}{\left|\mathbf{r}_i^0 - \mathbf{r}_j^n\right|}. \tag{16.69}$$

In this equation, indices i and j independently match k, and that splits the right-hand side in Equation 16.69 into two sums:

$$\mathbf{F}_k = \frac{1}{2} \sum_{n=-\infty}^{\infty} \sum_{j=1}^{N} \frac{d}{d\mathbf{r}_k^0} \frac{q_k^0 q_j^n}{\left|\mathbf{r}_k^0 - \mathbf{r}_j^n\right|} + \frac{1}{2} \sum_{n=-\infty}^{\infty} \sum_{i=1}^{N} \frac{d}{d\mathbf{r}_k^n} \frac{q_i^0 q_k^n}{\left|\mathbf{r}_i^0 - \mathbf{r}_k^n\right|}, \tag{16.70}$$

where $k = i$ and $k = j$, respectively. Since it now becomes necessary to distinguish the location of particle k in the CUC, when $n = 0$, from its location in an image cell, when $n \neq 0$, the derivative $d/d\mathbf{r}_k$ assumes the form $d/d\mathbf{r}_k^0$ and $d/d\mathbf{r}_k^n$, respectively.

The two sums in Equation 16.70 represent both possible types of summation of the particle-particle interactions. Figure 16.8 visually assists in this interpretation. The first sum in Equation 16.70 represents the interaction of particle k in the CUC with every other particle in the periodic system that matches the left panel in Figure 16.8. The second sum in Equation 16.70 represents the interaction of every image of particle k with all particles located in the CUC that corresponds to the right panel in Figure 13.8.

A visual inspection of Figure 16.8 shows that for every pair 0-0, 0-1, 0-2, 0-3, 0-4, 0-5, 0-6, 0-7, and 0-8 of interacting cells in the left panel there exists an identical pair 0-0, 5-0, 6-0, 7-0, 8-0, 1-0, 2-0, 3-0, and 4-0 of interacting cells in the right panel that carries the same relative position of the particle in black

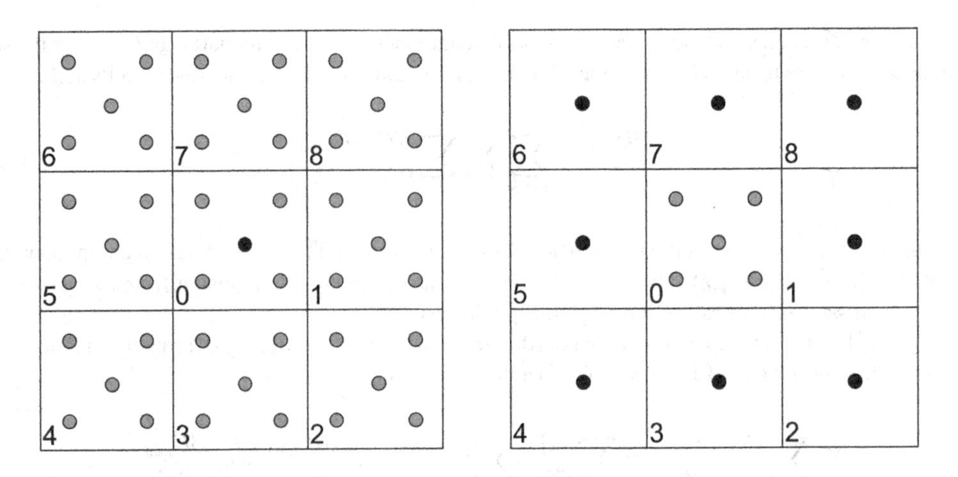

FIGURE 16.8 Left panel: Interaction of a particle (in black) with every other particle (in gray). Right panel: Interaction of every black particle with all particles in the central cell. The number in the lower left corner of each cell is the cell id. In the right panel, all particles not involved in the summation are omitted for visual clarity.

from an image cell to the particles from the central cell. Indeed, the pair of boxes 0-0 in the left panel has an identical particle arrangement to that of the pair 0-0 in the right panel; the pair 0-1 in the left panel matches the pair 5-0 in the right panel, and so on. Because of this symmetry both sums in Equation 16.70 are identical, and that leads to:

$$\mathbf{F}_k = \sum_{n=-\infty}^{\infty} \sum_{j=1}^{N} \frac{d}{d\mathbf{r}_k^0} \frac{q_k^0 q_j^n}{|\mathbf{r}_k^0 - \mathbf{r}_j^n|} = \sum_{n=-\infty}^{\infty} \sum_{i=1}^{N} \frac{d}{d\mathbf{r}_k^n} \frac{q_i^0 q_k^n}{|\mathbf{r}_i^0 - \mathbf{r}_k^n|}. \tag{16.71}$$

This symmetry is an intrinsic property of PBC, and applies to all periodic systems.

Having obtained the equation of force for PBC in a particle-particle formalism, work can now continue on Equation 16.68. Substituting the expression for energy from Equation 16.2 into Equation 16.68 gives:

$$\frac{\partial U_{total}}{\partial \varepsilon_{\alpha\beta}} = \frac{1}{2} \sum_{p=-\infty}^{\infty} \sum_{k=1}^{N} r_{k\beta}^p \frac{\partial}{\partial r_{k\alpha}^p} \left(\sum_{n=-\infty}^{\infty} \sum_{i=1}^{N} \sum_{j=1}^{N} \frac{q_i^0 q_j^n}{|\mathbf{r}_i^0 - \mathbf{r}_j^n|} \right). \tag{16.72}$$

An essential first step in dealing with this equation is to reduce the number of summations. Due to differentiation being performed over cell index p and particle index k, only those summation elements that satisfy the conditions ($p = 0$ and $k = i$) for the CUC or ($p = n$ and $k = j$) for the image cells are non-zero. Taking that requirement into account leads to:

$$\frac{\partial U_{total}}{\partial \varepsilon_{\alpha\beta}} = \frac{1}{2} \sum_{k=1}^{N} r_{k\beta}^0 \frac{\partial}{\partial r_{k\alpha}^0} \sum_{n=-\infty}^{\infty} \sum_{j=1}^{N} \frac{q_k^0 q_j^n}{|\mathbf{r}_k^0 - \mathbf{r}_j^n|} + \frac{1}{2} \sum_{n=-\infty}^{\infty} \sum_{k=1}^{N} r_{k\beta}^n \frac{\partial}{\partial r_{k\alpha}^n} \sum_{i=1}^{N} \frac{q_i^0 q_k^n}{|\mathbf{r}_i^0 - \mathbf{r}_k^n|}. \tag{16.73}$$

As expected, half of the contributing terms come from the CUC, and the other half comes from the image cells. Unfortunately, unlike Equation 16.71, the presence of additional multipliers $r_{k\beta}^0$ and $r_{k\beta}^n$ makes the two contributing groups numerically non-equivalent.

The one-dimension periodicity illustrated in Figure 16.9 provides some background for further work involving Equation 16.73. In Equation 16.73, the triple sum n along translation indices a, b, and c reduces to a single sum over cell index a in a 1D periodic system. In a 1D system, the translation vector $\mathbf{t}^a = \left(t_x^a, t_y^a, t_z^a\right)$ translates the particle coordinate, r_k according to:

$$r_{k\beta}^a = r_{k\beta}^0 + a t_\beta^a, \tag{16.74}$$

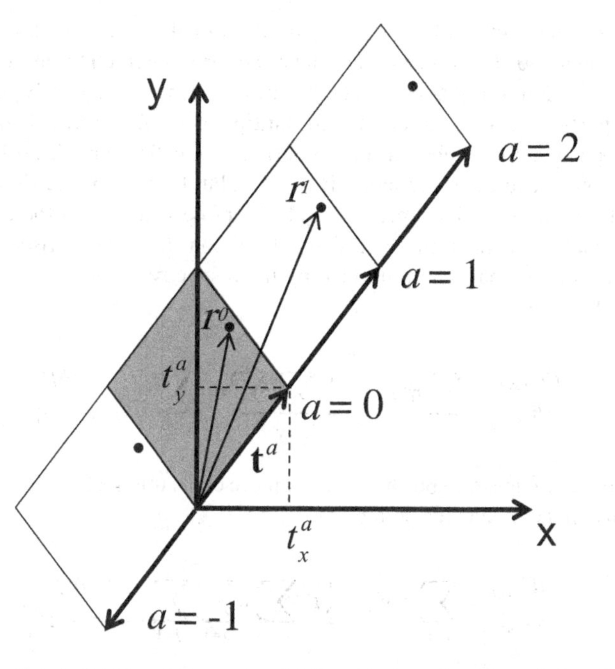

FIGURE 16.9 Periodic system in 1D. Translation vector $\mathbf{t}^a$ defines the direction of translation. Cell id $a = 0$ matches the CUC shown in gray.

where t_β^a is a constant, and a is an integer variable. Inserting Equation 16.74 into Equation 16.73 gives:

$$\frac{\partial U_{total}}{\partial \varepsilon_{\alpha\beta}} = \frac{1}{2}\sum_{k=1}^{N} r_{k\beta}^0 \frac{\partial}{\partial r_{k\alpha}^0} \sum_{a=-\infty}^{\infty}\sum_{j=1}^{N} \frac{q_k^0 q_j^a}{\left|\mathbf{r}_k^0 - \mathbf{r}_j^a\right|} + \frac{1}{2}\sum_{a=-\infty}^{\infty}\sum_{k=1}^{N}\left(r_{k\beta}^0 + a t_\beta^a\right)\frac{\partial}{\partial r_{k\alpha}^a}\sum_{i=1}^{N}\frac{q_i^0 q_k^a}{\left|\mathbf{r}_i^0 - \mathbf{r}_k^a\right|}. \tag{16.75}$$

Combining like terms for $r_{k\beta}^0$ transforms this equation to:

$$\begin{aligned}\frac{\partial U_{total}}{\partial \varepsilon_{\alpha\beta}} = &\sum_{k=1}^{N} r_{k\beta}^0 \left(\frac{1}{2}\sum_{a=-\infty}^{\infty}\sum_{j=1}^{N} \frac{\partial}{\partial r_{k\alpha}^0}\frac{q_k^0 q_j^a}{\left|\mathbf{r}_k^0 - \mathbf{r}_j^a\right|} + \frac{1}{2}\sum_{a=-\infty}^{\infty}\sum_{i=1}^{N}\frac{\partial}{\partial r_{k\alpha}^a}\frac{q_i^0 q_k^a}{\left|\mathbf{r}_i^0 - \mathbf{r}_k^a\right|}\right) \\ &+ \frac{1}{2}\sum_{a=-\infty}^{\infty}\sum_{k=1}^{N} a t_\beta^a \sum_{i=1}^{N}\frac{\partial}{\partial r_{k\alpha}^a}\frac{q_i^0 q_k^a}{\left|\mathbf{r}_i^0 - \mathbf{r}_k^a\right|}.\end{aligned} \tag{16.76}$$

In this equation, a comparison of the portion in brackets to Equation 16.70 shows that it represents the Cartesian component of force, $F_{k\alpha}$, along axis α. Substituting Equation 16.70 into Equation 16.76 gives:

$$\frac{\partial U_{total}}{\partial \varepsilon_{\alpha\beta}} = \sum_{k=1}^{N} r_{k\beta}^0 F_{k\alpha} + \frac{1}{2}\sum_{a=-\infty}^{\infty}\sum_{k=1}^{N} a t_\beta^a \sum_{i=1}^{N}\frac{\partial}{\partial r_{k\alpha}^a}\frac{q_i^0 q_k^a}{\left|\mathbf{r}_i^0 - \mathbf{r}_k^a\right|}. \tag{16.77}$$

The final task is to replace the inconvenient derivative in Equation 16.77 over the coordinate of an image particle with a derivative over the coordinate of a real particle in the CUC. This can be accomplished using the intrinsic PBC symmetry of the position of particles in a periodic system. Recall that the rightmost sum in Equation 16.77 represents the interaction of an image particle k with all particles i in the CUC. Pair 1-0 in the right panel of Figure 16.8 is an example of such an arrangement. Exactly the same mutual orientation of the interacting particles takes place in the pair 0-5 in the left panel of Figure 16.8, where a particle k in the CUC interacts with all image particles of cell 5. In both cases, the interacting pairs, 1-0 from the right panel and 0-5 from the left, have the same interaction energy. This symmetry was used earlier in the derivation of Equation 16.71. This time, when working on Equation 16.77 where each interacting pair of particles carries an additional multiplication factor a, it helps to notice that if the image cells were appropriately numbered for 3D periodicity, cells 1 and 5, positioned on the opposite sides of the CUC, would have the same numerical value, but be of opposite sign, that is, a would have the values -1 and 1, respectively. This means that for every derivative over the coordinates of a particle image multiplied by factor a on the right-hand side of Equation 16.77 there exists an equivalent derivative over the coordinates of a real particle multiplied by the respective factor $-a$. This symmetry property simplifies Equation 16.77 to:

$$\frac{\partial U_{total}}{\partial \varepsilon_{\alpha\beta}} = \sum_{k=1}^{N} r_{k\beta}^{0} F_{k\alpha} - \frac{1}{2} \sum_{a=-\infty}^{\infty} \sum_{k=1}^{N} a t_{\beta}^{a} \sum_{i=1}^{N} \frac{\partial}{\partial r_{k\alpha}^{0}} \frac{q_k^0 q_i^a}{\left|\mathbf{r}_k^0 - \mathbf{r}_i^a\right|}. \tag{16.78}$$

At this point, the particle indices can be changed to a more conventional form. Deleting the superscript 0, and moving the constant t_{β}^{a} outside the sum gives:

$$\frac{\partial U_{total}}{\partial \varepsilon_{\alpha\beta}} = \sum_{i=1}^{N} r_{i\beta} F_{i\alpha} - \frac{1}{2} t_{\beta}^{a} \sum_{a=-\infty}^{\infty} a \sum_{i=1}^{N} \sum_{j=1}^{N} \frac{\partial}{\partial r_{i\alpha}} \frac{q_i q_j^a}{\left|\mathbf{r}_i - \mathbf{r}_j^a\right|}. \tag{16.79}$$

This equation provides the stress tensor for a system that has one-dimensional periodicity.

In contrast to Equation 16.70, the general case of 3D periodicity for stress tensor requires accounting for three translation vectors $\mathbf{t}^a = \left(t_x^a, t_y^a, t_z^a\right)$, $\mathbf{t}^b = \left(t_x^b, t_y^b, t_z^b\right)$, and $\mathbf{t}^c = \left(t_x^c, t_y^c, t_z^c\right)$, accompanied by three translation factors a, b, and c, respectively. Since each translation happens independently, the stress tensor component for a system periodic in 3D becomes:

$$\begin{aligned} \frac{\partial U_{total}}{\partial \varepsilon_{\alpha\beta}} = \sum_{i=1}^{N} r_{i\beta} F_{i\alpha} &- \frac{1}{2} t_{\beta}^{a} \sum_{a=-\infty}^{\infty} a \sum_{i=1}^{N} \sum_{j=1}^{N} \frac{\partial}{\partial r_{i\alpha}} \frac{q_i q_j^a}{\left|\mathbf{r}_i - \mathbf{r}_j^a\right|} \\ &- \frac{1}{2} t_{\beta}^{b} \sum_{b=-\infty}^{\infty} b \sum_{i=1}^{N} \sum_{j=1}^{N} \frac{\partial}{\partial r_{i\alpha}} \frac{q_i q_j^b}{\left|\mathbf{r}_i - \mathbf{r}_j^b\right|} \\ &- \frac{1}{2} t_{\beta}^{c} \sum_{c=-\infty}^{\infty} c \sum_{i=1}^{N} \sum_{j=1}^{N} \frac{\partial}{\partial r_{i\alpha}} \frac{q_i q_j^c}{\left|\mathbf{r}_i - \mathbf{r}_j^c\right|} = \sum_{i=1}^{N} r_{i\beta} F_{i\alpha} + P_{\alpha\beta}. \end{aligned} \tag{16.80}$$

This equation illustrates the two different components that contribute to the stress. The first is the non-optimal position of particles in the CUC, which creates a force $\mathbf{F}$ acting on the particles. Optimizing the positions of the particles while holding cell vectors $\mathbf{t}$ fixed makes that term vanish. The second source of stress arises from the fixed cell dimensions. This is called the *rigid cell stress*, $P_{\alpha\beta}$, since it arises when the cell geometry is not relaxed. The change in cell vectors $\mathbf{t}^a$, $\mathbf{t}^b$, and $\mathbf{t}^c$ alters the cell geometry but leaves the distance between particles within the cell unchanged.

16.9 Lattice Sum for Stress Tensor

Computation of the stress tensor under PBC developed by Kudin and Scuseria follows a similar pattern to the computation of the electrostatic force. Both quantities include lattice summations of irregular multipole expansions of cell images from the periodic continuum. This involves an infinite number of image cells carrying a replica of multipole moments of the CUC, followed by differentiation of the accumulated potential. Given that, the algorithm for the computation of the lattice sum for the energy, as detailed in Section 16.4, provides a good starting point for obtaining the lattice sum for the stress tensor.

Since each image cell carries a replica of the multipole moments $\bar{M}^0$ of the CUC, computation of the central cell energy includes an M2L transformation of the image cell moments $\bar{M}^0$ into an irregular multipole expansion, $\tilde{L}(a,b,c) = \tilde{S}(a,b,c) \otimes \bar{M}^0$, that carries the electrostatic potential of an image cell $\{a, b, c\}$ into the CUC, where $\tilde{S}(a,b,c)$ is an irregular helper expansion of the radius vector connecting the center of the CUC with the center of the image cell $\{a, b, c\}$. Each image cell from the periodic continuum contributes in an identical manner, and only the expansion $\tilde{S}(a,b,c)$ accounts for the unique relative position of the image cell about the CUC.

Unlike the computation of electrostatic energy, determining the stress tensor requires three separate image cell contributions:

$$\begin{aligned} \tilde{L}^a(a,b,c) &= a\tilde{S}(a,b,c) \otimes \bar{M}^0, \\ \tilde{L}^b(a,b,c) &= b\tilde{S}(a,b,c) \otimes \bar{M}^0, \\ \tilde{L}^c(a,b,c) &= c\tilde{S}(a,b,c) \otimes \bar{M}^0, \end{aligned} \tag{16.81}$$

where each cell contribution needs to be scaled by the cell index, a, b, or c, as defined by Equation 16.80. The superscripts a, b, and c in $\tilde{L}^a$, $\tilde{L}^b$ and $\tilde{L}^c$ indicate the respective periodicity dimension. Summation over all image cells in the periodic continuum leads to three lattice moments $\tilde{T}^x$:

$$\tilde{T}^x = \sum_{a,b,c}^{\infty} x\tilde{S}(a,b,c). \tag{16.82}$$

Multipole operation $\tilde{T}^x \otimes \bar{M}^0$ gives the cumulative electrostatic contribution of all image cells into the CUC, scaled by the cell index $x = a$, b, or c along the periodicity dimension. The minor difference that exists in the summation of unscaled and scaled image cell contributions for energy and stress tensor will be addressed shortly.

As usual, the work on deriving the expression for lattice sum for stress tensor begins with analyzing the contribution from the first few hierarchy levels. According to the standard principles of FMM, the cumulative irregular multipole expansion of all image cells from the *FF* at the hierarchy level $h = 0$ that carries the electrostatic potential of the image cells into the central cell is:

$$\tilde{L}^{a,h=0} = \tilde{S}^{a,0} \otimes \bar{M}^0, \quad \text{where } \tilde{S}^{a,0} = \sum_{a,b,c \in FF^0} a\tilde{S}(a,b,c). \tag{16.83}$$

In obtaining the image cell contribution, each translation moment $\tilde{S}(a,b,c)$ associated with cell $\{a, b, c\}$ becomes multiplied by the value of cell index a to produce a cumulative irregular translation moment $\tilde{S}^{a,0}$. Figure 16.10 shows the *FF* zone at $h = 0$ for periodic dimension a. The *FF* zone becomes split into two disjoint segments by cells having index $a = 0$. The latter cells contribute to neither the *FF* nor the *NF* zone.

Equation 16.83 resolves the first contributing term, $\tilde{T}^{a,0}$, in the far-field portion of the lattice sum for stress tensor, $\tilde{T}^a$, along the periodic dimension a:

$$\tilde{T}^{a,0} = \tilde{S}^{a,0}. \tag{16.84}$$

This term automatically includes the proper scaling of the cell contribution to the stress tensor by the cell index a, which comes with $\tilde{S}^{a,0}$.

Further work on deriving the equation for the lattice sum for the stress tensor makes use of its similarity to the lattice sum for energy. The lattice moments $\tilde{C}$ for energy obey the recurrence relation:

$$\tilde{C}^{k+1} = \tilde{S}^0 + (XC^k) \otimes \bar{R}^{*0}, \tag{16.54}$$

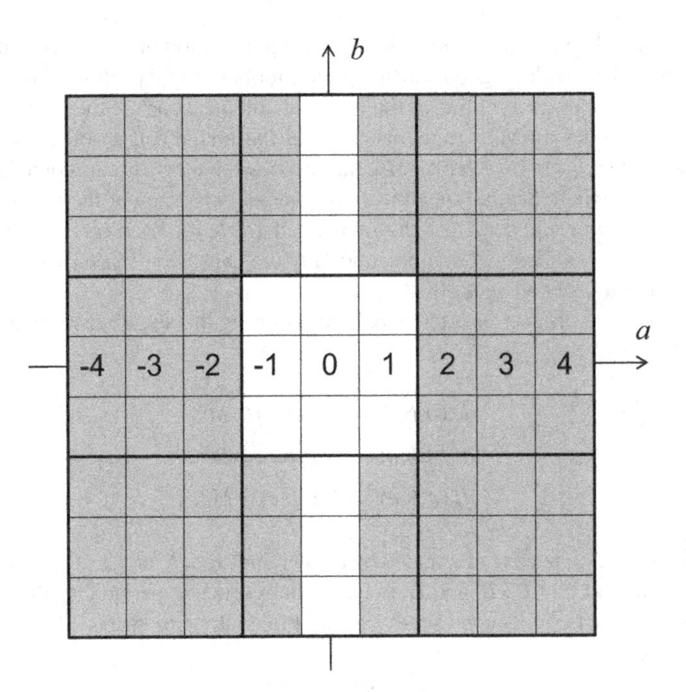

FIGURE 16.10 Cells contributing to the lattice sum along the periodic dimension a for hierarchy level $h = 0$ and $ws = 1$. The CUC has cell index 0; *FF* cells are in gray. The index a scales the image cell contributions along the respective periodicity dimension.

where $\tilde{C}^0 = \tilde{S}^0$, and X is the stretching operator, and $\bar{R}^{*0}$ and $\tilde{S}^0$ are translation moments:

$$\bar{R}^{*0} = \sum_{a,b,c \in FF^0} \bar{R}^{*0}(a,b,c), \quad \text{and} \quad \tilde{S}^0 = \sum_{a,b,c \in FF^0} \tilde{S}(a,b,c). \tag{16.85}$$

Converting Equation 16.54 to the lattice sum for the stress tensor requires performing a summation along each periodicity dimension. This results in three recurrence relations for the lattice sum for the stress tensor, where moments $\tilde{T}^x$ take the place of moments $\tilde{C}$, and $x = a$, b, or c. The term $\tilde{T}^0$ includes the initial scaling of the *FF* contribution by the cell index that is required for the stress tensor. An additional modification to Equation 16.54 will be needed in order to expand the scaling mechanism to higher hierarchy levels. Details and implementation of this modification will now be described.

Choosing the periodic dimension a to guide the implementation, and, based on analogy with Equation 16.54, while taking into account that $\tilde{T}^{a,0} = \tilde{S}^{a,0}$, the cumulative lattice moments for stress tensor for $h = 1$ that include the contribution from $h = 0$ would be $\tilde{T}^{a,1} \approx \tilde{S}^{a,0} + (2ws+1)(X\tilde{T}^{a,0}) \otimes \bar{R}^{*0} = \tilde{S}^{a,0} + (2ws+1)(X\tilde{S}^{a,0}) \otimes \bar{R}^{*0}$, where multiplier $(2\ ws + 1)$ provides an initial attempt to convert the scaling explicitly present in $\tilde{S}^{a,0}$ to the scaling of stretched super-cells for $h = 1$. Assuming, for illustration, that $ws = 1$ and, because $\tilde{S}^{a,0} = \sum_{a,b,c \in FF} a\tilde{S}(a,b,c)$, the factor $(2\ ws + 1)$ applied to the cell indices $a = (-4, -3, -2, -1, 0, 1, 2, 3, 4)$ yields a new set of numbers $(-12, -9, -6, -3, 0, 3, 6, 9, 12)$. The M2L operation $(X\tilde{S}^{a,0}) \otimes \bar{R}^{*0}$ attaches each of those new scale factors to the super-cells. Since each super-cell has the width of $2\ ws + 1$ elementary cells along the periodicity dimension, and, as before assuming that $ws = 1$, the contribution of every cell in *FF* zone at $h = 1$ now becomes scaled with the factor $(-12, -12, -12, \ldots, -3, -3, -3, 0, 0, 0, 3, 3, 3, \ldots, 12, 12, 12)$ along the dimension a, as shown at the bottom of the plot in Figure 16.11. Scaling the contribution of a super-cell equates to scaling the contribution of every elementary cell within that super-cell resulting in the $2\ ws + 1$ cells being multiplied by the same factor as that applied to their super-cell.

The resulting set of multipliers, $(-12, -12, -12, \ldots, -3, -3, -3, 0, 0, 0, 3, 3, 3, \ldots, 12, 12, 12)$, represents a close match to the actual coefficients needed, $(-13, -12, -11, \ldots, -4, -3, -2, -1, 0, 1, 2,$

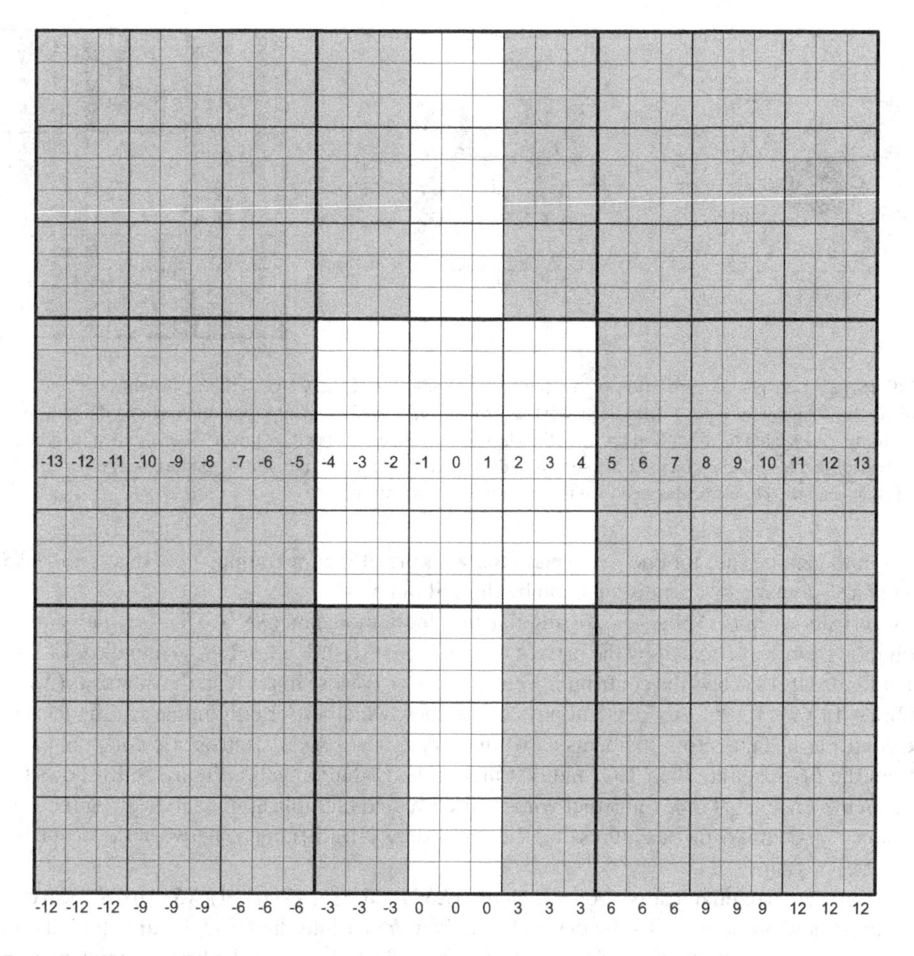

FIGURE 16.11 Cells contributing to the stress tensor for periodic dimension a at the hierarchy level $h = 1$ when $ws = 1$ due to the term $(2ws+1)(X\tilde{S}^{a,0}) \otimes \bar{R}^{*0}$. Contributing cells in the FF are shown in gray and non-contributing cells are shown in white. The numbers in the central row represent the actual cell index a. The numbers in the bottom come from the term $(2ws+1)(X\tilde{S}^{a,0}) \otimes \bar{R}^{*0}$.

3, 4,.., 11, 12, 13), for the stress tensor. Conversion from the multipliers to the coefficients is performed by the addition of irregular translation moments with the weights (−1, 0, 1, …, −1, 0, 1, −1, 0, 1, −1, 0, 1, …, −1, 0, 1) to the term $(2ws+1)(X\tilde{S}^{a,0}) \otimes \bar{R}^{*0}$. Conveniently, the pattern of the type (−1, 0, 1) exists in the cumulative regular translation moments, $\bar{R}^{*a,0}$, scaled by the cell index a:

$$\bar{R}^{*a,0} = \sum_{a,b,c \in NF^0} a\bar{R}^*(a,b,c). \tag{16.86}$$

These moments translate the regular multipole moments $\bar{M}^0$ of the CUC to the NF boxes at $h = 0$, by the means of M2M operation $\bar{R}^{*a,0} \oplus \bar{M}^0$ while simultaneously scaling each cell contribution by the cell index. The results are scaled multipole moments, $\bar{M}^{1,a}$, of the central super-cell at $h = 1$. Figure 16.12 illustrates the M2M operation and its result for $ws = 1$ in the left and middle panels, respectively.

Next, the pattern −1, 0, 1 obtained in the central super-cell at $h = 1$ is recast onto the FF contribution at the present hierarchy level. Collecting FF contributions scaled by the repeating pattern of type (−1, 0, 1) into the CUC requires replicating the regular multipole moments $\bar{M}^{1,a}$ of the central super-cell at $h = 1$ to the respective super-cells in the FF zone. This is accomplished using the M2L operation $(X\tilde{S}^0) \otimes \bar{M}^{a,1} = (X\tilde{S}^0) \otimes (\bar{R}^{*a,0} \oplus \bar{M}^0) = \left[(X\tilde{S}^0) \otimes \bar{R}^{*a,0}\right] \otimes \bar{M}^0$. Steps leading to Equation 16.35, which

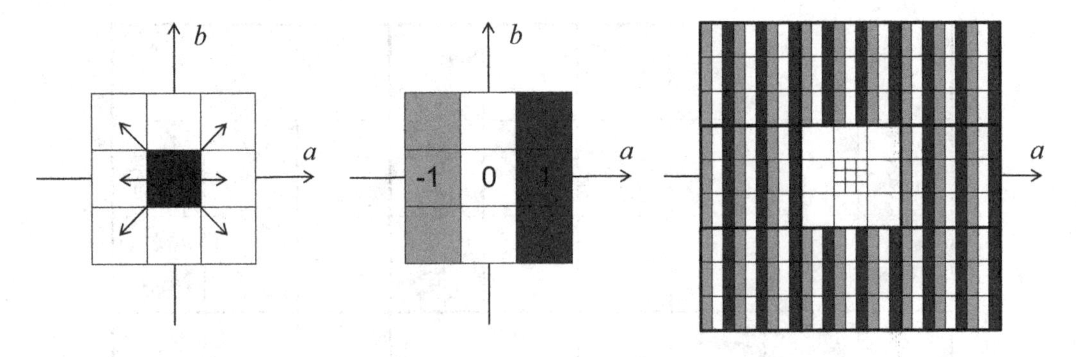

FIGURE 16.12 Left panel: translation of the multipole expansion of the CUC to the near-field boxes at $h = 0$ along periodicity dimension a for $ws = 1$. Middle panel: the result of that translation; multipole moments of the central super-cell at $h = 1$ for $ws = 1$ scaled by the cell index a. Cells whose contribution is scaled by factor -1 are shown in light gray. Cells whose contribution is scaled by factor 1 are shown in dark gray. Right panel: the scaling pattern (-1, 0, 1, ..., -1, 0, 1, ..., -1, 0, 1) that is applied to elementary cells in the *FF*-zone of $h = 1$ for $ws = 1$.

appears in the lattice sum for energy, explain the working of that operation. Note that, unlike $X\tilde{S}^{a,0}$, the moments $X\tilde{S}^0$ do not carry a multiplication by the cell index.

This derivation leads to the scaled irregular multipole moments $(X\tilde{S}^0) \otimes \bar{R}^{*a,0}$, which carry the *FF* contribution from $h = 1$ scaled by the pattern (-1, 0, 1, ..., -1, 0, 1, ..., -1, 0, 1) into the CUC. The right panel of Figure 16.12 shows the contributing cells for $ws = 1$. More important, the moments $(X\tilde{S}^0) \otimes \bar{R}^{*a,0}$ and $(2ws+1)(X\tilde{S}^{a,0}) \otimes \bar{R}^{*0}$ are identical physical entities, which differ only in the scaling factor applied to the contributing cells. Both moments assist in carrying the scaled electrostatic potential of an image cell from the *FF* zone into the CUC, and that makes their addition valid. Figure 16.13 shows the list of cells from the *FF* zone at $h = 1$ together with their corrected multiplication factors along the periodicity dimension a. A comparison of Figures 16.11 to 16.13 shows the difference between the uncorrected and corrected FF regions.

The corrected irregular translation moments $(X\tilde{S}^0) \otimes \bar{R}^{*a,0} + (2ws+1)(X\tilde{S}^{a,0}) \otimes \bar{R}^{*0}$ carry the *FF* contribution, now properly scaled by cell index a, from $h = 1$ into the CUC. Figure 16.13 shows where in the periodic continuum that contribution comes from. The portion of the lattice sum that comes from the central super-cell, shown in white in Figure 16.13, corresponds to the term $\tilde{T}^{a,0} = \tilde{S}^{a,0}$. With that, the lattice sum for the stress tensor that includes the contributions from $h = 0$ and $h = 1$ along the periodic dimension a becomes:

$$\tilde{T}^{a,1} = \tilde{S}^{a,0} + (X\tilde{S}^0) \otimes \bar{R}^{*a,0} + (2ws+1)(X\tilde{S}^{a,0}) \otimes \bar{R}^{*0}. \tag{16.87}$$

Construction of the explicit form of the lattice sum for the stress tensor for $h = 2$ repeats the pattern of transition from $h = 0$ to $h = 1$. A similarity between the lattice sum for the stress tensor and that for the energy suggests a possible solution in the form of $\tilde{T}^{a,2} \approx \tilde{S}^{a,0} + (2ws+1)(X\tilde{T}^{a,1}) \otimes \bar{R}^{*0}$, which comes from Equation 16.54. Scale factor (2 ws + 1) provides an initial attempt to scale the contribution of every elementary cell from the *FF* zone at $h = 2$ by the cell index that is required for the lattice sum of the stress tensor. When $ws = 1$, the multiplier (2 ws + 1) changes the existing 27 scale factors (-13, -12, -11, ..., -1, 0, 1, ..., 11, 12, 13) carried by $T^{a,1}$ to (-39, -36, -33, ..., -3, 0, 3, ..., 33, 36, 39). Application of the M2L operation $(X\tilde{T}^{a,1}) \otimes \bar{R}^{*0}$ then applies each scale factor to the corresponding super-cells. Since each super-cell has the width of 2 ws + 1 elementary cells along the periodicity dimension, and assuming that $ws = 1$, each of the 81 elementary cells determining the width of the hierarchy level $h = 2$ along the periodicity dimension a becomes scaled with the factor (-39, -39, -39, -36, -36, -36, ..., -3, -3, -3, 0, 0, 0, 3, 3, 3, ..., 36, 36, 36, 39, 39, 39). Once again, reaching the target scaling pattern of (-40, -39, -38, ..., -1, 0, 1, ..., 38, 39, 40) requires applying a correction term with the weights (-1, 0, 1, ..., -1, 0, 1, ..., -1, 0, 1) to the cell contributions.

According to the technique developed for $h = 1$, the correction needed would be of type $\tilde{S}^2 \otimes \bar{M}^{a,2}$, where $\bar{M}^{a,2}$ are the regular multipole moments of the central super-cell at $h = 2$, which carry the scale

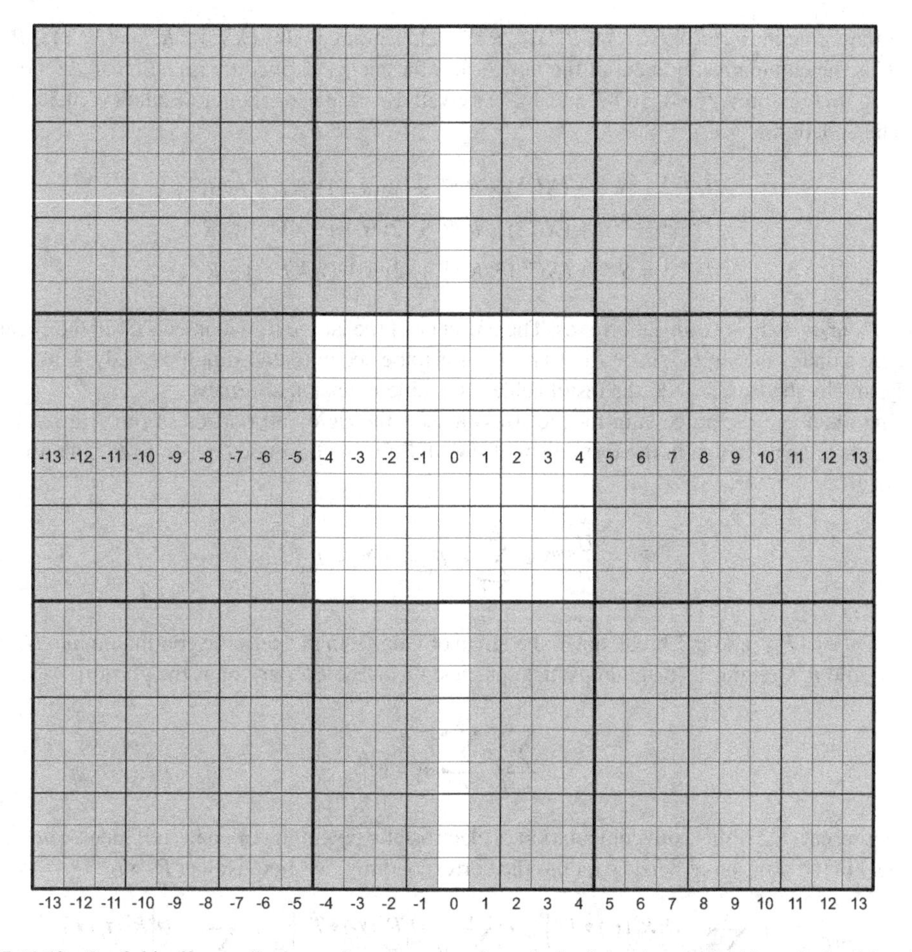

FIGURE 16.13 Far-field cells contributing to the stress tensor for periodic dimension a at the hierarchy level $h = 1$ when $ws = 1$. Contributing cells are gray and non-contributing cells are white. The numbers in the central row represent the cell index a and those in the bottom row show the corrected scale factors that now match the cell index a.

pattern (−1, 0, 1, …, −1, 0, 1, …, −1, 0, 1) along the periodic dimension a in its elementary cells, and $\tilde{S}^2$ are the irregular translation moments, which map the scaling pattern from the central super-cell to the elementary cells in the *FF* zone at $h = 2$. The non-elementary moments $\bar{M}^{a,2}$ expand down to the already known moments $\bar{R}^{*a,0}$ and $\bar{M}^0$:

$$\tilde{S}^2 \otimes \bar{M}^{a,2} = \tilde{S}^2 \otimes (\bar{R}^{*1} \oplus \bar{M}^{a,1}) = \tilde{S}^2 \otimes \left(\bar{R}^{*1} \oplus (\bar{R}^{*a,0} \oplus \bar{M}^0)\right). \tag{16.88}$$

The transformation steps detailed in the derivation of Equation 16.47 reduce Equation 16.88 to:

$$\tilde{S}^2 \otimes \bar{M}^{a,2} = \left((\tilde{S}^2 \otimes \bar{R}^{*1}) \otimes \bar{R}^{*a,0}\right) \otimes \bar{M}^0 = \left[\left(X\left((X\tilde{S}^0) \otimes \bar{R}^{*0}\right)\right) \otimes \bar{R}^{*a,0}\right] \otimes \bar{M}^0, \tag{16.89}$$

where the irregular translation moments $\left(X\left((X\tilde{S}^0) \otimes \bar{R}^{*0}\right)\right) \otimes \bar{R}^{*a,0}$ provide the necessary correction to the *FF* contribution. With that, the cumulative contribution to the stress tensor from the *FF* zone up to and including $h = 2$ becomes:

$$\tilde{T}^{a,2} = \tilde{S}^{a,0} + \left(X\left((X\tilde{S}^0) \otimes \bar{R}^{*0}\right)\right) \otimes \bar{R}^{*a,0} + (2ws + 1)(X\tilde{T}^{a,1}) \otimes \bar{R}^{*0}. \tag{16.90}$$

Here, the correction term $(X((X\tilde{S}^{0})\otimes\bar{R}^{*0}))\otimes\bar{R}^{*a,0}$ includes the equality $(X\tilde{S}^{0})\otimes\bar{R}^{*0}=(X\tilde{C}^{0})\otimes\bar{R}^{*0}=\tilde{C}^{1}$, where $\tilde{C}$ is the cumulative moment of the lattice sum for energy defined by Equation 16.54. Given that, and lining up Equations 16.84, 16.87, and 16.90 as well as comparing them to Equation 16.54 yields the recurrence relations:

$$\begin{aligned}\tilde{T}^{a,k+1}&=\tilde{S}^{a,0}+(X\tilde{C}^{k})\otimes\bar{R}^{*a,0}+(2ws+1)(X\tilde{T}^{a,k})\otimes\bar{R}^{*0},\\ \tilde{T}^{b,k+1}&=\tilde{S}^{b,0}+(X\tilde{C}^{k})\otimes\bar{R}^{*b,0}+(2ws+1)(X\tilde{T}^{b,k})\otimes\bar{R}^{*0},\\ \tilde{T}^{c,k+1}&=\tilde{S}^{c,0}+(X\tilde{C}^{k})\otimes\bar{R}^{*c,0}+(2ws+1)(X\tilde{T}^{c,k})\otimes\bar{R}^{*0},\end{aligned}\tag{16.91}$$

where a, b, and c indicate translation axes. The presence of the lattice sum moments $\tilde{C}^{k}$ in the equation for the lattice sum for the stress tensor allows both sums to be accumulated simultaneously. This converges quickly, and by the time $k=20$ the result achieves double precision accuracy.

Having obtained the lattice sum for the stress tensor, the rigid cell portion in the original Equation 16.80 can now be rewritten in the formalism of FMM. Splitting the rigid cell stress $P_{\alpha\beta}$ into the NF and FF parts gives:

$$\frac{\partial U_{total}}{\partial\varepsilon_{\alpha\beta}}=\sum_{i=1}^{N}r_{i\beta}F_{i\alpha}+P_{\alpha\beta}^{NF}+P_{\alpha\beta}^{FF},\tag{16.92}$$

where the terms $P_{\alpha\beta}^{NF}$ and $P_{\alpha\beta}^{FF}$ break down the range of index a in the periodic continuum into NF and FF zones. While $P_{\alpha\beta}^{NF}$ retains its original particle-particle form, the FF part maps the portion

$$\sum_{a\in FF}a\sum_{j=1}^{N}\frac{q_j^a}{\left|\mathbf{r}_i-\mathbf{r}_j^a\right|},$$

which represents the direct sum computation of electrostatic potential scaled by cell index a arising from particles j in the position of particle i, onto the corresponding FMM equivalent $\bar{R}^{*}\bullet\tilde{T}^{a}$:

$$P_{\alpha\beta}^{FF}=-\frac{t_\beta^a}{2}\sum_{i=i}^{N}q_i\frac{\partial\left[\bar{R}^{*}(\mathbf{r}_i)\bullet\tilde{T}^{a}\right]}{\partial r_{i\alpha}}-\frac{t_\beta^b}{2}\sum_{i=1}^{N}q_i\frac{\partial\left[\bar{R}^{*}(\mathbf{r}_i)\bullet\tilde{T}^{b}\right]}{\partial r_{i\alpha}}-\frac{t_\beta^c}{2}\sum_{i=1}^{N}q_i\frac{\partial\left[\bar{R}^{*}(\mathbf{r}_i)\bullet\tilde{T}^{c}\right]}{\partial r_{i\alpha}},\tag{16.93}$$

where $\tilde{T}^{a}$, $\tilde{T}^{b}$, and $\tilde{T}^{c}$ are lattice sum components along the periodicity dimension a, b, and c, respectively; $\bar{R}^{*}(\mathbf{r}_i)\cdot\tilde{T}^{a}\equiv\sum_{l=0}^{\infty}\sum_{m=-l}^{l}\bar{R}_{l,m}^{*}\tilde{T}_{l,m}^{a}$ is the scaled electrostatic potential at the position of particle i due to particles from the FF; and $\bar{R}^{*}(\mathbf{r}_i)$ are regular multipole moments of the radius vector $\mathbf{r}_i$ of particle i. Since differentiation applies only to $\bar{R}^{*}(\mathbf{r}_i)$, the lattice sum components of $\tilde{T}$ can be moved outside the differentiation block, to be separately summed up leading to:

$$P_{\alpha\beta}^{FF}=-\frac{1}{2}\sum_{i=i}^{N}q_i\frac{\partial}{\partial r_{i\alpha}}\left[\bar{R}^{*}(\mathbf{r}_i)\bullet\left(t_\beta^a\tilde{T}^{a}+t_\beta^b\tilde{T}^{b}+t_\beta^c\tilde{T}^{c}\right)\right].\tag{16.94}$$

Differentiation of the term $\bar{R}^{*}\bullet\tilde{T}$ in Equation 16.94 proceeds in the manner that previously led to the derivation of Equation 10.67, and has the following outcome:

$$\begin{aligned}\frac{\partial\left[\bar{R}^{*}(\mathbf{r}_i)\bullet\tilde{T}\right]}{\partial x_i}&=-\sum_{l=0}^{\infty}\sum_{m=-l}^{l}\left(\frac{lx_i+imy_i}{x_i^2+y_i^2}\bar{R}_{l,m}^{*}-\frac{x_iz_i}{x_i^2+y_i^2}\bar{R}_{l-1,m}^{*}\right)\tilde{T}_{l,m},\\ \frac{\partial\left[\bar{R}^{*}(\mathbf{r}_i)\bullet\tilde{T}\right]}{\partial y_i}&=-\sum_{l=0}^{\infty}\sum_{m=-l}^{l}\left(\frac{ly_i-imx_i}{x_i^2+y_i^2}\bar{R}_{l,m}^{*}-\frac{y_iz_i}{x_i^2+y_i^2}\bar{R}_{l-1,m}^{*}\right)\tilde{T}_{l,m},\\ \frac{\partial\left[\bar{R}^{*}(\mathbf{r}_i)\bullet\tilde{T}\right]}{\partial z_i}&=-\sum_{l=0}^{\infty}\sum_{m=-l}^{l}\bar{R}_{l-1,m}^{*}\tilde{T}_{l,m},\end{aligned}\tag{16.95}$$

where $\mathbf{r}_i = (x_i, y_i, z_i)$. The right-hand side of Equation 16.95 omits the argument $\mathbf{r}_i$ in $\bar{R}^*_{l,m}$ for the sake of brevity. Substituting the partial derivatives from Equation 16.95 into Equation 16.94 while consecutively replacing symbol α with x, y, and z leads to:

$$\begin{aligned} P^{FF}_{x\beta} &= \frac{1}{2}\sum_{i=i}^{N} q_i \sum_{l=0}^{\infty}\sum_{m=-l}^{l}\left(\frac{lx_i + imy_i}{x_i^2 + y_i^2}\bar{R}^*_{l,m} - \frac{x_i z_i}{x_i^2 + y_i^2}\bar{R}^*_{l-1,m}\right)\left(t^a_\beta \tilde{T}^a_{l,m} + t^b_\beta \tilde{T}^b_{l,m} + t^c_\beta \tilde{T}^c_{l,m}\right), \\ P^{FF}_{y\beta} &= \frac{1}{2}\sum_{i=i}^{N} q_i \sum_{l=0}^{\infty}\sum_{m=-l}^{l}\left(\frac{ly_i - imx_i}{x_i^2 + y_i^2}\bar{R}^*_{l,m} - \frac{y_i z_i}{x_i^2 + y_i^2}\bar{R}^*_{l-1,m}\right)\left(t^a_\beta \tilde{T}^a_{l,m} + t^b_\beta \tilde{T}^b_{l,m} + t^c_\beta \tilde{T}^c_{l,m}\right), \\ P^{FF}_{z\beta} &= \frac{1}{2}\sum_{i=i}^{N} q_i \sum_{l=0}^{\infty}\sum_{m=-l}^{l}\bar{R}^*_{l-1,m}\left(t^a_\beta \tilde{T}^a_{l,m} + t^b_\beta \tilde{T}^b_{l,m} + t^c_\beta \tilde{T}^c_{l,m}\right). \end{aligned} \tag{16.96}$$

Six more equations emerge from Equation 16.96 upon substitution of the symbol β with coordinates x, y, and z, thereby completing the nine-equation set for the *FF* portion of the rigid cell stress tensor.

All that remains is to resolve the *NF* part of the rigid cell stress tensor:

$$\begin{aligned} P^{NF}_{\alpha\beta} = &-\frac{1}{2}t^a_\beta \sum_{a\in NF} a \sum_{i=1}^{N}\sum_{j=1}^{N}\frac{\partial}{\partial r_{i\alpha}}\frac{q_i q^a_j}{\left|\mathbf{r}_i - \mathbf{r}^a_j\right|} - \frac{1}{2}t^b_\beta \sum_{b\in NF} b \sum_{i=1}^{N}\sum_{j=1}^{N}\frac{\partial}{\partial r_{i\alpha}}\frac{q_i q^b_j}{\left|\mathbf{r}_i - \mathbf{r}^b_j\right|} \\ &-\frac{1}{2}t^c_\beta \sum_{c\in NF} c \sum_{i=1}^{N}\sum_{j=1}^{N}\frac{\partial}{\partial r_{i\alpha}}\frac{q_i q^c_j}{\left|\mathbf{r}_i - \mathbf{r}^c_j\right|}. \end{aligned} \tag{16.97}$$

Differentiation of inverse-distance function has the outcome shown in Equation 9.9. Application of those steps to Equation 16.97 leads to:

$$\begin{aligned} P^{NF}_{x\beta} = &\frac{t^a_\beta}{2}\sum_{a\in NF} a \sum_{i=1}^{N}\sum_{j=1}^{N}\frac{q_i q^a_j\left(x_i - x^a_j\right)}{\left|\mathbf{r}_i - \mathbf{r}^a_j\right|^3} + \frac{t^b_\beta}{2}\sum_{b\in NF} b \sum_{i=1}^{N}\sum_{j=1}^{N}\frac{q_i q^b_j\left(x_i - x^b_j\right)}{\left|\mathbf{r}_i - \mathbf{r}^b_j\right|^3} \\ &+\frac{t^c_\beta}{2}\sum_{c\in NF} c \sum_{i=1}^{N}\sum_{j=1}^{N}\frac{\partial}{\partial r_{i\alpha}}\frac{q_i q^c_j\left(x_i - x^c_j\right)}{\left|\mathbf{r}_i - \mathbf{r}^c_j\right|^3}. \end{aligned} \tag{16.98}$$

Equation 16.98 refers to particle coordinates in the image cells. Since those cells are exact replicas of the CUC, their particle coordinates relate to those of the CUC via the translation operations $\mathbf{r}^a_j = \mathbf{r}_j + a\mathbf{t}^a$, $\mathbf{r}^b_j = \mathbf{r}_j + b\mathbf{t}^b$, and $\mathbf{r}^c_j = \mathbf{r}_j + c\mathbf{t}^c$, where $\mathbf{t}^a = \left(t^a_x, t^a_y, t^a_z\right)$, $\mathbf{t}^b = \left(t^b_x, t^b_y, t^b_z\right)$, and $\mathbf{t}^c = \left(t^c_x, t^c_y, t^c_z\right)$ are cell vectors. Using these translations to redefine particle coordinates of the image cells over those of the CUC, along with rearranging the sums to keep the equation compact, and considering all three possible values x, y, and z for index α in $P^{NF}_{\alpha\beta}$ leads to:

$$\begin{aligned} P^{NF}_{x\beta} &= \sum_{i=1}^{N}\sum_{j=1}^{N} q_i q_j \left[\frac{t^a_\beta}{2}\sum_{a\in NF}\frac{a\left(x_i - x_j - at^a_x\right)}{\left|\mathbf{r}_i - \mathbf{r}_j - a\mathbf{t}^a\right|^3} + \frac{t^b_\beta}{2}\sum_{b\in NF}\frac{b\left(x_i - x_j - bt^b_x\right)}{\left|\mathbf{r}_i - \mathbf{r}_j - b\mathbf{t}^b\right|^3} + \frac{t^c_\beta}{2}\sum_{c\in NF}\frac{c\left(x_i - x_j - ct^c_x\right)}{\left|\mathbf{r}_i - \mathbf{r}_j - c\mathbf{t}^c\right|^3}\right], \\ P^{NF}_{y\beta} &= \sum_{i=1}^{N}\sum_{j=1}^{N} q_i q_j \left[\frac{t^a_\beta}{2}\sum_{a\in NF}\frac{a\left(y_i - y_j - at^a_y\right)}{\left|\mathbf{r}_i - \mathbf{r}_j - a\mathbf{t}^a\right|^3} + \frac{t^b_\beta}{2}\sum_{b\in NF}\frac{b\left(y_i - y_j - bt^b_y\right)}{\left|\mathbf{r}_i - \mathbf{r}_j - b\mathbf{t}^b\right|^3} + \frac{t^c_\beta}{2}\sum_{c\in NF}\frac{c\left(y_i - y_j - ct^c_y\right)}{\left|\mathbf{r}_i - \mathbf{r}_j - c\mathbf{t}^c\right|^3}\right], \\ P^{NF}_{z\beta} &= \sum_{i=1}^{N}\sum_{j=1}^{N} q_i q_j \left[\frac{t^a_\beta}{2}\sum_{a\in NF}\frac{a\left(z_i - z_j - at^a_z\right)}{\left|\mathbf{r}_i - \mathbf{r}_j - a\mathbf{t}^a\right|^3} + \frac{t^b_\beta}{2}\sum_{b\in NF}\frac{b\left(z_i - z_j - bt^b_z\right)}{\left|\mathbf{r}_i - \mathbf{r}_j - b\mathbf{t}^b\right|^3} + \frac{t^c_\beta}{2}\sum_{c\in NF}\frac{c\left(z_i - z_j - ct^c_z\right)}{\left|\mathbf{r}_i - \mathbf{r}_j - c\mathbf{t}^c\right|^3}\right]. \end{aligned} \tag{16.99}$$

In Equation 16.99, replacing symbol β with coordinates x, y, and z generates six additional equations, and that completes the nine-equation set for the *NF* portion of the rigid cell stress tensor.

Note that, in the completely assembled stress tensor, only six of the nine derivatives are truly independent; the other three degrees of freedom correspond to the three rotations, and, in principle, could be eliminated.

Appendix

A.1 Spherical Polar Coordinates

Within the spherical polar coordinate system, shown in Figure A.1, any point in 3D space can be defined by coordinates r, θ, and ϕ, where r is the radial coordinate, which represents the distance to the point from the center of the coordinate system, θ is the polar angle between the z-axis and the radius vector of the point, and ϕ is the azimuthal angle created by the projection of a particle's radius vector onto the xy-plane. Polar angle is measured from the positive direction of the z-axis toward the particle radius vector, and takes values in the range [0, 180]. Azimuthal angle is measured from the positive direction of x-axis toward the projection in a counterclockwise direction around the z-axis when looking at the xy-plane from the tip of the z-axis, and takes values in the range [0, 360].

Cartesian and spherical polar coordinates are related by the equations:

$$\begin{aligned} x &= r\sin\theta\cos\phi, \\ y &= r\sin\theta\sin\phi, \\ z &= r\cos\theta. \end{aligned} \tag{A1.1}$$

From these relations, it immediately follows that:

$$\begin{aligned} r &= \sqrt{x^2+y^2+z^2}, \\ \theta &= \arccos\left(\frac{z}{r}\right), \\ \phi &= \arctan\left(\frac{y}{x}\right), \\ x+iy &= r\sin\theta(\cos\phi + i\sin\phi) = r\sin\theta e^{i\phi}. \end{aligned} \tag{A1.2}$$

Because spherical polar coordinates are functions of Cartesian coordinates, their derivatives over Cartesian coordinates need to be evaluated as derivatives of the type $\partial\Phi(r,\theta,\phi)/\partial x$. In this particular case the derivative is:

$$\frac{\partial\Phi(r,\theta,\phi)}{\partial x} = \frac{\partial\Phi}{\partial r}\frac{\partial r}{\partial x} + \frac{\partial\Phi}{\partial\theta}\frac{\partial\theta}{\partial x} + \frac{\partial\Phi}{\partial\phi}\frac{\partial\phi}{\partial x}. \tag{A1.3}$$

Derivatives of $\Phi(r,\theta,\phi)$ over the other two Cartesian coordinates, y and z, are similar.

While derivatives of Φ over Cartesian coordinates are specific to that function, partial derivatives of the type $\partial r/\partial x$, $\partial\theta/\partial x$, and $\partial\phi/\partial x$ are independent of Φ, and describe the spherical polar coordinate system itself. These derivatives are obtained by differentiation of the equations that establish the relationship between spherical polar and Cartesian coordinate systems.

Differentiation of the polar distance r over Cartesian coordinate x gives:

$$\frac{\partial r}{\partial x} = \frac{\partial}{\partial x}(x^2+y^2+z^2)^{\frac{1}{2}} = \frac{x}{(x^2+y^2+z^2)^{\frac{1}{2}}} = \frac{x}{r} = \frac{r\sin\theta\cos\phi}{r} = \sin\theta\cos\phi. \tag{A1.4}$$

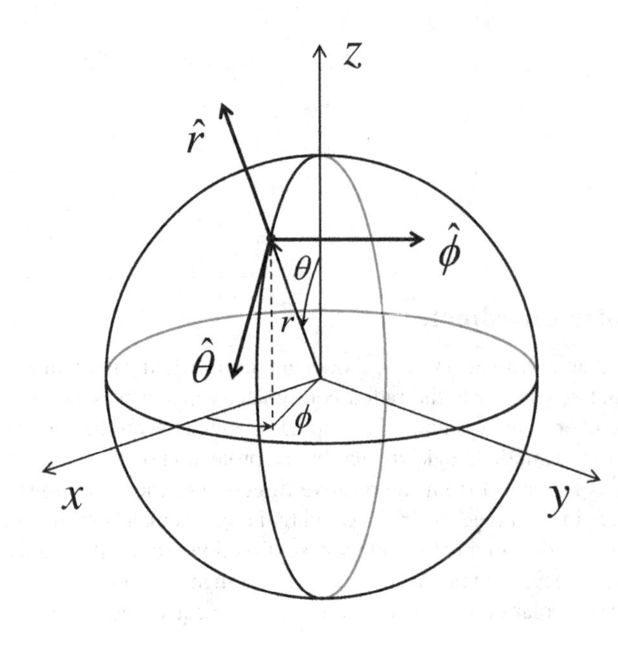

FIGURE A.1 Spherical polar coordinates r, θ, ϕ and their unit vectors $\hat{r}, \hat{\theta}, \hat{\phi}$.

Likewise, partial derivatives of r over the other two Cartesian coordinates are:

$$\frac{\partial r}{\partial y} = \frac{y}{r} = \frac{r \sin\theta \sin\phi}{r} = \sin\theta \sin\phi. \tag{A1.5}$$

and

$$\frac{\partial r}{\partial z} = \frac{z}{r} = \frac{r \cos\theta}{r} = \cos\theta. \tag{A1.6}$$

Differentiation of the polar angle θ over Cartesian coordinates requires application of the chain rule. This gives:

$$\frac{\partial \theta}{\partial x} = \frac{\partial\left[\arccos(z/r)\right]}{\partial[z/r]} \frac{\partial[z/r]}{\partial x}. \tag{A1.7}$$

Finding the derivative of arccos function over its argument requires a few additional steps. For a function $f = \arccos(g)$, its inverse function is $g = \cos(f)$. The differential of g is:

$$dg = d\left[\cos(f)\right] = -\sin(f)df. \tag{A1.8}$$

Replacing sine by cosine, to eliminate the unknown function $\sin(f)$, leads to:

$$dg = -\sqrt{1 - \cos^2(f)}df. \tag{A1.9}$$

From that, obtaining the derivative of f over g is now a straightforward operation of dividing the increments:

$$\frac{df}{dg} = -\frac{1}{\sqrt{1 - \cos^2(f)}} = -\frac{1}{\sqrt{1 - g^2}}. \tag{A1.10}$$

Replacing function *f* by arccos provides the derivative being sought:

$$\frac{d}{dg}\left[\arccos(g)\right]=-\frac{1}{\sqrt{1-g^2}}. \tag{A1.11}$$

Going back to $\partial\theta/\partial x$, the last part of the derivative to find is:

$$\frac{\partial}{\partial x}\left[\frac{z}{r}\right]=\frac{\partial}{\partial x}\left[z(x^2+y^2+z^2)^{-\frac{1}{2}}\right]=-zx(x^2+y^2+z^2)^{-\frac{3}{2}}=-\frac{zx}{r^3}. \tag{A1.12}$$

Now that both parts of the derivative $\partial\theta/\partial x$ are known, they can be combined to give:

$$\frac{\partial\theta}{\partial x}=-\frac{1}{\sqrt{1-(z^2/r^2)}}\left(-\frac{zx}{r^3}\right)=\frac{r}{\sqrt{r^2-z^2}}\frac{zx}{r^3}=\frac{zx}{r^2\sqrt{x^2+y^2}}. \tag{A1.13}$$

Substituting Cartesian coordinates with their spherical polar equivalents leads to:

$$\frac{\partial\theta}{\partial x}=\frac{r\cos\theta\, r\sin\theta\cos\phi}{r^2\sqrt{r^2\sin^2\theta\cos^2\phi+r^2\sin^2\theta\sin^2\phi}}=\frac{\sin\theta\cos\theta\cos\phi}{r\sin\theta}=\frac{\cos\theta\cos\phi}{r}. \tag{A1.14}$$

Determination of the derivative $\partial\theta/\partial y$ uses the same approach. The derivative of arccos() is already known, and the derivative of the second part that appears after application of the chain rule to $\partial\theta/\partial y$ is:

$$\frac{\partial}{\partial y}\left[\frac{z}{r}\right]=\frac{\partial}{\partial y}\left[z(x^2+y^2+z^2)^{-\frac{1}{2}}\right]=-zy(x^2+y^2+z^2)^{-\frac{3}{2}}=-\frac{zy}{r^3}, \tag{A1.15}$$

so combining the two parts gives:

$$\frac{\partial\theta}{\partial y}=-\frac{1}{\sqrt{1-(z^2/r^2)}}\left(-\frac{zy}{r^3}\right)=\frac{r}{\sqrt{r^2-z^2}}\frac{zy}{r^3}=\frac{zy}{r^2\sqrt{x^2+y^2}}. \tag{A1.16}$$

Replacing Cartesian coordinates by their spherical polar equivalents leads to the final form of the derivative of polar angle θ over coordinate y:

$$\frac{\partial\theta}{\partial y}=\frac{r\cos\theta\, r\sin\theta\sin\phi}{r^2\sqrt{r^2\sin^2\theta\cos^2\phi+r^2\sin^2\theta\sin^2\phi}}=\frac{\sin\theta\cos\theta\sin\phi}{r\sin\theta}=\frac{\cos\theta\sin\phi}{r}. \tag{A1.17}$$

The third derivative, $\partial\theta/\partial z$, uses the chain rule in a similar manner to give:

$$\begin{aligned}\frac{\partial}{\partial z}\left[\frac{z}{r}\right]&=\frac{\partial}{\partial z}\left[z(x^2+y^2+z^2)^{-\frac{1}{2}}\right]=(x^2+y^2+z^2)^{-\frac{1}{2}}-z^2(x^2+y^2+z^2)^{-\frac{3}{2}}=\frac{1}{r}-\frac{z^2}{r^3}\\&=\frac{x^2+y^2}{r^3}.\end{aligned} \tag{A1.18}$$

Combining both parts into the target equation, and substituting the Cartesian coordinates with their spherical polar counterparts leads to:

$$\frac{\partial\theta}{\partial z}=-\frac{r}{\sqrt{x^2+y^2}}\frac{x^2+y^2}{r^3}=-\frac{\sqrt{x^2+y^2}}{r^2}=-\frac{r\sin\theta}{r^2}=-\frac{\sin\theta}{r}. \tag{A1.19}$$

This completes the differentiation of polar angle θ over Cartesian coordinates.

The partial derivatives of azimuthal angle ϕ over each of the Cartesian coordinates is somewhat more complicated. Since ϕ has no explicit dependence on the Cartesian coordinates, its differentiation has to be performed using the chain rule, leading to:

$$\frac{\partial \phi}{\partial x} = \frac{\partial \left[\arctan\left(\frac{y}{x}\right)\right]}{\partial \left[\frac{y}{x}\right]} \frac{\partial \left[\frac{y}{x}\right]}{\partial x}. \tag{A1.20}$$

This composite derivative includes a derivative of the arctan() function that will need to be determined.
The inverse function of $f = \arctan(g)$ is $g = \tan(f)$, which is easy to differentiate using the well-known differentiation rules for the sin() and cos() trigonometric functions. The differential of g is:

$$dg = d\left[\tan(f)\right] = d\left[\frac{\sin(f)}{\cos(f)}\right] = \frac{\cos^2(f)+\sin^2(f)}{\cos^2(f)} df = \left[1+\tan^2(f)\right] df = (1+g^2)\, df. \tag{A1.21}$$

Rearranging the terms on the left- and right-hand sides leads to:

$$\frac{df}{dg} = \frac{1}{1+g^2}. \tag{A1.22}$$

Replacing $f = \arctan(g)$ gives the formula for differentiation of the arctan() function:

$$\frac{d}{dg}\arctan(g) = \frac{1}{1+g^2}. \tag{A1.23}$$

This can now be substituted into the equation for $\partial\phi/\partial x$ to give:

$$\frac{\partial \phi}{\partial x} = \frac{1}{1+(y^2/x^2)} \frac{\partial (y/x)}{\partial x} = \frac{x^2}{x^2+y^2}\left(-\frac{y}{x^2}\right) = -\frac{y}{x^2+y^2} = -\frac{r\sin\theta\sin\phi}{r^2\sin^2\theta} = -\frac{\sin\phi}{r\sin\theta}. \tag{A1.24}$$

Using a similar procedure, the partial derivative of azimuthal angle ϕ over the Cartesian coordinate y is:

$$\frac{\partial \phi}{\partial y} = \frac{x^2}{x^2+y^2} \frac{\partial (y/x)}{\partial y} = \frac{x^2}{x^2+y^2}\left(\frac{1}{x}\right) = \frac{x}{x^2+y^2} = \frac{r\sin\theta\cos\phi}{r^2\sin^2\theta} = \frac{\cos\phi}{r\sin\theta}. \tag{A1.25}$$

The remaining derivative of azimuthal angle ϕ over coordinate z is zero, because ϕ does not depend on coordinate z:

$$\frac{\partial \phi}{\partial z} = 0. \tag{A1.26}$$

A.2 Unit Vectors in Spherical Polar Coordinates

The unit vectors in spherical polar coordinates form a set of mutually orthogonal vectors $(\hat{r}, \hat{\theta}, \hat{\phi})$, as shown in Figure A.1. These unit vectors are orthogonal tangent lines on the surface of the sphere at the position of the point. The positive direction of the unit vectors corresponds to the direction in which the value of the corresponding spherical polar coordinate increases.

Unlike Cartesian unit vectors $\hat{x}, \hat{y}, \hat{z}$, which are constant, the direction of spherical polar unit vectors $\hat{r}, \hat{\theta}, \hat{\phi}$ varies with the change in position of the point in 3D space. Each unit vector in spherical polar coordinates can be expressed in terms of unit vectors in Cartesian coordinates:

$$\hat{\mathbf{r}} = \sin\theta\cos\phi\,\hat{\mathbf{x}} + \sin\theta\sin\phi\,\hat{\mathbf{y}} + \cos\theta\,\hat{\mathbf{z}},$$
$$\hat{\theta} = \cos\theta\cos\phi\,\hat{\mathbf{x}} + \cos\theta\sin\phi\,\hat{\mathbf{y}} - \sin\theta\,\hat{\mathbf{z}}, \tag{A2.1}$$
$$\hat{\phi} = -\sin\phi\,\hat{\mathbf{x}} + \cos\phi\,\hat{\mathbf{y}}.$$

Likewise, the direction vectors in Cartesian coordinates can be expressed in terms of unit vectors in spherical polar coordinates:

$$\hat{\mathbf{x}} = \sin\theta\cos\phi\,\hat{\mathbf{r}} + \cos\theta\cos\phi\,\hat{\theta} - \sin\phi\,\hat{\phi},$$
$$\hat{\mathbf{y}} = \sin\theta\sin\phi\,\hat{\mathbf{r}} + \cos\theta\sin\phi\,\hat{\theta} + \cos\phi\,\hat{\phi}, \tag{A2.2}$$
$$\hat{\mathbf{z}} = \cos\theta\,\hat{\mathbf{r}} - \sin\theta\,\hat{\theta}.$$

In order to derive the relationship between unit vectors in spherical polar coordinates and those in Cartesian coordinates, applying the following starting conditions, which originate from the definition of spherical polar coordinates, is useful:

$$\mathbf{r} = (x,y,z) = (r\sin\theta\cos\phi, r\sin\theta\sin\phi, r\cos\theta) = r\sin\theta\cos\phi\,\hat{\mathbf{x}} + r\sin\theta\sin\phi\,\hat{\mathbf{y}} + r\cos\theta\,\hat{\mathbf{z}},$$
$$|\mathbf{r}| = r = \sqrt{x^2+y^2+z^2}, \tag{A2.3}$$
$$\hat{\mathbf{r}} = \frac{\mathbf{r}}{r}.$$

Where $\mathbf{r}$ is a radius vector of a particle in 3D that has Cartesian coordinates (x, y, z), r is the length of the radius-vector, and $\hat{\mathbf{r}}$ is a radial unit vector of unit length pointing in the same direction as vector $\mathbf{r}$. The angles θ and ϕ are polar and azimuthal angles of the particle, respectively.

From the definitions of the radius vector of a particle and that of the length of the vector, it immediately follows that the unit vector for the radius vector is:

$$\hat{\mathbf{r}} = (\sin\theta\cos\phi, \sin\theta\sin\phi, \cos\theta) = \sin\theta\cos\phi\,\hat{\mathbf{x}} + \sin\theta\sin\phi\,\hat{\mathbf{y}} + \cos\theta\,\hat{\mathbf{z}}. \tag{A2.4}$$

The other two unit vectors can be found from the relationships that exist between the previously defined vectors. Given that, by definition, vectors $\hat{\mathbf{r}}$ and $\hat{\mathbf{z}}$ are coplanar and that vector $\hat{\phi}$ is orthogonal to both of them, it follows that:

$$\hat{\phi} = \frac{\hat{\mathbf{z}}\times\hat{\mathbf{r}}}{|\hat{\mathbf{z}}\times\hat{\mathbf{r}}|}. \tag{A2.5}$$

The cross product of vectors $\hat{\mathbf{z}}$ and $\hat{\mathbf{r}}$ is:

$$\hat{\mathbf{z}}\times\hat{\mathbf{r}} = \begin{vmatrix} \hat{\mathbf{x}} & \hat{\mathbf{y}} & \hat{\mathbf{z}} \\ 0 & 0 & 1 \\ \sin\theta\cos\phi & \sin\theta\sin\phi & \cos\theta \end{vmatrix} = -\sin\theta\sin\phi\,\hat{\mathbf{x}} + \sin\theta\cos\phi\,\hat{\mathbf{y}}, \tag{A2.6}$$

and has the length:

$$|\hat{\mathbf{z}}\times\hat{\mathbf{r}}| = \sqrt{\sin^2\theta\sin^2\phi + \sin^2\theta\cos^2\phi} = \sin\theta. \tag{A2.7}$$

Combining these relations gives the equation for the unit vector of azimuthal angle:

$$\hat{\phi} = -\sin\phi\,\hat{\mathbf{x}} + \cos\phi\,\hat{\mathbf{y}}. \tag{A2.8}$$

As soon as two unit vectors are known, the third one, orthogonal to the first two, appears as a result of a vector cross product:

$$\hat{\theta}=\hat{\phi}\times\hat{\mathbf{r}}=\begin{vmatrix}\hat{\mathbf{x}} & \hat{\mathbf{y}} & \hat{\mathbf{z}}\\ -\sin\phi & \cos\phi & 0\\ \sin\theta\cos\phi & \sin\theta\sin\phi & \cos\theta\end{vmatrix}$$
$$=\cos\theta\cos\phi\,\hat{\mathbf{x}}+\cos\theta\sin\phi\,\hat{\mathbf{y}}-\left(\sin\theta\sin^2\phi+\sin\theta\cos^2\phi\right)\hat{\mathbf{z}}. \quad \text{(A2.9)}$$

This leads to the expression for the unit vector of polar angle:

$$\hat{\theta}=\cos\theta\cos\phi\,\hat{\mathbf{x}}+\cos\theta\sin\phi\,\hat{\mathbf{y}}-\sin\theta\,\hat{\mathbf{z}}. \quad \text{(A2.10)}$$

To obtain the inverse relations, the expressions for Cartesian unit vectors over spherical polar unit vectors, it is helpful to observe that the unit vectors $\hat{\mathbf{r}}$ and $\hat{\theta}$ are in the same plane with unit vector $\hat{\mathbf{z}}$. This leads to the conclusion that $\hat{\mathbf{z}}$ is a linear combination of $\hat{\mathbf{r}}$ and $\hat{\theta}$. A suitable combination appears from the multiplication of $\hat{\mathbf{r}}$ by $\cos\theta$ and from the multiplication of $\hat{\theta}$ by $\sin\theta$. This leads to:

$$\begin{aligned}\cos\theta\,\hat{\mathbf{r}}&=\sin\theta\cos\theta\cos\phi\hat{\mathbf{x}}+\sin\theta\cos\theta\sin\phi\,\hat{\mathbf{y}}+\cos^2\theta\,\hat{\mathbf{z}},\\ \sin\theta\,\hat{\theta}&=\sin\theta\cos\theta\cos\phi\hat{\mathbf{x}}+\sin\theta\cos\theta\sin\phi\,\hat{\mathbf{y}}-\sin^2\theta\,\hat{\mathbf{z}}.\end{aligned} \quad \text{(A2.11)}$$

Subtracting the second relation from the first one cancels out the terms for $\hat{\mathbf{x}}$ and $\hat{\mathbf{y}}$, giving the desired expression for unit vector $\hat{\mathbf{z}}$:

$$\hat{\mathbf{z}}=\cos\theta\,\hat{\mathbf{r}}-\sin\theta\,\hat{\theta}. \quad \text{(A2.12)}$$

The strategy to get the expression for the next Cartesian unit vector is to construct a linear combination of vectors $\hat{\mathbf{r}}$ and $\hat{\theta}$ so that it does not contain the component $\hat{\mathbf{z}}$. This is achieved by multiplying the expressions for $\hat{\mathbf{r}}$ and $\hat{\theta}$ by $\sin\theta$ and $\cos\theta$, respectively. This leads to:

$$\begin{aligned}\sin\theta\,\hat{\mathbf{r}}&=\sin^2\theta\cos\phi\hat{\mathbf{x}}+\sin^2\theta\sin\phi\,\hat{\mathbf{y}}+\sin\theta\cos\theta\,\hat{\mathbf{z}},\\ \cos\theta\,\hat{\theta}&=\cos^2\theta\cos\phi\,\hat{\mathbf{x}}+\cos^2\theta\sin\phi\,\hat{\mathbf{y}}-\sin\theta\cos\theta\,\hat{\mathbf{z}}.\end{aligned} \quad \text{(A2.13)}$$

Adding these expressions together gives:

$$\sin\theta\,\hat{\mathbf{r}}+\cos\theta\,\hat{\theta}=\cos\phi\,\hat{\mathbf{x}}+\sin\phi\,\hat{\mathbf{y}}, \quad \text{(A2.14)}$$

with the resulting vector having no dependence on the component $\hat{\mathbf{z}}$. Recall that the unit vector $\hat{\phi}$ also has no dependence on the component $\hat{\mathbf{z}}$. This leads to:

$$\hat{\phi}=-\sin\phi\,\hat{\mathbf{x}}+\cos\phi\,\hat{\mathbf{y}}. \quad \text{(A2.15)}$$

Since vectors $\sin\theta\,\hat{\mathbf{r}}+\cos\theta\,\hat{\theta}$ and $\hat{\phi}$ lie in the same plane, it is possible to construct a linear combination that will produce the vectors $\hat{\mathbf{x}}$ and $\hat{\mathbf{y}}$. Multiplying the above two vectors by $\sin\phi$ and $\cos\phi$, respectively leads to:

$$\begin{aligned}\sin\theta\sin\phi\hat{\mathbf{r}}+\cos\theta\sin\phi\,\hat{\theta}&=\sin\phi\cos\phi\,\hat{\mathbf{x}}+\sin^2\phi\,\hat{\mathbf{y}},\\ \cos\phi\,\hat{\phi}&=-\sin\phi\cos\phi\,\hat{\mathbf{x}}+\cos^2\phi\,\hat{\mathbf{y}}.\end{aligned} \quad \text{(A2.16)}$$

Adding these two equations cancels out the component $\hat{\mathbf{x}}$ and gives the desired expression for $\hat{\mathbf{y}}$:

$$\hat{\mathbf{y}} = \sin\theta\sin\phi\,\hat{\mathbf{r}} + \cos\theta\sin\phi\,\hat{\theta} + \cos\phi\,\hat{\phi}. \tag{A2.17}$$

The final relation for component $\hat{\mathbf{x}}$ employs a similar strategy of constructing an appropriate linear combination. In this case, it is necessary to cancel out the component $\hat{\mathbf{y}}$. Multiplying vectors $\sin\theta\,\hat{\mathbf{r}} + \cos\theta\,\hat{\theta}$ and $\hat{\phi}$ by $\cos\phi$ and $\sin\phi$, respectively, leads to:

$$\begin{aligned} &\sin\theta\cos\phi\,\hat{\mathbf{r}} + \cos\theta\cos\phi\,\hat{\theta} = \cos^2\phi\,\hat{\mathbf{x}} + \sin\phi\cos\phi\,\hat{\mathbf{y}}, \\ &\sin\phi\,\hat{\phi} = -\sin^2\phi\,\hat{\mathbf{x}} + \sin\phi\cos\phi\,\hat{\mathbf{y}}. \end{aligned} \tag{A2.18}$$

Subtracting the second relation from the first one cancels out the terms containing component $\hat{\mathbf{y}}$ and produces the desired expression for component $\hat{\mathbf{x}}$:

$$\hat{\mathbf{x}} = \sin\theta\cos\phi\,\hat{\mathbf{r}} + \cos\theta\cos\phi\,\hat{\theta} - \sin\phi\,\hat{\phi}. \tag{A2.19}$$

One unusual property of the spherical polar coordinate system is that the values of unit vectors $\hat{\mathbf{r}}$, $\hat{\theta}$, $\hat{\phi}$ depend on the position of the point of their evaluation. For comparison, in the Cartesian coordinate system there is only one set of unit vectors $\hat{\mathbf{x}}$, $\hat{\mathbf{y}}$, $\hat{\mathbf{z}}$, which are the same for every point in 3D space. Since $\hat{\mathbf{r}}$, $\hat{\theta}$, $\hat{\phi}$ are functions of the position of the point, in order to completely characterize the spherical polar coordinate system, it is necessary to obtain derivatives for the corresponding unit vectors.

A convenient property of the spherical polar coordinate system is that none of its unit vectors depend on the radial coordinate, that is,

$$\frac{\partial\hat{\mathbf{r}}}{\partial r} = 0, \quad \frac{\partial\hat{\theta}}{\partial r} = 0, \quad \frac{\partial\hat{\phi}}{\partial r} = 0. \tag{A2.20}$$

With that, the only non-zero derivatives are those that have angular dependence. The derivatives of the unit vectors over polar angle θ are:

$$\begin{aligned} &\frac{\partial\,\hat{\mathbf{r}}}{\partial\theta} = \cos\theta\cos\phi\,\hat{\mathbf{x}} + \cos\theta\sin\phi\,\hat{\mathbf{y}} - \sin\theta\,\hat{\mathbf{z}} = \hat{\theta}, \\ &\frac{\partial\,\hat{\theta}}{\partial\theta} = -\sin\theta\cos\phi\,\hat{\mathbf{x}} - \sin\theta\sin\phi\,\hat{\mathbf{y}} - \cos\theta\,\hat{\mathbf{z}} = -\hat{\mathbf{r}}, \\ &\frac{\partial\,\hat{\phi}}{\partial\theta} = 0. \end{aligned} \tag{A2.21}$$

Using these results, the derivatives of the unit spherical polar vectors are easy to construct. For the azimuthal angle ϕ, they are:

$$\begin{aligned} &\frac{\partial\,\hat{\mathbf{r}}}{\partial\phi} = -\sin\theta\sin\phi\,\hat{\mathbf{x}} + \sin\theta\cos\phi\,\hat{\mathbf{y}} = \sin\theta\,\hat{\phi}, \\ &\frac{\partial\hat{\theta}}{\partial\phi} = -\cos\theta\sin\phi\,\hat{\mathbf{x}} + \cos\theta\cos\phi\,\hat{\mathbf{y}} = \cos\theta\,\hat{\phi}, \\ &\frac{\partial\hat{\phi}}{\partial\phi} = -\cos\phi\,\hat{\mathbf{x}} - \sin\phi\,\hat{\mathbf{y}}. \end{aligned} \tag{A2.22}$$

At this point, the result of the differentiation of $\partial\hat{\phi}/\partial\phi$ is not in terms of spherical polar unit vectors, so the next step is to replace the Cartesian unit vectors by their spherical polar equivalents. Performing this substitution gives:

$$\begin{aligned} \frac{\partial\hat{\phi}}{\partial\phi} = &-\sin\theta\cos^2\phi\,\hat{\mathbf{r}} - \cos\theta\cos^2\phi\,\hat{\theta} + \sin\phi\cos\phi\,\hat{\phi} \\ &-\sin\theta\sin^2\phi\,\hat{\mathbf{r}} - \cos\theta\sin^2\phi\,\hat{\theta} - \sin\phi\cos\phi\,\hat{\phi}. \end{aligned} \tag{A2.23}$$

Simplifying and canceling out common terms leads to the final expression for the derivative $\partial\hat{\phi}/\partial\phi$ in spherical polar coordinates:

$$\frac{\partial\hat{\phi}}{\partial\phi} = -\sin\theta\,\hat{\mathbf{r}} - \cos\theta\,\hat{\theta}. \tag{A2.24}$$

These derivatives of unit vectors in spherical polar coordinates will become useful in the derivation of the square of the angular momentum operator.

A.3 Trigonometric Functions of the Half-Angle

Half-angle formulae express trigonometric functions of an angle $\theta/2$ in terms of functions of angle θ. A good starting point for the derivation of the various relationships is the cosine function:

$$\cos 2\theta = \cos^2\theta - \sin^2\theta = 2\cos^2\theta - 1. \tag{A3.1}$$

Applying the substitution $\theta \to \theta/2$ leads to:

$$\cos\theta = 2\cos^2\left(\frac{\theta}{2}\right) - 1, \tag{A3.2}$$

and, from that, to:

$$\cos\left(\frac{\theta}{2}\right) = \pm\sqrt{\frac{1+\cos\theta}{2}}. \tag{A3.3}$$

The sign in front of the square root is determined by the quadrant that θ is in. The equation for $\sin(\theta/2)$ begins from almost the same starting point as that for $\cos(\theta/2)$:

$$\cos 2\theta = \cos^2\theta - \sin^2\theta = 1 - 2\sin^2\theta. \tag{A3.4}$$

Once again, applying the substitution $\theta \to \theta/2$ gives:

$$\cos\theta = 1 - 2\sin^2\left(\frac{\theta}{2}\right). \tag{A3.5}$$

Rearranging the terms leads to the half-angle formula for the sine function:

$$\sin\left(\frac{\theta}{2}\right) = \pm\sqrt{\frac{1-\cos\theta}{2}}. \tag{A3.6}$$

The half-angle expression for the tangent can be obtained using tan = sin/cos. Substituting the corresponding expressions for sine and cosine functions, and ignoring the uncertainty in sign, gives:

$$\tan\left(\frac{\theta}{2}\right) = \sin\left(\frac{\theta}{2}\right)\Big/\cos\left(\frac{\theta}{2}\right) = \sqrt{\frac{1-\cos\theta}{1+\cos\theta}}. \tag{A3.7}$$

This equation may be rearranged in two different ways. One way involves multiplying both the numerator and the denominator by $(1 - \cos\theta)$, and simplifying, to give:

$$\tan\left(\frac{\theta}{2}\right)=\sqrt{\frac{(1-\cos\theta)^2}{(1+\cos\theta)(1-\cos\theta)}}=\frac{1-\cos\theta}{\sin\theta}. \tag{A3.8}$$

The other way involves multiplying both the numerator and the denominator by $(1+\cos\theta)$. This leads to:

$$\tan\left(\frac{\theta}{2}\right)=\sqrt{\frac{(1-\cos\theta)(1+\cos\theta)}{(1+\cos\theta)^2}}=\frac{\sin\theta}{1+\cos\theta}. \tag{A3.9}$$

Taking into account the periodicity of trigonometric functions leads to the following relations:

$$\begin{aligned}\cos\left(\frac{\theta}{2}\right)&=(-1)^{(\theta+\pi)/2\pi}\sqrt{\frac{1+\cos\theta}{2}},\\ \sin\left(\frac{\theta}{2}\right)&=(-1)^{\theta/2\pi}\sqrt{\frac{1-\cos\theta}{2}},\\ \tan\left(\frac{\theta}{2}\right)&=(-1)^{\theta/\pi}\sqrt{\frac{1-\cos\theta}{1+\cos\theta}}.\end{aligned} \tag{A3.10}$$

In these expressions, the power to which (-1) is raised should be rounded down to the nearest integer.

A.4 Arithmetic Series

An arithmetic series is a sum in the form:

$$S_n=a_1+a_2+a_3+\cdots+a_n, \tag{A4.1}$$

in which each successive element a_k can be computed from the previous one by adding or subtracting a constant d, so that the series may be rewritten as

$$S_n=a_1+(a_1+d)+(a_1+2d)+\cdots+(a_1+(n-1)\,d). \tag{A4.2}$$

Finding the sum S_n of an arithmetic series rests on a unique property of the arithmetic series. This property says that if the elements of the series are arranged in the opposite order, that is, starting from the last element and going to the first one as in $S_n=a_n+a_{n-1}+a_{n-2}+\cdots+a_1$, then the element-wise addition of the original and the reversed series produces a new series that has all elements of the series having the same value. Indeed,

$$\begin{aligned}S_4&=3+7+11+15,\\ S_4&=15+11+7+3,\\ 2S_4&=18+18+18+18.\end{aligned} \tag{A4.3}$$

For a general case,

$$\begin{aligned}S_n&=a_1+a_2+a_3+\cdots+a_n,\\ S_n&=a_n+a_{n-1}+a_{n-2}+\cdots+a_1,\\ 2S_n&=(a_1+a_n)+(a_2+a_{n-1})+(a_3+a_{n-2})+\cdots+(a_n+a_1),\end{aligned} \tag{A4.4}$$

in order to make it evident that all elements of $2S_n$ are of equal value, it is necessary to express each element of the series *via* the element a_1 and the increment d. This leads to:

$$\begin{aligned}
S_n &= a_1 + (a_1+d) + (a_1+2d) + \ldots + (a_1+[n-1]d),\\
S_n &= (a_1+[n-1]d) + (a_1+[n-2]d) + (a_1+[n-3]d) + \ldots + a_1,\\
2S_n &= (2a_1+[n-1]d) + (2a_1+[n-1]d) + (2a_1+[n-1]d) + \ldots + (2a_1+[n-1]d).
\end{aligned} \tag{A4.5}$$

Since $2a_1 + [n - 1]d = a_1 + a_n$ the sum of general series S_n is simply one-half of the product of the sum of the first and last elements $(a_1 + a_n)$ of the series and the number of elements n in the series:

$$S_n = n\left(\frac{a_1 + a_n}{2}\right). \tag{A4.6}$$

This sum is valid only for finite arithmetic series; all infinite arithmetic series sum to infinity.

A.5 Geometric Series

Geometric series are power series and have the general form:

$$S_n = \sum_{k=l}^{n} x^k = x^l + x^{l+1} + \cdots + x^n, \tag{A5.1}$$

where S_n is the sum of the series up to its n-th term, and $l \geq 0$. For infinite series, that is, $n \to \infty$, a power series can be either convergent or divergent.

Finding the sum of a geometric series is straightforward. Indeed, the sum for $n + 1$ elements, S_{n+1}, is equal to the sum, S_n, for n elements plus the new element:

$$S_{n+1} = S_n + x^{n+1}. \tag{A5.2}$$

Because of the nature of power series, where each successive element is x-times the preceding element, the sum S_{n+1} can also be obtained by multiplying S_n by x and adding the first element in the series, x^l:

$$S_{n+1} = xS_n + x^l. \tag{A5.3}$$

Combining Equations A5.2 and A5.3 gives the sum for $n + 1$ elements:

$$S_n + x^{n+1} = xS_n + x^l, \tag{A5.4}$$

which, after rearrangement, becomes:

$$S_n = \frac{x^l - x^{n+1}}{1-x} = \frac{x^{n+1} - x^l}{x-1}. \tag{A5.5}$$

When $|x| < 1$ and $n \to \infty$, this series converges to:

$$\lim_{n\to\infty} S_n = \lim_{n\to\infty} \frac{x^l - x^{n+1}}{1-x} = \frac{x^l}{1-x}. \tag{A5.6}$$

If $x = 1/s$, $|s| > 1$, and $l = 1$, the limit becomes:

$$\lim_{n\to\infty} S_n = \lim_{n\to\infty} \frac{(1/s)-(1/s^{n+1})}{1-(1/s)} = \frac{1/s}{(s-1)/s} = \frac{1}{s-1}. \tag{A5.7}$$

One popular power series is the exponential function:

$$e^x = \sum_{k=0}^{\infty} \frac{x^k}{k!}. \tag{A5.8}$$

For complex exponential functions, the sum is:

$$e^{ix} = 1 + ix + \frac{(ix)^2}{2!} + \frac{(ix)^3}{3!} + \cdots = 1 + ix - \frac{x^2}{2!} - i\frac{x^3}{3!} + \frac{x^4}{4!} + i\frac{x^5}{5!} + \cdots. \tag{A5.9}$$

Rearranging by grouping real and complex terms separately leads to:

$$e^{ix} = \left(1 - \frac{x^2}{2!} + \frac{x^4}{4!} - \ldots\right) + i\left(x - \frac{x^3}{3!} + \frac{x^5}{5!} - \ldots\right) = \cos x + i \sin x. \tag{A5.10}$$

In the sum, one can recognize the power series for trigonometric functions as Euler's formula for a complex exponent:

$$\begin{aligned} e^{ix} &= \cos x + i \sin x, \\ e^{-ix} &= \cos(-x) + i \sin(-x) = \cos x - i \sin x. \end{aligned} \tag{A5.11}$$

A.6 Taylor Series

An infinitely differentiable function $y(x)$ can be expanded as a power series in the neighborhood of $x = a$. Such an expansion is called a Taylor series. In its general form, this series is:

$$y(x) = \sum_{k=0}^{\infty} \frac{1}{k!}(x-a)^k y^{(k)}(a) = y(a) + \frac{(x-a)}{1!} y^{(1)}(a) + \frac{(x-a)^2}{2!} y^{(2)}(a) + \cdots. \tag{A6.1}$$

where $y^{(k)}$ denotes the k-th derivative of function y over argument x. When the expansion is made in the neighborhood of another function, it can be regarded as a form of perturbation:

$$y(x+a) = \sum_{k=0}^{\infty} \frac{1}{k!} a^k y^{(k)}(x) = y(x) + \frac{a}{1!} y^{(1)}(x) + \frac{a^2}{2!} y^{(2)}(x) + \cdots. \tag{A6.2}$$

This particular expansion is known as the Taylor theorem. Both expansions are completely equivalent. The series is infinite only if the function $y(x)$ is infinitely differentiable; otherwise, it is finite.

A.7 Binomial Coefficient

Binomial coefficient $\binom{a}{k}$ is the combinatorial number of ways of selecting k members from the set of a members. For integer numbers $a \geq 0$ and $0 \leq k \leq a$, the binomial coefficient is

$$\binom{a}{k} = \frac{a!}{k!(a-k)!}. \tag{A7.1}$$

For $k > a$, the binomial coefficient is zero because the factorial $(a - k)!$ in the denominator turns to infinity. This outcome is consistent with the combinatorial definition of a binomial coefficient that provides zero ways of selecting more members than are available in the set.

Binomial coefficients readily extend to negative arguments. For a negative number $n = -a$, where $a > 0$, the infinite tail of the factorials in the numerator and denominator cancel each other out, leading to:

$$\binom{n}{k} = \binom{-a}{k} = \frac{(-a)!}{k!(-a-k)!} = \frac{(-a)(-a-1)\cdots(-a-k+1)}{k!}. \tag{A7.2}$$

Separating the negative sign from each multiplier and rearranging the remaining terms in descending order provides the final result:

$$\binom{-a}{k} = (-1)^k \frac{(a+k-1)\cdots(a+1)(a)}{k!} = (-1)^k \frac{(a+k-1)!}{k!(a-1)!} = (-1)^k \binom{a+k-1}{k}, \tag{A7.3}$$

which defines the binomial coefficient $\binom{-a}{k}$ for the negative component $-a$ in terms of a binomial coefficient $\binom{a+k-1}{k}$ that includes all positive numbers $a > 0$ and $k \geq 0$.

For two negative arguments $n = -a$, and $u = -k$, where $a > 0$ and $k \geq a$, cancelation of the infinite tails in the numerator and denominator leads to:

$$\binom{n}{u} = \binom{-a}{-k} = \frac{(-a)!}{(-k)!(-a+k)!} = \frac{(-a)(-a-1)\cdots(-k+1)}{(k-a)!}. \tag{A7.4}$$

Separating the negative sign from each multiplier in the numerator, and rearranging the sequence in descending order provides:

$$\binom{-a}{-k} = (-1)^{k-a} \frac{(k-1)\cdots(a+1)(a)}{(k-a)!} = (-1)^{k-a} \frac{(k-1)!}{(a-1)!(k-a)!} = (-1)^{k-a} \binom{k-1}{a-1}. \tag{A7.5}$$

When $k < a$, there are two factorials from the negative numbers in the denominator, and only one similar factorial in the numerator. Therefore, one infinite tail remains in the denominator, and that zeroes out the factorial:

$$\binom{-u}{-k} = \frac{(-u)!}{(-k)!(-a+k)!} = \frac{1}{(-k)\cdots(-a+1)(k-a)!} = 0. \tag{A7.6}$$

Binomial coefficients occur frequently in the algebra of series expansion. For example, the binomial theorem expands the two-component function $(x + y)^a$ raised to the integer non-negative power a into the finite series:

$$(x+y)^a = \sum_{k=0}^{a} \frac{a!}{k!(a-k)!} x^k y^{a-k} = \sum_{k=0}^{a} \binom{a}{k} x^k y^{a-k}. \tag{A7.7}$$

The binomial coefficient can also be used in a series expansion of the function $(1 + x)^a$:

$$(1+x)^a = \sum_{k=0}^{a} \frac{a!}{k!(a-k)!} x^k = \sum_{k=0}^{a} \binom{a}{k} x^k, \tag{A7.8}$$

where $\binom{a}{k}$ is a binomial coefficient of x^k.

The binomial theorem also holds for negative exponents, where it produces an infinite series:

$$(x+y)^{-a}=\sum_{k=0}^{\infty}\frac{(-a)!}{k!(-a-k)!}x^k y^{-a-k}=\sum_{k=0}^{\infty}\binom{-a}{k}x^k y^{-a-k}. \tag{A7.9}$$

As before, choosing $y = 1$ produces a more elegant series expansion:

$$\left(1+x\right)^{-a}=\sum_{k=0}^{\infty}\frac{(-a)!}{k!(-a-k)!}x^k=\sum_{k=0}^{\infty}\binom{-a}{k}x^k, \tag{A7.10}$$

where $\binom{-a}{k}$ is the binomial coefficient of x^k.

Binomial expansions are not limited to two-component functions $(x + y)^a$. When the binomial expansion is applied to multi-component functions, it is known as the multinomial theorem:

$$(x_1+x_2+\dots+x_b)^a=\sum_{k_1+k_2+\dots k_b=a}\frac{a!}{k_1!k_2!\dots k_b!}x_1^{k_1}x_2^{k_2}\dots x_b^{k_b}=\sum_{k_1+k_2+\dots k_b=a}\binom{a}{k_1,k_2,\dots,k_b}x_1^{k_1}x_2^{k_2}\dots x_b^{k_b}. \tag{A7.11}$$

In a multinomial expansion, the indices of each summation term must satisfy the requirement $k_1 + k_2 + \dots + k_b = a$.

Binomial coefficients have a number of useful properties. Among those, Vandermonde's identity establishes some important relationships between various binomial coefficients:

$$\sum_{k=0}^{c}\binom{a}{k}\binom{b}{c-k}=\binom{a+b}{c}, \tag{A7.12}$$

where $a > k$, $b > c - k$, $a + b > c$, and where a, b and c are positive integer numbers. The proof of this follows from the binomial expansion of the function $(1 + x)^{a+b}$:

$$(1+x)^{a+b}=\sum_{c=0}^{a+b}\frac{(a+b)!}{c!(a+b-c)!}x^c. \tag{A7.13}$$

Alternatively, representing the function $(1 + x)^{a+b}$ as a product of its two components $(1 + x)^a$ and $(1 + x)^b$, and separately expanding each of them provides a complementary relation:

$$(1+x)^{a+b}=(1+x)^a(1+x)^b=\left[\sum_{k=0}^{a}\frac{a!}{k!(a-k)!}x^k\right]\left[\sum_{s=0}^{b}\frac{b!}{s!(b-s)!}x^s\right]. \tag{A7.14}$$

Since the left-hand sides of these two equations are equivalent, both right-hand sides must be equal as well. Therefore,

$$\left[\sum_{k=0}^{a}\frac{a!}{k!(a-k)!}x^k\right]\left[\sum_{s=0}^{b}\frac{b!}{s!(b-s)!}x^s\right]=\sum_{c=0}^{a+b}\frac{(a+b)!}{c!(a+b-c)!}x^c. \tag{A7.15}$$

Coefficients of the same power of x in the left- and right-hand sides of the equation may be equated when $c = k + s$. Obviously, many combinations of the indices k and s on the left-hand side will satisfy this requirement, therefore their coefficients can be summed up. This can be achieved by maintaining a sum over index k, and substituting the other index s by the relation $s = c - k$. This leads to:

$$\sum_{k=0}^{c} \frac{a!}{k!(a-k)!} \frac{b!}{(c-k)!(b-c+k)!} = \frac{(a+b)!}{c!(a+b-c)!}, \tag{A7.16}$$

which represents Vandermonde's expansion in its factorial form.

During the transition from Equation A7.15 and A7.16, the imposed condition $c = k + s$ dictates the choice of the lower and upper boundaries for index k in the sum. Since c, k, and s are non-negative numbers, the possibility of $c = 0$ defines $k = 0$ as the lower boundary for index k. Similarly, the option of $s = 0$ requires the upper boundary for index k to be $k = c$. However, not all values of index k from the established range contribute to the sum. For example, c might potentially be greater than a, but index k cannot exceed the value of a. The presence of the factorial $(a - k)!$ in the denominator gracefully handles that condition; effectively zeroing out the summation term for all $k > a$ by going to infinity. Therefore, it is safe to have c as the upper boundary for index k. Also, for all $c > b$ it is necessary that index k starts from the values $k \geq c - b$. Once again the presence of the factorial $(b - c + k)!$ in the denominator zeroes all summation terms that have $k < c - b$ by going to infinity. Therefore, it is appropriate to have $k = 0$ as the lower boundary for index k.

The utility of Vandermonde's identity is that the product of binomial coefficients that appear in the separate expansion of functions $(1 + x)^a$ and $(1 + x)^b$ sums up to a single binomial coefficient from the expansion of function $(1 + x)^{a+b}$. In the proof given here, the assumption is made that the exponents a, b, and c are non-negative numbers. However, decomposition of function $(1 + x)^{a+b}$ into a product of functions $(1 + x)^a$ and $(1 + x)^b$ does hold for negative exponents. The only difference it would make to Equation A3.1 is to change the upper boundary of the sums from a finite number to infinity because of the presence of the binomial expansion of a function raised to negative exponent. This turns Equation A7.15 to:

$$\left[\sum_{k=0}^{\infty} \frac{(-a)!}{k!(-a-k)!} x^k\right]\left[\sum_{s=0}^{\infty} \frac{(-b)!}{s!(-b-s)!} x^s\right] = \sum_{c=0}^{\infty} \frac{(-a-b)!}{c!(-a-b-c)!} x^c, \tag{A7.17}$$

where indices k, s, and c are non-negative numbers.

As before, the coefficients in front of argument x^c in the left- and right-hand sides of Equation A7.17 can be equated when $c = k + s$. Therefore, Equation A7.16 and Vandermonde's identity remain valid for values of a and b in the full integer range encompassing positive and negative numbers.

For the sake of clarity and consistency in notation, it is preferable to keep a and b as non-negative numbers, and explicitly show the negative sign in front of the number. This turns Equation A7.16 to

$$\sum_{k=0}^{c} \frac{(-a)!}{k!(-a-k)!} \frac{(-b)!}{(c-k)!(-b-c+k)!} = \frac{(-a-b)!}{c!(-a-b-c)!}, \tag{A7.18}$$

and rewrites Vandermonde's identity as:

$$\sum_{k=0}^{c} \binom{-a}{k}\binom{-b}{c-k} = \binom{-a-b}{c}. \tag{A7.19}$$

The fact that a binomial coefficient of negative argument has a finite value, and that it may be rewritten in terms of a non-negative argument provides an opportunity to simplify Equation A7.19. According to Equation A7.3, conversion of all three factorials from Equation A7.19 to those using non-negative arguments produces:

$$\binom{-a}{k} = (-1)^k \binom{a+k-1}{k},$$
$$\binom{-b}{c-k} = (-1)^{c-k} \binom{b+c-k-1}{c-k}, \quad \text{(A7.20)}$$
$$\binom{-a-b}{c} = (-1)^c \binom{a+b+c-1}{c}.$$

Substitution of these equations back to Equation A7.19 gives:

$$\sum_{k=0}^{c} (-1)^k \binom{a+k-1}{k} (-1)^{c-k} \binom{b+c-k-1}{c-k} = (-1)^c \binom{a+b+c-1}{c}. \quad \text{(A7.21)}$$

Canceling out the powers of minus one in the left- and right-hand sides of the equation leads to:

$$\sum_{k=0}^{c} \binom{a+k-1}{k} \binom{b+c-k-1}{c-k} = \binom{a+b+c-1}{c}, \quad \text{(A7.22)}$$

which is a rewrite of Vandermonde's identity given by Equation A7.19 for negative arguments $-a$ and $-b$ in terms of non-negative arguments a and b.

Substitution of the binomial coefficients in Equation A7.22 by factorials simplifies that equation to:

$$\sum_{k=0}^{c} \frac{(a+k-1)!(b+c-k-1)!}{k!(a-1)!(c-k)!(b-1)!} = \frac{(a+b+c-1)!}{c!(a+b-1)!}. \quad \text{(A7.23)}$$

Keeping the terms that depend on index k in the left-hand side, and moving all other terms to the right-hand side, and after some rearrangement, gives:

$$\sum_{k=0}^{c} \frac{(b+c-k-1)!(a+k-1)!}{k!(c-k)!} = \frac{(a+b+c-1)!(b-1)!(a-1)!}{(a+b-1)!c!}. \quad \text{(A7.24)}$$

The present set of parameters a, b, c, and k may be expressed by another set of parameters j_1, j_2, j, and m_1, which appear in the context of angular momentum. Specifically, applying the substitution $k = j_2 - j + m_1$, $a - 1 = j_1 - j_2 + j$, $b - 1 = j_2 - j_1 + j$, and $c = j_1 + j_2 - j$ to Equation A7.24 changes that equation to:

$$\sum_{k=0}^{c} \frac{(j_2 - j_1 + j + j_1 + j_2 - j - j_2 + j - m_1)!(j_1 - j_2 + j + j_2 - j + m_1)!}{(j_2 - j + m_1)!(j_1 + j_2 - j - j_2 + j - m_1)!}$$
$$= \frac{(j_1 - j_2 + j + j_2 - j_1 + j + 1 + j_1 + j_2 - j)!(j_2 - j_1 + j)!(j_1 - j_2 + j)!}{(j_1 - j_2 + j + j_2 - j_1 + j + 1)!(j_1 + j_2 - j)!}. \quad \text{(A7.25)}$$

This simplifies to:

$$\sum_{k=0}^{c} \frac{(j_2 + j - m_1)!(j_1 + m_1)!}{(j_2 - j + m_1)!(j_1 - m_1)!} = \frac{(j_1 + j_2 + j + 1)!(j_2 - j_1 + j)!(j_1 - j_2 + j)!}{(2j+1)!(j_1 + j_2 - j)!}. \quad \text{(A7.26)}$$

At this point, the lower and upper boundaries in the summation symbol in Equation A7.26 are not yet properly defined. The new summation index is going to be m_1. For the upper boundary, the condition $k = c$ leads to the equality $j_2 - j + m_1 = j_1 + j_2 - j$, therefore it resolves to $m_1 = j_1$. For the lower boundary, the condition $k = 0$ leads to $j_2 - j + m_1 = 0$ or to $m_1 = j - j_2$. In angular momentum theory, the lowest allowable value of azimuthal number m_1 is $-j_1$. However, using any value of m_1 smaller than $j - j_2$ results in $(j_2 - j + m_1)!$ becoming infinite because of the negative argument. This would effectively annihilate the corresponding summation element on the left-hand side of Equation A7.26. Therefore, the lower boundary for index m_1 in Equation A7.26 can be set to the lowest possible value of $-j_1$, resulting in a more conventional–looking equation:

$$\sum_{m_1=-j_1}^{j_1} \frac{(j_2+j-m_1)!(j_1+m_1)!}{(j_2-j+m_1)!(j_1-m_1)!} = \frac{(j_1+j_2+j+1)!(j_2-j_1+j)!(j_1-j_2+j)!}{(2j+1)!(j_1+j_2-j)!}. \tag{A7.27}$$

B.1 Leibnitz Formula for the Derivative of a Product

This Leibnitz formula defines the n-th derivative of a product of two functions, denoted as u and v, stating that:

$$\frac{d^n}{dx^n}(u \cdot v) = \sum_{a=0}^{n} \frac{n!}{a!(n-a)!}\left[\frac{d^a}{dx^a}u\right]\left[\frac{d^{n-a}}{dx^{n-a}}v\right]. \tag{B1.1}$$

This equation can be proved as follows. The cases when $n = 0$ and $n = 1$ are trivial:

$$\begin{aligned} &\frac{d^0}{dx^0}(u \cdot v) = u \cdot v, \\ &\frac{d}{dx}(u \cdot v) = \frac{du}{dx} \cdot v + u \cdot \frac{dv}{dx} = u^{(1)}v + uv^{(1)}. \end{aligned} \tag{B1.2}$$

Proof of the general case involves using the method of induction. Assume that the following is true for $n \geq 1$:

$$\frac{d^{n-1}}{dx^{n-1}}(u \cdot v) = \sum_{a=0}^{n-1} \frac{(n-1)!}{a!(n-1-a)!} u^{(a)} v^{(n-1-a)}. \tag{B1.3}$$

Then,

$$\frac{d^n}{dx^n}(u \cdot v) - \frac{d}{dx}\left[\frac{d^{n-1}}{dx^{n-1}}(u \cdot v)\right] = \sum_{a=0}^{n-1} \frac{(n-1)!}{a!(n-1-a)!}\left[u^{(a)}v^{(n-a)} + u^{(a+1)}v^{(n-1-a)}\right]. \tag{B1.4}$$

Next, taking the zeroth order derivatives out of the sum leads to:

$$\frac{d^n}{dx^n}(u \cdot v) = uv^{(n)} + \sum_{a=1}^{n-1} \frac{(n-1)!}{a!(n-1-a)!} u^{(a)} v^{(n-a)} + \sum_{a=0}^{n-2} \frac{(n-1)!}{a!(n-1-a)!} u^{(a+1)} v^{(n-1-a)} + u^{(n)} v. \tag{B1.5}$$

The boundaries of the second sum can then be shifted by substituting b for $a + 1$, to give:

$$\frac{d^n}{dx^n}(u \cdot v) = uv^{(n)} + \sum_{a=1}^{n-1} \frac{(n-1)!}{a!(n-1-a)!} u^{(a)} v^{(n-a)} + \sum_{b=1}^{n-1} \frac{(n-1)!}{(b-1)!(n-b)!} u^{(b)} v^{(n-b)} + u^{(n)} v. \tag{B1.6}$$

Since both sums now have the same boundaries they can be merged into a single sum:

$$\frac{d^n}{dx^n}(u \cdot v) = uv^{(n)} + \sum_{a=1}^{n-1}\left[\frac{(n-1)!}{a!(n-1-a)!} + \frac{(n-1)!}{(a-1)!(n-a)!}\right] u^{(a)} v^{(n-a)} + u^{(n)} v. \tag{B1.7}$$

A simple transformation to obtain the same denominator gives:

$$\frac{d^n}{dx^n}(u\cdot v)=uv^{(n)}+\sum_{a=1}^{n-1}\left[\frac{(n-1)!(n-a)}{a!(n-a)!}+\frac{(n-1)!a}{a!(n-a)!}\right]u^{(a)}v^{(n-a)}+u^{(n)}v. \tag{B1.8}$$

On combining the terms in square brackets:

$$\frac{d^n}{dx^n}(u\cdot v)=uv^{(n)}+\sum_{a=1}^{n-1}\frac{n!}{a!(n-a)!}u^{(a)}v^{(n-a)}+u^{(n)}v=\sum_{a=0}^{n}\frac{n!}{a!(n-a)!}u^{(a)}v^{(n-a)}. \tag{B1.9}$$

This completes the proof of Equation B1.1.

B.2 Factorization of Associated Legendre Functions

Hobson provided a useful factorization of the associated Legendre function to be explained here. Associated Legendre functions exist in the form:

$$P_l^m(\mu)=(1-\mu^2)^{m/2}\sum_{k=0}^{(l-\text{mod}(l,2))/2}(-1)^{m+k}\frac{(2l-2k)!}{2^l k!(l-k)!(l-2k-m)!}\mu^{l-2k-m}. \tag{B2.1}$$

The coefficients change with change in the value of index k as follows:

$$k=0:\quad \frac{(2l)!}{2^l l!(l-m)!}, \tag{B2.2}$$

$$\begin{aligned} k=1:\quad -\frac{(2l-2)!}{2^l(l-1)!(l-m-2)!}&=-\frac{1}{2^l}\frac{(2l)!}{2l(2l-1)}\frac{l}{l!}\frac{(l-m)(l-m-1)}{(l-m)!}\\ &=-\frac{(2l)!}{2^l l!(l-m)!}\frac{(l-m)(l-m-1)}{2(2l-1)}, \end{aligned} \tag{B2.3}$$

and separating a common multiplier leads to the factorization:

$$\begin{aligned} P_l^m(x)=\frac{(2l)!}{2^l l!(l-m)!}(-1)^m(1-x^2)^{m/2}\Bigg\{&x^{l-m}-\frac{(l-m)(l-m-1)}{2(2l-1)}x^{l-m-2}\\ &+\frac{(l-m)(l-m-1)(l-m-2)(l-m-3)}{4\cdot 2\cdot(2l-1)(2l-3)}x^{l-m-4}-\cdots\Bigg\}. \end{aligned} \tag{B2.4}$$

Taking into account the following equality (for $k>0$)

$$\begin{aligned} &\frac{(2l-2k)!}{2^l k!(l-k)!(l-m-2k)}\\ &=\frac{1}{2^l k!}\frac{(2l)!}{(2l)(2l-1)\cdots(2l-2k+1)}\frac{l(l-1)\cdots(l-k+1)}{l!}\frac{(l-m)(l-m-1)\cdots(l-m-2k+1)}{(l-m)!}\\ &=\frac{(2l)!}{2^l l!(l-m)!}\frac{1}{k!}\frac{l(l-1)\cdots(l-k+1)\cdot(l-m)(l-m-1)\cdots(l-m-2k+1)}{(2l)(2l-1)(2l-2)\cdots(2l-2k+2)(2l-2k+1)}\\ &=\frac{(2l)!}{2^l l!(l-m)!}\frac{(l-m)(l-m-1)\cdots(l-m-2k+1)}{2^k k!(2l-1)(2l-3)\cdots(2l-2k+1)}, \end{aligned} \tag{B2.5}$$

Equation B2.4 can be rewritten more compactly as:

$$P_l^m(x)=\frac{(2l)!}{2^l l!(l-m)!}(-1)^m(1-x^2)^{m/2}\left\{x^{l-m}+\sum_{k=1}^{(l-m)/2}(-1)^k\frac{(l-m)(l-m-1)\cdots(l-m-2k+1)}{2^k k!(2l-1)(2l-3)\cdots(2l-2k+1)}x^{l-m-2k}\right\}. \tag{B2.6}$$

This is one of the many possible factorizations of the associated Legendre function, and is valid for both positive and negative values of m.

Equation B2.6 can be rearranged to eliminate the summation. Denoting the sum in curly brackets by $f(l,m,x)$ gives:

$$P_l^m(x)=\frac{(2l)!}{2^l l!(l-m)!}(-1)^m(1-x^2)^{m/2}f(l,m,x). \tag{B2.7}$$

The objective is to find an expansion for the series f in the form

$$f=b_0x^{l-m}+b_1x^{l-m-2}y^2+b_2x^{l-m-4}y^4+\cdots, \tag{B2.8}$$

where $y^2=1-x^2$. To aid in this effort, it is useful to rewrite the original expansion, Equation B2.4, in the form:

$$f=a_0x^{l-m}+a_1x^{l-m-2}+a_2x^{l-m-4}+\cdots=x^{l-m}\left\{1+a_1(1+t^2)+a_2(1+t^2)^2+a_3(1+t^2)^3\cdots\right\}, \tag{B2.9}$$

where $t^2=(1-x^2)/x^2$. Substituting $1+t^2=x^{-2}$ confirms the validity of Equation B2.9. In this form, the a coefficients can be readily associated with the coefficients of the original expansion in Equation B2.4. The equality $t^2=y^2/x^2$ makes it possible to equate the coefficients of t^{2k} (representing terms y^{2k} in Equation B2.8), of the same power k, to the coefficients b_k. In order to do that, it is necessary to sum up the coefficients of t^{2k} in Equation B2.9. It is easy to recognize the Pascal triangle, representing the coefficients of the binomial expansion, in this expansion. Left-justified, Pascal's triangle has the following form:

	$k=0$	$k=1$	$k=2$	$k=3$	$k=4$	$k=5$	...
$n=0$	1						
$n=1$	1	1					
$n=2$	1	2	1				
$n=3$	1	3	3	1			
$n=4$	1	4	6	4	1		
$n=5$	1	5	10	10	5	1	
$n=6$	1	6	15	20	15	6	1

(B2.10)

The coefficients that need to be summed are those in the k-th column. Their sum is defined by the binomial expansion:

$$b_k=\sum_{n=k}^{(l-m)/2}\frac{n!}{k!(n-k)!}a_n. \tag{B2.11}$$

Substituting $n=k+i$ gives:

$$\begin{aligned} b_k&=\sum_{i=0}^{(l-m)/2}\frac{(k+i)!}{k!i!}a_{k+i}=\sum_{i=0}^{(l-m)/2}\frac{(k+1)\cdots(k+i)}{i!}a_{k+i}\\ &=a_k+(k+1)a_{k+1}+\frac{(k+1)(k+2)}{2!}a_{k+2}+\cdots. \end{aligned} \tag{B2.12}$$

This establishes the relationship between coefficients b and a. An examination of Equation B2.6 shows that there is also a relationship between coefficients a_k and a_{k+1}:

$$\begin{aligned}
a_k &= \frac{(l-m)\cdots(l-m-2k+1)}{2^k k!(2l-1)(2l-3)\cdots(2l-2k+1)}, \\
a_{k+1} &= \frac{(l-m)\cdots(l-m-2k-1)}{2^{k+1}(k+1)!(2l-1)(2l-3)\cdots(2l-2k-1)}, \\
a_{k+1} &= a_k \frac{(l-m-2k)(l-m-2k-1)}{2(k+1)(2l-2k-1)}, \\
a_{k+2} &= a_{k+1} \frac{(l-m-2k-2)(l-m-2k-3)}{2(k+2)(2l-2k-3)} \\
&= a_k \frac{(l-m-2k)(l-m-2k-1)(l-m-2k-2)(l-m-2k-3)}{2^2(k+1)(k+2)(2l-2k-1)(2l-2k-3)},
\end{aligned} \tag{B2.13}$$

which can then be used to derive a recurrence relation for b_k, leading to the following expression:

$$b_k = a_k \left\{ 1 - \frac{(l-m-2k)(l-m-2k-1)}{2(2l-2k-1)} + \frac{(l-m-2k)(l-m-2k-1)(l-m-2k-2)(l-m-2k-3)}{2\cdot 4(2l-2k-1)(2l-2k-3)} - \ldots \right\} \tag{B2.14}$$

Comparing the expression in curly brackets with Equation B2.4 leads to the equality $b_k = a_k f(l-k, m+k, 1)$, where x is set to 1. Now that the coefficients a_k are known, the form of the expansion of f can be developed. With $P_1(x)$ being the coefficient r^l in the expansion of $(1-2xr+r^2)^{-1/2}$, $P_l^m(x) = (d^m/dx^m)P_l(x)$ is the r^l-th coefficient in the Taylor expansion of the derivative

$$\frac{d^m}{dx^m}(1-2xr+r^2)^{-\frac{1}{2}} = \frac{1}{2}\frac{3}{2}\frac{5}{2}\cdots\frac{2m-1}{2}(2r)^m(1-2xr+r^2)^{-m-\frac{1}{2}}. \tag{B2.15}$$

In this equation, differentiating m times produces m multiplication terms, with the largest coefficient $(2m-1)$ in the numerator terminating the multiplication series. At the terminus, parameter r is elevated to the power m, as $(2r)^m$, and another r^{l-m}-th order term can be obtained from the Taylor expansion. That, in combination with the differentiation part, will produce coefficients of the order r^l. Using the substitution $t = r(r-2x)$ gives:

$$\left(1-2xr+r^2\right)^{-m-\frac{1}{2}} = (1+t)^{-m-\frac{1}{2}}. \tag{B2.16}$$

A Taylor expansion around $t = 0$ has the following general form:

$$g(x) = \sum_{k=0}^{\infty} \frac{1}{k!} x^k g^{(k)}(0). \tag{B2.17}$$

The first few derivatives of this expansion are:

$$g^{(0)}(0) = g(0) = 1, \tag{B2.18}$$

$$g^{(1)}(0) = -\frac{2m+1}{2}, \tag{B2.19}$$

$$g^{(2)}(0) = \frac{2m+1}{2}\frac{2m+3}{2}, \tag{B2.20}$$

$$g^{(3)}(0) = -\frac{2m+1}{2}\frac{2m+3}{2}\frac{2m+5}{2}. \tag{B2.21}$$

Putting these terms together gives the expansion:

$$\begin{aligned}(1+t)^{-m-\frac{1}{2}} &= 1 - \frac{2m+1}{2}t + \frac{1}{2}\frac{(2m+1)(2m+3)}{2\cdot 2}t^2 - \frac{1}{3\cdot 2}\frac{(2m+1)(2m+3)(2m+5)}{2\cdot 2\cdot 2}t^3 + \cdots \\ &= 1 - \frac{2m+1}{2}(r^2 - 2xr) + \frac{1}{2}\frac{(2m+1)(2m+3)}{2\cdot 2}(r^4 - 4xr^3 + 4x^2r^2) \\ &\quad - \frac{1}{3\cdot 2}\frac{(2m+1)(2m+3)(2m+5)}{2\cdot 2\cdot 2}(r^6 - 6xr^5 + 12x^2r^4 - 8x^3r^3) + \cdots \end{aligned} \tag{B2.22}$$

Combining the terms with same power of r gives:

$$\begin{aligned}(1+t)^{-m-\frac{1}{2}} &= 1 + (2m+1)\,xr + \frac{1}{2}(2m+1)(2m+3)\,x^2r^2 - \frac{1}{2}(2m+1)\,r^2 \\ &\quad + \frac{1}{3\cdot 2}(2m+1)(2m+3)(2m+5)\,x^3r^3 - \frac{1}{2}(2m+1)(2m+3)\,xr^3 + \cdots \end{aligned} \tag{B2.23}$$

Setting $x = 1$ leads to:

$$(1+t)^{-m-\frac{1}{2}} = 1 + (2m+1)\,r + \frac{1}{2}(2m+1)(2m+2)\,r^2 + \frac{1}{3\cdot 2}(2m+1)(2m+2)(2m+3)\,r^3 + \cdots \tag{B2.24}$$

Finally, the r^l-th coefficient of expansion of $P_l^m(x)$ for $x = 1$ becomes:

$$\frac{1}{2}\frac{3}{2}\frac{5}{2}\cdots\frac{2m-1}{2}2^m\frac{(2m+1)(2m+2)\cdots(l+m)}{(l-m)!}. \tag{B2.25}$$

On the left-hand side relative to 2^m, there are m multiplication terms coming from the differentiation d^m/dx^m, and, on the right-hand side, there are $(l - m)$ multiplication terms coming as a result of the $(l - m)$-th order Taylor expansion. This order is reflected in the $(l - m)!$ term in the denominator. The largest multiplication term from the Taylor expansion is $(2m + l - m) = (l + m)$.

This expression can be simplified by canceling the 2^m term in the numerator and denominator, and simultaneously multiplying and dividing by $(2m)$. After a few additional transformations the equation reduces to:

$$\begin{aligned}&\frac{1\cdot 3\cdot 5\cdots(2m-1)(2m)(2m+1)(2m+2)\cdots(l+m)}{(2m)(l-m)!} \\ &= \frac{1\cdot 2\cdot 3\cdot 4\cdot 5\cdot 6\cdots(2m-2)}{2^{m-1}(m-1)!}\cdot\frac{(2m-1)(2m)(2m+1)(2m+2)\cdots(l+m)}{(2m)(l-m)!} \\ &= \frac{(l+m)!}{2^{m-1}(m-1)!(2m)(l-m)!} = \frac{(l+m)!}{2^m m!(l-m)!},\end{aligned} \tag{B2.26}$$

or:

$$\frac{(2l)!}{2^l l!(l-m)!}f(l,m,1) = \frac{(l+m)!}{2^m m!(l-m)!}, \tag{B2.27}$$

and

$$f(l,m,1) = \frac{2^{l-m}l!(l+m)!}{m!(2l)!}. \tag{B2.28}$$

Turning now to the coefficients b_k:

$$
\begin{aligned}
b_0 &= f(l,m,1)\cdot a_0 = \frac{2^{l-m}l!(l+m)!}{m!(2l)!},\\
b_1 &= f(l-1,m+1,1)\cdot a_1 = -\frac{2^{l-m-2}(l-1)!(l+m)!}{(m+1)!(2l-2)!}\cdot\frac{(l-m)(l-m-1)}{2(2l-1)}\\
&= -b_0\frac{2l(2l-1)}{4l(m+1)}\cdot\frac{(l-m)(l-m-1)}{2(2l-1)} = -b_0\frac{(l-m)(l-m-1)}{4(m+1)},\\
b_2 &= f(l-2,m+2,1)\cdot a_2 = \frac{2^{l-m-4}(l-2)!(l+m)!}{(m+2)!(2l-4)!}\cdot\frac{(l-m)(l-m-1)(l-m-2)(l-m-3)}{2\cdot 4\cdot(2l-1)(2l-3)}\\
&= b_0\frac{(2l)(2l-1)(2l-2)(2l-3)}{4^2 l(l-1)(m+2)(m+1)}\cdot\frac{(l-m)(l-m-1)(l-m-2)(l-m-3)}{2\cdot 4\cdot(2l-1)(2l-3)}\\
&= b_0\frac{(l-m)(l-m-1)(l-m-2)(l-m-3)}{2\cdot 4^2(m+2)(m+1)}.
\end{aligned}
\tag{B2.29}
$$

With that, the expression for $f(l,m,x)$ becomes:

$$
\begin{aligned}
f(l,m,x) = \frac{2^{l-m}l!(l+m)!}{m!(2l)!}\Bigg\{& x^{l-m} - \frac{(l-m)(l-m-1)}{4(m+1)}x^{l-m-2}y^2\\
&+ \frac{(l-m)(l-m-1)(l-m-2)(l-m-3)}{2\cdot 4^2(m+1)(m+2)}x^{l-m-4}y^4 - \cdots\Bigg\}.
\end{aligned}
\tag{B2.30}
$$

Finally, making the substitution $x = \cos\theta$ gives the final expression for the associated Legendre function:

$$
\begin{aligned}
P_l^m(\cos\theta) = \frac{(l+m)!}{2^m m!(l-m)!}(-1)^m\sin^m\theta\Bigg\{&\cos^{l-m}\theta - \frac{(l-m)(l-m-1)}{2\cdot(2m+2)}\cos^{l-m-2}\theta\sin^2\theta\\
&+ \frac{(l-m)(l-m-1)(l-m-2)(l-m-3)}{2\cdot 4\cdot(2m+2)(2m+4)}\cos^{l-m-4}\theta\sin^4\theta - \cdots\Bigg\}.
\end{aligned}
\tag{B2.31}
$$

The factorization given by Equation B2.31 applies to non-negative values of m only. That is, $l \geq m \geq 0$. The expansion for P_l^{-m} follows from the symmetry relation:

$$
P_l^{-m} = (-1)^m\frac{(l-m)!}{(l+m)!}P_l^m. \tag{B2.32}
$$

B.3 Recurrence Relation for Associated Legendre Functions

In order to derive the recurrence relations for the associated Legendre functions it is necessary to develop the recurrence relations for the Legendre polynomials first. Legendre polynomials may be expressed in terms of a generating function,

$$
\sum_{l=0}^{\infty} P_l(x)\,t^l = (1-2tx+t^2)^{-1/2}, \tag{B3.1}
$$

which originates from

$$
\frac{1}{|\mathbf{r}-\mathbf{r}'|} = (r^2 - 2rr'\cos\gamma + r'^2)^{-\frac{1}{2}} = \frac{1}{r}(1-2tx+t^2)^{-\frac{1}{2}} = \sum_{l=0}^{\infty}\frac{r'^l}{r^{l+1}}P_l(\cos\gamma) = \frac{1}{r}\sum_{l=0}^{\infty}P_l(x)\,t^l, \tag{B3.2}
$$

where $t = r'/r$ and $x = \cos\gamma$. Differentiating both parts of Equation B3.1 with respect to t gives:

$$\sum_{l=0}^{\infty} l\, P_l(x)\, t^{l-1} = (1-2tx+t^2)^{-\frac{3}{2}}(x-t). \tag{B3.3}$$

Multiplying both sides of Equation B3.3 by $(1 - 2tx + t^2)$ and substituting in Equation B3.1 gives:

$$(1-2tx+t^2)\sum_{l=0}^{\infty} l\, P_l(x)\, t^{l-1} = (x-t)\sum_{l=0}^{\infty} P_l(x)\, t^l. \tag{B3.4}$$

Equating coefficients at t^l leads to:

$$(l+1)\, P_{l+1}(x) - 2lxP_l(x) + (l-1)\, P_{l-1}(x) = xP_l(x) - P_{l-1}(x). \tag{B3.5}$$

Rearrangement of the terms of this equation then gives the recurrence relation for the Legendre polynomials:

$$(l+1)\, P_{l+1}(x) - (2l+1)\, xP_l(x) + lP_{l-1}(x) = 0. \tag{B3.6}$$

This allows the generating function in Equation B3.1 to be differentiated with respect to x, leading to:

$$\sum_{l=0}^{\infty} P_l'(x)\, t^l = (1-2tx+t^2)^{-\frac{3}{2}}\, t. \tag{B3.7}$$

Multiplying both sides of this equation by $(1 - 2tx + t^2)$ gives:

$$(1-2tx+t^2)\sum_{l=0}^{\infty} P_l'(x)\, t^l = \sum_{l=0}^{\infty} P_l(x)\, t^{l+1}. \tag{B3.8}$$

Equating the coefficients at t^l gives:

$$P_l'(x) - 2xP_{l-1}'(x) + P_{l-2}'(x) = P_{l-1}(x),$$

or

$$P_l'(x) - 2xP_{l-1}'(x) + P_{l-2}'(x) - P_{l-1}(x) = 0. \tag{B3.9}$$

In this form, substituting the index $l - 1$ by l gives a recurrence relation for the derivatives of the Legendre polynomials:

$$P_{l+1}'(x) - 2xP_l'(x) + P_{l-1}'(x) - P_l(x) = 0. \tag{B3.10}$$

Substituting $xP_l'(x) = P_{l+1}'(x) - (l+1)P_l(x)$ into Equation B3.10 leads to another recurrence relation:

$$(2l+1)\, P_l(x) = P_{l+1}'(x) - P_{l-1}'(x). \tag{B3.11}$$

The generating function for associated Legendre functions can be found by differentiating Equation B3.1 m-times (assuming $m > 0$) with respect to x, and multiplying both sides by $(-1)^m(1-x^2)^{m/2}$. This leads to:

$$\sum_{l=m}^{\infty} P_l^m(x)\,t^l = (-1)^m (1-x^2)^{\frac{m}{2}} \left[\frac{d^m}{dx^m}(1-2tx+t^2)^{-\frac{1}{2}}\right]. \tag{B3.12}$$

According to Appendix B2:

$$\frac{d^m}{dx^m}(1-2tx+t^2)^{-\frac{1}{2}} = (2m-1)!!\,t^m(1-2tx+t^2)^{-m-\frac{1}{2}}, \tag{B3.13}$$

where $(2m-1)!! = (2m-1)(2m-3)...5\cdot 3\cdot 1 = \frac{(2m)!}{2^m m!}$. Note, $1!! = 1$ and $0!! = 1$. This gives:

$$\sum_{l=m}^{\infty} P_l^m(x)\,t^l = (-1)^m \frac{(2m)!}{2^m m!}(1-x^2)^{\frac{m}{2}} t^m (1-2tx+t^2)^{-m-\frac{1}{2}}. \tag{B3.14}$$

Equation B3.14 is a generating function for the associated Legendre functions. Differentiating both parts of this equation with respect to t leads to:

$$\sum_{l=m}^{\infty} l\,P_l^m(x)\,t^{l-1} = m\sum_{l=m}^{\infty} P_l^m(x)\,t^{l-1} + (-1)^m \frac{(2m)!}{2^m m!}(1-x^2)^{\frac{m}{2}} t^m (1-2tx+t^2)^{-m-\frac{1}{2}-1}(2m+1)(x-t). \tag{B3.15}$$

Next, multiplying both parts of this equation by $(1 - 2tx + t^2)$ and substituting in Equation B3.14 gives:

$$(1-2tx+t^2)\sum_{l=m}^{\infty} l\,P_l^m(x)\,t^{l-1} = m\,(1-2tx+t^2)\sum_{l=m}^{\infty} P_l^m(x)\,t^{l-1} + (2m+1)(x-t)\sum_{l=m}^{\infty} P_l^m(x)\,t^l. \tag{B3.16}$$

Rearranging this equation by combining the terms t^{l-1} on the left side leads to:

$$(1-2tx+t^2)\sum_{l=m}^{\infty}(l-m)\,P_l^m(x)\,t^{l-1} = (2m+1)(x-t)\sum_{l=m}^{\infty} P_l^m(x)\,t^l. \tag{B3.17}$$

Equating coefficients of the type t^l gives:

$$(l+1-m)\,P_{l+1}^m(x) - 2(l-m)\,x\,P_l^m(x) + (l-1-m)\,P_{l-1}^m(x) = (2m+1)\,x\,P_l^m(x) - (2m+1)\,P_{l-1}^m(x). \tag{B3.18}$$

Rearranging the terms in this equation leads to the following recurrence relation in index l for the associated Legendre functions:

$$(l-m+1)\,P_{l+1}^m(x) - (2l+1)\,x\,P_l^m(x) + (l+m)\,P_{l-1}^m(x) = 0. \tag{B3.19}$$

To find a recurrence relation in index m, it is necessary to differentiate Equation B3.11 m-times to give:

$$(2l+1)\frac{d^m}{dx^m}P_l(x) = \frac{d^{m+1}}{dx^{m+1}}P_{l+1}(x) - \frac{d^{m+1}}{dx^{m+1}}P_{l-1}(x). \tag{B3.20}$$

Multiplying both sides by $(-1)^{m+1}(1-x^2)^{(m+1)/2}$ produces:

$$-(2l+1)\sqrt{1-x^2}\,P_l^m(x) = P_{l+1}^{m+1}(x) - P_{l-1}^{m+1}(x), \tag{B3.21}$$

or

$$(2l+1)\sqrt{1-x^2}\,P_l^m(x) = P_{l-1}^{m+1}(x) - P_{l+1}^{m+1}(x). \tag{B3.22}$$

This recurrence relation for associated Legendre functions is instrumental in solving Gaunt's integral over a triple product of spherical harmonics.

B.4 Hypergeometric Functions

Hypergeometric functions are series in which the successive terms have the same ratio. They are solutions of the hypergeometric differential equation:

$$x(1-x)y''+\left[c-(a+b+1)x\right]y'-aby=0, \tag{B4.1}$$

where $y = f(x)$ and a, b, and c are constants. The ratio of the successive terms having indexes k and $k + 1$ is:

$$\frac{t_{k+1}}{t_k}=\frac{(a_1+k)(a_2+k)\cdots(a_p+k)}{(k+1)(b_1+k)(b_2+k)\cdots(b_q+k)}x. \tag{B4.2}$$

Hypergeometric functions can be expressed using a compact notation:

$${}_pF_q(a_1,\ldots,a_p;b_1,\ldots,b_q;x). \tag{B4.3}$$

In this, index p indicates the number of terms in the numerator, index q specifies the number of terms in the denominator, and the last element in the brackets, argument x, can be a function. Often the ratio in a hypergeometric function has two coefficients in the numerator and one in the denominator: such a function is designated ${}_2F_1$ or, simply, $F = (a,b;c;x)$. In such cases, the ratio of successive terms is:

$$\frac{t_{k+1}}{t_k}=\frac{(a+k)(b+k)}{(k+1)(c+k)}x. \tag{B4.4}$$

The term $(k + 1)$ gives rise to a factorial in the denominator in each term:

$$F(a,b;c;x)=1+\frac{ab}{1!c}x+\frac{a(a+1)b(b+1)}{2!c(c+1)}x^2+\cdots. \tag{B4.5}$$

The utility of hypergeometric function, as shown by Hobson, is that it provides the mechanism to factorize the associated Legendre function. For that purpose, rewriting $(x^2 - 1)^l$, which is present in the Legendre polynomial, in the form $2^l(x-1)^l(1+(x-1)/2)^l$ and expanding it in powers of $x - 1$ *via* the binomial expansion gives:

$$(x^2-1)^l=2^l2^l\left(\frac{x-1}{2}\right)^l\sum_{k=0}^{l}\frac{l!}{k!(l-k)!}\left(\frac{x-1}{2}\right)^k=2^l2^l\sum_{k=0}^{l}\frac{l!}{k!(l-k)!}\left(\frac{x-1}{2}\right)^{l+k}. \tag{B4.6}$$

Differentiating Equation B4.6 $l + m$-times and converting both its sides into an associated Legendre function gives:

$$\begin{aligned}
P_l^m(x)&=\frac{1}{2^l l!}(-1)^m(1-x^2)^{\frac{m}{2}}\frac{d^{l+m}}{dx^{l+m}}(x^2-1)^l\\
&=\frac{1}{2^l l!}(-1)^m(1-x^2)^{\frac{m}{2}}2^l2^l\frac{d^{l+m}}{dx^{l+m}}\sum_{k=0}^{l}\frac{l!}{k!(l-k)!}\left(\frac{x-1}{2}\right)^{l+k}\\
&=(-1)^m(1-x^2)^{\frac{m}{2}}2^l\sum_{k=0}^{l-m}\frac{l!}{k!(l-k)!}\frac{(l+k)!}{(l+k-l-m)!}\left(\frac{1}{2}\right)^{l+m}\left(\frac{x-1}{2}\right)^{l+k-l-m}\\
&=(-1)^m(1-x^2)^{\frac{m}{2}}2^m\sum_{k=m}^{l}\frac{(l+k)!}{k!(l-k)!(k-m)!}\left(\frac{x-1}{2}\right)^{k-m}.
\end{aligned} \tag{B4.7}$$

For $m \geq 0$, it is necessary that $k - m \geq 0$, otherwise the summation term vanishes. Given that, the sum in the last equation can be changed so that it starts with $k = m$. Substituting variables $a = k - m$ and $k = m + a$ leads to:

$$P_l^m(x) = (-1)^m (1-x^2)^{\frac{m}{2}} 2^m \sum_{a=0}^{l-m} \frac{(l+m+a)!}{(m+a)!(l-m-a)!a!} \left(\frac{x-1}{2}\right)^a. \tag{B4.8}$$

The factorials in this equation can be written in an alternative form as:

$$\begin{aligned}
(l+m+a)! &= (l+m)!(l+m+1)\cdots(l+m+a), \\
(m+a)! &= m!(m+1)\cdots(m+a), \\
(l-m-a)! &= \frac{(l-m)!}{(l-m)(l-m-1)\cdots(l-m-a+1)} \\
&= (-1)^a \frac{(l-m)!}{(-l+m)(-l+m+1)\cdots(-l+m+a-1)}.
\end{aligned} \tag{B4.9}$$

The denominator in the last factorial contains terms involving a. With that, inverting the sign in each bracket in the denominator produces the multiplier $(-1)^a$. Explicitly writing out first few terms in the sum gives:

$$\begin{aligned}
a=0: &\quad \frac{(l+m+a)!}{(m+a)!(l-m-a)!a!}\left(\frac{x-1}{2}\right)^a = \frac{(l+m)!}{m!(l-m)!}\cdot 1, \\
a=1: &\quad \frac{(l+m)!}{m!(l-m)!}\cdot(-1)^1 \frac{(l+m+1)\cdot(-l+m)}{(m+1)\cdot 1!}\left(\frac{x-1}{2}\right)^1 \\
&= \frac{(l+m)!}{m!(l-m)!}\cdot\frac{(l+m+1)\cdot(-l+m)}{(m+1)\cdot 1!}\left(\frac{1-x}{2}\right)^1, \\
a=2: &\quad \frac{(l+m)!}{m!(l-m)!}\cdot(-1)^2 \frac{(l+m+1)(l+m+2)\cdot(-l+m)(-l+m+1)}{(m+1)(m+2)\cdot 2!}\left(\frac{x-1}{2}\right)^2 \\
&= \frac{(l+m)!}{m!(l-m)!}\cdot\frac{(l+m+1)(l+m+2)\cdot(-l+m)(-l+m+1)}{(m+1)(m+2)\cdot 2!}\left(\frac{1-x}{2}\right)^2.
\end{aligned} \tag{B4.10}$$

Since the ratio of successive elements in this sum satisfies

$$\frac{t_{a+1}}{t_a} = \frac{(l+m+1+a)(-l+m+a)}{(a+1)(m+1+a)}\left(\frac{1-x}{2}\right), \tag{B4.11}$$

the sum is in fact a hypergeometric function, $F\left(l+m+1,-l+m;m+1;(1-x)/2\right)$. Therefore,

$$P_l^m(x) = \frac{(l+m)!}{2^m m!(l-m)!}(-1)^m (1-x^2)^{\frac{m}{2}} F\left(l+m+1,-l+m;m+1;\frac{1-x}{2}\right). \tag{B4.12}$$

This factorization of P_l^m is valid for all values of m in the interval $[-l; +l]$.

Another useful factorization of P_l^m follows from the relation

$$(x^2-1)^l = (x-1)^l (x+1)^l. \tag{B4.13}$$

Differentiating this using the Leibniz formula gives:

$$\begin{aligned}\frac{d^{l+m}}{dx^{l+m}}(x+1)^l(x-1)^l &= \sum_{k=0}^{l+m}\frac{(l+m)!}{k!(l+m-k)!}\left[\frac{d^k}{dx^k}(x+1)^l\right]\left[\frac{d^{l+m-k}}{dx^{l+m-k}}(x-1)^l\right] \\ &= \sum_{k=m}^{l}\frac{(l+m)!}{k!(l+m-k)!}\frac{l!}{(l-k)!}(x+1)^{l-k}\frac{l!}{(k-m)!}(x-1)^{k-m}.\end{aligned} \tag{B4.14}$$

Eliminating all zero-value terms by enforcing the conditions $k \le l$ and $l+m-k \le l$, which simplify to $m \le k \le l$, leads to constraints on the summation limits, then applying the variable substitution, either $t = k - m$ or $k = t + m$, and observing that

$$(x+1)^{l-m-t}(x-1)^t = \frac{(x+1)^l}{(x+1)^m}\left(\frac{x-1}{x+1}\right)^t, \tag{B4.15}$$

leads to:

$$\frac{d^{l+m}}{dx^{l+m}}(x+1)^l(x-1)^l = \sum_{t=0}^{l-m}\frac{(l+m)!}{(m+t)!(l-t)!}\frac{l!}{(l-m-t)!}\frac{l!}{t!}\frac{(x+1)^l}{(x+1)^m}\left(\frac{x-1}{x+1}\right)^t. \tag{B4.16}$$

Substituting this into the expression for the associated Legendre function gives:

$$\begin{aligned}P_l^m &= \frac{(-1)^m}{2^l l!}(1-x^2)^{\frac{m}{2}}\sum_{t=0}^{l-m}\frac{(l+m)!}{(m+t)!(l-t)!}\frac{l!}{\left(l-m-t\right)!}\frac{l!}{t!}\frac{(x+1)^l}{(x+1)^m}\left(\frac{x-1}{x+1}\right)^t \\ &= \frac{(-1)^m i^m}{2^l}(x-1)^{\frac{m}{2}}(x+1)^{\frac{m}{2}}\frac{(x+1)^l}{(x+1)^m}(l+m)!\sum_{t=0}^{l-m}\frac{(l+m)!}{t!(m+t)!(l-t)!}\frac{l!}{(l-m-t)!}\left(\frac{x-1}{x+1}\right)^t \\ &= (-1)^m i^m\left(\frac{x-1}{x+1}\right)^{\frac{m}{2}}\left(\frac{x+1}{2}\right)^l\frac{(l+m)!}{m!(l-m)!}\sum_{t=0}^{l-m}\frac{l!}{t!(l-t)!}\frac{m!}{(m+t)!}\frac{(l-m)!}{(l-m-t)!}\left(\frac{x-1}{x+1}\right)^t.\end{aligned} \tag{B4.17}$$

For $t > 1$, the following takes place:

$$\begin{aligned}\frac{l!}{(l-t)!} &= l\cdots(l-t+1), \\ \frac{m!}{(m+t)!} &= \frac{1}{(m+t)\cdots(m+1)}, \\ \frac{(l-m)!}{(l-m-t)!} &= (l-m)\cdots(l-m-t+1).\end{aligned} \tag{B4.18}$$

For $t = 0$, the first term in the sum is unity. For $t = 1$, the summation term becomes:

$$\frac{-l(m-l)}{1!(m+1)}\left(\frac{x-1}{x+1}\right)^1. \tag{B4.19}$$

For $t = 2$, the summation term is:

$$\frac{(-l)(-l+1)(m-l)(m-l+1)}{2!(m+1)(m+2)}\left(\frac{x-1}{x+1}\right)^2. \tag{B4.20}$$

Recognizing a hypergeometric function, $F(-l,m-l;m+1;(x-1)/(x+1))$, in this series reduces the expansion to:

$$P_l^m=(-1)^m i^m\left(\frac{x-1}{x+1}\right)^{\frac{m}{2}}\left(\frac{x+1}{2}\right)^l\frac{(l+m)!}{m!(l-m)!}F\left(-l,m-l;m+1;\frac{x-1}{x+1}\right). \quad \text{(B4.21)}$$

Hypergeometric functions transform according to:

$$F(a,b;c;x)=(1-x)^{c-a-b}F(c-a,c-b;c;x), \quad \text{(B4.22)}$$

or

$$F(c-a,c-b;c;x)=(1-x)^{-c+a+b}F(a,b;c;x). \quad \text{(B4.23)}$$

With that,

$$F\left(-l,m-l;m+1;\frac{x-1}{x+1}\right)=F\left([m+1]-[l+m+1],[m+1]-[l+1];m+1;\frac{x-1}{x+1}\right). \quad \text{(B4.24)}$$

Given that $a=l+m+1$, $b=l+1$, and $c=m+1$, gives:

$$F\left(-l,m-l;m+1;\frac{x-1}{x+1}\right)=\left(\frac{2}{x+1}\right)^{2l+1}F\left(l+m+1,l+1;m+1;\frac{x-1}{x+1}\right). \quad \text{(B4.25)}$$

This leads to:

$$P_l^m=(-1)^m i^m\left(\frac{x-1}{x+1}\right)^{\frac{m}{2}}\left(\frac{2}{x+1}\right)^{l+1}\frac{(l+m)!}{m!(l-m)!}F\left(l+m+1,l+1;m+1;\frac{x-1}{x+1}\right). \quad \text{(B4.26)}$$

Finally, values of P_l^{-m} can be obtained using the symmetry relationship:

$$P_l^{-m}=(-1)^m\frac{(l-m)!}{(l+m)!}P_l^m=i^m\left(\frac{x-1}{x+1}\right)^{\frac{m}{2}}\left(\frac{2}{x+1}\right)^{l+1}\frac{1}{m!}F\left(l+m+1,l+1;m+1;\frac{x-1}{x+1}\right). \quad \text{(B4.27)}$$

B.5 Generating Function for Irregular Solid Harmonics

The generating function for irregular solid harmonics obtained by Hobson has the following form:

$$(z+ix)^{-l-1}=\sum_{m=-l}^{l}i^m\frac{(l-m)!}{l!}S_{l,m}+2r^{-l-1}\sum_{m=l+1}^{\infty}i^{-m}\frac{(l+m)!}{l!}P_l^{-m}\cos(m\phi). \quad \text{(B5.1)}$$

Derivation of this equation employs the following relations:

$$\begin{aligned} x&=r\sin\theta\cos\phi,\\ y&=r\sin\theta\sin\phi,\\ z&=r\cos\theta, \end{aligned} \quad \text{(B5.2)}$$

$$e^{\pm i\phi}=\cos\phi\pm i\sin\phi, \quad \text{(B5.3)}$$

$$(x \pm iy) = r\sin\theta\cos\phi \pm ir\sin\theta\sin\phi = r\sin\theta e^{\pm i\phi}. \tag{B5.4}$$

Substituting z and x by their spherical polar counterparts gives:

$$(z+ix)^{l-1} = r^{l-1}(\cos\theta + i\sin\theta\cos\phi)^{l-1}. \tag{B5.5}$$

Denoting $\mu = \cos\theta$ and $i\sin\theta = \sqrt{\mu^2-1}$ gives:

$$\begin{aligned}(z+ix)^{-l-1} &= r^{-l-1}\left(\mu+\sqrt{\mu^2-1}\cos\phi\right)^{-l-1} = \left(\frac{2}{r}\right)^{l+1}\left[2\mu+\sqrt{\mu^2-1}\,(e^{i\phi}+e^{-i\phi})\right]^{-l-1} \\ &= \left(\frac{2\sqrt{\mu^2-1}}{r}\right)^{l+1}\left[2\mu\sqrt{\mu^2-1}+(\mu^2-1)(e^{i\phi}+e^{-i\phi})\right]^{-l-1} \\ &= \left(\frac{2\sqrt{\mu^2-1}\,e^{i\phi}}{r}\right)^{l+1}\left[2\mu\sqrt{\mu^2-1}\,e^{i\phi}+(\mu^2-1)\,e^{2i\phi}+\mu^2-1\right]^{-l-1} \\ &= \left(\frac{2\sqrt{\mu^2-1}\,e^{i\phi}}{r}\right)^{l+1}\left[\left(\mu+\sqrt{\mu^2-1}\,e^{i\phi}\right)^2-1\right]^{-l-1}.\end{aligned} \tag{B5.6}$$

Additionally, denoting $w = \sqrt{\mu^2-1}\,e^{i\phi}$ leads to:

$$\begin{aligned}(z+ix)^{l-1} &= r^{-l-1}(2w)^{l+1}\left[(\mu+w)^2-1\right]^{-l-1} = r^{-l-1}(2w)^{l+1}\{\mu+w-1\}^{-l-1}\{\mu+w+1\}^{-l-1} \\ &= r^{-l-1}2^{l+1}\left\{1+\frac{\mu-1}{w}\right\}^{-l-1}\{\mu+w+1\}^{-l-1} = r^{-l-1}2^{l+1}(\mu+1)^{-l-1}\left\{1+\frac{\mu-1}{w}\right\}^{-l-1}\left\{1+\frac{w}{\mu-1}\right\}^{-l-1}.\end{aligned} \tag{B5.7}$$

The first task is to find a series expansion of the terms in curly brackets, which involves finding the absolute value of a complex number. For $z = \alpha + i\beta$, the absolute value is $|z| = \sqrt{\alpha^2+\beta^2}$. Given that, $|e^{\pm i\phi}| = |\cos\phi \pm i\sin\phi| = 1$.

The following derivation assumes that μ is positive, that is, that $0 < \mu \leq 1$. This leads to $|(\mu-1)/(\mu+1)| < 1$. This relation also holds when μ is complex. To find the absolute value of w, it is helpful to expand its expression to:

$$w = i\sqrt{1-\mu^2}\left(\cos\phi + i\sin\phi\right) = -\sqrt{1-\mu^2}\sin\phi + i\sqrt{1-\mu^2}\cos\phi. \tag{B5.8}$$

Then, $|w| = (1-\mu^2)\sin^2\phi + (1-\mu^2)\cos^2\phi = 1-\mu^2$. From this, it follows that values of $|w|$ are in the range $|\mu-1| \leq |w| \leq |\mu+1|$.

Applying a binomial expansion to the terms in curly brackets leads to two series denoted A and B:

$$\begin{aligned}A &= \left\{1+\frac{\mu-1}{w}\right\}^{-l-1} = \sum_{k=0}^{\infty}\frac{(-l-1)!}{k!(-l-1-k)!}\left(\frac{\mu-1}{w}\right)^k \\ &= 1+\sum_{k=1}^{\infty}\frac{(-l-1)(-l-2)\cdots(-l-k)}{k!}\left(\frac{\mu-1}{w}\right)^k \\ &= 1+\sum_{k=1}^{\infty}(-1)^k\frac{(l+1)(l+2)\cdots(l+k)}{k!}\left(\frac{\mu-1}{w}\right)^k = a_0+a_1+a_2+\cdots\end{aligned} \tag{B5.9}$$

$$B=\left\{1+\frac{w}{\mu+1}\right\}^{-l-1}=\sum_{k=0}^{\infty}\frac{(-l-1)!}{k!(-l-1-k)!}\left(\frac{w}{\mu+1}\right)^{k}$$
$$=1+\sum_{k=1}^{\infty}\frac{(-l-1)(-l-2)\cdots(-l-k)}{k!}\left(\frac{w}{\mu+1}\right)^{k} \tag{B5.10}$$
$$=1+\sum_{k=1}^{\infty}(-1)^{k}\frac{(l+1)(l+2)\cdots(l+k)}{k!}\left(\frac{w}{\mu+1}\right)^{k}=b_0+b_1+b_2+\cdots$$

so that,

$$AB=\left\{1+\frac{\mu-1}{w}\right\}^{-l-1}\left\{1+\frac{w}{\mu+1}\right\}^{-l-1}=\left\{1+\sum_{k=1}^{\infty}(-1)^{k}\frac{(l+k)!}{l!k!}\left(\frac{\mu-1}{w}\right)^{k}\right\}\left\{1+\sum_{t=1}^{\infty}(-1)^{t}\frac{(l+t)!}{l!t!}\left(\frac{w}{\mu+1}\right)^{t}\right\}. \tag{B5.11}$$

The product of series A and B corresponds to a direct multiplication of their terms:

$$AB=(a_0b_0+a_1b_1+\cdots)+\cdots+(a_0b_k+a_1b_{k+1}+\cdots)+(a_kb_0+a_{k+1}b_1+\cdots)+\cdots. \tag{B5.12}$$

In this sum, vector **a** multiplies vector **b**; index i in a_i goes from 0 to ∞; index j in b_j goes from k to ∞; in turn, index k goes from 0 to ∞ upon transitioning from one vector product to the next. With every vector product except the first one, the indices of vectors **a** and **b** flip, so that the rule described for **a** gets applied to **b** and vice versa. With that, for every k, there are two products in the sum, except for $k=0$, which corresponds to a single vector product.

Series A and B are alternating series in the sign of the terms. Convergence of the alternating series is determined by the absolute value of their terms:

$$\frac{\mu-1}{w}=\left(\frac{\mu-1}{\mu+1}\right)^{\frac{1}{2}}e^{-i\phi},$$
$$\left|\frac{\mu-1}{w}\right|=\left|\frac{\mu-1}{\mu+1}\right|^{\frac{1}{2}}, \tag{B5.13}$$

$$\frac{w}{\mu+1}=\left(\frac{\mu-1}{\mu+1}\right)^{\frac{1}{2}}e^{i\phi},$$
$$\left|\frac{w}{\mu+1}\right|=\left|\frac{\mu-1}{\mu+1}\right|^{\frac{1}{2}}. \tag{B5.14}$$

According to d'Alembert's test for an alternating or complex series, a series is convergent when its elements satisfy the limit of the ratio:

$$\lim_{k\to\infty}\frac{|s_{k+1}|}{|s_k|}<1. \tag{B5.15}$$

A series satisfying this test is known as absolutely convergent. Such series have special properties in that the order of its terms can be rearranged arbitrarily without altering the series sum, and can be manipulated as regular numbers, that is, they can be added and multiplied.

The series A and B have the same absolute value of their corresponding terms, so they have equivalent convergence properties. The application of the d'Alembert's test shows that

$$\lim_{k\to\infty}\frac{|s_{k+1}|}{|s_k|}=\lim_{k\to\infty}\frac{((l+1)\cdots(l+k+1))/(k+1)!}{((l+1)\cdots(l+k))/k!}\left|\frac{\mu-1}{\mu+1}\right|^{\frac{1}{2}}=\lim_{k\to\infty}\frac{l+k+1}{k+1}\left|\frac{\mu-1}{\mu+1}\right|^{\frac{1}{2}}<1. \quad \text{(B5.16)}$$

Although the ratio of the binomial coefficients, $\lim_{k\to\infty}(l+k+1)/(k+1)=1$, on its own, does not satisfy the ratio test, because both $0<\mu\leq 1$, and $|(\mu-1)/(\mu+1)|^{1/2}<1$, for any μ arbitrarily close to 0, the following condition does hold: $\lim_{k\to\infty}(l+k+1)/(k+1)\,|(\mu-1)/(\mu+1)|^{1/2}<1$. Since a series composed of positive terms converges, both series, A and B, are absolutely convergent.

Next, consider a helper expansion:

$$C=\left(1-\left|\frac{\mu-1}{\mu+1}\right|^{\frac{1}{2}}\right)^{-l-1}=\sum_{k=0}^{\infty}\frac{(-l-1)!(-1)^k}{k!(-l-1-k)!}\left|\frac{\mu-1}{\mu+1}\right|^{\frac{k}{2}}=1+\sum_{k=1}^{\infty}\frac{(l+1)...(l+k)}{k!}\left|\frac{\mu-1}{\mu+1}\right|^{\frac{k}{2}}=c_0+c_1+c_2+\cdots$$

(B5.17)

From the properties of the series A and B, it is known that $|a_k|=|b_k|=c_k$ for all values of k. Since the series A and B are convergent, the series C is also absolutely convergent. For absolutely convergent series A and B,

$$\sum_{k=0}^{\infty}a_k=A,\quad\text{and}\quad\sum_{k=0}^{\infty}b_k=B, \quad \text{(B5.18)}$$

the sum of their product, $s_n=\sum_{k=0}^{\infty}a_kb_{n-k}$, converges to the product of their sums, $\sum_{n=0}^{\infty}s_n=A\cdot B$. Similarly, the square of series $C\cdot C$ converges to the sum $(1-|(\mu-1)/(\mu+1)|^{1/2})^{-2l-2}$ and the terms in the double sum $\sum_{k=0}^{\infty}\sum_{t=0}^{\infty}c_kc_t$ can therefore be rearranged in a series of any type without altering its sum. By definition, the resulting series

$$D=C\cdot C=(c_0^2+c_1^2+\cdots)+2(c_0c_1+c_1c_2+\cdots)+\cdots+2(c_0c_k+c_1c_{k+1}+\cdots)+\cdots, \quad \text{(B5.19)}$$

is convergent. Based on that, and, since $|a_kb_t|=c_kc_t$, the product of series $\sum_{k=0}^{\infty}\sum_{t=0}^{\infty}a_kb_t$ is absolutely convergent. Therefore, it can be rearranged into a series of any type which will also be convergent. Correspondingly, the series

$$AB=(a_0b_0+a_1b_1+\cdots)+\cdots+(a_0b_k+a_1b_{k+1}+\cdots)+(a_kb_0+a_{k+1}b_1+\cdots)+\cdots, \quad \text{(B5.20)}$$

converges to the product of the series A and B, which is

$$\left\{1+\frac{\mu-1}{w}\right\}^{-l-1}\left\{1+\frac{w}{\mu+1}\right\}^{-l-1}. \quad \text{(B5.21)}$$

Next, it is necessary to clarify the character of the convergence of the series AB depending on the value of azimuthal angle ϕ. According to the Weierstrass' test, if, for the series $\sum_{k=0}^{\infty}f_k(\phi)$, one can find positive numbers X_k for all ranges of ϕ, such that $|f_k(\phi)|\leq X_k$ and if the series $\sum_{k=0}^{\infty}X_k$ is convergent, then the series $\sum_{k=0}^{\infty}f_k(\phi)$ converges uniformly for all values of ϕ. A comparison of the series AB with series D, which is convergent and consists of only positive terms, and remembering that $|\sin\phi|\leq 1$ and $|\cos\phi|\leq 1$, shows that the term

$$(a_0b_k+a_1b_{k+1}+\cdots)+(a_kb_0+a_{k+1}b_1+\cdots), \quad \text{(B5.22)}$$

from the series AB is not greater than $2(c_0c_k + c_1c_{k+1} + \cdots)$ in the series D. Thus the Weierstrass' test confirms that the product of the series A and B converges uniformly for all values of ϕ. Additionally, this series is absolutely convergent because the series D is absolutely convergent. This allows the terms in the product AB to be rearranged in powers of w^m *and* w^{-m} in the derivation. In the demonstration that follows, index k denotes terms of the expansion A, and index t denotes terms of the expansion B. The objective is to construct the following series:

$$\begin{aligned}
&AB(w^0)=\sum_{k=0}^{\infty}a_kb_k,\\
&AB(w^1)=\sum_{k=0}^{\infty}a_kb_{k+1},\quad AB\left(w^{-1}\right)=\sum_{k=0}^{\infty}a_{k+1}b_k,\\
&\ldots\\
&AB(w^m)=\sum_{k=0}^{\infty}a_kb_{k+m},\quad AB\left(w^{-m}\right)=\sum_{k=0}^{\infty}a_{k+m}b_k,\\
&\ldots.
\end{aligned} \tag{B5.23}$$

For $m = 0$ and w^0, the result is:

$$\begin{aligned}
\sum_{k=t=0}^{\infty}AB &= 1+\frac{(l+1)}{1!}\left(\frac{\mu-1}{w}\right)^1\cdot\frac{(l+1)}{1!}\left(\frac{w}{\mu+1}\right)^1+\frac{(l+1)(l+2)}{2!}\left(\frac{\mu-1}{w}\right)^2\cdot\frac{(l+1)(l+2)}{2!}\left(\frac{w}{\mu+1}\right)^2\\
&+\frac{(l+1)(l+2)(l+3)}{3!}\left(\frac{\mu-1}{w}\right)^3\cdot\frac{(l+1)(l+2)(l+3)}{3!}\left(\frac{w}{\mu+1}\right)^3+\cdots=F\left(l+1,l+1;1;\frac{\mu-1}{\mu+1}\right),
\end{aligned} \tag{B5.24}$$

which is a hypergeometric function.

For $m = 1$ and w^1, the sum goes over indexes $t = k + 1$, or $t = k + m$ in general, starting from $k = 0$:

$$\begin{aligned}
\sum_{k,t}^{\infty}AB &= -(l+1)\left(\frac{w}{\mu+1}\right)^1-\frac{(l+1)}{1!}\left(\frac{\mu-1}{w}\right)^1\cdot\frac{(l+1)(l+2)}{2!}\left(\frac{w}{\mu+1}\right)^2\\
&-\frac{(l+1)(l+2)}{2!}\left(\frac{\mu-1}{w}\right)^2\cdot\frac{(l+1)(l+2)(l+3)}{3!}\left(\frac{w}{\mu+1}\right)^3+\cdots
\end{aligned} \tag{B5.25}$$

Based on:

$$\begin{aligned}
&\frac{\mu-1}{w}\cdot\left(\frac{w}{\mu+1}\right)^2=\frac{w}{\mu+1}\cdot\left(\frac{\mu-1}{\mu+1}\right)^1,\\
&\text{and}\\
&\left(\frac{\mu-1}{w}\right)^2\cdot\left(\frac{w}{\mu+1}\right)^3=\frac{w}{\mu+1}\cdot\left(\frac{\mu-1}{\mu+1}\right)^2,
\end{aligned} \tag{B5.26}$$

and, after separating the common multiplier, the sum turns to:

$$\begin{aligned}
\sum_{k,t}^{\infty}AB &= (-1)^m\frac{(l+m)!}{l!m!}\left(\frac{w}{\mu+1}\right)^m\left\{1+\frac{(l+1)\cdot(l+m+1)}{1!(m+1)}\left(\frac{\mu-1}{\mu+1}\right)^1\right.\\
&\left.+\frac{(l+1)(l+2)\cdot(l+m+1)(l+m+2)}{2!(m+1)(m+2)}\left(\frac{\mu-1}{\mu+1}\right)^2+\cdots\right\}.
\end{aligned} \tag{B5.27}$$

For $m = 2$ and w^2, the sum goes over indexes $t = k + 2$, or $t = k + m$ in general, starting from $k = 0$:

$$\begin{aligned}\sum_{k,t}^{\infty} AB &= \frac{(l+1)(l+2)}{2!}\left(\frac{w}{\mu+1}\right)^2 + \frac{(l+1)}{1!}\left(\frac{\mu-1}{w}\right)^1 \cdot \frac{(l+1)(l+2)(l+3)}{3!}\left(\frac{w}{\mu+1}\right)^3 \\ &+ \frac{(l+1)(l+2)}{2!}\left(\frac{\mu-1}{w}\right)^2 \cdot \frac{(l+1)(l+2)(l+3)(l+4)}{4!}\left(\frac{w}{\mu+1}\right)^4 + \cdots \\ &= (-1)^m \frac{(l+m)!}{l!m!}\left(\frac{w}{\mu+1}\right)^m \left\{1 + \frac{(l+1)\cdot(l+m+1)}{1!(m+1)}\left(\frac{\mu-1}{\mu+1}\right)^1\right. \\ &\left.+ \frac{(l+1)(l+2)\cdot(l+m+1)(l+m+2)}{2!(m+1)(m+2)}\left(\frac{\mu-1}{\mu+1}\right)^2 + \cdots\right\}.\end{aligned} \quad \text{(B5.28)}$$

For negative powers of w^{-m}, it is useful to explicitly show the negative sign outside the parameter m, thus keeping m formally positive. For w^{-1}, the sum in the product of series goes over indexes $t = k - 1$, or $t = k - m$ in general, starting from $k = 1$:

$$\begin{aligned}\sum_{k,t}^{\infty} AB &= -\frac{(l+1)}{1!}\left(\frac{\mu-1}{w}\right)^2 - \frac{(l+1)(l+2)}{2!}\left(\frac{\mu-1}{w}\right)^2 \cdot \frac{(l+1)}{1!}\left(\frac{w}{\mu+1}\right)^1 \\ &- \frac{(l+1)(l+2)(l+3)}{3!}\left(\frac{\mu-1}{w}\right)^3 \cdot \frac{(l+1)(l+2)}{2!}\left(\frac{w}{\mu+1}\right)^2 - \cdots \\ &= (-1)^m \frac{(l+m)!}{l!m!}\left(\frac{\mu-1}{w}\right)^m \left\{1 + \frac{(l+1)\cdot(l+m+1)}{1!(m+1)}\left(\frac{\mu-1}{\mu+1}\right)^1\right. \\ &\left.+ \frac{(l+1)(l+2)\cdot(l+m+1)(l+m+2)}{2!(m+1)(m+2)}\left(\frac{\mu-1}{\mu+1}\right)^2 + \cdots\right\}.\end{aligned} \quad \text{(B5.29)}$$

The series construction can now be generalized. For non-negative ($m \geq 0$) powers of w^m, the product

$$r^{-l-1}2^{l+1}(\mu+1)^{-l-1}\left(1+\frac{\mu-1}{w}\right)^{-l-1} \cdot \left(1+\frac{w}{\mu+1}\right)^{-l-1} \quad \text{(B5.30)}$$

sums up to:

$$\begin{aligned}&2^{l+1}r^{-l-1}(\mu+1)^{-l-1}\sum_{m=0}^{\infty}(-1)^m \frac{(l+m)!}{l!m!}\left(\frac{w}{\mu+1}\right)^m F\left(l+1, l+m+1; m+1; \frac{\mu-1}{\mu+1}\right) \\ &= 2^{l+1}r^{-l-1}(\mu+1)^{-l-m-1}\sum_{m=0}^{\infty}(-1)^m \frac{(l+m)!}{l!m!}(\mu^2-1)^{\frac{m}{2}} e^{im\phi} F\left(l+1, l+m+1; m+1; \frac{\mu-1}{\mu+1}\right).\end{aligned} \quad \text{(B5.31)}$$

For negative powers of w^{-m}, the product sums up to:

$$\begin{aligned}&2^{l+1}r^{-l-1}(\mu+1)^{-l-1}\sum_{m=0}^{\infty}(-1)^m \frac{(l+m)!}{l!m!}\left(\frac{\mu-1}{w}\right)^m F\left(l+1, l+m+1; m+1; \frac{\mu-1}{\mu+1}\right) \\ &= 2^{l+1}r^{-l-1}(\mu+1)^{-l-1}\sum_{m=0}^{\infty}(-1)^m \frac{(l+m)!}{l!m!}(\mu-1)^m(\mu^2-1)^{-\frac{m}{2}} e^{-im\phi} F\left(l+1, l+m+1; m+1; \frac{\mu-1}{\mu+1}\right).\end{aligned} \quad \text{(B5.32)}$$

Due to

$$P_l^m = (-1)^m \frac{2^{l+1}}{m!} \frac{(l+m)!}{(l-m)!} \left(\frac{1-\mu}{1+\mu}\right)^{\frac{m}{2}} (1+\mu)^{-l-1} F\left(l+1, l+m+1; m+1; \frac{\mu-1}{\mu+1}\right), \tag{B5.33}$$

and

$$P_l^{-m} = \frac{2^{l+1}}{m!} \left(\frac{1-\mu}{1+\mu}\right)^{\frac{m}{2}} (1+\mu)^{-l-1} F\left(l+1, l+m+1; m+1; \frac{\mu-1}{\mu+1}\right), \tag{B5.34}$$

the series for w^m is:

$$\begin{aligned} & r^{-l-1} 2^{l+1} (\mu+1)^{-l-m-1} (-1)^m \frac{(l+m)!}{l!m!} (\mu^2-1)^{\frac{m}{2}} e^{im\phi} F\left(l+1, l+m+1; m+1; \frac{\mu-1}{\mu+1}\right) \\ &= (-1)^m r^{-l-1} \frac{(l+m)!}{l!m!} \left(\frac{2}{\mu+1}\right)^{l+1} \left(\frac{\mu-1}{\mu+1}\right)^{\frac{m}{2}} e^{im\phi} F\left(l+1, l+m+1; m+1; \frac{\mu-1}{\mu+1}\right) \\ &= i^m r^{-l-1} \frac{(l-m)!}{l!} P_l^m e^{im\phi} \Big|_{0 \le m \le l} + i^{-m} r^{-l-1} \frac{(l+m)!}{l!} P_l^{-m} e^{im\phi} \Big|_{m>l} . \end{aligned} \tag{B5.35}$$

Similarly, for w^{-m}:

$$\begin{aligned} & r^{-l-1} 2^{l+1} (\mu+1)^{-l-1} (-1)^m \frac{(l+m)!}{l!m!} (\mu-1)^m (\mu^2-1)^{-\frac{m}{2}} e^{-im\phi} F\left(l+1, l+m+1; m+1; \frac{\mu-1}{\mu+1}\right) \\ &= (-1)^m r^{-l-1} \frac{(l+m)!}{l!m!} \left(\frac{2}{\mu+1}\right)^{l+1} \left(\frac{\mu-1}{\mu+1}\right)^{\frac{m}{2}} e^{-im\phi} F\left(l+1, l+m+1; m+1; \frac{\mu-1}{\mu+1}\right) \\ &= i^{-m} r^{-l-1} \frac{(l+m)!}{l!} P_l^{-m} e^{-im\phi}. \end{aligned} \tag{B5.36}$$

Note that $(-1)^m i^m = i^{-m}$.

Combining the parts leads to:

$$\begin{aligned} (z+ix)^{-l-1} &= r^{-l-1} \sum_{m=0}^{l} i^m \frac{(l-m)!}{l!} e^{im\phi} P_l^m + r^{-l-1} \sum_{m=1}^{l} i^{-m} \frac{(l+m)!}{l!} e^{-im\phi} P_l^{-m} \\ &\quad + r^{-l-1} \sum_{m=l+1}^{\infty} i^{-m} \frac{(l+m)!}{l!} \left[e^{im\phi} + e^{-im\phi}\right] P_l^{-m} \\ &= \sum_{m=-l}^{l} i^m \frac{(l-m)!}{l!} S_{l,m} + 2r^{-l-1} \sum_{m=l+1}^{\infty} i^{-m} \frac{(l+m)!}{l!} P_l^{-m} \cos(m\phi). \end{aligned} \tag{B5.37}$$

The right-most sum in Equation B5.37 is the residual term. Unlike positive values of m, which cannot exceed l, negative values of m are allowed to go to infinity, so that $|-m| > l$ is a valid range, in which case differentiation inside P_l^{-m} gets replaced by an integral.

Numerical computation of the residual term in Equation B5.37 involves expressing associated Legendre functions *via* hypergeometric functions, as shown above, and replacing the infinite series by a finite polynomial using an Euler transformation:

$$F\left(l+1, l+m+1; m+1; \frac{\mu-1}{\mu+1}\right) = \left(\frac{\mu+1}{2}\right)^{2l+1} F\left(-l, m-l; m+1; \frac{\mu-1}{\mu+1}\right). \tag{B5.38}$$

Bibliography

1. Arfken G. B., Weber H. J. 2005. *Mathematical Methods for Physicists*. 6th edition. Burlington, MA: Elsevier Academic Press.
2. Biedenharn L. C., Louck J. D. 1981. *Angular Momentum in Quantum Physics. Theory and Application*. Reading, MA: Addison-Wesley Publishing Company, Inc, Series: Encyclopedia of mathematics and its applications, Vol. 8. Editor Gian-Carlo Rota.
3. Caola M. J. 1978. Solid harmonics and their addition theorems. *J. Phys. A: Math. Gen.* 11: L23–L25.
4. Cartan E. 1966. *The Theory of Spinors*. Cambridge, MA: The M.I.T. Press.
5. Edmonds A. R. 1960. Angular momentum in quantum mechanics. In *Series: Investigations in Physics*, Edited by E. P. Wigner and R. Hofstadter. Princeton, NJ: Princeton University Press.
6. Fano U., Racah G. 1959. Irreducible tensorial sets. In *Pure and Applied Physics. A series of monographs and textbooks*, Vol. 4, Consulting Editor H. S. W. Massey. New York: Academic Press.
7. Gimbutas Z., Greengard L. 2009. A fast and stable method for rotating spherical harmonic expansions. *J. Comp. Phys.* 228: 5621–5627.
8. Greengard L., Rokhlin V. 1987. A fast algorithm for particle simulations. *J. Comp. Phys.* 73: 325–348.
9. Hobson E. W. 1931. *The Theory of Spherical and Elliptical Harmonics*. London: Cambridge University Press.
10. Kudin K. N., Scuseria G. E. 2000. Analytic stress tensor with the periodic fast multipole method. *Phys. Rev. B* 61: 5141–5146.
11. Kudin K. N., Scuseria G. E. 2004. Revisiting infinite lattice sums with the periodic fast multipole method. *J. Chem. Phys.* 121: 2886–2890.
12. Noyan I. C., Cohen J. B. 1987. *Residual Stress. Measurement by Diffraction and Interpretation*. New York: Springer-Verlag.
13. Nye J. F. 1957. *Physical Properties of Crystals. Their Representation by Tensors and Matrices*. London: Oxford University Press.
14. Rose M. E. 1957. Elementary theory of angular momentum. In *Structure of Matter Series*, Advisory Editor M. G. Mayer. New York: John Wiley and Sons, Inc.
15. White C. A., Head-Gordon M. 1994. Derivation and efficient implementation of the Fast Multipole Method. *J. Chem. Phys.* 101: 6593–6605.
16. White C. A., Head-Gordon M. 1996. Rotating around the quartic angular momentum barrier in fast multipole method calculations. *J. Chem. Phys.* 105: 5061–5067.
17. Wigner E. P. 1959. Group theory and its application to the quantum mechanics of atomic spectra. In *Pure and Applied Physics. A series of monographs and textbooks*, Vol. 8, Consulting Editor H. S. W. Massey. New York: Academic Press. Translated from German by J. J. Griffin.

Index

Taylor & Francis Group
an informa business

Taylor & Francis eBooks

www.taylorfrancis.com

A single destination for eBooks from Taylor & Francis with increased functionality and an improved user experience to meet the needs of our customers.

90,000+ eBooks of award-winning academic content in Humanities, Social Science, Science, Technology, Engineering, and Medical written by a global network of editors and authors.

TAYLOR & FRANCIS EBOOKS OFFERS:

A streamlined experience for our library customers

A single point of discovery for all of our eBook content

Improved search and discovery of content at both book and chapter level

REQUEST A FREE TRIAL

support@taylorfrancis.com

Routledge Taylor & Francis Group

CRC Press Taylor & Francis Group

9781032337401